The Application of

Modern Physics to

The Earth and Planetary Interiors

NATO ADVANCED STUDY INSTITUTE

Conference organized by

the School of Physics at the University

of Newcastle upon Tyne, England

29 March–4 April 1967

Director of Institute PROFESSOR S. K. RUNCORN, F.R.S.

Organizing Committee MRS. J. ROBERTS

MR. F. WIDDAS

The Application of Modern Physics to The Earth and Planetary Interiors

Edited by S. K. RUNCORN, F.R.S.

School of Physics, University of Newcastle upon Tyne, England

1969

WILEY—INTERSCIENCE

a division of John Wiley & Sons

LONDON NEW YORK SYDNEY TORONTO

First published 1969 John Wiley & Sons Ltd. All rights reserved. No part of this book may be reproduced by any means, nor transmitted, nor translated into a machine language without the written permission of the publisher.

Library of Congress catalog card number 70-81232

SBN 471 74505 7

QC
806
N2
1967

Printed in Great Britain by Robert MacLehose and Co. Ltd, The University Press, Glasgow

Contributors

PROFESSOR C. O. ALLEY Department of Physics and Astronomy, University of Maryland, College Park, Maryland, U.S.A.

PROFESSOR T. J. AHRENS Division of Geological Sciences, California Institute of Technology, 220 N. San Rafael Avenue, Pasadena, California 91105, U.S.A.

DR. D. W. ALLAN Department of Mathematics, King's College, The Strand, London W.C.2 England

MR. S. I. VAN ANDEL Geologisch Instituut, Nieuwe Prinsengracht 130, Amsterdam-C, Netherlands

PROFESSOR D. L. ANDERSON Seismological Laboratory, California Institute of Technology, 220 N. San Rafael Avenue, Pasadena, California 91105, U.S.A.

PROFESSOR O. L. ANDERSON Lamont Geological Observatory, Palisades, New York 10964, U.S.A.

PROFESSOR A. E. BECK Department of Geophysics, University of Western Ontario, London, Ontario, Canada

P. L. BENDER Joint Inst. for Laboratory Astrophysics, National Bureau of Standards, University of Colorado, Colorado 80304, U.S.A.

PROFESSOR C. A. BERG Department of Mechanical Engineering, Rooms 3-354, Massachusetts Institute of Technology, Cambridge, Massachusetts 02139, U.S.A.

PROFESSOR F. BIRCH Hoffman Laboratory, Harvard University, Cambridge, Massachusetts 02138, U.S.A.

PROFESSOR K. E. BULLEN Department of Applied Mathematics, University of Sydney, Sydney, New South Wales, Australia

DR. R. G. BURNS Department of Chemistry, Victoria University of Wellington, New Zealand

DR. B. CANER Victoria Magnetic Observatory, Dominion Astrophysical Observatory, Victoria, British Columbia, Canada

DR. S. CHILDRESS New York University, Courant Institute of Mathematical Sciences, 251 Mercer Street, New York, New York 10012, U.S.A.

DR. A. H. COOK Division of Quantum Metrology, National Physical Laboratory, Teddington, Middlesex, England

DR. D. G. CURRIE Department of Physics and Astronomy, University of Maryland, College Park, Maryland, U.S.A.

DR. D. DAVIES Department of Geodesy and Geophysics, University of Cambridge, Madingley Rise, Madingley Road, Cambridge, England

DR. R. DEARNLEY, Institute of Geological Sciences, Exhibition Road, London S.W.7, England

PROFESSOR R. H. DICKE Department of Physics, Princeton University, Princeton, New Jersey 08540, U.S.A.

PROFESSOR L. EGYED Geophysical Institute, Eötvös University, Budapest, Hungary

PROFESSOR W. M. ELSASSER Institute for Fluid Dynamics, University of Maryland, College Park, Maryland 20742, U.S.A.

PROFESSOR J. E. FALLER, PH.D. Department of Physics, Wesleyan University, Middletown, Connecticut 06457, U.S.A.

DR. R. L. FLEISCHER General Electric Research and Development Center, Building K-1, Schenectady, New York 12301, U.S.A.

PROFESSOR F. C. FRANK Department of Physics, The University, Bristol 2, England

MR. R. D. GIBSON   School of Mathematics, University of Newcastle upon Tyne, England
DR. C. GILBERT   School of Mathematics, University of Newcastle upon Tyne, England
DR. J. J. GILVARRY   The Rand Corporation, 1700 Main Street, Santa Monica, California 90406, U.S.A.
PROFESSOR D. I. GOUGH   Department of Physics, University of Alberta, Edmonton, Alberta, Canada
DR. T. HERCZEG   Hamburg Observatory, 205 Hamburg 80, Cojenbergsweg 112, West Germany
PROFESSOR R. HIDE   Meterological Office, London Road, Bracknell, Berkshire, England
PROFESSOR J. HOSPERS   Geologisch Instituut, Nieuwe Prinsegracht 130, Amsterdam-C, Netherlands
PROFESSOR E. IRVING   Dominion Observatory, Ottawa, Ontario, Canada
DR. C. E. JOHNSON   Solid State Physics 8, Atomic Energy Research Establishment, Harwell, Didcot, Berkshire, England
PROFESSOR P. JORDAN   Universität Hamburg, Hamburg 13, Den Isestrasse 123, West Germany
PROFESSOR E. R. KANASEWICH   Department of Physics, University of Alberta, Edmonton, Alberta, Canada
MR. G. C. P. KING   Department of Geodesy and Geophysics, University of Cambridge, Madingley Rise, Madingley Road, Cambridge, England
DR. H. H. KOLM   Massachusetts Institute of Technology, Francis Bitter National Magnet Laboratory, Cambridge, Massachusetts 02139, U.S.A.
MR. R. C. LIEBERMAN   Lamont Geological Observatory, Palisades, New York 10964, U.S.A.
PROFESSOR G. C. MCVITTIE   University of Illinois Observatory, Urbana, Illinois 61801, U.S.A.
PROFESSOR N. H. MARCH   Department of Physics, The University, Sheffield 10, England
PROFESSOR F. R. N. NABARRO   Metal Physics Group, Cavendish Laboratory, Free School Lane, Cambridge, England
DR. E. R. NIBLETT   Division of Geomagnetism, Dominion Observatory, Ottawa, Ontario, Canada
DR. N. D. OPDYKE   Lamont Geological Observatory, Columbia University, Palisades, New York 10964, U.S.A.
DR. W. O'REILLY   Department of Geophysics and Planetary Physics, School of Physics, University of Newcastle upon Tyne, England
MR. C. F. PETERSEN   Stanford Research Institute, Menlo Park, California 94025, U.S.A.
DR. R. A. PHINNEY   Department of Geological and Geophysical Sciences, Princeton University, Princeton, New Jersey 08540, U.S.A.
DR. P. B. PRICE   General Electric Research and Development Center, Building K-1, Schenectady, New York 12301, U.S.A.
DR. J. S. REITZEL   Southwest Center for Advanced Studies, Dallas, Texas, U.S.A.
MR. G. O. ROBERTS   School of Mathematics, University of Newcastle upon Tyne, England
PROFESSOR P. H. ROBERTS   School of Mathematics, University of Newcastle upon Tyne, England

PROFESSOR L. ROSENFELD   Nordita, Blegdasvej 17, Copenhagen, Denmark

PROFESSOR I. W. ROXBURGH   Department of Mathematics, Queen Mary College, Mile End Road, London E.1, England

PROFESSOR S. K. RUNCORN   Department of Geophysics and Planetary Physics, School of Physics, University of Newcastle upon Tyne, England

MR. C. G. SAMMIS   Seismological Laboratory, California Institute of Technology, 220 N. San Rafael Avenue, Pasadena, California 91105, U.S.A.

PROFESSOR J. C. SAVAGE   Department of Physics, University of Toronto, Toronto, Ontario, Canada

MR. S. SCOTT   Department of Mathematics, Rutherford College of Technology, Newcastle upon Tyne, England

DR. U. SCHMUCKER, 34 Göttingen, Herzberger Landstrasse 42, West Germany

DR. T. J. SHANKLAND   Department of Geological Sciences, Harvard University, Cambridge, Massachusetts 02138, U.S.A.

PROFESSOR K. STEWARTSON   Department of Applied Mathematics, University College, Gower Street, London W.C.1, England

DR. R. G. J. STRENS   School of Physics, University of Newcastle upon Tyne, England

MR. G. SUFFOLK   Department of Mathematics, King's College, The Strand, London W.C.2, England

PROFESSOR and MADAME H. TERMIER   Université de Paris, Laboratoire de Géologie, Générale à la Sorbonne, 1 Rue Victor-Cousin, Paris V^e., France

MR. J. G. TOUGH   School of Mathematics, University of Newcastle upon Tyne, England

PROFESSOR R. M. WALKER   McDonnell Professor of Physics, Washington University, St. Louis 30, Missouri, U.S.A.

DR. K. WHITHAM   Division of Geomagnetism, Dominion Observatory, Ottawa, Ontario, Canada

MR. V. R. WILMARTH   National Aeronautics and Space Administration, FOB6, Room 52034, 400 MD Avenue S.W., Washington D.C., U.S.A.

Preface

The subject of geophysics has in the past largely been based on the application of classical physics. Seismology and other studies of the mechanics of the Earth have derived from classical elasticity, the Earth's gravitational and magnetic fields have been studied by the application of potential theory, and the theory of the geomagnetic field and its variations has been based upon electromagnetism. It has been by the application of these three branches of physics that our knowledge of the Earth's interior has been so greatly developed in the last 60 years. The next important advance will undoubtedly be in the application of the ideas of modern physics to the study of geophysics and during the last 10 years a beginning has been made. The theory of semiconduction has been used to give a satisfactory explanation of the distribution of electrical conductivity in the Earth's mantle. The modern theory of solids has also been used in attempts to predict the laws of the mechanical behaviour of the Earth's mantle over long periods of time, especially the relation of creep with continental drift and convection currents in the Earth's mantle, for which there is now strong evidence from palaeomagnetism and satellite gravity observations respectively. The thermal conductivity of the mantle and its elastic constants are also being interpreted by simple models of solids based on quantum theory and this work has been given great impetus by the development of experimentation at high pressures. New methods of dating the Earth and its rocks have been developed recently in nuclear physics laboratories. Moreover, cosmological theories involving changes in the gravitational constant have been applied to the Earth and may have relevance in view of the fact that the Earth's age is about half that of the Universe.

Again it is a pleasure to express my thanks to the Organizing Committee and other members of the staff of the School of Physics for their help in making this NATO Advanced Study Institute possible. The Vice-Chancellor of the University of Newcastle upon Tyne, Dr. C. I. C. Bosanquet, and other members of the University, and of our sister University at Durham, made our visitors most welcome.

Professors K. M. Creer and P. H. Roberts contributed greatly to the planning of the programmes of sections A and E respectively.

Much of the work in preparing this Proceedings has been borne by Mrs. Joyce Roberts to whom much of any success attributed to this exchange of our information and theories is due.

S. K. RUNCORN

Contents

Preface S. K. Runcorn

Celestial and terrestrial physics in historical perspective 1
L. Rosenfeld

A. Cosmology and Geophysics

I Theoretical considerations concerning a change in G

1. Considerations from cosmology and galactic structure which suggest a secular change in G 9
C. Gilbert
2. A cosmologist looks at the hypothesis of a change in the gravitational constant 19
G. C. McVittie
3. The oblateness of the sun 29
I. W. Roxburgh

II Geophysical methods of detecting a change in G

1. Dirac's principle of cosmology and radioactive dating 35
E. R. Kanasewich and J. C. Savage
2. A palaeontological method of testing the hypothesis of a varying gravitational constant 47
S. K. Runcorn

III Theoretical aspects of the hypothesis of Earth Expansion

1. On the possibility of avoiding Ramsey's hypothesis in formulating a theory of Earth Expansion 55
P. Jordan
2. The slow expansion hypothesis 65
L. Egyed
3. Energy changes in an expanding Earth 77
A. E. Beck

IV Geological and geophysical evidence related to the hypothesis of Earth expansion

1. Global palaeogeography and Earth expansion 87
H. Termier and G. Termier
2. Crustal tectonic evidence for the Earth expansion 103
R. Dearnley
3. Outline of the palaeomagnetic method of determining ancient Earth radii 111
E. Irving

4. New determinations of ancient Earth radii from palaeomagnetic data . 113
      S. I. Van Andel and J. Hospers

   B. SOLID STATE PHYSICS AND GEOPHYSICS

I New developments in the study of the electrical conductivity of the Earth's mantle
   1. Conductivity anomalies with special reference to the Andes . . . 125
      U. Schmucker
   2. Magnetic deep sounding and local conductivity anomalies . . . 139
      D. I. Gough and J. S. Reitzel
   3. Electrical conductivity anomalies in the mantle and crust in Canada . 155
      E. R. Niblett, K. Witham and B. Caner

II Laboratory measurements
   1. Transport properties of olivines. 175
      T. J. Shankland
   2. Optical absorption in silicates 191
      R. G. Burns
   3. The nature and geophysical importance of spin pairing in minerals of Ion (II) 213
      R. G. J. Strens

III Theory of creep applied to the Earth's mantle
   1. Convection and stress propagation in the upper mantle . . . 223
      W. M. Elsasser
   2. Diamonds and deep fluids in the upper mantle 247
      F. C. Frank
   3. Steady-state diffusional creep 251
      F. R. Nabarro
   4. The diffusion of boundary disturbances through a non-Newtonian mantle 253
      C. A. Berg
   5. The formation of oceanic trenches and the mechanism of permanent deformation in the mantle 273
      C. A. Berg

   C. HIGH-PRESSURE PHYSICS AND THE EARTH'S INTERIOR

I Chemical composition of the mantle
   1. Seismic and related evidence on compressibility in the Earth . . . 287
      K. E. Bullen
   2. Composition and state of the Earth's interior 299
      F. Birch
   3. Cosmogomical aspects in the theory of planetary interiors . . . 301
      T. Herczeg

Contents

II Equations of state

1. Equations of state at high-pressure from the Thomas–Fermi atom model . 313
   J. J. Gilvarry
2. High-pressure behaviour of solids 405
   N. H. March

III Laboratory data

1. Elastic constants of oxide compounds used to estimate the properties of the Earth's interior 425
   O. L. Anderson and R. C. Lieberman
2. Shock wave data and the study of the Earth 449
   T. J. Ahrens and C. F. Petersen

D. Developments in Techniques

I Application of techniques of modern physics to rock magnetism

1. Brillouin scattering — a new geophysical tool 465
   D. L. Anderson, C. G. Sammis and R. A. Phinney
2. Application of neutron diffraction and Mössbauer studies to rock magnetism 479
   W. O'Reilly
3. Mössbauer effect studies of magnetic minerals 485
   C. E. Johnson

II New applications of atomic and nuclear physics to geophysical measurements

1. Fission track dating and processes in the Earth's interior . . . 499
   R. L. Fleischer P. B. Price and R. M. Walker
2. Gas lasers and the measurement of long period and secular changes . 505
   A. H. Cook
3. A laser seismometer 513
   G. C. P. King and D. Davies
4. Some implications for physics and geophysics of laser range measurements from Earth to a lunar retro-reflector 523
   C. O. Alley, P. L. Bender, D. G. Currie, R. H. Dicke and J. E. Faller

III Space Technology

1. Apollo lunar sample analysis program — a progress report . . . 533
   V. R. Wilmarth

E. Magnetohydrodynamics

I Geomagnetic reversals

1. The Jaramillo event as detected in oceanic cores 549
   N. D. Opdyke

II The dynamo problem

1. The Braginskii dynamo 555
   J. G. Tough and R. D. Gibson
2. The Herzenberg dynamo 571
   R. D. Gibson
3. The Bullard-Gellman dynamo 577
   R. D. Gibson and P. H. Roberts
4. Dynamo waves 603
   G. O. Roberts
5. A class of solutions of the magnetohydrodynamic dynamo problem . 629
   S. Childress

III Aspects of the geomagnetic secular variation

1. On the dynamics of the Earth's deep interior 651
   R. Hide
2. Planetary magnetohydrodynamic waves as a perturbation of dynamo solutions 653
   G. Suffolk and D. W. Allan
3. Second-class magneto-Rossby waves 657
   K. Stewartson

F. THE EARTH AS A FUNDAMENTAL PHYSICS LABORATORY

1. Search for magnetic monopoles 661
   H. H. Kolm

Author Index 673

Subject Index 683

1

L. ROSENFELD

Nordita
Copenhagen, Denmark

Celestial and terrestrial physics in historical perspective

Ingenious, but inconclusive, arguments about the hardness of the Moon's surface have recently been settled in a very simple way: by scraping it. It is curious that this experiment has been accepted as the most natural thing to do, not only by the small community of scientists, but by the public at large. It has been hailed as a technological feat, but its whole conceptual background has just been taken for granted. Nobody doubts that the Moon is made of the same material as the Earth and that it is just a question of finding out how soft its soil is, much as one would do on the Earth. This sounds trivial enough, but think how revealing it is for the change of attitude undergone by, at any rate, a large part of mankind in the short span of two centuries under the impact of technological progress. I say 'technological' and not 'scientific', since the true spirit of science, which is of course the prime mover, has not yet taken hold of the minds of men; this will be, perhaps, the next stage on the path to happiness. Even in the late 18th century, while philosophers were discoursing about the 'progress of enlightenment', the peasants of western Europe held beliefs about the celestial phenomena hardly different from those entertained by the peasants of Egypt and Sumer four thousand years earlier.

In those remote times, when nature appeared to men in all its unanalysed complexity, there was no reason to assume any similarity between the celestial and terrestrial phenomena; on the contrary, it was natural to ascribe to them, as to all others, a quality of their own. In fact, they were just regarded as a part of a terrestrial environment that was identified with the Universe. Like all the forces of nature that escaped human control, they were popularly regarded as manifestations of divine powers. On this narrow scene, the first stage of sophistication of which we have documented evidence is a succession of imaginative attempts to reduce the variety of appearances to some underlying simplicity. In such an endeavour it is pretty obvious to associate the Sun, the Moon, and the stars with the fires that burn on the Earth, and even to regard the fire emanating from them as more perfect than the earthly ones, which are always mixed with other elements. But how is it, now, that this celestial fire appears to us in the shape of Sun, Moon, and stars and that these shapes move about as they do?

The answers given to these questions by the ancient cosmogonies illustrate the formidable difficulty of drawing analogies from terrestrial experience. The idea of bodies moving freely through space was the last to occur. Indeed, no arrow or stone hurled by whatever kind of earthly force could be imagined to move on in such regular and never-ending revolutions. It was easier to give a material form to the vault of the heavens and to ascribe to it a continual rotation. In one early cosmogony, the Sun, the Moon, and the stars were interpreted as holes in a revolving spherical screen, through which one had glimpses of the celestial fire burning behind. In others, the stars were thought to be hanging from the heavenly vault like lamps. Eventually, the necessity for giving a more accurate account of the celestial motions—of 'saving the phenomena'—led to a multiplication of the spheres carrying the Sun, Moon, planets, and stars and to an increasing complication of their motions; but the idea that consisted in interpreting these fiery appearances as bodies moving through space without support was truly revolutionary.

Firstly, it was one of those moves towards the unification of apparently heterogeneous phenomena, which are so decisive for the development of a scientific view of the Universe. Moreover, it was a step towards the substitution of a rational explanation of the celestial phenomena for the mystical features attached to them by popular belief. As such, it was a threat to the stability of the social institutions which were sanctified by being related to the heavenly powers. For expressing the view that the Moon and the other celestial bodies were made of the same kind of matter as the Earth, Anaxagoras was accused of impiety by the Athenians: such an offence to the divinities of the heavens could induce them to withdraw the protective influence they exerted on the affairs of the city. This involuntary opposition to popular religion remained for a long time a formidable obstacle to the progress of scientific cosmology—the rational attitude of a handful of philosophers was no match for the social forces bent on maintaining a divine sanction for institutions which it was deemed too hazardous to abandon to the will of men. Whenever these social inhibitions were weakened, as happened in hellenistic times, and later during the Renaissance in Western Europe, a more tolerant climate prevailed in scientific pursuits. From the former period, we obtain a glimpse of the degree of sophistication attained by the leisured class of society through Plutarch's curious dialogue *On the Face of the Moon*. Here, all the conflicting concepts, including that of the identical nature of the Moon and Earth, are freely discussed, but significantly the question is left undecided: debating it was a pleasant exercise for the dilettanti, but there was no possibility of either testing the ideas or drawing from them consequences of any use.

Likewise, the problem of 'saving the phenomena' by perfecting the machinery of rotating spheres was bogged down in sterile formalism, completely divorced as it was from the mechanics of terrestrial bodies. The latter was itself in too rudimentary a state to be of any help even if this help had been sought, but the rift was deeper: the celestial motions were regarded as radically different from those of terrestrial bodies, and the idea of any relation between the two was beyond the conceptual horizon. This was as true of Copernicus as of the scholastic philosophers. There was indeed a Copernican revolution, but it was not made by Copernicus: it came much later. Copernicus wanted to simplify the Ptolemaic system, and he saw that one did not need so many spheres if they were arranged around the Sun rather than around the Earth, and if the Earth rotated around its axis; as a humanist, he referred these purely kinematic ideas to their source in the Pythagorean tradition, but he could, no more than Plutarch, produce compelling arguments for regarding the Earth's mobility as a dynamical reality. Indeed, by proposing the alternative system in which only the Sun revolved around the Earth, while all the planets revolved in the Copernican way around the Sun,

Tycho Brahe made it clear that the Copernican reform did not imply the assimilation of the Earth to a planet. Tycho is a good example of those clever people who apply their ingenuity to preventing revolutions.

The issue had to be faced, however, and it was Tycho's own disciple who, by his remarkably unprejudiced approach to the problem of planetary motions, initiated the radical change of outlook that gave birth to modern science. By putting his trust in the accuracy of Tycho's observations, which showed that the heliocentric motion of the planet Mars could not be circular, Kepler brushed away the whole concept of the celestial spheres and opened up the heavens. The concept of the planets as bodies moving along definite paths around the Sun acquired the character of a law of nature directly derived from observation, and by the same token the whole Copernican system could no longer be denied substantial reality. The ultimate consequences of the Copernican view had received a poetic formulation in Giordano Bruno's vision of an infinite Universe with innumerable suns and planetary systems—a vision so destructive to the very foundations of social order that its prophet had to pay for his imprudence with his life. But how much more forceful were the rigorous arguments by which scientists like Kepler or Otto von Guericke (who allied an equally powerful imagination with a severely critical judgment) established these consequences as facts of observation! Of special importance to our theme is von Guericke's contribution; his concept of the heavenly expanse as an unlimited empty space in which suns and planets were moving was directly inspired by his experiments on the production of such limited empty volumes on Earth. The complete identification of the Earth and the other heavenly bodies is illustrated by Kepler's essay *Somnium*— a 'dream' in which he imagines himself on the Moon and goes on to describe the lunar landscape and the appearance of the heavens from this extra-terrestrial vantage point, exactly as he would have done for some remote region of the surface of the Earth.

In the immensely widened picture of the Universe as it emerged from Kepler's life-work, one essential element was still missing. Kepler realized that the planetary motions must be ascribed to some influence emanating from the Sun, but he did not know how to attack this dynamical problem, because no sound method was yet available for the analysis of the motion of ordinary terrestrial bodies. It is just on this last point that Galileo concentrated his efforts, and, although he did not arrive at a complete solution, he laid the foundations of the dynamics which eventually enabled Newton to find the answer to Kepler's question. It is an irony of history that Galileo himself was unaware of the problem of celestial mechanics to the solution of which his own work contributed. He completely ignored Kepler's laws of planetary motion (if indeed he ever knew of them—even Newton did not appreciate Kepler's law of areas until he found that it was a consequence of the general laws of motion). On the contrary, he clung to the Platonic idea of the perfection of circular motion, and it was circular motion, and not motion in a straight line, that he regarded as inertial (as we would call it); when he applied his principle of inertia to the rectilinear component of the motion of projectiles on Earth, it was as an approximation to the actual circular displacement along the surface of the Earth. Kepler was also engrossed in Platonic philosophy, but the strict demand he imposed on his theories of accurately accounting for the observations led him gradually to abandon it. Galileo had no need of accurate experiments to establish the law of falling bodies on Earth, because he took it for granted that it should be some simple law of proportionality, and rough measurements sufficed to guide him to the right formulation. In his astronomical work, Galileo's intention was quite different from Kepler's: he looked for physical arguments in support of the reality of the Copernican system, and in his quest he proceeded with a surer intuition of physical phenomena (though not without some blundering) than the more mathematically minded Kepler. The latter was a better logician than the sanguine Italian, but it was Galileo's brilliant dialectics that won the field for the Copernican system. Kepler was better than Galileo at understanding optics, but it was Galileo's practical sense that prompted him to take the telescope from the hands of the craftsmen and make of it a working tool for disclosing the physical aspects of the celestial bodies, and thus definitely clinch the argument for the identity of matter in the whole Universe.

For Newton this identity—which he called 'the analogy of nature'—became a heuristic principle of paramount importance. As an instance of the guidance he always sought in this 'analogy', I shall just mention his refutation of a hypothesis proposed by Flamsteed in his correspondence with Newton about the comet of 1680. Flamsteed had correctly interpreted the observations as showing that the comet, first seen moving towards the Sun and later seen receding from it, had been deflected in its neighbourhood (where it was not visible); he went on to suggest that this deflection might be due to magnetic forces exerted by the Sun on the comet, the magnetic axis of which he naïvely imagined would remain parallel to the trajectory, so that the comet, after an initial deflection on entering the vortex of the Sun, would gradually turn towards the Sun the pole opposite to that initially facing it. Newton did not only point out this last mistake in Flamsteed's argument; he observed that the Sun could not possibly exert magnetic forces, because it was too hot: for we know from experience with magnets on Earth that they lose their magnetism on being heated.

Newton's discovery of the law of universal gravitation made such an impression on the scientific community that the celestial mechanics he founded upon this law became the model for a purely dynamical concept of the whole physical Universe: all the phenomena (including those of light and heat) would in principle be explained by the motions of material particles interacting by central forces. This ambitious programme, which received its most comprehensive elaboration in Laplace's great work, had an extraordinarily stimulating influence on the development of our modern theoretical physics. Taking their inspiration from the problems of celestial mechanics, Laplace and his successors—Gauss in Germany, George Green, William Thomson, and Maxwell in Britain—created the refined mathematical methods which found application not only in terrestrial mechanics, but in optics, electricity, and magnetism as well.

A parallel line of development that led to momentous results was the study of radiation from both stellar and terrestrial sources, initiated in the 18th century by Bouguer and Lambert. Its culmination was Kirchhoff's creation of the method of spectral analysis, which provided an additional proof of the universal identity of matter, this time from the point of view of its chemical character. In this domain we have again numerous examples of fruitful influences in both directions, one of the most famous being the spectroscopic discovery, in the solar chromosphere, of a new element, helium, which was then also detected in terrestrial locations.

As the recognition of the uniformity of the material constitution of the Universe thus, step by step, was taking shape, speculation never ceased about the fascinating problems of cosmogony, which likewise passed from the mythical to the scientific level. There is indeed no difference of principle between the study of the constitution of the Universe and that of its formation: in the former we are testing theoretical predictions about the results of experiments; in the latter, we are working backwards from the present evidence, formulating 'retrodictions' instead of predictions. The real difficulty here, apart from the vastness of the object of study, is in fact our impatience to reach definite conclusions: cosmogony may just be retrodiction, but it is mostly retrodiction on insufficient evidence. Hence its chequered history reflects a steady increase in our basic knowledge and a succession of ephemeral inferences rashly drawn from it. Still, from Thomas Wright's first vision of a Universe of separate galaxies—a static vision that inspired Kant's dynamical hypothesis about the formation of galaxies and planetary systems by condensation of nebular matter—to our present sophisticated cosmogonies, a broadening of the World picture and a deepening of its foundations has taken place that completely dwarfs the Copernican revolution.

One might say that this revolution was extended to the time dimension, and the obstacle of the traditional myths with their hopelessly inadequate time scale proved as difficult to overcome as the geocentric prejudice. In this respect, history tells of a continual struggle to push back the origins beyond the artificial barriers raised by timid imagination. In Newton's time, cosmogony was still discussed in terms of the biblical account. Burnet's *Theory of the Earth* is an attempt to interpret this account as a symbolical description of actual events; in his correspondence with Burnet, Newton himself tries to defend a more literal interpretation of it, as giving a description of the process as it would have appeared to a human observer. Only the demand from expanding mining activity for better methods of prospection freed geological science from the straitjacket of biblical mythology, and the same fight had to be fought again on the question of the antiquity of man and of organic evolution. As regards the time scale of celestial phenomena, it offers a still greater challenge to imagination, and there have been quite a few hurdles raised by the physicists themselves. Let us only recall Lord Kelvin's stubborn opposition to the estimates of the age of the Earth arrived at by the geologists, on the ground that they conflicted with his own estimate, based on the theory that the Sun's heat originated from gravitational contraction—a theory regarded as being beyond doubt.

After putting such emphasis, in the course of this survey of the relations between celestial and terrestrial physics, on the realization of the essential unity of constitution of the Universe, I shall conclude by raising a question which in a sense could suggest a limitation of this general conclusion. It concerns the force of gravitation itself, which was historically the first example of a universal agency, and which indeed dominates all phenomena on the cosmic scale. The temptation is great to treat gravitation on the same basis as the other universal forces of nature—the trinity of strong, electromagnetic, and weak interactions, whose domain of action is primarily that of the phenomena on the atomic and nuclear scale. Is this assimilation justified, however? In particular, is the gravitation field quantized like the others? By its identification with the metrical field, gravitation is ascribed a unique function in the description of nature, a function which it does not seem easy to reconcile with the quantal features of the field. Even in its dynamical action, gravitation differs from other forces in one important respect (Professor Prigogine recently drew my attention to this): a system of particles interacting only by gravitation does not show any tendency towards a state of equilibrium; being of infinite range and always attractive, the force of gravitation does not lead to a statistical distribution of the energy among the particles within any finite average time. Thus, purely gravitating systems are outside the range of application of the second law of thermodynamics—a circumstance well worth remembering in cosmological studies.

A

COSMOLOGY AND GEOPHYSICS

I. Theoretical considerations concerning a change in G

1

C. GILBERT

School of Mathematics
University of Newcastle upon Tyne
England

Considerations from cosmology and galactic structure which suggest a secular change in G

Introduction

The ratio of the electric force to the gravitational force between a proton and an electron is a number of the order of magnitude 10^{40}. Eddington long ago proposed that this number was the square root of the number of protons and electrons in the Universe[1,2]. Since then there has been continued speculation about the reason for the occurrence of this very large number in nature, but it has usually been along different lines. It is now known that there are other combinations of the physical and astrophysical constants which yield dimensionless numbers of this order of magnitude[3-5]. Dirac assumed that these coincidences do not occur purely by chance and stated, as a fundamental principle, 'any two of the very large dimensionless numbers occurring in nature are connected by a simple mathematical relation in which the coefficients are of the order of magnitude unity'. Since the age of the Universe T, measured in terms of an atomic unit of time, gives a number of order of magnitude 10^{40}, Dirac's principle led to the result that the gravitational constant G was in fact not a constant, but varied inversely as T. By a similar argument he deduced that Hubble's constant H and the mean density of matter in the Universe were quantities which varied as T^{-1}. In the theories of Jordan[6] and Brans and Dicke[7], G is also a quantity which varies with the epoch, although these theories are not so specific about the form of the dependence on T.

The general theory of relativity is the most satisfactory theory of gravitation which has so far been formulated. It is a dualistic theory in the sense that the fields and the sources of the fields are different entities. The mass particles which are the sources of the gravitational field are not, in general, part of the field and are defined independently of the field. The field equations thus contain an energy tensor T_{ij} which accounts for the stress, energy, and momentum of the sources of the field. There is, however, an important exception. Many results of the theory are not dependent on the use of an energy tensor, and then the general theory of relativity has the nature of a unitary theory, with the sources of the field described by the singularities of the field. The Schwarzschild field of a mass particle is a solution of the field equations in the latter form, and thus the experimental tests of the theory of relativity, which are based on this solution, do not serve as a test of the 'energy-tensor form' of the field equations. The experimental tests are essentially tests of (i) the vacuum field equations, and (ii) the hypothesis concerning the propagation of light. The 'energy-tensor form' of the field equations must be used to determine the solutions of relativistic cosmology, with T_{ij} describing the stress, energy, and momentum of a continuous distribution of matter. Since one of Dirac's relations concerns the value of the mean density of matter in the Universe, any attempt to derive the relations from the general theory of relativity is dependent on using this form of the field equations. The 'energy-tensor form' of the field equations, however, contains a constant κ which can be shown to have the value $8\pi G/c^4$, which implies that G is a constant. This points strongly to the need to reformulate the field equations of the general theory of relativity if the results deduced from Dirac's principle are to be within the scope of that theory. This, however, is not sufficient because it is necessary also to modify the principle of equivalence. The latter principle has been discussed fully by Dicke[5] in this connection, and need not be considered here. It is sufficient to note that a theory which can account for varying G must have an essentially different structure from the general theory of relativity.

The theory of Jordan[6] is based on projective relativity. Conformal transformations of the metric are permitted, and thus a scalar function, directly related to the gravitational power of matter, enters into the field equations in addition to the usual metric tensor components and their first and second derivatives. The theory can account for varying G in accordance with Dirac's hypothesis, but leads to non-conservation of mass. The theory of Brans and Dicke[7] is based on a weak form of the principle of equivalence and is related to the theory of Jordan, but has some differences, notably that mass is conserved. Although these theories permit a varying G, the form of the variation with T does not necessarily approximate to that derived from Dirac's hypothesis. The theory of Brans and Dicke is in fact more intimately concerned with an attempt to construct a theory which incorporates Mach's principle. According to this principle the inertial properties of matter are determined by the distant masses in the Universe. Einstein hoped, when he formulated the general theory of relativity, that it would be found to incorporate this principle[8]. It has since been found that Mach's principle has only a limited expression in the general theory of relativity[7,9]. This was also indicated by the work of Einstein and Strauss[10], who showed that a local gravitating system would not appear to be affected to any appreciable extent by the distant matter of an expanding Universe. A diminishing value of G in an expanding Universe would be in accordance with the principle that the inertial masses in a local system depend on the distribution of matter in the Universe.

A conformal theory of gravitation has recently been proposed by Thomas[11] in order to account for certain discretization effects found in the structure of galaxies by Wilson[12]. An example of this type of effect is that the mass of a galaxy can assume only discrete values for a given size. Although not directly connected with the problem of accounting for a possible variation of G over the ages, this is quoted as an indication of the way in which it may be necessary to modify the general theory of relativity in order that it shall be capable of accounting for certain observed phenomena.

In the next section the equations of Newtonian cosmology are derived and on pp. 13–14 some of the results obtained from relativistic cosmology are stated. Although the equations of relativistic cosmology are analogous to those of Newtonian cosmology, the analogy only exists in general relativity in the local coordinate systems of observers.

On pp. 14–17 an outline is given of a relativistic theory for a cosmology having equations analogous to those of Newtonian cosmology, but with the standard measure of length based on the unit of the spatial coordinates, when the usual 'cosmic coordinates' are used. In this theory the value of G varies with the time.

Newtonian Cosmology

It was shown by Milne[13] and Milne and McCrea[14] that, with a suitable interpretation of Newtonian concepts, it is possible to describe Newtonian Universes which are in most respects completely equivalent to those of relativistic cosmology. There is, however, a difference of interpretation, which has been aptly described by Milne[13] as follows: "The phenomenon of the expansion of the Universe has usually been discussed by students of relativity by means of the concept of 'expanding space'. This concept though mathematically significant has by itself no physical content; it is merely the choice of a particular mathematical apparatus for describing and analysing phenomena. An alternative procedure is to choose a static space, as in ordinary physics, and analyse the expansion phenomena as actual motion in this space. Moving particles in a static space will give the same observable phenómena as stationary particles in expanding space." The mathematical basis for the last statement is given on pp. 14–17. In this section the equations of Newtonian cosmology are derived in a manner closely resembling the account by Bondi[15].

We assume that ordinary Euclidean space is filled with a medium of uniform density at any time, which we shall call the substratum. Any observer of the system is assumed to be moving with the medium in his immediate neighbourhood, and to measure coordinates of other elements of the medium relative to a Euclidean frame in which he is at rest. The latter Euclidean frame, rather than one which is in uniform motion relative to absolute space, will be said to define a Newtonian inertial frame. This modification of Newtonian theory is necessary in order to be able to satisfy the cosmological principle, which may be stated in the form that to any observer making measurements in his own inertial frame the system has the same appearance as it has to any other observer, provided they carry out their observations at the same time. It is necessary to make two further assumptions in order to be able to satisfy the cosmological principle: (i) the medium extends to infinity in all directions, and (ii) the gravitational intensity in the inertial frame of any observer O at a point P, distant ρ from O, is equal to the field intensity of the material contained in a sphere of radius ρ surrounding O. The implication of (ii) is that the distribution of matter at infinity is obtained as the limiting case of a spherical distribution, centred on O, as the radius tends to infinity.

Let O' be another observer and let the vectors $\vec{OO'}$, $\vec{OP}$, $\vec{O'P}$ be represented by ρ_0, ρ, ρ' respectively. Also let us denote the velocity of O' relative to O, of P relative to O, and of P relative to O' by $V(\rho_0)$, $V(\rho)$, and $V'(\rho')$ respectively. From the cosmological principle $V'(\rho')$ must be the same function of ρ' as $V(\rho)$ is of ρ. Hence

$$\rho = \rho' + \rho_0 \tag{1}$$
$$V(\rho) = V'(\rho') + V(\rho_0) \tag{2}$$
$$V'(\rho') = V(\rho') \tag{3}$$

From (1), (2), and (3)

$$\mathbf{V}(\rho-\rho_0) = \mathbf{V}(\rho) - \mathbf{V}(\rho_0) \tag{4}$$

It follows from (4) that $\mathbf{V}(\rho)$ is a linear function and can be written

$$\mathbf{V}(\rho) = F(t)\rho \tag{5}$$

where $F(t)$ is an arbitrary function of Newtonian time t, which is the same function for all observers. If $\mathbf{P}(\rho_0)$, $\mathbf{P}(\rho)$, $\mathbf{P}'(\rho')$ are the gravitational intensities at the points O' in O's frame, P in O's frame, and P in O'''s frame respectively, we find by a similar argument that

$$\mathbf{P}(\rho) = \phi(t)\rho \tag{6}$$

where $\phi(t)$ is an arbitrary function which is the same for all observers. Also, since density and pressure of the medium are scalars having the same value for all observers at any time, they must be constants, or functions of t only. Let $\omega(t)$ be the density and $p(t)$ the pressure at time t.

In the inertial frame of O the Newtonian equations of motion give

$$\frac{\partial}{\partial t}\omega(t) + \mathrm{div}\,\{\omega(t)\mathbf{V}(\rho)\} = 0 \tag{7}$$

$$\frac{D\mathbf{V}(\rho)}{Dt} + \frac{1}{\omega(t)}\,\mathrm{grad}\,p(t) = \mathbf{P}(\rho) \tag{8}$$

and, since the second term in equation (8) is zero, the motion is independent of the pressure. Defining a function $R(t) = \exp(\int F(t)\,dt)$, we find from equations (5) and (7)

$$\rho = R(t)\mathbf{r} \tag{9}$$

$$\mathbf{V} = \frac{\dot{R}(t)}{R(t)}\rho \tag{10}$$

$$\omega(t) = \frac{A^2}{R^3(t)} \tag{11}$$

where A is a constant of integration, the dot denotes differentiation with respect to t, and $\mathbf{r}$ is a constant vector, which is different for each element of the fluid. Also from equations (6), (8), and (10)

$$\ddot{R}(t) = \phi(t)R(t) \tag{12}$$

The gravitational intensity $\mathbf{P}(\rho)$ is of the form (6), and may be expressed as the sum of a term representing the resultant intensity coming from the mutual gravitational attractions of the elements of the fluid, and a cosmological term. Thus

$$\mathbf{P}(\rho) = -\frac{G(t)M\rho}{\rho^3} + \frac{\Lambda(t)c^2}{3}\rho \tag{13}$$

where M is the mass of material contained in a sphere of radius ρ surrounding O.

The usual gravitational constant G and cosmological constant Λ are here replaced, by functions of the time, which is permissible, since then the law of force (13) still has the form (6) required by the cosmological principle. We find in fact from (6), (11), (12), and (13) that

$$\phi(t) = -\frac{4\pi G(t)A^2}{3R^3(t)} + \frac{\Lambda(t)c^2}{3} \tag{14}$$

which gives, by (12), the differential equation for $R(t)$

$$\frac{\ddot{R}(t)}{R(t)} = -\frac{4\pi G(t)A^2}{3R^3(t)} + \frac{\Lambda(t)c^2}{3} \tag{15}$$

The solutions of equation (15) determine infinitely many Newtonian Universes compatible with the cosmological principle, corresponding to different choices of the functions $G(t)$, $\Lambda(t)$.

Relativistic Cosmology

Although the concepts of general relativity require the use of considerably more specialized mathematical techniques than those used in Newtonian theory, the Universes of relativistic cosmology were in fact discovered before those of Newtonian cosmology. The connection with Newtonian theory was first given by Milne[13] and Milne and McCrea[14], who showed in effect that the equations of Newtonian theory, (9), (10), (11), and (15), with $G(t)$ and $\Lambda(t)$ replaced by constants, are the same as those defining relativistic Universes in which the pressure of matter is zero. In this section some of the concepts of general relativity will be briefly discussed. The reader is referred to the standard work of Tolman[16] and to the excellent review of relativistic cosmology by Robertson[17] for further details.

According to the general theory of relativity the space–time of our experience is a four-dimensional manifold in which an invariant interval between neighbouring events is defined by an indefinite Riemannian metric. Introducing coordinates x^i, $i = 1, 2, 3, 4$, with x^0 a time coordinate and x^α, $\alpha = 1, 2, 3$, spatial coordinates, the measure of interval between x^i and $x^i + dx^i$ is $(\varepsilon\, ds^2)^{1/2}$, where

$$ds^2 = g_{ij}\, dx^i\, dx^j \tag{16}$$

is the space–time metric of signature -2, and ε has the value $+1$ for a time-like, and -1 for a space-like, interval.

The space–time metric depends on the material energy content of the Universe and this dependence finds its expression in Einstein's field equations

$$-G_{ij} - \Lambda g_{ij} = \kappa T_{ij} \tag{17}$$

where G_{ij} is the Einstein tensor, T_{ij} is the stress–energy–momentum tensor of matter and radiation, and Λ, κ are constants. It is a consequence of these equations that a free test particle will describe a path which is a non-null geodesic of (16). The path of a beam of light is assumed to be a null geodesic, for which $ds = 0$.

The Universes of relativistic cosmology are based on the Robertson–Walker line element

$$ds^2 = c^2 dt^2 - R^2(t) du^2 \qquad (18)$$

where $R(t)$ is a function of t only and

$$du^2 = \frac{(dx^1)^2 + (dx^2)^2 + (dx^3)^2}{(1 + \tfrac{1}{4} k r^2)^2} \qquad (19)$$

$$r^2 = (x^1)^2 + (x^2)^2 + (x^3)^2$$

is a space of constant curvature $k = +1, 0,$ or -1. A class of privileged observers exists whose paths are geodesics of (18) and (19) given by $x^\alpha = $ constant ($\alpha = 1, 2, 3$). The cosmological principle is satisfied for these observers who are supposed to move with the material of a uniform distribution of matter forming the substratum of the Universe. The energy tensor of the substratum is that of a perfect fluid of proper density $\omega(t)$ and pressure $p(t)$, having the form

$$T^i_j = \{\omega(t) c^2 + p(t)\} \delta^i_0 \delta^0_j - p(t) \delta^i_j \qquad (20)$$

where $T^i_j = g^{ik} T_{kj}$, g^{ij} being the tensor associated with g_{ij} and defined by the equation $g^{ik} g_{jk} = \delta^i_j$. From (17), (18), (19), and (20), it is found that[18]

$$\kappa \omega(t) c^2 = -\Lambda + \frac{3\{k + \dot{R}^2(t)/c^2\}}{R^2(t)} \qquad (21)$$

$$\kappa p(t) = \Lambda - \frac{2\ddot{R}(t)}{R(t) c^2} - \frac{\{k + \dot{R}^2(t)/c^2\}}{R^2(t)} \qquad (22)$$

In the case $p(t) = 0$, equations (21) and (22) give

$$\omega(t) = \frac{A^2}{R^3(t)} \qquad (23)$$

$$\frac{\ddot{R}(t)}{R(t)} = -\frac{4\pi G A^2}{3 R^3(t)} + \frac{\Lambda c^2}{3} \qquad (24)$$

where A is a constant of integration.

Thus the field equations are exactly equivalent to the equations (11) and (15) of Newtonian theory when $G(t)$, $\Lambda(t)$ have the constant values G, Λ.

A Basis for a Relativistic Theory in which G Varies with the Time

In the general theory of relativity a comparison with Newtonian theory is normally made by choosing a coordinate system such that the metric tensor g_{ij} has the approximate components $(c^2, -1, -1, -1)$ in a restricted region. In the case of cosmology, these coordinates can be defined relative to a tetrad of orthogonal vectors, which is taken along the world-line of an observer by parallel transport. Let us consider a tetrad of vectors e_i at an event on the world-line $x^\alpha = 0$ of an observer in a Universe having metric (18), in the case $k = 0$. Let e_0 be tangent to the world-line and pointing into the future, and let e_α be tangent to the coordinate curves of parameter x^α. (Latin suffixes have values 0, 1, 2, 3 and Greek suffixes have values 1, 2, 3, and a summation is implied for repeated suffixes. Suffixes which are used as labels and have no tensor significance will be enclosed in brackets.) As the tetrad is transported along the world-line, which is a geodesic of space–time, e_0 remains tangent to the path, and the changes of the tetrad components $e_{i(a)}$ for the displacement $dx^k = dt\,\delta_0^k$ are given by

$$de_{i(a)} = \Gamma^j_{i0} e_{j(a)} dt \tag{25}$$

where $\Gamma^i_{jk} = \{^i_{jk}\}$ are Christoffel symbols of the second kind for the metric (18). When the equations (25) are integrated along the world-line, it is found that

$$e_{i(0)} = c\delta_{i(0)}, \qquad e_{i(\alpha)} = R(t)\delta_{i(\alpha)} \tag{26}$$

where the constants of integration have been chosen to make the components c and unity at the event for which $R(t) = 1$. ($\delta_{i(j)}$ has the value $+1$ if $i = j$, and 0 if $i \neq j$.) From the second equation of (26), the vectors e_α have unit *metric* length. Let us define vectors $e_{(\beta)}$ by the equation $e_\alpha = e_{\alpha(\beta)} e_{(\beta)}$. Then any event (t, x^α) can be represented in the observer's Euclidean 3-space by the position vector

$$\rho = x^\alpha e_\alpha = X_{(\alpha)} e_{(\alpha)} \tag{27}$$

where $X_{(\alpha)} = R(t)x^\beta \delta_{\beta(\alpha)}$. The $e_{(\beta)}$ will be regarded as vectors of unit length in terms of a *physical* standard of measure.

The equivalence of Newtonian and relativistic cosmology comes from (9) and (27) on setting $\mathbf{r} = x^\alpha \delta_{\alpha(\beta)} e_{(\beta)}$. No physical significance should be attached to the coordinates x^α, because the vectors e_α do not define an inertial frame. The unit of the $X_{(\alpha)}$ coordinates gives the standard measure for distance both in the 'local field' of a mass particle and in the cosmological field.

Whilst the general theory of relativity allows freedom in the choice of coordinate systems, the standard measures of interval are assumed to have the same fixed values at different events. The idea behind the theory discussed below is to attach greater physical significance to the 'cosmic coordinates' (t, x^α), and to permit changes of the standard measures of interval. According to the theory of Weyl[18], a Riemannian space–time is completely specified by a quadratic form and its calibration ratio. The quadratic form (16) gives the measure of length $(\varepsilon\,ds^2)^{1/2}$, corresponding to coordinate intervals dx^i, and the calibration ratio specifies how the standard measures of length vary from one space–time event to another. The comparison of standards is made by taking a vector from a fixed event P to some other event Q by parallel transport according to (25), or by its generalization for other paths. If the length is l_0 at P and l at Q, the ratio l^2/l_0^2 gives the calibration ratio $e^{2\sigma}$, and the coefficients of connection are given by

$$\Gamma^i_{jk} = \{^i_{jk}\} + \delta^i_j \sigma_{,k} + \delta^i_k \sigma_{,j} - g_{jk} g^{il} \sigma_{,l} \tag{28}$$

the comma denoting partial differentiation. In the case when $\sigma = 0$, the space–time is said to be normally calibrated.

It was shown by the author[19] that results in agreement with the deductions from Dirac's hypothesis could be obtained, on the basis of Weyl's theory, when the cosmological space–time had a calibration ratio which was a function of the time. However, the junction of the local field and the cosmological field had to be made when the fields had different calibration ratios and this was unsatisfactory, as regards both the interpretation of physical standards of measure and the determination of free paths in the fields. In the following investigation it will be assumed that the coordinates are always cosmic coordinates, and it will be shown that the unit of the distance coordinates has the same constant length at different events. The physical standards of length are based on this unit, which will be assumed to be of the order of magnitude of the 'electron radius'. The unit of time will be assumed to be $1/c \times$ unit of length, where c is a constant in the metric (18).

The local field of a mass particle will also be described in cosmic coordinates and continuity of the two fields on a given hypersurface will ensure consistency of measurements of distances. The continuity of geodesic paths is obtained by taking over some results from affine field theory. It has been found in the affine theory[20] that, when there is no skew field, the coefficients of connection

$$\Gamma^i_{jk} = \{^i_{jk}\} + \delta^i_j \sigma_{,k} + \delta^i_k \sigma_{,j} - 3g_{jk} g^{il} \sigma_{,l} \tag{29}$$

are associated with the field equations

$$R_{ij} + 6\sigma_i \sigma_j = \Lambda g_{ij} \tag{30}$$

$$\frac{\partial}{\partial x^i}(\sigma^i \sqrt{-g}) = 0 \tag{31}$$

where R_{ij} is the Ricci tensor formed from the components g_{ij} of the metric tensor and $g = |g_{ij}|$. The difference between (28) and (29) should be noted. It is the factor 3 in (29) which makes it possible to obtain continuity of the geodesic paths at the hypersurface separating the fields. The equations (30) are the generalization in the affine theory of Einstein's free-space field equations, which are given by the special case $\sigma = 0$. (The equations (29), (30), and (31), when $\Lambda = 0$, are given by case (e) is Schrödinger's general survey of affine field theories[20].)

Let us consider the metric (18) for cosmology, and let σ be a function of t only. From the field equations (30) and (31), it can be shown that[19]

$$\frac{\ddot{R}(t)}{R(t)} = -2\frac{A^2}{R^6(t)} + \frac{\Lambda c^2}{3} \tag{32}$$

$$\dot{\sigma} = \pm \frac{A}{R^3(t)} \tag{33}$$

Since the units of the x^α coordinates give the physical standard of length, the equation (32) can be compared with the equations (15) of Newtonian cosmology. We find $\Lambda(t) = \Lambda$ and

$$G(t) = 3/2\pi R^3(t) \tag{34}$$

It will be shown, however, that this is only strictly true in the case when $k = 0$, $\Lambda = 0$, because otherwise the unit of the coordinates is not a constant standard of measure.

Let the tetrad $\mathbf{e}_a$ be carried along the world-line of the observer $x^\alpha = 0$. From equation (25), using the coefficients (29) for the metric (18), it is found that

$$e_{i(0)} = ce^{-\sigma}\delta_{i(0)}, \qquad e_{i(\alpha)} = e^\sigma R(t)\delta_{i(\alpha)} \tag{35}$$

Although we shall continue to regard t as Newtonian time, it is interesting to note that, in terms of the components on the tetrad, $dT = e^{-\sigma}\,dt$, $X_{(\alpha)} = e^\sigma R(t)u_{(\alpha)}$ (where $u_{(\alpha)}$ are the components of distance for an event with coordinates x^α in the subspace (19)), the equations of motion of a particle in the substratum have the form

$$\frac{d^2 X_{(\alpha)}}{dT^2} = \tfrac{1}{3}e^{2\sigma}X_{(\alpha)}\Lambda c^2 \tag{36}$$

Thus the local frames may be considered as being inertial, with a cosmical force proportional to the distance acting at every point.

Case when $k = 0$, $\Lambda = 0$

The equations (32) and (33) have the integrals $R(t) = (3At)^{1/3}$, $\sigma = \pm\tfrac{1}{3}\log t + C$. Let the constants A and C be given the values $\tfrac{1}{3}$ and 0. Then

$$R(t) = t^{1/3}, \quad e^\sigma = t^{1/3} \text{ or } t^{-1/3} \tag{37}$$

Both values of e^σ are consistent with the same metric, but they give different coefficients of connection. Taking $e^\sigma = t^{-1/3}$, it follows from (35) that $e_{i(\alpha)} = \delta_{i(\alpha)}$, showing that the unit of the spatial coordinates can be taken as a constant standard of measure. It is easily seen from (32), (33), and (35) that this is the only case when this result holds good. For the cosmological field take $e^\sigma = t^{1/3}$, then the standard measure of space-like intervals, given by $\{-e^i_{(\alpha)}e_{i(\alpha)}\}^{1/2}$, increases as $t^{1/3}$. From the metric this means that the coordinate interval between the ends of a standard rod is constant. The measure of space-like intervals $(-ds^2)^{1/2}$ is therefore made in terms of the constant physical standard based on the unit of coordinate length.

When each observer has set up a reference frame for Euclidean 3-space using a constant unit of coordinate length, there exists an exact analogy with the system of Newtonian cosmology. The previous investigation of the junction conditions, where the field of a mass particle is joined to the field of the substratum[19], showed that the particles of the substratum had constant masses, and therefore the density of matter must be proportional to t^{-1}. Also from (34) and the first of equations (37), $G(t) = 3/2\pi t$, in agreement with the result deduced from Dirac's hypothesis. On account of the modified form of connection (29), the conclusions reached previously, concerning the path of a free test particle, have to be modified, but it does not seem likely that this will make it necessary to alter the conclusion, that gravitational forces are capable of accounting for the spiral arms in the neighbourhood of a nebular nucleus.

References

1. A. S. Eddington, *The Mathematical Theory of Relativity*, 2nd ed., Cambridge University Press, London, 1930, p. 167.
2. A. S. Eddington, *The Expanding Universe*, Cambridge University Press, London, 1933, pp. 106–19.

3. P. Jordan, *Nature*, **164,** 637 (1949).
4. P. A. M. Dirac, *Proc. Roy. Soc.* (*London*), *Ser. A*, **165,** 199 (1938).
5. R. H. Dicke, *Rev. Mod. Phys.*, **29,** 355 (1957).
6. P. Jordan, *Schwerkraft und Weltall*, Vieweg, Braunschweig, 1955.
7. C. Brans and R. H. Dicke, *Phys. Rev.*, **124,** 925 (1961).
8. A. Einstein, *The Meaning of Relativity*, 4th ed., Methuen, London, 1950.
9. D. W. Sciama, *Monthly Notices Roy. Astron. Soc.*, **113,** 34 (1953).
10. A. Einstein and E. G. Strauss, *Rev. Mod. Phys.*, **17,** 120 (1945).
11. T. Y. Thomas, *Perspectives in Geometry*, Indiana University Press, Bloomington, 1966, pp. 389–412.
12. A. G. Wilson, *Astron. J.*, **68,** 547 (1963).
13. E. A. Milne, *Quart. J. Math.*, **5,** 64 (1934).
14. E. A. Milne and W. H. McCrea, *Quart. J. Math.*, **5,** 73 (1934).
15. H. Bondi, *Cosmology*, Cambridge University Press, London, 1950, pp. 75–89.
16. R. C. Tolman, *Relativity Thermodynamics and Cosmology*, Clarendon Press, Oxford, 1949.
17. H. P. Robertson, *Rev. Mod. Phys.*, **5,** 62 (1933).
18. H. Weyl, *Space–Time–Matter*, 4th ed. (reprint), Dover Publications, New York, 1950.
19. C. Gilbert, *Monthly Notices Roy. Astron. Soc.*, **120,** 367 (1960).
20. E. Schrödinger, *Proc. Roy. Irish Acad.*, *Sect. A*, **51,** 205 (1948).

2

G. C. McVITTIE

University of Illinois Observatory
Urbana, Illinois, U.S.A.

A cosmologist looks at the hypothesis of a change in the gravitational constant

It should be made quite clear at the outset that these remarks are to be regarded as a commentary on certain theories which contain the hypothesis that the 'constant of gravitation' G is in fact a variable. It must be insisted that anyone who wishes to study these theories should read the original papers by the authors of the theories first and only then look at this commentary.

I shall begin with a point of terminology. The expression 'variable constant' of gravitation seems to be a contradiction in terms. The expression 'scalar component' of gravitation has, for me, a misleading connotation that a vector with components is involved. As far as I can judge, the intention is to assert that $G = G_0 F$, where F is a scalar function of position and time. Its value is to be unity 'here' and 'now' and F is a dimensionless function. The name that I shall give to F is the 'gravitation function'. The constant of gravitation is G_0 and it has the usual physical dimensions.

Quasi-Newtonian Gravitational Theory

The time is an absolute invariant in Newtonian mechanics. Therefore it is meaningful to assume that F is a function of absolute Newtonian time T alone. But, since F is to be dimensionless, it is necessary to invoke the existence of a fixed time interval T_a also and to say that $F = F(T/T_a)$. If T_0 is 'now', then it must also be the case that $F(T_0/T_a) = 1$.

The magnitude of the force of gravitation between two particles of masses M and m is now $G_0 F(T/T_a) Mm/r^2$, where r is the absolute Euclidean distance between the particles at the instant T of absolute Newtonian time. Consider then the simplest, and yet basic, problem of celestial mechanics, namely the motion of a small particle m about a fixed massive particle M. The reader is reminded that Newton's second law of motion states that the rate of change of the linear momentum of the particle is proportional to the force. Let **r** be the position vector of m relative to M as origin; then the vector equation of motion m is

$$\frac{\mathrm{d}}{\mathrm{d}T}(m\dot{\mathbf{r}}) = -G_0 F\left(\frac{T}{T_a}\right)\frac{Mm}{r^2}\mathbf{i}, \tag{1}$$

where **i** is a unit vector that is always directed from M to m. There are two ways in which it is now possible to proceed: (i) when M and m are regarded as constants the vector equation of motion (1) becomes

$$\frac{d}{dT}(\dot{\mathbf{r}}) = -G_0 F\left(\frac{T}{T_a}\right)\frac{M}{r^2}\mathbf{i}; \qquad (2)$$

but (ii) when M and m are regarded as functions of the time such that $M' = MF^{1/2}$ and $m' = mF^{1/2}$ are constants then (1) becomes

$$\frac{d}{dT}\left(\frac{\dot{\mathbf{r}}}{F^{1/2}}\right) = -\frac{G_0 M'}{r^2}\mathbf{i}. \qquad (3)$$

The vector equations of motion (2) and (3) are not identical and therefore the motion of m would be different according as the first or the second assumption was made. Therefore the hypothesis of a variable gravitation function F, which leads to (2), is different from the hypothesis that the masses of all particles are undergoing a secular change, which leads to (3). The two hypotheses would be equivalent if Newton's second law always reduced to the statement that mass times acceleration was proportional to force. But this is true only if the mass of the particle involved is a constant.

It is easy to show from equation (2), in the first approximation at least, that the orbit of m is not a closed curve but a kind of spiral of continually increasing or decreasing size. This can be established by assuming that F varies slowly with T and that the orbit is quasi-elliptic. Therefore the orbits of all solar system planets and their satellites, all the orbits of double stars, the orbit of the Sun around the centre of the Galaxy, etc., are no longer closed elliptic curves but are open spiral-like curves. I have been unable to find in the textbooks on celestial mechanics that this effect is taken into consideration or that it has been observed and measured. The appeal to the argument from ignorance, namely the statement that 'F varies so slowly with the time that the spiralling effect eludes observation and so the secular change in F can be accepted', I discard as totally unsatisfactory. In a matter of such fundamental importance, direct measurement of the effect, if it exists, must be provided.

Quasi-Newtonian Cosmology and the Identification of T_a

It was pointed out thirty years ago by Milne and McCrea[1] that there was a formal analogy between the formulae of general relativity cosmology (in which $F = 1$) and the solutions of the hydrodynamical equations combined with Poisson's equation, with $F = 1$ also. These are the equations (A1) to (A3) of the appendix with $F = 1$. In these cosmological problems the universe is represented by a hydrodynamical fluid whose density, pressure, and velocity have the very special forms

$$\rho = \rho(T), \qquad p = p(T), \qquad u_i = \frac{dR}{dT}\frac{x_i}{R} \qquad (i = 1, 2, 3). \qquad (4)$$

The function $R(T)$ corresponds to the scale factor of the general relativity model universes.

We accept this idealized representation of the material content of the universe and we assert arbitrarily, in addition, that

$$F = \left(\frac{T_a}{T}\right)^n, \quad n \text{ positive.} \tag{5}$$

This means that G was infinite at the instant $T = 0$ and that it decreases in magnitude as time proceeds. Moreover, if 'now' is $T = T_0$, then clearly $T_0 = T_a$. When these rules for F, ρ, p, and the u_i are introduced into equations (A1) to (A3), the result (A13) is obtained. This is not the only solution but it is the simplest one and it corresponds, in the relativistic theories, to the case of zero space curvature. The general relativity analogue is given in (A14) and it is unique. T_a is called the 'age of the universe' which means the time that has elapsed since $T = 0$, when R was zero and G was infinite. But this calculation implies that the highly idealized assumptions regarding ρ, p, and the u_i are valid even when the volume occupied by the material content of the universe was so small that all galaxies interpenetrated each other. That the assumptions should still then be valid is most unlikely. I think that T_a measures the period during which conditions roughly similar to those at present observed held sway. The singularities in the density, the scale factor, and the 'constant' of gravitation are merely consequences of the over-idealization of the problem. At any rate, it is thought at the present time that T_a is equal to about 10^{10} years.

It is to be noticed that the equations (A1) to (A3) are no longer consistent with a condition of equilibrium, if by that is meant that all dependent variables do not involve the time. The presence of F, an explicit function of the time, in the gravitational potential V, means that all the dependent variables are necessarily explicit functions of the time, as well as of the spatial coordinates. But no doubt, quasi-equilibrium conditions could be set up in which the velocity components were negligibly small and the density and pressure changed very slowly with time.

Relativistic Theories

The time and the space coordinates must now be treated on a level with one another and no one coordinate is an invariant. Therefore the gravitation function F must now be regarded as a scalar function of all four coordinates. It is still, of course, dimensionless. Some investigators of the problem are the following.

(i) *Milne*[2]: he sought to make F depend on invariant combinations of the time and space coordinates of the space–time of special relativity.

(b) *Jordan*[3]: his field equations (p. 141, equations (9)) are reproduced in our notation as equations (A7), (A8), and (A9). The g_{mn}, R_{mn}, and T_{mn} are the components of the metrical tensor, the Ricci tensor, and the energy tensor, respectively. The scalar curvature is denoted by R^*. A comma denotes a covariant derivative. The presence of an arbitrary pure number ζ in the field equations and the form of the conservation law (A9) should be noted. It differs from the general relativity law (A5) and it leads to a kind of 'creation of matter' situation.

(c) *Hoffman*[4]: this is probably the most general theory of them all, but it has not been applied in any detail to the gravitational problem.

(d) *Brans and Dicke*[5], *Dicke*[6]: the field equations of the Brans–Dicke version (equations (9), (11), and (7) of their paper) are reproduced in (A10), (A11), and (A12). But let us remember that Brans and Dicke use the conventions of Landau and Lifshitz[7] regarding the signs of the metrical and Ricci tensors. Thus (A10), (A11), and (A12) cannot be compared term by term with Jordan's equations, because Jordan uses the convention of Einstein and of most other writers on relativity. The presence of an arbitrary pure number ω and the conservation law (A12), which is the same as that used in general relativity, should be noticed. This is no longer true in Dicke's[6] version and is one of the reasons why I prefer the earlier one.

Dependence of F on the Physical Situation

The most striking feature of these relativistic theories—except Milne's—is that there is now no universal function F, as there is in the quasi-Newtonian theory. Every physical situation has its own F.

Example (i): in the gravitational field for a vacuous region outside a spherically symmetric body, F is shown to depend on r alone, where r is the radial coordinate measured from the centre of the body.

Example (ii): if the universe is represented by a hydrodynamical fluid of zero pressure and whose density varies with the cosmic time t alone, then F also depends on t alone. In the simplest case of zero space curvature, F and the corresponding scale factor R are exhibited in equations (A15) and (A16). The 'age of the universe' is t_a in both cases. It should be noted also how the index ω now depends on either ζ or ω. Apart from the connection between n and ω, the Brans–Dicke cosmological F and R in (A16) have the same forms as in the quasi-Newtonian solution (A13). This is not true in Jordan's case (A15) because of the modified conservation law (A9).

The field equations are nonlinear so that different solutions cannot be superposed and still yield a solution. This has an important bearing on the theory of the internal constitution of the Earth, of the Sun, and of all stars. Inside these bodies there are spatial gradients of density and pressure so that the components T_{mn} of the energy tensor vary primarily with the distance r from the centre. They may, of course, vary with the time also, if nonequilibrium conditions are imposed. Similarly the gravitational potentials g_{mn}, and their derived tensors R_{mn} and R^*, would be expected to vary primarily with r, and with t also if there is no equilibrium. It is therefore not self-evident that the purely time-dependent cosmological $F (\propto 1/\phi)$ of equation (A16) can be introduced into the equations (A10) to (A12) and produce a consistent solution. In fact the 'external' solution found by Brans and Dicke—referred to in example (i) above—suggests that the basic solution of the Brans–Dicke field equations for the interior of any spherically symmetric body would be one in which *F and all tensors depended on r alone*. At any rate, it would have to be carefully proved that an F dependent on the time alone could be combined with expressions for the tensors that were mainly dependent on r.

I suspect that this point has been overlooked by Pochoda and Schwarzschild[8] and by Roeder and Demarque[9] in their studies of solar evolution with a time-dependent G. Both sets of authors imply that they are using the Brans–Dicke theory but they do not reveal what basic equations they have employed in their calculations. In particular, they do not state what equation they have derived from the Brans–Dicke theory for the dependence of the pressure gradient on the gravitational force. However, Dr. Roeder informs me that he and Dr. Demarque employed the equations (7) to (10c) of Larson and Demarque[10], with the addition of a secular variation in G. Therefore they have used the quasi-equilibrium case of equations (A1) to (A3) with the 'cosmological' expression for F given in (A13). In this function F the index n satisfies only the inequality $2 \geq n \geq 0$; the Brans–Dicke relation between n and ω exhibited in (A16) is absent. As far as I can judge, Pochoda and Schwarzschild have also used the quasi-Newtonian equations. Therefore these investigations throw no light on the validity or otherwise of the Brans–Dicke theory; they are applications of the quasi-Newtonian theory.

The same type of consideration applies in the 'motion of the perihelion of Mercury' problem. The Brans–Dicke theory combined with the numbers furnished by Duncombe[11] leads to the following:

$$\text{Bran–Dicke theoretical result} \quad \frac{4+3\omega}{6+3\omega} \, 43{\cdot}03''$$

$$\text{observation} \quad 43{\cdot}11'' \pm 0{\cdot}45''$$

It is important to remember that the theoretical result depends on the assumption that the Sun is spherically symmetric and is not oblate. Three times the formal probable error of the observational result is $1{\cdot}35''$. Thus the Brans–Dicke theory might be regarded as in agreement with the observed perihelion motion of Mercury if $\omega \geq 30$. Such a value of ω would also secure agreement in the 'bending of light rays' problem. But Brans and Dicke, for reasons that are not clear to me, want ω to be of the order of unity and they say that $\omega \geq 6$. This produces a defect in the predicted, as compared with the observed, perihelion motion, which Dicke wants to cure by attributing an oblateness to the Sun. But such an oblateness would destroy the theory on which the theoretical result is based, in particular by altering the expression for the gravitation function. The general relativity perihelion motion about an oblate Sun can also be worked out[12], though it is a matter of considerable complexity. In any case, it is this 'oblate Sun' general relativity result that must be compared with the corresponding 'oblate Sun' Brans–Dicke prediction, when this is worked out. It is only after this comparison has been made that the solar oblateness effect might be accepted, even if one has been convinced that Dicke has detected the effect observationally.

In any case it is clear that the constants ω and ζ of the Brans–Dicke and the Jordan theories are parameters introduced for an *ad hoc* reason, namely to secure as much agreement as possible with local effects in the solar system. It is doubtful that limitations on their values can be imposed by *a priori* appeals to 'reason'.

The Brans–Dicke 'Empty' Case

The special case in which the energy tensor is zero everywhere is worth examining in some detail. This may be done with the aid of the space–times whose metrics are

$$ds^2 = -dt^2 + a^2(t)\left(\frac{dr^2}{1-kr^2} + r^2\, d\Omega^2\right),$$

$$d\Omega^2 = d\theta^2 + \sin^2\theta\, d\phi^2, \tag{6}$$

where the scale factor is $a(=R)$ and the space curvature constant is k. The second symbol is replaced by λ in the work of Brans and Dicke. It is well known that the space–time of special relativity, expressed in terms of spatial spherical polar coordinates, may be obtained from (6) in two ways as follows. If a is constant and k is zero, the metric (6) obviously reduces to that of special relativity; but also, if $a = t$ and $k = -1$, then the coordinate transformation

$$u = rt, \quad \tau = t(1+r^2)$$

again reduces (6) to the formula for special relativity, with τ as a Lorentz time and u as radial coordinate.

In the equation (A11) it is assumed that ϕ is also a function of t alone and the components of the tensors R_{kl} and R^*g_{kl} are calculated for the metric (6). Four equations result, namely,

$$-3\left(\frac{\dot{a}^2}{a^2}+\frac{k}{a^2}\right) = \frac{8\pi}{\phi c^4}T_4^4 - \frac{\frac{1}{2}\omega\dot{\phi}^2}{\phi^2} + 3\frac{\dot{a}\dot{\phi}}{a\phi}, \qquad (7)$$

$$-\left(\frac{2\ddot{a}}{a}+\frac{\dot{a}^2}{a^2}+\frac{k}{a^2}\right) = \frac{8\pi}{\phi c^4}T_i^i + \frac{\frac{1}{2}\omega\dot{\phi}^2}{\phi^2} + \frac{\ddot{\phi}}{\phi} + 2\frac{\dot{a}\dot{\phi}}{a\phi}, \qquad (8)$$

where i can take the values 1, 2, and 3 and it is convenient to use the mixed forms of the tensors. Brans and Dicke have also shown that an equation equivalent to (A10) is

$$-a^{-3}\frac{d}{dt}(a^3\dot{\phi}) = \frac{8\pi}{(3+2\omega)c^4}T, \qquad (9)$$

while, of course, (A12) is also true.

It is obvious that, if the two independent assumptions are made, namely that the components of the energy tensor, and therefore also the invariant density T, are everywhere zero and also that ϕ is a constant, then either $a = $ constant, $k = 0$ or $a = t$, $k = -1$. In either case the space–time is that of special relativity. But this conclusion does not follow if only one of the assumptions is made, namely that the energy tensor is zero. This may be seen from the following example. Let us suppose that the energy tensor is zero and that $a = t^n$. Then (A12) is identically satisfied while (9) yields

$$\dot{\phi} = +At^{-3n}, \qquad \phi = +A(1-3n)^{-1}t^{1-3n}, \qquad (10)$$

where A is the constant of integration. Substitution of these expressions for a and ϕ into (7) and (8) produces, after some calculation, the two equations

$$-3kt^{2-2n} = 3n(1-2n)-\frac{\omega}{2}(1-3n)^2,$$

$$-kt^{2-2n} = -3n(1-2n)+\frac{\omega}{2}(1-3n)^2. \qquad (11)$$

From these equations it follows immediately that $k = 0$ and that

$$n = \frac{1+\omega\pm(1+2\omega/3)^{1/2}}{4+3\omega}. \qquad (13)$$

Hence n is positive for positive ω, whichever sign is chosen. Thus a is zero at $t = 0$. Now G is proportional to $1/\phi$, and so to $t^{-(1-3n)}$. Therefore, if G is to decrease as t increases and is to be infinite at $t = 0$, it must be the case that $1 - 3n$ is positive. Hence the negative sign must be chosen in (13) and the desired solution is

$$n = \frac{1+\omega-(1+2\omega/3)^{1/2}}{4+3\omega}$$
$$1-3n = \frac{1+3(1+2\omega/3)^{1/2}}{4+3\omega}.$$
(14)

But $a = t^n$, with n given by (14), combined with $k = 0$, does not reduce (6) to the space–time of special relativity. Hence the Brans–Dicke field equations possess solutions in which space–time is curved and yet in which there is no matter or energy present. This arises because of the presence of the gravitation function which need not have a constant value, as shown in equation (10), even if the scalar density T in equation (9) is zero.

Comparison with Einstein's General Relativity

In some ways the Brans–Dicke and Jordan theories are more comprehensive than general relativity. But it should be noticed that they reduce to a particular case of Einstein's theory only. If in their field equations F or ϕ is taken to be a constant, it follows from (A7) or from (A10) that $R^* = 0$. Space–times with a zero scalar curvature do occur in Einstein's theory but they form a restricted class only. In general, R^* is proportional to the scalar density T, as in (A6), and the scalar density need not be zero.

The customary deduction from Einstein's equations (A4) is that, if the energy tensor is zero, then the space–time is that of special relativity.* The converse is certainly true. Thus special relativity is the particular case of general relativity in which gravitational effects are negligible. However, the true parallel with the Brans–Dicke theory occurs when Einstein's equations are modified by the inclusion of the cosmical constant Λ. Instead of (A4), the form of Einstein's equation is

$$R_i^k - \tfrac{1}{2}(R^* - 2\Lambda)\delta_i^k = -\frac{8\pi G_0}{c^4} T_i^k \tag{15}$$

where the mixed forms of the tensors have been used. It is well known[13] that there are solutions of the equations (15), with the form (6) for the metric, in which the energy tensor vanishes identically, but the space–time is not flat. The most famous of these cases is that of the de Sitter universe, which was discovered some fifty years ago. There is no difficulty in interpreting such 'empty' cases in general relativity: the cosmical constant represents, in Newtonian language, a universal 'force' additional to gravitation. It is the presence of this force which accounts for the curvature of space–time even when no matter or energy is present. It has been shown in equation (14) that the Brans–Dicke theory also includes curved space–times in which there is no matter or energy. However, the gravitation function ϕ is now intimately bound up with gravitation, it does not represent an independent 'force' as does the cosmical constant. On

---

* But see McVittie[13] (p. 77) for a discussion of this point.

the face of it, therefore, the Brans–Dicke theory appears to predict that a universe would be possible in which no gravitating matter or energy was present but in which gravitational effects would still occur and, presumably, be observable by examining the motions of free particles following geodesic paths. But perhaps all such 'empty' cases are meant to be excluded by an appeal to Mach's principle[14].

Summary

Newtonian gravitational theory is mathematically compatible with a universal gravitation function F, which is a function of absolute Newtonian time alone. But there is no prescription for selecting the functional form of F. Hence the theory of orbits in the solar system and of double stars, the dynamics of globular clusters and of the Galaxy, the theory of stellar structure, etc., must be recalculated for an infinity of functions F. In each case the predictions so obtained would have to be shown to be observable and also to be confirmed by observation; in my view it would be insufficient to argue that a variable F was acceptable because it produced only certain effects. The relativistic theories make F a function of the four coordinates, the time and the three space coordinates. I put forward the conjecture that a different law of gravitation (i.e. a different function of F) is predicted for the interior of the Earth, for the Earth's orbital motion round the Sun, for the motion of the Sun in the Galaxy, for the cosmological problem, etc. These various laws of gravitation are not superposable because of the nonlinearity of the field equations. It seems to me that it is the responsibility of the propounders of the theories to show whether this conjecture is correct or not. They would also have to explain why such a remarkable degree of agreement with observation had been obtained during the past three hundred years through the assumption that there was one universal, and constant, function F. The reader is reminded that the path of development of relativity theory during the past 50 years is strewn with the corpses of modified forms of general relativity. The demise in each case was largely due to the fact that the inventor of the modification drew from his theory only one or two conclusions favorable to his point of view. When a more comprehensive analysis was carried out, the theory was found to be unacceptable and no more was heard of it.

Appendix

Quasi-Newtonian theory

$F = F(T/T_a)$, index values 1, 2, 3 only.

$$\frac{\partial \rho}{\partial T} + \frac{\partial \rho u_j}{\partial x_j} = 0, \tag{A1}$$

$$\frac{\partial \rho u_i}{\partial T} + \frac{\partial}{\partial x_j}(\rho u_i u_j + \delta_{ij} p) = \rho \frac{\partial V}{\partial x_i}, \tag{A2}$$

$$\nabla^2 V = -4\pi G_0 F\left(\frac{T}{T_a}\right)\rho. \tag{A3}$$

The energy tensor is given by
$$T_{44} = \rho, \qquad T_{i4} = \rho u_i, \qquad T_{ij} = \rho u_i u_j + \delta_{ij} p.$$

Einstein theory

$F=1$, index values 4, 1, 2, 3,

$$R_{mn} - \tfrac{1}{2} R^* g_{mn} = -\frac{8\pi G_0}{c^4} T_{mn}, \tag{A4}$$

$$T^{mn}_{,n} = 0, \qquad T^{mn} = g^{im} g^{jn} T_{ij}, \tag{A5}$$

$$R^* \propto T = T^n_n. \tag{A6}$$

Jordan theory

$$\kappa = \frac{8\pi G_0}{c^2} F(x^4, x^1, x^2, x^3), \qquad F_k = \frac{\partial F}{\partial x^k},$$

$$R^* + \zeta g^{mn} \left(\frac{2 F_{m,n}}{F} - \frac{F_m F_n}{F^2} \right) = 0, \tag{A7}$$

$$R_{mn} - \tfrac{1}{2} R^* g_{mn} + \left\{ \frac{F_{m,n}}{F} - g_{mn} g^{ij} \frac{F_{i,j}}{F} - \zeta \left(\frac{F_m F_n}{F^2} - \tfrac{1}{2} g_{mn} g^{ij} \frac{F_i F_j}{F^2} \right) \right\} = -\frac{8\pi G_0}{c^4} F T_{mn}, \tag{A8}$$

$$(F^2 T^{mn})_{,n} = 0. \tag{A9}$$

Brans–Dicke theory

$$\phi \propto \frac{1}{F}, \qquad \phi_k = \frac{\partial \phi}{\partial x^k},$$

$$R^* + \omega g^{mn} \left(\frac{2\phi_{m,n}}{\phi} - \frac{\phi_m \phi_n}{\phi^2} \right) = 0, \tag{A10}$$

$$R_{mn} - \tfrac{1}{2} R \left\{ \frac{\phi_{m,n}}{\phi} - g_{mn} g^{ij} \frac{\phi_{i,j}}{\phi^2} + \omega \left(\frac{\phi_m \phi_n}{\phi^2} - \tfrac{1}{2} g_{mn} g^{ij} \frac{\phi_i \phi_j}{\phi^2} \right) \right\} = +\frac{8\pi}{c^4} \frac{1}{\phi} T_{mn}, \tag{A11}$$

$$T^{mn}_{,n} = 0. \tag{A12}$$

Simplest cosmological solutions

Newton:

$$\left. \begin{array}{l} R \propto \left(\dfrac{T}{T_0} \right)^{(2-n)/3}, \qquad G \propto \left(\dfrac{T_0}{T} \right)^n, \qquad 2 \geq n \geq 0 \\[2mm] \rho = \rho_0 \dfrac{R_0^3}{R^3}, \qquad p(T) \text{ undetermined.} \end{array} \right\} \tag{A13}$$

Einstein:

$$\left. \begin{array}{l} R \propto \left(\dfrac{t}{t_0} \right)^{2/3}, \qquad G = \text{constant} \\[2mm] \rho = \rho_0 \dfrac{R_0^3}{R^3}, \qquad p = 0. \end{array} \right\} \tag{A14}$$

Jordan:

$$\left.\begin{aligned} &R \propto \left(\frac{t}{t_0}\right)^{(2+n)/3}, \quad G \propto \left(\frac{t_0}{t}\right)^n, \quad n = \frac{2}{3\zeta - 4} \\ &\rho = \rho_0 \frac{R_0^3}{R^3}, \quad p = 0. \end{aligned}\right\} \quad \text{(A15)}$$

Brans–Dicke:

$$\left.\begin{aligned} &R \propto \left(\frac{t}{t_0}\right)^{(2-n)/3}, \quad G \propto \left(\frac{t_0}{t}\right)^n, \quad n = \frac{2}{3\omega + 4} \\ &\rho = \rho_0 \frac{R_0^3}{R^3}, \quad p = 0. \end{aligned}\right\} \quad \text{(A16)}$$

References

1. E. A. Milne and W. H. McCrea, *Quart. J. Math.*, **5**, 73 (1934).
2. E. A. Milne, *Kinematic Relativity*, Clarendon Press, Oxford, 1948.
3. P. Jordan, *Schwerkraft und Weltall*, Vieweg, Braunschweig, 1952.
4. B. Hoffman, *Phys. Rev.*, **89**, 49, 52; **91**, 751 (1953).
5. C. Brans and R. H. Dicke, *Phys. Rev.*, **124**, 925 (1961).
6. R. H. Dicke, 1962, *Phys. Rev.*, **125**, 2163 (1962).
7. L. D. Landau and E. M. Lifshitz, *The Classical Theory of Fields*, 2nd ed., Pergamon Press, London, 1962.
8. P. Pochoda and M. Schwarzschild, 1964, *Astrophys. J.*, **139**, 587 (1964).
9. R. C. Roeder and P. R. Demarque, 1966, *Astrophys. J.*, **144**, 1016 (1966).
10. R. B. Larson and P. R. Demarque, *Astrophys. J.*, **140**, 524 (1964).
11. R. L. Duncombe, *Astron. J.*, **61**, 174 (1962).
12. P. A. Geisler and G. C. McVittie, *Monthly Notices Roy. Astron. Soc.*, **131**, 483 (1966).
13. G. C. McVittie, *General Relativity and Cosmology*, 2nd ed., Chapman and Hall, London, 1965, pp. 223–5.
14. R. H. Dicke, *Phys. Today*, **20**, 55 (1967).

3

I. W. ROXBURGH*

Astronomy Centre
University of Sussex
Brighton, Sussex, England

The oblateness of the Sun

Dicke and Goldenberg[1] have measured the oblateness of the Sun and found a difference between the equatorial and polar radii of 35 km, which amounts to 5 parts in 10^5 of the radius of the Sun. In principle a measurement of the oblateness can reveal the quadrupole moment of the Sun, and, because a distorted Sun produces a perihelion advance of the planets, the contribution to the perihelion advance of Mercury from an oblate Sun can be determined. From the measured oblateness they deduced a contribution to the perihelion advance of Mercury of 4 seconds of arc per century, which implies a mean rotation period for the inside of the Sun of 1·8 days.

The spirit of this chapter is to enquire whether we could reasonably expect any other effect than a rapidly spinning core to produce the observed oblateness, and in particular whether a slow uniform rotation could be responsible for so large an oblateness.

The effect of rotation is usually estimated by considering the ratio of centrifugal force to gravity

$$\alpha = \frac{1}{3}\frac{\Omega^2 r^3}{GM}$$

For the Sun this is small, of the order of 10^{-5}. However, the parameter α need not be the correct one for estimating rotational effects; in the convective layers of a star we have another small parameter

$$\varepsilon = \frac{\beta}{dT/dr}$$

where β is the difference between the actual and adiabatic temperature gradients. ε is of the order unity in radiative zones and very small, of the order of 10^{-5}, in convective zones. The parameter α/ε therefore is small in radiative zones but large in convective zones; thus it is not unreasonable to expect a slow rotation to have a large effect on a convective layer, and a detailed analysis is called for.

Unfortunately no such detailed analysis is available; indeed we still have no satisfactory theory of turbulent convection without rotation. Our first approach therefore is to look at the theory of the onset of convection. In this case it is found that instability occurs when the critical Rayleigh number is related to the Taylor number by a relation of the approximate form

$$R_c = 657 + 8\tau^{2/3}, \quad R_c = \frac{g\beta d^4}{T\kappa\nu}, \quad \tau = \frac{4\Omega^2 \cos^2\theta \, d^4}{\nu^2}$$

* Now at Mathematics Department, Queen Mary College, University of London, England.

where g is the value of gravity, T the temperature, d the depth of the layer, v the viscosity, κ the conductivity, and θ the angle between the rotation axis and gravity. The difference in the superadiabatic gradient between pole and equator is then

$$\Delta\beta = 8\left(\frac{4\Omega^2 d^4}{v^2}\right)^{2/3} \frac{T\kappa v}{g\, d^4}$$

In the outer layers of the Sun with $\kappa \sim 10^{10}$, $v \sim 10^{-1}$ this gives $\Delta\beta \sim 10^{-8}$.

However, a more reasonable estimate of $\Delta\beta$ would be obtained by replacing κ and v by the eddy conductivity and viscosity in the convective layer. This is very difficult to estimate. However, if we assume that elements formed at the top of the Sun rapidly become isolated from their surroundings and therefore fall freely under gravity, then

$$v^2 = 2GM\left(\frac{1}{R} - \frac{1}{r}\right) = 5 \times 10^{14} \text{ cm}^2/\text{sec}^2$$

hence $v \sim 2 \times 10^7$ cm/sec. The depth of the convective zone is of the order of 10^{10} cm; hence

$$\kappa_e = v_e = v\,d = 2 \times 10^{17} \text{ cm}^2/\text{sec}$$

This is an upper limit on the value of κ, and v. However, with this value

$$\Delta\beta = 2\cdot 5 \times 10^{-7} \text{ degK/cm}$$

A more realistic value can be obtained from detailed integrations of the solar outer layers such as those given by Faulkner, Griffiths, and Hoyle[2]. Typical convective velocities deep in the convective zone are of the order of 3×10^5 and hence the eddy viscosity is 3×10^{15} cm^2/sec and the corresponding

$$\Delta\beta \simeq 2 \times 10^{-8} \text{ degK/cm}$$

Of course this value of $\Delta\beta$ cannot be considered exact; it could easily be wrong by a factor 2 and maybe by a factor of 10; such is our lack of knowledge of turbulent convection.

However, if we assume that the above calculation gives the correct order of magnitude for the difference in temperature gradient between pole and equator, we can calculate the difference in polar and equatorial radii, Δr. The actual temperature gradient inside the solar convective zone is of the order of 10^{-4} degK/cm since the temperature drops from some $1\cdot 5 \times 10^6$ °K to essentially zero in $1\cdot 5 \times 10^{10}$ cm. If the temperature gradient up the pole is $10^{-4} + 2 \times 10^{-8}$, as opposed to 10^{-4} along the equator, then the temperature drops to zero from $1\cdot 5 \times 10^6$ °K in a distance

$$R_e = \frac{1\cdot 5 \times 10^6}{10^{-4} + 2 \times 10^{-8}} = 1\cdot 5 \times 10^{10} - 3 \times 10^6 \text{ cm}$$

as opposed to $1\cdot 5 \times 10^{10}$ cm in the equatorial regions. The difference between polar and equatorial radius is then

$$r = 30 \text{ km}$$

Thus an oblateness of the order of 30 km may well be produced by the interaction of rotation with turbulent convection. This is all that is necessary to explain Dicke and Goldenberg's observations.

Of course the different temperature gradient will produce a temperature difference, and hence a pressure difference, over an equidensity surface. This will drive motions in the convective zone which may well manifest themselves as the solar equatorial acceleration. Kippenhahn[3] has shown that meridional motions in a viscous layer could maintain an equatorial acceleration.

References

1. R. Dicke and R. Goldenberg, *Phys. Rev. Letters*, in press (1967).
2. J. Faulkner, K. Griffiths, and F. Hoyle, *Monthly Notices Roy. Astron. Soc.*, **129,** 363–93 (1965).
3. R. Kippenhahn, in *Atti del Convegno Sui Campi Magnetica Solari, Firenze* (Ed. G. Barbera), 1966.

A

COSMOLOGY AND GEOPHYSICS

II. Geophysical methods of detecting a change in G

1

E. R. KANASEWICH

Department of Physics
University of Alberta
Edmonton, Canada

and

J. C. SAVAGE

Department of Physics
University of Toronto
Toronto, Canada

Dirac's principle of cosmology and radioactive dating

Introduction

Dirac[1] has proposed as a fundamental principle of cosmology that 'any two of the very large dimensionless numbers occurring in nature are connected by a simple mathematical relation in which the coefficients are of the order of unity'. A hypothesis of Blackett[2] and Dirac's principle were used by Houtermans and Jordan[3] to relate the beta-decay constant to the reciprocal square root of the age of the Universe. They noted that the theory could be tested by a comparison of the beta-decay and alpha-decay ages of naturally occurring minerals. Dicke[4-6] has replaced Blackett's hypothesis with more recent theories of nuclear decay. Using Dirac's principle he found that the alpha-decay constant should be essentially independent of the age of the Universe (T), but that the beta-decay constant should be proportional to T^{-n}, where n is a number between $\frac{1}{4}$ and $\frac{1}{2}$. Dicke then attempted to test Dirac's hypothesis by comparing the Rb–Sr (beta-decay) and Pb–Pb (alpha-decay) ages of meteorites. The measured values of the decay constant of ^{87}Rb varied[7] from 1·19 to 1·47 × 10^{-11} years^{-1} and this large uncertainty dominated the comparison so that no conclusion could be made on the validity of the hypothesis. This difficulty may be overcome to some extent by using the radioactive ages of terrestrial rocks as well as meteorites. Such an analysis was described by Kanasewich and Savage[8] and is discussed further in this chapter.

Theory

In order to render the various quantities dimensionless, we express them in terms of the fundamental atomic units $\hbar$, c and the mass of an elementary particle such as the proton, pion, or electron. There seems to be no theoretical basis on which to choose which if any of the elementary particles should serve as a fundamental unit of mass, so it is necessary to try each one in turn. The square of the strong coupling constant ($f_{str}^2/\hbar c$) is then about one in an order-of-magnitude calculation. The alpha-decay constant depends upon the square of the strong coupling constant and, since it is not a large number, the alpha-decay constant should be essentially independent of T. The weak coupling constant[9] is $g = 1\cdot 99 \times 10^{-49}$ erg cm^3 and has dimensions of energy times volume, so it is necessary to specify a unit of mass. The reciprocal square of the weak coupling constant is of the order of 10^{23} if the electron is the unit of mass and 10^{10} if the neutron or proton had been used. On occasion[10] the pion mass is used, in which case the dimensionless reciprocal square of the weak coupling constant is between 10^{13} and 10^{14}. This may be compared with the dimensionless expression for the reciprocal of the gravitational interaction ($\hbar c/Gm^2$) between two pions which is of the order of 10^{40}. An alternate comparison is with the dimensionless reciprocal of the Hubble constant ($H^{-1}/\hbar/m_\pi c = 10^{41}$) which may be related to the age of the Universe by various cosmological theories. If 10^{14} is accepted as a large number, then Dirac's principle implies that the beta-decay constant is proportional to T^{-n}. With the selection of the electron or the proton as the unit of mass, n would be $\frac{1}{2}$ or $\frac{1}{4}$ respectively; the choice of a pion for the unit of mass leads to the intermediate value of $\frac{3}{8}$.

Using the convention that time is measured positively into the past the beta-decay constant at any time t is given in terms of the decay constant λ as measured at present:

$$\bar{\lambda} = \lambda\left(\frac{T}{T-t}\right)^n \qquad (1)$$

The modified form of Rutherford's equation of radioactive decay is then

$$dN = \bar{\lambda}N\,dt \qquad (2)$$

where N is the number of parent atoms present at any time. The solution of the equation of radioactive decay is then

$$\ln\left(\frac{D+N_0}{N_0}\right) = \frac{\lambda T}{1-n}\left\{1-\left(1-\frac{\tau_\alpha}{T}\right)^{1-n}\right\} \qquad (3)$$

N_0 and D are the number of parent and daughter atoms as measured at the present time and τ_α is the time since the sample was last differentiated chemically. The expression in (3) may be compared with the results obtained from Rutherford theory:

$$\ln\left(\frac{D+N_0}{N_0}\right) = \lambda_e \tau_\beta \qquad (4)$$

where $\lambda_e = 0\cdot 0139 \times 10^{-9}$ years^{-1}, the value of the ^{87}Rb-decay constant presently used in the literature and τ_β (called the Rb–Sr age) is calculated to fit the measured abundance. It should be noted that the value of λ_e cited was determined by Aldrich and others[11] in the same manner as the decay constants are determined in this chapter. The observable quantity for any sample is given by the left-hand sides of equations (3) and (4). Equating these expressions we have

$$\lambda_e \tau_\beta = \frac{\lambda T}{1-n}\left\{1-\left(1-\frac{\tau_\alpha}{T}\right)^{1-n}\right\} \qquad (5)$$

Letting

$$x = \frac{T}{1-n}\left\{1 - \left(1 - \frac{\tau_\alpha}{T}\right)^{1-n}\right\} \qquad (6)$$

we have

$$\tau_\beta = \frac{\lambda}{\lambda_e} x \qquad (7)$$

Values of both τ_β and τ_α (and hence x) are available for many terrestrial rocks and some meteorites. Thus we have a collection of observed values of τ_β and x which should lie on the straight line represented by equation (7). A least-squares analysis, in which the sum of the squares of the perpendicular distances of the data points from a straight line is minimized, may be used to determine the value of λ in equation (7) which best fits the data for each of the possible pairs of n and T. In order to judge how well the data fit a particular set of values of n and T we shall also calculate the standard deviation σ_θ of the angle θ, where $\tan \theta$ is the slope of the best straight line representing equation (7). Clearly a small value of σ_θ indicates a good fit.

We propose to find which of two theories gives the smaller value of σ_θ. The first theory (hereafter referred to as the Rutherford theory) requires that beta decay be independent of the age of the Universe, i.e. $n = 0$. The second theory (hereafter referred to as the Dirac theory), being based upon Dirac's principle, requires that the beta-decay constant should depend upon the age of the Universe as T^{-n}; the precise value of n is not known but it is unlikely that it lies outside of the range $\frac{1}{4}$ to $\frac{1}{2}$. In the Dirac theory[1] there is also a restriction upon the age of the Universe, namely that

$$T = \frac{1}{3H} \qquad (8)$$

The Hubble age[12] H^{-1} is 13×10^9 years with an uncertainty of a factor 2.

Data

Isotopic analysis from eleven terrestrial localities have been found which yield concordant Pb–U dates and also have Rb–Sr data in closely related geologic units (see table 1). A lead–uranium determination was judged to be concordant if the ages from the ratios ^{206}Pb/^{238}U, ^{207}Pb/^{235}U, and ^{206}Pb/^{207}Pb were within 10% of the mean of these three determinations. For these samples the lead–lead age was used in the calculations. This is the most reliable index, even if the mineral has suffered a small amount of lead loss either by diffusion or by geochemical processes. As may be seen in Tilton's[43] tables 1 and 2 and also his Concordia–Discordia plots, the lead–lead ages agree more closely with the upper intersection of the Concordia which is the best estimate of the time of mineral formation. The lead–lead age is also most directly comparable with the model lead age which was used for the meteorites.

Most areas of the Earth have suffered some degree of metamorphism subsequent to a major period of mineral formations. Both Rb–Sr and K–Ar ages are particularly sensitive to the thermal history of the area. In the areas from which the samples were

TABLE 1 Table of age determinations

Area	Reference	Mineral	K-A	Age ($\times 10^9$ years)[a]				Rb-Sr
				$\frac{^{206}Pb}{^{238}U}$	$\frac{^{207}Pb}{^{235}U}$	$\frac{^{208}Pb}{^{232}Th}$	$\frac{^{206}Pb}{^{232}Pb}$	
1. Spinelli Quarry, Connecticut	13	Samarsk.		0·253	0·255	0·266	0·280	
	14	U. mineral		0·257	0·258	0·257	0·292	
	15	Samarsk.		0·253		0·266	0·280	
	16	muscovite	0·248					0·258
		microcline						0·286
Mean							0·284	0·272
2. Spruce Pine, North Carolina	17	uraninite		0·385	0·390	0·400		
	18	uraninite		0·370	0·375	0·420		
	18	muscovite	0·335					0·375
	18	microline						0·380
Mean							0·410	0·378
3. Quanah Mts., Oklahoma	19	zircon		0·520	0·525	0·505	0·550	
	19	zircon		0·515	0·520	0·495	0·550	
	19	biotite	0·510					0·535
	19	feldspar						0·520
Mean							0·550	0·528
4. Wilberforce, Ontario	14	U. mineral		1·015	1·038	0·990	1·060	
	20	zircon		0·900	0·930	1·130	1·000	
	21	uraninite		1·150	1·110	1·000	1·030	
	22	uraninite		1·060	1·055	0·995	1·040	
	18	uraninite		1·020	1·020		1·020	
	23	uraninite		1·077			1·035	
	24	uraninite		1·000	1·010	0·870	1·030	
	11	muscovite						0·990
	18	biotite	0·920					1·000
	23	lepidolite						0·960
	18	biotite	0·960					1·030
Mean							1·030	0·995

Dirac's principle of cosmology and radioactive dating

Location		Mineral					
5. Llano, Texas	20	zircon		0·950	0·990	1·070	
	25	zircon		0·970	1·020	1·120	
	18	biotite	1·060				1·100
	26	biotite	1·060				1·110
	27	micas	1·060				1·060
Mean						$\overline{1·095}$	$\overline{1·090}$
6. Pump Station Hills, Texas	25	zircon		1·065	1·105	1·175	
	25	zircon		1·005	1·050	1·140	
	25	feldspar					1·060
	25	whole rock					1·020
Mean						$\overline{1·158}$	$\overline{1·040}$
7. Bob Ingersoll Mine, South Dakota	28	uraninite		1·580	1·600	1·440	1·630
	11	lepidolite					1·680
	28	muscovite	1·515				1·730
	11	microcline					1·600
Mean						$\overline{1·630}$	$\overline{1·670}$
8. Viking Lake, Saskatchewan	29	uraninite		1·850	1·880	1·670	1·910
	11	biotite–chl. mica	1·830				1·880
	29						1·930
Mean						$\overline{1·910}$	$\overline{1·905}$
9. Inari, Finland	30	zircon		1·820	1·870	1·960	1·930
	30	biotite	1·880				1·860
Mean						$\overline{1·930}$	$\overline{1·860}$

TABLE 1 (contd.) Table of age determinations

Area	Reference	Mineral	K–A	$\frac{^{206}\text{Pb}}{^{238}\text{U}}$	Age ($\times 10^9$ years)[a] $\frac{^{207}\text{Pb}}{^{235}\text{U}}$	$\frac{^{208}\text{Pb}}{^{232}\text{Th}}$	$\frac{^{206}\text{Pb}}{^{232}\text{Pb}}$	Rb–Sr
10. Ebonite claim, Bikita Quarry, South Rhodesia	31	monzonite		2·660	2·660	2·640	2·680	
	31	monzonite		2·640	2·610	2·640	2·620	
	32	monzonite		2·675	2·680	2·645	2·680	
	31	mica	2·480					2·530
	31	mica	2·340					2·710
	31	mica						2·685
	11	lepidolite						2·640
	11	lepidolite						2·680
	33	lepidolite						2·540
	34	lepidolite						2·625
	35	lepidolite						2·600
	36	lepidolite	2·430					2·625
Mean							2·660	2·626
11. Cooke City and Beartooth Mountains, Montana	37	uraninite		2·600	2·650		2·700	
	37	micas	2·470					2·800
	37	micas						2·770
	37	micas	2·530					2·700
	37	micas	2·510					2·530
	37	micas						2·730
	37	micas	2·360					2·750
Mean							2·700	2·713
12. Mean of meteoritic data	38							4·620
	39							4·440
	40						4·550	
	41							4·740
	42							4·450
Mean							4·550	4·560

[a] The decay constants used for the age determinations are as follows: ^{238}U, 1.54×10^{-10} years^{-1}; ^{235}U, 9.72×10^{-10} years^{-1}; ^{232}Th, 4.99×10^{-11} years^{-1}; ^{87}Rb, 1.39×10^{-11} years^{-1}; ^{40}K (λ_β), 4.72×10^{-10} years^{-1}; ^{40}K (λ_α), 0.585×10^{-10} years^{-1}.

taken, the present surface represents such a deep erosional level that what we now sample are the old mountain roots. Such samples must have been at a fairly high temperature during much of geologic time. Consequently, K–Ar ages may be used as an index to check on both episodic loss of daughter products and the effect of continuous diffusion. The K–Ar ages are all lower than the lead–lead ages, but the percentage difference is small and no subsequent major chemical fractionation is indicated for the eleven terrestrial samples.

The Rb–Sr age of meteorites was obtained from four recent studies. Gast's[38] model I age of $4 \cdot 62 \times 10^9$ years ($\lambda = 0 \cdot 0139 \times 10^{-9}$ years^{-1}) included analyses from eight stone meteorites (Beardsley, Forest City, Modoc, Moore County, Nuevo Laredo, Pasamonte, Richardton, and Sioux County). Pinson and Schnetzler[39] obtained a value of $4 \cdot 44 \times 10^{-9}$ years for superior analyses of six stone meteorites (Bath, Estacada, Farmington, Homestead, Nakhla, and Estherville). Rama Murthy and Compston[41] obtained an age of $(4 \cdot 74 \pm 0 \cdot 15) \times 10^9$ years from four carbonaceous chondrites (Orgueil, Murray, Mokoia, and Lancé) if the mean initial value is assumed from achondritic samples. Shields[42] has taken the greatest care in the chemical technique and reduced the contamination level for Sr by an order of magnitude and for Rb by a factor of 2. He obtained an age of $(4 \cdot 45 \pm 0 \cdot 03) \times 10^9$ years for six stony meteorites (Pasamonte, Estherville, Nakhla, Murray, Bath, and Bruderheim). The

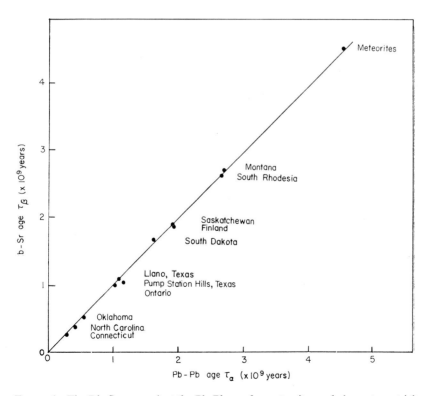

FIGURE 1 The Rb–Sr age against the Pb–Pb age for meteorites and eleven terrestrial rocks. The full line represents the straight line passing through the origin which best fits the data. The full circles indicate the data used in computations (see table 1)

mean of these four sets of data is $4\cdot56 \times 10^9$ years. Anders[44] has discussed the evidence for the significance and reliability of both Rb–Sr and model Pb ages of meteorites and the subject will not be reviewed here. Since 1953 over fifty lead isotopic determinations on stone and iron meteorites have been reported. Although widely varying uranium-to-lead ratios occur in meteorites, the isotopic compositions of stone and iron meteorites are, within the experimental error, linearly related. This indicates that they represent closed systems isolated since the time at which they underwent chemical differentiation. Rama Murthy and Patterson[40] have obtained the slope of the meteoritic isochron and from this the mean model age of meteorites may be calculated as $(4\cdot55 \pm 0\cdot03) \times 10^9$ years. The standard deviation is given with the assumption that there is no error in the decay constants.

Figure 1 shows a plot of the Rb–Sr ages against the Pb–Pb ages for the samples studied here. The full line represents the straight line which best fits the data. It is clear that the scatter is small. Figure 1 is included merely to present the data. The fact that the straight line, which represents the Rutherford theory, fits the data quite well does not imply that the Dirac theory does not. In fact, the Dirac theory with $T = 13 \times 10^9$ years, $\lambda = 0\cdot0132 \times 10^{-9}$ years^{-1}, $n = \frac{1}{4}$ would be an equally good fit; we shall show, however, that such a large value of T is unlikely in the Dirac theory and the predicted decay constant does not agree with recent experimental determinations.

Results

A least-squares analysis was used to determine the value of λ most consistent with the data for several values of n and many values of T. The analysis also yielded the associated value of σ_θ, which is a measure of how well the particular theory fits the data. In performing the analysis it would be desirable to weight each of the entries in table 1 according to its standard error. However, we do not have reliable estimates of the standard errors for most of the data points. We have, therefore, assumed that the standard error for each of the data points is the same. In other words, no extraneous weights have been introduced into the adjustment.

Table 2 presents the present value of the Rb–Sr decay constant λ which best fits each of several pairs of n and T. It is notable that these values all fall below the best laboratory measurements of the decay constant which are $0\cdot0147 \times 10^{-9}$ years^{-1}, $0\cdot0145 \times 10^{-9}$ years^{-1} by the liquid scintillation method[45, 46], and $0\cdot0147 \times 10^{-9}$ years^{-1} by a measure of decay products[47].

TABLE 2 Present value of decay constant of ^{87}Rb for various values of n and T

T ($\times 10^9$ years)	λ ($\times 10^9$ years)$^{-1}$		
	$n = \frac{1}{2}$	$n = \frac{1}{4}$	$n = 0$
5	0·0104	0·0122	0·0138
6	0·0114	0·0126	0·0138
7	0·0118	0·0128	0·0138
8	0·0122	0·0130	0·0138
9	0·0124	0·0130	0·0138

Dirac's principle of cosmology and radioactive dating

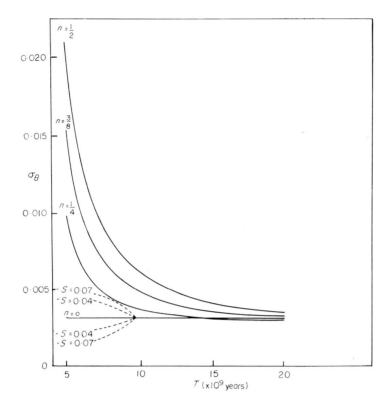

FIGURE 2 Graph of σ_θ as a function of the age of the Universe (T) for the hypothesis that the rate of β decay is proportional to T^{-n}. A small value of σ_θ indicates the data fit the theory closely. The broken curves represent confidence limits

Curves representing σ_θ as a function of T are shown in figure 2 as full lines for each of four values of n. The curve for $n = 0$ represents the Rutherford theory; the curves for $n = \frac{1}{4}, \frac{1}{2}$, and $\frac{3}{8}$ represent the Dirac theory. Inasmuch as σ_θ is a measure of how well the theory fits the data, an inspection of figure 2 shows that the Rutherford theory fits the data better than the Dirac theory for $T < 13 \times 10^9$ years. If one recalls that on the basis of the Dirac theory T should be one-third of the Hubble age, it is seen that it is very unlikely that T is as large as 13×10^9 years. A judgement of whether the agreement is satisfactory involves a knowledge of the standard error in σ_θ itself as well as an estimate of the likelihood that T can exceed a given value.

The standard deviation of σ_θ is, of course, determined by the standard deviations of the individual data points of table 1. In fact, the parameter which determines the standard deviation of σ_θ is the standard deviation of $\{(\tau_\alpha^2 + \tau_\beta^2)/2\}^{1/2}$. We shall call this latter standard deviation S. We should have a value of S for each of the entries in table 1. However, in the least-squares analysis we have assumed a common standard deviation (and hence a common S) for each of the entries. The analysis then yields a value $S = 0 \cdot 04 \times 10^9$ years. We may take an independent estimate of S from the meteorite data alone. From these data we estimate the uncertainty in τ_α as $0 \cdot 1 \times 10^9$ years ($0 \cdot 03 \times 10^9$ years scatter in analyses and perhaps $0 \cdot 07 \times 10^9$ years caused by uncertainty in the decay constant) and the uncertainty in τ_β as $0\cdot1 \times 10^9$ years (all scatter in analyses since the decay constant is not involved). Thus S is unlikely to exceed $0\cdot07 \times 10^9$ years.

The likelihood that T may exceed a given value has been estimated by assuming that the Hubble age (13×10^9 years with an uncertainty of a factor of 2) may be approximated by $(16 \pm 10) \times 10^9$ years, where $\pm 10 \times 10^9$ years will be treated as a standard error. This would appear to be a generous allowance in favour of the Dirac theory; however, the fact must be kept in mind that the present estimate of the Hubble constant differs by a factor of 6 from that used a decade ago. Combining the likelihood (measured by the standard deviation of σ_θ) that σ_θ differs from the curve given in figure 2 with the likelihood (measured by the standard deviation of the Hubble age) that T exceeds certain values, we may draw confidence limits (95% level) which bound the region of the (σ_θ, T) plane in which the Dirac theory would not differ significantly from the Rutherford theory. These limits are shown as broken curves in figure 2 for two different values of S, $0\cdot04$ and $0\cdot07 \times 10^9$ years. It is seen that the curves representing the Dirac theory lie outside of these limits for all T.

Conclusions

A test of Dirac's principle is very difficult. The year-to-year effects of the principle are so small, at least at the present epoch, that no experiment has been proposed which will detect them. On the other hand, our knowledge of the remote post is so imperfect that identification of discrepancies which might be produced by Dirac's principle is exceedingly difficult. We think that the recent determinations by McMullen, Fritz, and Tomlinson[47] of the ^{87}Rb-decay constant by mass spectrometric measurement of the amount of ^{87}Sr formed during a seven-year period is critical to the discussion here. It forms an independent verification of Flynn and Glendenin's[45] value and is strong evidence in support of the Rutherford theory in which λ is constant.

It is very likely that the Rb–Sr ages of the terrestrial samples systematically tend to be small owing to the loss of radiogenic strontium by diffusion. There is reason to believe that the Rb–Sr age of meteorites is less affected by this error. If these presumptions are correct, a curvature would be introduced into the line representing τ_β against τ_α which would be in the same sense as that introduced by Dirac's principle. It would follow that the analysis presented here would be biased in favour of the Dirac theory. In spite of this the analysis favours the Rutherford theory. Conversely, the result that the Rutherford theory is favoured by the analysis suggests that the systematic errors are rather minor. It is this latter consequence which gives us some confidence in this analysis.

It should be noted that the analysis presented here follows only the most straightforward development of Dirac's principle. It is, of course, possible that the four coupling constants (strong, electromagnetic, weak, and gravitational) are not all independent. In that case other forms of Dirac's theory become possible (e.g. Dicke[6], Dicke and Peebles[48]). A critical review of this approach is given in section VI of a review article by Zeldovich[49].

The present analysis offers no support for the Dirac theory of beta decay. However, the use of the theory of errors (with the implicit assumption of a normal distribution for the errors) in an analysis in which many of the parameters are highly uncertain (e.g. Hubble constant) cannot be accepted without serious reservation. In particular, the 95% confidence limits must not be taken too seriously. All we can conclude is that the Rutherford theory appears to be more consistent with the data than the Dirac theory and that this condition obtains in spite of a suspected systematic error which should favour the latter theory.

References

1. P. A. M. Dirac, *Proc. Roy. Soc. (London), Ser. A*, **165**, 199 (1939).
2. P. M. S. Blackett, *Nature*, **144**, 30 (1939).
3. F. G. Houtermans and P. Jordan, *Z. Naturforsch.*, **1**, 125 (1946).
4. R. H. Dicke, *Rev. Mod. Phys.*, **29**, 355 (1957).
5. R. H. Dicke, *Nature*, **183**, 170 (1959).
6. R. H. Dicke, *Phys. Rev.*, **128**, 2006 (1962).
7. H. Leutz, H. Wenniger, and K. Ziegler, *Z. Physik*, **169**, 409 (1962).
8. E. R. Kanasewich and J. C. Savage, *Can. J. Phys.*, **41**, 1911 (1963).
9. P. Roman, *Theory of Elementary Particles*, Interscience, New York, 1960, p. 403.
10. T. Fulton, G. Källén, J. D. Jackson, and C. Fronsdal, *Elementary Particle Physics and Field Theory*, Benjamin, New York, 1963, p. 276.
11. L. T. Aldrich, G. W. Wetherill, G. L. Davis, and G. R. Tilton, *Phys. Rev.*, **103**, 1045 (1956).
12. A. R. Sandage, *Astrophys. J.*, **127**, 513 (1958).
13. J. Rogers, *Am. J. Sci.*, **250**, 411 (1952).
14. W. R. Eckelmann, *Bull. Geol. Soc. Am.*, **68**, 1117 (1957).
15. A. O. Nier, R. W. Thompson, and B. F. Murphey, *Phys. Rev.*, **60**, 112 (1941).
16. P. M. Hurley and M. I. T. Staff, *6th Ann. Prog. Rept., Dept. Geol. Geophys., Mass. Inst. Technol.*, No. NYO-3939, 28 (1958).
17. G. L. Davis, G. R. Tilton, and G. W. Wetherill, *J. Geophys. Res.*, **67**, 1987 (1962).
18. L. T. Aldrich, G. W. Wetherill, G. L. Davis, and G. R. Tilton, *Trans. Am. Geophys. Union*, **39**, 1124 (1958).
19. G. R. Tilton, G. W. Wetherill, and G. L. Davis, *J. Geophys. Res.*, **67**, 4011 (1962).
20. G. R. Tilton, G. L. Davis, G. W. Wetherill, and L. T. Aldrich, *Trans. Am. Geophys. Union*, **38**, 360 (1957).
21. C. B. Collins, R. M. Farquhar, and R. D. Russell, *Bull. Geol. Soc. Am.*, **65**, 1 (1954).
22. A. O. Nier, *Phys. Rev.*, **55**, 153 (1939).
23. G. W. Wetherill, G. R. Tilton, G. L. Davis, and L. T. Aldrich, *Trans. Am. Geophys. Union*, **36**, 533 (1955).
24. G. J. Wasserburg and R. J. Hayden, *Geochim. Cosmochim. Acta*, **7**, 51 (1955).
25. G. J. Wasserburg, G. W. Wetherill, L. T. Silver, and P. T. Flawn, *J. Geophys. Res.*, **67**, 4021 (1962).
26. S. S. Goldich, A. O. Nier, H. Baadsgaard, J. H. Hoffman, and H. W. Krueger, *Minn. Geol. Surv., Bull.*, No. 41 (1961).
27. R. E. Zartman and G. J. Wasserburg, *J. Geophys. Res.*, **67**, 1664 (1962).
28. G. W. Wetherill, G. R. Tilton, G. L. Davis, and L. T. Aldrich, *Geochim. Cosmochim. Acta*, **9**, 292 (1956).
29. G. W. Wetherill, G. J. Wasserburg, L. T. Aldrich, G. R. Tilton, and R. J. Hayden, *Phys. Rev.*, **103**, 987 (1956).
30. G. W. Wetherill, O. Kouvo, G. R. Tilton, and P. W. Gast, *J. Geol.*, **70**, 74 (1962).
31. A. Holmes and L. Cahen, *Acad. Roy. Sci. Coloniales (Brussels), Classe Sci. Tech., Mem., Collection in 8°*, **5**, No. 1 (1957).
32. L. H. Ahrens, *Rept. Prog. Phys.*, **19**, 80 (1956).
33. L. F. Herzog, *Natl. Acad. Sci.—Natl. Res. Council, Publ.*, No. 400, 114 (1956).
34. L. O. Nicolaysen, L. T. Aldrich, and J. B. Dosk, *Trans. Am. Geophys. Union*, **34**, 342 (1953).
35. R. K. Webster, W. J. Morgan, and A. A. Smales, *Trans. Am. Geophys. Union*, **38**, 543 (1957).

36. G. W. Wetherill, *Geochim. Cosmochim. Acta*, **9**, 290 (1956).
37. P. W. Gast, *Geochim. Cosmochim. Acta*, **26**, 927 (1962).
38. P. W. Gast, J. L. Kulp, and L. E. Long, *Trans. Am. Geophys. Union*, **39**, 322 (1958).
39. W. H. Pinson, Jr., and C. C. Schnetzler, *10th Ann. Prog. Rept., Dept. Geol. Geophys., Mass. Inst. Technol.*, No. NYO-3943, 19 (1962).
40. V. Rama Murthy and C. C. Patterson, *J. Geophys. Res.*, **67**, 1161 (1962).
41. V. Rama Murthy and W. Compston, *J. Geophys. Res.*, **70**, 5297 (1965).
42. R. M. Shields, *12th Ann. Prog. Rept., Mass. Inst. Technol.*, No. M.I.T. 138-12, 1 (1964).
43. G. R. Tilton, *J. Geophys. Res.*, **65**, 2933 (1960).
44. E. Anders, *Rev. Mod. Phys.*, **34**, 298 (1962).
45. K. F. Flynn and L. E. Glendenin, *Phys. Rev.*, **116**, 744 (1959).
46. A. Kovach, *Acta Phys. Acad. Sci. Hung.*, **17**, 341 (1964).
47. C. C. McMullen, K. Fritze, and R. H. Tomlinson, *Can. J. Phys.*, **44**, 3033 (1966).
48. R. H. Dicke and P. J. Peebles, *Space Sci. Rev.*, **4**, 419 (1965).
49. Ya. B. Zeldovich, *Advan. Astron. Astrophys.*, **3**, 242 (1965).

2

S. K. RUNCORN
Department of Geophysics and Planetary Physics
School of Physics
University of Newcastle upon Tyne
England

A palaeontological method of testing the hypothesis of a varying gravitational constant

The hypothesis that the gravitational constant—one of the fundamental constants of nature—has changed with time is at first sight self-contradictory. However, a history of physics might almost be written around the successive discoveries that quantities which had at first been assumed constant, e.g. mass, the length of the day, but which, for a variety of reasons, were later found not to be so. In the general theory of relativity, the gravitational constant (G) features as a constant which could logically be dependent on time but, as nothing within this theory necessitates this, the principle of the economy of hypothesis has led most physicists to accept its independence of time. If, therefore, one wishes to examine such speculations, one must find another theoretical basis. Dicke[1] and Jordan[2] have argued that Einstein's theory is incomplete and that the necessary modification requires the gravitational constant to decrease inversely with time since the origin of the Universe.

The argument can be put very generally if Mach's principle, as well as the equivalence of gravitational and inertial accelerations, is accepted. Supposing there is no aether, the rotation of the Earth, which gives rise to clearly measurable effects both geophysically on the figure of the Earth and on motions in the atmosphere and core and in laboratory experiments, is a meaningless concept unless it is a rotation relative to other masses in the Universe. It then follows by the principle of equivalence that the Coriolis force and the centrifugal force in such a rotating frame of reference arise as gravitational attractions from distant masses of the Universe which are assumed isotropic in their distribution. Order-of-magnitude calculations have shown that the attraction of masses in the solar system and galaxy, which is not isotropic, are of negligible importance compared with the distant masses, which give rise to effects of the right order. Apart from the steady state theory, cosmological theories postulate that the mass distribution of the Universe changes with time. The 'big bang' theory of its origin and the law of expansion of the Universe, based on the observed ratio of the red shift to distance, predict, on this interpretation, that the gravitational constant will decrease with the age of the Universe. Bondi[3] has recently summarized this line of thought clearly.

This conclusion was reached by a different argument by Dirac[4], who drew attention to the clustering (to a few powers of 10) of the non-dimensional constants of nature around 1, 10^{40}, and 10^{80}. He argued that to consider this an accident was implausible.

The size and age of the Universe expressed in terms of atomic units of distance and time are around 10^{40}, but so also is the ratio of the electrical to gravitational attraction of elementary particles; thus G must vary inversely as the age of the Universe.

Changes in the neighbouring heavenly bodies, let alone cosmological changes, have rarely been considered as of concern to geologists and geophysicists. However, it is interesting to note here that, before continental drift was suggested, the discoverer of the Permo-Carboniferous glaciation of India was bold enough to suggest, as an explanation of such a cold climate close to the equator, that the Sun was a variable star[5]. Egyed[6] and Creer[7] have, by directing attention to the possibility that there exist geological and geophysical evidence for a variable G, made a very original contribution to Earth science. The search for such critical observations is of great importance.

Creer[7] has carefully evaluated a previous observation that the continental blocks, now covering about a quarter of the Earth's surface, would fit together rather well as a complete spherical shell on a globe of radius of about a half of the present one. Some objection could be raised to the method by which the blocks are bent to accommodate a smaller radius of curvature and, of course, there is an arbitrary element in the choice of the edge of the continental shelf, although there is no doubt that the 500 fathom isobath is a better boundary to the continental blocks than the present coastlines. The fit of Europe and North America and South America and Africa is so good that it has been accepted as among the most convincing arguments for the drifting apart of these continents and the goodness of the fit—the exceptions, Iceland and the Niger delta, are post-drift—accords with the recent time (5% of the Earth's life) of this continental drift episode. The fit Creer finds of the other continental boundaries is less good but, as he supposes this shell to have been disrupted soon after the Earth's origin, this reconstruction of the globe 4000 million years ago seems plausible. To suppose this fit corresponds to the situation an order of magnitude more recent in time, as Carey[8] and Heezen[9] have argued, results in a very rapid rate of expansion (0·5 to 1 cm/year) and raises insuperable difficulties about the Pre-Cambrian history of the Earth. On the other hand, to suppose the present shapes of the continental blocks have been substantially unchanged for 4–5 eons is equivalent to rejecting the idea of the gradual separation of sialic material from the mantle, implicit in the theory of the growth of continents. It would be a view easier to accommodate if the Earth had once been molten and its differentiation a brief episode at its origin. Viewing Creer's reconstruction purely as an observation, unencumbered by questions of the effect of its acceptance on theories of the Earth's evolution, it appears striking but, as always in the case of such qualitative evidence, hard to evaluate.

Other arguments for the hypothesis of an expanding Earth based on detailed geological observations are set out in this section, but they raise the same difficulties as the rather similar lines of evidence brought forward in the earlier discussions on continental drift. The uncertainties of geological dating and the gaps in the geological column (for times of deposition only are represented—the much longer periods of no deposition or erosion being recorded only by bedding planes) both increase as one goes to earlier geological periods. The observations of Termier and Termier (see section AIV, chapter 1 of this book) that greater areas of the continents were covered by sea in the earlier geological periods could perhaps be accounted for in this way, without supposing a change in the Earth's surface area.

Geophysical measurements have the advantage over geological observations in

The hypothesis of a varying gravitational constant

being quantitative and therefore in one important respect easier to evaluate. Unfortunately, most geophysical quantities are known only by measurements over historical times. The observations of ancient eclipses giving values over the last 2000 years of the secular accelerations of the Sun and Moon, interpreted as changes in the length of the day and month, do in principle allow changes in the Earth's moment of inertia (I) to be determined. Such a change alters the length of the day and not the month: the latter is changed by the action of the lunar tides and the consequent changes in the day can thus be calculated and allowed for.

However, 2000 years is not any greater than the periods in the geomagnetic secular variation and therefore in the varying torque which the Earth's core exerts on the mantle (as the lower mantle is a semiconductor). Thus, while the irregular changes in the length of the day since 1650 which are clearly present in the observatory data can be explained by interchange of angular momentum between the core and mantle, there may be present longer periods with the same cause which cannot be distinguished from those arising from a slow expansion (or contraction) of the Earth altering its moments of inertia. Dicke's attempts[1] to use this astronomical data to study changes of G seem therefore doomed to failure.

If, however, data on the length of the day or the month can be recovered from the geological record (as the geomagnetic field record has been recovered from the remanent magnetization of rocks), then this method can be used, for in the long run we can expect the relative rotation rates of the core and mantle to be constant and probably zero. This important extension of palaeogeophysics now seems possible for Wells[10] and Scrutton[11] have discovered growth increments on Devonian rugose corals, which appear to be daily and monthly, as well as the annual rings previously studied by Ma[12]. If one accepts the palaeontological evidence two pieces of data for each geological period are available: the number of days in the month (s) and the number of days in the year (w).

If, as seems permissible, one neglects changes in the eccentricity of the Moon's orbit in the last half eon, Kepler's third law applied to the Earth's orbit around the Sun and the Moon's around the Earth yields (see Runcorn[13])

$$\frac{w/s}{13 \cdot 37} = \left(\frac{L_0}{L}\right)^3$$

where L_0 and L are the present and past orbital angular momenta of the Moon. As G is eliminated in deriving this relation, it is true even if G changes. The loss of angular momentum from the Earth by tides, the rate of which may vary, can be found. It equals $(1+\beta)(L_0-L)$, where β is the mean ratio of the solar and lunar tidal torques, which in the absence of a complete theory of tidal friction is not known with great accuracy but lies between $\frac{1}{3}$ and $\frac{1}{5}$. From Runcorn[13] it is easy to show that

$$1 + \frac{(1+\beta)(L_0-L)}{H_0} = \frac{I}{I_0}\frac{w}{365 \cdot 2}\left(\frac{G}{G_0}\right)^2$$

where H_0 is the present spin angular momentum of the Earth, the suffix 0 indicating present values.

A variety of possible processes in the Earth's evolution could contribute to a change in I, other than the increase to be expected if G decreases with time. Runcorn[13] shows that virtually all the competing theories of the Earth's evolution imply changes of the same order of magnitude as the cosmological one.

Though all these are of the right order to be in principle detectable by the coral growth ring method, it is not clear whether there is any practical method of distinguishing between changes in I and in G. A linear decrease in G both lengthens the year and causes the Earth to retreat from the Sun and the energy generation in the Sun to change. Biological processes on the Earth set limits which were first set out by Teller[14], whose conclusions, when modified to allow for the new estimates of the age of the Universe, are not in contradiction with the Dirac hypothesis, provided one is prepared to suppose that the recent discoveries of fossils in the earlier Pre-Cambrian are forms of life which were rather tolerant of higher temperatures than those of Phanerozoic times. If they were not so, Gamov's view[15] is that even such a slow change in G as that of Dirac's hypothesis must be excluded (unless the age of the Universe is again underestimated by say 50%).

Can, therefore, we look forward—even speculatively—to a measurement which allows the separation of changes in I and G? This is possible only if there are other growth increments dependent on another astronomical period of time. Marine life, corals, stromatalites, etc., is influenced in its growth by the great periodic changes in environment, the diurnal and seasonal variations in sunlight and temperature, and the monthly and bimonthly tides. These periodicities give explanations of the three growth increments found by Wells and Scrutton. Is it too fanciful to think of the variations in the tides produced by the next longest astronomical period, that of the 18·6 year precession of the nodes of the lunar orbit? Even supposing the inclination of the lunar orbit to the ecliptic changes, this period is directly proportional to the orbital angular momentum of the Moon (L) and inversely to the torque exerted by the gravitational field gradient of the Sun. The latter is directly proportional to G and to the square of the Earth–Moon distance and inversely proportional to the cube of the Earth–Sun distance. The variation of L is determined from the ratio of the palaeontological counts w and s, and the Earth–Moon distance varies as L^2/G. This period is therefore inversely proportional to $G^2 L^3$. The number of years in this period is inversely proportional to L^3. Even if this could be determined, the variation of G cannot be calculated independently of the variation of I, unless this latter change arose wholly or mainly from a change in G.

Considerable improvement in the accuracy of determination of the quantities L/L_0 and $(I/I_0)(G/G_0)^2$ for each geological period should be possible by meaning the results from large numbers of specimens. The tidal frictional torque dL/dt may only depend inversely on the sixth power of the Earth–Moon distance, in which case it is proportional to G^6/L^{12}. To test whether $(dL/dt)L^{12}$ is constant throughout geological time would take these discussions a step further.

This possibility may seem entirely visionary: however, stromatolites occur throughout the whole geological column since the early Pre-Cambrian and corals in every geological period since the Ordovician. Palaeontology as a tool in palaeogeophysics is only at a very early stage of investigation, and it is of interest to consider to which problems it might be applied.

References

1. R. H. Dicke, *The Theoretical Significance of Experimental Reativity*, Blackie, Glasgow, 1964.
2. P. Jordan, *Die expansion der Erde*, Vieweg, Braunschweig, 1966.
3. H. Bondi, *Assumption and Myth in Physical Theory*, Cambridge University Press, Cambridge, 1967.
4. P. A. M. Dirac, 1938, *Proc. Roy. Soc. (London), Ser. A*, **165,** 199–208 (1938).
5. H. F. Blanford, *Quart. J. Geol. Soc. London*, **31,** 519 (1875).
6. L. Egyed, *Nature*, **178,** 534 (1956).
7. K. M. Creer, *Nature*, **204,** 1115–20 (1964).
8. S. W. Carey, *Symp. on Continental Drift, University of Tasmania, 1958.*
9. B. C. Heezen, *Intern. Symp. on Oceanography, New York, 1959.*
10. J. W. Wells, *Nature*, **197,** 948 (1963).
11. C. T. Scrutton, *Paleontology*, **7,** 552 (1964).
12. T. Y. H. Ma, 'Research on the past climate and continental drift', *Taiwan XIV* (1958).
13. S. K. Runcorn, *Nature*, **204,** 823–5 (1964).
14. E. Teller, *Phys. Rev.*, **73,** 801 (1948).
15. G. Gamov, *Phys. Rev. Letters*, **19,** 759–61 (1967).

A

COSMOLOGY AND GEOPHYSICS

III. Theoretical aspects of the hypothesis of Earth expansion

P. JORDAN

Hamburg University
Hamburg, West Germany

On the possibility of avoiding Ramsey's hypothesis in formulating a theory of Earth expansion

In the year 1937 Dirac, motivated by considerations regarding cosmology on the one hand, and the physics of elementary particles on the other, put forward a hypothesis expressing the idea that the 'gravitational constant' G might be in reality *not* a constant but a physical value slowly decreasing in the course of cosmological evolution. Having been fascinated by this idea of linking the Universe with the electron, the present author devoted several years to a study of the question: how might Einstein's theory of gravitation be revised or generalized if Dirac's hypothesis is to be correct?

A series of physicists, astronomers, and mathematicians joined me in this endeavour. The French mathematician Thiry independently found that this question is closely related to the so-called five-dimensional formulation of general relativity. Dicke from other considerations arrived at a similar theory. But though mathematically beautiful relations were revealed by these studies—summarized in a book by Jordan[1]—it remained unsatisfactory that the extreme slowness of the surmised decrease seemed to leave scarcely any hope of connecting these mathematical speculations with empirical facts.

Therefore the author again became interested when his late friend Fisher in New York made the remark that Dirac's decrease of G, if it existed, must have caused a marked expansion of the Earth in the course of its history. The author believed that here could be seen a possible answer[1] to one of the great problems of Earth research: why is there a division of the surface of the Earth into two different parts, continental areas and deep sea?

Thus the present author began to read a little about the sciences concerning the Earth and Moon, and he realized that there are many theories which cannot be reconciled with Dirac's hypothesis (for instance the interpretation of the *great* craters of the Moon as volcanoes). But there were no proved *facts* which could be used in order to *discard* the hypothesis. Selecting a set of ideas, every one of which has been put forward and discussed by at least *one* specialist already (though remaining controversial), the author obtained a theoretical picture which seems to be in agreement with what we really know, and with what we should have to infer if Dirac's hypothesis were to be correct.

During the year since the publication of the book by Jordan[2] about this topic, rapid progress has been made in all the fields of science that were touched upon, and the results seem to be encouraging.

Pochoda and Schwarzschild[3] (and in a similar manner also Gamow[4]) discussed, as an argument *against* Dirac's hypothesis, that, in the case of a reciprocity of G with the age of the Universe, the Sun in the course of its history would have burnt so much of its hydrogen that today it would have to belong not to the main sequence of stars, but to the red giants. This very valuable contribution to the discussion makes explicit use not only of the hypothesis of slow decrease of G, but also of the proportionality with the reciprocal of the age of the Universe; this latter point is not supposed (and not acknowledged) in the author's considerations. He (the present author) thinks that this question leads us to problems not yet solvable, concerning the origin of galaxies. Considering the unorthodox ideas of the late Ambarzumian about the creation of galaxies by *explosive* processes, one can surmise that in the phase of formation of our galaxy the value of G may have been far from that proportionality, and also far from agreement with the spatial mean value, of G in the whole Universe. It seems that these difficult questions cannot be decided today, but that the above-mentioned investigations will later become very useful for attacking problems which now better remain separate from those questions which concern only the last 500 million years.

Discussions with Elsasser convinced the author that a certain detail of his book must certainly be corrected: at one point use was made of a hypothesis of Ramsey and, though such a famous geophysicist as Bullen has taken this hypothesis seriously too, Elsasser convinced the author that it must necessarily be discarded, because it is in contrast with modern experimental and theoretical knowledge in high-pressure physics and its application to the interior of the Earth, as performed especially by Birch.

In this book an idea of Fisher was also mentioned, i.e. the doubt that the fluid state of the outer core would be definitively proved by the existence of only *one* kind of wave in it; the author felt a certain amount of reluctance about acknowledging Elsasser's theory of geomagnetism as the definitive solution of this problem. On both points the present author is now sure of the definitive correctness of Elsasser's opinion.

From Dirac's hypothesis it can be inferred by a very simple mathematical calculation[2] that *ephemeris time T*—definable also in the case of correctness of this hypothesis—cannot be identical with the exact inertial time t. We have theoretically the relation

$$T = t(1 - \varepsilon t) \tag{1}$$

with

$$\varepsilon = -\frac{\dot{G}}{G}, \quad \varepsilon^2 \simeq 0 \tag{2}$$

For nearly ten years Nicholson and Sadler[5] measured and registered the relation between T and t, using sharply defined values of T (given by the occultations of stars by the Moon) and measuring the corresponding inertial times by an atomic clock. Becker and Fischer[6] recalculated their results with the help of a computer, comparing them with the theoretical formula (1). Ten years of registering results were not enough to put the results beyond every doubt. But it *seems* that the most probable empirical value fulfils the relation

$$10^{-10} \text{ year}^{-1} < \varepsilon < 10^{-9} \text{ year}^{-1} \tag{3}$$

From this basis, further noting of results may probably soon lead to a considerable diminution of the probable error and to a real decision about Dirac's hypothesis.

The relation (1) is what can be measured—investigating the planetary system—from the following more complete theoretical statement: the solution of the equations of motion for a Dirac n-body problem, with

$$G = G_0(1-\varepsilon t)$$

with masses M_k, spatial coordinates $\mathbf{r}_k$, and with $\dot{\varepsilon} \neq 0$ (but $\varepsilon^2 = 0$), can be inferred from a Newtonian n-body problem with G_0 instead of G, with masses M_k, spatial coordinates $\mathbf{R}_k$, and time T instead of t, where

$$\begin{aligned} \mathbf{R}_k &= \mathbf{r}_k(1-\varepsilon t) \\ T &= t(1-\varepsilon t) \end{aligned} \tag{4}$$

This mathematical result (verified by a trivial calculation) shows that the effect of Dirac's decrease of G is a slow *expansion* in every purely gravitational n-body problem, associated with a corresponding slowing down of the movements[7].

One may be inclined to ask whether the Hubble effect is also contained in this formula (4), so that ε would be identical with the Hubble constant α. Indeed, ε and α have the same order of magnitude; but whether they are exactly equal must be decided theoretically by further investigation of cosmological models in the theory of gravitation with $\dot{G} \neq 0$.

Classical planetary astronomy—without radio echos—obtains precision measurements of *angular* coordinates with deviations of only about 10^{-7}; in *radial* coordinates there remained relative uncertainties of 10^{-4}. Using radio echos these uncertainties can now be reduced to about 10^{-6}. Only in the case of the Moon, by the application of orthogonal reflectors transported to the Moon and by the application of lasers, may corresponding uncertainties be reduced to 10^{-8} or even 10^{-9}, so that here is a further possibility to test Dirac's hypothesis by precision measurements—as mentioned in Jordan's book, and fully discussed elsewhere in this book.

Let us now consider the consequences of Dirac's hypothesis on the expansion of the Earth. The work of Bullen and other geophysical investigators gives a basis for evaluating the *elastic* response of the Earth to $\dot{G} < 0$: Hess and Murphy, two coworkers of Dicke's, found

$$-\dot{R}/R = 0{\cdot}1 \dot{G}/G \tag{5}$$

This leads us to expect a very small slow expansion of the Earth, not sufficient to allow us to see in it a source of really interesting geophysical consequences. But the situation becomes more interesting if we try to analyse some empirical facts.

Let us remember the sensational confirmation of Heezen's concept about the role of the *oceanic rifts* in the process of the *spreading of ocean floors*, by the investigation of magnetic anomalies of the ocean bottom parallel to these rifts (cf. Vine[8], and also

Cox, Dalrymple, and Doell[9]). We can say now that this much discussed and much doubted spreading has ceased to be any hypothetical process: it is a *fact*.

Two different interpretational hypotheses are possible now.

(i) If there is no expansion ($\dot{R} = 0$), then there must exist some compensating compressional feature.

(ii) If there is not sufficient compensation, then expansion $\dot{R} > 0$ must unavoidably be inferred from oceanic spreading.

We shall discuss here only hypothesis (ii), assuming not more than perhaps 10% compensation, which may be neglected totally in a first rough approximation.

According to this, the growth of oceans must be (approximately) equal to the growth (caused by expansion) of the total surface $4\pi R^2$ of the surface of the Earth. From this an *estimation* of the growth of $4\pi R^2$ since the Carboniferous Age is possible. From geological arguments, not discussed here, the author is convinced that the separation of Africa and South America (surmised by Wegener and proved by Carey) began in the Carboniferous Age—though Wegener and also some modern authors believed that they were together still in the Triassic Age. Summarizing similar other estimations we think that the Earth's radius R in the Carboniferous Age may have been about 80% of its present value.

This 80%, discussed in Jordan's[2] book, is in accord with van Hilten's[10] attempt to determine from palaeomagnetism the change of R in the course of the Earth's development, and the similar attempt of Khramov and Komissarova[11]. Now we have heard in the course of this Conference severe criticism about van Hilten's method, and certainly one cannot deny that this criticism is justified to a certain degree. From geometrical grounds there cannot be any general and exact law connecting the measurable palaeomagnetic quantities of today with the value of the former curvature of continental lumps. Taking this into consideration radically, one has to discard van Hilten's results, but one has to discard too the results of similar investigations which tried to prove that *no* appreciable expansion took place. But one cannot, on the other hand, deny that the method may allow *some restricted* applications, under certain conditions, in which the calculations of van Hilten, though only *approximately* correct, can give some information of approximate value. Especially, one must avoid its application to regions with mountain folding—orogenic activity being caused, according to the expansion theory, as a result of the adaptation of continental lumps to the decreasing curvature of the Earth's surface. (This interpretation, discussed in the book by Jordan in only a very sketchy manner, has successfully been made precise in a paper by Glashoff[12].) The author is inclined to believe that, in spite of the imperfections of the method, van Hilten's results are probably not very far from veracity. But the estimation 80% was obtained by the present author *before* he knew about the work of van Hilten and of Khramov and Komissarova.

The decisive empirical fact leading us inevitably to the correct interpretation of the existence of areas *with or without* a sialic layer lies in the evidence that the sialic layer, where it exists, *has in all cases (approximately) the same thickness*.

All speculation about the growth of continents (which, according to the expansion theory, is only the growth of dry land areas), the formation of the Moon out of the sialic material of the Earth, the formation of the Earth's sialic layer out of the material of a captured Moon, and the decrease of continental areas in consequence of the spreading of ocean floors gives us no indication as to an explanation of the existence of *two preferred hypsographic levels* on the Earth, meaning the existence of a *spatially constant* thickness of sial.

No attempt at all has been made to explain this fundamental empirical fact—one has to state simply that in a large amount of literature even *mention* of this fact is lacking. From this lack of any attempt to explain the two-level situation we have to infer the *impossibility* of any explanation in such a manner that any plausible process might be invented which could *generate* this splitting of hypsographic levels—all known processes can only contribute to *destroying* it; no process has been invented in any form of hypothesis which would be able to *restore* it, after destruction, or to *maintain* it against damage; *erosion* especially would be totally unable to do so. Therefore also any *change* of this thickness in the course of geological time cannot have been performed: this thickness must be interpreted as a characteristic value of the structure of the Earth, determined in the primary steps of the Earth's formation, and never changed in later ages.

The author does not see any possibility of explaining the two-level situation, other than the following one. Just after the Earth's formation the sial layer—then still fluid—covered the whole surface of the body of the Earth in spherical symmetry and equal thickness. After solidification no essential change of this thickness could be made; expansion of the Earth could only *tear* this sialic skin into separate lumps—oceanic areas originating between them.

(The word 'essential' means that a certain amount of decrease of the total sum of continental areas must be acknowledged, caused by orogenic processes. But this amount must have remained relatively small—otherwise the empirical fact of approximate constancy of sial thickness (equivalent to the two-level phenomenon) would have been destroyed.)

Therefore to a first approximation the sum of all the sialic areas had to remain *constant*. The continental lumps, covering in the beginning the whole surface, left no room for the oceans between them. The radius R then must have been about 65% of that of today.

Already in Wegener's old concepts there appeared a division of the history of the Earth into two clearly distinct parts. During the greatest part of this history there existed—according to him—only *one* ocean and *one* huge continent. But at a certain age of the Earth (in the Carboniferous Age, according to the author's opinion, though *Wegener* himself preferred another dating) there began the great processes of continental disruption and drift.

Now we see clearly the meaning of these two chief ages in the history of the crust: in the first one R grew from 65% to about 80%, needing a duration of about 4000 million years. Obviously this expansion process may be interpreted as resulting from the *elastic* response of the Earth's material.

But our considerations, if they have anything to do with reality, force us now to acknowledge that approximately

$$\dot{R}/R = 10^{-9} \text{ year}^{-1}, \qquad \dot{R} = 5 \text{ mm/year} \tag{6}$$

since the Carboniferous Period.

Today this statement seems no longer so embarassing as two years ago, when the author—reluctantly—wrote it down in his book. For it resulted from the study of magnetic anomalies at the bottom of the oceans[8] that the spreading along the expansion rifts there gives at some points *several* cm/year breadth of new oceanic bottom, so that $\dot{R}$ surely cannot be still smaller than 5 mm/year.

Therefore apart from the elastic response, as had already taken place *before* the Carboniferous Period, still another additional cause for expansion must have been action in later ages.

It is a very remarkable fact that the spreading of the oceans and the corresponding formation of 'Klaffspalten' in the grabens of continents show the expansion of the *surface* of the Earth's crust to be wholly concentrated along a few long lines—the active parts of the system of rifts discovered by Ewing, Heezen, and Tharp. These facts show that there is a strong nucleus of truth in Runcorn's tendency to believe in far-reaching slow movements in the mantle material. But in spite of this the author is reluctant to believe that the mobility of the mantle material would be great enough to make possible (i) polar migration, and (ii) such movements of continents which are *not* caused directly by the expansion.

Let it be said openheartedly that all theories about polar migration seem to me to be pure fairy tales.

The concept of polar migration resulted from the fact that many cases of old glaciation are known from places lying a long way from the recent poles. This difficulty led also to the attempt to doubt the reality of some reported cases of old glaciation. Wegener himself thought that, for instance, the facts known about the Congo domain were not really a case of glaciation; he was inclined to consider them as pseudoglacial phenomena. But for instance Kummel's[13] book (see the figure on p. 334) shows that at least several of the modern geologists are convinced that the Congo glaciation was a reality; in spite of the fact that, among all the numerous contradictory attempts to reconstruct the path of a migrating North Pole, there was never a reconstruction making the Congo domain a part of a polar region.

But the idea of polar migration seemed to be unavoidable only because it had been made an axiom of palaeoclimatology that also the older glaciations—including the Permo-Carboniferous one—must have been *polar* glaciations like the Diluvian one. The Dirac hypothesis on this point gives a totally different picture, discussed in a certain amount of detail in Jordan's[2] book. In this connection the author arrived at the concept of a closed layer of clouds around the Earth still in the Carboniferous. Meteorological objections to such a picture can be overcome[14].

The author's late coworker Binge put forward an interpretation of *volcanism* on the basis of Dirac's hypothesis, claiming that the hitherto unknown (or misinterpreted) primary processes of volcanic activity (including intrusions) must be *explosive phase transformations*—these are caused by the fact that materials of the lower crust are brought by decreasing pressure to conditions in which their original high-pressure phases cease to be the most stable ones, the materials therefore gaining a tendency to go over into less-dense low-pressure phases. It seems that Binge indeed found the key to the great problem of volcanism.

It has been inferred[2] from his concept that on the *Moon* there could not be any real volcanism with the outflow of lava, though certainly there is a lively activity of gaseous outbreaks, which caused the many well-known chains of very small craters. A valuable contribution to an empirical confirmation of this conclusion has been given (by Hibbs) by the fact that the many already existing satellite-made photographs of the Moon did *not* give any evidence of a volcanic character of the central peak in the crater Alphonsus, where Kozyrew observed two cases of gaseous outbreaks (misinterpreted as proof for real volcanic lava eruptions). This recent result is of fundamental interest because authors, believing that the *great* craters would also be volcanoes instead of collision marks, claimed such central peaks as strong proof against the impact theory (cf. von Bülow[15] and also his other numerous publications).

At last we come back to the *second* part of the history of the crust. What cause may have been in action since the Carboniferous Era giving, in addition to the *elastic* expansion, another about ten times stronger contribution to the expansion process? It may now become understandable that the author felt at this point a strong temptation to succumb to Ramsey's hypothesis, though several years ago Teller had already warned against this hypothesis. According to Ramsey, the material of the outer core is not chemically different from that of the mantle; it is another denser *phase* of the same material as the mantle itself. Therefore it seemed an attractive idea to say that perhaps at the beginning of the Earth's history there was scarcely any mantle beneath the sial layer, and this absence or approximate absence of the mantle continued during several thousand million years; the expansion during this part of time was only an *elastic* one. But afterwards, in the second part of the crust development, the chief part of the expansion was performed by the phase change: the core material was transformed to mantle material.

After having learnt from Elsasser that Ramsey's hypothesis really must be abandoned totally (the arguments by which the author had hoped to escape from this conclusion were insufficient), the question arose whether the nucleus of the idea indicated above can be maintained anyhow *without* using Ramsey's concept. The nucleus of what was said above seems to lie in the idea that (since about the Carboniferous Age) material from the core began to separate itself from the Fe/Ni, and reinforce the mantle. That would mean that in the primordial core there had been *dissolved* considerable amounts of elements other than Fe and Ni. Surely these could *not* be elements corresponding to the elementary composition of olivine—we know that O has a very low solubility[16] in the core material. In order to correspond to what our idea requires, this material must have had a strong portion of elements possessing an atomic volume markedly greater than the corresponding ion volume, so that these elements after emerging from the core needed more space than before, in solution in the core material. If these requirements can be fulfilled, we should come possibly to understand, not only without Ramsey's hypothesis, but also in a more plausible form, that for a certain time the mantle could have grown by crystallization of matter originally dissolved in the fluid core.

But has the mantle, according to our modern knowledge, really got a structure which can be thought to be in agreement with these tentative conclusions?

Our understanding of the interior of the Earth won considerable refinement since Bullen in the third edition of his famous book[17] defined a certain parameter η, calculable from seismologically measurable quantities, which has the value $\eta = 1$ in the case of chemical homogeneity, and values $\eta > 1$ in the case of chemically inhomogeneous matter. Bullen obtained the result that the old hypothesis of a chemically homogeneous mantle can scarcely be maintained. Recently, Anderson[18] obtained still more precision by making use of some simple, though only approximate, laws given by the quantum theory of crystal lattices.

It is well known already from older investigations that the mantle shows several spherical limits in different parts. The Byerly sphere, 400 km deep, and the Repetti sphere, 1000 km deep, are examples of these. Anderson calculated the values (given in table 1) of η for several zones of the mantle.

TABLE 1   Values of η for several zones of the mantle

Limits	Depths (km)		Interpretation
Byerly sphere	33– 400	1·0	old
	400–1000	1·8	
Repetti sphere	1000–2500	1·4	younger
	2500–2900	3·0	

Therefore it seems allowable to interpret (tentatively) the matter above the Byerly sphere as the *old* mantle, already formed at the beginning of the history of Earth, and all matter below the Byerly sphere as the *young* mantle, formed *afterwards* by crystallization of the originally dissolved matter from the Fe/Ni fluid. The *strong* expansion of the geological ages since the Carboniferous would then be attributed to the segregation of dissolved material from the core.

Finally, we emphasize that this picture of the interior of the Earth and its historical development seems to be in best accord with a remark of Elsasser[19] about the origin of the Earth–Moon system. The content of the remark is as follows: in order to understand this origin *without* taking recourse to any capture hypothesis, one has to assume that the differentiation between core and mantle had already taken place in the formation process of this sytem.

Though MacDonald showed, in a highly interesting manner, that the capture hypothesis can be formulated without violation of the laws of mechanics (so often neglected in speculations about capture processes), *if* strong *tidal friction* is assumed, the author prefers to believe that the origin of the Earth–Moon system took place without capture. Then a picture of this origin can scarcely be drawn in any other manner than the following one. In the planetoidal state there existed a mass consisting of small and larger pieces of matter, from atoms and molecules to meteoritic and asteroidal bodies. Among these were ones of a chondritic character and ones similar to iron meteorites—though it is well known that the asteroid belt must be thought to have originated from the destruction of a greater body; on the other hand, the occasional occurrence of iron monocrystals among meteorites shows that direct formation of meteorites containing practically only Fe is also possible. During the process of uniting the greater part of these masses in a central mass, strong friction in the planetoid matter caused pieces of denser matter to fall to the centre more rapidly than less dense ones. As the rotational impulse of the system allowed no concentration in *one* small body, the Moon or its predecessors were formed from less dense material; but in the centre the Fe/Ni accumulated, and became fluid in consequence of the impact energy. Considerable contributions to the formation of this fluid centre originated also from non-metallic elements; if they contained oxygen, they became separated soon afterwards from the Fe/Ni fluid. But other materials became dissolved in the fluid core, so that a long time afterwards their separation from the core could begin, after large changes of G, pressure, and temperature.

References

1. P. Jordan, *Schwerkraft und Weltall*, 2nd ed., Vieweg, Braunschweig, 1955.
2. P. Jordan, *Die Expansion der Erde*, Vieweg, Braunschweig, 1966.
3. P. Pochoda and M. Schwarzschild, *Astrophys. J.*, **139**, 587 (1964).
4. G. Gamow, *Proc. Natl. Acad. Sci. U.S.*, **57**, 187 (1967).
5. W. Nicholson and D. H. Sadler, *Nature*, **210**, 187 (1966).
6. G. Becker and B. Fischer, *Mitt. Physik.-Tech. Bundesanstalt*, **77**, 15 (1967).
7. P. Jordan, *Z. Physik*, **201**, 394 (1967).
8. F. J. Vine, *Science*, **154**, 1405 (1966).
9. A. Cox, G. B. Dalrymple, and R. R. Doell, *Sci. Am.*, **216**, 44 (1967).
10. D. van Hilten, *Geophys. J.*, **9**, 279 (1965).
11. A. N. Khramov and R. A. Komissarova, in *Rock Magnetism and Palaeomagnetism* (Ed. A. Ya. Vlasov), Siberian Division of the Academy of Sciences of USSR, Krasnojarsk, 1963.
12. H. Glashoff, *Akad. Wiss. Lit. (Mainz), Abhandl. Math.-Nat. Kl.*, **1966**, 671.
13. B. Kummel, *History of the Earth*, Freeman, San Francisco, London, 1961.
14. P. Jordan, *Akad. Wiss. Lit. (Mainz), Abhandl. Math.-Nat. Kl.*, **1967**, 43.
15. K. von Bülow, *Ann. N. Y. Acad. Sci.*, **123**, 528 (1965).
16. B. J. Alder, *J. Geophys. Res.*, **71**, 4973 (1966).
17. K. E. Bullen, *An Introduction to the Theory of Seismology*, 3rd ed., Cambridge University Press, London, 1963.
18. O. L. Anderson, *Trans. N. Y. Acad. Sci.*, **27**, 288 (1965).
19. W. Elsasser, private communication.

2

L. EGYED

*Eötvös University
Budapest, Hungary
and
IISEE
Tokyo, Japan*

The slow expansion hypothesis

All the hypotheses have their empirical basis and can be supported by observations as well as by theoretical considerations. In the following we shall try to summarize the observations pointing towards an expansion of the Earth and amounting to an increase of 0·65±0·15 mm/year in radius. Furthermore, some theoretical basis of the expansion is suggested resulting in the same rate of radius increase and some consequences will be considered.

Let us consider first the dimension of water-covered continental areas during the geological past[1]. These values are represented in figures 1(a) and 1(b) based on the palaeogeographic maps of the Termiers and Strahow, respectively. If we take into evidence that, during the past 600 million years, the mass of the hydrosphere increased by about[2] 4% and that the level of the ocean bottom has been controlled first of all by isostasy, the decreasing trend exhibited in these diagrams definitely supports an expanding Earth. From the diagrams and the actual hypsometric curve, it is possible to determine a probable minimum value of the yearly radius increase. The result is 0·54 mm/year from the first, and 0·76 mm/year from the second, series of data. Hence the probable minimum rate of yearly radius increase based on palaeogeographic maps amounts to 0·65±0·11 mm/year.

The expansion of the Earth offers the most simple explanation of the formation of the continental crust and oceanic basins. On this explanation, the primaeval Earth became a layered one by spherically symmetrical differentiation, and a uniform continental crust covering the whole surface was formed. The deviations of this surface from the level surface would show a Gaussian distribution. This primaeval uniform crust disrupted under the effect of internal expansion, and along the fracture lines the emerging denser material of the deeper parts controlled by isostasy occupied a level about 5 km deeper than the original crust, providing the base of the first oceanic basins (figure 2).

Many properties characterizing the continental and oceanic areas are plausible consequences of the above simplified mechanism, e.g. the double maxima of the hypsometric curve, reflected also in the frequency of the depth distribution of the Moho-surface data (figure 3); the possibility of matching all the continents on a globe with a total surface area nearly equal to that of the continental areas[3]; etc.

Fracture lines very similar to those observed along the coast lines of continents can be produced by very simple model experiments. If a layer of plaster applied to the

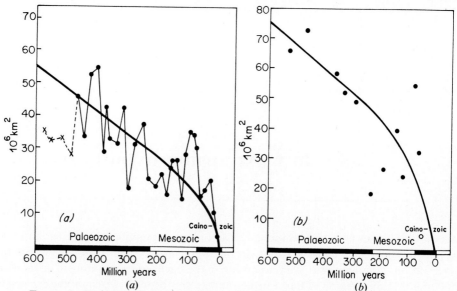

FIGURE 1 Variation of water-covered continental areas during the Earth's history computed from the data of (a) Termiers' and (b) Strahov's palaeogeographic maps

surface of a blown-up football bladder is subjected after hardening to an internal pressure, the consequent fracture lines show surprising similarities to known features of continental margins (figures 4(a), (b)).

This mechanism is reliable for estimating also the rate of expansion. According to the mechanism the first surface of the Earth coincided with that of the present continents, so that the primaeval radius can be computed.

An upper limit for the surface of continents can be regarded as the area above the isobath 2000 m below sea level. This area is $2 \cdot 07 \times 10^{18}$ cm^2 and the corresponding radius is $4 \cdot 06 \times 10^8$ cm. The most probable value of the age of the crust has been estimated as $4 \cdot 3 \pm 0 \cdot 2 \times 10^9$ years. Hence a lower limit to the average yearly radius increase is $0 \cdot 51$ mm/year.

The second limit can be obtained by assuming the continental component of the hypsometric curve (figure 5) to be equal to the area of the first crust. This area is $1 \cdot 12 \times 10^{18}$ cm^2 and the corresponding radius is $3 \cdot 0 \times 10^8$ cm. Thus an upper limit to the average yearly expansion derived from the area of the continents is $0 \cdot 83$ mm/year.

The suggested explanation of the formation of continents and ocean basins results in an average yearly radius increase of $0 \cdot 67 \pm 0 \cdot 16$ mm/year, a value coinciding exactly with that derived from palaeogeographic maps.

This coincidence supports not only the reality of the expansion but suggests that, since Cambrian times, the distribution of elevations on the Earth's surface (hypsometric curve) has not been essentially different from that of the recent distribution.

It is very easy to show that the recent slowing down of the speed of rotation of the Earth is consistent with the above rate of radius increase if the effect of tidal friction is excluded. As a matter of fact the rate of yearly radius increase $\dot{R}$ may be derived from the formula

$$\dot{R} = -\frac{R}{2}\frac{\dot{\omega}}{\omega}$$

The slow expansion hypothesis

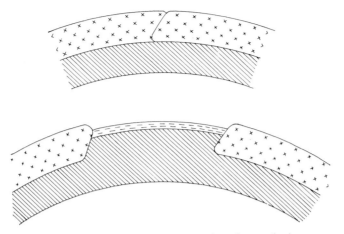

FIGURE 2 Mechanism of the formation of ocean basins

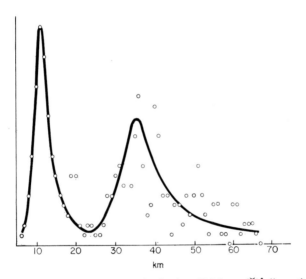

FIGURE 3 Frequency diagram of the depths of Mohorovičić discontinuity

where R is the Earth's radius and ω the angular velocity. According to the most recent data[4]

$$\dot{\omega} = -4\cdot 81 \times 10^{-22} \text{ rad/sec}^2 = 1\cdot 44 \times 10^{-14} \text{ rad/sec year}$$

$$\omega = 7\cdot 29 \times 10^{-5} \text{ sec}^{-1}, R = 6\cdot 37 \times 10^8$$

These data give $\dot{R} = 0\cdot 63$ mm/year, a value in excellent agreement with those obtained above.

This result suggests that, at least at present, no significant effect of tidal forces is involved in the slowing down of the Earth, if the stated yearly radius increase is correct.

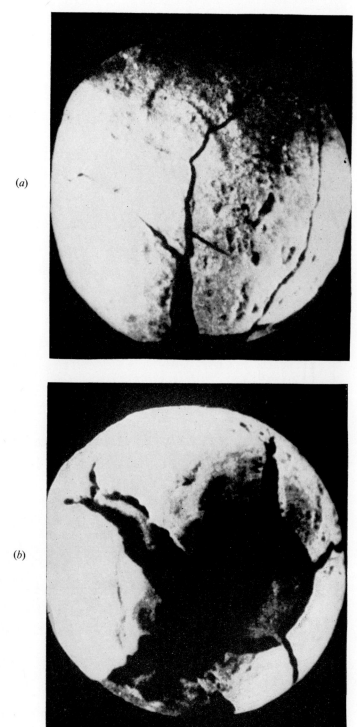

FIGURE 4(a), (b) Fracture patterns, caused by the increase of internal pressure in layers of plaster solidified on an inflated football bladder

The slow expansion hypothesis

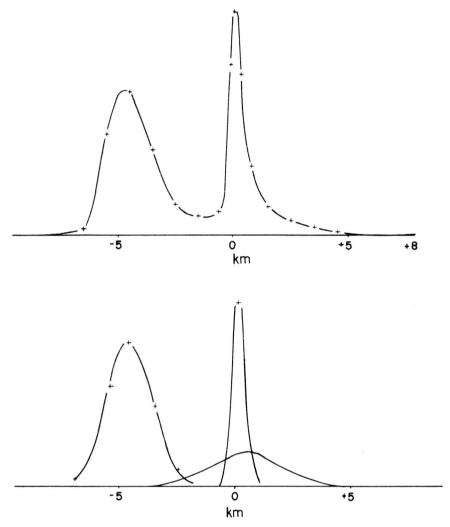

FIGURE 5 The hypsometric curve decomposed into (a) oceanic, (b) continental, and (c) sedimentary parts with Gaussian distributions

It was shown by Dearnley[5] that the orogenic belts (up to 2750 million years) show a regular distribution with respect to the actual equator if an expansion of 0.65 ± 0.25 mm/year is supposed.

It was also suggested by the author that palaeomagnetic data be used to determine the ancient radii of the Earth[6, 7]. However, the scattering of the data is too large for this purpose, although the results did not contradict the rate of expansion derived above[8].

The following detail is relevant to the theoretical background of the expansion hypothesis. The yearly energy necessary to lift the mantle by the expansion exceeds 2.6×10^{29} erg/year. On the other hand, the heat energy released in radioactive decay by an Earth with a composition like that of the basalt, for which the rate is 50 erg/year g-rock, amounts to 3×10^{29} erg/year. This suggests an expansion caused by thermal dilatation. Such an explanation was suggested, e.g. by Eder[9]. A detailed computation shows, however, that this effect is not sufficient. A result consistent with the rate of radius increase estimated from different observation data can be obtained only

TABLE 1 $\dot{R}$ and time interval

	Rate of radius increase $\dot{R}$ (mm/year)	Time interval (million years)
Palaeogeographic maps	0.65 ± 0.11	600
Continental crust	0.67 ± 0.16	4300 ± 200
Slowing down of the speed of the Earth's rotation	0.63	recent
Orogenic belts (Dearnley)	0.65 ± 0.25	3000

if we suppose (i) a Dirac[10] cosmology in which the 'gravity constant' is a quantity decreasing inversely with a time parameter comparable with the age of the solar system, (ii) the existence of high-pressure phases of the type proposed by Ramsey[11], and (iii) the average radioactivity for the whole Earth's interior corresponds to that of the basalt, the heat generated by radioactive decay thus amounting to 50 erg/year g-rock.

On these assumptions the yearly radius increase can be computed using the following relations based on the first law of thermodynamics:

$$\delta \dot{p} = \delta p \left(\frac{4\dot{r}}{r} + \frac{1}{t} \right)$$

$$\dot{p} = \sum_R^r \delta \dot{p}$$

$$\beta = \left(\alpha \dot{T} + \frac{\dot{p}}{K_T} \right) \Delta$$

$$\gamma = \frac{2\dot{r}}{r} \Delta$$

$$\dot{T} = \left(\dot{Q} - \frac{p\dot{p}}{\rho K_T} \right) \Big/ \left(C_V + \frac{\alpha p}{\rho} \right)$$

then

$$\dot{R} = \sum_R^0 (\beta - \gamma) + \frac{\lambda_h p_h}{g_h} \left(\frac{1}{\rho_{h-0}} - \frac{1}{\rho_{h+0}} \right)$$

where

$$\lambda_h = \begin{cases} 0 & \text{if } h \neq 2900 \text{ km} \\ 1 & \text{if } h = 2900 \text{ km} \end{cases}$$

The symbols used are as follows: δp, partial pressure increment of the layer considered; $\delta \dot{p}$, yearly change of the partial pressure in the layer considered; r, distance of the layer from the centre of the Earth; t, time parameter of the Dirac relation, estimated for the present as 4.5×10^9 years; γ, geometrical decrease of the thickness of the layer caused by expansion; β, increase of the thickness of the layer caused by elastic and thermal expansion; $\dot{Q}$, yearly heat generation, amounting to 50 erg/g; $\dot{T}$, yearly temperature increase; K_T, isothermal incompressibility coefficient; α, volume thermal expansion; C_V, heat capacity at constant volume; ρ, density; g_h, gravity intensity at depth h; $\dot{R}$, the yearly expansion of the Earth.

It is interesting that in the interior of the mantle $C_V + \alpha p/\rho$ is nearly a constant, equal to 14.6×10^6 erg/degC.

The value of this expression is supposed to be constant also in the core, the most probable being derived as 7.3×10^6 erg/degC.

The method was applied to the Landisman–Satô–Nafe[12] model M_1 and gave the rate of radius increase as 0.72 mm/year.

It is interesting to note the result of the yearly temperature increase which is plotted against the depth in figure 6. If we take into account the low heat conduction, this result can be regarded as a first approximation to the components affecting the temperature distribution in the Earth's interior. This suggests, moreover, a surprisingly simple explanation of the circumstances at the core–mantle boundary entailing a low temperature at the bottom of the mantle in contact with a high temperature surface of the core. The estimated temperature of the outer core, namely a value around 8000 °C, is consistent with the fluid character of the outer core, while the temperature in the inner core does not exceed the melting point at the given pressure thus the inner core can be in a solid state.

The reality of continental drift cannot be questioned any more, at least in the sense that the Americas, Africa, and Europe formed a single continent in the Palaeozoic.

The palaeomagnetic data enable us to draw the conclusion that Gondwanaland existed during the Palaeozoic and started to break up during the Permo-Triassic, i.e. about 200 million years ago[13].

The Earth's expansion seems to give a reliable explanation of continental drift, i.e. continental drift can be regarded as an episode of ocean-basin formation. But the most probable rate of expansion, 0.65 ± 0.15 mm/year, results in an increase of $2.1 \pm 0.5 \times 10^{17}$ cm^2 of the oceanic areas from the beginning of the Mesozoic. This amounts to 10 to 15% of area of the Atlantic, Indian, and Arctic Oceans regarded as a result of continental drift, but it is nearly exactly equal to the area of the main parts of the midoceanic ridges of these areas.

A quantitative consideration of the rate of expansion points to the following consequence.

If continental drift resulted in the evolution of the Atlantic, Indian, and Arctic ocean basins, the surface increase due to expansion must have been restricted to the area of midoceanic ridges and the continents must have moved away from the residual part of the actual oceanic areas and overridden parts of the contemporaneous oceanic areas.

The whole phenomenon can be considered, however, as a result of internal expansion. The mechanism suggested is the following.

The episode of continental drift started with the fracture system separating the more or less contiguous Gondwanaland and Laurasia into the main blocks of the actual continents. A deep zone of diminished strength and rigidity was brought about along this new fracture system. Along the deep fracture zone a steady convection of material started as a consequence of the expansion of the Earth's interior, the mass transport

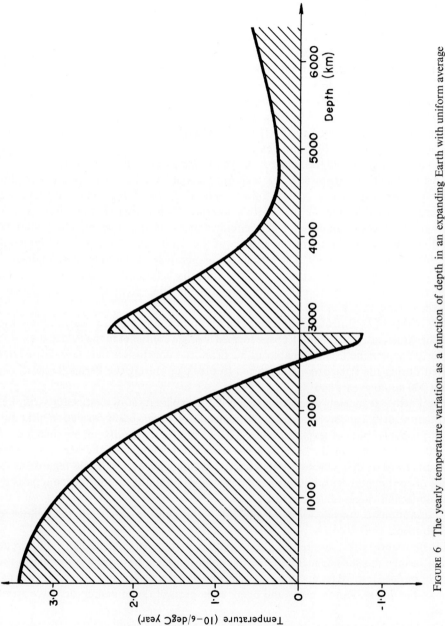

FIGURE 6 The yearly temperature variation as a function of depth in an expanding Earth with uniform average distribution of radioactive elements

The slow expansion hypothesis 73

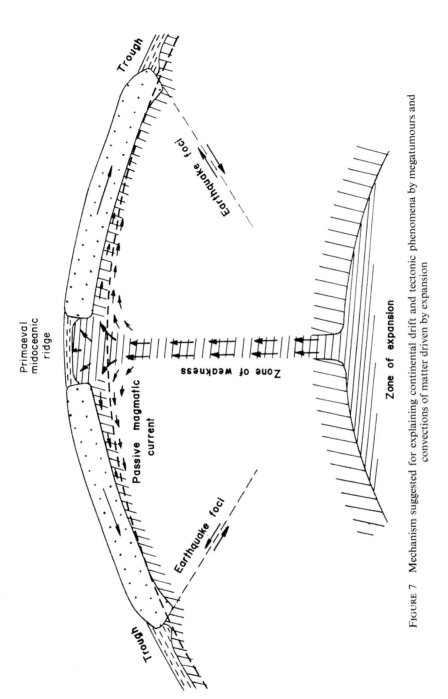

FIGURE 7 Mechanism suggested for explaining continental drift and tectonic phenomena by megatumours and convections of matter driven by expansion

adjusting the space deficiency arising from the density decrease, and near the surface a megatumour has formed. The broken-up pieces of the original continent Pangaea found their way onto the flanks of the megaundulations (figure 7). Under the effect of the relevant gravity component the continental blocks started to slide down, partly together, with the material forming the megatumour. This movement was hindered by internal friction of the subcrustal material only, as the sliding-down phenomenon was accompanied by an outflow of the masses of the megaundulation, i.e. the expansion driven from the interior brought about a magmatic current, giving a continuous supply of new material for the megatumour.

This explanation, derived from strict physical considerations, can be regarded as a combination of the expanding Earth theory, the megaundulation theory (e.g. van Bemmelen[14, 15]) and the passive convective current theory. On this concept the megaundulations and the magmatic currents are physical consequences of expansion. The above mechanism can be regarded as effective if the resistance of subcrustal masses does not surpass the acting gravity component.

The distribution of the earthquake foci in areas where deep earthquakes occur shows the direction of drift. The mechanism sketched above seems to be the only acceptable explanation for the fact that the zone of epicentres of these belts runs below the continents.

The above mechanism fits all the requirements of convection currents suggested by Runcorn[16], except only that the heat field should be replaced by a stress field. In any case a convection driven by expansion seems to be more realistic than any other mechanism attributed to temperature differences.

References

1. L. Egyed, 'Determination of changes in the dimensions of the Earth from palaeogeographical data', *Nature*, **173**, 534 (1956).
2. P. H. Kuenen, *Marine Geology*, Wiley, New York, London, 1950.
3. K. M. Creer, 'An expanding Earth?', *Nature*, **205**, 539–44 (1965).
4. W. H. Munk and G. J. F. MacDonald, *The Rotation of the Earth*, Cambridge University Press, London, 1960.
5. R. Dearnley, 'Orogenic fold belts and a hypothesis of Earth evolution', *Phys. Chem. Earth*, **7**, 1–117 (1966).
6. L. Egyed, 'Some remarks on continental drift', *Geofis. Pura Appl..*, **45**, 115–16 (1960).
7. L. Egyed, 'Palaeomagnetism and the ancient radii of the Earth', *Nature*, **190**, 1097–8 (1961).
8. M. A. Ward, 'On detecting changes in the Earth's radius', *Geophys. J.*, **8**, 217–25 (1963).
9. G. Eder, 'Der Zuwachs des Erdradius', *Z. Geophysik*, **31**, 206–11 (1965).
10. P. A. M. Dirac, 'A new basis for cosmology', *Proc. Roy. Soc. (London), Ser. A*, **165**, 199–208 (1938).
11. W. H. Ramsey, 'On the nature of the Earth's core', *Monthly Notices Roy. Astron. Soc., Geophys. Suppl.*, **5**, 409–26 (1949).
12. M. Landisman, Y. Satô, and F. E. Nafe, 'Free vibrations of the Earth and the properties of its deep interior regions: 1. Density', *Geophys. J.*, **9**, 439–502 (1965).
13. K. M. Creer, 'Palaeomagnetic data from the Gondwanic continents', *Phil. Trans. Roy. Soc. London, Ser. A*, **258**, 27–40 (1965).
14. R. W. van Bemmelen, 'The evolution of the Atlantic megaundation (causing the American continental drift)', *Tectonophysics*, **1**, 385–430 (1964).
15. R. W. van Bemmelen, 'The evolution of Indian Ocean megaundation', *Tectonophysics*, **2**, 29–57 (1965).

16. S. K. Runcorn, Palaeomagnetic evidence for continental drift and its geophysical cause', in *Continental Drift* (Ed. S. K. Runcorn), Academic Press, New York, London, 1962, pp. 1–40.

A. E. BECK

Department of Geophysics
University of Western Ontario
Ontario, Canada

Energy changes in an expanding Earth

Introduction

When a celestial body undergoes a change in dimension, its gravitational potential energy also changes. Thus, if a body such as the Earth changes in such a way as to decrease its gravitational potential energy, the difference in energy is available for compressing materials, heating, changes of state, chemical reactions, changes in angular rotation, etc. To determine the evolutionary history of the Earth not only should the initial and final conditions of the above interacting quantities and density distribution be known, but there should also be some knowledge of these data at intermediate times.

Because of the complexity of the problem, and of our lack of knowledge of the behaviour of Earth materials in the deep interior, it appears to be a rather futile exercise to try to prove or disprove a particular theory of Earth expansion or contraction. However, by taking a very general approach it should be possible to set limits, albeit very wide limits, on what is plausible and what is not. Then, as further information becomes available, these limits can be narrowed.

The approach taken will be to consider possible changes in the gravitational potential energy of the Earth as a whole and to neglect the problem of how this energy is used in the various regions of the Earth.

Although the subject of an expanding Earth has been discussed for several years, it is still often assumed that for the Earth to expand some internal or external source of energy must be available. This assumption is based upon the well-known equation for a sphere of uniform density:

$$\text{gravitational potential energy} = -0.6GM^2/a \qquad (1)$$

where G is the universal constant of gravitation, M is the mass of the sphere, and a is its radius. It can be seen from equation (1) that, if M is kept constant, then an increase in the radius requires the injection of energy. That this is not necessarily so for a sphere of non-uniform density can be seen from figure 1. Figure 1(a) represents a sphere of uniform density ρ, radius a, and mass $8M$. If the sphere is partly formed out to the radius $b \, (= \tfrac{1}{2}a)$, then the gravitational potential energy of that portion is

$$-0.6GM^2/b$$

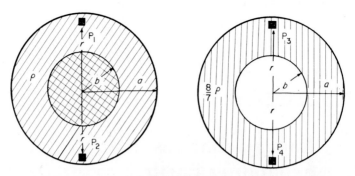

FIGURE 1 Illustration of two models of same mass and radius but different density distribution and gravitational potential energy: (a) $0 < r < a$ uniform density ρ, gravitational potential energy $= -1$ (say); (b) $0 < r < b$ density $= 0$, $b < r < a$ uniform density $= 8/7\rho$, gravitational potential energy $= -43/49$ (referred to model in (a))

If we now continue forming the main sphere by bringing up particles P_1 and P_2, which for the sake of simplicity we assume to be of unit volume, to the positions shown in figure 1(a), then the gravitational potential energy of the system thus far is equal to

$$-\left(\frac{0.6GM^2}{b} + \frac{2GM\rho}{r} + \frac{G\rho^2}{2r}\right)$$

Referring to figure 1(b), this represents a sphere of the same mass and radius as figure 1(a) but with a hollow core out to radius b; the density in the region $b < r < a$ is therefore $8\rho/7$. The gravitational potential energy in assembling the system of the core, radius b, and the two particles P_3 and P_4, again assumed to be of unit volume, is

$$-32G\rho^2/49r$$

Clearly the gravitational potential energy of this system is much less than the gravitational potential energy of the similar system which occupies the same volume in figure 1(a). If the process is carried to completion, it can readily be shown that the non-uniform body of figure 1(b) has a higher gravitational potential energy than that of figure 1(a). Thus there are two bodies of the same mass and radius but with different gravitational potential energies. The body of figure 1(b) could be shrunk until it has the same gravitational potential energy as that of figure 1(a), producing two bodies of the same mass and gravitational potential energy but of different radii.

With the principle established it is now necessary to discuss the real Earth, that is, Earth models without hollow cores.

The basic assumptions are that the Earth has at all times been spherically symmetrical, and that the mass of the Earth and G have remained essentially constant throughout geological time. Initially, the effect of both internal and external energy sources or sinks and of the changes in the moments of inertia (I) are neglected; these will be discussed later in the chapter.

Units used will be erg/10^{39} for the gravitational potential energy, g cm^2/10^{44} for moments of inertia, and g/cm^3 for density.

The Gravitational Potential Energy

The first requirement is a value for the gravitational potential energy of the present Earth using a suitable model. Unfortunately, there appear to be several such models all fitting the seismic data fairly well. A value has therefore been obtained for two models which represent extremes of central density—namely, the Birch model[1] 1 with a central density of 12·62 g/cm^3 and the Bullen[2] model B with a central density of 17·90. Nearly all other acceptable models appear to have central densities lying between these two. In an earlier paper[3] only the Bullen model B was used. Density distributions of the form

$$\rho_r = \rho_i (1 - k_i r^n) \qquad (2)$$

where ρ_r is the density at radius r, ρ_i and k_i are constants for the ith layer, and $n=1$ were fitted to the six well-defined regions of the model, and used in the equation developed for a layered Earth (Beck[3], equation (4)). The method has been refined by using $n=2$ in some layers to give a better fit and yields a gravitational potential energy of $-2·482$. A similar treatment for the Birch model 1 gives a figure of $-2·492$. Thus it seems reasonable to assume that the gravitational potential energy of the present Earth will lie between $-2·48$ and $-2·50$ no matter which reasonable model is chosen; for the purposes of the present chapter the figure of $-2·49$ is used. Table 1 summarizes the relevant data and also shows data for the Birch[4] primitive model.

TABLE 1 Data for various Earth models

Model	^{a}No. of regions	Mass (g/10^{27})	I (g cm^2/10^{44})	Gravitational potential energy (erg/10^{39})
Bullen B	6	5·977	8·11	$-2·482$
Birch 1	6	5·976	8·02	$-2·492$
Birch primitive	4	5·985	9·06	$-2·332$
Birch primitive	1 ($n=2·7$)	5·973	9·03	$-2·324$
Mean present		5·977	8·05	$-2·490$

^{a} This is the number of regions into which the Earth was divided for the purposes of curve fitting.

The Gravitational Potential Energy of Primitive Earth Models

The density distribution within the Earth at the time of its formation and for some time afterwards was very probably continuous, and can be represented by

$$\rho_r = \rho_0 (1 - k r^n) \qquad (3)$$

where ρ_0 is the central density, k is a constant, and ρ_r is the density at radius r. With the present knowledge of the Earth's interior this form is preferable to a straight power series since, to evaluate more than two constants, too many unreliable assumptions would have to be made.

If the equation of state for the primitive Earth was known, a reliable primitive Earth model could be used. Unfortunately, equations of state seem to vary with the author (and, occasionally, with time, keeping author constant), and the above approach using equation (3) seems to be better since it can be made to fit practically any present or future Earth model by the simple adjustment of n. Thus, for instance $n = 2 \cdot 7$ for a single curve fit to the Birch primitive model.

The gravitational potential energy of a primitive Earth model of radius a and the density distribution of the form of equation (3) is given by

$$\text{gravitational potential energy} = -\frac{16}{15}\pi^2 G \rho_0^2 a^5 \left\{ 1 - \frac{5(n+6)ka^n}{(n+3)(n+5)} + \frac{15(ka^n)^2}{(n+3)(2n+5)} \right\} \quad (4)$$

To evaluate ρ_0 and k for a known value of a, the primitive radius, it is necessary to know, in addition to the mass, a value of the density at some point on or within the Earth. If it is assumed that the Earth was formed rapidly by the aggregation of meteoritic material of average density σ, then the surface density ρ_a will be σ for all models. Varying n simply redistributes the mass throughout the Earth in such a way that for the lower values of n there is more mass in the central regions of the Earth than for the higher values of n. Thus, for a given value of a, the higher the value of n the higher will be the gravitational potential energy and I. There is some uncertainty as to the mean value of meteoritic material, the extremes being about $2 \cdot 8$ and $4 \cdot 2$. In general, for a given radius a and power index n, higher values of σ will be associated with higher values of the gravitational potential energy and I. This is shown in figure 2 for $n=2$ and $n=10$; the number by the side of a point is the value of the central density for that model.

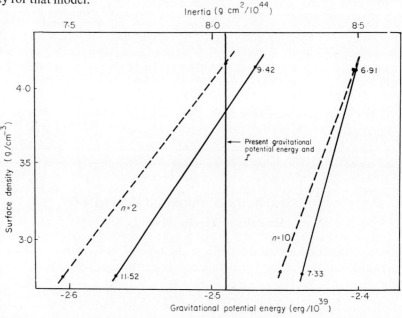

FIGURE 2 Gravitational potential energy (full lines) and moments of inertia (broken lines) against surface density for a model of radius 6100 km and density distribution

$$\rho_r = \rho_0(1 - kr^n)$$

for $n=2$ and $n=10$. $a=6100$ km. Smaller numbers by the side of points are central densities ρ_0. Full vertical line is present gravitational potential energy

Energy changes in an expanding Earth 81

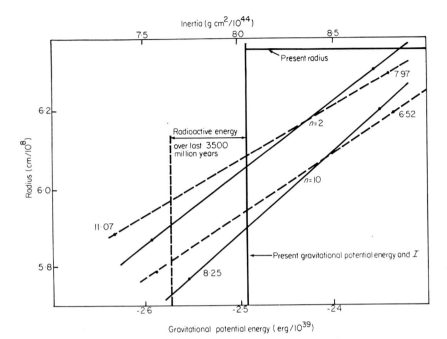

FIGURE 3 Gravitational potential energy (full lines) and moments of inertia (broken lines) against radius for surface density of 4·2 g/cm² and density distribution

$$\rho_r = \rho_0(1 - kr^2)$$

for $n=2$ and $n=10$. $\rho_a = 4·2$. Smaller numbers by the side of points are central densities ρ_0. Full vertical line is present gravitational potential energy and I; broken vertical line indicates extra energy available from radioactive decay over last $3·5 \times 10^2$ years

Figure 3 shows a plot of primitive radius against gravitational potential energy and I using $\sigma = 4·2$; figure 4 shows a similar plot for $\sigma = 2·8$. The full vertical lines are the gravitational potential energy and I of the present Earth, and the number by the side of a point represents the central density for that particular model.

Any point lying within the rectangle on the right bounded by full lines represents an Earth model with a higher gravitational potential energy and smaller radius than the present Earth. Some of these models have I higher, and some lower, than the present Earth. It can be seen from figures 3 and 4 that there are a large number of models to choose from, many of them having reasonable central densities. Even the Birch primitive model, with a central density of 7·12 and $I = 9·06$, must expand a small amount to reach the diameter of present Earth.

It appears that, in terms of energy of the Earth as a whole, there is no problem for expansion of the radius of the order of 100 km. As with many things concerning the Earth it is difficult to draw a line between the probable and improbable. However, a lower limit of probability, for a smaller primitive Earth, can be effectively set by the Earth which has the same gravitational potential energy and mass as the present Earth but is of uniform density. Such an Earth has a radius of 5769 km and a density of 7·42. For a doubling of the Earth's surface area, as required by some theories of Earth expansion, densities of 25 to 40 and large internal sources of energy are required.

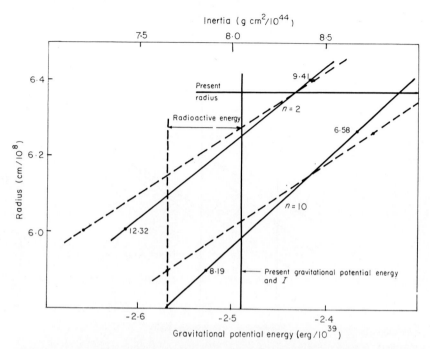

FIGURE 4 Gravitational potential energy (full lines) and moments of inertia (broken lines) against radius for surface density of 2·8 g/cm² and density distribution

$$\rho_r = \rho_0(1 - kr^2)$$

for $n=2$ and $n=10$. $\rho_a = 2·8$. Smaller numbers by the side of points are central densities ρ_0. Full vertical line is present gravitational potential energy and I; broken vertical line indicates extra energy available from radioactive decay over last $3·5 \times 10^9$ years

Energy Sources and Sinks

If the Earth was formed by chondritic aggregation, the total amount of radioactive energy generated over the last $3·5 \times 10^9$ years is $0·082 \times 10^{39}$ erg; over the last $4·5 \times 10^9$ years it is $0·146 \times 10^{39}$ erg[3,5]. This energy must be added to the gravitational potential energy; in other words, to take this energy into account in all the figures, the line representing the gravitational potential energy of the present Earth must be moved to the left; this is indicated for the $3·5 \times 10^9$ year quantity in both figures 3 and 4. This effectively gives much more latitude in the amount of allowable expansion based on energy alone.

It is difficult to estimate the effect of energy sinks. For instance, the surface flow of heat at the present is approximately 10^{28} erg/year. Summed over $3·5 \times 10^9$ years this means that $0·035 \times 10^{11}$ erg has been lost since the formation of the crust. However, it is not known how constant this heat flow has been with time and it is doubtful whether it can be considered lost to the system since such energy is presumably trapped in the atmosphere. Again, energy lost or gained from phase transformation and changes in angular rotation is not lost from the system although it does, of course, effect the distribution within the system. Possibly the only effective energy sink is loss by radiation. This is most likely to have been significant during the accretion stages of the Earth when very high temperatures could be expected at radii greater than about 1000 km. However, even if all accretion energy is lost owing to radiation, this does not affect the general argument since we are considering changes of energy since accretion and, more specifically, since the formation of the crust about 3.5×10^9 years ago. The proportion of accretion energy lost by radiation will, of course, effect the internal distribution of energy and hence the subsequent history of the Earth. The thermal energy required for heating the Earth would come from energy changes associated with gravitational reorganization energy. In this case there would be a maximum in the curve of energy density against radius in the vicinity of the core–mantle boundary[6]; a more sophisticated approach than that used by Beck[6], using the Birch 1 and Birch primitive models, results in a much sharper maximum than was obtained with the rather elementary models used in the earlier paper. Thus, although it has often been stated that there is sufficient energy from gravitational reorganization to melt the whole of the Earth, this is not necessarily available in the right quantities at every point in the Earth, particularly in the region of the inner core.

Conclusions

From considerations of gravitational potential energy alone, neglecting energy sources and sinks, primitive Earth models with plausible density distributions can be found which allow expansion of the radius of about 400 km; if all radioactive energy is put into expansion, this could be used to allow a further 200- or 300-km expansion. However, when certain possible restrictions are taken into account this range may be reduced. In particular, if the moment of inertia of the primitive Earth was much higher than the present Earth, for instance, about 9·0 as suggested by the Birch primitive model and by Runcorn[7], this immediately restricts the possible radial increase, using reasonable primitive density distributions, to less than 100 km.

Thus on present-day knowledge radial increases of the order of 1000 km seem improbable but radial increases of the order of 100 km cannot yet be ruled out.

Acknowledgements

Some of this work was carried out while the author was in receipt of a National Research Council of Canada Senior Fellowship at the Department of Geophysics and Geochemistry, Australian National University. Grateful acknowledgements are made to both organizations.

References

1. F. Birch, 'Density and composition of mantle and core', *J. Geophys. Res.*, **69**, 4377–88 (1964).
2. K. E. Bullen, 'An Earth model based on a compressibility–pressure hypothesis', *Monthly Notices Roy. Astron. Soc., Geophys. Suppl.*, **6**, 50–9 (1950).
3. A. E. Beck, 'Energy requirements of an expanding Earth', *J. Geophys. Res.*, **66**, 1485–90 (1961).
4. F. Birch, 'Energetics of core formation', *J. Geophys. Res.*, **70**, 6217–21 (1965).
5. G. J. F. MacDonald, 'Calculations on the thermal history of the Earth', *J. Geophys. Res.*, **64**, 1967–2000 (1959).
6. A. E. Beck, 'A note on the thermal history of the Earth and the possible origin of a solid inner core', *Can. J. Phys.*, **42**, 825–9 (1964).
7. S. K. Runcorn, 'Changes in the Earth's moment of inertia', *Nature*, **204**, 823–5 (1964).

A

COSMOLOGY AND GEOPHYSICS

IV. Geological and geophysical evidence related to the hypothesis of Earth expansion

1

H. TERMIER
and

G. TERMIER
University of Paris
Paris, France

Global palaeogeography and Earth expansion

Possibilities and Impossibilities which Geologists Use to Explain All Palaeogeographic Facts in a Present Geographical Sense

All palaeogeographic arguments that are, in general, in favour of continental drift are also in favour of the hypothesis of the Earth's expansion.

However, when attempting to reconstruct the primitive continental assemblage for Pre-Mesozoic Periods, it is relatively easy to reassemble Gondwanaland or to fit North America with Greenland and western Europe, or Alaska with the Tchouktchen Peninsula, or, of course, South America with Africa. But we meet many difficulties with the Tethys, some of them being inherent in the notable difference in the relative orientation between Eurasian and African axes with regard to their present ones. For example, Asia–Australian relationships are difficult to imagine before Mesozoic times. While Audley-Charles[1] reconstructs the south part of that linkage since Permian times, the supposed distance from Indonesia to Australasia is too large during the preceding periods if the Earth's diameter was constant. But perhaps the diameter growth was not important enough to supply a valuable explanation here.

Palaeogeographic Clues in Favour of the Earth's Expansion

Pre-Cambrian times

The first possibility for the palaeogeographer to reconstruct the Earth's face is given by orogenesis and megatectonics. As first reconstructed by Choubert[2] and more recently and more precisely by Dearnley[3, 4], palaeogeographic frames constitute successive networks that are different from the modern ones. We shall not add anything to the frame reconstruction by Dearnley for the Earth during the 'Superior regime' (named from the province of Lake Superior in the Canadian Shield).

The 'Banded Iron Ores' Era. In the first phase of the palaeogeographic story of the Earth, we have the appearance of living organisms. These are generally linked with a special facies of these times, the 'Banded Iron Ores', which are associated with microorganisms and stromatolites, as well as with graphitic rocks (Madagascar). We must say that these 'Banded Iron Ores' are almost uniformly dispersed upon the surface of the different continents.

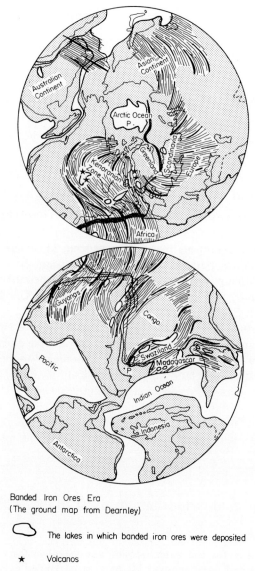

Banded Iron Ores Era
(The ground map from Dearnley)

⌒ The lakes in which banded iron ores were deposited

★ Volcanos

FIGURE 1 Banded Iron Ores Era (the ground map from Dearnley). Superior regime

If we compare the orogenetic partition of Pre-Cambrian times, for example, from the Dearnley classification, with that which is implied by the great stages of biological development, in the light of the most modern knowledge, we notice a surprising fit. Calvin's experiments and discoveries show that life arose about 3100 million years ago (Banded Iron Ores in Swaziland). Then, during the following 1500 million years (between 3100 and 1600 million years ago), occurred what we consider, with regard to palaeogeography, as the 'Banded Iron Ores Era'. It includes the 'Superior regime' of Dearnley up until 1950 million years ago: we notice that the richest microflora in that period is 1900 million years old.

The Riphean Era (*1600 to 700 million years ago*). Available data for the 'Hudsonian regime', which is the second period of Dearnley's classification, are too poor to be discussed here. Therefore, we prefer to use the lower limit of the Riphean Era defined by Soviet authors—the last part of the Pre-Cambrian times prior to the appearance of any Coelomata, i.e. the Ediacarian Era (between 680 and 600 million years ago). The major biochemical feature at the beginning of the Riphean Era was the existence of an oxidizing atmosphere. The characteristics of life were, in those days, the expansion of lagoonal seas which allowed the spreading out of the stromatolites associated with Protocaryota, among which the genus *Conophyton* is almost ubiquitous (U.S.S.R., Spitzbergen, eastern Congo, Australia, Labrador).

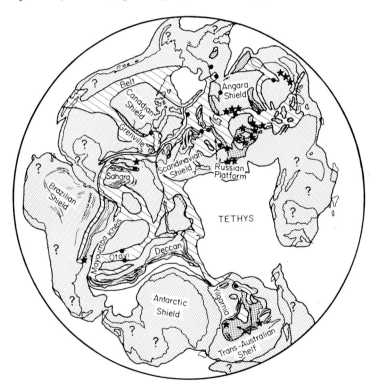

★ Volcanos

● Stromatolites (*Conophyton etc.*)

FIGURE 2. Riphean

Epicontinental seas were very poor in geosynclines and basins, but it seems that they were rather shallow, possibly lagoonal, favouring cyanoschizophytic development.

Eocambrian and Ediacarian Eras (680 to 600 million years ago). The period that corresponds to the Eocambrian and Ediacarian Eras (approximately equivalent to the Soviet Vendian Era) takes place about the end of Pre-Cambrian times or, alternatively, at the beginning of fossiliferous times. Since here we commence Biokinesis (organic evolution in its dynamic sense), we assume that it is the first part of the Cambrian System (Cambrian I). Palaeogeographically, Cambrian I is a period of large continentalization, i.e. emerged lands are widely distributed, much more than during the Riphean Era.

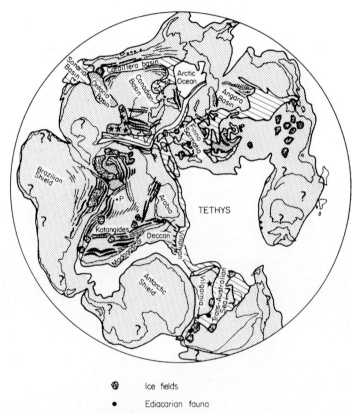

FIGURE 3. Eocambrian

Cambrian I is also a period of nearly homogeneous repartition of some sediments, such as tillites which, at the base of this stage, seem to mean a rather uniformly glacial climate. The map (figure 3) shows tillites (spread out around the polar area in Africa, as well as in Australia, Greenland, Scandinavia, and central Europe) in areas that should normally be temperate or even tropical. The same uniform repartition may be assigned to the Ediacarian fauna, which is known in South Australia, South Africa, England (Charnian), the Russian Platform, Siberia, and is probably represented by Medusae in the Beltian area.

Conclusions regarding Pre-Cambrian palaeogeography. Palaeogeographic maps supply evidence of conditions very different from modern ones. First, they give evidence of the wideness of orogenic belts mainly in the 'Superior regime': it seems there were neither defined orogens nor high mountains but mobile regions and rich petrogenesis (volcanism, granitization). No true deep sea can be detected and it seems likely that even ocean bottoms did not demonstrate any gross unevenness. 'Banded Iron Ores' within the same frame seem to be linked to these belts as very shallow lagoonal seas or lacustrine deposits (of lightly salted and acid water) fed in part by sluggish rivers.

Neither climate nor anything else that discriminates today the diverse regions from the proximity of their poles, oceans, or high lands can be reconstructed. The same can be said for Riphean and Cambrian I Eras.

Whatever the geophysical hypothesis for the beginnings of the evolution of the Earth's crust, it is obvious that climates and seasons were not yet differentiated.

Let us now look at the characteristics of different types of life. Pure biological conclusions pertaining to the low stages of evolution are difficult to distinguish from geophysical imperatives. Thus, Protocaryota are very tolerant to heat, salinity, and acidity. The fact that plankton first spread out suggests that light, floating organisms were favoured on a denser Earth, as we can imagine it according to the expansion theory.

The same conditions seem to have been carried through all Pre-Cambrian time with, however, some variations. This is the case for stromatolites: these are not fossil skeletons but concretions lining living 'algae' (or Protocaryotes), these concretions being the product of photosynthesis. Now, stromatolites are the only calcareous fossils of Pre-Cambrian times. It seems that their calcareous crust formation was like a shroud and almost adverse to living conditions.

The Ediacarian fauna contrasts with Protocaryota and plankton from the preceding times. Representatives of nektobenthonic (Medusae) and maybe also of benthonic, (Rangeides and Pennatulides) Cnidarians, but also various Coelomata (Annelids and Nymphoids), have been found here. The term Nymphoids needs some explanation. We have used it for larva-shaped and relatively large-dimensioned organisms that do not display any imaginal stage that is morphologically different. They are like larvae developed from other larvae without more highly structured ancestors. No Nymphoids evolved a mineralized skeleton. They seem to have lived half-way between plankton and benthos, leading a semipelagic existence.

There were also some burrows of unknown organisms (Annelids or other Coelomata, probably) even before the Ediacarian (from 1000 million years ago).

The Ediacarian fauna was very different from its successor. Let us recall that Soviet geologists now set the limit between the Pre-Cambrian and Cambrian as after the Vendian (Ediacarian) because the Ediacarian fauna lacks a mineralized skeleton. Now, everyone emphasizes the similarities of Ediacarian fauna to deep-sea benthos. Perhaps we may see in these forms the realization of the utmost development permitted in those times, and we might consider that one of the causes of restricted structural development in Metazoa would be in relation to the actual gravitational field. Indeed, the pH could not have been so low that calcareous skeletons were non-existent, since calcareous stromatolites abounded contemporaneously.

Palaeozoic times

As we have just seen, the Ediacarian establishes something new in the global palaeogeographical record. We are able to characterize it as the first marine transition of fossiliferous times. What kind of a transition was it? We are struck by the fact that it immediately followed the large Eocambrian glaciation. We know that, after the last Würmian glaciation, a wide eustatic transgression took place, the Flandrian transgression. Accordingly, it seems likely that glacial control began to act at the dawn of fossiliferous times at least.

While Ediacarian fauna shows a general climatic uniformity, the Cambrian II fauna gives the first evidence for climatic differentiation. Meanwhile the Cambrian (from 600 to 550 million years ago) marine spreading grew larger and larger, especially over Laurasia; east and west North Ameria, northwest Europe, north and south Asia, South America, central Antarctica and Transaustralian Sea, north India, south Iran, Israel, south Spain, and southwest Morocco are flooded.

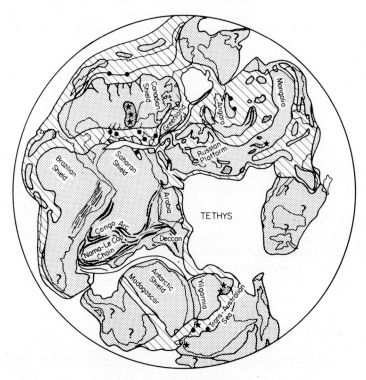

● Principal archaeocyatha bioherms
★ Volcanos

FIGURE 4. Lower Cambrian

Logically this picture suggests that part of these seas seeped into weakness lines and separated the continents, the main separation being the present-day North Atlantic. The evidence of a tropical climate is displayed by the repartition of Archaeocyathid bioherms in North America, Morocco, Siberia, and Australia. Biological provinces characterized by Trilobites were being established: this is one of the main implications of the benthonic Coelomata development.

Palaeogeography seems to display a rather similar repartition of sea and land during the Lower Palaeozoic, associated with some climatic variations. These variations

Global palaeogeograghy and Earth expansion

induced a real rhythm, according to which biohermal periods alternated with non-biohermal ones. As Archaeocyathid reefs were confined to the Lower Cambrian, all the rest of Cambrian and Tremadocian was practically devoid of any bioherm up to Ordovician.

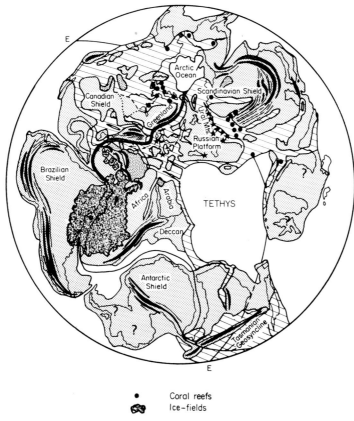

●   Coral reefs
🍥   Ice-fields

FIGURE 5. Ordovician

The Upper Ordovician epoch (XIII) (450 million years ago) offers the first coralliferous tropical belt (in the Arctic American archipelago, Scandinavia, Siberia), related to the most controversial map since the beginning of geological times, owing to the fact that many glacial indications have just been found in Africa, near the supposed position of one of the poles. There was here a true glaciation (Eglab, Ajjer, Tibesti) which left striated rocks, U-shaped valleys, and 'roches moutonnées'.

During the Silurian and, finally, the Siluronian epoch (XVI), i.e. the lower limit of Devonian Era (400 million years ago), the end of the great Caledonian orogenesis led to land emergence in North America and north Europe, so that a large part of north Eurasia became dry while shallow seas invaded South America and South Africa. The wide continental emergence during the Siluronian is on a par with the terrestrial biological development, i.e. vertebrates (Agnathids and fishes) and continental plants (Tracheophytes) which mark the Lower Old Red Sandstone.

Climatic belting and its seasonal implications for some countries were then fairly well established. We find the best example of this in the palaeogeographic map of the

Pennsylvanian System (XXVI) (260 million years ago), i.e. during the Permo-Carboniferous glaciation that affects essentially Gondwanian lands, after the climax of the Variscan orogenesis.

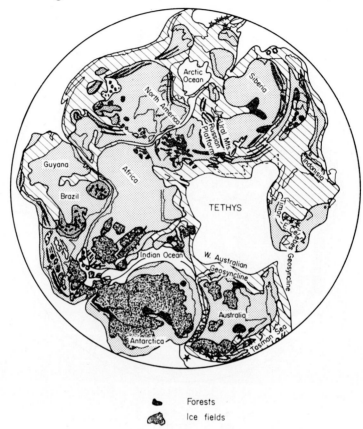

FIGURE 6. Upper Carboniferous

Conclusions regarding Palaeozoic palaeogeography. During Palaeozoic times, it seems likely that the relations between continents did not undergo great changes. A continuous pattern of continents favours connections that would be unlikely in the recent pattern, for example the Scandinavia–Australian marine linkage, attested by identical marine fossil zonation (Trilobites) during Cambrian times at least. We see the successive appearance of climates and alternation of tropical biohermal phases with ahermal phases.

There is good evidence to prove that the circumpolar area (the pole 'running' between the middle of Africa and South Africa) was almost unceasingly glaciated from Cambrian I to Pennsylvanian XXVI, with peaks during epeirogenesis, for instance at the end of Ordovician times; after this occurred the eustatic transition of the graptolite sea. It is necessary to introduce the volume of glaciated water if we wish to compute the amount of water covering the continents during these times.

On the other hand, it may be important for expansion purposes to notice that, several times in the course of the Palaeozoic Era, not only orogenetic belts but also nearly all continents were uplifted. For this upheaval we use the name of epeirogenesis. The most remarkable ones were located at the very end of the Ordovician (Taconic epeirogenesis), at the end of Silurian (Caledonian or Ardennic epeirogenesis), and at the boundary between Mississipian and Pennsylvanian (Mid-Carboniferous epeirogenesis). As already pointed out, these epeirogenetic periods favoured continental life, probably since it was then possible (but the highly continentalized Eocambrian does not display any subaerial organism). However, for the above-mentioned epeirogenesis, and all those that followed in Mesozoic and Tertiary times it is obvious that no return to the former position is normally detected after epeirogenesis. We shall see that this phenomenon occurs again several times—this is registered by the successive planation surfaces contained in record of every old platform stabilized from Pre-Cambrian times onwards. Therefore the oldest surface is also the highest and so on.

However, continentalization grew larger and larger. Struck by this feature, Egyed computed from the authors'[5-7] and Strakhow's[8] palaeogeographic maps a lowering of sea level which fits with expansion theory. He found that the sea level had fallen by 275 m since the Middle Ordovician, and by 550 m since the early Cambrian. From the graphic data obtained from both of these independent sources, Egyed inferred that the Earth's radius was increasing from 0·4 to 0·66 mm/year—these figures fit the theoretical ones (0·58 mm/year) calculated from an initial Earth whose surface exactly coincides with that of recent continents (including the continental shelf and slope down to −2000 m).

Mesozoic times

During the larger part of Mesozoic times, conditions were not very different from Palaeozoic ones. Continentalization had increased through epeirogenesis and the accompanying Variscan orogenesis, so that the Permian Era (XXVII) (240 million years ago) and the early Trias System (XXVIII) displayed large continents with the new accompanying development of terrestrial fauna and flora. Climates becoming dry and organisms which had long-living eggs or seeds and water-free young stages were favoured by natural selection.

Sea water was then essentially restricted to some ancient structures such as circum-Pacific geosynclines and the Tethys, a large oceanic basin communicating with the Pacific but largely penetrating between Laurasia and Gondwana. Since the Middle Trias, a double orogenetic girdle originating from the east Tethys followed south Europe and North Africa. During the same time (Upper Permian), while South Africa was the seat of an important volcanic activity (Karroo), another part of the Tethys was coming to an end by being swallowed up between Madagascar and southeast Africa. In Toarcian times (Lias epoch) (XXXIII), a real sea was born, the *Bouleiceras* Sea (167 million years ago), related to the Tuwaiq Gulf[9], the future Transerythrean Channel. It is the origin not only of Mozambique Channel but also of the severing of the peninsular of India (Deccan) from the African realm.

We have to wait until the Middle Cretaceous Era (95 million years ago) before any new events in mobilistic palaeogeography occurred. First, India began to be displaced somewhat rapidly through the Tethys: Madagascar was approximately in the same place as nowadays; secondly, the sea seeped slowly between South Africa and South America: primarily lagoonal basins settled on both of the opposite sides[10]; afterwards

FIGURE 7. Permian

brackish or neritic populations invaded these basins (The joining of both continents up until the Middle Cretaceous is guaranteed by the amphi-Atlantic earth-worm basins which cross the severing line.)

Conclusions regarding Mesozoic palaeogeography. Mesozoic palaeogeography displays the first signs of the independence marked by some continents in the frame of the general phenomenon of continental drift.

Alternation of transgressions and regressions is noticeable. We have shown that Tethysian transgressions started from the Tethys and spread tropical faunas throughout most of the seas, while Arctic transgressions spread temperate faunas from the Arctic Ocean, and, on the other side, circum-Pacific transgressions were connected with regressions in the Tethys area. Beginning in Palaeozoic times, this alternation is more discernible in Mesozoic times. From our maps Egyed[11] drew a diagram which shows $8\frac{1}{2}$ regular oscillations, each of them including a transgression and a regression. The duration of each oscillation was 47 million years, and the total duration of all oscillation was 400 million years. The period of 47 million years is very close to 50 million years, the time calculated by the same author for the stress accumulation into

Global palaeogeography and Earth expansion 97

shallow seismic belts. According to Egyed[12], epeirogenetic movements were transmitted to the Earth's crust by accumulation of tectonic energies, whilst orogenetic ones resulted from relaxation of the same energies. As accumulation times are very much longer than relaxation ones, which are nearly instantaneous, epeirogenesis is practically continuous and orogenesis very short. Palaeogeography and expansion of the Earth seems to be in harmony on that point.

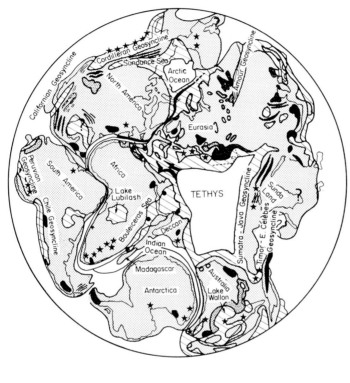

- • Bouleiceras fauna
- ★ Volcanos

FIGURE 8. Upper Lias

Cainozoic and recent times

From the Middle Cretaceous the continental separation increased at the same time that the rate of drift of some lands increased. Since the explorations of Ewing and his colleagues in oceanic basins, we know the shock lines where disjunction originated (most of them are midoceanic ridges) and through the pattern of fracture zones we can find the tracks of the displacement itself.

Continentalization reached its climax at the end of Cainozoic times: the Alpine chain took place by the joining of Africa and India with Eurasia on the Tethys emplacement. The Indian Ocean replaced a part of the Tethys.

At last, the Red Sea appeared 11 million years ago, African rift valleys 2 or 3 million years ago. These features might be the origin of new oceanic structures.

FIGURE 9. Cretaceous

★ Volcanoes

FIGURE 10. Oligocene

Comparing Transgressions–Regressions, Epeirogenesis, Bathygenesis, and Glaciations in the Light of Earth Expansion

Some great periodic events stand put as landmarks in the record of the Earth, when Earth expansion is established as a non-reversible phenomenon. We are going to compare both sets of data, our target being to find how they relate to each other. This will help us to understand some anomalies displayed by inferences drawn from our palaeogeographic maps for the early Cambrian times.

Let us first present the problem. From Ordovician times the emerging continental areas grow regularly and epicontinental seas reduce reciprocally; continentalization appears very important during Pre-Cambrian and Cambrian times, as we have already seen. If expansion is the sole parameter for epicontinental seas spreading during the whole of geological times, it would then be that more continental areas were inundated in the early Palaeozoic. Since they were in geosynclinal conditions and later on metamorphized, it might not be possible to find any fossils in them (although they can be found in other marine zones): this is Holmes' interpretation[13]. However, it is doubtful whether true geosynclinal marine series developed in the countries quoted by this author (western and south Australia, the Antarctic continent, Madagascar, east and west Africa, south Deccan and Ceylon, Arabia and eastern Brazil). We can only assert that granites, of 600 to 500 million years age and more recent ones, were formed in these areas, but the last marine series was the Riphean (Grenville about 1000 million years ago). As mentioned above, most of these areas were flooded and have supplied chemicofossils during a large part of Pre-Cambrian times ('Superior regime' and Riphean). But it seems that epicontinental waters were then very shallow and hardly deserved the name of seas, as they were rather lagoonal, belonging to the phytal or euphotic zone.

Kuenen computed that, since the beginning of Cambrian, the increase of the hydrosphere did not exceed 4%. On the other hand, within the same period (i) epeirogenesis was almost continuous, (ii) the folded belts were more and more restricted and higher, and (iii) oceanic basins enlarged and deepened through bathygenesis.

If we consider the record of depth in oceanic basins, palaeontological knowledge suggests that, up to the Upper Cretaceous, there were no ocean bottoms deeper than 2000 ($1\frac{1}{4}$ mile), i.e. the lower part of the continental slope. The same conclusion was reached by the oceanographers and biologists among Ewing's colleagues (Lamont Observatory). Bathygenesis displayed now by sea mounts and guyots has reached 1000 m since Cretaceous times (1650 m for the Mid-Pacific Ridge), as against about 2000 m for epeirogenesis during Mesozoic times. During the Pliocene alone, bathygenesis of the Pacific bottom (Bikini) was 700 m. This might be a sufficient explanation for the more rapid continental formation shown since that time.

From before that critical epoch of high bathygenesis, some oceanic deposits by seismic soundings have been investigated: deposition in the southern part of the abyssal Hatteras plain (North Atlantic) dates from 2780 million years ago, that of the Indian Ocean, south Bengal Bay, from 670 million years, that of the east Atlantic from 412 million years, and that of the eastern equatorial Pacific from 100 million years.

General Conclusions

We assume three great stages for the palaeogeographical evolution of the Earth's crust.

(i) During early times, corresponding to the pre-Superior regime (pre-geologic era), there were not any differentiated orogenic belts or consolidated shields: the very unstable crust displayed numerous volcanic phenomena and finally reached the eddying pattern of the Superior regime folding on a global scale. Crustal differentiation trended to granitization (sial) from the basaltic initial crust while volcanic fissures appeared, producing basalt, and exploded during the Earth's degassing. The predominant aspect of the Earth seems to have been a total lack of atmosphere. It is very likely that volcanic gases primitively present were water, carbon dioxide, ammonia, and methane, and this state extended over not less than 500 million years. Hydrogen (a part of which was lost in space) and water, which had escaped from the crust, were progressively incorporated into the gaseous shell—and probably also hydrogen sulphide and carbon dioxide. About the same time the hydrosphere also formed. At first, the primitive gases were probably associated with light elements such as Na, K, Mg, and perhaps also with heavier elements such as Si, Ca, Fe. This gaseous ionized shell was termed the pneumatosphere[14].

(ii) Next the separation of the hydrosphere (the water associated with the heaviest ions of the pneumatosphere) and atmosphere happened. From then on, the conditions presiding regarding erosional and sedimentary processes are comparable with present ones. For the first pre-biological ocean about 3800 million years ago, a water content of 5 to 10% of the modern one is given.

An important stage in the history of the hydrosphere was the persistence of sheets of water on the crustal surfaces. It is possible that they were materialized by the settling of 'Banded Iron Ores', more than 3000 million years ago. Lagoonal or lacustrine, these deposits were laid down in shallow waters, randomly scattered; their banding suggests periodic phenomenon, 'seasons' of an odd type (or maybe, better, a wide tidal phenomenon). That stage is attended by the appearance of life.

(iii) The third stage of palaeogeographical evolution, corresponding to the Riphean, is characterized by its oxidizing atmosphere and by the presence of epicontinental seas, where life is now represented, over every known outcrop, by Protocaryotes in benthos and plankton, i.e. very shallow indices. We are able to assume that the Ediacarian marks the end of this stage and the beginning of the fourth one: it is a period of high continentalization, favoured by many folded belts filling the place left by the flooded areas of the Riphean.

During the Riphean, the Pacific Ocean at least obtained, on the whole, its present areal importance so that it constitutes an oceanic hemisphere, whilst continents, still close-packed, cover the other hemisphere. There were not yet any seasons, polar areas being probably in the centre of each hemisphere.

(iv) The fourth stage of palaeogeographical evolution corresponds approximately to fossiliferous times. After Cambrian I, the first stages of biokinesis and climatic diversity appear and the first marine transgression of these times takes place. We consider that, with the beginning of the Cambrian transgression in late Cambrian I, new conditions affected our planet, nearer to present ones from every point of view. In this respect, the oceanic mass had approximately its present value, notwithstanding its 4% growth since then.

Polar wandering, that is global continental drift and finally continental disruption, happened at the time when the primitive equilibration between an oceanic hemisphere and a continental one ceased. During these vicissitudes, the Earth's crust underwent periodic transgressions and regressions and almost unceasing epeirogenesis and bathygenesis.

At all events global palaeogeography does not display any argument against Earth expansion.

References

1. M. G. Audley-Charles, 'Permian palaeogeography of the northern Australia–Timor region', *Palaeogeography, Palaeoclimatology, and Palaeoecology*, **1**, 297–305 (1965).
2. B. Choubert, 'Recherches sur la genèse des chaînes paléozoïques et antécambriennes', *Rev. Geograph. Phys. Geol. Dyn.*, **8**, 1 (1935).
3. R. Dearnley, 'Orogenic fold belts, convection and expansion of the Earth', *Nature*, **206**, 1284–90 (1965).
4. R. Dearnley, 'Orogenic fold belts and continental drift', *Nature*, **206**, 1083–6 (1965).
5. H. Termier and G. Termier, *Histoire Géologique de la Biosphère*, Masson, Paris, 1952.
6. H. Termier and G. Termier, *L'Evolution de la Lithosphère*, Vol. III: *Glyptogénèse*, Masson, Paris, 1961.
7. H. Termier and G. Termier, *Evolution et Biocinèse*, Masson, Paris, 1968.
8. N. M. Strakhow, *Outlines of Historical Geology*, in Russian, Moscow, 1948.
9. W. J. Arkell, R. A. Bramkamp, and M. Steineke, 'Jurassic ammonites from Jebel Tuwaiq, central Arabia', *Phil. Trans. Roy. Soc. London, Ser. B*, **236**, 241 (1952).
10. P. Hirtz, 'Les bassins salifères du Gabon et du Congo (Brazzaville): Essai de reconstitution paléogéographique', *Congr. Natl. Petrole, 5th, Le Touquet, 1965*, Vol. 2.
11. L. Egyed, 'The change of the Earth's dimensions determined from palaeogeographical data', *Geofis. Pura Appl.*, **33**, 42–8 (1956).
12. L. Egyed, 'A new dynamic conception of the internal constitution of the Earth', *Geol. Rundschau*, **46**, 101–21 (1957).
13. A. Holmes, *Principles of Physical Geology*, 2nd ed., Nelson, London, 1965, p. 969.
14. A. Rittmann, 'Die prägeologische Pneumatosphäre und ihre Bedentung für die geologischen Probleme der Gegenwart', *Experientia*, **3**, 310–15 (1947).

2

R. DEARNLEY

Institute of Geological Sciences
London, England

Crustal tectonic evidence for Earth expansion

Over the past decade there have been a number of proponents of the hypothesis of an expanding Earth (e.g. Egyed[1-3], Carey[4], Dicke[5], Heezen[6]), although there seems not to have been much agreement on the possible causes or the actual amount or rate of expansion. Expansion has been variously postulated as the cause of continental drift or as a process which has been steadily continuing in the background independently of continental drift. A recent appraisal of the problem has been presented by Holmes[7], as a result of which he came to the conclusion that 'there is now ample evidence for seriously considering the hypothesis that expansion of the interior of our planet may have played a dominant role in geological history and the evolution of surface features' (Holmes[7], p. 965).

Some Pre-Cambrian crustal geotectonic data are outlined here which lend considerable support to previously suggested continental drift reconstructions and which could be interpreted also as evidence for expansion.

It is only comparatively recently that enough data have become available, in the form of age determinations, to provide any means of correlation within the Pre-Cambrian and therefore these formations figure little, if at all, in the continental drift reconstructions of Wegener[8], Taylor[9], du Toit[10], Carey[4], King[11,12], and others. Thus, the application of these data covering the major portion of the readable geological history of the crust may be regarded as a completely independent check on the validity of previous suggestions of the fit of continents.

The most important method of geotectonic analysis which can be applied to the Pre-Cambrian Period is an investigation into the distribution of dated geosynclinal orogenic fold belts. Considerable metamorphic and igneous activity is associated with these orogenic belts and, if the incidence of this activity is correctly represented by the frequency distribution of the available age determinations, then variations of activity throughout geological time may be deduced. A previous investigation by Gastil[13], using about 400 such age determinations, has been repeated with a much larger sample of about 3400 age determinations[14]. This new compilation is by no means exhaustive; but the sample is large enough to avoid any undue overall place or age bias and, as the results do not differ significantly from the earlier study, the histogram of figure 1 may be considered to reflect the actual variations of igneous, metamorphic, and orogenic activity.

The cumulative curve derived from the age determination frequency histogram of figure 1 shows three particularly well-defined changes of slope at 1950 million years,

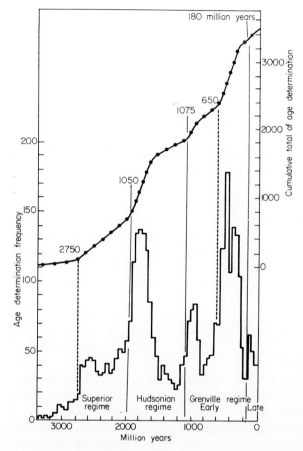

FIGURE 1 Frequency histogram and cumulative curve of igneous and metamorphic age determinations plotted against geological time, showing division into Superior, Hudsonian, and Grenville regimes. [Reproduced from Dearnley[15]]

1075 million years, and 180 million years, and two other less abrupt changes at 2750 million years and 650 million years. The changes of slope at 2750 million years, 1950 million years and 1075 million years correspond closely to the Katarchaean–Archaean (2700±150 million years), Archaean–Lower Proterozoic (1900±100 million years) and Lower Proterozoic–Upper Proterozoic (1000±100 million years) boundaries in the classification of the Pre-Cambrian suggested by Vinogradov and Tugarinov[16]. These major changes of slope correspond closely to the periods suggested by Sutton[17] as marking the beginnings of evolution of the three major 'chelogenic' provinces, each comprising a number of essentially parallel or subparallel fold belts which truncate the older provinces and which in turn are cut by the succeeding major clusters of fold belts. The major peaks of igneous and metamorphic activity fall within the periods covered by these provinces, whereas the transitions between the provinces are marked by pronounced decreases of activity.

The numerous individual fold belts of different ages which comprise the *Superior regime* (> 2750 to 1950 million years ago) are dominantly concordant in any given region and they make up the earliest recognizable structural systems of the continents forming semicontinuous and relatively uniform trends for long distances, up to 2000 km in Africa and North America. A series of maps have been prepared to show the structural trends of the fold belts of the Superior regime where these are known[15]. These compilations, which are based on published structural and tectonic maps where accompanied by adequate coverages of age determinations, are combined in figure 2 to show the continental distributions of orogenic fold belts > 1950 million years old.

The continental reconstructions used in figure 2 are based on the morphological fit of the continental shelves in the Atlantic Ocean, and on the fit of the Lomonosov Ridge against the Siberian continental shelf in the Arctic Ocean; the Gondwanaland reconstruction used and the relative position of this to the Laurasian continental reconstruction are consistent with the palaeomagnetic evidence and are similar to those proposed by du Toit[10].

FIGURE 2 Superior regime fold belts (> 2750–1950 million years) on a continental reconstruction based on the morphological fit of the Atlantic Ocean continental shelves and on a Gondwanaland reconstruction similar to that proposed by du Toit[10]. Approximate trend lines are shown between the isolated provinces. Convergence of trends towards the Arctic and the eastward convergence along latitude 30° of this projection should be noted. [Reproduced from Dearnley[14]]

All the continental regions of figure 2 consist of fold belts within the same broad time span and, in many of the provinces, fold belts of different ages within this time span occur in an essentially parallel arrangement. Thus, within the limits of the available data, approximate trend lines may be drawn between the isolated provinces to obtain the overall structural pattern. The major structure is relatively simple and, although the evidence of the Superior regime is relatively fragmentary, there are two significant features of the overall arrangement of the fold belts on the continental reconstruction. First, there is a centre of convergence of the fold belt trends towards the region of the present Arctic Ocean and, secondly, a line of symmetry can be drawn to bisect the eastward convergence of the fold belts which corresponds approximately with latitude 30° of the projection shown. The major structural patterns of figure 2 tend to confirm the general predrift 'fit' of the continents as suggested by the morphological features of the Atlantic continental shelves. But, in addition, since any fold belt pattern must be regarded as a reflection of subcrustal activity, and since on present knowledge there is no apparent mechanism for the generation of this large-scale activity other than that of convection in the mantle, it is reasonable to consider the possible relationships on this basis.

The fold belt convergence in the region of the present Arctic Ocean can be related only either to an upwelling 'source' or to the downward 'sink' of a zonal convection cell centred in that region. Since fold belts would tend to form perpendicular to the direction in which the convection currents in the mantle flow, it is possible to deduce the flow pattern necessary to generate crustal fold belts of the Superior regime shown in figure 2. This would take the form of a system of spiral flow lines centred on the same region as the convergence of the fold belts, and everywhere perpendicular to the fold belt trends. The fact that the pattern of flow converges towards the west could be due to the deflection caused by the force of the Earth's rotation.

On this basis, only two possible arrangements of convective flow could give rise to the pattern of fold belts observed for the Superior regime—a three-cell zonal system, with converging currents moving downwards along latitude 30° N, or a two-cell symmetrical system with currents converging along the equator of a smaller-radius Earth. A three-cell system of convection is unlikely, since the structural evidence of the Grenville regime, indicating that a four-cell system operated at that time, shows that the transition from three to four cells took place between the Hudsonian and Grenville regimes, 1075 million years ago, and not between the Superior and Hudsonian regimes 1950 million years ago. The only alternative consistent with the differences in the patterns of fold belts between the Superior, the Hudsonian, and the Grenville regimes is to fit the pole and the line of symmetry of the Superior regime fold belt pattern onto the pole and equator of a smaller-radius Earth. In this way the fold belts would correspond to a two-cell pattern of convection modified by the force of the Earth's rotation[15, 18].

On this basis, the apparent Earth radius for the Superior regime can be determined to be about 4400 km. North and south polar stereographic projections of this reconstruction are shown in figure 3 to indicate the relative simplicity of the overall structural pattern of the orogenic fold belts. This pattern would suggest a two-cell system of convection in the mantle with currents rising at the poles and falling at the equator—the eastward convergence of the fold belts would require a westwardly directed convergence of the convective flow below the crust consistent with the effects which would be produced by the Coriolis force of the Earth's rotation.

For the Coriolis force to be effective in producing such a deflection the Taylor number[19] must be about 10^3, and therefore the viscosity of the mantle would require to be much lower than the usually derived figure[20] of 10^{20}–10^{23} c.g.s. or 10^{26} c.g.s.

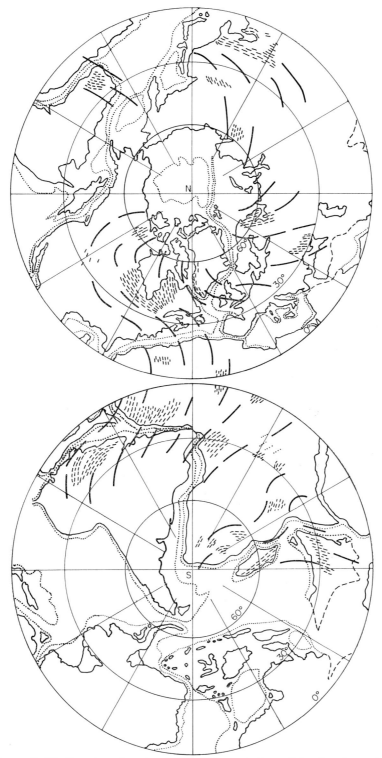

FIGURE 3 North and south polar stereographic projections of the Superior regime (> 2750–1950 million years) fold belts and trend lines. An Earth radius of 4400 km has been determined by fitting the Superior regime fold belt pattern to a line of equatorial symmetry and to a polar convergence. [Reproduced from Dearnley[18]]

The typical response of crystalline materials and metals to stress at elevated temperatures is non-Newtonian plastic flow[21, 22], and it has been shown by Griggs, Turner, and Heard[23] that crystalline rocks behave in the same way.

The effects of large changes in strain rate in the experimental deformation of marble have been studied by Heard[24] over the range 4×10^{-1} to 3×10^{-8} sec^{-1} and at 0.19–0.67 T_m (where T_m is the ratio of the experimental temperature to the melting temperature in °K). Experimental studies on deformation of Cr–Mo, Mo–V, and Cr–Ni–Nb steels[25-27] from 10^{-7} sec^{-1} to 10^{-12} sec^{-1} over the range 0.45–0.51 T_m carry the experimental data to strain rates and temperatures nearer to those operating during tectonic deformations in the upper part of the mantle. For example, according to the temperature–depth curves of Ringwood, MacGregor, and Boyd[28], at a depth of 35 km below the oceanic crust and about 60 km below the continental crust the temperatures are in the region of 900 °K. At this temperature the viscosity of even a high-strength steel for strain rates less than about 10^{-12} sec^{-1} is no more than about 10^{19} c.g.s. units and this would set an upper limit to the viscosity of the mantle at these depths. At greater depths, for instance at about 150 km, where the temperature ranges between about 1500 and 1800 °K, the viscosity would be many orders of magnitude less. According to a T_m-stress (strength) graph which may be constructed from the published data, the viscosity of steel at this temperature for strain rates of 10^{-12} sec^{-1} or less would be about 10^{10} or 10^{11} c.g.s. units. These values are close to those derived by Cook[29] and reported by Caloi[30] for the mantle. On this basis the Taylor number is

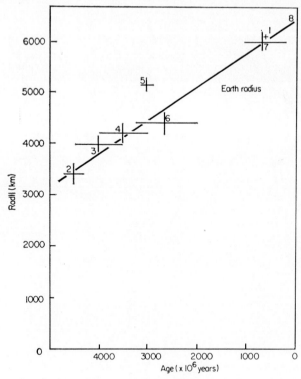

FIGURE 4 Estimates of Earth expansion—average gives 0·65 mm/year increase in radius. See text for explanation of numbers referring to the various different methods of estimation. [Reproduced from Dearnley[18]

10^3 at a depth of less than 150 km, and thus the Coriolis force could be effective in causing deflection of convective flow in the mantle.

The rate of expansion if the Superior regime fold belts are fitted to an Earth radius of 4400 km, with the simple two-cell flow pattern shown on figure 3, is compared with other estimates in the graph of figure 4. One quantitative estimate has been given by Egyed[2], based on areas of maring sedimentation since Cambrian times; a pronounced trend is apparent, indicating that progressively smaller continental areas are covered by water, clearly suggesting expansion of the Earth (by oceanic rifting) over this period amounting to an increase in the Earth's radius averaging between 0·40 and 0·66 mm/year (position 1). McDougall and others[31] have pointed out the 'remarkably close agreement between the rate of increase in the Earth's radius and that of the Universe according to Hubble's law. Using the at present accepted value of Hubble's constant, $H = 100$ km/sec Mparsec, which is $1·65 \times 10^{-4}$ mm/year mile and substituting the value of the Earth's radius in the Hubble equation $V = RH$ we obtain the radial expansion for the Earth of 0·66 mm/year'; although 'this agreement may be fortuitous it may suggest a fundamental concordance between the expansion processes in the Earth's core and those responsible for the expansion of the Universe'. The position corresponding to this estimate for 4500 million years is shown in position 2.

This rate is similar to that calculated from the radius of the Earth after differentiation of the crust on the assumption that the Earth's surface was then entirely composed of the granitic continental layer. The present area of the continental sialic material, taken as extending from an average depth of 1500 m on the continental slope is approximately 200×10^6 km^2, which gives an average annual increase in radius of $0·60 \pm 0·07$ mm, using the value of 4000 ± 500 million years as the estimated age of the differentiation of the crust (position 3).

If the shapes of the continents are to be preserved without marked distortion, then the smallest diameter for which this can be achieved is about 4200 km and, since the oldest rocks on the continents are about 3500 million years, the continents in approximately their present shapes are probably about this age (position 4).

Since the age of the Earth is about 4500 million years but the oldest sediments are only about 3500 million years, it may be that prior to this the sialic areas were not exposed to erosion above sea level. According to Egyed[32], Holmes has calculated on the foregoing assumption an average annual increase in radius of 0·40 mm/year since this period. No details of the method of calculation have been given, and for this reason, although the estimate is marked as position 5, it has not been used in the determination of the average rate of expansion.

An approximate estimate of the average Earth radius during the Grenville regime[18], obtained by fitting the fold belts younger than 1075 million years to an equator, results in a value of 6000 km (position 7).

The above estimates are closely comparable with the average annual increase of radius of $0·75 \pm 0·20$ mm/year calculated from the Superior regime fold trends, assuming that they were equatorially symmetrical (position 6) and based on an Earth radius of 4400 km at 2750 million years.

The combined results of each of these entirely different and independent methods of estimation suggest a relatively uniform rate of expansion of the Earth's radius of about $0·65 \pm 0·15$ mm/year as far back as 4500 million years.

References

1. L. Egyed, 'Determinations of changes in the dimensions of the Earth from palaeographical data', *Nature*, **178**, 534(1956).
2. L. Egyed, 'A new dynamic conception of the internal constitution of the Earth', *Geol. Rundschau*, **46**, 101 (1957).
3. L. Egyed, 'The expanding Earth?', *Nature*, **197**, 1058 (1963).
4. S. W. Carey, 'The tectonic approach to continental drift', in *Continental Drift, A Symposium, 1958*, University of Tasmania, Hobart, 1958, p. 177.
5. R. H. Dicke, 'New research on old gravitation', *Science*, **129**, 621 (1959).
6. B. C. Heezen, 'The rift in the ocean floor', *Sci. Am.*, **203**, 98 (1960).
7. A. Holmes, *Principles of Physical Geology*, Nelson, London, 1965.
8. A. Wegener, *The Origin of Continents and Oceans*, translation of 3rd ed., Methuen London, 1924.
9. F. B. Taylor, 'Gliding continents and tidal rotational forces, theory of continental drift,' *Tulsa Geol. Soc. Digest* (1928).
10. A. L. du Toit, *Our Wandering Continents*, Oliver and Boyd, Edinburgh, London, 1937.
11. L. C. King, 'Basic palaeogeography of Gondwanaland during the late Palaeozoic and Mesozoic Eras', *Quart. J. Geol. Soc. London*, **114**, 47 (1958).
12. L. C. King, *'The Morphology of Earth'*, Oliver and Boyd, Edinburgh, London, 1962.
13. G. Gastil, 'The distribution of mineral dates in time and space', *Am. J. Sci.*, **258**, 1 (1960).
14. R. Dearnley, 'Orogenic fold belts and continental drift', *Nature*, **206**, 1083 (1965).
15. R. Dearnley, 'Orogenic fold belts and a hypothesis of Earth evolution', *Phys. Chem. Earth*, **7**, 1 (1966).
16. A. P. Vinogradov and A. I. Tugarinov, 'Geochronology of the Pre-Cambrian', *Geochemistry (USSR) (English Transl.)*, **9**, 787 (1961).
17. J. Sutton, 'Long-term cycles in the evolution of the continents', *Nature*, **198**, 731 (1963).
18. R. Dearnley, 'Orogenic fold belts, convection and expansion of the Earth', *Nature*, **206**, 1284 (1965).
19. R. Hide, 'Hydrodynamics of the Earth's core', *Phys. Chem. Earth*, **1**, 94 (1956).
20. G. J. F. MacDonald, 'The deep structure of continents', *Rev. Geophys.*, **1**, 587 (1963).
21. E. Orowan, 'Mechanism of seismic faulting', *Geol. Soc. Am., Mem.*, **79**, 323 (1960).
22. E. Orowan, 'Continental drift and the origin of mountains', *Science*, **146**, 1003 (1964).
23. D. T. Griggs, F. J. Turner, and H. C. Heard, 'Deformation of rocks at 500 to 800 °C', *Geol. Soc. Am., Mem.*, **79**, 39 (1960).
24. H. C. Heard, 'Effect of large changes in strain rate in the experimental deformation of Yule marble', *J. Geol.*, **71**, 162 (1963).
25. J. D. Murray, J. S. Blair, G. G. Foster, H. W. Kirkby, and J. Blackhurst, 'The creep and nature properties of $2\frac{1}{4}\%$ chromium–1% molybdenum quality steel', *J. Iron Steel Inst. (London)*, **193**, 354 (1959).
26. A. I. Smith, E. A. Jenkinson, D. S. Armstrong, and L. M. T. Hopkin, 'Creep, mechanical, and metallurgical properties of an experimental molybdenum–vanadium steel rotor forging', *J. Iron Steel Inst. (London)*, **196**, 117 (1960).
27. E. A. Jenkinson, M. F. Day, A. I. Smith, and L. M. T. Hopkin, 'The long-time creep properties of an 18% Cr–12% Ni–1% Nb steel steampipe and super heater tube', *J. Iron Steel Inst. (London)*, **200**, 1011 (1962).
28. A. E. Ringwood, I. D. MacGregor, and F. R. Boyd, 'Petrological constitution of the upper mantle', *Ann. Rept. Dir. Geophys. Lab., Carnegie Inst., Wash.*, 147 (1964).
29. M. A. Cook, 'Viscosity–depth profiles according to the Ree–Eyring viscosity relations', *J. Geophys. Res.*, **68**, 3515 (1963).
30. P. Caloi, 'On the upper mantle', *Advan. Geophys.*, **12**, 79 (1967).
31. J. McDougall, R. Butler, P. Kronberg, and A. Sandqvist, 'A comparison of terrestrial and universal expansion', *Nature*, **199**, 1080 (1963).
32. L. Egyed, 'On the origin and constitution of the upper part of the Earth's mantle', *Geol. Rundschau*, **50**, 251 (1960).

3

E. IRVING

Department of Earth Sciences
The University
Leeds, England

Outline of the palaeomagnetic method of determining ancient Earth radii

Abstract

The Egyed–Ward method is outlined. In calculating the Earth's palaeoradius from palaeomagnetism it is assumed that the geomagnetic field has, on average, been that of a geocentric dipole and that the continental block from which the observations are obtained has remained constant in size. As the radius increases (decreases) the angle subtended at the Earth's centre by a crustal block decreases (increases) so that the rate of change of a geocentric dipole field over the block changes with time and this can be measured palaeomagnetically. The method is crude because the palaeomagnetic observations are subject to statistical fluctuations of $10°$ or so and because it is a change in the rate of change which is to be measured. Only large changes (20% or so) can be detected. The radius (expressed as a ratio to the present radius) for the Permian Period is calculated to be 0·94 and for the Triassic Period 0·99. Therefore no significant change in radius can be detected by this method since the Palaeozoic Era.

4

S. I. VAN ANDEL

and

J. HOSPERS

Division of Geophysics
Geological Institute of the University of Amsterdam
Amsterdam, Netherlands

New determinations of ancient Earth radii from palaeomagnetic data

Introduction

Three different methods have been developed until now to calculate ancient Earth radii from palaeomagnetic data. These methods will be referred to as the palaeomeridian method[1], the triangulation method[2], and Ward's minimum dispersion method[3].

The palaeomeridian method was invented by Egyed[1], who pointed out that, if two widely separated palaeomagnetic sampling sites are available, pertaining to the same geological period and situated on the same continent, the ancient Earth's radius may be computed if these sites are situated on the same palaeomeridian. The fundamental point here is that the observed palaeomagnetic inclinations allow a difference in palaeolatitude to be computed, whereas the unchanged linear distance between the sampling sites is known from their position on the present Earth's surface.

Ward's minimum dispersion method[3] makes use of all palaeomagnetic data for a given continent. Ward computes palaeomagnetic pole positions for a number of values of the radius of the ancient Earth and accepts the value of the ancient radius for which the scatter in the computed pole positions reaches a minimum as the most probable one.

These two methods will be discussed in detail by Irving (see section A IV, chapter 3 of this book). This paper will be mainly concerned with the triangulation method. In what follows, R_a stands for the radius of the ancient Earth and R_p for the present radius of the Earth (6370 km). Results will usually be expressed as R_a/R_p ratios.

Earlier Calculations of R_a/R_p Ratios Based on the Triangulation Method

The triangulation method uses palaeomagnetic sampling sites situated on different palaeomeridians. It was proposed by Egyed[2] and was first used in a much modified form by van Hilten[4] for the actual calculation of R_a/R_p ratios.

114   Geological and geophysical evidence related to the hypothesis of Earth expansion

van Hilten[4] assumes that the Earth expanded in such a way that the area of the continents remained substantially unaltered. The increase in area of the Earth's surface therefore primarily occurred in the oceans. The continents continually adapted themselves to an Earth of larger radius by the formation of radial tears near their edges.

It is clear that a great-circle arc, connecting two sampling sites situated on the stable part of a continent and not cut by any such tears, will have a constant linear length on an expanding Earth. Furthermore, it is implied in van Hilten's model that at each site the angle between the horizontal component of the palaeomagnetic direction and the connecting great-circle arc remains constant during expansion[5].

Accordingly, the problem of what happens to the palaeomagnetic pole positions pertaining to the two sampling sites A and B reduces to the problem of what happens to a spherical triangle ABC (figure 1) when the radius of the Earth increases.

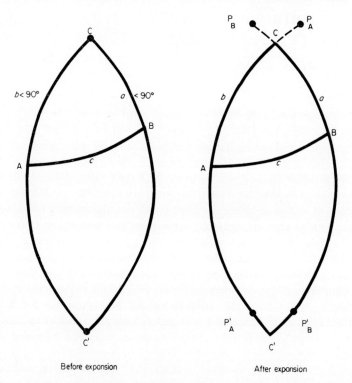

Before expansion   After expansion

FIGURE 1   Diagram illustrating the relative positions of the palaeomagnetic poles P_A and P_B (pertaining to sites A and B respectively) and the intersection C of the palaeomeridians before and after expansion. Unprimed symbols refer to the nearer pole, primed ones to the further pole

From the model requirements it follows that the linear length of the side c and the angles A and B must remain constant. Accordingly, when the radius of the Earth (R) changes, the side c expressed in angular measure will change in such a way that for both expansion and contraction

$$\frac{dc}{dR} < 0$$

New determinations of ancient Earth radii

If the standard formulae of spherical trigonometry which relate A, B, and c in the spherical triangle ABC to a and b are differentiated with respect to R, it can be shown[5] that

$$\frac{da}{dR} < 0, \quad \frac{db}{dR} < 0, \quad \text{if} \quad a < 90° \text{ and } b < 90°$$

and

$$\frac{da}{dR} > 0, \quad \frac{db}{dR} > 0, \quad \text{if} \quad a > 90° \text{ and } b > 90°$$

This means that on an expanding Earth the intersections of the palaeomeridians must lie between the sampling sites and the nearer palaeomagnetic poles P_A and P_B and beyond the further palaeomagnetic poles P_A' and P_B' (figure 1). This situation arises because, expressed in angular measure, the sides a, b, and c change as shown above, whereas the site-to-pole distances, expressed in angular measure, remain constant as they are derived from the palaeomagnetic inclinations at A and B.

These conditions are not satisfied by van Hilten's[4] Carboniferous data, and his Carboniferous ratio $R_a/R_p = 0.87$ must hence be rejected. This was also concluded by Ward[6] in his comments on van Hilten's calculations. However, the line of argument followed by Ward cannot be correct, since Ward also concluded that, if van Hilten had used the opposite Carboniferous poles, he would have found a contraction. This is not so, since the configuration of the opposite poles and the intersection is also at variance with the model.

It has further been shown by the present authors[5] that, even if the configuration of poles and intersections is in agreement with the requirements stated above, van Hilten's[4] method still can only yield approximate values of the R_a/R_p ratios corresponding to his data.

Since van Hilten's version of Egyed's triangulation method does not yield proper results, we shall now briefly refer to the original method[2]. In this method, the R_a/R_p ratio is computed from the equation $R_a/R_p = c_p/c_a$, where c_a and c_p represent the great-circle arc AB on the ancient and the present Earth, expressed in angular measure. In order to calculate c_a Egyed constructs the triangle ABC (figure 1) on the ancient Earth, where C represents the coinciding pole positions as seen from A and B before expansion. However, this construction can only be successful in the ideal case illustrated here. This is so because every spherical triangle must satisfy the following three conditions:

(i) if $A+B \gtreqless 180°$, then $a+b \gtreqless 180°$
(ii) if $A \gtreqless B$, then $a \gtreqless b$
(iii) $\sin A/\sin a = \sin B/\sin b = \sin C/\sin c$

Egyed has realized the importance of this last condition, since in his method of calculating c_a he uses the geometric mean of $\sin A/\sin a$ and $\sin B/\sin b$. However, these three conditions are not satisfied by any set of actual palaeomagnetic data. This must be explained by the fact that even on the ancient Earth before expansion the pole positions corresponding to the data at A and B would not have coincided because of unavoidable imperfections in the data. As a result, Egyed's calculation method, though mathematically sound, does not yield results when applied to the actual palaeomagnetic data. Accordingly, a new method was developed which will now be discussed.

A New Triangulation Method and its Results

The new method may be explained by referring to figure 2. The angles A and B may be calculated on the present Earth and are consequently known on the ancient Earth. For every value of R_a one wishes to assume, the magnitude of c_a in degrees may be calculated from $c_a/c_p = R_p/R_a$. From these quantities the sides a and b and the angle C_1 may be calculated.

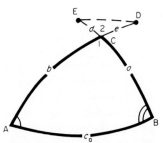

FIGURE 2 Spherical triangle illustrating the calculation of the angular magnitude of the arc DE, by means of which the separation of the palaeomagnetic poles D and E is measured, as a function of a variable ancient Earth's radius

As the arcs AD and BE are known from the palaeomagnetic inclinations at A and B respectively, the quantities d and e, as well as of course the angle $C_2 = C_1$, are known. As a consequence, the great-circle arc DE can be computed. Thus, varying R_a/R_p from 0·5 to 1·5, one finds the corresponding values of DE.

One may assume that the value of R_a for which DE reaches a minimum is the most probable value of the Earth's radius at the time of the formation of the rocks at A and B. The method is therefore a method of minimum scatter, in some respects similar to the one used by Ward[3] (for a further discussion the reader is referred to van Andel and Hospers[7]).

New values of R_a/R_p have been determined using this method. Figure 3 shows the values of DE when R_a varies from $0·5 R_p$ to $1·5 R_p$. One sees that for the Carboniferous, Permian, and Triassic data of North America there are no minimum values of DE, so that a value of R_a cannot be determined for these data.

The Cretaceous data from North America, however, do yield minimum values of DE, for which the corresponding mean value of R_a/R_p is 0·98. The Permian data from Europe and Russia similarly yield $R_a/R_p = 1·03$, the Permian data from Siberia yield $R_a/R_p = 0·70$, and the Triassic data from Europe and Russia yield $R_a/R_p = 0·89$. These four results are incorporated in table 2 which will be discussed later.

New Calculations with the Palaeomeridian Method

In addition, we also calculated new R_a/R_p ratios for the Devonian, Carboniferous, Permian, and Triassic data of Eurasia using Egyed's[1] palaeomeridian method[8].

For the Devonian, Carboniferous, and Permian data we calculated not only the arithmetic mean, but also the harmonic mean R_a/R_p ratio from the available data. This was done because it is clear that, in the ratios from which R_a/R_p is calculated, the errors in the denominator are very much larger than those in the numerator.

New determinations of ancient Earth radii

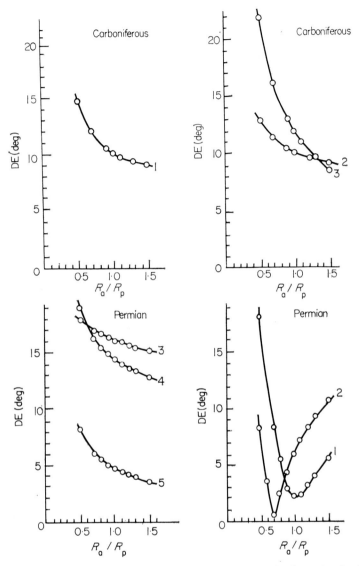

FIGURE 3 Graphs showing the calculated separation DE (in degrees) of pairs of palaeomagnetic poles as a function of R_a/R_p. The key explains by means of numbers to which continent each graph pertains (for key see next page)

118 *Geological and geophysical evidence related to the hypothesis of Earth expansion*

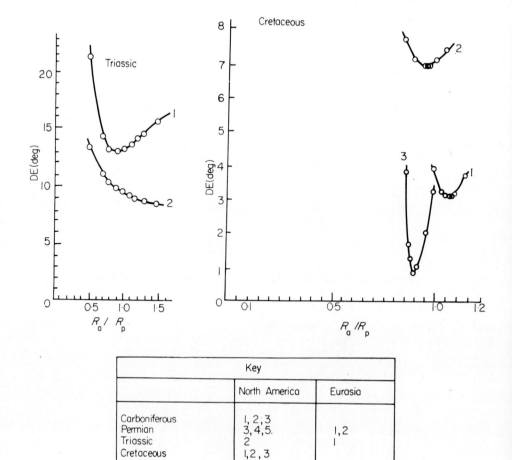

FIGURE 3 (contd.)

Each of these two mean ratios was calculated using two different weighting procedures. The first gives the individual ratios unit weight; the other gives them a relative weight equal to the number of entries in Irving's[9] catalogue of palaeomagnetic data they represent. Accordingly, four different mean ratios were obtained for each of these three geological periods. They are listed in table 1 and incorporated in table 2.

TABLE 1 Mean R_q/R_p ratios calculated by means of the palaeomeridian method

	Devonian	Carboniferous	Permian
Arithmetic mean	1·34	1·73	0·90
Weighted arithmetic mean	0·89	1·49	0·91
Harmonic mean	0·83	0·85	0·90
Weighted harmonic mean	0·71	0·80	0·91

TABLE 2 Accepted R_a/R_p ratios

Period	Reference number in Irving[9]	Author[b]	Method[c]	Weight	R_a/R_p[d] (1)	(2)	(3)	(4)
Devonian	507, 08, 09, 17, 18–22, 24–30, 39, 40, 42–45	B	P	22	1·34	0·83	0·89	0·71
Devonian	502, 05–09, 11–15, 18–22, 24–29	C	W	22	1·12	1·12	1·12	1·12
Carboniferous	601, 02, 03, 40, 41, 43, 44, 45, 47–54, 69, 70	B	P	18	1·73	0·85	1·49	0·80
Permian	707–11, 13–18, 22, 23, 24, 26, 27, 28, 30, 32, 33, 35, 36, 38, 39, 54a	B	P	27	0·90	0·90	0·91	0·91
Permian	705, 07–11, 13–18, 22, 23, 24, 26, 27, 28, 32, 33, 35, 36, 38	C	W	23	0·94	0·94	0·94	0·94
Permian	704, 05, 07–11, 13–18, 22, 23, 24, 26, 27, 28, 30, 32, 33, 35, 36, 38, 54	A	T	26	1·03	1·03	1·03	1·03
Permian	739, 57	A	T	2	0·70	0·70	0·70	0·70
Triassic	803, 04, 06, 07, 08, 15–19, 21–24	B	P	14	1·12	1·12	1·12	1·12
Triassic	803, 04, 06–13, 161–9, 21–24	C	W	18	0·99	0·99	0·99	0·99
Triassic	803, 04, 06–13	A	T	10	0·89	0·89	0·89	0·89
Cretaceous	1011, 12, 17, 18	A	T	4	0·98	0·98	0·99	0·99

[a] Also de Magnée and Nairn[10], Kruseman[11].
[b] A, van Andel and Hospers[7]; B, van Andel and Hospers[8]; C, Ward[3].
[c] P, palaeomeridian method; T, triangulation method; W, Ward's method.
[d] (1) arithmetic mean, (2) harmonic mean, (3) weighted arithmetic mean, and (4) weighted harmonic mean where applicable.

For the Triassic data it was not possible to calculate these different mean ratios, since only one combination of sampling sites could be found, yielding only one ratio of $R_a/R_b = 1·12$.

Combination of All Accepted R_a/R_p Ratios

In table 2 the new R_a/R_b ratios we calculated using our own triangulation method and Egyed's palaeomeridian method are shown, together with the ratios Ward[3] calculated using his method.

Since for three geological periods we calculated four different mean R_a/R_b ratios we now have four different groups of ratios, shown in the columns labelled (1)–(4) in table 2. The ratios within one group cannot, at present, be properly evaluated statistically, because (i) the ratios within one geological period are mutually dependent as they are partly based on the same palaeomagnetic data, (ii) it is not possible to calculate the standard deviation of each ratio, (iii) it is therefore not possible to give these ratios their proper relative weights, and (iv) the ratios may not be normally distributed.

Accordingly, we combined the ratios within one group with one another using the principle of least squares and assuming a linear relation between R_a and geological age. We did this in two different ways. In the first one we gave each ratio in table 2 unit weight. In the second one we gave each ratio a relative weight equal to the number of entries in Irving's[9] catalogue upon which each ratio is based. (These weights are listed in the fifth column of table 2.)

Each straight line fitted to the appropriate date is calculated as follows. As a straight line through the origin is required (that is, through the point representing the present-day radius R_p of 6370 km), the equation of the line is $y = ax$, where x is the geological age expressed in units of 10^9 years and $y = R_a/R_p - 1$. If w_i represents the weight assigned to the ratio used, the slope a is found from

$$a = \frac{\sum w_i x_i y_i}{\sum w_i x_i^2}$$

These slopes are listed in table 3. This table may also be used to identify the corresponding straight lines in figure 4. The straight lines in figure 4 give some idea of the large statistical uncertainties which are inherent in the use of palaeomagnetic data for the present purpose.

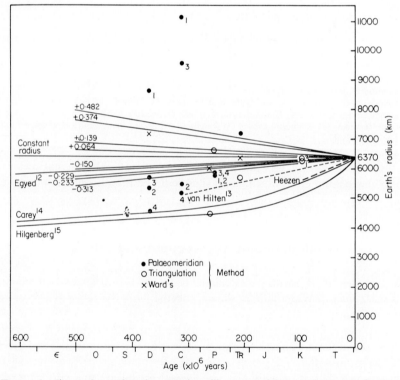

FIGURE 4 Comparison of ancient Earth radii computed from palaeomagnetic data with various proposed expansion hypotheses. The numbers 1, 2, 3, and 4 denote arithmetic, harmonic, weighted arithmetic, and weighted harmonic mean ratios respectively, where this applies. The fitted straight lines are identified by the value of their slopes (table 3)

Also shown in figure 4 are the ancient radii corresponding to the R_a/R_p ratios in table 2, as well as lines indicating the expansion rates assumed by Egyed[12], van Hilten[13], Heezen, Carey[14], and Hilgenberg[15]. From the position of the points representing the calculated R_a/R_p ratios and of the eight straight lines we fitted to these

New determinations of ancient Earth radii

TABLE 3 Values of the slopes of the fitted straight lines

Weighting procedure used	Column in table 2			
	(1) (sixth column)	(2) (seventh column)	(3) (eighth column)	(4) (ninth column)
each ratio unit weight	+0·374	−0·229	+0·064	−0·313
each ratio the relative weight shown in the fifth column of table 2	+0·482	−0·150	+0·139	−0·233

points it will be clear that an expansion rate as advocated by Egyed, a constant radius of the Earth, or a slow contraction, are more plausible than the fast expansion rates advocated by Hilgenberg, Carey, van Hilten, and Heezen. These fast expansion rates must hence be considered as implausible.

References

1. L. Egyed, *Geofis. Pura Appl.*, **45**, 115 (1960).
2. L. Egyed, *Nature*, **190**, 1097 (1961).
3. M. A. Ward, *Geophys. J.*, **8**, 217 (1963).
4. D. van Hilten, *Nature*, **200**, 1277 (1963).
5. J. Hospers and S. I. van Andel, *Tectonophysics*, **5**, 5 (1967).
6. M. A. Ward, *Geophys. J.*, **10**, 445 (1966).
7. S. I. van Andel and J. Hospers, 'Palaeomagnetism and the hypothesis of an expanding Earth: a new calculation method and its results', *Tectonophysics*, in press (1968).
8. S. I. van Andel and J. Hospers, 'Palaeomagnetic calculations of the radius of the ancient Earth by means of the palaeomeridian method', *Phys. Earth Planetary Interiors*, in press (1968).
9. E. Irving, *Palaeomagnetism*, Wiley, New York, 1964.
10. I. de Magnée and A. E. M. Nairn, *Bull. Soc. Belge Geol.*, **71**, 551 (1962).
11. G. P. Kruseman, *Geol. Ultriectina*, **9**, 1 (1962).
12. L. Egyed, *Nature*, **197**, 1059 (1963).
13. D. van Hilten, *Tectonophysics*, **1**, 3 (1964).
14. S. W. Carey, *Continental Drift, A Symposium, 1958*, University of Tasmania, Hobart, 1958.
15. O. C. Hilgenberg, *Neues Jahrb. Geol. Palaeontol.*, **116**, 1 (1962).

B

SOLID STATE PHYSICS AND GEOPHYSICS

I. New developments in the study of the electrical conductivity of the Earth's mantle

1

U. SCHMUCKER

Göttingen
West Germany

Conductivity anomalies, with special reference to the Andes

Internal and External Parts of Geomagnetic Variations

Transient fluctuations of the Earth's magnetic field are commonly known as geomagnetic variations. They arise from shifting current systems in the ionosphere or beyond and diffuse through the conductive layers of the Earth's interior with amplitude reduction and phase rotation whilst inducing currents. The depth of these currents, and thereby the depth of penetration of the incident variation field, increases from a few kilometres for fast pulsations (60 cycles/hour) to hundreds of kilometres for the slow diurnal variations ($\frac{1}{24}$ cycle/hour).

The decisive frequency–conductivity parameter is the skin-depth value

$$p = \frac{1}{(2\pi\omega\sigma\mu)^{1/2}} \qquad (1)$$

where $\omega = 2\pi f$ is the angular frequency of the incident field, μ the magnetic permeability (usually set to unity), and σ the electrical conductivity of the subterranean matter. All quantities are to be measured in electromagnetic c.g.s. units (e.m.u.). In the case of a uniform conductor, p is the depth beneath its surface where the amplitude of an incident electromagnetic field is attenuated to 1/e of its surface value, assuming that this depth is small compared with the lateral non-uniformity of the incident field. Rewriting (1) in convenient units for geomagnetic induction problems, namely in cycles/hour for f and $\Omega^{-1}\,\text{m}^{-1}$ for σ, gives

$$p = \frac{30 \cdot 2}{(f\sigma)^{1/2}} \text{ km} \qquad (1a)$$

The surface field of the internal eddy currents is superimposed upon the primary source field from above, and we distinguish accordingly between the *internal* (induced) and the *external* (inducing) part of geomagnetic variations as observed at the Earth's surface. Both parts can be separated when the spatial distribution of the variation field is known.

We are concerned here with local anomalies of the internal part, which are caused by an unequal distribution of subterranean conductivity, involving large gradients of σ in the horizontal direction. They produce characteristic differences of simultaneously recorded variations at adjacent sites, say less than 100 km apart, which usually cannot be attributed to the smoothly varying primary field from above. Hence, a closely spaced network of temporary magnetic recording stations is needed to detect such internal conductivity anomalies.

Two inherent limitations of this method of *geomagnetic depth sounding* should be mentioned. Since the observations are made within small areas (small in comparison with the spatial extent of the primary field) a complete separation of internal and external parts is not possible. As a consequence, the average change of conductivity with depth in the surveyed area remains unknown and any local anomalies of geomagnetic variations have to be interpreted by lateral conductivity changes within a preconceived normal, i.e. stratified conductivity distribution which must be inferred from other sources of information (cf. p. 127).

Secondly, oceans and continental surface layers form a thin conducting cover of great complexity. The flow of superficial eddy currents is therefore highly distorted. This may lead to local anomalies of the internal part, in particular near coast lines because of the outstanding conductivity contrast of sea water and rock formations on land. Such surface effects have to be taken into account before conclusions are drawn about possible anomalies at greater depth.

Conductivity and Temperature

At first sight the electrical conductivity σ of subterranean matter may not seem to be a very noteworthy parameter. There is, however, its close relation to temperature, following the general theory of semiconduction, and even small changes of temperature can cause drastic changes of conductivity. Olivine, for instance, doubles its conductivity when the temperature is raised by just 50 degC in the range[1] from 1000 to 1250°C.

We have to bear in mind, however, that semiconduction in non-metallic solids is primarily an impurity effect. Thus, minute changes of composition, in particular of the iron content, can have an equally strong effect upon σ, not counting the largely unknown influence of pressure. This limits the effective use of the conductivity as *absolute thermometer* for the Earth's interior.

There remains, however, the important aspect to use σ as *relative thermometer*, namely to infer deep-seated lateral gradients of conductivity and possibly temperature from their distorting effect upon the internal part of geomagnetic variations. Such thermal imbalances in the upper mantle could be connected with ascending and descending branches of convection cells or with local concentrations of radioactive heat sources, which may be the underlying cause for the diversified tectonic and magmatic history of the Earth's outermost layers.

Conductivity Distribution in the Upper Mantle

The electric conduction in surface rocks is mainly electrolytic through salty solutions filling pores and cracks. Their conductivity varies accordingly between $1\ \Omega^{-1}\ m^{-1}$ for unconsolidated clastic sediments and 0·001 for dense igneous rocks. Sea water in comparison has an average conductivity of 4, copper a conductivity of $10^8\ \Omega^{-1}\ m^{-1}$.

Rocks become insulating under pressure when their pores and cracks are closed, and the Earth's crust and uppermost mantle must be indeed very poor conductors. There is clear evidence, however, that the conductivity rises again in the upper mantle. and it is not unreasonable to relate this rise to the downward increase of ambient temperature (see the previous section).

Two methods have been in use to infer the change of conductivity with depth by means of natural electromagnetic fields. The first and classical method is based on magnetic observations alone and uses the surface ratio of internal to external parts of geomagnetic variations, averaged on a global scale. Disregarding in this way regional differences Lahiri and Price[2] gave two possible distributions, representing limiting cases, which are compatible with the internal parts of semidiurnal S_q variations and smoothed storm-time D_{st} variations. In the first model 'e', an insulating intermediate layer extends downwards from the surface to 600 km depth, where the conductivity rises abruptly to infinity. The whole model is surrounded by a thin outermost shell which has an integrated total conductivity of $5 \cdot 1 \times 10^{-6}$ e.m.u. cm, equivalent to 1280 m sea water (cf. equation (5)). In the alternative model 'd', the conductivity rises smoothly with depth beneath a surface shell of 500 m sea water. Starting with $0 \cdot 004 \ \Omega^{-1} \ m^{-1}$ beneath this shell the conductivity reaches $0 \cdot 1$ at 500 km depth and unity at 900 km depth.

The second method, introduced by Tichonov and Cagniard, uses the surface impedance, i.e. the ratio of tangential electric to orthogonal magnetic field fluctuations, as observed at one site over a wide frequency range. Lateral conductivity variations are excluded and it is assumed that the primary field is of great lateral uniformity in comparison with its depth of penetration. The magnetotelluric method yields in this way estimates for the mean layered conductivity distribution on a regional scale. The analysis of pertinent observations at various places proved the existence of a high-resistivity zone between surface layers of great complexity and highly conducting matter in the upper mantle.

Returning to the first mentioned magnetic method we observe that the slow diurnal variations propagate with negligible attenuation through the upper mantle above 500 km and cannot yield more than an upper limit for the conductivity existing here. Detailed information about this depth range comes therefore mainly from fast variations around 1 cycle/hour with a reduced depth of penetration. Rikitake[3] was the first to attempt a World-wide analysis of bays and other short-period events. It became soon evident, however, that their internal part is subject to numerous local anomalies and that the upper mantle must be extremely non-uniform as far as its conductivity is concerned. In particular, standard magnetic observatories seemed to have the tendency to lie close to anomalous zones, which facilitated their detection[4].

Since then interest has focused on these *induction anomalies* of fast variations. They have been found at many places around the World, even though the depth of their origin and their significance for the upper-mantle structure are not always clear. The prominent coastal anomalies near large and deep oceans[5], for instance, coincide with an outstanding superficial conductivity contrast and can be interpreted—at least partially —as surface effect (see p. 129). The Rio Grande anomaly in the southwestern United States, on the other hand, is presumably of deep origin, since the overall surface conductivity is rather low here and without marked changes within the zone of anomalous variations[6].

At times it appeared as if such induction anomalies could be found everywhere, and clearly, when the anomalous becomes the norm, its significance for the unusual diminishes. But this is not so and there are large areas where the normal behaviour of the variations indicates a stratified internal conductivity structure. This applies for southern Arizona and New Mexico between the Colorado river and the Rio Grande anomaly, where we observe exceedingly small but uniform Z variations[6]. Another example is Bavaria, where a north–south profile from Upper Palatinate across the Bavarian 'Molasse' into the Alps failed to give indications for internal conductivity anomalies[7]. Hence, the well-known north German anomaly does not seem to have a counterpart in southern Germany.

Outline of the Data Reduction

Considering the internal conductivity σ at the level z beneath the Earth's surface we distinguish between its constant normal part $\bar{\sigma}$ and its variable anomalous part σ_a:

$$\sigma = \bar{\sigma} + \sigma_a \tag{2}$$

The transient magnetic field vector $\mathbf{F}(t, P)$ is accordingly the sum of a normal plus anomalous part:

$$\mathbf{F}(t, P) = \bar{\mathbf{F}}(t, P) + \mathbf{F}_a(t, P) \tag{3}$$

$\bar{\mathbf{F}}(t, P)$ represents the smoothly varying external plus internal surface field above the averaged conductivity distribution $\bar{\sigma}(z)$, while the induction anomaly $\mathbf{F}_a(t, P)$ is, by definition, of internal origin alone. We have to find the perturbation σ_a as a function of depth from an observed induction anomaly F_a on the basis of a presumed normal distribution $\bar{\sigma}(z)$.

The flow of eddy currents in a stratified substratum is parallel to its surface. Hence, the internal and external parts of $\bar{F}$ have matching distributions at the Earth's surface, provided that the depth of penetration of the normal variation field is small in comparison with its lateral non-uniformity. This is a justified assumption in the case of fast variations and in the absence of overhead current concentrations (jets). We obtain then the normal part of the tangential H (northward) and D (eastward) variations by smoothing the observed variation field within the surveyed area. Local deviations from this smoothed level are considered as anomalous parts of H and D.

The normal part of the observed vertical Z variations is given by the overall depth of the eddy currents in relation to the lateral non-uniformity of the primary field. Let their mean depth be represented by a perfect conductor at the frequency-dependent depth h. (This substitution accounts of course only for the in-phase component of the induced surface field.) Then

$$\bar{Z} = h(\bar{H}_x + \bar{D}_y) \tag{4}$$

where $\bar{H}_x$ denotes the northward gradient of $\bar{H}$ and $\bar{D}_y$ the eastward gradient of $\bar{D}$ ($\bar{H}, \bar{D}, \bar{Z}$ are the components of the normal variation vector $\bar{\mathbf{F}}$).

The midlatitude bay field, for instance, has in H a relative northward gradient of 5% per hundred kilometres and a negligible eastward gradient in D. Thus, by setting

Conductivity anomalies

$h = 200$ km, we obtain $\bar{Z}/\bar{H} = 0.1$ as normal ratio of vertical to horizontal variations. This reflects the well-known fact that the inducing and induced fields above a conductive substratum supplement each other in the tangential components, but oppose each other in the vertical component, yielding a nearly tangential transient surface field under normal conditions. Consequently, internal conductivity anomalies which disturb this sensitive balance between external and internal Z variations are more obvious in Z than in H and D, where the anomalous parts are superimposed upon substantial normal parts (cf. figure 2).

The second step of the data reduction is a statistical correlation analysis between the thus separated anomalous and normal parts, involving numerous magnetic disturbances of the same general type (e.g. bays) but of different form and intensity. This postulated correlation is necessarily linear, since the governing equations, Maxwell's field equations, establish linear relations between the electromagnetic field components and their time and space derivatives. We obtain as result for each survey station a 3×3 matrix of transfer functions connecting the components of $\bar{\mathbf{F}}$ and $\mathbf{F}_a$ in the frequency domain. They describe the induction anomaly as a function of frequency and location in a statistically condensed form and provide the proper basis for the subsequent interpretation.

It remains to verify the truly internal origin of the anomalous surface field, normalized in this way. This can be done by applying appropriate separation methods to its spatial distribution for each resolved frequency component, thereby eliminating unwanted contaminations of external origin. Siebert and Kertz[8] proposed a convenient method for two-dimensional fields which Hartmann[9] and Weaver[10] extended to three-dimensional distributions. Price and Wilkins[11] used in their treatise on the S_q field a somewhat different, but also very suitable, separation technique. The separation involves in either case elaborate numerical calculations and it is carried out preferably as final step of the data reduction. We may presume that the statistical treatment of numerous events help to minimize random contributions of external origin to $\mathbf{F}_a$.

Interpretation of Induction Anomalies

At the outset we have to estimate the possible effect of near-surface conductivity variations upon the internal part. Following Price[12] it is convenient and, for the frequencies around 1 cycle/hour which are considered here, also permissible to treat the outermost layers (oceans and geological strata on land) as a thin surface sheet of variable *total conductivity:*

$$\tau = \int_0^d \sigma(z)\,dz$$

separated by an insulating intermediate zone from highly conductive matter further down. The integration is carried out from the outer to the inner face of the sheet, d denoting its thickness.

Let $\bar{\tau}$ be, in analogy to $\bar{\sigma}$, the averaged total conductivity of the surface layers in the surveyed area and h the depth of a substitute perfect conductor in reference to the mean depth of the deeply induced currents. Theoretical considerations show that

$$\eta_\varepsilon = 4\pi\omega\bar{\tau}h$$

controls as dimensionless *induction parameter* the relative strength of those eddy currents which are induced in the surface layers. Their contribution to the internal part is greater than the contribution of deeply induced currents when $\eta_s > 1$, and vice versa.

The inclusion of h accounts for the dampening effect of the inductive couple between superficial and deep eddy currents, assuming again that their depth is small when compared with the spatial wave length of the primary field.

With regard to bays we may insert $\omega = 2\pi$ cycles/hour and $h = 200$ km. This gives $\eta_s = 7$ for $\bar{\tau} = 16 \times 10^{-6}$ e.m.u. cm (= 4 km of sea water, $\sigma = 4 \, \Omega^{-1} \, m^{-1}$) and $\eta_s = 0.18$ for $\bar{\tau} = 4 \times 10^{-7}$ e.m.u. cm (= 4 km of rock formations, $\sigma = 0.1 \, \Omega^{-1} \, m^{-1}$). We see that bay disturbances penetrate with little attenuation by eddy currents through continental surface layers of the indicated conductivity but not through large and deep oceans, at least not for the value of h postulated here. This discrepancy would explain the anomalous behaviour of bays near coast lines, but it becomes obvious at the same time that prominent inland anomalies of bays could hardly arise from superficial conductivity contrasts alone. It may be added that the proper mean value $\bar{\tau}$ is not the arithmetic but the harmonic mean over a variable total conductivity.

Observations with a self-contained D variometer, lowered to the bottom of the Pacific Ocean offshore from California, revealed that the D amplitude of bays is reduced indeed to one-quarter of its surface value beneath 4 km of sea water[13]. This implies that about three-quarters of the internal part above the ocean comes from eddy currents induced in the ocean.

Corresponding observations beneath continental surface layers, say in deep bore holes, have not been made yet, but a preliminary estimate for their shielding effect upon geomagnetic variations can be derived from magnetotelluric measurements. Let E/H be the surface impedance and $\bar{\tau}$ the total conductivity of surface layers at a given site. The difference between the tangential variations above (H) and below (H^-) these layers is equal to the integrated sheet-current density $E\tau$, multiplied by 4π. Hence,

$$\frac{H^-}{H} = 1 - 4\pi\tau\frac{E}{H} \qquad (7)$$

Wiese[14] reported, for instance, as impedance of bays $E_{EW}/H = 0.14$ mv/km $\gamma = 1.4 \times 10^4$ e.m.u. for the observatory Niemegk near Potsdam, situated above the highly conducting sediments of northern Germany. Inserting $\tau = 1.6 \times 10^{-6}$ e.m.u. cm (= 4 km of sediments, $\sigma = 0.4 \, \Omega^{-1} \, m^{-1}$) yields $H^-/H = 0.72$. Thus, only a small part of the internal bay field would be due to eddy currents in the sediments, indicating a deep-seated cause for the anomaly of bays in northern Germany.

Induction anomalies which are of truly deep origin can be interpreted on the basis of two basic models.

(*a*) We approximate the internal conductivity distribution by a stratified substratum with undulating interfaces between various layers of uniform but different conductivity.

(*b*) We use a stratified substratum with plane interfaces but assume that one or more layers are non-uniform as indicated in equation (2).

Both models merge at some distance from the anomaly into a preconceived normal distribution $\bar{\sigma}(z)$ and to be determined are either the undulations or the perturbations σ_a from an induction anomaly at the surface of the substratum. The greatest possible

Conductivity anomalies

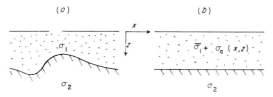

FIGURE 1 Two-layer models to illustrate the proposed types of internal conductivity anomalies

depth of these non-uniformities is given by the depth of penetration of the normal variation field which of course must reach the undulating interfaces or non-uniform layers, respectively.

Figure 1 shows these basic concepts of interpretation applied to a simple two-layer model for the Earth's interior. A straightforward treatment of the model 1(a) is possible in the limiting case that $\sigma_1 = 0$ and $\sigma_2 = \infty$. The boundary condition for a transient field in the upper non-conducting half-space requires that its magnetic vector is tangential to the surface of the underlying perfect conductor. Hence, this surface can be found by deriving the internal field-line pattern for an anomalous plus normal variation field which is given at the surface of the upper non-conducting layer and extended downwards in one way or the other. It is of course presumed that the anomalous and normal parts of the observed variations are roughly in phase.

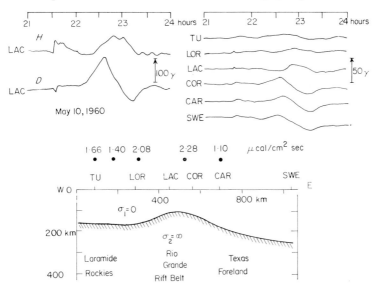

FIGURE 2 Rio Grande anomaly of fast variations in southern Arizona, southern New Mexico, and west Texas, shown for a typical bay. Subdued Z amplitudes west of the Rio Grande (Tucson, Lordsburg) suggest high mantle conductivities at shallow depth beneath the Laramide Rockies. Their conspicuous increase east of the Rio Grande (Cornudas, Carlsbad, Sweetwater) reflects in comparison, the low conductivity of the upper mantle beneath the Texas foreland. The Z reversal between Las Cruces and Cornudas can be explained by an additional rise of highly conductive matter under the Rio Grande rift belt. The horizontal variations of D and H are shown only for Las Cruces. Notice that the Z variations east of the Rio Grande have the same form as the D variations at Las Cruces, indicating a north–south trend of the conductivity structure shown below (see text). [Heat flow values from Warren[15] and Herrin and Clark[16]]

Each field line which does not intersect the Earth's surface is one possible interface between non-conducting and perfectly conducting matter. From the family of curves we choose that which merges at some distance from the anomaly into a postulated normal depth h of a substitute perfect conductor as introduced above. Each particular frequency component of the induction anomaly yields undulations around another normal depth h, indicating the varying depth of penetration of the incident field as a function of location and frequency.

The application of this field-line method to the Rio Grande anomaly of bays is shown in figure 2. In accordance with the width of the coastal anomaly in southern California we chose $h = 160$ km as the normal depth of the substitute perfect conductor under southern Arizona (Tucson) for 1 cycle/hour. Its surface then rises to $h = 100$ km under the Rio Grande rift belt and sinks to more than 200 km under the Texas foreland.

This contrast of low mantle conductivities under west Texas to high conductivities under southern Arizona and New Mexico affects even the internal part of the deeply penetrating diurnal variations as seen in figure 3. We observe that the centre of the northern S_q vortex passes during the equinoxes more or less overhead the east–west line of survey stations, which lie therefore in the range of maximum external Z variations. Hence, the observed diurnal Z amplitude is, as the sum of the external plus internal parts, a sensitive measure for the depth to the highly conducting part of the upper mantle.

It may be suggestive that the Rio Grande anomaly lies at the border of two structural provinces, the Laramide Rockies to the west and the Texas foreland to the east, the

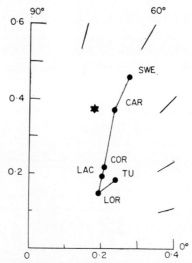

FIGURE 3 Rio Grande anomaly of the slow diurnal variations (cf. figure 2). The dots show the Z/Y ratio of the third time harmonic in polar coordinates, derived from hourly means in Z and Y (= true east component) of four quiet days (April 19–22, 1960). The angle indicates a phase lead of Z relative to Y. The star shows the global average of the Z/Y ratio for the magnetic latitude of the survey stations (40° N), calculated from Chapman's equinoctial ratio of internal to external parts for the P_3^1 term (cf. Lahiri and Price[2], table 1). The reduced Z/Y ratio in southern Arizona and its gradual increase toward Texas conform with the anomalous behaviour of bays along the same profile, indicating a deep-seated cause of the anomaly

latter being a region of great tectonic stability and magmatic inactivity since Pre-Cambrian times. Furthermore, the postulated rise of highly conductive and probably hot mantle material under the Rio Grande valley coincides with a belt of intense vulcanism in recent times, high terrestrial heat flow[15-17], and unusual attenuation of seismic waves[18].

Conductivity Anomaly in the Andes of Peru and Bolivia

In 1957 the Carnegie Institution of Washington sent a seismic expedition to the Andes in South America[19]. One of its conspicuous results was the discovery of an extremely high attenuation of seismic waves which travel across the mountain range. It was suggested at that time that 'this attenuation may have some connection with the volcanic structure of the Andes', involving hot and perhaps even molten material at shallow depth.

Would this zone of high seismic attenuation appear as a zone of high conductivity in the internal part of geomagnetic variations? A field programme to test this hypothesis began in 1963 with a net of nine magnetic recording stations (Askania variographs). It has been in progress since that time as a joint venture of the Instituto Geofisico del Peru, the Instituto Geofisico Boliviano, and the Department of Terrestrial Magnetism (Carnegie Institution of Washington).

The first reconnaissance survey of 1963 revealed that the expected coastal anomaly of bays is not only missing in southern Peru but even reversed in its sign[20]. In other words, the superficial conductivity contrast between ocean and continent is more than compensated by an internal conductivity gradient in the opposite direction, bringing deep induction currents close to the surface beneath the Andes. After the second survey 1965–6 it became clear that this postulated zone of high internal conductivity ends near the eastern slope of the mountain range[21].

Let a few introductory remarks precede the detailed discussion of selected magnetograms. Peru and Bolivia are unusual countries in various aspects, but their geomagnetic distinction is the presence of the dip equator of the main field as shown in figure 4.

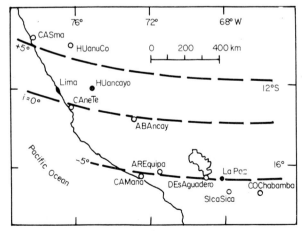

FIGURE 4   Magnetic stations during the 1965–6 survey in Peru and Bolivia, shown in relation to lines of equal dip i

This line of zero Z component exerts a pinching effect upon ionospheric currents on the day-lit side of the Earth, leading to an overhead current concentration which is known as *equatorial electrojet*. The surface field of this jet has a half-width of about 250 km. It may be visualized as the field of a line current, flowing 2–300 km above the line of zero dip.

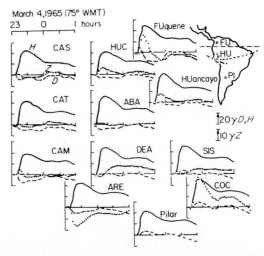

FIGURE 5 Equatorial night event (bay) as recorded at the stations of the 1965–6 survey (see figure 4) and the permanent observatories Fuquene (Columbia), Pilar (Argentina). Slight enhancement of H amplitudes at mountain stations relative to those at coastal stations and concurrent reversal of Z amplitudes (CAS–HUC, CAT–HU, ARE–COC). Both observations suggest that deep induction currents are brought close to the surface in a high conductivity channel beneath the Andes. [From Schmucker and others[21]]

The electrojet is absent during the night hours and the equatorial bay field, for example, is indeed of remarkable uniformity (figure 5). Thus, we have two distinct source fields at our disposal: the spatially smooth field of night events and the highly non-uniform jet field of day events.

Local differences of night events are undoubtedly of internal origin alone and due to subterranean conductivity anomalies. Local differences of day events, on the other hand, reflect not only the distorting effect of these internal anomalies σ_a but also the non-uniformity of the external plus internal jet field above a normal stratified distribution $\bar{\sigma}(z)$. In short, the combined anlaysis of day and night events gives us in equatorial regions the unusual opportunity to investigate concurrently the anomalous and normal conductivity distribution with observations in a limited area.

Beginning with a typical night event we infer from the traces of figure 5 that the H amplitude of the equatorial bay field hardly changes over 37° in latitude. Hence, we may expect minute Z amplitudes under normal conditions (cf. equation (4)). There is a slight anomalous increase of the maximum H deflection at mountain stations relative to those at the coast. We notice also some irregular differences in D, even though substantial changes of the compass deviation from station to station obscure their significance. Nevertheless, the overall D amplitude is small and the horizontal disturbance vector points northwards to the high-latitude centre of the ionospheric bay vortex.

Conductivity anomalies

A preliminary evaluation of numerous night events showed that their normal parts in X (= true north component) and Y (= true east component) are well represented by the horizontal variations at Arequipa, the capital of southern Peru half-way between the coast and the high Andes. We subtract the thus-defined normal part from the observed X and Y amplitudes and obtain for each survey station the respective components of the anomalous horizontal vector $\mathbf{B}_a$ as shown in figure 6.

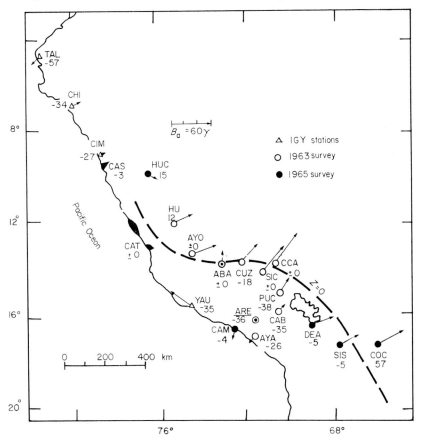

FIGURE 6 Andean anomaly of bays, normalized with the true north amplitude X of Arequipa and averaged over two events for each station. The arrows show the anomalous part B_a of the horizontal disturbance vector:

$$B_a = (X - X_{ARE})i + (Y - Y_{ARE})j$$
$$X_{ARE} = 100 \, \gamma$$

The numbers give the Z amplitude. Arrows of maximum length are found in the high mountains where they are more or less perpendicular to a line of zero Z amplitude. This line indicates the trend of the postulated high-conductivity zone in the Andes (see text). [From Schmucker and others[21]]

These vectors indicate, when rotated anticlockwise by 90°, strength and direction of the anomalous internal current field which is superimposed upon the normal east–west flow of subterranean eddy currents. We see that these currents are channelled into a high-conductivity zone which follows the general trend of the mountain range. The resultant current concentration beneath the Andes explains at the same time the reversed anomalous Z amplitudes along the eastern and western slope which is so strikingly demonstrated by the opposite Z deflections at Arequipa and Cochabamba.

Turning now to day-time fluctuations of comparable frequency we observe that their jet field should be uniform along lines of equal dip and without variations in D when internal anomalies are absent. (The horizontal force of the main field is nearly perpendicular to the dip equator in Peru.) Hence, induction anomalies of the jet field are characterized by different day-time variations at stations of the same dip and by D variations in general, provided of course that the trend of the anomaly is not parallel to the dip equator.

The equatorial day-time fluctuations in Peru and Bolivia show indeed these criterions for internal conductivity anomalies (figure 7). The rugged Z trace of Cochabamba stands in sharp contrast with the small Z amplitudes of Sicasica, Desaguadero, and Arequipa (not shown), even though these stations lie more or less on the same isocline, namely on the southern isocline of maximum Z amplitude of the normal jet field. We conclude that Cochabamba is located above the edge of an extremely shallow concentration of internal eddy currents to the south, while the other stations are on top of it. No explanation can be offered yet for the anomalous behaviour of Z at the coastal station Camanà. Clearly visible are also anomalous D variations along the southern isocline, indicating a northward deflection of the internal jet current by the high-conductivity zone under the Andes.

The depth of this zone has to be small in comparison to the half-width of the jet field, i.e. of the order of 50 km or less. Otherwise, this narrow-spaced source field would not reach the anomalous zone at all. The skin-depth value for 1 cycle/hour of

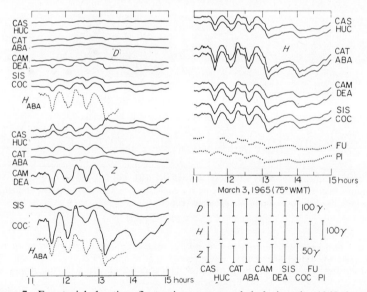

FIGURE 7 Equatorial day-time fluctuations as recorded during the 1965–6 survey (see figure 4). Maximum H amplitudes near the dip equator (CAT, ABA) reflect the electrojet effect upon the ionospheric source field. Subdued Z amplitudes at the northern stations CAS and HUC indicate that the Z component of the external jet field is nearly compensated by the field of subterranean eddy currents near 100 km depth. The large Z amplitude at COC, CAM and the D fluctuations at DEA, SIS, COC are anomalous and can be related to high internal conductivities under the Andes. [From Schmucker and others[21]]

the material existing here must be also small as against the half-width in order to permit a significant attentuation of incident day-time fluctuations by eddy currents. Hence, in virtue of equation (1a) we have to postulate a conductivity of at least $0.1\,\Omega^{-1}\,m^{-1}$ in contrast with much lower values under the adjacent Brazilian shield and the offshore Peruvian trench.

A careful intercomparison of the slow diurnal variations at the same survey stations revealed that their relatively small internal parts are not affected by the Andean anomaly. Hence, the jet field penetrates for $f = 1/12$ cycle/hour through the high-conductivity zone with negligible attenuation, which establishes $0.1\,\Omega^{-1}\,m^{-1}$ also as an upper permissible conductivity in this zone.

We recall from p. 126 that conductivities of this magnitude are expected only deep within the mantle under normal conditions. Their presence at shallow depth beneath the Andes (0–50 km) is clear evidence for the unusual thermal state or composition of the existing mantle material, reflecting perhaps the remarkably intense tectonic and magmatic history of this mountain range.

A complete review of this type of work can be found in the book by Rikitake[22].

References

1. D. C. Tozer, 'The electrical properties of the Earth's interior', *Phys. Chem. Earth*, **3**, 414–36 (1959).
2. B. N. Lahiri and A. T. Price, 'Electromagnetic induction in non-uniform conductors, and the determination of the conductivity of the Earth from terrestrial magnetic variations', *Phil. Trans. Roy. Soc. London, Ser. A*, **237**, 509–40 (1939).
3. T. Rikitake, 'Electromagnetic induction within the Earth and its relation to the electrical state of the Earth's interior', *Bull. Earthquake Res. Inst., Tokyo Univ.*, **28**, 45–100, 219–283 (1950).
4. O. Meyer, Über eine besondere Art von erdmagnetischen Bay-Störungen', *Deut. Hydrograph. Z.*, **4**, 61 (1951).
5. W. D. Parkinson, 'The influence of continents and oceans on geomagnetic variations', *Geophys. J.*, **6**, 441–9 (1962).
6. U. Schmucker, 'Anomalies of geomagnetic variations in the southwestern United States', *J. Geomagnetism Geolectricity*, **15**, 193–221 (1964).
7. A. Berktold, 'Erste Auswertung von Messungen des zeitlich variablen erdmagnetischen Feldes entlang eines Profiles vom Oberpfälzer Wald bis zu den Kitzbüheler Alpen', *Z. Geophysik*, **32**, 492–501 (1966).
8. M. Siebert and W. Kertz, 'Zur Zerlegung eines lokalen erdmagnetischen Feldes in äusseren und inneren Anteil', *Nachr. Akad. Wiss. Goettingen, Math.-Physik. Kl.*, **1957**, No. 5, 87–112.
9. O. Hartmann, 'Behandlung lokaler erdmagnetischer Felder als Randwertaufgabe der Potentialtheorie', *Abhandl. Akad. Wiss. Goettingen, Math.-Physik. Kl. Beiträge zum Internationalen Geophysikalischen Jahr*, **9** (1963).
10. J. T. Weaver, 'On the separation of local geomagnetic fields into external and internal parts', *Z. Geophysik*, **30**, 29–36 (1963).
11. A. T. Price and G. A. Wilkins, 'New method for the analysis of geomagnetic fields and their application to the S_q field of 1932-3', *Phil. Trans. Roy. Soc. London, Ser. A*, **256**, 31–98 (1963).
12. A. T. Price, 'The induction of electric currents in non-uniform thin sheets and shells', *Quart. J. Mech. Appl. Math.*, **2**, 283–310 (1949).
13. J. H. Filloux and C. S. Cox, 'Magnetotelluric sounding under the sea floor', *Trans. Am. Geophys. Union*, **47**, 52 (1966).

14. H. Wiese, 'Geomagnetische Tiefentellurik Teil I: Die elektrische Leitfähigkeit der Erdkruste und des oberen Erdmantels', *Geofis. Pura Appl.*, **51**, 59–78 (1962).
15. R. E. Warren, 'Heat flow measurements in southern Arizona and New Mexico', unpublished.
16. E. Herrin and S. P. Clark, 'Heat flow in west Texas and eastern New Mexico', *Geophysics*, **21**, 1087–99 (1956).
17. E. R. Decker, 'Crustal heat flow in Colorado and New Mexico', *Trans. Am. Geophys. Union*, **47**, 180 (1966).
18. J. Jordan, R. Black, and C. C. Bates, 'Patterns of maximum amplitudes of P_n and P waves over regional and continental areas', *Bull. Seismol. Soc. Am.*, **55**, 693–720 (1965).
19. 'Seismic studies in the Andes', *Trans. Am. Geophys. Union*, **39**, 580–2 (1958).
20. U. Schmucker, O. Hartmann, A. A. Giesecke, M. Casaverde, and S. E. Forbush, 'Electrical conductivity anomalies in the Earth's crust in Peru', *Carnegie Inst. Wash. Yearbook*, **63**, 354–62 (1964).
21. U. Schmucker, S. E. Forbush, O. Hartmann, A. A. Giesecke, M. Casaverde, J. Castillo, R. Salgueiro, and S. del Pozo, 'Electrical conductivity anomaly under the Andes', *Carnegie Inst. Wash. Yearbook*, **65**, 11–28 (1966).
22. T. Rikitake, *Electromagnetism and the Earth's Interior*, Elsevier, Amsterdam, London, New York, 1966.

2

D. I. GOUGH

University of Alberta
Edmonton, Canada

and

J. S. REITZEL*

Southwest Center for Advanced Studies
Dallas, Texas, U.S.A.

Magnetic deep sounding and local conductivity anomalies

Introduction

The electrical conductivity within the Earth varies principally with radius; in particular, there is a rapid rise in conductivity to values greater than 1 Ω^{-1} m^{-1} at depth 700 to 800 km from much lower values (typically 10^{-4} Ω^{-1} m^{-1} in the crust below sedimentary layers). This increase is associated with the general rise of temperature with depth, through excitation of ionic and intrinsic semiconduction in the silicates of the mantle. The time-varying part of the Earth's magnetic field with periods from minutes to days is strongly influenced by currents induced in the conducting regions at depth; analysis of the World-wide fields of daily variations and magnetic storms has provided the most direct evidence for the general increase of conductivity with depth. The state of knowledge of these topics is reviewed by Rikitake[1].

In many places there are remarkable differences between the time-varying magnetic fields recorded at stations a few tens of kilometres apart, especially in the vertical component and for periods from a few minutes to a few hours. These differences appear to arise from currents induced in local regions of enhanced conductivity at depths less than 700 km; for the most part, these are probably regions of anomalously high temperature in the upper mantle. In some cases these hot regions may possibly be equivalent to irregularities on the upper surface of the Gutenberg low-velocity layer of seismology. The delineation and study of such regions by simultaneous recording of three components of magnetic variations on a line or array of stations is called *magnetic deep sounding* (Erdmagnetische Tiefensondierung). Its interest lies largely in the probable association of variation anomalies with temperature anomalies. Unlike the study of surface heat flow, this method provides information on the present deep temperatures, uncomplicated by the relaxation times of many millions of years which characterize thermal conduction in the upper mantle.

A natural magnetic variation is the vector sum, at each instant, of an external field and an internal field. *The external field* is that of an electric current system in or beyond

* Present address: The University, Leeds, England.

the ionosphere, generated by various interactions of the solar radiation and plasma flux with the Earth's magnetosphere and ionosphere. *The internal field* is that of a current system in the solid Earth, electromagnetically induced by the external field.

If the potential and the normal component of a variation field (either an instantaneous value or a Fourier component) are known over the whole of a closed surface, the internal and external parts of the field can be separated by methods of potential theory. The configuration of the internal field and its relations to the inducing external field may then be compared with those derived from model conductivity distributions within the surface. For the study of local variation anomalies, where V and $\partial V/\partial r$ may be known only over a limited region of the Earth's surface, a separation may still be carried out provided the region of observation extends well beyond the region of strongly anomalous variations. Pioneer studies by Siebert and Kertz[2] and Schmucker[3] have shown a way to the development of methods for separating the fields from observations over a limited area. Apart from these, existing studies of local conductivity anomalies have been made without benefit of quantitative separation of the observed field. This is hardly to be avoided, since the small numbers of instruments used (commonly four to eight) cannot adequately describe a time-varying field over an area of surface. The more or less intuitive separation of the field made in such studies can be meaningful though, since in midlatitudes the external field of many disturbances is nearly uniform over the variometer array, when the concentrated parts of the source currents are thousands of kilometres distant in the auroral zones. Local changes of the fields can thus safely be associated with local internal currents, in regions far from concentrated ionospheric currents.

Among the known local conductivity anomalies are the *continental edge anomalies*, the *intracontinental anomalies*, and others which do not belong unambiguously to either of these groups. Continental edge anomalies have been studied by Parkinson[4] in Australia, by Schmucker[5] in California, and by Lambert and Caner[6] in British Columbia. They generally show evidence of internal currents parallel to the continental edge, induced by external fields normal to that edge. The currents flow partly in the ocean and partly in the upper mantle; at the longer periods the mantle currents may predominate, and suggest that the surface of the highly conducting part of the mantle, like the seismic M discontinuity, rises offshore. Intracontinental anomalies have been studied by Schmucker and others in North Germany[3] and in New Mexico[5], and by Whitham and his collaborators in the Arctic islands of Canada[7]. Complicated anomalies which do not belong clearly in either class include the Japanese anomaly intensively studied by Rikitake and others[1], and an anomaly near the southern coast of Peru studied by Schmucker and others[8].

A New Variometer

The authors believe that real advances may be made if variation fields can be recorded simultaneously at fifteen or twenty stations over an area. In particular, it may then prove possible to make a valid separation of the internal and external fields without restrictive assumptions about the nature of the external field. As a first step a new three-component recording variometer has been designed. The prime consideration has been to keep the cost low enough so that a set of twenty instruments may be constructed in a research laboratory within the limits of a reasonable budget. Tests of the prototype and of eight similar variometers have shown that the variometer is a reliable instrument for field studies of local conductivity anomalies. The cost for materials and time is about $1500 each. Construction in Dallas and Edmonton will produce a total of forty-six variometers in 1967.

A detailed description of the variometer has been published elsewhere[9]; here we sketch its principal features. The design is classical, and has three magnets suspended on torsion wires, with optical magnification and photographic recording of the small rotations of the magnets in response to the variation field. The arrangement of magnets and optics is shown schematically in figure 1(a). A cardinal point of the design is the mounting of the instrument in a long narrow tube of aluminium (figure 1(d)). In use it stands vertically in a hole drilled in soil so that the suspended magnets are effectively thermostatted. In addition, first-order thermal compensation is provided. Creep and mechanical hysteresis of the suspensions are minimized by the mounting design and by careful handling during assembly. Much of the success of this variometer is due to the skill of Mr. J. A. Keiller in the detailed design of parts.

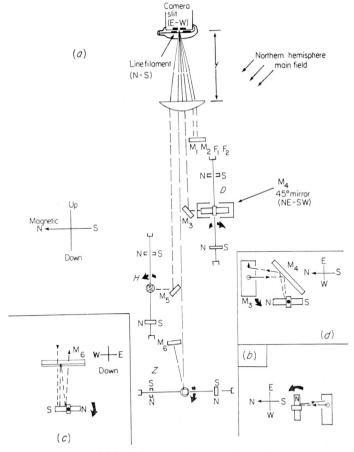

FIGURE 1(a), (b), (c), and (d) Schematic diagram of variometer, not to scale: (a) looking eastwards; (b) H, looking downwards; (c) Z, magnet looking northwards; (d) D, magnet looking downwards. [Reproduced from Gough and Reitzel[9]]

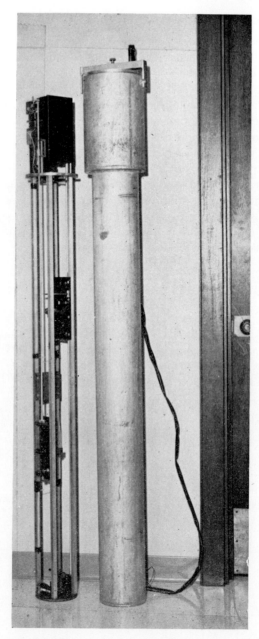

FIGURE 1(e)   Variometer and case: camera, lamphouse, and switching circuits at top; D, H, and Z suspensions spaced down the frame; case is 2 m long

The variometer records the magnetic field every ten seconds, and has good baseline stability over many days, unaffected by surface temperature changes, so that variations with periods from one minute to several days can be reliably recorded. Figure 2 shows a 42-minute sequence of a record, directly printed from the 35-mm film. Resolution is better than one gamma. The instrument records unattended for three weeks on power from modest batteries, and is convenient to install and maintain in the field.

Magnetic deep sounding and local conductivity anomalies

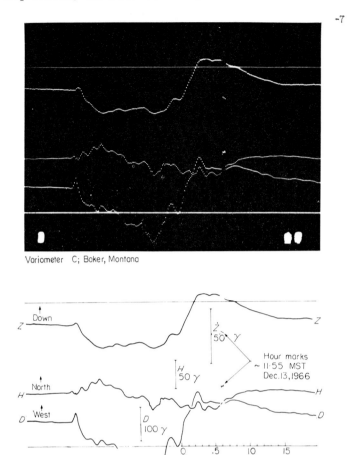

FIGURE 2 Print of 42-minute section of a variometer record, with explanatory key. [Reproduced from Gough and Reitzel[9]]

The flow of data which will come from an array of these variometers, recording three field components 360 times per hour for several weeks at forty stations, demands modern data-handling methods as a necessary condition of extracting any result. Work is proceeding on the design of a high-speed reading and analogue to digital conversion system.

A New Anomaly in Colorado

In October 1966 we made a field test with eight of the new variometers along the Arkansas and Gunnison Rivers in the state of Colorado. This line was chosen to look for a possible northward continuation of the anomaly found by Schmucker[5] in southern New Mexico (named by him the 'Texas anomaly').

Schmucker observed a change in the character of variations with periods in the range 15 min to 2 hours at stations along an east–west line from Tucson, Arizona, to Sweetwater, Texas. West of Las Cruces, New Mexico, the vertical component Z of these variation fields was rather small; east of Las Cruces it was about three times larger. The horizontal components H (magnetic N–S) and D (magnetic E–W) were relatively uniform along the line. As functions of time, Z at the eastern stations resembled D rather than H. Evidence from other observations, particularly from temporary International Geophysical Year observatories, suggested that small Z variations might be generally characteristic of the Cordillera and large Z of the plains. Taking the horizontal field components to be mainly external and the vertical component mainly internal, Schmucker suggested that the internal currents were induced mainly by D, since $Z(t)$ resembled $D(t)$, and were therefore more or less north–south in direction. He suggested that horizontal structures existed throughout the region, but that these differed between the mountains and the plains, with highly conducting layers nearer the surface under the mountains. The boundary between the two might thus correspond to a north–south aligned step structure in a highly conductive layer, about 100 km west of the eastern front of the Cordillera. Figure 3 reproduces a bay event from Schmucker[5] (figure 2) and illustrates the change in Z along the line of stations (W–E from top to bottom) and the resemblance of Z to D.

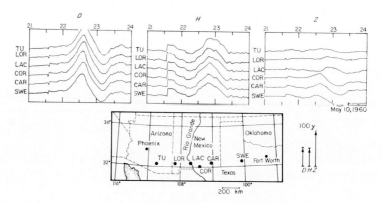

FIGURE 3 Geomagnetic bay recorded by Schmucker[5] at stations in Arizona, New Mexico, and Texas. [Reproduced from Schmucker[5]]

Figure 4 illustrates the main result of our own exploratory line in Colorado. Here are shown the three components D, H, and Z recorded at Cimarron during a bay disturbance in October 1966 and the Z component at six stations progressively further east. D is recorded in opposite senses in the new variometers and in the Askania variometers used by Schmucker. Remembering this, the resemblance between figures 3 and 4 is evident. Along the Colorado line Z again increases by a factor of about 3 from west to east; Z as a function of time resembles D much more than H; the transition occurs near Salida, which is about 100 km west of the front range.

After the discovery of the anomaly here reported, the authors learned that Caner, Cannon, and Livingstone[10] had used variometers along a line running about N 75° E across northern New Mexico and the Texas Panhandle, i.e. between Schmucker's profile and that reported here, in a search for a northward continuation of Schmucker's anomaly. They found that Z variations continued small across the combined width of New Mexico and the Texas Panhandle, and Z increased only at their easternmost station east of the Texas–Oklahoma border (some 500 km east of the mountain front).

It is clear that more observations are needed to establish the number of variation anomalies and their limits in this large region. The authors plan to make observations with about forty variometers in the summer of 1967 over much of the area between 36° N and 42° N, and between 102° W and 116° W.

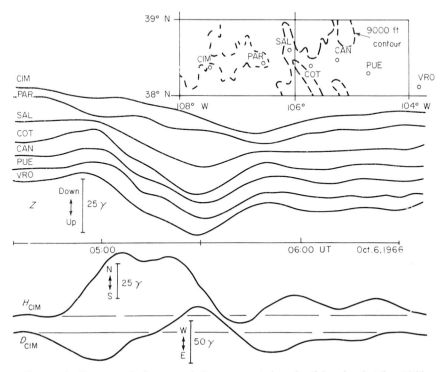

FIGURE 4 Geomagnetic bay recorded at seven stations in Colorado, October 1966

Separation of External and Internal Fields

The time-honored way of separating a magnetic field observed on the Earth's surface into parts of external and internal origin is to fit the observed values to sets of orthogonal harmonic functions taken in pairs, with one of each pair corresponding to an upward-increasing field and the other to a downward-increasing field. This method has weaknesses when applied to a field observed only over a part of the Earth small enough to be treated as plane. For example, fitting the observations to double Fourier series in x and y would express a field observed only within a rectangle as a field repeated periodically over an infinite plane. For most local anomalies, almost any smooth and reasonable extrapolation to great distances would be less misleading than such a wallpaper pattern.

A different method of separation, derived by Vestine[11] and discussed and used by Price and Wilkins[12], is likely to be more suited to local field anomalies. This method does not rely on special functions, but uses surface integrals (over a closed surface or an infinite plane) that may be derived most directly from theorems due to Green, as is shown in the appendix. For separation at any chosen origin on an infinite plane, the integral formulae are as follows:

$$2\pi(V_e - V_i)_0 = \int_P \frac{Z(x,y)\,dS}{r} + 2\pi V' \tag{1}$$

where V' is an arbitrary constant in the potential,

and
$$2\pi(X_e - X_i)_0 = -\int_P \frac{xZ(x,y)\,dS}{r^3}$$
$$2\pi(Y_e - Y_i)_0 = -\int_P \frac{yZ(x,y)\,dS}{r^3} \tag{2}$$

$$2\pi(Z_e - Z_i)_0 = -\int_P \frac{\{V(x,y) - V_0\}\,dS}{r^3} \tag{3}$$

$$2\pi(Z_e - Z_i)_0 = -\int_P \frac{xX(x,y)\,dS}{ry^2} = -\int_P \frac{yY(x,y)\,dS}{rx^2} \tag{4}$$

Here r is the distance on the plane $z = 0$ from the origin where the separation is being made, to a surface element dS at (x, y). The customary coordinates are used, with distance x and field component X positive northwards, y and Y positive eastwards, and z and Z positive downwards.* The suffix e (for external) refers to fields from sources in $z < 0$, above the plane; i refers to fields from sources in $z > 0$.

A case likely to be important in practice is the two-dimensional anomaly, in which Y and Z are functions of y and z only and $X = $ constant. Continental edge anomalies and the anomaly discussed above may approximate to this type. The formulae (2) and (4) reduce, in this case, to

$$\pi(Y_e - Y_i)_0 = -\int_{-\infty}^{\infty} \frac{Z(y)\,dy}{y} \tag{5}$$

and

$$\pi(Z_e - Z_i)_0 = \int_{-\infty}^{\infty} \frac{Y(y)\,dy}{y} \tag{6}$$

Siebert and Kertz[2] have shown that these two-dimensional formulae may be used to separate three-dimensional fields along special straight profiles, if any such can be found, where the contours of the component perpendicular to the profile are themselves perpendicular to the profile.

Figure 5 shows a simple example of the separation of a two-dimensional field by means of formula (6). The points on the left represent observations of Y and Z at nine stations on a plane lying between the overhead wire 1 and the underground wire 2, which carry equal currents in the same direction. The curves connecting these points are no more than interpolations by eye. With equally sketchy extrapolation of Y to $\pm \infty$, and crude numerical evaluations of the integral (6), the curves for Z_1 and Z_2

---

* One could, of course, work with the magnetically oriented components H and D instead of X and Y over an area in which the magnetic declination was constant.

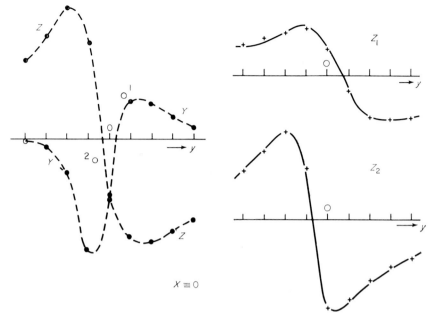

FIGURE 5 An example of field separation in a two-dimensional case
$$(Z_1-Z_2)_A = \frac{1}{\pi}\int_{-\infty}^{\infty} \frac{Y-Y_A}{y-y_A}\,dy$$

were found, which agree within 10% or less with the actual separate fields, shown by crosses. Y may be separated similarly by formula (5), but in this case the regions where the field is invented, rather than observed, will have more influence on the integral, and the separation will be so much the less reliable.

In general, the separation of fields observed over a three-dimensional anomaly by use of the surface integrals (1)–(4) will be more informative and more objective than two-dimensional methods. The use of (1) and (3), after calculating the potential from the observed X and Y, will employ all the information from X and Y together with equal weight to each. The use of (2) and (4) separately for each horizontal component, on the other hand, makes it unnecessary to introduce the potential at all and provides two independent separations of Z.

In any case, reduction of the observations will begin with interpolation and extrapolation of the observed field. For the horizontal field, these may be improved if the condition $\partial X/\partial y = \partial Y/\partial x$ is kept in mind. Price and Wilkins[12] have used this condition to interpolate the global S_q field, in the form

$$\int_c \mathbf{H}\cdot d\mathbf{s} = 0$$

around any closed loop. It may be applied in a more localized way by drawing separate trial contours of X and Y and adjusting them so that the superposed contours form a net of rhombuses, rather than unequal parallelograms, while still satisfying the observations. The best result obtainable is still not, of course, unique.

The convergence of the integrals (1)–(4) at the origin, and at great distances, is discussed in the appendix. It is clear, both physically and from study of the integrals, that a uniform field is not separable, and that some apparently uniform or slowly changing part of the field will often have to be removed from the observed field to ensure convergence of the integrals. The removed part will generally contain an internal field reflecting the World-wide conductivity of the mantle at depth. To find the true relation of the total or the anomalous internal field to the total external field, the removed part could ideally be separated itself with the aid of records from distant observatories. Rikitake[1, 13] (pp. 138–9 of Rikitake[1]), for instance, has found a simple form to represent the general field of bay disturbances, which might suggest the proper approximation of the global field to subtract from the locally observed field of a bay, and might further allow its internal and external parts to be estimated.

Interpretation of the Separated Variation Fields

The ultimate aim of magnetic deep sounding is to deduce internal conductivity distributions; a semifinal stage of interpretation is to delimit the range of internal current systems that can produce the observed internal field. The usual ambiguity of all geophysical methods involving fields is present here: there are many current systems able to produce a given field on a surface. In this case, however, we have an important source of extra information that is not available in the interpretation of gravity or static magnetic anomaly maps. This is the relationship between the external and internal fields. Having isolated the external part, we should not incontinently discard it.

A current element flowing near the boundary of a conducting region will only be induced by fields normal to the boundary, and will itself produce a field at the boundary with no component in the direction of flow, so the correlations between different external and internal components will often be indicative of the current geometry. This idea has been more or less formally applied for some time to the well-known correspondence on many observatory records between Z and a particular azimuth of horizontal variation, characteristic of the station. Parkinson[4, 14] sharpened this notion by pointing out that the variation vector at many stations consistently lies close to a preferred plane; the projection of the downward normal to this plane on the horizontal is 'Parkinson's arrow', which may be supposed to point in the direction from the station toward the region of highest conductivity in the surrounding. Schmucker[5], in analysing the fields of bay disturbances, has used a least-squares method to find the direction of 'normal' horizontal variation, recorded far away from an anomaly, that is best correlated with upward 'anomalous' Z variation (Z minus 'normal' Z). He has applied the method to Fourier sine and cosine components of individual events, to explore the frequency dependence and phase lags of the induced fields. This sort of attack will be still more powerful if used upon fields that have been formally separated over an area.

One similar approach would be to deal with individual events or groups of events as time series, and to compute the horizontal component of the external field in chosen azimuths at intervals around a semicircle. Then one could compute cross-products of each of these with the *internal vertical* field at a single station, with a range of time lags in each case. If the internal current has a well-defined azimuthal direction, a maximum in the correlation should appear when the orthogonal external horizontal field is chosen. Two estimates could thus be made of the direction of a horizontal internal current from each of the inside stations of an array (separation might be impracticable for some edge stations). One estimate would come from cross-correlation between the external horizontal and internal vertical fields, the other simply from the internal horizontal field, which may be nearly orthogonal to the local current. The $2n$ estimates so found from n stations might go far towards defining the configuration of the current. It is worth noting that the internal field may have very small horizontal components at some stations; here the cross-correlation method would still indicate the current direction.

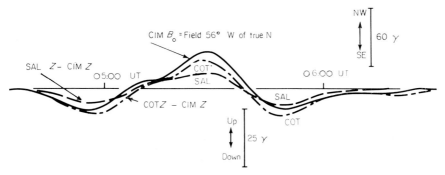

FIGURE 6   Cross-correlation between vertical field differences (Cotopaxi–Cimarron and Salida–Cimarron) and horizontal field N 56° W at Cimarron

A rough application of the cross-correlation idea, to the event recorded in Colorado and illustrated in figure 4, is shown in figure 6. Assuming baldly that the variations at Cimarron, the westernmost station, are 'free of anomaly', the course of Z at Cotopaxi minus Z at Cimarron was compared with the horizontal variation at Cimarron in various azimuths. The best fit was found for a horizontal field positive toward 304° east of north, for which a good coherence with upward Z (without phase lag) held through the whole course of the event. The azimuth indicated for the current is 34° east of true north. Since no separation was made, this conclusion is no better than the implicit assumption that the residual Z field (Cotopaxi minus Cimarron) was 'all internal' and the horizontal field 'all external'. The vertical field difference Salida minus Cimarron is also shown in the figure.

The tracking of internal Z with the external horizontal field in some azimuth over the whole of a complicated sequence of variation, as suggested here and in many observations elsewhere, indicates that the induction process involves only small phase lags not strongly dependent on frequency. This in turn suggests that the induction takes place in a thick region of high conductivity, probably with conductivity increasing downwards; presumably in the upper mantle and perhaps at the depth of the seismic low-velocity layer. When strong internal fields that do not track in this way are evident, induction in a thin region of limited total conductivity, such as a body of sediment saturated with water, is to be suspected. Since fields of higher frequency are less well able to penetrate a shallow conducting region and so to induce fields in the mantle beneath, study of the frequency spectrum of the ratio Z_i/H_e, say, may serve (as it does in the magnetotelluric method) to detect and characterize shallow conductors. The same sort of cross-correlation analysis may be used for this, applied to data in frequency bands isolated by digital filtering; one might in this way detect currents flowing in different directions at different depths, with more sensitivity than is given by the magnetotelluric study of anisotropies in conductivity.

Acknowledgements

We are supported in the study here described by the United States National Science Foundation, the Southwest Center for Advanced Studies, the National Research Council of Canada, the Defence Research Board of Canada, and the University of Alberta. To these organizations we express appreciation.

Appendix. Integral Formulae for Separation of Fields

Let R be a closed region, with boundary S; let r be the distance from any point A to the surface element dS at any point B in S; let n be the outward normal to dS, and let $d\Omega$ be the solid angle subtended at A by dS. Applying the divergence theorem to the vector $(1/r)\,\text{grad}\, V$, V being a scalar function, we obtain the result that for any point A not in R or S

$$I_s(V) = \int_S \left(\frac{1}{r}\frac{\partial V}{\partial n} - V\frac{\partial}{\partial n}\frac{1}{r}\right) dS = \int_R \frac{1}{r}\nabla^2 V\, dv$$

Other forms for I_s are

$$I_s = \int_S \left(\frac{1}{r}\frac{\partial V}{\partial n} + \frac{V}{r^2}\frac{\partial r}{\partial n}\right) dS = \int_S \frac{\partial}{\partial n}(rV)\frac{dS}{r^2} = \int_S \frac{\partial}{\partial r}(rV)\,d\Omega = \int_S \left(V + r\frac{\partial V}{\partial r}\right) d\Omega$$

The last form given shows that any constant V' may be added to V without changing I_s, since

$$\int_S V'\, d\Omega = 0$$

when A is not in R or S. This last form is also convenient for deriving similar identities when A is in R or S. For any point A in R, we may construct a new region R' that excludes A by removing from R a sphere of radius ε round A. As $\varepsilon \to 0$, R' goes to R, S' goes to S, and the contribution to I_s' from the inner spherical surface goes to $-4\pi V_A$. So, when A is in R,

$$\int_R \frac{1}{r}\nabla^2 V\, dv = I_s - 4\pi V_A$$

This is Green's third identity. For any A in the boundary S, we may similarly construct a new region R'' to exclude A by removing from R a hemispherical dimple of radius ε, centered on A. Now as $\varepsilon \to 0$ we obtain

$$\int_R \frac{1}{r}\nabla^2 V\, dv = I_s - 2\pi V_A$$

for A in S. (The same result is obtained if we construct instead a region R''' with a pimple instead of a dimple, so that A is within R'''.)

Now let us consider a magnetic field whose sources (currents or dipoles) are all outside R and S, so that in R and S it will have a scalar potential V_e, for which we may say $\nabla^2 V_e \equiv 0$ in R. Then, at a point A in S, $2\pi V_e = I_s(V_e)$. Next, taking V_i to be the scalar potential of a field due to sources within R, let us enclose R by a new region R^* whose inner surface is S and whose outer surface is arbitrarily far removed from any point of S. Then, if V_i vanishes as r^{-1} or faster at infinity, $2\pi V_i = -I_s(V_i)$; I_s appears here with negative sign because the outward normal from R^* into R is $-n$. Combining these results, and noting that $V_e + V_i = V$, the total potential, we have

$$2\pi(V_e - V_i) = I_s(V_e) + I_s(V_i) = I_s(V)$$

or

$$2\pi(V_e - V_i) = \int_s \left(\frac{1}{r}\frac{\partial V}{\partial n} + \frac{V}{r^2}\frac{\partial r}{\partial n}\right) dS$$

Since V_i must vanish at infinity, it can contain no constant part, and any arbitrary constant added to V must also be included in V_e alone, as is shown directly by Price and Wilkins[12].

The field component F in a direction x_A at A is given by $-\partial V_A/\partial x_A$. Taking the derivative of I_s with respect to x_A, and observing that r is the only quantity in the integrand that is a function of x_A, we obtain

$$2\pi(F_e - F_i)_A = -\int_s \left\{\frac{(x_B - x_A)}{r^3}\frac{\partial V_B}{\partial n} - \frac{V_B}{r^3}\frac{\partial}{\partial n}(x_B - x_A) + \frac{3V}{r^4}(x_B - x_A)\frac{\partial r}{\partial n}\right\} dS_B$$

When S is taken, as a useful limiting case, to be the infinite plane $z = 0$, with z positive downwards and all $z < 0$ considered to be outside R, the expression for $V_e - V_i$ at the origin takes the form (1) given before, while the expression for $F_e - F_i$ takes on the forms (2) or (3), as the direction of F lies in the plane or normal to it.

If the potential in (3) is such that $V(x, y) + V(-x, y)$ goes to zero for all y at large x, (3) may be integrated by parts with respect to x or y to produce the formulae of (4). These, however, will not converge unless we exclude from the plane a small region about the origin, for which special numerical treatments must be devised in the separation of actual fields.

The other formulae (1)–(3) do converge over the whole plane for reasonably extrapolated fields. Cartesian coordinates are most convenient for numerical work with observed fields, but polar coordinates give a quick view of the requirements for extrapolating the field to infinity, and quick estimates of the contribution to each integral from a small region about the origin.

Let us take (2) as an example, and consider

$$-\int_{P'} \frac{xZ \, dS}{r^3} = -\int_a^b \frac{dr}{r} \int_0^{2\pi} Z(r, \theta) \cos \theta \, d\theta$$

For bounded Z, this will converge as $b \to \infty$ if

$$\frac{1}{r}\int_0^{2\pi} Z(r, \theta) \cos \theta \, d\theta$$

goes to zero for large r; it is seen that Z may be extrapolated to any constant value at great distances, so that (2) may equally well be written in terms of $Z-Z_0$ for any arbitrary origin. To see what happens as $a \to 0$, let us expand Z as a double Taylor series about the origin in powers of $r \cos \theta$ and $r \sin \theta$, giving

$$Z(r,\theta) = Z_0 + r\cos\theta \left.\frac{\partial Z}{\partial x}\right|_0 + r\sin\theta \left.\frac{\partial Z}{\partial y}\right|_0 + \ldots$$

to put into the expression

$$J = -\int_0^a \frac{dr}{r} \int_0^{2\pi} Z(r,\theta) \cos\theta \, d\theta$$

The result is the series

$$J = -\pi a \left.\frac{\partial Z}{\partial x}\right|_0 + O(a^3)$$

all of whose terms go to zero as $a \to 0$. The leading term approximates the contribution to the integral (2) from a small circular region of radius a about the origin. The contribution from a small square region of side $2a$, centered on the origin, is more useful to know, and may be shown to be

$$-4a \ln\left\{\frac{\sqrt{(2)}+1}{\sqrt{(2)}-1}\right\} \left.\frac{\partial Z}{\partial x}\right|_0$$

Integrals (1) and (3) may be treated in the same way. The corresponding contributions from the small square of side $2a$ are as follows:

for (1)

$$4a \ln\left\{\frac{\sqrt{(2)}+1}{\sqrt{(2)}-1}\right\} Z_0$$

and for (3)

$$2a \ln\left\{\frac{\sqrt{(2)}+1}{\sqrt{(2)}-1}\right\} \left(\frac{\partial X}{\partial x} + \frac{\partial Y}{\partial y}\right)_0$$

References

1. T. Rikitake, *Electromagnetism and the Earth's Interior*, Elsevier, Amsterdam, 1966.
2. M. Siebert and W. Kertz, 'Zur Zerlegung eines lokalen erdmagnetischen Feldes in äusseren und inneren Anteil', *Nachr. Akad. Wiss. Goettingen, Math.-Physik. Kl.*, **1957**, No. 5, 87–112.
3. U. Schmucker, 'Erdmagnetische Tiefensondierung in Deutschland 1957/59: Magnetogramme und erste Auswertung', *Abhandl. Akad. Wiss. Goettingen, Math.-Physik. Kl.*, **5**, 1–51 (1959).
4. W. D. Parkinson, 'Conductivity anomalies in Australia and the ocean effect', *J. Geomagnetism Geoelectricity*, **15**, 222–6 (1964).
5. U. Schmucker, 'Anomalies of geomagnetic variations in the southwestern United States', *J. Geomagnetism Geoelectricity*, **15**, 193–221 (1964).
6. A. Lambert and B. Caner, 'Geomagnetic depth-sounding and the coast effect in western Canada', *Can. J. Earth Sci.*, **2**, 485–509 (1965).

7. K. Whitham and F. Andersen, 'Magnetotelluric experiments in northern Ellesmere Island', *Geophys. J.*, **10**, 317–45 (1965).
8. U. Schmucker, S. E. Forbush, O. Hartmann, A. A. Giesecke, Jr., M. Casaverde, J. Castillo, R. Salgueiro, and S. del Pozo, 'Electrical conductivity anomaly under the Andes', *Ann. Rept.*, 1965–1966, *Dept. Terrest. Magnetism, Carnegie Inst., Wash.*, 11–28 (1967).
9. D. I. Gough and J. S. Reitzel, 'A portable three-component magnetic variometer', *J. Geomagnetism Geoelectricity*, **19**, 203–15 (1967).
10. B. Caner, W. H. Cannon, and C. E. Livingstone, 'Geomagnetic depth-sounding and upper mantle structure in the Cordillera region of western North America', *J. Geophys. Res.*, **72**, 6335–51 (1967).
11. E. H. Vestine, 'On the analysis of surface magnetic fields by integrals, 1', *Terrest. Magnetism Atmos. Elec.*, **46**, 27–41 (1941).
12. A. T. Price and G. A. Wilkins, 'New methods for the analysis of geomagnetic fields and their application to the S_q field of 1932–3', *Phil. Trans. Roy. Soc. London, Ser. A*, **256**, 31–98 (1963).
13. T. Rikitake, 'Electromagnetic induction within the Earth and its relation to the electrical state of the Earth's interior, 1 (1)', *Bull. Earthquake Res. Inst., Tokyo Univ.*, **28**, 45–100 (1950).
14. W. D. Parkinson, 'Directions of rapid geomagnetic fluctuations', *Geophys. J.*, **2**, 1–14 (1959).

3

E. R. NIBLETT
and
K. WHITHAM
Dominion Observatory
Ottawa, Canada

B. CANER
Victoria Magnetic Observatory
Dominion Astrophysical Observatory
Victoria, British Columbia, Canada
and
Department of Geophysics
University of British Columbia
Vancouver, British Columbia, Canada

Electrical conductivity anomalies in the mantle and crust in Canada*

Introduction

Three prominent magnetic variation anomalies have been found in Canada and subjected to detailed study. Two of these are located in the Arctic Archipelago and the third lies in the Cordillera in southern British Columbia.

The first anomaly, at Alert on the northern end of Ellesmere Island, was discovered by Whitham, Loomer, and Niblett[1] from the records of a temporary magnetometer station established there for the IGY. Higher-quality data became available after a permanent magnetic observatory was set up late in 1961. The standard magnetograms at Alert (recording speed 2 cm/hour) display an abnormally high level of irregular magnetic activity and also a persistent tendency for the horizontal variation vector to be confined to a single direction. The main features of the anomaly have been explained[2-5] by postulating the presence of a massive highly conducting body in the lower part of the crust. Field experiments in the vicinity have indicated that the structure may be several hundred kilometres long with a cross-section scale length of roughly 100 km.

The second Arctic anomaly is at Mould Bay on Prince Patrick Island and was discovered after the establishment of a permanent magnetic observatory there in 1961. Here the outstanding feature of the magnetograms is the absence of shorter-period fluctuations in the vertical component. The existence of a large subterranean conductor—a horizontal slab perhaps a few hundred kilometres in extent—is again suggested in order to explain the effect. At this location supporting evidence for an anomalous structure deep in the crust or in the upper mantle is available from seismic observations.

In southern British Columbia a series of field observations by the University of British Columbia and the Dominion Observatory[6-9] have shown that along east–west profiles across the Cordillera from the coast to the Alberta plains, the vertical force variations are consistently smaller west of longitude 118° (approximately) than they

---

* Canadian Contribution No. 161 to the International Upper Mantle Project.

are to the east of this meridian. Similar effects have been observed on east–west profiles farther to the south in the United States[9, 10] and it has been found that the western Cordillera region may be generally characterized by small vertical force variations and low seismic P_n velocities, while the eastern Cordillera and plains area is characterized by higher values of these parameters. These observations suggest that the physical properties of the deep crust or upper mantle under the eastern and western Cordillera are not the same, and that two broad structural regimes may be separated by a narrow transition zone.

The Alert Anomaly

The high level of irregular magnetic activity found at Alert is illustrated in figure 1 which shows, for Canadian stations, the latitude variation in the mean hourly range

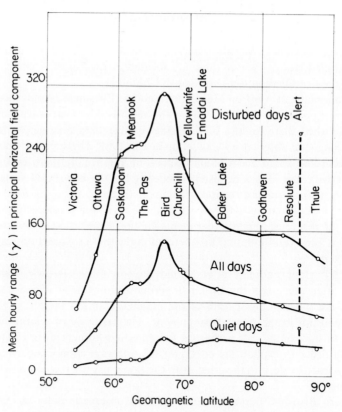

FIGURE 1 Irregular magnetic activity in Canada during August and September 1957 as a function of geomagnetic latitude. [After Whitham and Andersen[2]]

Electrical conductivity anomalies

of the principal horizontal component (at southern stations the principal horizontal component is H; at northern stations it is either X or Y, whichever is largest). Another very unusual feature of magnetic variations at this location is a strong tendency for the X-component and Y-component traces to mirror each other, particularly for bay-like events with a duration of 2 hours or less. The effect is illustrated in figure 2 which shows the magnetograms for August 6, 1961, from Alert along with those from Lake Hazen 150 km to the southwest, Eureka 500 km to the southwest, and Resolute Bay 1100 km distant in a similar direction. The 'mirror image' effect has been found to be a persistent phenomenon, prominent at Alert, less pronounced at Lake Hazen and Eureka, and not at all evident at Resolute.

After detailed study of records from Alert, Resolute Bay, Thule, and temporary field stations on Ellesmere Island and Axel Heiberg Island, Whitham and Andersen[2] have been unable to find evidence that Alert is located in an inner zone of enhanced

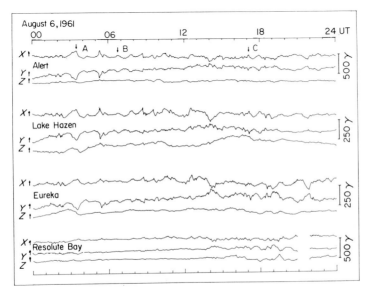

FIGURE 2   Magnetograms for August 6, 1961, for Alert, Lake Hazen, Eureka, and Resolute Bay. The sensitivity of the Alert and Resolute Bay records is one-half that of the records from Lake Hazen and Eureka. [After Whitham and Andersen[2]]

magnetic activity. Instead, the data indicate that the high activity is caused by an extraordinarily large induced component in the horizontal field fluctuations in a certain period range. The 'mirror image' effect between X and Y components reveals a strong confinement of the horizontal change vector in the NW–SE direction for events of 2 hours' duration or less. The only hypothesis that seems able to account for these observations is that a large highly conducting body exists within the Earth in which large induced currents are able to flow. The induced currents must be confined in the NE–SW direction in the vicinity of Alert, and recordings at a line of satellite stations across the strike of the anomaly[3] confirm this. Since the anomalous effects are still evident at Lake Hazen 150 km downstrike from Alert it is inferred that an elongated conductor, extending for perhaps several hundred kilometres and with a width of 50–100 km, must lie in the crust or upper mantle beneath the immediate vicinity of these stations. The results of a potential analysis, in which anomalous internal contributions to total measured disturbances were estimated from simultaneous records at Alert and the satellite stations, confirmed the presence of such a conductor.

Law and others[3] have examined the frequency response of the anomaly using the method of power spectrum analysis. Energy density estimates from simultaneous recordings at Alert and Greenland station G1 120 km to the southeast were computed for the horizontal component H_1 in the direction perpendicular to strike, and for the Z component. G1 was assumed to be sufficiently remote to be unaffected by the anomaly. The energy density ratios for horizontal force and for vertical force at the two stations are plotted as a function of frequency in figure 3. For the horizontal component H_1 the energy density ratio at long periods is less than or close to unity (the two long-period points are low as a consequence of temperature drifts in the magnetometers), but at a period close to 100 min the ratio at first rises rapidly and then continues a gradual increase as the period reduces to about 5 min. The vertical force ratio falls rapidly as period decreases to about 60 min and then rises again at still shorter periods.

For the horizontal component perpendicular to strike the anomalous variations are restricted to periodicities of 2 hours and less. The theoretical H_1 and Z responses for

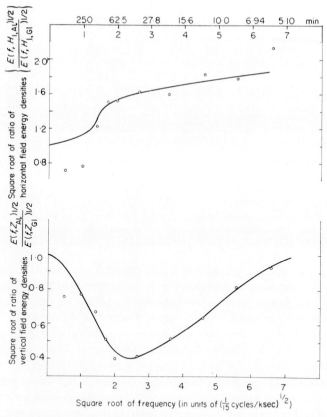

FIGURE 3 Frequency dependence of the ratio of the square root of power spectral densities at stations Alert and G1. Periods in minutes which correspond to the plotted frequencies are shown at the top of the diagram. [After Law and others[3]]

Electrical conductivity anomalies

an infinitely long conducting cylinder of radius ρ and conductivity σ in a uniform inducing field have been calculated. A reasonable fit to the observed H_1 response was obtained with $\sigma = 3 \times 10^{-11}$ e.m.u., $\rho = 51$ km and with the top of the cylinder only 10 km below the surface. The observed Z response does not agree with that of the model and it is not known why the Z ratio rises again at shorter periods.

If the high conductivity is interpreted in terms of temperature, the cylindrical model just referred to implies that the 1400 °C isotherm could be upheaved to a depth of about 10 km from the Earth's surface, which appears absurd. However, Rikitake and Whitham[4] have shown that, when electromagnetic coupling between an infinite circular cylinder and the underlying conducting mantle is taken into account, it is possible to obtain theoretical agreement with the experimental observations with the top of the cylindrical body lowered to a more reasonable depth of 25–30 km.

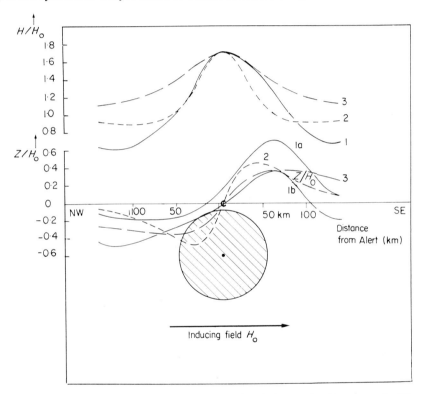

FIGURE 4 Observed and theoretical responses in relation to the Alert anomaly. The theoretical curves include the response of a line current at 60 km depth and a cylindrical model as shown with $\sigma = 10^{-11}$ e.m.u. Curve 1a, actual response of typically 10-min periods; curve 1b, actual response of typically 20-min periods; curve 2, cylindrical response; curve 3, line current response. [After Whitham[11], courtesy of *Journal of Geomagnetism and Geoelectricity*]

The horizontal and vertical field responses (figure 4) along a line perpendicular to the strike of the conductor, for a uniform horizontal inducing field in the same direction, have also been studied by Whitham[11], using range measurements scaled off the records at both 10-min and 20-min intervals. The measured horizontal response appears to be quite symmetrically distributed on either side of the centre of the anomaly, but the vertical force response does not possess the symmetry to be expected for a cylindrical model. Rikitake and Whitham[4] have derived response curves for various models with elliptical and triangular cross section in order to try and explain this asymmetry. However, more and better quality field data are required before a quantitative discussion of the asymmetry of the underground structure can be reasonably attempted.

Whitham and Andersen[5] and Whitham[12] have given an account of magnetotelluric experiments conducted at and near Alert in 1963. The electric field was found to be very strongly confined to a north–south direction and to be in phase with the magnetic fields between periods of 4 and 100 min. The modulus ratio $|E/H|$ was substantially independent of period in the same range and was abnormally low (~ 0.13 mv/km γ). The apparent resistivity values in a direction close to the principal conductivity axis of the anisotropy were extremely low and indicate that the conductivity for the shortest-period measurements (~ 4 min) may be greater than 10^{-11} e.m.u.

In a qualitative sense the magnetic and magnetotelluric results are compatible since both methods reveal highly anomalous induction in the crustal layers near Alert. Quantitatively, however, the two types of data lead to inconsistencies when interpreted with an approximate theory. Thus, for a cylindrical structure with parameters appropriate for the magnetic anomaly, Whitham and Andersen[5] found that (i) the north–south electric field confinement is unexplained, (ii) the $|E/H|$ ratio should decrease by a factor of about 10 in the period range 4 to 400 min, together with a phase advance of the electric field of about 70°, and (iii) the magnetotelluric data imply a depth for the conductor much less than 25–30 km. A more advanced theory taking into account induction over a wide area, coupling between the anomalous structure and underlying conducting layers, and rigorous treatment of boundaries will probably be required for meaningful interpretation of all the available data.

Apart from the magnetotelluric data, no geological or geophysical evidence for an extensive near-surface conductor has been found. Recent airborne magnetometer flights by Serson, Hannaford, and Haines of the Dominion Observatory show that there is a striking reduction in the amplitude of total field anomalies over the large portion of Ellesmere Island covered by the Franklinian geosyncline with particularly smooth profiles over that part of the syncline covered by the younger rocks of the Sverdrup Basin. This absence of relief in the magnetic profiles is possibly due to great depths of sediments and/or to a basement complex which is unusually hot. However, there is no indication from these data of massive graphite or sulphide mineralization in the Alert–Lake Hazen area.

Bouguer anomaly profiles shown by Law and others[3] give highs of 40–50 mgal directly over the zone where the anomalous conductor is thought to exist. Calculations have shown that there is no gross inconsistency between the experimentally observed gravity anomalies and the hypothesis of a large hot body in equilibrium with 10–20 km of crustal thinning.

The Mould Bay Anomaly

Geomagnetic effects and their interpretation

In 1961 a joint magnetic and seismic observatory was established at Mould Bay on Prince Patrick Island. All magnetograms from this station show an unusually strong

Electrical conductivity anomalies 161

attenuation of the shorter-period vertical force fluctuations. The horizontal force activity, on the other hand, appears to have a more or less normal frequency distribution. Power spectrum analysis has shown that for periods less than two hours the vertical force amplitudes decrease very rapidly with increasing frequency. Consequently the Z traces have a very smoothed appearance when compared with those of the horizontal components (figure 5).

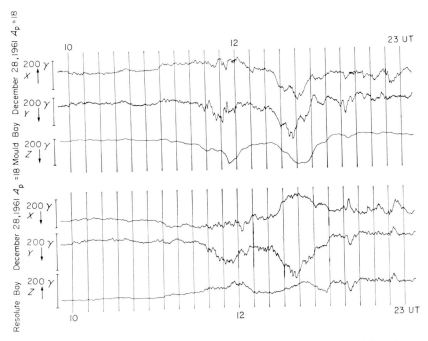

FIGURE 5   Magnetograms for December 28, 1961, for Mould Bay and Resolute. [After Whitham[13]]

This unusual magnetic behaviour has been investigated by Whitham[11-13]. Since the Z attenuation is always present, and since the horizontal variations do not appear to be associated with any particular preferred direction, Whitham[13] suggested that the effect could be caused by a highly conducting horizontal slab located at some depth beneath Mould Bay. Using the technique of power spectrum analysis, energy densities were computed for all three components from a number of simultaneous magnetograms from Mould Bay and Resolute Bay. The spectrum calculations ranged over periods of 10 to 1000 min. The energy density ratios $E(Z)/E(X)$ and $E(Z)/E(Y)$ were found to depend on frequency in a characteristic manner at Mould Bay quite different from that at Resolute. Typical results are shown in figure 6 illustrating energy density ratios at Mould Bay and Resolute. Using an extension of Price's[14] magnetotelluric

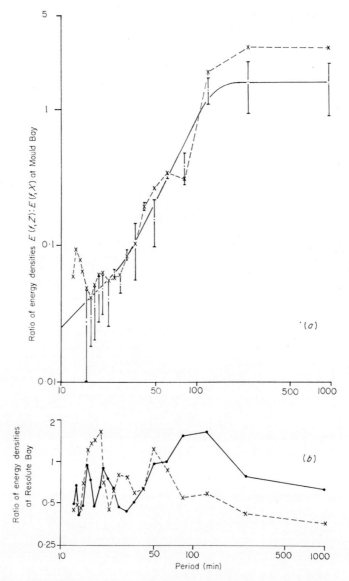

FIGURE 6 Comparison of energy density ratios at Mould Bay and Resolute Bay for February 5, 1962. (a) —·— $E(f, Z)/E(f, X)$ mean of all available data; -- × -- $E(f, Z)/E(f, X)$ February 5, 1962. (b) —·— $E(f, Z)/E(f, X)$, February 5, 1962; -- × -- $E(f, Z)/E(f, Y)$, February 5, 1962. [After Whitham[13]]

theory, Whitham showed that the frequency characteristics of the Mould Bay magnetograms could be reasonably well explained by induction in a horizontal conducting layer about 10–20 km thick, having a conductivity of 10^{-11} e.m.u., and located near the bottom of the crust. Approximately 100 different models were examined in which the thickness of the conductor was varied between 2 and 20 km, the conductivity between 10^{-10} and 10^{-11} e.m.u., and the depth to the top of the conductor from 0 to 100 km. The source field wave number v was varied between 10^{-7} and 10^{-8} cm^{-1}

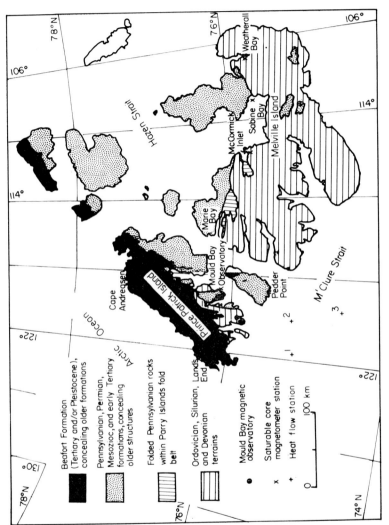

FIGURE 7 Map showing positions of temporary variometer stations occupied in the western Arctic Archipelago during 1963 and 1964. [After Whitham[12], courtesy of *Journal of Geomagnetism and Geoelectricity*]

for these model calculations. It was assumed in the calculations that (i) the horizontal conducting slab is large in extent compared with typical scale lengths of the inducing field, (ii) the energy density against frequency characteristic at Mould Bay is time invariant, and (iii) the scale lengths of the inducing fields are both time and frequency independent.

During the summers of 1963 and 1964 a number of temporary magnetic and telluric recording stations were set up to investigate the spatial extent of the Mould Bay anomaly (figure 7). An analysis of hourly range data from these locations[12] has shown that the anomalous effects are appreciably reduced 100 km to the north and 100 km to the south of Mould Bay. The anomalous zone appears to extend about 200 km to the east of the observatory and an unknown distance (> 100 km) to the west.

Power spectrum analysis was done from magnetograms recorded over two 24 hour periods (July 21, 1964, and July 30, 1964) at Mould Bay, McCormick Inlet, Sabine, Weatherall Bay, and Resolute Bay. The energy density ratio $E(Z)/E(Y)$ is plotted against period for each station in figures 8 and 9. The responses at Mould Bay and McCormick are quite similar, and both stations are clearly well within the zone of anomalous Z variations. At Sabine, Weatherall Bay, and Resolute rapid attenuation of the Z variations with increasing frequency is no longer evident and the energy density plots are no longer diagnostic of a subterranean layer of high conductivity. Either this layer has disappeared altogether somewhere between McCormick and Sabine, or it has dipped very steeply in this area so that under Weatherall Bay and Resolute it exists at a much greater depth. Induction vectors of the Parkinson type have also been estimated (for a 20-min interval) from the aforementioned data and the results are of interest. At Mould Bay and McCormick the vectors are very small

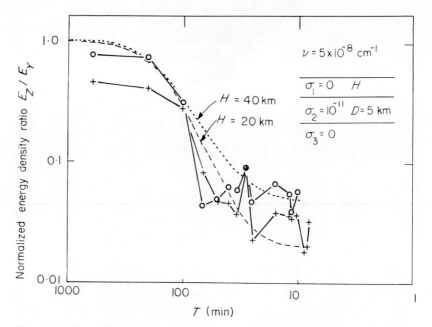

FIGURE 8 Energy density ratios at Mould Bay (+) and McCormick Inlet (○) [from De Laurier[15]]. The dotted curves show the effect of a highly conducting layer ($\sigma^2 = 10^{-11}$ e.m.u.) 5 km thick at a depth H km beneath the surface

Electrical conductivity anomalies

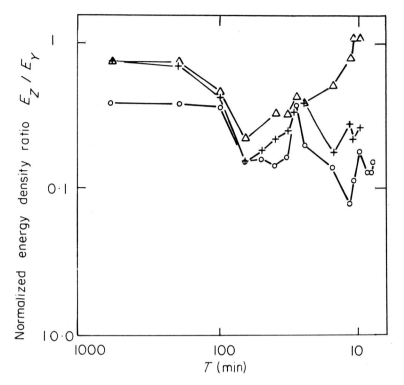

FIGURE 9 Energy density ratios at Sabine Bay, Weatherall Bay, and Resolute Bay: o Resolute Bay; + Weatherall Bay; △ Sabine. [After DeLaurier[15]]

as would be expected. At Sabine Bay and Weatherall Bay the vectors are large and indicate that positive correlation exists between Z variations and those in the horizontal plane. The presence of conductivity inhomogeneities (edge effects?) is therefore indicated in the vicinity of these stations. At Resolute Bay no systematic correlation between Z and H variations was found. These observations are qualitatively consistent with the interpretation that the gross conductor under Mould Bay must either vanish or plunge to greater depths somewhere in the vicinity of Sabine–Weatherall Bay.

The telluric data collected at Mould Bay and the temporary sites during 1964 have not yet been analysed in detail. However the records from the eastern part of Melville Island give normal electric to magnetic field ratios (1 mv/km γ) whereas at Mould Bay the ratio is about ten times smaller at periods of about 30 min.

Significance of other geophysical and geological data

Whitham[13] pointed out the possibility that the Mould Bay anomaly could be produced by large regional upwarping of the 1400 °C isotherm since at this temperature ionic semiconduction in olivine could lead to a conductivity of about 10^{-11} e.m.u. If this were the case, high heat flows could be expected in the zone of anomalous magnetic activity. To test this idea, Law, Paterson, and Whitham[16] and Paterson and Law[17] measured heat flows from the sea ice at ten sites near southern Prince Patrick Island in 1964 and 1965. Water depths varied between 200 and 500 m. At two stations on the continental shelf about 100 km northwest of Prince Patrick Island the mean heat flow was found to be 0.46 ± 0.09 $\mu cal/cm^2$ sec (30% of the World continental average). At five stations in Crozier Channel and Kellett Strait, all lying less than 100 km east of Mould bay, the mean heat flow was 1.46 ± 0.16 $\mu cal/cm^2$ sec (close to World average). At three stations in M'Clure Strait roughly 130 km south of Mould Bay the mean value was 0.84 ± 0.09 $\mu cal/cm^2$ sec (60% of World average). Thus the heat flow in the area is far from uniform and the significance of the observations is difficult to assess. In addition, there is still the unresolved problem of whether the heat flows measured in the comparatively shallow channels are equilibrium flows. The water temperatures at depths where the measurements were made were assumed to be unaffected by annual or long-term variations. However, the influence of such disturbing effects plus instrumental errors should not exceed 30% of the estimated heat flow values. It is also possible that the Mould Bay anomaly is sufficiently young that the isotherms have not yet diffused to the surface. This would be the case for a thermal upheaval which occurred in the late Quaternary Period less than a million years ago.

The surface geology in the vicinity of Mould Bay is complex and does not appear to provide any clue relating to the source of the magnetic anomaly. Palaeozoic sediments about 5 km thick are believed to overlie the basement rocks of the Franklinian geosyncline. Neither aeromagnetic profiles nor gravity data from the area provide evidence for extensive mineralization in the crust. As at Alert, the aeromagnetic profiles in the region show remarkably little relief and suggest a basement complex which is quite non-magnetic.

Recent seismic results provide additional evidence for an anomalous structure in the vicinity of Mould Bay. Overton[18] has described seismic refraction surveys in the neighbourhood of the western Arctic Achipelago and has deduced a most complex structure with a velocity inversion problem in the basement complex. The best estimate of the upper-mantle P-wave velocity in the area surrounding the induction anomaly is 8.18 ± 0.04 km/sec, which is normal. The average crustal thickness is 36 km. However, Overton has outlined an anomaly in the upper mantle defined by P_n time delays (normally ± 0.2 sec) from 1.2 to 3.0 sec. This anomaly covers Eglington Island and the western part of Melville Island and appears to correlate reasonably well in spatial extent with the magnetic anomaly. No self-consistent explanation has been given for the time delays in terms of either a locally smaller P_n velocity or an unusual crust.

In the spring of 1965 a unique sequence of more than 2000 microearthquakes occurred near Mould Bay over an interval of nearly 2 months. This swarm occurred 15 km southeast of Mould Bay at a depth of about 6 km. Many studies are in progress by Smith, Piché, and Whitham including particle motion, energy studies, and magnitude–frequency relationships. The magnitude–frequency plot has a slope of -0.7 and therefore suggests that the origin is tectonic, not volcanic.

Wickens and Pěč[19] have studied Love wave dispersion on a line from Edmonton to Mould Bay using reversed data. The usual mantle S-wave low-velocity channel appears to rise between Coppermine and Mould Bay and there is substantial evidence for a shallow (second) low S-wave velocity layer in the crust under Mould Bay.

Anomalous Effects in Western Canada and United States

Geomagnetic effects

In 1963 Hyndman[6] recorded magnetic variations in three components at twelve field locations along an east–west profile from Westham Island near Vancouver to Lethbridge in Alberta. The permanent magnetic observatory at Victoria added a thirteenth station to this long chain across the Cordillera. The sites were spaced about 80–100 km apart and recording speeds and sensitivities were similar to those of standard observatory magnetograms. At the seven stations on this profile lying west of the 118th meridian, Hyndman found that the Z variations with durations between 5 and 120 min were two or three times smaller than those recorded at the remaining stations to the east. The horizontal components D and H, however, showed little change over the entire profile

In discussing Hyndman's data Whitham[12] pointed out that the small Z variations at the western stations could be explained if highly conducting layers of the upper mantle lie closer to the surface in this region. He suggested that a rise of conducting material with $\sigma = 10^{-11}$ e.m.u. to within 200 km of the surface might be sufficient to account for the observed attenuation.

Caner, Cannon, and Livingstone[9] have reported further field studies during 1965–1966, conducted jointly by the Department of Geophysics at the University of British Columbia and the Dominion Observatory establishment at Victoria. Two additional profiles of variometer stations were completed using Askania GV-3 three-component recording variographs with a paper speed of 2 cm/hour. The first of these is profile B in figure 10 near latitude 35° N between Horse Springs (HOR), New Mexico, and the permanent observatory at Norman, Oklahoma. The second is profile D in figure 10 at latitude 51° N from Cache Creek (CAC) in British Columbia to Calgary (CAL), Alberta. Hyndman's 1963 profile is labelled C, and an earlier one by Schmucker[10] at latitude 33° N is labelled A. Typical magnetogram pairs from profiles B, C, and D, are shown in figure 11. Profile D revealed effects similar to those found by Hyndman with a transition zone close to longitude 117° W. On profile B the transition was found near the Texas–Oklahoma border (longitude 100° W), while further south on profile A Schmucker found the discontinuity nearly 400 km further to the west at Cornudas, New Mexico. Observations were also made during the IGY at a chain of widely spaced magnetic recording stations near latitude 39° N. Schmucker[10] has pointed out that the records from these stations display the same effect—to west of longitude 106° W the Z variations are attenuated, while to the east they are not.

Thus an abrupt discontinuity in geomagnetic variations appears to extend from southern British Columbia to the Mexican border. In Canada it lies well within the Cordillera while in the southwestern United States it has been found both inside the Rocky Mountains and in the plains area further east. Though the main trend is parallel to the Cordillera, the direction of strike appears to vary widely and does not correlate closely with surface geology.

For quantitative analysis Caner, Cannon, and Livingstone[9] computed power spectra from the variometer data from profiles B, C, and D. The attenuation of the vertical component at the western stations was found to depend on frequency in essentially the same way on all three profiles. For structural interpretation the frequency dependence of the dimensionless number

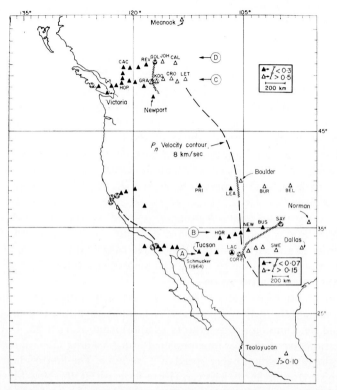

FIGURE 10 Location of geomagnetic recording stations and of transition zones between regions of high and low Z variations. The full triangles ▲ indicate recording stations where the ratio

$$I = \frac{\Delta Z}{\{(\Delta D)^2 + (\Delta H)^2\}^{1/2}}$$

is low and the open triangles △ indicate stations where I is large. Anomalous stations near the transition zones and on the coast are circled ⊚, ⊛. Paths of the geomagnetic transition zones are indicated by wavy lines. [After Caner, Cannon, and Livingstone[9]]

$$M = \frac{(E_Z/E_H)_{\text{west}}}{(E_Z/E_H)_{\text{east}}}$$

was computed for pairs of stations on a given profile. The numerator represents the energy density ratio of vertical to horizontal components at a station west of the transition zone, and the denominator represents the same ratio for a station to the east. M was found to be strongly dependent on frequency between 0·1 cycle/hour and 1·25 cycle/hour, falling from a value of nearly unity to about 0·2 in this range. At frequencies higher than 1·25 cycle/hour (48 min period) the attenuation curve levels off at $M \simeq 0·1$. The variographs used in the field were too insensitive to permit interpretation at periods less than about 15 min.

Several horizontally layered models were computed for comparison with the experimental attenuation curve. Good agreement was obtained for a conducting layer ($\sigma \simeq 10^{-11}$ e.m.u.) about 20 km thick lying at or near the bottom of the crust underneath the western stations. Reliable data at periods less than 15 min are required to confirm the existence of such a shallow conductor.

Electrical conductivity anomalies 169

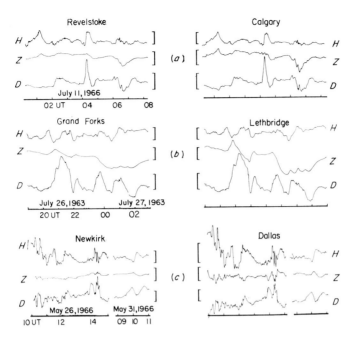

FIGURE 11 Magnetogram sections from pairs of stations at (*a*) latitude 51° N (profile D), (*b*) latitude 49·5° N (profile C), and (*c*) latitude 33–35° N (profile B). Scale bars are 50 γ and time marks are at hourly intervals. [After Caner, Cannon, and Livingstone[9]]

Significance of other geophysical data

Caner and Cannon[7] and Caner, Cannon, and Livingstone[9] have studied the available data from seismic refraction work and from heat flow observations in the Cordillera regions of western North America. Figure 10 shows 8·0 km/sec contours of upper-mantle P_n-wave velocity. Pakiser and Zietz[20] noted that these contours appear to separate a low-velocity region ($P_n \simeq 7\cdot8$ km/sec) between the Rocky Mountains and the California coast from regions of normal velocity ($P_n \simeq 8\cdot2$ km/sec) to the east of the Rockies and to the west of the Pacific continental margin.

In general, therefore, a large part of western United States seems to be associated with low upper-mantle *P*-wave velocities and small geomagnetic *Z* variations. Because the seismic results are rather sparse, and because horizontal resolution is inherently poor for refraction methods, the position of the 8·0 km/sec contour may be uncertain by more than 100 km. In particular the extension into Canada is uncertain. White, Bone, and Milne[21] have discussed the results available to date and shown that the best estimates for the Canadian Cordillera indicate a P_n velocity of 8·0 km/sec as far east as longitude 118° W, and an unusually thick crust under Vancouver Island and the Coast Ranges. Under the Great Plains the P_n velocity rises to 8·1 or 8·2 km/sec but the transition is not clearly defined. However, Caner and Cannon[7] have pointed out that in the United States the P_n velocity transition at least follows the main trend of the geomagnetic transition near the eastern side of the Cordillera. On the Pacific Coast, Schmucker[10] and Lambert and Caner[8] have demonstrated that magnetic variations are influenced by the well-known 'coast effect' in which variations in the horizontal plane are positively correlated with vertical force changes. This effect could be due simply to the conductivity contrast along the land–sea interface or it could be caused by a deeper structure in the crust or upper mantle parallel to the continental margin. Schmucker[10] attributed his results at coastal stations in southern California to the effect of the ocean but Lambert and Caner[8] have pointed to evidence favouring the existence of a structural change at depth along the Pacific coast.

Caner and Cannon[7] and Caner, Cannon, and Livingstone[9] have concluded that upper-mantle material of high electrical conductivity (10^{-11} e.m.u.) must lie near the base of the crust underneath the whole region from southern British Columbia to Mexico between the continental margin and the transition zone for seismic and magnetic effects. Along these two boundaries the conducting layers either terminate or drop to much greater depths. As at Mould Bay, this hypothesis implies that upper-mantle temperatures should be high inside the region and that high heat flows are to be expected at the surface. Unfortunately heat flow data in the Cordillera are scarce. Very high values have been observed in the Gulf of California and Menard[22] has suggested that these are associated with a continuation of the East Pacific Rise through the Gulf into western United States. Lee and Uyeda[23] quote an average for nine heat flow measurements in the Cordillera of $1 \cdot 73 \pm 0 \cdot 53$ μcal/cm^2 sec which is perhaps 25–50% higher than average values in other parts of North America. Preliminary unpublished measurements of the Dominion Observatory in the Canadian Cordillera give one value of $2 \cdot 0$ μcal/cm^2 sec along with a number of geothermal gradients which are certainly higher than in shield areas.

Pakiser and Zietz[20] have drawn attention to a striking difference in character between aeromagnetic profiles west of the Rocky Mountains and those to the east. Across the Cordillera the profiles are generally flat and featureless, while to the east strong local anomalies and broad regional anomalies are common. Thus in a qualitative sense the aeromagnetic data are also consistent with the hypothesis that high temperature underlie the western region.

Discussion

The two Arctic anomalies and the western anomaly have certain features in common. All suggest that high electrical conductivities and high temperatures characteristic of the upper mantle exist at unusually shallow depths; all are associated with regions in which the aeromagnetic profiles are, for the most part, exceptionally flat; all lie adjacent to the continental margin; and none show any correlation with surface geology.

The spatial extent of each of these anomalies has so far been only partially determined. At Alert the effect is confined to a narrow belt striking NE–SW but the extension along strike is unknown in either direction. At Lake Hazen, 150 km to the southwest anomalous effects are still evident, though less pronounced, for both magnetic and electric fields. Other locations occupied briefly during the summer of 1961 were Isachsen on Ellef Rignes Island, Meighen Island, Axel Heiberg Island, and Eureka on the west side of Ellesmere Island. The high hourly range activity that is characteristic of Alert was not found at any of these stations, but each demonstrated horizontal field disturbances appreciably higher than at Resolute Bay. Whitham and Andersen[2] show a magnetogram from Eureka in which the antiphase character of X and Y variations is clearly evident (figure 2).

More field investigations in the northern part of the Arctic Archipelago are required to determine the extent of the anomalies at both Mould Bay and Alert. While the data collected so far indicate that both are probably local rather than regional effects, the interesting possibility of a major feature striking NE–SW from Alert to Prince Patrick Island along the northern margin of the Archipelago has not been eliminated. Further field work on Ellesmere Island is planned for the summer of 1967.

In western Canada and the United States the anomalous zone appears to be related to an upper mantle feature of major proportions. Its extension south of latitude 32° N and north of latitude 51° N is still undetermined, but between these parallels we have a structure which underlies nearly the whole of the Cordillera. Menard[22] has discussed topography, seismicity, and heat flow in the East Pacific Rise. The crest of the rise is about 800 km wide, is seismically active, has surface heat flows much higher than the oceanic average, and has upper-mantle seismic velocities which are much less than normal mantle velocities. Furthermore, the topographic relief of the rise is matched by a very similar bulge in the upper mantle. Near the equator the rise swings eastwards towards the edge of the Pacific Basin and its eastern flank and crest appear to pass under western North America in the neighbourhood of Mexico and the Gulf of California. Thereafter the main trend of the rise is roughly parallel to the continental margin as far north as Alaska, though there is some evidence that the crest reenters the Pacific in the vicinity of Washington State or southern British Columbia.

Caner and Cannon[7] have noted that the upper-mantle structure they propose to explain the zone of low P_n velocities and reduced Z variations in western North America could well represent that part of the East Pacific Rise which lies beneath the continent. The low P_n velocities appear to be directly continuous across the continental boundary. Many more heat flow data on the land side are required, but the high values in the Gulf of California and the moderately high average west of the Rockies lend qualitative support to the hypothesis.

Speculation on the origin and character of the induction anomalies is severely hampered because adequate correlations between electromagnetic effects and other aspects of geophysics and structural geology are still lacking. At Alert seismic activity is low and indicates that failure is not occurring at the present time in the region of nothern Ellesmere Island. However, faulting is known to have taken place during the Mesozoic in a direction roughly parallel to the anomaly axis. We can therefore speculate that tensional stresses due to the presence of a hot subterranean body could not have exceeded crustal breaking strength during the past hundred million years and that the body is therefore cooling and now exists in a state which is more or less stress free. The observed gravity anomalies in the area provide reasonable support for this hypothesis. Upper-mantle P-wave velocities are unknown in the anomalous region and their measurement in such a restricted zone presents formidable difficulties. No heat flow measurements have yet been attempted.

At Mould Bay and in the Cordillera, seismicity can be generally associated with regions of unusual induction, though correlations are quite uncertain in their quantitative aspects. Such uncertainties are even more pronounced in the case of terrestrial heat flow, while in neither region do the available gravity data appear to offer much hope for meaningful comparisons.

At present one can merely speculate on the structural and tectonic implications of these interesting phenomena. Much more field experimentation and theoretical development are required before we can understand their proper relation to the distribution of physical and chemical properties in the interior and to events in geological time.

References

1. K. Whitham, E. I. Loomer, and E. R. Niblett, *J. Geophys. Res.*, **65**, 3961 (1960).
2. K. Whitham and F. Andersen, *Geophys. J.*, **7**, 220 (1962).
3. L. K. Law, J. DeLaurier, F. Andersen, and K. Whitham, *Can. J. Phys.*, **41**, 1868 (1963).
4. T. Rikitake and K. Whitham, *Can. J. Earth Sci.*, **1**, 35 (1964).
5. K. Whitham and F. Andersen, *Geophys. J.*, **10**, 317 (1965).
6. D. H. Hyndman, Thesis, University of British Columbia (1963).
7. B. Caner and W. H. Cannon, *Nature*, **207**, 927 (1965).
8. A. Lambert and B. Caner, *Can. J. Earth Sci.*, **2**, 485 (1965).
9. B. Caner, W. H. Cannon, and C. E. Livingstone, *J. Geophys. Res.*, **72**, 6335 (1967).
10. U. Schmucker, *J. Geomagnetism Geolectricity*, **15**, 193 (1964).
11. K. Whitham, *J. Geomagnetism Geoelectricity*, **15**, 227 (1964).
12. K. Whitham, *J. Geomagnetism Geoelectricity*, **17**, 481 (1965).
13. K. Whitham, *Geophys. J.*, **8**, 26 (1963).
14. A. T. Price, *J. Geophys. Res.*, **67**, 1907 (1962).
15. J. M. DeLaurier, private communication (1967).
16. L. K. Law, W. S. B. Paterson, and K. Whitham, *Can. J. Earth Sci.*, **2**, 59 (1965).
17. W. S. B. Paterson and L. K. Law, *Can. J. Earth Sci.*, **3**, 237 (1966).
18. A. Overton, *Can. J. Earth Sci.*, **5**, in press (1968).
19. A. J. Wickens and K. Pěč, *Bull. Seismol. Soc. Am.*, in press (1968).
20. L. C. Pakiser and I. Zietz, *Rev. Geophys.*, **3**, 505 (1965).
21. W. R. H. White, M. N. Bone, and W. G. Milne, 'Seismic refraction surveys in British Columbia, 1964–1966: a preliminary interpreatation', *Proceedings of the Tokyo Upper Mantle Meeting, Am. Geophys. Monograph No. 12*, in press (1967).
22. H. W. Menard, *Science*, **132**, 1737 (1960).
23. W. H. K. Lee and S. Uyeda, 'Review of heat flow data', in *Terrestrial Heat Flow* (Ed. W. H. K. Lee), *Am. Geophys. Monograph No. 8*, 87 (1965).

B

SOLID STATE PHYSICS AND GEOPHYSICS

II. Laboratory measurements

1

T. J. SHANKLAND

Department of Geophysics and Planetary Physics
School of Physics
University of Newcastle upon Tyne
England

Transport properties of olivines*

Introduction

The combination of information gained from seismic properties of the Earth, together with laboratory measurements on rocks and minerals and with solid-state physical theory of elastic properties, has provided the greater part of our information about the composition and state of the Earth's interior. Similarly, it is desirable to use the growing body of knowledge of electronic properties of materials to narrow the range of speculation on electrical, magnetic, and thermal conditions inside the Earth.

In discussing electronic properties of solids, we greatly complicate the physics and chemistry of minerals owing to the fact, now well appreciated, that very small amounts of impurities in a host lattice can radically alter electrical and optical properties. This chapter discusses some effects of iron as an impurity in a silicate lattice, both to emphasize the similarity of silicates to other oxides, and to demonstrate the dominance by transition-metal ions, principally iron, of electrical and optical properties of the materials which probably constitute the Earth's mantle.

Measurements have been made of optical absorption and electrical transport in synthetic single crystals of forsterite grown by the Verneuil or flame fusion method[1]. In the complete solid solution of the olivine series, of which forsterite is the magnesium-rich end member, ferrous iron can substitute for magnesium in the formula

$$(Mg_{1-x}Fe_x)_2SiO_4$$

where x can vary from 0 to 1. Although in geophysical problems the most interesting value of x is 10–20%, the synthetic crystals used had $x = Fe/Mg$ in the range 0·007–1%. However, within this range of variation it was possible to separate effects due to iron from those intrinsic to pure Mg_2SiO_4. While detailed description of the synthetic crystals is given elsewhere[1], it should be mentioned that measured density was equal to that calculated from the known X-ray lattice parameters, and the stoichiometry was better than that tabulated for natural olivines. One result of growth in a rather oxidizing atmosphere is that the ratio of Fe^{3+}/Fe^{2+} is much higher than it would be in a natural crystal. This result is apparent both in chemical analyses[1] and in the absorption spectra discussed below.

---

* This work was supported in part by the Office of Naval Research under Contract Nonr-1866 (10) and by the Division of Engineering and Applied Physics, Harvard University, U.S.A.

Band Structure of an Insulator

There are two common ways of looking at the energy levels of electrons in semiconductors and insulators. The methods which have produced the most useful and elegant descriptions of the properties of semiconductors emphasize the periodic nature of a crystal lattice in which the electron wave function is a travelling wave of wavelength λ very similar to a free electron. This method is most useful for materials of high mobility in which the electrons have a long travel time between collisions; consequently they can traverse a large number of lattice distances and have a mean path greater than their wavelength λ.

In the other approach, the tight-binding method, calculations start from the wave functions of the free atom. The energy levels of the free ions are broadened as the ions are brought closer together. However, the wave functions are still considered as principally localized on an individual ion rather than being spread out through the crystal as in the first picture.

With both pictures we have a forbidden gap E_g, in which no electronic energy states exist, between a filled valence band and an empty conduction band. While the nearly free electron picture affords a detailed description of many materials, especially the electronic semiconductors such as Si or Ge which have small E_g, of the order of 1 ev, it is frequently less appropriate for materials which we consider insulators, those with a wide band gap. In particular, there are simpler methods for investigating impurity energy levels in the centre of a large band gap. Because some impurity levels—such as those which arise from the unfilled d shell of an iron ion—are widely separated in energy from the band edges, it would take an enormous number of travelling wave functions to describe the impurity energies. With impurity levels which are remote from the band edge of the crystal a tight-binding approach gives an adequate description, and it is sufficient for many purposes to use wave functions localized on the impurity ion and affected only by the ions in the immediate neighbourhood. In a completely localized description we can regard the transition-metal ion as being taken from infinity and surrounded by a symmetrical arrangement of oxygen ions (six in the case of iron in MgO, Al_2O_3, or Mg_2SiO_4).

Two localized calculations of the possible energy states of a d electron can then be made. If we calculate the relative movement of the d levels as the ion is surrounded by the electrostatic field of the nearby ions (sometimes called ligands), then we have a crystal-field calculation. If we try to include the combined levels of the impurity d electrons plus the electrons on the ligands, then we have the more general molecular orbital theory.

Figure 1 qualitatively depicts the energy levels to be expected for the molecular orbital states of a metal ion with a single $3d$ electron when it is surrounded by an octahedral array of six oxygen ions. The molecular orbital states are combined from the free ion states on either side. The states are labelled according to the representations appropriate for octahedral symmetry. Those labelled with a superscript 'b' are filled bonding states which merge with the filled valence band of the host crystal, and those labelled with a superscript 'a' are empty antibonding states which merge with the empty conduction band of the entire crystal. The separation of bonding and antibonding levels, or valence and conduction bands, gives an energy gap within which are the crystal-field impurity levels labelled t_{2g} and e_g.

Transport properties of olivines

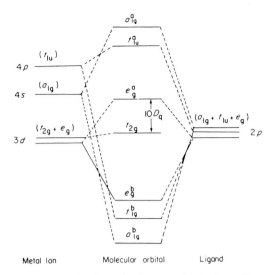

FIGURE 1 Formation of molecular orbital energy levels from free on evels. The states are labelled according to their representations in octahedral symmetry; superscript 'a' implies an antibonding state and 'b' a bonding state

Transitions among the d levels, t_{2g} to e in figure 1, are crystal-field transitions. Those between the d levels and the molecular orbital state or the band states are called charge-transfer transitions because an electron leaves a state mainly on the metal ion for a state mainly on a ligand, or vice versa. Finally, transitions between valence and conduction bands—or between bonding and antibonding states in the molecular orbital case—are the fundamental absorptions intrinsic to the lattice.

Measurements of optical absorption can reveal relative energies of the various transitions, and the details of transition-metal spectra have been extensively discussed[2,3]. Monochromatic light of an energy corresponding to the energy separations of the levels is absorbed, weakly by the crystal-field transitions in the middle of the band gap and more strongly by the charge-transfer and fundamental transitions at higher energy. In a plot of optical transmission, which is the ratio of light intensity I through a sample to light intensity I_0 in a reference beam, low transmission T corresponds to high absorption according to the formula

$$\frac{I}{I_0} = T = (1-R)^2 \exp(-\alpha d) \tag{1}$$

R is the fraction of energy reflected at the surface of the sample, d is the thickness of the sample, and α, the absorption constant, is directly related to the microscopic properties of the material.

Absorption Spectra of Forsterite

Fundamental transitions

As can be seen in figure 2, the optical transmission of a very pure synthetic sample measured into the vacuum ultraviolet shows that the pure material is transparent out to 7·5 ev before an absorption, too strong to be due to impurities, sets in. Thus, since fundamental transitions from the valence band to the conduction band are not found at lower energies than 7·5 ev, there is a forbidden gap of at least this magnitude in pure magnesium orthosilicate. Electrical insulators are characterized by a large band gap, and we shall not be surprised to find that pure forsterite falls into this class.

The apparent width of the band gap will be reduced if energy levels can be introduced into the forbidden gap. One observed source of such levels is an exciton, an effect intrinsic to a given lattice in which a hole in the valence band is bound to an electron in the conduction band, giving rise to one or more stable bound states whose energy of formation is less than the band gap. It is not clear whether the 7·5 ev edge is the true value of E_g or is due to excitonic effects just below the gap. We can begin a comparison with the spectra of MgO by noting that MgO shows an exciton peak[4] at 7·60 ev and that the band gap[5] has been estimated to occur at about 8·7 ev. Should the comparison hold true, the fundamental absorption at 7·5 ev in Mg_2SiO_4 would be the slope of an exciton band, and the band gap would then occur at a higher value.

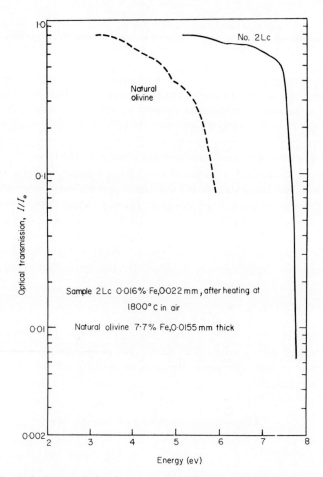

FIGURE 2 Optical transmission in a natural olivine compared with that in synthetic forsterite (sample No. 2Lc). Because of its low iron concentration the synthetic crystal is transparent out to the onset of fundamental lattice transitions

Strong impurity transitions

The apparent absorption edge in the olivine sample of figure 2, which led some early experimenters to hypothesize that the band gap was 3·3 ev, can be shown to be the result of impurity iron in the lattice rather than an intrinsic effect. This conclusion is clearly demonstrated in figure 3, where an increase of iron causes a corresponding

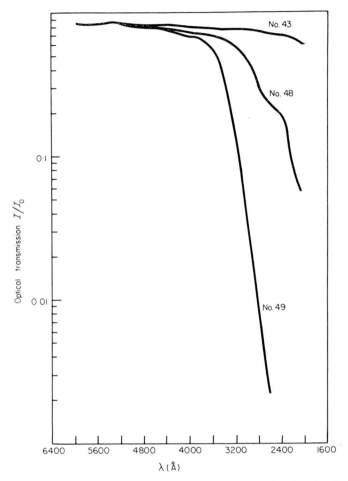

FIGURE 3   Optical transmission in three synthetic samples showing the increased absorption in the ultraviolet caused by increased iron concentration

Sample	Fe/Mg (%)	Thickness (mm)
43	0·06	0·889
48	0·20	0·719
49	0·26	0·978

increase in absorption. Since, as we have seen, the true absorption edge is greater than 7·5 ev and is not the 3·3 ev edge, it follows that olivines are not apt to be intrinsic electronic semiconductors at temperatures below their melting points because of the fact that much higher temperatures are needed to excite electrons across a forbidden gap of close to 8 ev.

As figures 4 and 5 show, the absorptions at 4·6 and 6·2 ev involve the presence of Fe^{3+} for they are reduced in intensity by a reducing or neutral atmosphere. Since the

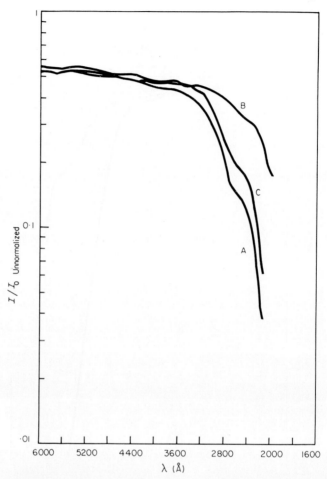

FIGURE 4 Transmission in forsterite, sample No. 48, Fe/Mg=0·20%. Reversible reduction of ultraviolet absorption which occurs after annealing in a neutral atmosphere. Curve A, as grown; curve B, after 70 hours at 1300 °C in N_2; curve C, after 90 hours at 1300 °C in air. The ratio Fe^{3+}/Fe^{2+} is the value as grown

absorptions are really too high in energy and far too strong to be crystal-field transitions, and since they are not intrinsic to pure forsterite, they can plausibly be attributed to charge transfer. However, another possibility is absorption at defects which are associated with compensation for the extra charge of Fe^{3+}. Clark[6] has previously identified these absorptions in the near ultraviolet as charge-transfer effects, although in Fe^{2+}. The present work does not prove that Fe^{2+} is not slightly responsible but does show that Fe^{3+} has far stronger transitions. Only about 0·6 mol. % Fe^{3+} would be needed to produce the absorption intensity of the olivine sample of figure 2, and more than this amount of Fe^{3+} has been observed in analyses[7] of natural olivines. Despite such strong indications, though, Fe^{3+} may not be the only source of charge-transfer absorptions in natural crystals. Firstly, the shoulders of the olivine absorption curve

Transport properties of olivines

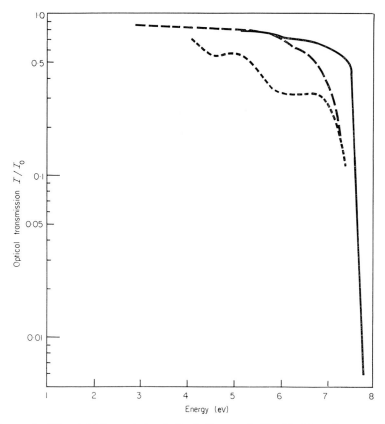

FIGURE 5 Ultraviolet transmission in forsterite, sample No. 2L. Near-disappearance of ultraviolet absorptions after heating in a strongly reducing atmosphere. – – – – sample No. 2Lb, 1·02 mm thick, after heating at 1800 °C in air; —— —— sample No. 2Lb, 0·978 mm thick after heating at 1400 °C in H_2; ——— sample No. 2Lc, 0·022 mm thick, after heating at 1800 °C in air. The very thin sample No. 2Lc is included to show the location of the absorption edge

TABLE 1 Absorption bands in forsterite

	Material		
Mg_2SiO_2	MgO	Others	Source
>7·5	7·6[4]		exciton
6·2	5·7		
4·6 {4·9 / 4·4} at 4·2 °K	4·8[8, 9, 10, 11]		charge transfer involving Fe^{3+}
2·9–3·1	3·1[9, 10]	3·2[15] in Al_2O_3	crystal field in Fe^{3+} $^6A_{1g}$–$^4A_{1g}$, ^4E_g
0·95–1·44	1·21–1·44[9, 14]	1·23–1·65[14] in spinel	crystal field in Fe^{2+} $^5T_{2g}$–^5E_g

Energies are in ev. The references are as given in the text. The crystal-field transitions have the standard notations for electron states in an octahedral symmetry[3].

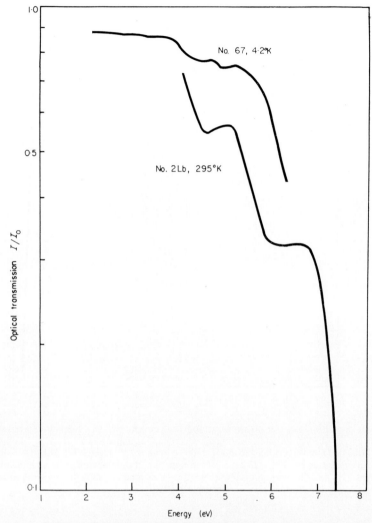

FIGURE 6 Splitting of the 4·6 ev charge-transfer band at low temperatures. Sample No. 2Lb: 1·02 mm thick, after heating in air at 1800 °C. Sample No. 67: 1·26 mm thick, as grown, Fe content not known

appear at different energies, 4·0 and 5·0 ev, than the shoulders of the Fe^{3+} curve, which indicates that additional transitions are taking place. Secondly, it is quite likely that there are other impurities, manganese and nickel, for example, which can contribute to the ultraviolet absorption in natural crystals.

Two further refinements of the absorption spectra should be noted. One is the fact that the 4·6 ev band splits into two bands at 4·4 and 4·9 ev when the sample is cooled to liquid helium temperature as shown in figure 6. The other is that it is necessary to take into account the fact that the olivine structure is orthorhombic, hence its absorption spectrum is pleochroic. Figure 7 shows the intensity, though not the energy, of the 4·60 ev line to depend quite strongly upon polarization.

As table 1 demonstrates, the change-transfer spectrum of Fe^{3+} in MgO is very similar to the Fe^{3+} spectrum in Mg_2SiO_4. Fe^{3+} lines have been seen [8-10] at 4·3 and

Transport properties of olivines

5·7 ev with an additional weak line at 4·8 ev inferred from line-shape analysis[11]. As with the forsterite lines of similar energy, they are removed by heating the sample in hydrogen or in vacuum.

Weak impurity transitions

As the iron doping in the crystal is increased, it becomes possible to see the crystal-field absorptions which take place among the impurity levels of the d functions in the band gap. The intensity of the transitions increases with iron content. The principal lines are those around 1·18 ev and those at 2·9–3·1 ev. These are shown in figure 8 for which only a single polarization direction was chosen.

It is helpful in interpreting the spectra to notice the changes in intensity which take place when the sample is heated in hydrogen. The intensity of the lines near 1·18 ev increases slightly. Ignoring the unexpected line at 3·3 ev which originated from furnace

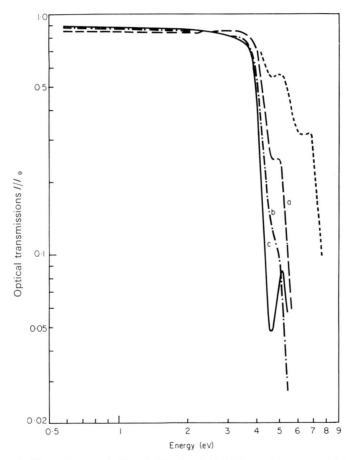

FIGURE 7 Forsterite, sample No. 2L, Fe/Mg=0·016%. Dependence upon light polarization of the 4·60 ev band

Polarization	Thickness (cm)	Symbol
E ∥ a	0·469	— —
E ∥ b	0·517	–·–·–
E ∥ c	0·517	——
unpolarized	0·0978	- - - -

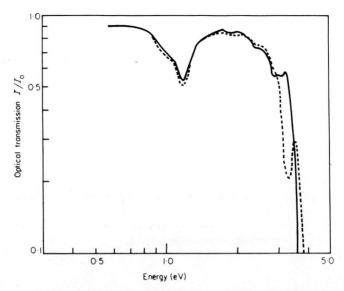

FIGURE 8 Transmission in forsterite, sample No. 3L, **E** ∥ **a**, beam ∥ **b**. Crystal-field transitions in forsterite. ——— 0·457 cm, after annealing in air at 1800 °C; — — — — 0·419 cm, after annealing in H_2 at 1400 °C. The intensity of the absorption increases on heating the sample in hydrogen, implying it is an Fe^{2+} transition, while the intensity of the 2·9–3·1 ev transition decreases, implying an Fe^{3+} transition. The sharp line at 3·3 ev is due to furnace contamination and occurred only in this particular sample

contamination when this sample was heated in hydrogen, one can see that the absorption at 2·9–3·1 ev decreases. The obvious inference is that the 1·18 ev absorption is due to Fe^{2+}, and the 2·9–3·1 ev group is due to Fe^{3+}. The 1·18 ev line has been previously identified as being the principal crystal-field transition of octahedral Fe^{2+} in olivines[6, 12, 13] as well as in a number of other oxides, for instance MgO[9, 14] or spinel[14]. Likewise the 2·9–3·1 ev group is well known in Fe^{3+} in different oxides. It occurs at 3·1 ev in[9] MgO and at 3·2 ev in[15] Al_2O_3. Only the low iron doping of the synthetic crystals makes it possible to oxidize sufficient Fe^{2+} to Fe^{3+} to make the dependence on Fe^{3+} apparent without damaging the crystal.

The details of the weaker absorptions and their interpretation will not be given here.

Conclusions from optical measurements

It is seen, then, that pure Mg_2SiO_4 has the wide band gap characteristic of such insulators as MgO, and that lower energy levels in the olivines are exclusively due to the presence of impurities, mainly iron. It is these lower levels which give rise to the electronic properties of interest in solid earth geophysics.

Aside from what they reveal about the electronic nature of a material, the most direct application of optical absorption measurements is to the estimation of the contribution to thermal conductivity of radiative transport. Significant heat conduction by radiative transport is expected to be effective in the region above 1000 °C, where its T^3 dependence would begin to add to the normal lattice conduction.

Radiative transport is blocked by the absorption of light; hence the measured absorption bands of minerals directly affect the possibility of radiative transport. Previous evaluations of radiative transport have assumed a 'grey body' having constant absorption with wavelength. It is time to allow for the fact that some minerals are quite transparent over certain energy ranges and that these ranges are functions of temperature and pressure. For instance, the crystal-field transition in Fe^{2+} near 1·2 ev has been observed to move up in energy with increased pressure, in accordance with the predictions of crystal-field theory[16]. On the other hand, the very strong charge-transfer bands of iron, those once thought to have been the band edge, move towards lower energies with pressure, in a direction to shut off thermal transport[17, 18].

Electrical Transport

In figure 9 is given the electrical conductivity of two synthetic samples of forsterite (not oriented) compared with some other measurements made on the olivine system which were selected for having particularly low values of conductivity. Four-probe measurements were made[1] using single-crystal samples in an atmosphere of argon which was buffered by being flowed over iron–iron oxide (rusty steel wool) to maintain a constant and low partial pressure of oxygen. While the results are only preliminary and the highest-temperature points somewhat doubtful, several interesting points emerge.

Firstly, it is seen that the conductivity of the synthetic samples, which are relatively iron free, is substantially lower than any conductivities reported for the olivine system. It is reasonable to attribute the very low conductivity of the single crystals of forsterite to their lack of impurities, contrasting with the samples of Hughes[19] and of Hamilton[20], and to the lack of grain boundaries, contrasting with the samples of Hamilton[20], of Jander and Stamm[21], and of Bradley, Jamil, and Munro[22]. Impurities are a source of charge carriers, and grain boundaries can be both a source of carriers and a relatively high-mobility channel for the movement of ions. Measurements on sintered samples of MgO have been made by Hamilton[22], and they show conductivities higher by orders of magnitude than in single-crystal MgO; Fe–Ni–spinels show the same behaviour[23]. Thus, grain boundaries can as often shunt bulk conductivity as block it, as Hutson[24] has discussed; hence, measurements on polycrystalline samples are not reliable guides to bulk properties.

Secondly, it is seen that the two samples of differing iron content have the same conductivity over at least five orders of magnitude. The overlap suggests that at this low iron content the conductivity mechanism does not involve iron. If conduction is not an impurity effect and since impurities are apt to be the only source of electrons or holes in a material with such a wide forbidden gap, we can infer that, in this temperature range, conduction must be ionic. Such behaviour is consistent with the previous lengthy parallel with electronic structure of MgO, where transport by ions is thought to be the principal conduction mechanism[25-27].

The fact of a long region of overlap in the curves of figure 9 is uncommon in transport measurements on oxides, and, while it may be fortuitous, it argues the need for measurements in a controlled atmosphere on single crystals of known, preferably high, purity.

In order to illustrate the similarity of forsterite to some refractory oxides, figure 10 compares the conductivity data of figure 9 with conductivities selected for being among the lowest for two oxides: for[28] MgO, for[29] Al_2O_3. While arguments from the magnitudes or slopes of conductivity curves are not reliable, it is apparent that forsterite behaves more like these oxides than like other materials, including a known ionic conductor[30] such as KCl, which have still higher conductivities.

Since pure forsterite is seen to be an excellent insulator, it is clear that electrical conduction in olivine will be due to the iron which substitutes for magnesium in the lattice. Electrical transport measurements in pure silicates are useful in geophysical problems as a background for studying the effects of addition of impurity ions such

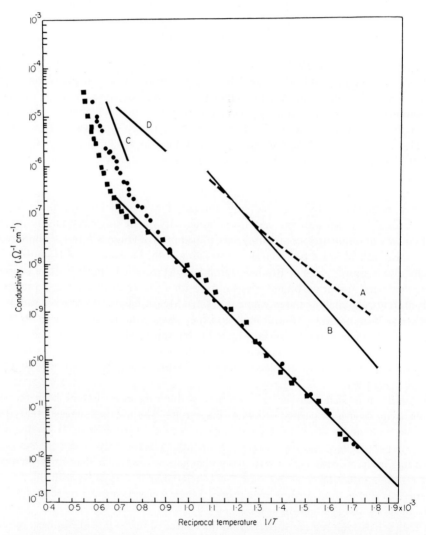

FIGURE 9 Electrical conductivity in two synthetic forsterite single crystals compared with conductivities in more iron-rich or sintered samples. Curve A, from Hamilton, sintered sample, 11·5 kb pressure, 0·75 and 0·91 ev; curve B, from Bradley, Jamil, and Munro, sintered sample, 11·8 kb pressure, 1·10 ev; curve C, from Hughes, peridot crystal, ~ 10% Fe (a.c.), 2·7 ev; curve D, from Jander and Stamm, sintered sample, 0·92 ev; ■ single-crystal forsterite, sample No. 48, Fe/Mg=0·2%, 1·00 ev; ● single-crystal forsterite, sample No 47, Fe/Mg=0·05%, 1·00 ev. The activation energies given are the slopes of each curve

Transport properties of olivines

as iron. Hamilton[20] and Bradley, Jamil, and Munro[22] showed that conductivity in sintered samples increases with increasing iron substitution. Further, there seems to be a qualitative change in conduction mechanism for Fe/Mg near 1%; the low-iron samples have a positive coefficient of activation energy with pressure, whereas the more iron-rich olivines show the opposite dependence. A similar qualitative change occurs in[31] MgO when the iron doping reaches 10%. It may be that the qualitative change comes at a lower iron concentration in olivines because of the anisotropy of its

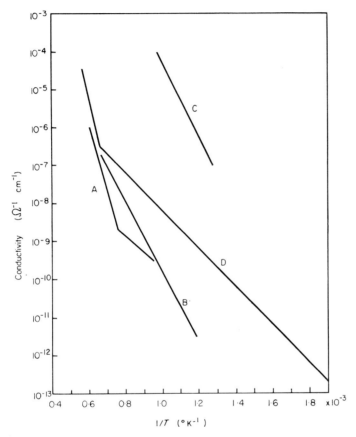

FIGURE 10 Electrical conductivities of several single-crystal insulators. Curve A, from Mitoff[28], MgO; curve B, from Rochow[29], Al_2O_3; curve C, from Kelting and Witt[30], KCl, intrinsic ionic; curve D, Mg_2SiO_4

lattice: along the crystal c axis the cations are much more closely spaced than in other directions. Hence, impurities might be expected to interact more readily in olivines than in MgO. Measurements on single crystals can indicate whether this hypothesis holds true.

It is clear that we are dealing with materials which are more difficult than either the electrolytic semiconductors like KCl in which ions are the charge carriers or the well-behaved materials like Si for which electronic semiconductor theory has been elaborated. 'Well-behaved' in this context is taken to mean that the material is described by a high mobility of electrons or holes. Mobility μ is the mean drift velocity per unit applied electric field, measured in units of cm²/v sec. It is related to electrical conductivity σ by

$$\sigma = ne\mu \qquad (2)$$

where n is the number of carriers per unit volume having a charge of magnitude e. It is possible to write the mobility as

$$\mu = \frac{e}{m^*}\tau \qquad (3)$$

Here m^* is the effective mass of the electron or hole in a crystal and τ, the relaxation time, is roughly the time for an electron of drift velocity v to travel a mean free path l. As is clearly explained by Ioffe[32], if l decreases to the order of the interatomic separation, then it is no longer meaningful to speak of free electrons moving in a conduction band or holes in a valence band, because to do so one must represent the carriers by travelling wave functions extending throughout the crystal. Below a certain mobility the formalisms of band theory do not apply; at room temperature the dividing line is of the order of 1 cm²/v sec and probably higher. For higher mobilities it is usually possible to perform those experiments which have revealed the band structure of solids, and at lower mobilities the carriers behave as if they do not perceive the long-range periodicity of the lattice. The low-mobility semiconductors are a large class of materials for which no single formalism has been found adequate.

Those materials below AgBr in table 2 have mobilities too low for the usual semiconductor band theory to have quantitative significance. Attempts to measure Hall

TABLE 2 Some representative mobilities compared with upper limits on mobilities in Mg_2SiO_4

Material	Electron mobility at 300 °K (cm²/v sec)
InSb	70000
Si	1900
PbS	600
AgBr	80
TiO_2	10^{-1}–1
Fe_2O_3	10^{-3}–10^{-1}
Mg_2SiO_4	$\begin{cases} <10\ (1000\ °K) \\ <10^{-2\,a} \end{cases}$

[a] As given by Wilson[33].

effect at high temperatures in the samples of figure 9 were unsuccessful, putting an upper limit of 10 cm²/v sec on mobility. Wilson's[33] measurements on his synthetic samples show the room-temperature mobility to be less than 10^{-2} cm²/v sec.

Thus, with the remotely possible exception of crystals far more perfect than those studied here, and certainly for natural crystals with their relatively high content of transition-metal ions, we must put aside the conduction theories of the well-behaved semiconductors. It is necessary to look at the much less complete models appropriate for simple oxide systems, particularly for the transition-metal oxides where models of the interaction of almost completely localized wave functions are being developed[34]. Such models, for instance, include the following: 'hopping conductivity', a term which implies that it is necessary to supply an activation energy for movement of electron from site to site, perhaps through intermediate oxygen states or through a very narrow band; or conduction by polarons (a polaron is an electron or hole surrounded by a region of lattice polarization which must be transported along with the electron). Without developing this suggestion into anything concrete, we can emphasize the need for stronger examination of the field of low-mobility semiconductors rather than of the better-understood high-mobility ones. Since the field is still developing and suffers from the same problems of dealing with materials of uncertain composition and impurity content as does the study of silicates, the elaboration of problems of electrical transport in the bulk of the Earth is going to require a still more sophisticated approach than in the past.

References

1. T. J. Shankland, *Synthesis and Optical Properties of Forsterite*, Thesis, Harvard Univ., and Div. Eng. Appl. Phys., Tech. Rept, No. HP-16 (1966).
2. D. S. McClure, 'Electronic spectra of molecules and ions in crystals: Part II. Spectra of ions in crystals', *Solid State Phys.*, **9**, 400 (1959).
3. J. S. Griffith, *The Theory of Transition Metal Ions*, Cambridge University Press, Cambridge, 1961.
4. D. M. Roessler and W. C. Walker, *Phys. Rev. Letters*, **17**, 319 (1966).
5. G. H. Reiling and E. B. Hensley, *Phys. Rev.*, **112**, 1106 (1958).
6. S. P. Clark, *Carnegie Inst. Wash. Yearbook*, **58**, 187 (1959).
7. W. A. Deer, R. A. Howie, and J. Zussman, *Rock-Forming Minerals*, Vol. 1, Longmans, Green, London, 1962, pp. 2–33.
8. W. T. Peria, *Phys. Rev.*, **112**, 423 (1958).
9. G. A. Slack, private communication.
10. J. E. Wertz, G. Saville, P. Auzins, and J. W. Orton, *J. Phys. Soc. Japan, Suppl. II*, **18** 305 (1963).
11. R. W. Soshea, A. J. Dekker, and J. P. Sturtz, *J. Phys. Chem. Solids*, **5**, 23 (1958).
12. E. F. Farrell and R. E. Newnham, *Am. Mineralogist*, **50**, 1972 (1965).
13. W. B. White and K. L. Keester, *Am. Mineralogist*, **51**, 774 (1966).
14. G. A. Slack, *Phys. Rev.*, **134**, A1268 (1964).
15. D. S. McClure, *J. Chem. Phys.*, **36**, 2757 (1962).
16. D. R. Stephens and H. G. Drickamer, *J. Chem. Phys.*, **35**, 424 (1961).
17. A. S. Balchan and H. G. Drickamer, *J. Appl. Phys.*, **30**, 1446 (1959).
18. S. K. Runcorn, *J. Appl. Phys.*, **27**, 598 (1956).
19. H. Hughes, *J. Geophys. Res.*, **60**, 187 (1955).
20. R. M. Hamilton, *J. Geophys. Res.*, **70**, 5679 (1965).
21. W. Jander and W. Stamm, *Z. Anorg. Allgem. Chem.*, **207**, 289 (1932).
22. R. S. Bradley, A. K. Jamil, and D. C. Munro, *Geochim. Cosmochim. Acta*, **28**, 1669 (1964).
23. A. A. Samokhvalov and A. G. Rustamov, *Fiz. Tverd. Tela*, **7**, 1198 (1965) (*Soviet Phys.— Solid State (English Transl.)*, **7**, 961 (1965)).
24. A. R. Hutson, 'Semiconducting properties of some oxides and sulfides', in *Semiconductors* (Ed. N. B. Hannay), Reinhold, New York, 1959, p. 541.
25. S. P. Mitoff, *J. Chem. Phys.*, **36**, 1383 (1962).
26. S. P. Mitoff, *J. Chem. Phys.*, **41**, 2561 (1964).
27. M. O. Davies, *J. Chem. Phys.*, **38**, 2047 (1963).

28. S. P. Mitoff, *J. Chem. Phys.*, **31,** 1261 (1959).
29. E. G. Rochow, *J. Appl. Phys.*, **9,** 664 (1938).
30. H. Kelting and H. Witt, *Z. Physik*, **126,** 697 (1949).
31. K. W. Hansen and I. B. Cutler, *J. Am. Ceram. Soc.*, **49,** 100 (1966).
32. A. F. Ioffe, *Physics of Semiconductors*, Infosearch, London, 1960.
33. J. Wilson, *The Optical and Electrical Properties of Synthetic Olivine Crystals*, Thesis, University of Newcastle upon Tyne (1965).
34. F. J. Morin, 'Oxides of the $3d$ transition metals', in *Semiconductors* (Ed. N. B. Hannay) Reinhold, New York, 1959, p. 600.

2

R. G. BURNS

Department of Geology and Mineralogy
University of Oxford
Oxford, England

Optical absorption in silicates

Introduction

Considerable interest centres around the thermal structure of the Earth and the importance of radiative transfer in the mantle[1-6]. The efficiency of the mantle to transmit heat to the Earth's surface is strongly dependent on absorption properties of mineral phases constituting the mantle in the visible and near infrared regions.

Present knowledge of the mineralogy of the mantle suggests that the upper mantle consists of some combination of the ferromagnesian silicates, olivine, orthopyroxene, garnet, clinopyroxene, and, perhaps, amphibole in restricted regions[7,8]. High-pressure polymorphs and breakdown products of these phases are believed to constitute the lower mantle.

The majority of silicate minerals owe the property of absorption in the visible and near infrared regions to the presence of transition-metal ions in the structure[9]. Iron is the most abundant transition metal in the Earth and the Fe^{2+} ion occurs in most common silicate minerals. Thus, it is of interest to know regions of the spectrum which are absorbed by iron in ferromagnesian silicates.

Comparatively few spectral measurements have been made of the common rock-forming silicate minerals because of manipulation difficulties[10-12]. A technique has been devised for measuring polarized spectra of small crystals in thin sections of rock[13]. In the present chapter absorption spectra of ferromagnesian silicates measured at atmospheric pressure and room temperature are described. The effects of pressure and temperature on the spectra are discussed and possible applications of the spectral data to processes occurring in the mantle are suggested.

Origins of Optical Absorption

Significant advances have been taking place in the understanding of the mechanisms of heat conduction in the mantle[5,6]. Estimates of the variation of thermal conductivity with depth indicate that heat conduction by radiative transfer is more important than the conventional lattice conductivity[3] below about 100 km. To evaluate radiative conductivities absorption coefficients obtained from spectral measurements have been used[10]. To facilitate calculations it has been generally assumed that the material of the mantle is a 'grey' body, and that absorption coefficients are independent of wavelength[6]. The veracity of this assumption may be examined by measuring positions and intensities of absorption bands in the visible and near infrared. In addition, thermodynamic data may be obtained from the positions of absorption bands and used to interpret element enrichments in crystal structures of minerals[9, 13-17].

The term optical absorption implies absorption in the visible region (4000–7000 Å) of the electromagnetic spectrum, usually through electronic processes. Absorption of this radiation produces coloured transmitted light. However, radiation outside the visible region may be similarly affected and not be apparent optically.

Electronic processes leading to absorption in the near ultraviolet, visible, and short-wave infrared regions (3000–20 000 Å or 33 000–5000 cm^{-1}) include the following: (i) electron transfer through crystal structure imperfections, (ii) interelement or charge-transfer transitions, (iii) transitions between bonding and antibonding π orbitals, and (iv) internal electron transitions within transition elements.

Electron transitions induced by crystal structure imperfections

Many crystals, particularly those of the alkali and alkaline-earth metal compounds, possess colour centres or lattice defects which absorb light[18–20]. There are several types of colour centres, but F centres (electrons filling anion vacancies) and F' centres (electrons filling interstices) are the most common. Transitions involving electrons in vacancies and interstices in crystal structures probably have widespread occurrence in nature, but are often obscured by other electronic processes. Well-known mineralogical examples are natural, deformed, and irradiated halites, fluorites, and calcites. Pleochroic haloes are also common. In this case excitation is induced in minerals such as biotite and cordierite by included radioactive element-bearing minerals such as zircon and allanite.

Interelement electron transitions or charge transfer

Charge transfer occurs when electrons migrate between neighbouring ions in a crystal structure. The process is induced by high-energy ultraviolet radiation, but absorption bands may extend into the visible. Charge-transfer processes are favoured when neighbouring elements in a crystal structure are able to exist in a variety of oxidation states (for example Fe^{2+} and Fe^{3+}, Ti^{3+} and Ti^{4+}, Mn^{2+} and Mn^{3+}). The process is facilitated whenever there is local misbalance of electrostatic charge in a crystal structure accompanying isomorphous substitution (for example replacement of Fe^{2+} and Mg^{2+} by Al^{3+} and Fe^{3+} in ferromagnesian silicates). These factors enable charge transfer to take place more readily along certain crystallographic directions than others, and lead to both absorption of radiation at lower energy (longer wavelength) and pleochroism in polarized light[21]. Mineralogical examples are the red–brown colours observed in hornblendes, augites, biotites, fayalite (figure 3(b)), staurolite, and 'hypersthene' (figure 5)[21].

Transitions between π orbitals

Closely related to charge transfer are the electronic transitions between bonding and antibonding π orbitals in compounds containing appreciable orbital overlap or covalent bonding. Common mineralogical examples are graphite and sulphides, arsenides, tellurides, etc., of the transition-metal and B-subgroup elements.

Internal electron transitions

Transitions involving d and f electrons within ions of the transition-metal and rare-earth elements, respectively, are the most common causes of colour and optical absorption in minerals. Absorption spectra of transition-metal ions may be explained semiquantitatively by crystal-field theory[9, 22–24].

Optical absorption in silicates

Crystal-field Spectra

The elements of the first transition series, scandium, titanium, vanadium, chromium, manganese, iron, cobalt, nickel, and copper, are characterized by incompletely filled $3d$ atomic orbitals. These five orbitals have different spatial configurations. Two of the orbitals, $d_{x^2-y^2}$ and d_{z^2}, have lobes directed along three Cartesian axes; the other three orbitals, d_{xy}, d_{yz}, and d_{xz}, have lobes projecting between these axes. When a transition-metal ion is in a crystal structure the $3d$ orbitals are separated into different energy levels owing to differences in electrostatic repulsion between electrons and surrounding anions. In octahedral coordination, with six anions located along the Cartesian axes, electrons in the $d_{x^2-y^2}$ and d_{z^2} orbitals are repelled to a greater extent than those in the d_{xy}, d_{yz}, and d_{xz} orbitals (figure 1(a)). The reverse situation occurs with ions in tetrahedral and cubic coordinations (figure 1(a)). The induced energy separation between the two groups of orbitals is termed crystal-field splitting and is

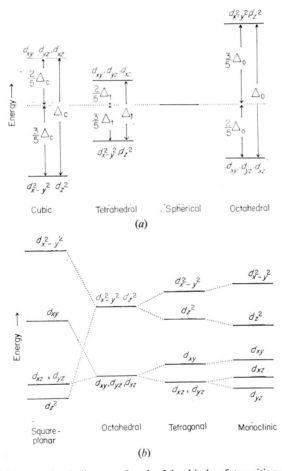

FIGURE 1 (a) Energy level diagrams for the $3d$ orbitals of transition-metal ions in octahedral, tetrahedral, and cubic coordinations. Δ_o, Δ_t, and Δ_c are the crystal-field splittings for octahedral, tetrahedral, and cubic coordinations, respectively. (b) Relative energies of the $3d$ orbitals of transition-metal ions in coordination sites with octahedral, tetrahedral, square-planar, and monoclinic symmetries designated by Δ or $10Dq$. The relative 3d-orbital energy levels for a transition-metal ion in undistorted octahedral, tetrahedral, and cubic coordination sites are shown in figure 1.

Measurements of electronic absorption spectra enable the crystal-field splitting Δ to be calculated. Observed crystal-field splittings for transition-metal ions in octahedral, cubic, and tetrahedral coordinations agree closely with those predicted from electrostatic theory[25]. Other factors which contribute to the magnitude of the crystal-field splitting are as follows[9, 24]: (i) the oxidation state of the transition-metal ion (values of Δ are higher for M^{3+} ions than M^{2+} ions), (ii) the nature and type of anion or dipolar group coordinated to the transition-metal ion (for example Δ_0 for 6 CN^- ions is considerably larger than that for six oxygen ions), and (iii) the interatomic distance R between metal and anion (according to crystal-field theory Δ should vary as R^{-5} for anions and R^{-6} for dipolar groups).

In structures of ferromagnesian silicates, coordination sites usually have symmetries lower than octahedral, tetrahedral, or cubic. This leads to further resolution of the 3d-orbital energy levels. For example the d_{xz} and d_{yz} orbitals are stabilized relative to d_{xy}, and the $d_{x^2-y^2}$ orbital destabilized relative to d_{z^2}, when a transition-metal ion is in a tetragonally distorted octahedral site elongated along one of the tetrad axes (for example the M_1 site in the olivine structure[26]). In square-planar coordination the d_{z^2} orbital may become the most stable[27]. In coordination sites of lower symmetries (for example the M_2 site of the orthopyroxene structure[28]) the 3d orbitals may become separated into five energy levels (figure 1(b)).

Additional electron transitions are possible between the resolved 3d-orbital energy levels of a transition-metal ion in a distorted coordination site. These transitions are usually polarization dependent and cause pleochroism, which may occur both inside and outside the visible region. Symmetry criteria and group theory are used to interpret polarized absorption spectra of transition-metal ions in low-symmetry crystal fields[27, 29]. One result stemming from such considerations concerns intensities of absorption bands arising from transition-metal ions in centrosymmetric and non-centrosymmetric sites. Electron transitions between 3d-orbital energy levels lead to relatively weak absorption bands in the spectra of transition-metal ions in centrosymmetric environments such as regular octahedral and square-planar coordinations. The absorption bands are intensified considerably when the transition-metal ion occurs in non-centrosymmetric environments such as many of the six-coordinate sites in ferromagnesian silicate crystal structures.

The width of absorption bands is related to varying metal–anion distances in an oscillating system and the resultant changes in Δ and energy separations between 3d orbitals. As transition-metal ions and surrounding anions vibrate about lattice points, Δ oscillates about a mean energy corresponding to the average positions of the atoms. In a static system electron transitions between energy levels would lead to absorption over a very narrow energy range. This energy range increases within a vibrating system and absorption bands become wider. The effect of rising temperature, which increases thermal vibrations, is to broaden absorption bands. Theoretically, rising pressure should inhibit vibrations and narrow absorption bands. Another factor contributing to the breadth of absorption bands is superposition of two or more absorption bands. Thus, separations may be small between energy levels of a transition-metal ion either located in a single low-symmetry coordination site, or distributed over two or more

Optical absorption in silicates

structurally similar sites. Electron transitions to these closely spaced levels may produce superimposed absorption bands.

Techniques for Measuring Spectra of Silicates

The majority of spectral measurements of silicates have been made on polished plates of large crystals[10, 12, 30-55]. Most of the common rock-forming silicates are too fine grained for this technique and various methods have been devised using powdered samples[56-59]. Microscopes have been used in conjunction with a spectrophotometer optical system to measure the spectra of small crystals[11, 21, 60-64]. To measure polarized spectra of fine-grained minerals in rock a technique has been developed which uses a polarizing microscope equipped with a universal stage, which is inserted in a Cary model 14 spectrophotometer[13]. The universal stage attachment enables a crystal to be orientated accurately and the calcite Nicol prism of the microscope provides polarized radiation in the region 4000–25 000 Å. Beyond these limits appreciable absorption by the glass optics of the microscope system takes place. Most of the spectra described in this chapter were measured with this microscope technique.

Apparatus has been described for measuring spectra at high pressures[65, 66]. For room-temperature measurements, cells useful to 55 kb and 200 kb have been used[65]. A high-temperature high-pressure cell has also been designed for spectral measurements[66] at 500 °C and 180 kb.

Mineralogy and Composition of the Mantle

On the basis of seismic velocity distributions[67, 68] the mantle is believed to consist of three distinct regions: the upper mantle (10–400 km), the transition zone (400–900 km), and the lower mantle (900–2900 km). Two rock types have been suggested for the upper mantle[7, 8]: peridotite-dunite, consisting dominantly of olivine, pyroxenes and garnet; eclogite, consisting mainly of garnet and clinopyroxene. Amphibole is also believed to occur in the uppermost levels of the mantle[7, 69].

In the transition zone a series of major phase transformations from normal silicate to dense oxide structures takes place[7, 70, 71]. Some of the transitions suggested at various depths in the mantle include the following:

enstatite → forsterite + stishovite at about 400 km
forsterite → spinel at about 500 km
spinel + stishovite → ilmenite below 500 km
spinel → ilmenite + periclase between 500 km and 900 km

Other possible transformations include the following:

garnet → perovskite
diopside → forsterite + stishovite + garnet
     (or spinel)   (or perovskite)
ilmenite → corundum

Thus common ferromagnesian silicate structures, containing Fe^{2+} ions in distorted six-coordinate sites, transform to oxide structures in which Fe^{2+} ions occur in more regular octahedral sites.

At 900 km the phases are considered to be too closely packed for further major changes[7]. Thus, the lower mantle is thought to be homogeneous and to consist of dense oxide phases with the periclase, corundum, ilmenite, perovskite, and rutile (stishovite) structures.

The chemical composition of the mantle, deduced from element abundances in carbonaceous chondrites[72, 73], indicates that Mg, Si, and Fe dominate the mantle with the molecular ratio FeO/(FeO+MgO) lying between 0·1 and 0·2. Smaller amounts of Al, Ca, and Na and lesser quantities of Cr, Ni, K, Ti, Mn, and P are thought to be present. The composition derived for the upper mantle[72] shows a significant reduction in the elements Fe, Co, Ni, Cu, Pt, and Au relative to their abundances in carbonaceous chondrites.

Spectra of Iron Silicates

Iron is probably the most abundant metal in the Earth and the mantle is believed to consist of oxides and silicates of iron and magnesium. Therefore, the electronic spectra of Fe^{2+} compounds measured at atmospheric pressure will be discussed in some detail in this section. These measurements, in conjunction with data from spectral measurements at elevated pressures and temperatures, enable one to evaluate energies and intensities of absorption bands at the high pressures and temperatures found in the lower mantle.

Spectra of Fe^{2+} ions in regular coordination sites

Iron in tetrahedral and octahedral coordinations in oxide structures gives rise to comparatively simple spectra. The spectra of Fe^{2+}-doped zincite[74] and spinel[75], in

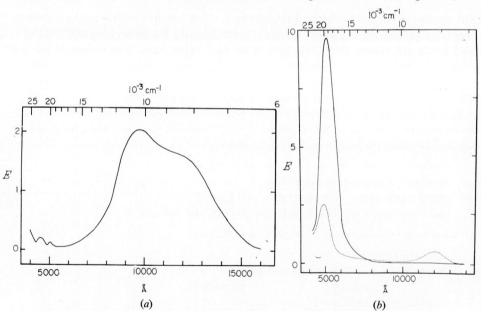

FIGURE 2 (a) Absorption spectrum of $Fe(H_2O)_6^{2+}$ in aqueous ferrous ammonium sulphate (extinction coefficient ε in l./mole cm). (b) Polarized absorption spectra of gillespite: dotted curve, ordinary ray; full curve, extraordinary ray. (Extinction coefficients ε, in l./mole cm, are calculated for the formula $BaFeSi_4O_{10}$)

Optical absorption in silicates

which iron occurs in tetrahedral coordination, consist of single absorption bands at 22 200 Å and 23 300 Å, respectively. Iron in octahedral coordination with H_2O in solution gives rise to a broad asymmetric band at about 10 000 Å (figure 2(a)). The spectrum of Fe^{2+}-doped periclase[76], containing iron in regular octahedral coordination, consists of a single band at 10 000 Å.

The spectra of Fe^{2+} ions in non-cubic environments are more complex. For example the polarized spectra of gillespite $BaFeSi_4O_{10}$, containing Fe^{2+} ions in square-planar coordination[77], are shown in figure 2(b). Two absorption bands occur at 5000 Å and 12 000 Å, and a third is predicted to lie[78] at about 35 000 Å. The spectra have been interpreted by group theory[27] and indicate that the order of increasing energy of the 3d orbitals is $d_{z^2} < d_{xz}, d_{yz} < d_{xy} < d_{x^2-y^2}$. The absorption band at 5000 Å indicates that the energy separation between 3d orbitals is about 20 000 cm^{-1} in the Fe^{2+} ion in gillespite.

Ferrous ions in distorted and asymmetric environments give rise to more complex absorption spectra, which is illustrated by the spectra of ferromagnesian silicate minerals.

Spectra of olivines $(Mg, Fe)_2SiO_4$

The polarized spectra of forsterite (Fa_{17}) and fayalite (Fa_{96}) are shown in figures 3(a) and 3(b). The spectra illustrate the pronounced pleochroism of olivine in the infrared which is not seen. Absorption maxima show compositional variations and shift to longer wavelengths with increasing iron content. Ferrous ions in the centrosymmetric tetragonally distorted M_1 site[26] give rise to the weak broad bands and

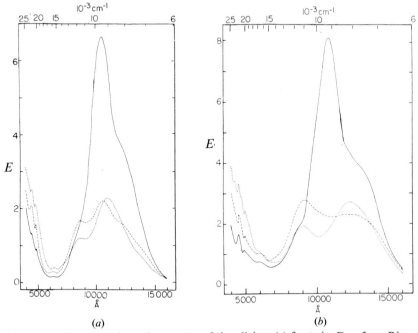

FIGURE 3 Polarized absorption spectra of the olivines (a) forsterite Fa_{17} from Rice, Arizona, and (b) fayalite Fa_{96} from Rockport, Massachusetts: broken curve, X polarization; dotted curve, Y polarization; full curve, Z polarization. (Extinction coefficients ε, in l./mole cm, are calculated for moles of Fe_2SiO_4)

shoulders at 8500–9200 Å and 10 450–12 800 Å. Intense absorption bands at 10 400–10 800 Å in the Z spectra are due to Fe^{2+} ions in the non-centrosymmetric M_2 site. The absorption band at 60 000 Å in the infrared spectrum of olivine[79] is probably due to electronic absorption in the Fe^{2+} ion.

Spectra of pyroxenes

Enstatite–orthoferrosilite series, $(Mg, Fe)SiO_3$. The polarized spectra of two orthopyroxenes, Fs_{14} and Fs_{86}, possessing low alumina contents, are shown in figures 4(a) and 4(b). The X spectra show one intense absorption band at 9100–9400 Å. The Y spectra consist of two bands at 8900–9350 Å and 18 500–21 000 Å, and a shoulder at about 11 500 Å, which becomes increasingly prominent in iron-rich orthopyroxenes.

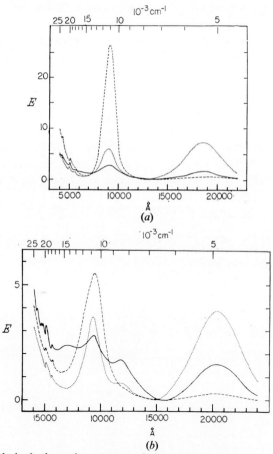

FIGURE 4 Polarized absorption spectra of the orthopyroxenes (a) bronzite Fs_{14} from Bamle, Norway, and (b) orthoferrosilite Fs_{86} from south west Manchuria: broken curve, X polarization; full curve, Z polarization. (Extinction coefficients ε, in l./mole cm, are calculated for moles of $FeSiO_3$)

The Z spectra of orthopyroxenes with less than about 50% $FeSiO_3$ consist of weak bands at approximately 9000 Å and 19 000 Å. In iron-rich orthopyroxenes these bands become more intense and occur at slightly longer wavelengths, and additional bands appear at 6000–6900 Å and 11 500–11 700 Å. An additional band has been observed in the unpolarized spectra of enstatite[12] at 32 000 Å. This band has an intensity comparable with that at 18 600 Å. Absorption maxima of all bands move to longer wavelengths with increasing iron content.

The spectra of aluminous orthopyroxenes have additional features (figure 5). The broad band at 6000–6900 Å in the Z spectra of iron-rich orthopyroxenes becomes prominent at 5500–6000 Å in aluminous orthopyroxenes with low iron content. In addition, absorption edges of charge-transfer bands in the ultraviolet extend further into the visible region in aluminous orthopyroxenes. The band associated with X polarized light absorbs more blue and green radiation than those for Y and Z polarizations, which accounts for the diagnostic pleochroic scheme of 'hypersthene', $X = $ pink, $Z = $ green[21].

The absorption bands at 32 000 Å, 18 500–21 000 Å, 9100–9400 Å, and 5500–6900 Å may be attributed to Fe^{2+} ions in the very distorted, non-centrosymmetric M_2 site of the orthopyroxene structure[28]. X-ray diffraction[28] and Mössbauer spectroscopy[80] measurements have shown that iron strongly favours this position in metamorphic orthopyroxenes. Iron in the more regular M_1 site gives rise to absorption bands at 9000–9400 Å and 11 500–11 700 Å.

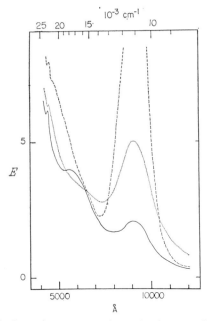

FIGURE 5 Polarized absorption spectra of an aluminous orthopyroxene—bronzite Fs_{26}, containing 4·35 wt.% Al_2O_3, from Madras, India: broken curve, X polarization; dotted curve, Y polarization; full curve, Z polarization

Diopside–hedenbergite series, $Ca(Mg, Fe)Si_2O_6$. The polarized spectra of a calcic clinopyroxene with 67% $CaFeSi_2O_6$ are shown in figure 6. The spectra contain three prominent bands or shoulders between 8000 Å and 12 000 Å. The unpolarized spectrum of diopside[12] shows absorption bands at 7350 Å, 10 300 Å, and 22 600 Å (ε about 15 l./mole cm).

The spectra of the diopside–hedenbergite series represent transitions within Fe^{2+}

ions in the distorted, non-centrosymmetric M_1 site of the calcic clinopyroxene structure[81].

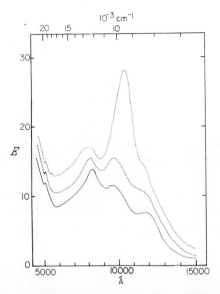

FIGURE 6 Polarized absorption spectra of a calcic clinopyroxene—ferrosalite $Wo_{50}En_{16}Fs_{24}$ from a metamorphosed iron formation in Quebec, Canada: broken curve, X polarization; dotted curve, Y polarization; full curve, Z polarization. (Extinction coefficients ε, in l./mole cm, are calculated for moles of $CaFeSi_2O_6$)

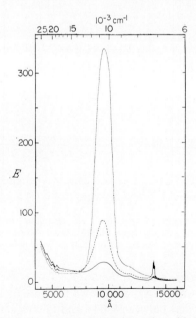

FIGURE 7 Polarized absorption spectra of cummingtonite, with 35·8% $Fe_7Si_8O_{22}(OH)_2$ from a metamorphosed iron formation in Quebec, Canada: broken curve, X polarization; dotted curve, Y polarization; full curve, Z polarization. (Extinction coefficients ε, in l./mole cm, are calculated for moles of $Fe_7Si_8O_{22}(OH)_2$)

Spectra of amphiboles

Cummingtonite–grunerite series, $(Mg, Fe)_7Si_8O_{22}(OH)_2$. The polarized spectra of a cummingtonite with 35·8% $Fe_7Si_8O_{22}(OH)_2$ are shown in figure 7. The X and Z spectra consist of broad bands around 9600 Å and prominent shoulders at 11 500 Å. The Y spectrum contains a very intense and broad band at about 9650 Å. All bands migrate to longer wavelengths with increasing iron content[11]. The intensities of the bands and shoulders in the X and Z spectra and the width of the intense band in the Y spectra increase in the spectra of grunerites[11].

The intense bands in the Y spectra are attributed to Fe^{2+} ions in the very distorted, non-centrosymmetric M_4 site of the cummingtonite structure[82,83]. X-ray crystallography[82, 83], Mössbauer[84], and infrared[85] measurements have shown that Fe^{2+} ions enter preferentially the M_4 position of the cummingtonite structure, discriminate against the M_2 position, and are randomly distributed over the M_1 and M_3 positions. The increased intensity and width of absorption bands in the spectra of grunerites is due to increased absorption by Fe^{2+} ions in the more regular octahedral M_1, M_2, and M_3 sites.

The weak, sharp peaks at 13 940 Å, 14 000 Å, 14 065 Å, and 14 150 Å in the X spectra are the first overtone of the hydroxyl stretching frequency[85] at 3619–3669 cm^{-1}. The positions of the peaks are not affected by iron content. However, the number and intensity of the peaks in each spectrum depend on the Fe^{2+} and Mg^{2+} contents of the M_1 and M_3 positions. The peaks have been used to estimate site populations in the cummingtonite–grunerite series[85]. For example, the spectra of the cummingtonite illustrated in figure 7 indicate that 34% of the iron is in the M_1 and M_3 positions, which agrees favourably with the estimate, 39·5%, from X-ray measurements of the same cummingtonite[83].

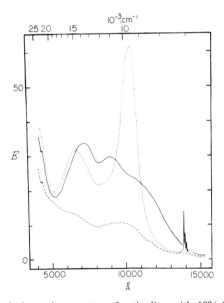

FIGURE 8 Polarized absorption spectra of actinolite, with 10% $Ca_2Fe_5Si_8O_{22}(OH)_2$, from glaucophane schist, Berkeley, California: broken curve, X polarization; dotted curve, Y polarization; full curve, Z polarization. (Extinction coefficients ε, in l./mole cm, are calculated for moles of $Ca_2Fe_5Si_8O_{22}(OH)_2$)

Tremolite–actinolite series, $Ca_2(Mg, Fe)_5Si_8O_{22}(OH)_2$. The polarized spectra of an actinolite with 10% $Ca_2Fe_5Si_8O_{22}(OH)_2$ are shown in figure 8. The X spectrum consists of weak, broad bands spanning 8000 Å to 11 000 Å. The Y spectrum contains an intense band at 10 350–9950 Å and a less intense band at 6600–6950 Å. The Z spectrum shows absorption bands at 7250–7150 Å and 8950–9200 Å, and a prominent inflection at about 10 000 Å. The spectra represent superimposed absorption bands due to Fe^{2+} ions in the distorted M_2 and more regular octahedral M_1 and M_3 sites in the actinolite structure[86].

The weak, sharp peaks at 13 925 Å, 13 985 Å, 14 050 Å, and 14 130 Å again reflect the occupancies of the M_1 and M_3 positions. Calculations based on the X spectrum shown in figure 8 indicate that 80% of the iron is in the M_1 and M_3 positions.

Spectra of garnets, $(Mg, Fe)_3Al_2(SiO_4)_3$

The spectrum of almandine with 70% $Fe_3Al_2(SiO_4)_3$ is shown in figure 9. Two weak, broad bands occur at 17 200 Å and 13 100 Å, and a third band occurs[10] at about 34 500 Å. Several weak, sharp peaks are present in the visible. The spectrum of pyrope with 20% $Fe_3Al_2(SiO_4)_3$ is similar and absorption bands occur at 16 500 Å and 12 600 Å.

Iron is in eightfold coordination in the pyrope structure[87]. The coordination site is distorted from cubic, and resolution of the 3d-orbital energy levels occurs in the Fe^{2+} ion. Electron transitions between these levels gives rise to the absorption bands in the infrared (figure 9).

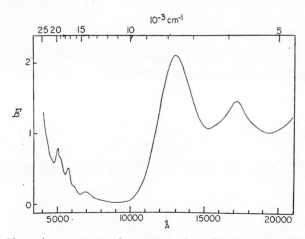

FIGURE 9 Absorption spectrum of a garnet—almandine $Pyr_{21}Alm_{70}Sp_3Gr_6$ from Fort Wrangle, Alaska. (Extincton coefficient ε, in l./mole cm, is calculated for moles of $Fe_3Al_2(SiO_4)_3$)

The Effects of Pressure and Temperature on Optical Spectra

Extensive measurements have been made of high-pressure spectroscopy of solids[88–91]. These include studies of the effect of pressure on absorption edges[30, 92–97], charge transfer[98–100], colour centres[101–105], and transition-metal[30, 106–113], and rare-earth[114] spectra. Some of the measurements have been made on minerals[30, 93, 106]. Important features of these studies may be summarized as follows.

In general, absorption bands in the spectra of transition-metal compounds shift to higher energies (shorter wavelength) and crystal-field splittings increase with increasing pressure. Many of the shifts are linear over the first 50 kb, but become curvilinear with decreasing splittings at pressures approaching 150 kb. Some of the pressure-induced shifts of transition-metal ions in oxide structures of interest to the present discussion are summarized in table 1. The absorption band at 12 900 Å in the spectrum of garnet moves to higher energies[30], and the absorption edge in the olivine spectrum moves into the visible region with increasing pressure and temperature[30].

The most general effect of pressure on the intensity of absorption bands is a small increase in intensity with increasing pressure[88, 89, 114]. This may be correlated with increased orbital overlap or covalent bonding with increased pressure[88, 89, 114].

Colour centres, which are induced by increased pressure[105] but bleached by rising temperature[115], gain intensity and shift to higher energies with increasing pressure[91, 105].

TABLE 1 Pressure-induced shifts of absorption bands in the spectra of transition-metal ions in oxide structures

Compound	Absorption band (cm^{-1})	Approximate pressure shift (cm^{-1})	Pressure range (kb)	Reference
MgO: Ti^{3+}	11360	2300	140	109
MgO: Cr^{3+}	16200	2100	140	109
	22700	2200	140	
MgO: Co^{2+}	8220	900	140	109
	18770	2200	140	
MgO: Ni^{2+}	8845	1400	140	109
	24500	2000	145	
Al$_2$O$_3$: Ti^{3+}	17720	1100	145	109
	17870	1300	145	
Al$_2$O$_3$: V^{3+}	17490	500	55	109
	25970	500	55	
Al$_2$O$_3$: Ni^{2+}	10050	300	55	109
Al$_2$O$_3$: Cr^{3+}	18300	1400	130	106
	25000	1300	130	
NiO: Ni^{2+}	8900	1000	130	108
	22900	400	45	
(NH$_4$)$_2$Cu(SO$_4$)$_2$.6H$_2$O: Cu^{2+}	11900	1300	90	107
FeSiF$_6$.6H$_2$O: Fe^{2+}	10480	2000	140	107
garnet: Fe^{2+}	7720	700	173	30
olivine: Fe^{2+}	31000 (absorption edge)	+3500	140 (250 °C)	30

N.B. All pressure shifts are negative (to higher energies), except olivine.

High-pressure spectroscopy has been used to study polymorphic transitions[116-118]. At the point where a first-order phase transformation occurs, there is a pronounced decrease in transmission of light, which can be traced to small particles scattering the light in the initial and final stages of a reaction.

Few spectral measurements of silicate minerals have been made at elevated temperatures[11, 30]. The intense band at 10 480 Å in the Z polarized spectrum of forsterite Fa_{17} was found to broaden, decrease in intensity, and shift to 9900 Å during measurements[11] from 25 °C to 750 °C. The same band at 10 770 Å in the spectrum of fayalite Fa_{96} moved to 10 350 Å at[11] 475 °C. A point at 31 000 cm^{-1} on the absorption edge of the spectrum of forsterite showed a linear shift of about 3·4 cm^{-1} per degree C towards the visible[30]. Extrapolation of the measured temperature- and pressure-induced shifts of this absorption edge indicates that it will have moved into the near infrared well before 1000 °C and 1000 kb are reached[91].

Applications of Spectral Data to Studies of the Mantle

Polymorphic transitions induced by spin pairing in transition-metal ions

Various phase changes have been postulated in the mantle[7, 70, 71]. Each transition is characterized by a decrease in molar volume and is favoured by increased pressure with depth in the mantle. Certain changes in electronic configuration of transition-metal ions can lead to contraction of metal–anion distances, and be induced by increased pressure[9, 119, 120].

For example, the electronic configuration of the Fe^{2+} ion in octahedral coordination in oxide structures on the Earth's surface is such that five of the $3d$ electrons occupy singly each of the $3d$ orbitals, and the sixth fills one of the low-energy orbitals d_{xy}, d_{yz}, and d_{xz} (figure 1). This sixth electron induces a crystal-field stabilization energy of $\frac{2}{5}\Delta_o$ on the Fe^{2+} ion, and the four unpaired electrons cause paramagnetism in iron (II) compounds. The resulting electronic configuration is referred to as the 'high-spin' or spin-free state. If the energy separation between the two groups of orbitals is very large, or if increased separation is induced by pressure, electrons may fill the low-energy orbitals and lead to the 'low-spin' or spin-paired state of iron with zero unpaired electrons. Iron (II) compounds containing 'low-spin' iron are diamagnetic (for example, pyrite, FeS_2), and the Fe^{2+} ion may acquire a relative crystal-field stabilization energy of $\frac{12}{5}\Delta_o$. In addition, since electrons no longer occupy orbitals projecting towards surrounding anions, metal–anion distances are about ten per cent shorter in 'low-spin' iron compounds[9].

It is possible to estimate the pressures at which 'high-spin' to 'low-spin' transitions may take place in the Fe^{2+} ion in various silicate and oxide phases believed to constitute the mantle. The basis for these calculations is as follows: (i) the energy separations between $3d$ orbitals of iron in gillespite; (ii) trends in the high-pressure spectral measurements of transition-metal compounds; (iii) the spectral measurements of ferromagnesian silicate and oxide minerals.

The maximum energy separation between the $3d$ orbitals of Fe^{2+} in gillespite was estimated from the spectra (figure 2(b)) to be approximately[27] 20 000 cm^{-1}. Experimental evidence[121] suggests that spin pairing may be induced in Fe^{2+} in gillespite at about 26 kb. This suggests that 20 000 cm^{-1} is close to the upper stability limit for

Optical absorption in silicates

3d-orbital energy separations for Fe^{2+} ions in silicate structures. We shall assume that energy separations in excess of 20 000 cm^{-1} lead to 'low-spin' configurations in Fe^{2+}.

The data summarized in table 1 indicate that the crystal-field splittings of Ni^{2+} and Co^{2+} ions in MgO, where divalent transition-metal ions replace Mg^{2+} ions in octahedral coordination, increase by approximately 10 cm^{-1} per kilobar. The data for $FeSiF_6 \cdot 6H_2O$, in which iron is surrounded by H_2O molecules at the vertices of a distorted octahedron, indicate that there is a 15 cm^{-1} per kilobar shift in the absorption band at 10 480 cm^{-1}. We shall assume that the maximum energy separations between 3d orbitals of Fe^{2+} in ferromagnesian silicates and oxides increase by 10 cm^{-1} per kilobar, and further assume that the increments are linear with pressure. As a result of these approximations the calculations summarized in table 2 probably represent minimum pressures or depths in the mantle at which spin pairing could take place in iron.

Since geochemical and geophysical evidence indicates that the molecular ratio FeO/(FeO+MgO) of the mantle is between[72, 73] 0·1 and 0·2, spectral data for ferromagnesian silicate and oxide phases with 15 mole % iron (II) component are used in the calculations. The results are summarized in table 2. The absorption band at highest energy in the spectrum of each mineral is used, and the pressure required to shift this band to 20 000 cm^{-1} is calculated for each mineral, assuming a pressure increment of 10 cm^{-1} per kilobar in most cases. The experimentally determined[30] shift of 700 cm^{-1} over 173 kb is used for pyrope garnet.

TABLE 2 Minimum pressures required to induce spin pairing in Fe ions in ferromagnesium silicates and oxides

Mineral	Absorption (Å)	Band (cm^{-1})	Pressure to shift band to 20000 cm^{-1} (kb)	Depth in mantle (km)	Reference
olivine (forsterite, $Fo_{85}Fa_{15}$)	8620	11600	840	1800	figure 3(a)
orthopyroxene (bronzite, $En_{85}Fs_{15}$)	5550	18000	200	500	figure 4(a) and figure 5
clinopyroxene (diopside, $Wo_{50}En_{42}Fs_8$)	7350	13600	640	1400	figure 6 12
garnet (pyrope, $Pyr_{85}Alm_{15}$)	12900	7720	1200	2500	figure 9 30
periclase (MgO: Fe^{2+})	10000	10000	1000	2100	76

The results in table 2 suggest that spin pairing within the Fe^{2+} ion in olivine, clinopyroxene, and garnet probably occurs at depths exceeding those at which these phases transform to denser oxide phases[7, 71]. The 'high-spin' to 'low-spin' transition of iron in the orthopyroxene structure might occur at a depth comparable with that at which orthopyroxene is considered to break down to olivine and stishovite[7]. Spin pairing within Fe^{2+} in orthopyroxene could catalyse this phase change in the transition zone. The calculations indicate that there is a strong possibility that spin pairing occurs in octahedrally coordinated Fe^{2+} ions in oxide phases of the lower mantle. These changes of electronic configuration could account for some of the minor irregularities found in the seismic velocity data from the lower mantle[122]. The decrease of magnetic susceptibility due to spin pairing in iron would affect the magnetic properties of the lower mantle.

Two alternative electronic configurations are also possible within transition-metal ions possessing four, five, six, and seven $3d$ electrons (9). These include the Mn^{2+} and Fe^{3+} ($3d^5$), Fe^{2+} and Co^{3+} ($3d^6$), and Co^{2+} ($3d^7$) ions. The Co^{3+} ion in oxide structures has the 'low-spin' configuration at atmospheric pressure, and the Fe^{2+} ion in close-packed oxide structures is postulated to revert to the 'low-spin' state in the lower mantle. The question arises whether spin pairing may be induced in Mn^{2+}, Fe^{3+}, and Co^{2+} in the lower mantle. Estimates from energy level diagrams and spectral data for the ions in octahedral coordination with oxygen[123] indicate that Fe^{3+} and Co^{2+}, but not Mn^{2+}, could acquire 'low-spin' configurations in the lower mantle.

Distributions of transition-metal ions in the mantle

Crystal-field stabilization energies have been used to account for relative enrichments of transition-metal ions during mineral formation[9, 14–17]. The stabilization energies are estimated from the spectra of transition-metal compounds. Ions receiving highest crystal-field stabilization energies in oxide structures include[9] Cr^{3+}, Ni^{2+}, and 'low-spin' Fe^{2+} and Co^{3+}.

The pressure-induced shifts of absorption bands to higher energies and the increased crystal-field splittings in the spectra of transition-metal ions in oxide structures (table 1) imply that crystal-field stabilization energies increase with depth in the mantle. We might expect those transition-metal ions with high crystal-field stabilization energies to be enriched in silicate and oxide phases of the lower mantle.

Evidence in support of this hypothesis comes from the measured element abundances in carbonaceous chondrites, representing the lower mantle, and the estimated composition of the model pyrolite for the upper mantle[72]. The elements Fe, Co, Ni, Cu, Pt, and Au are strongly depleted in the upper mantle compared with their concentrations in carbonaceous chondrites. The depletion of Fe, Co, Ni, and, to a smaller extent, Cu and Pt (belonging to the third transition series) may be explained by the increased crystal-field stabilization energies of the Fe^{2+}, Co^{3+}, and Ni^{2+} ions at depth in the lower mantle. The Cr^{3+} ion might also be expected to be enriched in the lower mantle. However, the chromium content of pyrolite is slightly larger than that of carbonaceous chondrites[72].

Absorption of radiation in the mantle

Since colour centres are destroyed by rising temperature, the two phenomena most likely to cause absorption of radiation in the mantle are electronic processes within the Fe^{2+} ion and charge-transfer transitions between neighbouring ions.

With the transformation of common ferromagnesian silicates, containing iron in distorted environments, to dense oxide phases with iron in more regular octahedral

Optical absorption in silicates 207

coordination, there will be a significant reduction in the amount of radiation absorbed in the infrared region. The pressure-induced shift and intensification of electronic absorption bands indicates that the broad, relatively weak band at 10 000 Å (characteristic of octahedral Fe^{2+} ions in oxide structures on the Earth's surface) will migrate towards the visible and become more intense with increasing depth in the mantle. Furthermore, should Fe^{2+} ions revert to the 'low-spin' state, absorption spectra analogous to those of Co (III) compounds[23, 124] would be anticipated with absorption bands in the visible and near ultraviolet. The high-pressure, high-temperature spectral measurements of olivine[30] suggest that the analogous intense charge-transfer bands in the ultraviolet in ferromagnesian oxide phases of the lower mantle will shift into the visible and near infrared with increasing temperature and pressure. Thus intense or continuous absorption of radiation shorter than at least 10 000 Å might be expected in the lower mantle. This will markedly affect the redistribution of heat by radiation towards the core–mantle boundary.

On the other hand, the presence of olivine, orthopyroxene, clinopyroxene, and garnet, which show prominent electronic absorption bands in the infrared, is expected to influence the heat balance of the upper mantle. The assumption that the material of the upper mantle is a 'grey' body, that is, a material in which absorption is independent of wavelength, is incorrect since electronic absorption bands in the infrared exist in olivine (60 000 Å), orthopyroxene (32 000 Å, $\varepsilon \simeq 4$ l./mole cm; 18 600 Å, $\varepsilon \simeq 8$ l./mole cm), clinopyroxene (22 600 Å, $\varepsilon \simeq 15$ l./mole cm), and garnet (17 000 Å and 13 000 Å, $\varepsilon \simeq 1-2$ l./mole cm). In addition, amphiboles, with intense O–H vibrational bands at about 27 500 Å and 14 000 Å, also absorb radiation in the infrared. These ferromagnesian silicates probably contribute to the efficiency of the upper mantle as a thermal insulator of the Earth's interior[6], and may bring about local and intermittent fusion and evolution of magma.

Finally it should be noted that, since transmission of radiation is reduced considerably during a first-order phase transformation, the transition zone, by its very nature, may shield the upper mantle from radiation from the lower mantle.

Summary

Measurements have been made of absorption spectra of the ferromagnesian silicates olivine, orthopyroxene, clinopyroxene, amphiboles, and garnet at atmospheric pressure and room temperature. The results, together with data from high-pressure and high-temperature spectroscopy of transition-metal compounds, have been used to interpret possible changes of electronic configurations in transition metals, enrichments of transition-metal ions and absorption of radiation in the mantle.

Calculations suggest that spin pairing will occur in the Fe^{2+} ion in oxide phases of the lower mantle, leading to contraction of iron–oxygen distances and a reduced magnetic susceptibility. 'High-spin' to 'low-spin' transitions may also occur in iron of orthopyroxenes in the transition zone, and could catalyse the breakdown of orthopyroxene to olivine (or spinel) plus stishovite. The Fe^{3+} and Co^{2+} ions are also postulated to occur in 'low-spin' states in the lower mantle.

The pressure-induced increases of crystal-field splittings indicate that crystal-field stabilization energies of transition-metal ions increase with depth in the mantle. The

Cr^{3+}, Ni^{2+}, and low-spin' Fe^{2+} and Co^{3+} ions, which have large crystal-field stabilization energies, are predicted to be enriched in the lower mantle.

Electronic absorption bands in the infrared indicate that the common ferromagnesian silicate minerals in the upper mantle may act as thermal insulators of radiation from the lower mantle. The pressure-induced shifts and intensifications of absorption bands suggest that Fe^{2+} ions in oxide phases will absorb visible radiation in the lower mantle. Intense absorption of radiation in the visible and near infrared by charge-transfer bands may seriously affect the heat balance near the core–mantle boundary.

The increased scattering of radiation associated with first-order phase transformations indicates that the transition zone may shield the upper mantle from radiation emitted from the lower mantle.

Acknowledgements

The author gratefully acknowledges scholarship support from the Royal Comission for the Exhibition of 1851 (London), the University of New Zealand, and the University of California. I wish to thank sincerely the following people who generously provided analysed specimens: Dr. S. O. Agrell, Professor R. A. Howie, Professor C. S. Hurlbut, Dr. C. Klein, Jr., Professor H. Kuno, Dr. B. Mason, Dr. R. F. Mueller, and Professor A. Pabst. I am indebted to Professor W. B. White and Professor A. E. Ringwood for the opportunity of seeing their papers in preprint form. Mrs. V. M. Burns gave invaluable assistance in editing and typing the manuscript. The work was supported by grants from the National Science Foundation and Petroleum Research Fund of the American Chemical Society.

References

1. S. P. Clark, Jr., *Bull. Geol. Soc. Am.*, **67**, 1123 (1956).
2. S. P. Clark, Jr., *Trans., Am. Geophys. Union* **38**, 931 (1957).
3. H. A. Lubimova, *Geophys. J.*, **1**, 115 (1958).
4. A. W. Lawson and J. C. Jamieson, *J. Geol.*, **66**, 540 (1958).
5. E. A. Lyubimova, in *Problems of Theoretical Seismology and Physics of the Earth's Interior* (Ed. V. A. Magnitskii), Israel Progressive Science Translation, Jerusalem, 1963, p. 60.
6. W. M. Elsasser, in *Advances in Earth Science* (Ed. P. M. Hurley), Massachusetts Institute of Technology Press, Cambridge, Massachusetts, 1966, p. 461.
7. A. E. Ringwood, in *Advances in Earth Science* (Ed. P. M. Hurley), Masachusetts Institute of Technology Press, Cambridge, Massachusetts, 1966, p. 357.
8. F. R. Boyd, *Science*, **145**, 13 (1964).
9. R. G. Burns and W. S. Fyfe, in *Researches in Geochemistry* (Ed. P. H. Abelson), Vol. 2, Wiley, New York, 1967, p. 259.
10. S. P. Clark, Jr., *Am. Mineralogist*, **42**, 732 (1957).
11. R. G. Burns, Ph.D. Dissertation, University of California, Berkeley, California (1965).
12. W. B. White and K. L. Keester, *Am. Mineralogist*, **51**, 774 (1966).
13. R. G. Burns, *J. Sci. Instr.*, **43**, 58 (1966).
14. R. J. P. Williams, *Nature*, **184**, 44 (1959).
15. R. G. Burns and W. S. Fyfe, *Science*, **144**, 1001 (1964).
16. C. D. Curtis, *Geochim. Cosmochim. Acta*, **28**, 389 (1964).
17. R. G. Burns, *Proc. Intern. Mineral. Assoc., 5th, Symp. II*, in press.
18. F. Seitz, *Rev. Mod. Phys.*, **26**, 7 (1954).
19. C. Kittel, *Introduction to Solid State Physics*, 2nd ed., Wiley, New York, 1956, p. 491.

20. J. H. Shulman and W. D. Compton, *Color Centers in Solids*, Macmillan, New York, 1962.
21. R. G. Burns, *Mineral. Mag.*, **35**, 715 (1966).
22. D. S. McClure, Ed., *Solid State. Phys.*, **9**, 400 (1959).
23. T. M. Dunn, in *Modern Coordination Chemistry* (Eds. J. Lewis and R. G. Wilkins), Interscience, New York, London, 1960, p. 229.
24. L. E. Orgel, *An Introduction to Transition-Metal Chemistry: Ligand-Field Theory*, 2nd ed., Methuen, London, 1966.
25. C. J. Ballhaussen, *Kgl. Danske Videnskab. Selskab, Mat.-Fys. Medd.*, **29**, 4 (1954).
26. K. Hanke, *Beitr. Mineral. Petrog.*, **11**, 535 (1965).
27. R. G. Burns, M. G. Clark, and A. J. Stone, *Inorg. Chem.*, **5**, 1268 (1966).
28. S. Ghose, *Z. Krist.*, **122**, 81 (1965).
29. F. A. Cotton, *Chemical Applications of Group Theory*, Interscience, New York, London, 1963.
30. A. S. Balchan and H. G. Drickamer, *J. Appl. Phys.*, **30**, 1446 (1959).
31. J. E. S. Bradley and O. Bradley, *Mineral. Mag.*, **30**, 26 (1953).
32. B. V. Chesnokov, *Dokl. Akad. Nauk SSSR*, **129**, 647 (1959).
33. B. V. Chesnokov, *Zap. Vses. Mineralog. Obshchestva*, **90**, 700 (1961).
34. E. F. Farrell and R. E. Newnham, *Am. Mineralogist*, **50**, 1972 (1965).
35. S. V. Grum-Grzhimailo, *Acta Physiochim.*, **20**, 933 (1945).
36. S. V. Grum-Grzhimailo, *Dokl. Akad. Nauk SSSR*, **60**, 1377 (1948).
37. S. V. Grum-Grzhimailo, *Zap. Vses. Mineralog. Obshchestva*, **82**, 142 (1953).
38. S. V. Grum-Grzhimailo, *Mineralog. Sb., L'vovsk. Geol. Obshchestvo pri L'vovsk. Gos. Univ.*, **8**, 281 (1954).
39. S. V. Grum-Grzhimailo, *Tr. Inst. Kristallogr., Akad. Nauk SSSR*, **12**, 79 (1956).
40. S. V. Grum-Grzhimailo, *Zap. Vses. Mineralog. Obshchestva*, **87**, 129 (1958).
41. S. V. Grum-Grzhimailo, *Materialy Vses. Nauchn.-Issled. Geol. Inst.*, **40**, 57 (1960).
42. S. V. Grum-Grzhimailo, *Kristallografiya*, **6**, 67 (1961) (*Soviet Phys.—Cryst.* (*English Transl.*), **6**, 54 (1961)).
43. S. V. Grum-Grzhimailo, *Zap. Vses. Mineralog. Obshchestva*, **91**, 86 (1962).
44. S. V. Grum-Grzhimailo, L. I. Anakina, E. N. Belova, and K. Tolstikhina, *Mineralog. Sb., L'vovsk. Geol. Obshchestvo pri L'vovsk. Gos. Univ.*, **9**, 90 (1955).
45. S. V. Grum-Grzhimailo, N. A. Brilliantov, D. T. Sviridov, and O. N. Sukhanova, *Opt. i Spektroskopiya*, **14**, 228 (1963) (*Opt. and Spectr.* (*USSR*) (*English Transl.*), **14**, 118 (1963)).
46. S. V. Grum-Grzhimailo and M. A. Enikova, *Kristallografiya*, **2**, 186 (1957).
47. S. V. Grum-Grzhimailo and L. A. Perneva, *Tr. Inst. Kristallogr. Akad. Nauk SSSR*, **12**, 85 (1956).
48. S. V. Grum-Grzhimailo, R. K. Sviridova, O. N. Sukhanova, and M. M. Kapitonova, *Opt. i Spektroskopiya*, **13**, 133 (1962) (*Opt. Spectr.* (*USSR*) (*English Transl.*), **13**, 72 (1962)).
49. E. Kolbe, *Neues Jahrb. Mineral., Abhandl.*, **69**, 183 (1935).
50. W. Low and M. Dvir, *Phys. Rev.*, **119**, 1587 (1960).
51. N. M. Melankholin, *Zap. Vses. Mineralog. Obshchestva*, **75**, 89 (1946).
52. A. Neuhaus, *Z. Krist.*, **113**, 195 (1960).
53. A. Neuhaus and W. Richartz, *Angew. Chem.*, **70**, 430 (1958).
54. O. Schmidt-DuMont and N. Moulin, *Z. Anorg. Allgem. Chem.*, **314**, 260 (1961).
55. K. H. Schüller, *Beitr. Mineral. Petrog.*, **6**, 112 (1958).
56. D. M. Gruen and M. Fred, *J. Am. Chem. Soc.*, **76**, 3850 (1954).
57. V. Stubican and R. Roy, *Am. Mineralogist*, **46**, 32 (1961).
58. D. S. McClure, *J. Phys. Chem. Solids*, **3**, 311 (1957).
59. R. G. Burns, R. H. Clark, and W. S. Fyfe, *Chemistry of the Earth's Crust*, Vol. 2, Science Press, Moscow, 1964, p. 88.
60. R. Tsuchida and M. Kobayashi, *Bull. Chem. Soc. Japan*, **13**, 618 (1937).
61. N. M. Melankholin, *Zap. Vses. Mineralog. Obshchestva*, **85**, 218 (1956).
62. V. N. Vishnevskii and L. K. Klimovskaya, *Ukr. Fiz. Zh.*, **3**, 239 (1958).
63. R. G. Burns and R. G. J. Strens, *Mineral. Mag.*, **36**, 204 (1967).
64. L. J. Robinson, M.Sc. Thesis, University of Alaska (1966).

65. R. A. Fitch, T. E. Slykhouse, and H. G. Drickamer, *J. Opt. Soc. Am.*, **47**, 1015 (1957).
66. A. S. Balchan and H. G. Drickamer, *Rev. Sci. Instr.*, **31**, 511 (1960).
67. K. E. Bullen, *Bull. Seismol. Soc. Am.*, **30**, 235 (1940).
68. K. E. Bullen, *An Introduction to the Theory of Seismology*, Cambridge University Press, Cambridge, 1953.
69. D. H. Green and A. E. Ringwood, *J. Geophys. Res.*, **68**, 937 (1963).
70. A. E. Ringwood, *Nature*, **178**, 1303 (1956).
71. A. E. Ringwood, *Earth Planetary Sci. Letters*, **2** (1967).
72. A. E. Ringwood, in *Advances in Earth Science* (Ed. P. M. Hurley), Massachusetts Institute of Technology Press, Cambridge, Massachusetts, 1966, p. 287.
73. B. Mason, *Nature*, **211**, 616 (1966).
74. C. Bates, W. B. White, and R. Roy, *J. Inorg. Nucl. Chem.*, **28**, 397 (1966).
75. G. A. Slack, *Phys. Rev.*, **134**, 1268 (1964).
76. W. Low and M. Weger, *Phys. Rev.*, **118**, 1130 (1960).
77. A. Pabst, *Am. Mineralogist*, **28**, 372 (1943).
78. M. G. Clark and R. G. Burns, *J. Chem. Soc.*, **1967**, 1034.
79. W. B. White and K. L. Keester, *Am. Mineralogist*, in press (1967).
80. G. M. Bancroft, R. G. Burns and R. A. Howie, *Nature*, **213**, 1221 (1967).
81. B. E. Warren and W. L. Bragg, *Z. Krist.*, **69**, 168 (1928).
82. S. Ghose and E. Hellner, *J. Geol.*, **67**, 691 (1959).
83. S. Ghose, *Acta Cryst.*, **14**, 622 (1961).
84. G. M. Bancroft, R. G. Burns, and A. G. Maddock, *Am. Mineralogist*, **52**, 1009 (1967).
85. R. G. Burns and R. G. J. Strens, *Science*, **153**, 890 (1966).
86. J. Zussman, *Acta Cryst.*, **8**, 301 (1955).
87. G. V. Gibbs and J. V. Smith, *Am. Mineralogist*, **50**, 2023 (1965).
88. H. G. Drickamer and J. C. Zahner, *Advan. Chem. Phys.*, **4**, 161 (1962).
89. H. G. Drickamer, in *Solids Under Pressure* (Eds. W. Paul and D. M. Warschauer), McGraw-Hill, New York, London, 1963, p. 357.
90. S. E. Babb, Jr., and W. W. Robertson, in *High Pressure Physics and Chemistry* (Ed. R. S. Bradley), Vol. 1, Academic Press, London, New York, 1963, p. 375.
91. H. G. Drickamer, *Solid State Phys.*, **17**, 1 (1966).
92. T. E. Slykhouse and H. G. Drickamer, *J. Phys. Chem. Solids*, **7**, 207 (1958).
93. H. L. Suchan, A. S. Balchan, and H. G. Drickamer, *J. Phys. Chem. Solids*, **10**, 343 (1959).
94. H. L. Suchan and H. G. Drickamer, *J. Phys. Chem. Solids*, **11**, 111 (1959).
95. J. C. Zahner and H. G. Drickamer, *J. Phys. Chem. Solids*, **11**, 92 (1959).
96. H. L. Suchan, S. Wiederhorn, and H. G. Drickamer, *J. Chem. Phys.*, **31**, 355 (1959).
97. A. S. Balchan and H. G. Drickamer, *J. Phys. Chem. Solids*, **19**, 261 (1961).
98. D. R. Stephens and H. G. Drickamer, *J. Chem. Phys.*, **30**, 1518 (1959).
99. W. H. Bentley and H. G. Drickamer, *J. Chem. Phys.*, **34**, 2200 (1961).
100. A. S. Balchan and H. G. Drickamer, *J. Chem. Phys.*, **35**, 356 (1961).
101. I. S. Jacobs, *Phys. Rev.*, **93**, 993 (1954).
102. E. Burstein, J. J. Oberley, and J. W. Davisson, *Phys. Rev.*, **85**, 729 (1952).
103. W. G. Maish and H. G. Drickamer, *J. Phys. Chem. Solids*, **5**, 328 (1958).
104. R. A. Eppler and H. G. Drickamer, *J. Chem. Phys.*, **32**, 1418 (1960).
105. S. Minomura and H. G. Drickamer, *J. Chem. Phys.*, **34**, 670 (1961).
106. D. R. Stephens and H. G. Drickamer, *J. Chem. Phys.*, **34**, 937 (1961).
107. D. R. Stephens and H. G. Drickamer, *J. Chem. Phys.*, **35**, 424 (1961).
108. D. R. Stephens and H. G. Drickamer, *J. Chem. Phys.*, **35**, 427 (1961).
109. S. Minomura and H. G. Drickamer, *J. Chem. Phys.*, **35**, 903 (1961).
110. J. C. Zahner and H. G. Drickamer, *J. Chem. Phys.*, **35**, 1483 (1961).
111. R. E. Tischer and H. G. Drickamer, *J. Chem. Phys.*, **37**, 1554 (1962).
112. R. W. Parsons and H. G. Drickamer, *J. Chem. Phys.*, **29**, 930 (1958).
113. D. R. Stephens and H. G. Drickamer, *J. Chem. Phys.*, **35**, 429 (1961).
114. K. B. Keating and H. G. Drickamer, *J. Chem. Phys.*, **34**, 140, 143 (1961).
115. R. Berman, *Am. Mineralogist*, **42**, 191 (1957).
116. S. Wiederhorn and H. G. Drickamer, *J. Appl. Phys.*, **31**, 1665 (1960).

117. T. E. Slykhouse and H. G. Drickamer, *Colloq. Intern. Centre. Natl. Rech. Sci.* (*Paris*), **1959**, 128.
118. M. Pagannone and H. G. Drickamer, *J. Chem. Phys.*, **43**, 4064 (1965).
119. W. S. Fyfe, *Geochim. Cosmochim. Acta*, **19**, 141 (1960).
120. R. G. J. Strens, in *The Application of Modern Physics to the Earth and Planetary Interiors* (Ed. S. K. Runcorn), Wiley, New York, London, to be published.
121. R. G. J. Strens, *Chem. Commun.*, **1966**, 777.
122. B. Gutenberg, *Trans. Am. Geophys. Union*, **39**, 486 (1958).
123. J. S. Griffith, *The Theory of Transition-Metal Ions*, Cambridge University Press, Cambridge, 1964, pp. 261, 262, 310.
124. C. J. Ballhaussen, *Introduction to Ligand Field Theory*, McGraw-Hill, New York, 1962, p. 259.

3

R. G. J. STRENS*

Department of Earth Sciences
The University
Leeds, England

The nature and geophysical importance of spin pairing in minerals of iron (II)

Introduction

In his book on the nature of the chemical bond, first published in 1939, Pauling[1] noted that certain transition-metal ions with incomplete d shells could exist in alternative electronic configurations, characterized by different numbers of unpaired electrons, and having different ionic radii, magnetic moments, and chemical properties. In particular, the 'low-spin' form of a d^6 ion such as Fe^{2+} or Co^{3+} (with no unpaired electrons) has a much smaller ionic radius than the 'high-spin' form.

Fyfe in 1960 argued[2] from this decrease in radius on pairing that the low-volume low-spin form of an iron mineral would be stabilized relative to the high-spin form by high pressures, and calculated that spin pairing would occur in (Fe,Mg) olivine (or its high-pressure equivalent) at moderate mantle pressures.

Strens in 1966 provided[3] experimental confirmation of the idea that high pressures would induce spin pairing in oxygen compounds of iron (II) by locating the spin-state transition in the mineral gillespite. The present chapter extends this work by considering the conditions under which such transitions will occur, by calculating the probable transition pressure at mantle temperatures, and by evaluating the geophysical consequences.

By far the commonest transition-metal ion in mantle materials is Fe^{2+}, which is characterized by an incomplete d shell containing six electrons distributed over five d-orbital energy levels. Since each such level can hold two electrons with opposed spins, the Fe^{2+} ion can exist in three alternative electronic configurations, possessing 4, 2, and 0 unpaired electrons. The stable configuration is determined by the energy separations of the various levels, and these in turn depend on the strength of the 'crystal field' in which the ion is placed.

For any given combination of metal and anion (in this case Fe^{2+} and O^{2-}), the nature and magnitude of the d-orbital splittings, and thus the stable electronic configuration, is determined by the nature of the coordination polyhedron about the metal ion, and by the Fe–O distance (figure 1). The field strength, and hence the

---

* Present address: Department of Geophysics and Planetary Physics, School of Physics, University of Newcastle upon Tyne, England.

splittings, normally decrease along the series square-planar > octahedral > cubic eightfold > tetrahedral coordination, and increase very rapidly, approximately as r^{-5}, with decreasing Fe–O distance (r). High pressures, by reducing r, increase the splittings and tend to stabilize the form of lowest spin multiplicity, which is also that of smallest radius.

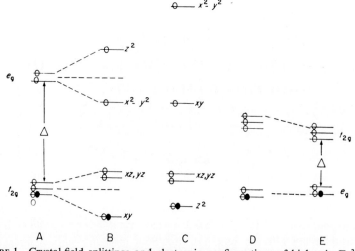

FIGURE 1 Crystal-field splittings and electronic configurations of high-spin Fe^{2+} in various coordinations as follows: A, regular octahedron; B, tetragonally distorted octahedron; C, square-planar coordination in gillespite; D, cubic eightfold; E, tetrahedral. Splittings approximately to scale, ignoring Jahn–Teller distortions in A, D, and E.

Under surface conditions of temperature and pressure, iron (II) in oxygen compounds is almost invariably in octahedral coordination, although examples of square-planar, tetrahedral, and eightfold coordination are also known. If the sequence of structure types in the mantle is that usually assumed, i.e.

olivine → spinel → ilmenite + NaCl ——→ rutile + NaCl

pyroxene ——————→ ilmenite ——————→ rutile + NaCl

then octahedral coordination of iron persists to the highest mantle pressures. Average metal–oxygen distances vary little between different iron (II) compounds under the same (P, T) conditions, implying approximate constancy of mean crystal-field splittings.

All known oxygen compounds of iron (II) exist in the high-spin form under surface conditions, with the possible exception of deerite. Carmichael, Fyfe, and Machin[4] have suggested, on the basis of its low magnetic susceptibility, that deerite contains a proportion of low-spin iron, but the alternative view that it is a ferrimagnetic with only short-range magnetic order is accepted here, as it yields a calculated susceptibility within 1% of the observed value, and accounts for the absence of both hyperfine splitting and recognizable low-spin iron (II) peaks in the Mössbauer spectrum.

Some Aspects of Ligand Field Theory

In field-free space the five d-electron orbitals of a transition-metal ion are all of equal energy, but in a coordination polyhedron those electrons in orbitals which point towards the negatively charged ligands are more strongly repelled than those in orbitals which project between the ligands, and the d orbitals split into high- and low-energy groups[5-7]. In an octahedron, these consist of a set of two high-energy (e_g) orbitals, sometimes designated d_{z^2} and $d_{x^2-y^2}$, separated by an energy Δ from a set of three low-energy (t_{2g}) orbitals, designated d_{xy}, d_{xz}, and d_{yz}. The former are directed along the tetrad axes (x, y, z) of the octahedron, i.e. towards the ligands, and the latter along diad axes, i.e. between the ligands. Spin pairing occurs when the crystal field splitting exceeds the energy required to pair the electrons in the low-energy orbitals (figure 2). Spin-state equilibria become important when the difference between

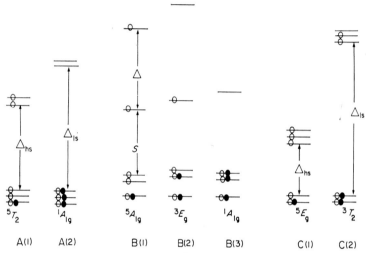

FIGURE 2 Electronic configurations of Fe^{2+} in various coordinations and spin states as follows: A, regular octahedral; B, square-planar; and C, cubic eightfold or tetrahedral coordination. The d-orbital sequence shown in B is that in gillespite, and $(\Delta+2S)/2$ is approximately equivalent to octahedral Δ for present purposes

the energies of high- and low-spin states becomes comparable with kT. The energies of d^6 ground states in an octahedral crystal field have been tabulated by Griffith[7] in terms of the Racah parameters B and C. The high-spin ($^5T_2$) and low-spin ($^1A_1$) states are found to have equal energy when

$$2\Delta_p = 5B+8C$$

It can be shown[7] that the intermediate-spin (^3E_g) state is unstable in all but the most distorted octahedral environments. Free-ion values of B and C are not accurately known, but taking $B = 1060$ cm^{-1} and $C = 3900$ cm^{-1} from the book by Griffith[7], we find $\Delta_p = 18\,250$ cm^{-1} in the free Fe^{2+} ion. This must be reduced by the nephelauxetic ratio, $\beta = B_{crystal}/B_{free\,ion}$, to find Δ_p in the crystal. The best value to adopt for β is itself uncertain but, taking 0·65 as a reasonable estimate, $\Delta_p = 12\,000$ cm^{-1}. This compares with the equivalent of 11 500 cm^{-1} for high-spin iron (II) in gillespite, and about 9100 cm^{-1} for high-spin iron (II) in octahedral coordination with oxygen. With Δ varying as r^{-5}, reasonable values of the compressibility then lead to spin pairing in square-planar and octahedral complexes at pressures below 0·1 and 0·5 mb respectively.

By contrast, the splitting required to stabilize the low-spin ($^3T_2$) form in tetrahedral or cubic eightfold coordination is very much larger:

$$\Delta_p = \beta(6B+5C) = 16\,800 \text{ cm}^{-1}$$

which compares with observed splittings of only 4500 cm^{-1} for tetrahedral iron (II) in staurolite[8], and just over 6000 cm^{-1} for eight-coordinated iron (II) in garnet[9]. No reasonable behaviour of the compressibilities will permit pairing in these minerals at pressures available in the mantle.

Experimental Data

The spin-state transition in gillespite

The unusually large crystal-field splitting resulting from the square-planar coordination of Fe^{2+} in the rare mineral gillespite ($BaFe^{II}_{hs}Si_4O_{10}$) made it particularly suitable for experiments on pressure-induced pairing in iron–oxygen systems[3]. A clearly visible, instantaneous, and reversible colour change (red to colourless) occurred when a (001) cleavage flake of gillespite was subjected to 26 kb mean applied pressure in a Van Valkenburg diamond-anvil apparatus. Under these conditions the high-pressure form covered about half the diameter of the anvil, and was separated from the rim of unaltered gillespite by a narrow strongly coloured zone, interpreted as a region of spin-state equilibrium. For a flake of the thickness used (30 μm) the true pressure at this point was probably near 50 kb, and the width of the transition zone certainly less than 5 kb.

Absorption spectra (3500 to 10 000 Å) of the low-pressure form show a single band ($\varepsilon = 2\cdot3$) which broadens, intensifies, and shifts to the blue with increasing pressure, disappearing at the transition point to give the featureless spectrum of the high-pressure form. No further changes occurred as the pressure was increased slowly to 80 kb, and the gillespite was recovered as the low-pressure form after the run.

These and other data[3,10] are entirely compatible with spin pairing $^5A_{1g}-{}^1A_{1g}$ as the cause of the transition, the ^3E_g (intermediate-spin) state apparently being unstable even in square-planar coordination in gillespite. Assuming a transition pressure of 50 kb, reasonable values of the compressibility suggest that pairing occurred at a point equivalent to an octahedral Δ of 11 900 cm^{-1}, in good agreement with the preceding calculation, and with experimental data for the organic complexes discussed below. As thick flakes could not be used in the Van Valkenburg apparatus, the weak spin-forbidden transitions in the low-spin form were not studied, and no value of Δ_{ls} could be obtained.

Spin-state transitions in $[Fe^{II}phen_2X_2]$ complexes

Complexes of the general type $[Fe^{II}phen_2X_2]$ may be prepared[11] in high- or low-spin forms, depending on the field strength produced by the ligand X. If X is a 'strong' ligand the complex is low spin, whilst if X is 'weak' a high-spin complex is formed. Ligands of intermediate strength, including (NCS)$^-$ and (NCSe)$^-$ yield complexes in which Δ is very close to the value at which pairing occurs.

Measurements[11] of the magnetic susceptibility, and of the infrared, visible, and Mössbauer spectra, indicate that both the (NCS) and the (NCSe) complexes are predominantly high spin at 300 °K, and low spin at 77 °K. Although the spin-state equilibrium results in anomalous magnetic behaviour over a wide temperature range, there is a well-defined transition temperature at which the magnetic moment of these complexes decreases from $\sim 4\cdot 9\ \mu_B$ to $\sim 1\cdot 7\ \mu_B$ (where μ_B is a Bohr magneton) within 3 degK, presumably owing to cooperative effects.

The spectral studies[11] show that $\Delta_{hs} = 11\ 900\ cm^{-1}$, in good agreement with both the calculated value of $12\ 000\ cm^{-1}$, and the estimated pairing value in gillespite; the low-spin forms have $\Delta_{ls} = 16\ 300\ cm^{-1}$.

(P, V) data for FeO and Fe_2SiO_4

The absence of any marked anomaly in the room-temperature compressibility data of Clendenen and Drickamer[12] sets a lower limit of 300 kb on the transition pressure in wustite. An upper limit of about 550 kb is implied by the shock wave studies of McQueen and Marsh[13] on fayalite, in which the Fe–O distance is close to that in wustite. Their density of 6·17 at 577 kb is so high as to imply that the transition has already occurred at that pressure.

Calculation of the Transition Pressure at Mantle Temperatures

The simplest compound containing Fe_{hs}^{2+} in octahedral coordination with oxygen for which extensive (P, V) data exist is wustite (FeO with the NaCl structure). Wustite prepared at 1 atm has a defect structure, with two Fe^{3+} ions and a vacancy replacing three Fe^{2+} ions to give a range of compositions $Fe_{1-x}O$, where $0\cdot 05 < x < 0\cdot 15$. High pressures will probably tend to reduce the defect concentration, and this calculation will be based on an idealized non-defect FeO, for which a short extrapolation[14] gives $r = a/2 = 2\cdot 166$ Å.

The data of Cotton and Meyer[15] suggest an average value of $9500\ cm^{-1}$ for the transition from the lowest t_{2g} level to the mean of the two e_g levels in several $[Fe_{hs}^{2+}(H_2O)_6]$ complexes. Applying an estimated correction of $400\ cm^{-1}$ for the splitting of the t_{2g} levels gives $\Delta = 9100\ cm^{-1}$ for Fe_{hs}^{2+} in octahedral coordination with (water) oxygen, compared with a value of $9120\ cm^{-1}$ found by Tischer and Drickamer[16] for a silicate glass. Mean Fe–O distances in silicates and oxides range from 2·14 to 2·18 Å, so that the convenient assumption that $\Delta = 9100\ cm^{-1}$ at $r = 2\cdot 166$ Å should not be far in error. Assuming now that Δ varies as r^{-5} over a wide range of values of r, it is found that spin pairing will occur in FeO when r has been reduced to 2·05 Å (and $\Delta_p = 11\ 900\ cm^{-1}$), at which point r is reduced discontinuously to 1·92 Å ($\Delta_{ls} = 16\ 300\ cm^{-1}$). A short extrapolation of the experimental (P, V) data[12] for $Fe_{0\cdot 95}O$ suggests that the pairing value of 2·05 Å will be reached in non-defect FeO at about 340 kb.

The change in r between high-spin iron at one atmosphere and low-spin iron at 340 kb is then 0·25 Å, of which half represents the contraction of the Fe–O bond between 1 atm and the transition pressure (P_t), and the other half the discontinuity in r at the transition. The overall change (Δr) can be calculated by an independent method, providing a check on the assumptions used:

$$\Delta r = (Mn–S – Fe–S) - (Mn–O – Fe–O) = 0\cdot 26\ \text{Å}$$

where Mn–S and Fe–S are the metal–sulphur distances in (high-spin) MnS_2 and (low-spin) FeS_2, and the second term is a correction for the difference between the radii of high-spin manganese and iron. The agreement is satisfactory.

On the basis of the calculated values of **r**, the volume change (ΔV_t) is $1\cdot77$ cm^3/mole. Calculation of the reaction slope, and thus of the pressure at which the transition will occur on the geotherm, requires also a knowledge of the entropy change. This may be divided into two parts, ΔS_v being that attributable to the change in vibrational frequencies resulting from the shortening and strengthening of the Fe–O bonds, and ΔS_{el} the change in the configurational entropy of the d-electron distribution. From classical thermodynamics

$$(\partial S/\partial V)_T = \alpha/\chi$$

whence $\Delta S_v = 1\cdot6$ entropy units.

The electronic entropy contribution is zero in low-spin FeO, in which the six d electrons are localized in the three t_{2g} levels, but it is large in high-spin FeO in which the 'sixth' d electron has a choice between three almost equienergetic t_{2g} orbitals, each of which will be occupied one-third of the time by an electron, and two-thirds of the time by a 'hole':

$$\Delta S_{el} = 3R(\tfrac{1}{3}\ln\tfrac{1}{3} + \tfrac{2}{3}\ln\tfrac{2}{3}) = 3\cdot8 \text{ entropy units}$$

Taking the values of ΔV_t and ΔS calculated above, the slope of the transition curve is found to be $0\cdot127$ kb/degC, i.e. the equation of the reaction is

$$P_t(\text{kb}) = 337 + 0\cdot127 T(°\text{C})$$

For geothermal gradients of the type usually assumed, this cuts the geotherm at about 590 kb, equivalent to a depth of 1400 km. Since average metal–oxygen distances and local compressibilities of Fe–O bonds vary little between different structures[16], this calculation should apply without much modification to other minerals containing iron (II) in octahedral coordination with oxygen. It has not been possible to estimate the limits of error of the calculation, or to find the width of the zone of spin-state equilibrium, although the experimental evidence of a narrow transition zone in gillespite, and of a well-defined transition temperature in the $[Fe^{II}\text{phen}_2X_2]$ complexes suggests that cooperative effects limit its width.

The agreement between the present results and Fyfe's[2] original calculation that the transition in olivine would occur at a depth of about 1400 km is purely coincidental. Fyfe equated $(P_t\Delta V_t)$ with the difference between twice the 1 atm Δ of olivine, and twice the free-ion value of Δ_p, thus ignoring the considerable work done in bringing the sample to the transition pressure, and the fact that $\beta \neq 1$. In addition, the effect of temperature on the transition pressure was ignored, and the inadequate experimental data available at that time did not permit accurate evaluation of the volume changes.

Digression on the Olivine–spinel Transition

The slopes of most solid–solid transitions are nearly linear, reflecting the approximate constancy of volume and entropy changes over wide (P, T) ranges. A curious feature of the fayalite–spinel transition noted by Akimoto, Komado, and Kushiro[17] is the marked change in slope of the transition curve above 1150 °C, which implies either a rather sudden increase in ΔV, or more likely a decrease in ΔS, with increasing temperature.

The importance of entropy terms arising from the distribution of the d electrons over the t_{2g} levels has already been noted in connection with the spin-pairing transition, but analogous effects are to be expected in all reactions involving iron (II) minerals. The $[FeO_6]$ octahedra in olivine are more strongly distorted and of lower symmetry than those in spinel, so that the splittings of the t_{2g} levels will be larger in olivine. This in turn implies that the difference between the electronic entropies of fayalite and spinel, and thus the overall entropy change of the reaction, will decrease with increasing temperature, accounting qualitatively for the observed curvature. Rough calculations indicate that an effect of the right magnitude would be obtained with t_{2g} splittings of the order of 1000 cm^{-1} in olivine, and rather less in spinel.

Geophysical Consequences of the Spin-state Transition

Depending on the phase involved, the spin-state transition should decrease the volume of a system with Fe:Mg = 20:80 by from 1 to 3%, causing corresponding changes in volume-dependent physical properties. A minor seismic discontinuity would thus be expected at a depth of 1400 km, although behaviour in this region might be complicated by the postulated increase in mean atomic weight (discussed below), and by the spin-state equilibrium.

At low pressures, spin-allowed d–d transitions between the t_{2g} and e_g levels of high-spin iron (II) lead to strong absorption between 9000 and 12 000 cm^{-1}. Increasing pressure and decreasing r should shift this absorption into the 12 000–15 000 cm^{-1} region at the transition pressure. At higher pressures the spectrum of low-spin iron (II) compounds displays only weak spin-forbidden peaks, so that the transition effectively eliminates absorption in the wide window between the charge-transfer edge at 4 to 7 ev (30 000–50 000 cm^{-1}) and the vibrational absorption at ~ 1500 cm^{-1}. Since this is the region in which the peak of the black-body radiation lies for inferred mantle temperatures, the transition would presumably increase the effectiveness of radiative transfer, and thus tend to decrease the temperature gradient within the lower mantle (> 1400 km).

It is often true that, of two isomorphous compounds A.X and B.X containing cations (A, B) of the same charge, that with the larger cation has the lower melting point[18]. Thus, at low pressures the iron end-members of common solid solution series have lower melting points than the magnesium end-members, and crystallization leads to enrichment of the solid in magnesium and of the liquid in iron. At pressures above the transition point, low-spin iron has a smaller ionic radius than magnesium, possibly leading to a reversal of the melting point relations, with consequent enrichment of the lower mantle in iron, and the production of a relatively magnesian upper mantle from the residual liquids.

In addition to these major effects, discontinuities (modified by the spin-state equilibrium) will exist in the magnetic susceptibility and electrical conductivity of iron (II) minerals at the transition pressure.

When combined with the results of detailed seismic work, experiments now in progress should lead to a much fuller understanding of spin-state transitions in planetary systems. In the meantime, it must be remembered that the calculations made here are based on many assumptions, some of which may be modified by future work: the results must therefore be used cautiously until they have been more fully tested.

Acknowledgements

In preparing this work, I have been particularly indebted to G. A. N. Connell, who provided the apparatus and gave generously of his time during the experiments on gillespite, to G. M. Bancroft, for unpublished Mössbauer data on deerite, and to R. G. Burns who supplied the gillespite.

References

1. L. Pauling, *Nature of the Chemical Bond*, 1st ed., Cornell University Press, New York, 1939.
2. W. S. Fyfe, *Geochim. Cosmochim. Acta*, **19**, 141 (1960).
3. R. G. J. Strens, *Chem. Commun.*, **1966**, 777.
4. I. S. E. Carmichael, W. S. Fyfe, and D. J. Machin, *Nature*, **211**, 1389 (1966).
5. L. E. Orgel, *Introduction to Transition-metal Chemistry: Ligand Field Theory*, Methuen, London, 1966.
6. B. N. Figgis, *Introduction to Ligand Fields*, Interscience, New York, 1966.
7. J. S. Griffith, *The Theory of Transition-metal Ions*, 1st ed., Cambridge University Press, London, 1964.
8. G. M. Bancroft and R. G. Burns, *Am. Mineralogist*, in press (1967).
9. P. Bloomfield, A. W. Lawson, and C. Rey, *J. Chem. Phys.*, **34**, 749 (1961).
10. R. G. Burns, M. G. Clark, and A. J. Stone, *Inorg. Chem.*, **5**, 1268 (1966).
11. E. Konig and K. Madeja, *Inorg. Chem.*, **6**, 48 (1967).
12. R. L. Clendenen and H. G. Drickamer, *J. Chem. Phys.*, **44**, 4223 (1966).
13. R. G. McQueen and S. P. Marsh, quoted by F. Birch, in *Handbook of Physical Constants* (Ed. S. P. Clark, Jr.) (*Geol. Soc. Am., Mem.*, **97** (1966)).
14. A. F. Wells, *Structural Inorganic Chemistry*, 3rd ed., Clarendon Press, Oxford, 1962.
15. F. A. Cotton and M. D. Meyer, *J. Am. Chem. Soc.*, **82**, 5023 (1960).
16. R. E. Tischer and H. G. Drickamer, *J. Chem. Phys.*, **37**, 1554 (1962).
17. S. Akimoto, E. Komada, and I. Kushiro, *J. Geophys. Res.*, **72**, 679 (1967).
18. V. M. Goldschmidt, *Geochemistry*, Clarendon Press, Oxford, 1954.

B

SOLID STATE PHYSICS AND GEOPHYSICS

III. Theory of creep applied to the Earth's mantle

1

W. M. ELSASSER*

*Department of Geology
Princeton University
Princeton, New Jersey, U.S.A.*

Convection and stress propagation in the upper mantle

Introduction and Summary

Continental topography and land-based geology have failed to exhibit quantitative relationships of large dimensions. In more recent times a number of such correlations have been brought to light by ocean-bottom topography and geology. Here belong the characteristic locations of the submarine ridges with respect to continental margins, the fairly uniform depths of the abyssal plains, and the indications that the ocean floor has moved and is still moving in a direction away from the active submarine ridges. These correlations present evidence for definite dynamical processes in the upper mantle; the latter are now often interpreted as due in the main to thermal convection currents.

The upper mantle is here taken as relatively shallow, of about 800 km depth. Following Birch we assume that the lower mantle has been drained of its radioactivity and possibly of much of its lighter constituents in a comparatively early phase of the Earth's history. This was accompanied by general convection in the lower mantle, leaving the latter with a low (adiabatic) temperature gradient and no heating, so that it is now convectively inert.

The most striking feature of upper-mantle convection is the preponderance of horizontal motions. This can only in part be attributed to the relative shallowness of the layer; it is mainly caused by a rather pronounced minimum of the 'viscosity' between about 100 and 200 km depth. The existence of such an 'asthenosphere' is readily explained by the fact that the closest approach to melting must occur in this region. It may be shown by mechanical arguments that horizontal sliding of the top layer, here called the 'tectosphere', can be more easily achieved than circulation in the material underneath.

Mechanical stresses are preferentially transmitted horizontally, along the tectosphere. This accounts for the otherwise quite unintelligible correlation between ridges, considered as upwellings in the mantle, and continental margins which are obviously rather shallow features on a mantle-wide scale. We conclude that the convective circulation in the upper mantle is largely controlled by irregularities of its free upper surface, in a manner inverse to the way in which thermal convection of a fluid is often controlled by the corrugations of its lower boundary.

* Present address: Institute for Fluid Dynamics and Applied Mathematics, University of Maryland, College Park, Maryland, U.S.A.

In the past, mantle convection was often dealt with by models drawn from fluid mechanics. But there are great differences between solids in creep and fluids. While not using calculations from the difficult theory of nonlinear creep, the author has tried to be extremely circumspect in applying fluid-mechanical concepts to the mantle. In this the author had the counsel of his colleague, E. Orowan, whose knowledge of the mechanics of solids proved repeatedly invaluable.

Thermal Stratification of the Mantle

The key to all models of convection in the Earth's mantle is the distribution of temperature with depth. Unfortunately, we have no quantitative knowledge of the temperature that would be even remotely comparable with our precise knowledge of many mechanical properties as obtained from seismic data, and we must be satisfied with semiquantitative estimates. If the temperature gradient as measured near the surface (some 20-30 degK/km) continued downwards, the mantle would be molten at a depth of less than 100 km. In order to avoid this conclusion the temperature gradient must be assumed to decrease rapidly with increasing depth: the temperature–depth curve must flatten out, that is, it must have a relatively pronounced bend at rather moderate depth (see figure 1, below). Even a qualitative study of the possible causes of this bend turns out to be instructive. We find that there are concurrently three such causes.

(i) In the first place, there is the gradual transition with rising temperature from ordinary conductive heat transport to radiative transport, which at higher temperatures is more efficient than lattice conduction; at sufficiently high temperatures, heat is carried almost entirely by radiation. Depending on the infrared transparency of the material, this change of mechanism occurs for silicates in general in the neighbourhood of 1600–2200 °K. Calculations of heat distribution in the Earth's interior which have been based on this phenomenon (e.g. Lubimova[1], MacDonald[2]) show very clearly the appearance of a rather sharp bend in the temperature curve owing to this effect alone.

(ii) Next, there is the concentration of the radioactive heat sources. Birch[3] has long since advocated the view, also propounded by Jeffreys, that the radioelements, owing to their large ionic volume, have tended to rise in the mantle and are now very strongly concentrated in the upper mantle. According to Birch this process of segregation should have taken place in a time of the order of a thousand million years from the beginning stage of the Earth (i.e. the stage in which the core was formed). Patterson and Tatsumoto[4] have interpreted their studies on the distribution of lead isotopes in an entirely similar fashion; they say, furthermore, that such a model is about the only one compatible with their experimental results. Another, independent argument pointing in the same direction comes from the estimates of the central temperature of the Earth that have been based on the existence of a solid inner core. A number of authors agree that this fixes the temperature at the bottom of the mantle as not much in excess of, say, 4000 °K. Since the mantle is nearly 3000 km thick and the temperature must certainly rise to 1500 °K rather close to the surface, the *average* temperature gradient over all but the topmost layers of the mantle should be below 1 degK/km; the gradient in the lowest 2000 km, say, should be appreciably less than 1 degK. Again, by a totally different approach, on the basis of electrical conductivity data, Tozer[5] has estimated the temperature of the lower mantle as of the order of 4500 °K, in fair agreement with the preceding value considering the crudeness of the arguments. Such relatively low temperatures and low gradients are hard to reconcile with the idea that the lower mantle has retained most of its original heat, as well as its radioactivity, and at the same time has never been convecting. If there was (solid-state) convection in the past, the lower mantle is also more likely than not to have undergone a degree of chemical segregation which would certainly include appreciable upward migration of radioelements—a point to which we shall revert presently. Furthermore, if one wanted to justify as low a temperature gradient as 1 degK/km in terms of radiative transport alone, the mantle would have to be highly transparent in the infrared (since at these temperatures the black-body curve reaches its maximum in the near infrared). Now it is a known fact that the prime absorber in the magnesium–iron silicates of the mantle is the Fe ion. The mantle will be transparent and a good radiative conductor only if it contains very little iron. On the basis of seismological data, Anderson[6] has estimated that the atomic ratio in the mantle is roughly as follows: above 400 km depth 1:4, below this depth 1:3, and a somewhat increasing iron fraction at still greater depth. With that much iron, it would be difficult to maintain a very low temperature gradient by radiation alone if there are strong heat sources at depth.

(iii) The most efficient way of reducing the temperature gradient is convection itself. Since convective transport is much more rapid than any transport by radiation or ordinary heat conduction, the temperature during vertical displacement changes along the adiabat. Using a thermodynamical identity, the adiabatic temperature slope as function of depth can be written as

$$\left(\frac{\partial T}{\partial z}\right)_S = \frac{g\alpha T}{c_p} \tag{1}$$

where α is the coefficient of thermal expansion at constant pressure and the other symbols have their usual meaning. Numerical estimates for the upper mantle lead to values in the range 0·25–0·40 degK/km. Thus in the presence of very active convection only small vertical temperature gradients can be expected. (This statement, however, will have to be greatly modified for convective patterns that are very shallow compared with their width, such as found in the upper mantle (see Elsasser[7]).)

Later on we shall discuss reasons why convection in the mantle must be considered from the viewpoint of solid-state *creep* which is a highly nonlinear phenomenon, and why a model in terms of a fluid with conventional linear viscosity might not be adequate. We shall use the term *ductility* to indicate qualitatively the ease with which a solid in creep can be deformed by a given stress. One thing which will be discussed more quantitatively later on is clear at the outset: ductility increases strongly with increasing temperature.

Now it seems rather senseless to discuss pronounced ductility of the mantle due to heating without also considering true melting. The two phenomena could only be separated in a strictly homogeneous material which remains crystalline and hard throughout, all the way to the melting point, and then turns liquid at a sharply defined temperature. But the mantle is quite heterogeneous chemically; some lesser components have lower melting points than its bulk material. As Clark[8] has pointed out, the melting point may be further lowered if these low-melting components form eutectics. We should note here how inhomogeneous the mantle probably is, since this is not always sufficiently emphasized in the geophysical literature. One can obtain a good idea from the measured values in meteoritic matter where the atomic abundance ratio of the four most abundant lighter ions to the magnesium ion is[9]

$$(Al+Ca+Na+K)/Mg = 0.21$$

If we multiply the Mg abundance value by 1·25 in order to account for the presence of iron in the upper mantle, we still find that, roughly, the lighter silicates should amount to over 15% (say, by volume). We might therefore readily encounter situations where the mechanical behavior of the mantle with respect to slow deformation such as creep is almost completely determined by its lighter components, while at the same time these light components may have only a modest influence upon the seismic properties.

A term such as 'melting temperature of the mantle' is quite ambiguous but the following general conclusions seem justified. From melting-point curves as a function of pressure determined in the laboratory one infers a rather regular rise of melting point with depth. For possible constituents of the upper mantle, the melting-point gradients as determined from laboratory data range[10] from 1·4 degK/km to about 3 degK/km. Since these gradients are considerably steeper than the adiabatic, equation (1), there follows the well-known (Adams–Williamson) result that a molten mantle will solidify from the bottom up. On using a model of a homogeneous mantle without creep the temperature curve in the lower part of the mantle should run close to the melting-point curve of the major silicates. (Such an argument is not affected by the idea that the Earth aggregated in a cold state and became molten only later on by the combination of two effects: first, the temperature rise due to the formation of the core, or by completion of the core formation if the core was partially formed already in the gas-dust stage and, secondly, by the comparatively high heat output of the radioelements at this early stage.) But the temperature gradient is radically altered when solid-state convection is present. In fact, if the mantle contains low-melting components and if, moreover, some components remain highly ductile even after solidification, solid-state convection can be expected to set in long before melting of a thick layer has occurred. Such convection will undoubtedly be associated with a partial upward segregation of the lighter constituents; the upward migration will be most pronounced for those ions which have the largest radii (even if they are not light); among them are of course the radioactive elements.

In this way we can give an answer to Urey's famous argument that the Earth can never have been molten, since an Earth once molten would thereafter be tectonically 'dead' because the light constituents and the radioelements would all be at the surface. In place of this we may now assume partial segregation in a mainly solid but convecting mantle. Such segregation may long since have reached a terminal stage in the lower mantle while still going on actively in the upper mantle. If the lower mantle is assumed to be some 2000 km thick, it contains nearly half of the entire volume of the mantle; hence the migration of any constituents into the upper mantle will not change relative concentrations there by more than a factor of about 2, usually by less.

Next, we should remember that the temperature distribution in a convecting medium tends towards the adiabatic. The adiabatic gradient is considerably smaller than the melting-point gradient. Thus, if the lower mantle was convecting in an earlier stage of the Earth's life, it should have assumed a temperature distribution close to the adiabatic. If this is combined with an efficient removal of radioactive heat sources from depth, it would lead to a condition where by now the lower mantle, if it has any convection at all, moves so slowly as to be effectively uncoupled from the upper mantle.

It is only realistic to think of the ductility of any layer of the mantle as time dependent on a geological scale. Such changes are especially to be expected after the lower mantle has ceased to convect and may have lost a fraction of its lighter constituents. It might then change its ductility gradually, becoming less ductile as time goes on. This occurs through recrystallization processes, thermally activated movements of dislocations, etc. Such phenomena are well known from ordinary life in the form of increasing brittleness of material with passing time. Of course brittleness is absent in the lower mantle but ductility can be expected to change in just the same way. Thus efforts at explaining the disequilibrium equatorial bulge of the Earth (amounting to about 60 m height) by pronounced 'hardness' of the lower mantle[11, 12] might point in the right direction. However, the assumption of a linear viscosity at low stresses must be questioned; if it does not hold, the numerical 'viscosities' of 10^{26} obtained may be irrelevant.

Thermochemistry of Convection

It will be convenient to divide the following analysis of thermal and chemical aspects of convection into four parts: cyclic and acyclic convection; phase transformations; thermal factors in creep; chemical factors in creep.

Cyclic and acyclic convection

There is no traditional, generally accepted definition of convection. The best thing we can say is that convection is motion which results from density differences in a gravitational field. Such density differences can be either thermal or chemical, and all intermediate cases are possible when complex phase transformations occur. Now these changes can be either *cyclic*, implying that the material runs through a thermodynamical cycle as it rises and then sinks to rise again eventually, or else they can be *acyclic* as exemplified by the selective rising of a lighter fraction which leaves a heavier residue behind. This may lead us to ask whether the slow falling of a heavy piece of matter, or else the rising of a light one, should be considered as a convective process. Our answer will be that verbal scholarship and science do not mix too well; after having recognized the ambiguities of a term we keep on using it as best we can. In dealing with the motions of the upper mantle we are clearly facing a highly complex situation: a cyclic phase transformation that involves expansion of some component on rising, and contraction on sinking, may become an acyclic process through the physical separation of the less dense component from the denser one. One cannot doubt that convective processes in the mantle are in this sense *always partly cyclic, partly acyclic*. We shall leave the degree of each of these open; this remains a matter of specific models and of specific geological situations which may vary both in space and in time.

While the acyclic component of convection is extremely important from the geological viewpoint, its contribution to the *energy* of the convective process seems to be relatively small. To show this, let us assume that during a given geological interval, say one thousand million years, 0·5% of upper-mantle material comes to the surface irreversibly. With an overall depth of, say, 800 km, a mean vertical travel of 400 km, and a density defect, $\Delta\rho = 0·5$, of the light material, we find the amount of gravitational energy released to be 4×10^8 cal/cm^2 of surface. This is to be compared with an aggregated heat flow of some 4×10^{10} cal/cm^2 which has left the Earth during the last thousand million years of geological history (that is, if the present heat flow were constant; actually the heat flow has been somewhat larger in the past). Not all of this heat is available to generate convective motion but only a fraction $\Delta T/T$, by well-known thermodynamical principles. It would seem, nevertheless, that the larger portion of the energy sources driving convective motions in the upper mantle is of purely thermal origin. Details will depend critically on the efficiency of the convective engine.

Note on the core. The following is a brief digression into the thermodynamical nature of convection in the core. We know that such convection is required to account for the generation of the Earth's magnetic field. Now and then efforts are resuscitated to account for this field by various other means, especially by generation of core motion through tides, or through equinoxial precession. However, the equations of magnetohydrodynamics say that magnetic fields destroy or create vorticity, as the case may be, and at a rapid rate if the field is sufficiently large (as observations show it to be). Hence a magnetohydrodynamic system requires a primary motion-generating mechanism that supplies more than small amounts of vorticity. No such mechanism other than convection is known.

Cyclic convection in the core should result from its overall cooling (there being very probably no radioelements in the core on chemical grounds). Heat transport in the adjacent mantle is essentially radiative; on the other hand, radiative transport in the core should be negligible because metals are highly opaque. Conventional heat conduction in the core should be of moderate magnitude but not large; this is seen as follows. The thermal and electrical conductivities are proportional to each other by the law of Dulong–Petit, but the core is not a very good conductor; its electrical conductivity is by a factor of at least 10^3 smaller than that of the well-conducting metals. Whether in these circumstances heat transport in the lower mantle exceeds that in the core sufficiently so that purely thermal convection of the core results is difficult to decide at present.

Acyclic convection should be expected in the core. Starting from the fact that the mechanical motions are dominated by the Coriolis force one can estimate the variations in density required to balance this force[13]. These turn out exceedingly small, of order $\Delta\rho/\rho = 10^{-9}$. Now there is no reason to assume that either the upper or the lower boundary of the liquid core should be in thermodynamical equilibrium. Studies of recent years (e.g. Knopoff and MacDonald[14]) have made it almost certain that the impurities dissolved in the outer core are of the general magnitude of 10%. From the dynamics of motions in the core one may infer that the turbulent mixing time will be of the order of 500 years or slightly larger; thus at most 10^7 'mixings' of the core have taken place in the lifetime of the Earth. With the above value of instantaneous density variations this would give a total density change of the core over the age of the Earth of order 1%.

Looking at this problem from the viewpoint of specific mechanisms, it seems clear that atoms or ions can enter the fluid downwards through the upper boundary and other atoms or ions can leave it by the reverse route. Similarly, if the inner core grows through overall cooling, there will be chemical disequilibria at the lower boundary which should contribute to convection. Such arguments show that convection in the fluid core which is demanded by the existence of the magnetic field, must also be considered altogether possible on the basis of the quite unrelated physicochemical evidence just given.

Phase transformations

A number of phase transformations have been attributed to the material of the mantle. They have been discussed for instance by Ringwood[15]. Of these, the breakdown of the main magnesium–iron silicates into corresponding oxides under pressure is the most spectacular. It is generally admitted to occur in the Birch transition zone around 800 km depth. We shall not deal with it since for reasons given in various other places of the present chapter we focus our attention here on the upper layers of the mantle, above this depth. In these layers, seismological analysis reveals some further stratification, probably attributable to phase changes. With exception of the M discontinuity which separates crust from mantle these seem to be of relatively minor importance. Verhoogen[16] has shown from thermodynamical arguments that a stream of convective motion in a medium of sufficient homogeneity of composition can cross a phase boundary; the material undergoes the phase transformation reversibly while it moves across the boundary. In this process, one may assume that a certain hysteresis occurs and that the phase boundary gets slightly deformed upstream, towards the moving current. Nevertheless, Verhoogen's result indicates that models of upper-mantle convection might not depend critically on a knowledge of all the possible phase transformations; it may be possible to construct such models along lines of more direct evidence as we shall do later on.

Bridgman commented early in his career on the multitude of pressure-induced phase transformations that he found. There is, however, a limit to phase transformations; as the pressure increases only the simplest configurations remain stable, and phase transformations become rare again. We may surmise that phase transformations are most common in the pressure field of kilobars to some tens of kilobars found in the upper layers of the mantle. Silicates are noted for the great variety of configurations which their space lattices can assume, but many of these arrangements are rather loose and can therefore only be realized at moderate pressures. We must therefore look towards the upper mantle as the prime locus of those transformations whose cyclic occurrence (leading to expansion on rising, and to contraction on sinking) will be the main driving force of convection. We need hardly say that phase transformations may be considered as an extreme form of more ordinary thermal expansion. Since the density changes under phase transformations are usually much larger than those due to thermal expansion, phase changes are very efficient in generating convection.

Another reason why the upper layers of the mantle are the region where phase changes as well as convection are most likely to occur, may be recognized from figure 1: near the top of the mantle the temperature gradient is far steeper than farther down. In the case of simple thermal expansion (where we have approximately $d\rho^2/dT^2 = 0$) the excess of the temperature gradient over the adiabatic gradient is a direct measure of the potential energy available in the stratification; this is the energy capable of release on convective overturn. An analysis of the thermomechanical principles that apply in the case of convection engendered by phase transformations is still lacking. Verhoogen's work deals with one aspect of it. Laboratory experiments with viscous fluids undergoing phase transformations would be of great interest to geophysics.

Let us consider now the submarine ridges as upwellings. If convection is confined to the top-most few hundred kilometers, the roots of the ridges cannot be explained by thermal expansion alone. (This would lead to very deep roots, well over 1000 km.) Instead, much of the buoyancy of these roots must come from phase transformations. Menard[17] in his review mentions two principal types of such transformations. The first of these has been proposed by Hess[18] and consists of the hydration of ultrabasic mantle material, peridotite, to form serpentine. According to Hess, serpentine becomes unstable at about 500 °C and above this temperature dissociates into peridotite and water. Owing to this low dissociation temperature, serpentine can only be formed at rather shallow depths; nevertheless, it might well make a contribution to the buoyancy of the ridges. Isotopic determinations indicate that the water is not virginal; it is presumably provided by the ocean in the vicinity.

The second major phase transformation of interest is that of basalt from its lighter gabbroic form into the denser eclogitic form. This was first suggested in connection with the mechanics of mantle movements by Dietz[19]. In the meantime much laboratory work has been done on the gabbro–eclogite transformation by Ringwood and Green[20] who assign to it a major role in convection of the upper mantle. Menard, however, is very sceptical about the importance of both of these types of transformations for convection. The present writer is not enough of a chemist to feel entitled to an opinion of any weight.

Thermal factors in creep

Although the slow deformation of solids known as creep has been known for ages, it was studied quantitatively only beginning with Andrade in 1913. The outstanding features of creep are its pronounced nonlinearity and its critical dependence on chemical parameters as well as temperature. Fluids tend to become homogeneous by virtue of rapid mixing but in solids thermal as well as chemical inhomogeneities persist for long times.

There is transient as well as steady-state creep. In the study of the mantle one may confine oneself to phenomena which are of a steady-state type overall but the longest intervals of time. Quantitatively, steady creep is described by relating the stress rate, say $\dot{e}$, to the stress σ. Below, $\dot{e}$ may be either a longitudinal or a shear strain, as the case may be, and the same holds for the stress. Observations show that the following expression gives an approximate description of creep:

$$\dot{e} = \begin{cases} C_1 \sigma^n \exp\left(\dfrac{-a}{T}\right) & (\sigma \text{ small}) \\ C_2 \exp\left(\dfrac{\sigma}{\sigma_0}\right) \exp\left(\dfrac{-e}{T}\right) & (\sigma \text{ large}) \end{cases} \quad (2)$$

It is apparent that the exponentials containing the temperature represent thermal activation of the atomic or molecular processes underlying creep. We also see that at large stresses the stress dependence is steeper (exponential) than at small stresses where it follows a power law. The border between large and small stresses depends on the material and on other conditions; as a rule it is of the general order of some tens to a hundred bars so that processes of mantle creep correspond more commonly to a small stress than to a large one in (2). There has been considerable variation in measurements bearing on the magnitude of the exponent n for small stresses. Apparently there is no general rule; the behavior depends on the molecular mechanisms which are in turn dependent on chemistry, lattice structure, graininess, etc. Most laboratory values of n are between 3 and 5 (e.g. Weertman[21], Jones[22]). Linear behaviour, $n = 1$, also seems to occur but, since it is not the rule in the laboratory, any rash assumption of linearity of creep in the mantle is a potential source of error. We have already introduced the term ductility to express qualitatively the dependence of strain rate upon stress. We shall postpone the discussion of the mechanical properties of creep and consider next its thermal and chemical characteristics.

Experience shows that substances begin to exhibit pronounced creep when the temperature exceeds about half the melting point. They continue to become more ductile with higher temperature but do not change their behavior qualitatively until the melting point is reached. The mantle, however, is a highly inhomogeneous mixture and cannot immediately be judged by the results of laboratory experiments on chemically homogeneous substances. Since the mantle contains comparatively large fractions of lower-melting constituents we may think of the high-melting ultrabasic materials as hard grains embedded in a more ductile matrix. Let us consider now figure 1. There is a region of closest approach to melting. In fact, there may be some substances that are at the melting point and fractionally molten in this zone, as indicated by the broken line in figure 1. In any event, we can expect that there will be a

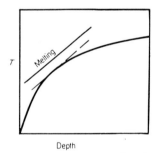

FIGURE 1 A plot of temperature T against depth in the mantle

layer of maximum ductility at some depth. It seems entirely legitimate to relate this layer to the zone of low seismic velocity discovered and so much emphasized by Gutenberg[23]. Much research on this zone has in recent years been done by Anderson[24] and others. It appears from Anderson's data that the low-velocity zone extends from about 75 to 225 km depth, being centered at about 150 km. Of course there is no compelling reason why seismic velocity should be proportional to creep rate; the molecular mechanisms are no doubt quite different. All we are entitled to infer is, qualitatively, a change in properties of the material. Geologists have often postulated the existence of a 'soft' layer at some depth, and the name *asthenosphere* coined by Daly is in common use. We shall employ this term here to designate a layer of greatly increased ductility whose extent, for want of more quantitative knowledge, we identify with the Gutenberg low-velocity layer.

There is one other significant index of stratification of the upper mantle which we can get from seismology. Anderson[24] has plotted the quantity K/μ obtainable directly from seismic velocities, as a function of depth (K is the bulk modulus, μ the shear modulus). This quantity rises first, then reaches a high level around 2–300 km depth (where $K/\mu \sim 2\cdot0$), falls thereafter uniformly to a value of $1\cdot81$ at 800 km depth, and at this point turns sharply upwards, continuing to increase at greater depth. Whatever the detailed and difficult interpretation of this behavior, depending no doubt on both temperature and pressure, it seems clear that there is a discontinuity near 800 km. In identifying it with the phase transition from silicates to oxides we may also surmise that radioactivity as well as materials enhancing ductility have to some extent been squeezed out of the denser oxide phases to congregate higher up.

Chemical factors in creep

Among the vast variety of chemical properties that may affect creep we shall consider here only two. The first of these concerns the difference between metallic and ionic crystals. For obvious reasons most of the earlier work on creep dealt with metals. It does not seem certain *a priori* that ionic crystals will show the same characteristics of creep as metals but all experience indicates that there are no major differences. In more recent years extensive creep studies have been made on ceramic materials, for instance MgO. Creep in rock (marble) under high-pressure confinement was investigated among others by Heard[25] who achieved lower creep rates than previous investigators; but his rates are still by a factor of order 10^7 above the creep rates of the mantle. On the whole, Heard's results are encouraging with regard to the use of formula (2) as a qualitative guide. But one must keep in mind that creep is found to be based on a variety of atomic mechanisms. The most important of these are as follows: first, sliding within the relative disordered boundary layers between crystal grains; secondly, the migration of dislocations across crystal grains (known as Nabarro–Herring creep). Since the material of the mantle has sojourned there for a long time, its crystals may be both quite large and also quite hard, owing to gradual loss of dislocations. It is not possible to judge crystals at depth from those found at the Earth's surface since in the process of rising they usually are violently deformed. (We owe this cautionary remark to Professor H. H. Hess.) Such deformation, by generating numerous new dislocations in the crystal grains, may radically change their mechanical behavior.

In sum, we can say rather little about the behavior of a system so far removed from laboratory conditions as the mantle in creep. We may vaguely surmise that creep is largely of the grain-boundary sliding type and this might possibly mean that creep is of the 'linear' type ($n = 1$ in equation (2)), but this conclusion is quite uncertain. We must largely rely on empirical arguments taken from the Earth in a more direct manner.

The second chemical aspect of creep to be discussed here is the capability of certain chemicals in even small concentrations to enhance the ductility vastly. These are designated as *fluxing* agents; water and the alkalis are the best known of these. So far as rocks are concerned, experiments on quartz under high-pressure confinement have had some rather spectacular results. The data obtained by Griggs and Blacic[26] show that the addition of small quantities of water softens quartz quite radically. In the same context the softness of serpentine, a hydrated ultrabasic rock, should be remembered.

Orowan[27] has suggested that the upper mantle including the ocean bottoms is perfused with a small concentration of water which acts as a fluxing agent and greatly increases the ductility as compared with the continents. The latter, conversely, have been desiccated in the course of time. They form therefore comparatively rigid rafts floating on the more ductile material of the upper mantle. Only in regions and during periods of intense orogeny will the continental crust become temporarily more ductile, owing to extensive magmatic intrusions which carry a great deal of water with them. On the whole, however, ocean-bottom and upper-mantle material may be assumed much more ductile than continental plates; hence the former can be expected to have suffered far more deformation than the continents. We have found this concept a most fertile working hypothesis and adopt it throughout.

Bullard, Maxwell, and Revelle[28] advanced the idea that the near-equality of heat flow from continents and oceans may be explained by the following assumption: the large heat flow coming from radioactivity embedded in continents is balanced on the ocean bottoms by a corresponding quantity of convectively supplied heat. This view can now be put on a much more plausible basis[29]. On the assumption of the Orowan hypothesis one can prove that the dynamics of the convection-generating forces is such that it tends to establish a near-equalization of surface heat flows (provided the thermal conductivities of the materials are approximately equal, which is plausible). While we cannot go into details here, it is clear that one is by no means entitled to infer unambiguously from the mean uniformity of heat flow a corresponding uniformity in the distribution of radioactive materials beneath.

Long-distance Correlations

We have seen that the upper-mantle forms, with respect to convection, a shallow tray whose depth, if estimated at 800 km, is only one-fiftieth of the Earth's circumference. There is every evidence that the surface expressions of convection involve distances very much larger than this depth. Hence the convective patterns appear to be horizontally extended and pancake shaped. Tozer[30] has carried out experiments which show that, when a fluid is heated internally (in his case by electric currents) rather than by contact with a bottom plate, the convective pattern will be distinctly broadened; its horizontal extension becomes considerably larger than the vertical one. This model of heat sources should apply to the upper mantle. The geophysical evidence indicates well beyond this that mantle material rises rather concentratedly under submarine ridges, moves away from these horizontally, and sinks along sloping layers (the Gutenberg fault zones) marked at the surface by trenches (figure 2). The very long horizontal legs of such a convective pattern are no doubt in the main due to the presence underneath of the asthenosphere, a layer of greatly enhanced ductility.

Convective activity is also reflected in the World-wide gravity data from satellite observations. Runcorn[31] has developed methods to correlate such data with the buoyancy forces inside the Earth. Qualitatively the correlation is good, but it is difficult to make it highly detailed. Owing to the narrowness of the zones of upwelling and sinking, the spherical-harmonic series of the geopotential, on whose lower harmonics the maps of satellite data are based, converges only very slowly.

Owing to the shallowness of the upper mantle it is profitable to study the surface topography and to look for indications of a convective pattern there. The nomenclature

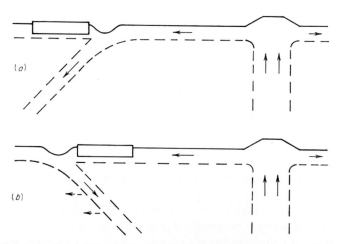

FIGURE 2 Convective pattern of mantle material under submarine ridges

with regard to behavior in depth (figure 3) used from now on is as follows: we divide the upper mantle into *tectosphere* (about 75 km deep), *asthenosphere* (from 75 km to 225 km, say), and *mesosphere* (from there to about 800 km). These are not well-defined layers in a material sense, of course, but only symbols for dynamical entities whose *depth may vary within large limits.*

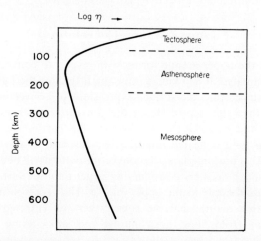

FIGURE 3 Division of upper mantle into tectosphere, asthenosphere, and mesosphere. A plot of log η against depth

The general pattern of a sliding tectosphere with rising and sinking branches is shown in figure 2. Hess[32] first assumed that submarine ridges do not just represent local swellings but loci of a steady upwelling with the material then moving off laterally to form the ocean bottom. This model has been amply confirmed. Among the corroborative results we mention only two. The recent seismic analysis of Sykes[33] shows directly some of the faults crossing the Mid-Atlantic Ridge to be transform faults in the sense of Wilson[34]. This makes the interpretation of the ridge quite unambiguous; it is incompatible with any known process other than rising which then turns into lateral spreading. Again, Oliver and Isacks[35] have given impressive seismological evidence to the effect that the slanting descent under about 45°, as shown in figure 2, is a realistic picture. Here, where we give a theoretical model rather than a survey of the literature, we are unable to mention numerous related observations.

We now confront an interesting question: while there is some asymmetry with regard to vertical components of the motion in figure 2. it is not remotely as strong as the asymmetry commonly found in the convective motion of stratified fluids. Let us take a very familiar form of convection, the thunderstorm; we find in it concentrated vertical updrafts with velocities reaching many meters per second; the compensating downdrafts are spread over large areas and correspondingly small, perhaps only a fraction of a millimeter per second. No similar asymmetry is visible in the upper mantle We see extensive upwellings, and the tremendous overall length of the ridges (over 7×10^4 km) makes it hardly necessary to look for any other places where mantle material rises—and none are apparent. There is now the question whether in addition to the downdrafts marked at the surface by trenches there exist regions of extensive but slow subsidence. On the basis of all the information we have, our answer is that there seem to be none.

We shall strengthen this conclusion by some examples. If there is any large and regular surface below which we could expect slow subsidence, it is the Pacific. The East Pacific Rise is an intensely active ridge and the material flowing from there eastwards is no doubt flowing down at the Andean Trench. To the west of the Rise there is a vast expanse of ocean but we do not see any generalized subsidence. Instead, we see the Tonga Trench, well defined, deep, and an active site of localized sinking. Let us take another example: the manner in which some continental blocks are surrounded by ridges leads to the inference that, if material does indeed travel away from the ridges, there should be subsidence below the continental plates[17, 19]. A place where this is most likely to occur is under Africa which is surrounded on three sides by well-marked ridges. Again, subsidence could not occur under eastern Africa, where there is the well-developed rift valley, certainly a rising not a sinking feature. Thus subsidence would almost certainly have to be centered on the Sahara. But Menard's map for Africa shows very clearly that the Mid-Atlantic Ridge in the region where it passes by north Africa is shifted to the west; this section of the ridge is set off by an east–west fault at each end (roughly at 10° and 30° northern latitude). The obvious qualitative explanation of this lies in a 'pull' by the Puerto Rico Trench to the west. No evidence of a downdraft under Africa is afforded by topography. Another example is the North American continent. Adherents of a continental drift model agree generally that the continent has drifted to the west and in the process has overridden the northernmost branch of the East Pacific Rise, whose extension is now under the Cordilleran Plateau. The eastern United States is therefore a natural place to look for evidence of deep and widespread subsidence under the continental plate, seeing that this region is roughly in the middle between two strips of upwelling which have been active in the fairly recent geological past. There is again no evidence.

If we speak of evidence for large-scale subsidence, we can only have in mind disturbances of the surface structure which are geologically marked. One might say that such disturbances are minimized by the relative hardness of the continental plates. It would be most surprising, however, if some such evidence could not be found on the ocean bottoms where one is so much closer to the more ductile material of the mantle.

The extensive explorations of the bottom of the Atlantic by means of echo soundings leave no serious doubt that the tectonic condition of the Atlantic Ocean bottom has not undergone any detectable disturbance since the Tertiary Period when sediments began to be deposited there (see in particular the work of the Lamont Observatory, e.g. Heezen[36], Ewing, Ewing, and Talwani[37]. Unquestionably, ocean-bottom sediments would be among the most sensitive indicators of such disturbances.

The Mid-Atlantic Ridge has been and still is at many places a very active structure. Failure to detect slow subsidence of material that has spread sideways from this ridge is highly significant; this induces us to suspend the idea that such subsidence exists until some favorable evidence, either observational or at the least theoretical, is provided. We are left then with the view that the only loci of sinking are the regions of the Gutenberg fault zones marked by trenches at the places where they intersect the ocean bottom. The crucial point of our model is therefore the sliding of the tectosphere over the more ductile asthenosphere, carrying material from the ridge to the trench.

This leads at once to the question whether there is a closer and dynamically motivated correlation between ridges and trenches. Surprisingly enough, the answer is, on the whole, no. The classical case, always cited, of the Andean Trench which runs parallel to the South Atlantic Ridge is an exception rather than the rule. Another example of a modest correlation is that of the East Pacific Rise with the Andean Trench to the east and with the Tonga Trench to the west. But here the ridge is vastly longer than these trenches and parallelism is limited. In the remaining parts of the ridge or trench system, no pairwise correlation is visible. From this we can legitimately conclude that horizontal displacements of the tectosphere or, rather, of pieces of it can take place only when there is a stretch of ridge at one end and a stretch of trench at the other. A rather spectacular example is as follows: the North Atlantic Ridge in the general neighbourhood of Iceland is very active at this time; but the material issuing from there and moving sideways cannot find a trench except very far away: to the west there is the Aleutian Trench and to the east we have to go all the way to the Pacific margin of the Asian continent to find trenches. Notably enough the Aleutian Trench is far from parallel to the Atlantic Ridge.

We may conclude this series of examples with the following: the Juan de Fuca Ridge is very short, only a few hundred kilometers long, situated in the Pacific off Vancouver Island and directed SW–NE. To understand that there can be any horizontal spreading from this ridge, one has to recall that it is about midway between two trenches, the Aleutian and the Puerto Rican. That this relationship is not just pure fancy is attested to by the San Andreas fault which begins at this ridge and runs from there over 2000 km to the southeast, through California and well into Mexico. The shear direction at the fault is compatible with a movement of the tectosphere from the region of the ridge toward the Caribbean.

Strange as these results may be, there is at least one major observational fact to support this trend of ideas. Menard[38] has been able to trace the Clipperton fault over 10 000 km along nearly a great circle, all the way across the Pacific. Now, if such a fault is indicative of sliding of pieces of the tectosphere past each other, as had been suggested by the early magnetic work[39], then such sliding is on a gigantic scale, 10 000 km being 90° on a great circle. Hence the idea that the mechanics of the tectosphere has some nearly World-wide aspects can be strongly supported.

The ridge system shows certain clear regularities of which the best known is the rather precise median position of the Atlantic Ridge between the continental margins. This is not the only case; the Carlsberg Ridge is median in the Indian Ocean, a ridge passes in the middle between Africa and Antarctica, etc. Menard[38] in his penetrating study of the regularities of submarine topography says that about half of the length of the ridge system is found in such median positions. He does not consider this as the most distinctive feature of ridges; according to him this feature is their tendency to be located at a fairly constant distance from the continental margins. A number of maps of continents in polar projection given in Menard's paper put this clearly in evidence. The most conspicuous case is Africa, which is surrounded by ridges on three sides at a fairly constant distance from the edge of the continent. Antarctica is fully surrounded by ridges but the distance is not quite so regular. South America is partially surrounded by ridges but the pattern is somewhat irregular there. Another very curious example is the following: at one place the East Pacific Rise shows a rather pronounced bend which carries it in an arc around the area once occupied by the now largely defunct Darwin Rise. It is quite likely that the Darwin Rise, having once been the locus of very intense volcanic activity, has become somewhat desiccated, and hence this region of the tectosphere might in hardness be closer to typical continental than to other oceanic regions.

On the whole, the existence of strong long-distance correlations in the position of ridges is a patent fact. It is obvious that such correlations must be the expression of a system of forces; in a solid this takes the form of stresses. These stresses act over very large distances such as the width of the Atlantic and apparently even across the Pacific. But stresses must somehow propagate. We ought to remember here that, since velocities in the mantle are so small, the mantle must be at any one time extremely close to equilibrium, meaning *elastic* equilibrium. We can hardly conceive of the stresses inferred as propagating through the deep interior of the mantle. In a homogeneous solid, stresses due to a point force decrease as $1/r^2$ with distance r and stresses due to a line force as $1/r$. No such impression of decrease with distance is gained from the topographic effects. Furthermore, the long-distance correlations are with the margins of continents which are clearly very shallow features compared with the depth of the mantle; the influence of continental plates should be quite small below the tectosphere. We may conclude safely that stresses are transmitted horizontally, along the Earth's curvature, in the topmost layers, that is, in the tectosphere. *The tectosphere acts as a stress guide.* This is possible only because it can slide readily on the underlying asthenosphere.

The correlations of the ridges to the continental margins are of two types. In the first place the ridges tend to keep a certain average distance from these margins; in the second place, when there are two roughly parallel continental margins and a ridge in the ocean basin between them, the ridge tends to appear in the median line between those two margins. There seems to be one way only in which these observations can be interpreted mechanically: the stresses in the tectosphere, whatever their nature, act in such a way as to move a ridge and a continental margin apart. Furthermore, the effect of the stresses must decrease somewhat with distance from the continents; otherwise the median position of the ridges would not be mechanically stable. It should be noted that the preceding propositions are not a hypothesis; so long as we are careful not to specify the stress-transmission mechanism prematurely, they are no more than a translation of the observed facts into the language of mechanics.

We can draw one other significant conclusion. The ridges could not show the kind of regularity we observe unless they were mobile within the material of the upper mantle. If the ridges are upwellings, the material which rises in them must come from an appreciable depth, probably from the mesosphere, which we estimated as extending from about 225 to about 800 km depth. In the language of hydrodynamics, the ridges are *dynamical*, not substantial, features. A dynamical pattern is one which moves across the fluid (e.g. a wave crest) keeping approximately its shape while the particles composing it change. This in turn makes it certain that the upper mantle must be rather homogeneous horizontally; otherwise dynamical features that propagate with little change of shape could not exist.

A closer look at figure 2 lets one recognize the difference between dynamical and substantial features. Figure 2(a) may be thought of, for instance, as representing the upwelling of mantle material in the East Pacific Rise, its migrating westwards and its going down at one of the trenches off southeastern Asia or Australia. This could be conceived of as a conveyor belt operation in which the main block of mantle in the center of the moving sheets remains untouched. But figure 2(b) has a quite different topological structure: the central mass cannot remain intact in the long run. We may think of this diagram as representing schematically material which rises in the South Atlantic Ridge, and then becomes tectosphere which moves to the west, carrying the South American continent with it. Room is made for this westward migration by material of the Pacific going down at the Andean Trench. The dynamic character of the pattern and the change of its constituent material particles is here quite clear.

Combining the previous arguments, we can draw a third major conclusion: the pattern of convection in the upper mantle is to a large extent controlled by the mechanical condition of the top layer, the tectosphere. We do not say that it is entirely so controlled because upwellings are found under continents also, and it would be difficult to justify this fact by any reasonably simple model of the tectosphere alone. But the fundamental role which the distribution of continents and oceans plays in the location of the ridges is clear enough. Again, this is less surprising than one might think. Convective flow is capable of a great variety of patterns, most of them not uniquely stable, and thus the influence of a boundary upon convective patterns is quite intelligible. The corrugations of the Earth's surface certainly have a great influence upon the convective patterns in the lower atmosphere. In the case of the mantle a similar influence seems to be exerted by its upper, free surface, which has become corrugated and modified through the effect of chemical and thermal variations in its material and by the existence of the crust.

So far we have emphasized the ridges and have said little about the trenches. These show even more definite correlations to continental margins. In fact all present-day trenches are located around the border of the Pacific Ocean or close to it. An occasional one is shifted towards the interior of the Pacific like the Tonga Trench; others, mainly the Java Trench and the Puerto Rico Trench, are found at a moderate distance from the Pacific. No trenches are found elsewhere in the World. Hence, if tectospheric motion is from ridge to trench, the corresponding transmission of stresses must be possible over a vast scale since about half of the Earth's surface has no trenches.

The distribution of trenches cannot always have been this way. One can hardly escape the conclusion that up to the Tertiary Period there was a trench in front of the Himalayas, and perhaps a whole trench system reaching from Spain to Szechuan, along the southern edge of the Eurasian continent. Without this assumption the orogenies along this edge would be hard to understand. Also one of the more conspicuous features of the continental drift hypothesis, the 'drifting' of India from a position near the southern end of Africa to its present location could not be understood without such an assumption. The topography of the bottom of the Indian Ocean shows rather distinct traces of an original north–south fault system dating from this period, overlain by east–west faults corresponding to a more recent drift direction.

Although correlation with continental margins is clear enough, none of the facts concerning trenches has as yet found a theoretical explanation. The only generalization we can seemingly make is one already mentioned, namely that a system of drift as described will not work unless the trenches as well as the ridges are dynamical and not substantial features.

To summarize, our main conclusions are as follows.

(i) There is a stress system (details as yet unknown) acting horizontally along the tectosphere which tends to move ridges away from continental margins.

(ii) Ridges and trenches are dynamical patterns, mobile within the material of the upper mantle.

(iii) The actual pattern of convection through the upper mantle is largely controlled by the mechanical–thermal condition of the tectosphere.

Sliding of the Tectosphere

From the preceding analysis there emerges at least a qualitative picture of mantle convection. We were led to assume that radioactive heating, and hence convection, is confined to the layer above the Birch transition zone (phase change from silicates to oxides) at about 800 km depth. There are four legs to the convective circulation, two long horizontal ones and two short vertical ones. The upper horizontal leg consists mainly in the sliding of the tectosphere over the asthenosphere; one of the consequences of this sliding is the presumptive spectacle of continental drift. The other near-horizontal leg must be some sort of 'return flow' in the mesosphere but the mechanism is altogether unknown. There is no reason to assume a one-to-one correspondence between a horizontal motion in the tectosphere and a transport in the opposite direction in the mesosphere. This would imply the picture of a 'cell'. While this concept has often been thought of, it lacks foundation in fact, and far more general patterns are readily possible.

Convection in the upper mantle differs from other convective processes in a variety of important characteristics; among them are the extreme thermomechanical stratification of the upper mantle, the highly superadiabatic temperature gradient which, as the evidence shows, is not wiped out by the convection itself, and nonlinear creep of a solid taking the place of ordinary viscosity of a fluid.

Still, it is becoming more and more clear that upper-mantle convection is the basic mechanism that underlies most of the large-scale tectonic phenomena observed on the Earth. The velocities of order one to several centimeters per year, which observations lead us to ascribe to convectively induced displacements, exceed by orders of magnitude those that would follow from the older concepts of a steadily contracting or expanding Earth. In a detailed paper, Orowan[40] has recently shown that both contraction and expansion could be ignored as causes of the observed amounts of tectonic deformation even in the absence of convection. Therefore, convection appears the more as the essential basic mechanism of tectonics.

We shall here concentrate on the horizontal displacements of the tectosphere as that branch of the convective upper-mantle circulation that is most readily accessible to some degree of theoretical analysis. Borrowing a term from oceanography and meteorology, the writer[7] has used the term 'advection' to describe such essentially horizontal motions. Ringwood and Green[20] have adopted this suggestion but use the term in an extremely broad sense, almost synonymous with convection. We would suggest, however, that the term be restricted to fully or almost horizontal motions so as to make its meaning the same in the various branches of geophysics.

In order to gain any insight into the mechanical conditions of upper-mantle convection, it is almost inevitable that one should begin with a model of a viscous fluid, that is, a substance with linear viscosity and incapable of purely elastic deformation. It must remain open how far this is a justifiable approximation. In terms of a viscous fluid, then, we represent the asthenosphere by a sharp decrease with depth of the viscosity η. A sketch is shown in figure 3 where $\log \eta$ is plotted against depth in a more or less arbitrary fashion. We assume here a rapid decrease of η from the surface to a minimum in the asthenosphere, by about a factor of 10^2–10^3, and then a much slower rise on going down into the mesosphere. The last-named, much slower trend seems indispensable if one is to account for a 'return flow' in the mesosphere. The latter is subjected to the severe constraint of having no free boundaries; but, since return flow must necessarily occur if the tectosphere moves, there must be some mechanical condition that facilitates flow. The only one that comes to mind for the mesosphere, in the absence of any evidence for a 'soft' layer farther down, is an increased ductility over an appreciable interval of depth.

To make numerical estimates about the sliding of the tectosphere over the asthenosphere, we require some further drastic simplifications. We represent the tectosphere by a *solid* plate of thickness h_1 and elastic modulus E. We represent the asthenosphere by a *fluid* layer of depth h_2 and viscosity η. In steady translational motion of the tectosphere the velocity distribution will be as shown in figure 4. We now assume that the

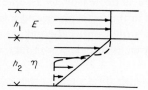

FIGURE 4 Velocity distribution of the tectosphere

shear stress exerted at the bottom of the plate is balanced by the total horizontal stress exerted at the end of the plate. The fluid is assumed at rest at the bottom of the asthenosphere. The velocity v_x of stationary sliding is readily found to be

$$v_x = \frac{\sigma_x}{\eta} \frac{h_1 h_2}{L} \qquad (3)$$

where L is the length of the sliding plate. (Calculations refer to unit thickness perpendicular to the plane of figure 4.) Now let us take $\eta = 10^{21}$, smaller than the often accepted value for the upper layers of the mantle but perhaps reasonable for the asthenosphere. Letting for simplicity $h_1 = h_2 = 100$ km and $L = 1000$ km, and assuming $v_x = 1$ cm/year, we obtain $\sigma_x = 30$ bars. This shows that the concept of a sliding tectosphere makes good mechanical sense, at least for a crude model.

Next, let us consider transient conditions. Under lateral transient stress the plate becomes locally compressed or extended. We assume these effects first to be purely elastic; then, if u is the displacement of a point on the plate, the local horizontal stress is $\sigma_x = E \, \partial u/\partial x$, and the horizontal force per unit slice of the plate is $h_1 \, \partial \sigma_x/\partial x$. On the other hand, the shear stress at the bottom is $\eta v_x/h_2 = (\eta/h_2)\partial u/\partial t$. (Here, all motion is assumed slow so that $d/dt = \partial/\partial t$.) Requiring again balance of the horizontal forces with the shear forces on the plate, we obtain finally

$$\frac{\partial u}{\partial t} = \kappa \frac{\partial^2 u}{\partial x^2} \tag{4a}$$

$$\kappa = \frac{h_1 h_2 E}{\eta} \tag{4b}$$

Equation (4) has the form of a heat-conduction or diffusion equation; its integration is standard. In particular, the mean distance x by which a disturbance propagates in time t is given approximately by

$$x \sim (4\kappa t)^{1/2} \tag{5}$$

Letting[41] $E = 1 \cdot 6 \times 10^6$ bars, the other quantities being as in (3), we find $\kappa = 1 \cdot 6 \times 10^5$; for $t = 10^4$ years, (5) gives $x = 1500$ km.

While these values are not radically different from those frequently quoted from the analysis of the Fennoscandian uplift, it is of importance here to remark that the definition (4b) of κ has no real physical meaning. Orowan[42] has pointed out that a plate of rock subjected to lateral forces will always react plastically, not elastically, provided the thickness of the plate is appreciably in excess of 10 m. This follows from standard formulae of the mechanics of solids and may therefore be safely accepted. The tectosphere will thicken on compression and be thinned under tension. Isostatic mechanisms will then come into play to restore equilibrium. If the deviations from equilibrium are small, one may tentatively assume that the restoring forces are linear in these deviations and (4a) will again hold, but with a different physical significance and numerical value of the coefficient κ. From the orders of magnitude observed in the postglacial rebound, it is rather safe, however, to assume that $\kappa = 10^4$–10^5 in order of magnitude, and hence horizontal stress propagation is rather fast on a geological time scale. For our purposes we need not specify the detailed mechanism of this propagation (see also Beloussov[43]). Of course the stress will diminish somewhat as it propagates. Still, this picture of horizontal propagation, where penetration into depth is buffered by the asthenosphere, is quite different from the stress pattern that a load would produce in a homogeneous spherical shell.

Next, let us compare sliding of the tectosphere with what would happen if material is moved *inside* the asthenosphere. If, for instance, the tectosphere buckles down, there should be a lateral outflow in the asthenosphere. The following simplified model will suffice[44]. The space between two parallel circular plates of radius R and mutual distance $h \ll R$ is filled with a viscous fluid. The plates are compressed with a force $\pi R^2 \sigma$, where σ is the normal stress. The mean velocity (averaged over h) at the rim is

$$v = \frac{\sigma}{3\eta} \frac{h^2}{R} \tag{6}$$

Dimensionally, (6) resembles (3) but the difference is readily apparent. We obtain large velocities in (6) only when the normal stress applies over *large areas*; if stresses act only in limited regions, the sliding motion is strongly preferred (provided of course the sliding plate is not stopped at the far end). On the other hand, we see from (6) that large h increases the velocity of internal motions; these motions then will tend to be more closely *isometric*, that is the horizontal dimension should be comparable with the vertical ones, both presumably of the order of a few hundred kilometers. Whether such internal localized circulations exist is not known at present. Some other examples of very slow viscous flow are calculated in Jaeger[45].

All arguments just given are based on a linear law of viscosity. If we have a nonlinear relationship between strain rate and stress (such as (2) with $n > 1$), the creep patterns will as a rule change in such a way that large strains become concentrated and enhanced and small strains tend to be reduced. Thus, while a viscous fluid flows through a tube with the well-known parabolic velocity profile, a plastic substance has its shear concentrated at the boundary, i.e. it moves like a plug. Orowan[27] has suggested that the ridges correspond to large rising blocks with the shears concentrated at the edges; however, seismic data point rather to a strong concentration of the shear in a narrow central band of the ridge. Since the material may undergo phase transformations as it rises, the elevation of the ridges would then be attributed to this less dense material having flowed sideways from the central current.

Nonlinearity of the type (2) is complicated by time delays in solids. There is strain hardening and strain softening, both well known in the laboratory. Earthquakes are due to nonlinear strain concentration together with strain softening of an as yet obscure type. It is remarkable that there are no earthquakes that would be indicative of a horizontal sliding of the tectosphere. If the creep law is highly nonlinear the layer of sliding will tend to be very concentrated as shown by the broken line in figure 4. The fact that no corresponding earthquakes are observed may indicate that the nonlinearity is not very pronounced—but definite conclusions seem premature.

Regarding earthquakes it is no doubt significant that practically all the deep-focus earthquakes (below about 200 km) are in regions where one would suspect material to sink. Perhaps the sinking material has first suffered loss of water or other fluxing agents fairly close to the surface; farther down it undergoes phase transformations to denser forms which makes it sink still farther. This material is likely to be far less ductile and therefore probably more subject to shear disruption than average mantle material at a similar depth. Earthquakes are indicators of motion but the reverse is not true: their absence does not prove absence of motions.

Analysis of observations shows directly[46] that large sections of the tectosphere slide or slide-and-rotate. This can be achieved by three kinds of stresses: (1a) local horizontal compression, (1b) local horizontal tension, and (2) shear drag from underneath. Clearly, (2) can coexist with either (1a) or (1b). Applicability of (2) would imply that the motion beneath the tectosphere is faster than the motion of the tectosphere itself. The discussion following equation (6) indicates that such flow underneath would be of limited horizontal extent, say a few hundred kilometers across. So far there is no clear evidence for the existence of such undercurrents.

Compressive stresses will occur at the flanks of ridges and tensile stresses adjacent to trenches; it is only when the two act simultaneously on a piece of the tectosphere that motion of the latter can occur. (The singular exception would be a circular piece of tectosphere which could rotate about its center without creating gaps or overlaps.) Little can be said about the tensile stresses since little is known. Material, which by various processes has been made denser and slides down under about 45° along the Gutenberg fault zones, can either be 'pushed' down or it can sink down on its own by having acquired a sufficiently greater density than surrounding mantle material. Several arguments suggest that the second of these alternatives is applicable. The motion along the 45° zones then leads to a tensile 'pull' in the adjacent horizontal part of the tectosphere by simple mechanical means which we shall not analyze here.

The lateral compression due to ridges offers some more interesting features which we now discuss. With a height of the ridge, say, 2 km above the sea bottom, the excess vertical pressure is around 600 bars. Now a piece of solid that is subjected to uniaxial compression has zero stress in a perpendicular plane only if laterally unconfined; if confined it has a lateral compressive stress. If stress relaxation occurs, this lateral stress grows asymptotically toward the full vertical stress (hydrostatic pressure). Very little is known either experimentally or theoretically about stress relaxation. Whether it occurs can sometimes depend on boundary conditions (figure 5). In figure 5(b) horizontal stresses will continue almost indefinitely as the material spreads to the sides. On the other hand, it is intuitively evident that, while stress relaxation in figure 5(a) must occur, it cannot depend on the particular geometry. One expects it to be an

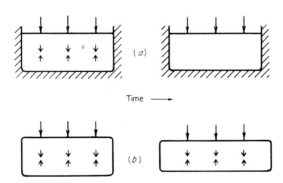

FIGURE 5 Effect of boundary conditions on stress relaxation

intrinsic property of the material. There is one quantity having the dimensions of a time which can be formed from the *macroscopic* parameters of the material, namely $t_r = \eta/E$, where we now interpret η as the (linearized) viscosity for slow creep, assuming that such a quantity can be meaningfully defined in order of magnitude. It should be noted that the often used value $\eta = 10^{23}$, derived from the postglacial uplift, assumes *a priori* a viscous fluid, that is, it assumes that stress relaxation has already occurred. With this value of η we find $t_r = 2000$ years. Thus, unless this value of η should be completely meaningless, stress relaxation is much faster than the postglacial uplift. It may be safe to assume that η is somewhat smaller on ridges than it is under Scandinavia, and the stress under ridges might thus not be far from hydrostatic at all times. Thus we get appreciable forces for a lateral push on the tectosphere.

By the nature of stress relaxation we do not mean the disappearance of the convection generating forces. We mean the conversion of a local stress tensor into an isotropic hydrostatic pressure. A tendency to relaxation always exists since relaxation diminishes the stress energy. There remain, after relaxation, the circulation-generating forces of hydrodynamics: such forces appear whenever the surfaces of constant pressure do not coincide with the surfaces of constant density. There is no implication in this that the viscous friction must be linear.

On the Earth, isostatic equilibrium, which means balance of vertical forces only, is often rather closely fulfilled. The uncompensated horizontal forces should lead to slow horizontal expansion or contraction, as the case may be. This was first emphasized by Benioff[47] for the continental plates but very clear evidence is lacking for these. Orowan[42] suggests that the breakup of northern Canada into an assembly of islands may be attributed to a 'Benioff spreading stress' of this kind. If continental plates are less ductile than ocean-bottom material, the spreading stresses of the latter, issuing from the ridges, should preponderate. In fact, the well-known double maximum of the terrestrial height–distribution curve (e.g. Scheidegger[48]) with a strong concentration of heights near sea level and another strong concentration at the level of the abyssal deeps (4–5 km) indicates decisively that there must be a lateral force holding the continents together and keeping them 'thick', as it were.

This would conclude our argument except for the need of avoiding one likely misconception. Sliding of the tectosphere as understood here is not identical with 'sea-floor spreading' as deduced from magnetic observations[49, 50]. At large distances from the center of a ridge the basaltic crustal layer, which may be assumed to contain the magnetization, is no doubt carried along passively by the moving tectosphere, and to this extent the two are equivalent. But, if the model of material rising under ridges and then spreading out to form new tectosphere is correct, there is a large discrepancy: a zone of stagnation should appear in the middle of the ridge where the horizontal velocity from being zero in the middle increases slowly as one goes outwards. One may estimate that the width of this zone should be comparable with the depth of the convective layer, thus at least 200–300 km. But the observed zone of stagnation for the magnetically observed spreading is only a few kilometers wide. To explain the difference, we must take note of the structure of the oceanic crust. Echo soundings show a thin upper layer of sediments, below it a layer of about 1 km thickness, presumably basalt, and below this a layer roughly 5 km thick resting on top of the mantle. According to Hess[18, 32], this lower crustal layer is largely made up of serpentine, that is, hydrated olivine. Serpentine is a notoriously highly ductile substance and retains high ductility under pressure[51]. So we are tempted to attribute the extreme narrowness of the zone of stagnation, and the gradual transition from there into motion synchronized with a sliding tectosphere, to the easy deformability of serpentine. The magnetically observed motion would be that of the crust which is not necessarily identical with that of the mantle. In our earlier analaysis we have therefore relied on features of macrotectonics rather than on specific magnetic data.

Acknowledgements

The research leading up to this chapter was supported by a grant from the National Science Foundation and by a grant from the National Aeronautics and Space Administration.

References

1. H. A. Lubimova, *Geophys. J.*, **1**, 115–34 (1958).
2. G. J. F. MacDonald, *J. Geophys. Res.*, **64**, 1967–2000 (1959).
3. F. Birch, *Bull. Geol. Soc. Am.*, **76**, 133–54 (1965).
4. C. C. Patterson and M. Tatsumoto, *Geochim. Cosmochim. Acta*, **28**, 1–22 (1964).
5. D. C. Tozer, *Phys. Chem. Earth*, **3**, 414–36 (1959).
6. D. L. Anderson, *Science*, **157**, 1165–73 (1967).
7. W. M. Elsasser, 'Thermal structure of the upper mantle and convection', in *Advances in Earth Science* (Ed. P. M. Hurley), Massachusetts Institute of Technology Press, Massachusetts, 1966, pp. 461–502.
8. S. P. Clark, Jr., 'Variation of density in the Earth and the melting curve in the mantle', in *The Earth Sciences* (Ed. T. W. Donnelly), Rice University Press, Houston, 1963, pp. 5–42.
9. B. Mason, *Principles of Geochemistry*, 2nd ed., Wiley, New York, 1958.
10. S. P. Clark, Jr., (Ed.), *Handbook of Physical Constants* (*Geol. Soc. Am., Mem.*, **97** (1966)).
11. G. J. F. MacDonald, *Rev. Geophys.*, **1**, 587–605 (1963).
12. D. F. McKenzie, *J. Geophys. Res.*, **71**, 3995–4010 (1966).
13. W. M. Elsasser, *Rev. Mod. Phys.*, **28**, 135–63 (1955).
14. L. Knopoff and G. J. F. MacDonald, *Geophys. J.*, **1**, 284–97 (1958).
15. A. E. Ringwood, 'The chemical composition and origin of the Earth', 'Mineralogy of the mantle', in *Advances in Earth Science* (Ed. P. M. Hurley), Massachusetts Institute of Technology Press, Massachusetts, 1966, pp. 287–356, 357–99.
16. J. Verhoogen, *Phil. Trans. Roy. Soc. London, Ser. A*, **258**, 276–83 (1965).
17. H. W. Menard, *Phys. Chem. Earth*, **6**, 317–64.
18. H. H. Hess, *J. Marine Res.* (*Sears Found. Marine Res.*), **14**, 423–39 (1955).
19. R. S. Dietz, *Nature*, **190**, 854–7 (1961).
20. A. E. Ringwood and D. H. Green, *Tectonophysics*, **3**, 383–427 (1966).
21. J. Weertman, *J. Geophys. Res.*, **67**, 1133–9 (1962).
22. R. B. Jones, *Nature*, **207**, 70–1 (1965).
23. B. Gutenberg, *Physics of the Earth's Interior*, Academic Press, New York, 1959.
24. D. L. Anderson, *Phys. Chem. Earth*, **6**, 1–131 (1966).
25. H. C. Heard, *J. Geol.*, **71**, 162–95 (1963).
26. D. T. Griggs and J. D. Blacic, *Science*, **147**, 292–5 (1965).
27. E. Orowan, *Phil. Trans. Roy. Soc. London, Ser. A*, **258**, 284–313 (1965).
28. E. C. Bullard, A. E. Maxwell, and R. Revelle, *Advan. Geophys.*, **3**, 153–81 (1956).
29. W. M. Elsasser, *J. Geophys. Res.*, **72**, 4768–70 (1967).
30. D. C. Tozer, in *The Earth's Mantle* (Ed. T. F. Gaskell), Academic Press, New York 1967, p. 325.
31. S. K. Runcorn, *Phil. Trans. Roy. Soc. London, Ser. A*, **258**, 228–51 (1965).
32. H. H. Hess, 'History of ocean basins', *Petrologic Studies: A Volume to Honor A. F. Buddington*, Geological Society of America, New York, 1962, pp. 599–620.
33. L. R. Sykes, *J. Geophys. Res.*, **72**, 2131–53 (1967).
34. J. T. Wilson, *Nature*, **207**, 343–7 (1965).
35. J. Oliver and B. Isacks, *J. Geophys. Res.*, **72**, 4259–75 (1967).
36. B. C. Heezen, 'The deep-sea floor', in *Continental Drift* (Ed. S. K. Runcorn), Academic Press, New York, 1962, pp. 235–88.
37. M. Ewing, J. I. Ewing, and M. Talwani, *Bull. Geol. Soc. Am.*, **75**, 17–35 (1964).
38. H. W. Menard, *Science*, **155**, 72–4 (1967).

39. V. Vacquier, A. D. Raff, and R. E. Warren, *Bull. Geol. Soc. Am.*, **72,** 1251–70.
40. E. Orowan, to be published (1967).
41. K. E. Bullen, *An Introduction to the Theory of Seismology*, 3rd ed., Cambridge University Press, London, 1963.
42. E. Orowan, *Mechanical Problems of Geology*, Lecture, California Institute of Technology (1958).
43. V. V. Beloussov, *Basic Problems in Geotectonics*, McGraw-Hill, New York, 1962.
44. L. D. Landau and E. M. Lifshitz, *Fluid Mechanics*, Academic Press, New York, 1959.
45. J. C. Jaeger, *Elasticity, Fracture and Flow*, Methuen, London, Wiley, New York, 1962.
46. W. J. Morgan, *J. Geophys. Res.*, **73,** 1959–82 (1968).
47. H. Benioff, *Bull. Geol. Soc. Am.*, **60,** 1837–56 (1949).
48. A. E. Scheidegger, *Principles of Geodynamics*, Academic Press, New York, 1963.
49. F. J. Vine, *Science*, **154,** 1405–15 (1966).
50. F. J. Vine and D. H. Matthews, *Nature*, **199,** 947–9 (1963).
51. C. B. Raleigh and M. S. Paterson, *J. Geophys. Res.*, **70,** 3965–85 (1965).

2

F. C. FRANK

H. H. Wills Physics Laboratory
University of Bristol
Bristol, England

Diamonds and deep fluids in the upper mantle*

'A snowflake is a letter from the sky.'—Ukichiro Nakaya

'Diamonds are letters
Still better worth the reading:
We can reach the sky.'—F.C.F.

Detailed study of the internal crystal defects of diamonds shows the following facts.

(i) Most diamonds, if not hardened by nitrogen precipitation, have suffered plastic deformation, generating dislocations arrayed on (111) slip bands, made visible by birefringence, or by X-ray topography or electron microscopy of the actual dislocations[2-4]. This implies that they have been subjected to high shear stress at high temperature (probably above 2000 °K)[5,6]. Others, with much smaller birefringence, are nevertheless dislocation rich, the dislocations appearing to have polygonized: this implies a further period at high temperature after deformation[7].

(ii) Most diamonds experienced numerous, sometimes abrupt, changes in chemical environment during growth. Among the most easily visible features are growth shells with widely different nitrogen content[1, 8-10].

(iii) Within this complex variation, a general sequence of three stages is recognizable for many diamonds.

(a) An early stage in which the growth habit was rounded, or mamillary, except for limited octahedral flat facets[1, 9-13]. During this phase the impurity content was different on the octahedral facets and other surfaces of the crystal growing at the same time. This implies a very low diffusion rate for the impurities concerned in the crystal at the time of growth[14]. Sectors of mamillary growth also incorporated minute specks of composition corresponding to single minerals (silica, corundum, and probably garnet)[15]. These must therefore have been present as solids, presumably finely divided, at the time of growth. Their non-incorporation in the growth sectors of the octahedral facets is explicable by considerations of surface energy. Stage (a) is missing from the growth history of some diamonds.

(b) A stage of growth in octahedral habit, with fluctuations in nitrogen uptake still occurring.

---

* This paper is substantially a compressed version of one presented to the Oxford Industrial Diamonds Conference 1966[1].

(c) A stage of re-solution, varying in amount from removal of a substantial fraction of the earlier weight (the commonest case) to a comparatively light etch[1, 16, 17].

The foregoing is applicable fairly generally to diamonds of South African, Tanganyikan, and Indian origin, and probably true of diamonds of much more widely various origin, but there are doubtless others to which it does not apply: in particular, the 'coated stones' of the Congo, with a surface layer of finely fissured diamond intermingled with small amounts of other matter, must have had a distinct final growth history.

(iv) The nitrogen in diamonds (in concentration up to about 0.3%[18]) is normally present as very thin platelets (? two atoms thick), on $\{100\}$ planes, of equant habit in the plane, and of equal frequency on each of the three equivalent planes, just like the Preston–Guinier zones of an age-hardening alloy. However, the boundary between a platelet-rich and platelet-free region is just as sharp as the average distance between platelets[19]. It does not seem possible to explain this situation otherwise than by supposing that nitrogen was first incorporated into the diamond, during growth, at a temperature at which diffusion of nitrogen in the crystal during the time of growth was negligible, and that the nitrogen atoms subsequently diffused to precipitate as platelets, either during hot storage for a time orders of magnitude longer than the growth time, or on a considerable rise of temperature after growth (it is almost certain that the effect of pressure change on diffusion would be insufficient)[1]. The evidence for a turbulent growth history makes one prefer the explanation in terms of temperature rise to that of prolonged isothermal storage. It is to be presumed that work on synthetic diamonds will in due course reveal the temperature required for the diffusion of nitrogen in diamond. 2000 °K may well be insufficient.

Diamond is unstable relative to graphite at all temperatures for pressures less than 13 kb. At 500 °K the equilibrium pressure is 20.2 kb, at 1000 °K 33.1 kb, at 2000 °K 61.2 kb, being closely approximated by $(30T)$ b except at low temperatures[20, 21]. Three times the pressure in kilobars is approximately equatable to depth in kilometres. Thus diamonds are only stable at subcrustal depths, exceeding 100 km for a temperature of 1000 °K, or 185 km for a temperature of 2000 °K. Unless the diamonds grew metastably (and human attempts at achieving this have so far been unsuccessful) they grew at depths such as this or greater. Since there is independent evidence that diamonds have experienced temperatures above 2000 °K, which imply a communication with great depth, the problem of the origin of diamonds is made no easier by the less probable assumption of metastable growth. Let us make the more straightforward assumption of thermodynamically stable growth.

Carbon should dissolve in silicates, especially molten silicates, under conditions forbidding the escape of CO_2 or CO, in the form of carbonate ions and heteropolysilicate–carbonate ions. At a suitably low oxidation–reduction potential (regulated by the Fe^{3+}/Fe^{2+} ratio) carbon should separate from such a system, and, at sufficient depth, in the form of diamond. Formation of diamond is thus no problem, but there is a real problem in bringing it to the surface, bearing in mind that it is unstable for the last 100 km of its rise from the depths. Very fast transport through the unstable regime, limiting the amount of re-solution by limitation of time, appears to offer the only way out. Only fluid flow could be fast enough. The following speculative picture of a sequence of events is thus formed.

A pocket of hot, perhaps water-rich, melt-fluids or magma, slowly rising from a depth of some hundreds of kilometres, pierced a channel through a continental crust, clearing a passage for accelerated flow of more material from below. The thermal changes in such a rising column depend notably on whether the work of expansion is expended elsewhere (as by lifting the surface of the Earth) or, frictionally, upon the rising material itself. In the former case, the expansion is adiabatic, and, without phase changes occurring, would amount to

$$\left(\frac{\partial T}{\partial p}\right)_S = \left(\frac{\partial V}{\partial S}\right)_p = \frac{\alpha T}{\rho c_p} \quad (1)$$

say, about 0·5 degK/kb, making a slow fall of temperature in rising material. However, in a magma the temperature would be regulated by progressive melting, and held to the pressure melting (or solution) point according to the law

$$\frac{dT}{dp} = \frac{\Delta V}{\Delta S} \quad (2)$$

the two latter quantities being volume and entropy changes respectively, for small increments of melting. This ratio may be estimated as about 2 degK/kb. This, being a more rapid temperature fall with rise of material than the rate of fall without phase change, would be accompanied by progressive melting. With the work expended internally, and without phase change, the law would be that of isenthalpic expansion:

$$\left(\frac{\partial T}{\partial p}\right)_H = -\frac{1}{c_p}\left\{V - T\left(\frac{\partial V}{\partial T}\right)_p\right\} = -\frac{1}{\rho c_p}(1 - \alpha T) \quad (3)$$

amounting to, say, −25 degK/km, the temperature rising relatively rapidly as the material rises. In a magma, this temperature rise (of about 8 degK/km) would be replaced by progressive melting, the temperature falling according to (2) during the melting of any major solid constituent, and rising as the more fusible constituents were exhausted, giving something like complete melting in 150 km, after which the temperature would rise rapidly.

The mean flow velocity for a fluid of density ρ, $\Delta\rho$ less than the density of its surroundings (which determines the pressure gradient), in a vertical pipe of radius r is[22]

$$\left(\frac{4rg\Delta\rho}{\rho\lambda}\right)^{1/2}$$

where the hydrodynamic resistance coefficient λ is not highly dependent on viscosity or wall roughness if the Reynolds number is sufficiently high for fully developed turbulence, and may be put equal to, say, 1/40. Putting $\Delta\rho/\rho$ at 1/20, and $r = 10^3$ cm, one obtains a velocity of about 100 km/hour. Vapour evolution would produce a higher velocity and a rapid fall in temperature in the last kilometre or so.

This model appears to provide the requisite conditions for diamond growth and ejection, with an initial stage of growth in magma, a later stage of growth in a significantly different environment, largely melted, with much turbulence in both stages, a subsequent important temperature rise, and stage of re-solution, and a rapid passage through the final 100 km in which diamond is unstable. It does not resemble any eruptive process happening at the present day.

One may speculate still further on a means of obtaining the requisite carbonaceous (and nitrogeneous) magma pocket at a depth of some hundreds of kilometres. Accepting the picture of the ocean bed underthrusting the continents, on surfaces indicated by the foci of deep earthquakes, and taking the view that the focal event is a fracture which can only occur in the presence of fluid, then 700 km, the limiting depth of earthquakes, marks the place where pressure finally drives all fluids into solid combination. If in its further travel the same material (which may carry with it carbonaceous and nitrogeneous matter of biological origin from the ocean bed) is brought to a depth of less than 700 km, then partial melting occurs in it once more, and the conditions precedent for the foregoing account are established.

References

1. F. C. Frank, *Proc. Intern. Industrial Diamond Conf., Oxford, 1966*, Industrial Diamond Information Bureau, London, 1967.
2. G. N. Ramachandran, *Proc. Indian Acad. Sci., Sect. A*, **24**, 65 (1946).
3. G. P. Freeman and H. A. van der Velden, *Physica*, **18**, 9 (1952).
4. T. Evans and C. Phaal, *Proc. Roy. Soc. (London), Ser. A*, **270**, 538 (1962).
5. T. Evans and R. K. Wild, *Phil. Mag.*, **12**, 479 (1965).
6. T. Evans and R. K. Wild, *Phil. Mag.*, **13**, 209 (1966).
7. A. R. Lang, *Nature*, **213**, 248 (1967).
8. C. V. Raman and G. R. Rendall, *Proc. Indian Acad. Sci., Sect. A*, **19**, 265 (1944).
9. M. Takagi and A. R. Lang, *Proc. Roy. Soc. (London), Ser. A*, **281**, 310 (1964).
10. F. C. Frank and A. R. Lang, in *Physical Properties of Diamonds* (Ed. R. Berman), Clarendon Press, Oxford, 1965, pp. 106–13.
11. E. R. Harrison and S. Tolansky, *Proc. Roy. Soc. (London), Ser. A*, **279**, 490 (1964).
12. M. Seal, *Proc. Intern. Congr. on Diamonds in Industry, 1st, Paris, 1962*, p. 361.
13. M. Seal, *Am. Mineralogist*, **50**, 105 (1965).
14. J. B. Mullin and K. F. Hulme, *Phys. Chem. Solids*, **17**, 1 (1960).
15. M. Seal, *Nature*, **212**, 1528 (1966).
16. F. C. Frank, K. E. Puttick, and E. Wilks, *Phil. Mag.*, **3**, 1262 (1958).
17. M. Seal, unpublished.
18. W. Kaiser and W. L. Bond, *Phys. Rev.*, **115**, 857 (1959).
19. T. Evans, in *Physical Properties of Diamonds* (Ed. R. Berman), Clarendon Press, Oxford, 1965, p. 131.
20. R. Berman and F. E. Simon, *Z. Elektrochem.*, **59**, 333 (1955).
21. R. Berman, in *Physical Properties of Diamond* (Ed. R. Berman), Clarendon Press, Oxford, 1965, p. 382.
22. L. Prandtl, *Essentials of Fluid Dynamics*, Blackie, London, 1952, p. 165.

3

F. R. N. NABARRO*

Cavendish Laboratory
Cambridge, England

Steady-state diffusional creep

Abstract

The theory of steady-state diffusional creep is developed for a material in which the grains are large and the dislocations form the principal sources and sinks for vacancies. Under a given applied shear stress p there is an equilibrium length of link in the dislocation network, only slightly longer than that link which provides a dislocation source under the applied stress. The dislocation density increases by the operation of Bardeen–Herring sources, and decreases as dislocations of opposite sign climb towards one another and annihilate. When diffusion occurs predominantly through the bulk of the crystal, the rate of strain is proportional to p^3. At lower temperatures, where diffusion occurs predominantly along the cores of dislocations, the rate of strain is proportional to p^5.

Approximate formulae for the two cases are

$$\dot{\varepsilon} = \frac{Dbp^3}{\pi kTG^2} \bigg/ \ln\left(\frac{4G}{\pi p}\right)$$

and

$$\dot{\varepsilon} = \frac{4D_c bp^5}{\pi^4 kTG^4}$$

Here $\dot{\varepsilon}$ is the rate of strain, b is the Burgers vector of the dislocations, p is the applied stress, G the shear modulus, and kT the Boltzmann factor. The self-diffusion coefficient, assumed to be produced by a vacancy mechanism, is D, and D_c is an enhanced self-diffusion coefficient in the cores of dislocations, which are taken to be of cross-sectional area b^2.

Details of the calculation are given in the paper by Nabarro[1].

It should be emphasized that these strain rates are achieved in the steady state, when there is no trace left of the original dislocation array, and the grain size has become very large.

References

1. F. R. N. Nabarro, *Phil. Mag.*, **16**, 231–7 (1967).

* On leave from the University of the Witwatersrand, Johannesburg, South Africa.

4

C. A. BERG

Department of Mechanical Engineering
Massachusetts Institute of Technology
Cambridge, Massachusetts, U.S.A.

The diffusion of boundary disturbances through a non-Newtonian mantle

Introduction

Classical quantitative work on the mechanics of deformation in the Earth's mantle is based on the assumption that the mantle can be treated as a Newtonian viscous fluid. With but a handful of exceptions, modern quantitative work on flow and convection in the mantle is also based upon the assumption of Newtonian viscosity. It has been pointed out by Orowan[1] that the material of the mantle is most likely a polycrystalline aggregate at an elevated temperature, and that therefore the main mechanism of permanent deformation in the mantle is probably either plastic deformation or Andradean creep. Orowan noted that the sharp curvatures of the vertical profiles of the oceanic ridges could not be provided by Newtonian viscous flow in the mantle because the radius of curvature of a cylindrical convection cell in a Newtonian fluid must be of the same order of magnitude as the diameter of the cell. In the same writing, Orowan investigaged the feasibility of convection via ideal plastic deformation entailing the rise of a hot dike of material under the oceanic ridge.

In addition to the observed sharp curvature of the oceanic ridge profiles, another experimental observation bears on the mechanism of permanent deformation in the mantle. According to Wilson[2] the ages of islands increase with distance from an oceanic ridge; recent potassium argon dating of rocks on the ocean floor also suggests that age is roughly proportional to this distance, all of which may be interpreted as the consequence of movement of the ocean floor with uniform velocity away from the crest of a ridge. A number of investigators have proposed convective velocity patterns in the vicinity of oceanic ridges which provide uniform velocities of spreading of the ocean floor away from the ridge. Figure 1(*a*) shows the velocity pattern adopted by Ewing for the purposes of calculating heat flow in the vicinity of a ridge; Orowan[3] has pointed out that the velocity field of figure 1(*a*) should be corrected as is shown in figure 1(*b*) for the width of the oceanic ridge, and finally that the velocity fields of figure 1(*a*) and 1(*b*) are incompatible with the mechanical properties of real materials and that the actual velocity field of a convective cell which provides for uniform spreading velocity of the ocean floor would look more like that given in figure 1(*c*). There are two common aspects to these velocity patterns. First, in all cases it is assumed that the convective

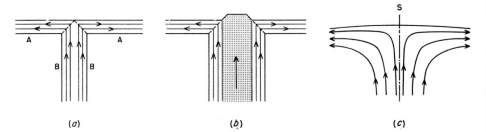

FIGURE 1 Velocity patterns which have been proposed to represent the ascending convective flow under an oceanic ridge. (*a*) shows the pattern proposed by Ewing. (*b*) shows Orowan's correction to Ewing's pattern, to allow for the width of the ridge. (*c*) shows the pattern which Orowan[3] proposes as the most likely to exist in view of both the width of the ridge and the actual mechanical properties of polycrystalline aggregates, such as the mantle

flow is plane. The midoceanic ridges are relatively straight and have rather large ratios of length to width, as is shown for example by Heezen's[4] map of the Mid-Atlantic Ridge given in figure 2. One should note that the radius of curvature of the ridge,

FIGURE 2 Heezen's[4] map of the Mid-Atlantic Ridge. Note that, except at points of concentrated curvature, the radius of curvature of the ridge is large compared with the width of the ridge

measured in a plane tangent to the Earth's surface, is large compared with the width of the ridge. The extent to which nonplanar flow might be significant in a convective upwelling under the ridge depends upon the ratio of width of the ridge to the radius of curvature of the ridge measured in the horizontal planes; as this ratio becomes smaller nonplanar effects become less significant, and would disappear entirely if the ratio were zero (i.e. if the ridge were straight). Assuming that the ridges are situated over ascending branches of convective flows then, because the ridges are relatively straight, one may, as a first approximation, treat the assumed upwelling convection under the ridges as planar flow. Second, in all of the assumed velocity patterns, the ascending branch of the convective motion is assumed to be relatively narrow (i.e. small compared with the width of the ocean). Therefore, there are assumed to be sharp horizontal gradients of the velocity as one moves across the ascending branch of convection and into the adjacent mantle material. These strong horizontal gradients of velocity are necessary if the velocity pattern is to provide uniform velocity of spreading of the ocean floor. Considering the behavior of the vertical component of velocity across any horizontal cut through the ascending branch at a level well below the ocean floor, one sees that the velocity passes from a maximum in the ascending branch to a vanishingly small magnitude in the adjacent, essentially nondeforming, mantle material. Thus there are both strong horizontal gradients and strong curvatures of the velocity profile. One of the major questions about the appropriate rheological model to be adopted in the study of mantle convection is just how the sharp gradients and curvatures of the velocity field which seem to be required for the building of the relatively narrow ridges may arise. Finally, it is noted that, according to Vine's[5] study of magnetic anomalies on the ocean floor, it can be inferred that uniform spreading of the ocean floor has been taking place for approximately 10^8 years. This is approximately the time interval over which continental drift is supposed to have occurred, and, if one attributes continental drift and spreading of the ocean floor to mantle convection, then from the observations of Wilson and Vine it may be inferred that convective motion in the mantle has been in steady state for approximately 10^8 years, that is during most of the period of continental drift. In the following, the possibilities that the main mechanism of permanent deformation is either Andradean viscous flow or ideal plastic deformation are considered, and the consequences of assuming either of these two types of material behavior in mantle convection are compared with the observations discussed above.

Mechanics of Flow in the Mantle

We shall first suppose that the main mechanism of permanent deformation in the mantle is Andradean flow. As noted above, the profile of the vertical component of velocity in a convective rise in the vicinity of a ridge should have a strong curvature in the horizontal direction. Now the pressure field in the apparently quiescent section of the mantle adjacent to the assumed convective rise is predominantly determined by gravity. In this case the strong curvature of the profile of the vertical velocity component, which one postulates to occur in passing from the convective rise into the quiescent material, can come about only by diffusion of momentum into the quiescent material. It is important to note here that the non-Newtonian flow which might occur in the vicinity of a convective rise in the mantle is qualitatively different from the non-Newtonian flow which one might observe in a pipe. In flow through a pipe the pressure gradient along the axis of the pipe is prescribed and the pressure gradient field is not balanced by a body force. To meet the conditions of equilibrium the prescribed axial pressure gradient in a pipe must be accompanied by a gradient of the shear stress in the radial direction across the pipe, and the radial gradient of shear stress causes corresponding curvatures in the profile of the velocity. In the case of flow in the mantle, the pressure gradients are, to close approximation, compensated by gravity and therefore the curvature of the profile of the vertical component of velocity, which occurs in the vicinity of an oceanic ridge, will be accompanied by strong horizontal gradients of the shear stress which are not balanced by a vertical pressure gradient, and will therefore cause local diffusion of momentum into the nearby quiescent material. In order to decide whether Andradean flow is an appropriate description of the mechanism of permanent deformation in the mantle, one should determine how rapidly momentum may diffuse through a non-Newtonian mantle.

In a Newtonian fluid, with kinematic viscosity v, the characteristic distance of diffusion (δ) of a boundary disturbance during the time t is given by $(vt)^{1/2}$. That $(vt)^{1/2}$ is the characteristic distance in the hydrodynamics of a Newtonian viscous fluid can be obtained via a number of different arguments—e.g. dimensional analysis—however, an explicit demonstration of this is given in Stokes'[6] solution for the flow in a Newtonian fluid near a flat plate which is suddenly accelerated from rest and moves in its own plane with constant velocity u_0. The velocity (u) in the fluid is everywhere parallel to the velocity of the plate and depends upon the coordinate (x_2) normal to the plate and time (t) according to

$$\frac{u}{u_0} = \mathrm{erfc}\left\{\frac{x_2}{2(vt)^{1/2}}\right\}$$

From this expression, the emergence of $(vt)^{1/2}$ as the characteristic distance of diffusion is clear.

In the case of polycrystalline materials at high temperatures, the connection between the shear stress (σ) and the shear-strain rate (ε) in a pure shear experiment is

$$\left(\frac{\sigma}{\sigma_0}\right)^n = \frac{\varepsilon}{\varepsilon_0} \qquad (1)*$$

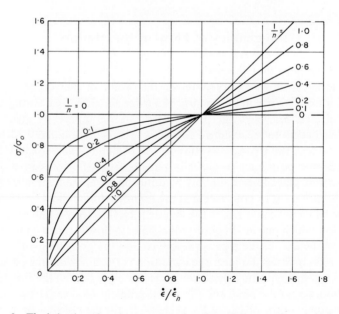

FIGURE 3 The behavior of strain–stress as a function of shear-strain rate (ϵ) for a fluid having a constitutive relation of the form $(\sigma/\sigma_0)^n = \varepsilon/\varepsilon_0$; these are the Andradean fluids. Note as $n \to \infty$ the behavior of the fluid approaches ideal plasticity

The (hardening) exponent (n) is greater than unity. Figure 3 shows the behavior of stress against strain rate for various values of the hardening exponent; it should be noted that with fixed values of σ_0 and ε_0 equation (1) passes in the limit as $n \to \infty$ to a description of ideal plastic deformation. The representation (1) for the steady non-Newtonian flow of crystalline matter was recognized, and separated from transient flow, by Andrade[7] and was confirmed experimentally by Norton[8]. Although experimentally determined values of n usually lie between 3·5 and 5·5 for glacial ice and common metals in temperature ranges of technical interest, the exponent n increases strongly as the temperature of the material is decreased below the melting point; for example, figure 4 shows steady creep data of stainless steel at 1100 °F (approximately 0·47 of the solidus temperature) and for this creep curve $n \sim 20$.

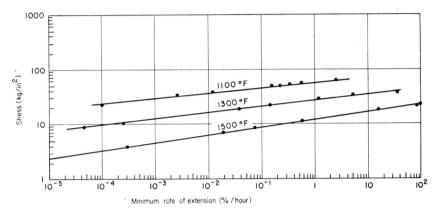

FIGURE 4 Data for creep of stainless steel at high temperatures. The shear-stress (σ)–shear-strain rate (ε) data can be represented by an Andradean law, and for the experiments at 1100 °F the Andradean exponent (n) is approximately 20

In the case of an Andradean fluid a characteristic distance of diffusion is not as readily evident as in the case of the Newtonian fluid; however, one can recognize that the local rate of diffusion through an Andradean fluid will depend upon the slope of the stress–strain rate curve for the shear rate at the point, and that as this slope decreases so does the rate of diffusion of momentum through the fluid. If one compared the rates of momentum diffusion in a Newtonian and an Andradean fluid having a common point on their stress–strain rate curves, as is shown in figure 5, then one would expect that at low shearing rates the Andradean fluid will diffuse momentum more rapidly than the Newtonian fluid, and at high shearing rates, where the slope of the stress–strain rate curve of the Andradean fluid decreases, the diffusion of momentum in the Newtonian fluid will be faster. One recognizes that the characteristic rate of diffusion of momentum in a non-Newtonian fluid depends upon the nature of the boundary disturbance as well as upon the properties of the fluid because the strength of the boundary disturbance determines whether the mean strain rate level propagated through the fluid is small (with high rates of diffusion of momentum) or large (with low rates of momentum diffusion). In order to obtain a quantitative estimate for the characteristic distance of diffusion of momentum through a non-Newtonian fluid, it is helpful to have an exact solution to examine. The most natural and least complicated problem to analyze for this purpose is the problem of the motion in a fluid adjacent to

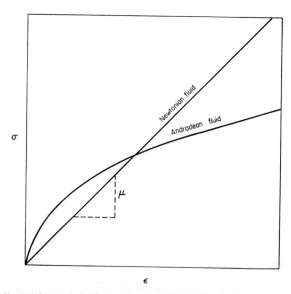

FIGURE 5 Comparison of Andradean and Newtonian fluid deformation. The local slope of the stress (σ)–strain rate (ε) curve determines the rate of diffusion of momentum through the fluid. At low levels of strain rate, momentum will diffuse faster through the Andradean fluid than through the Newtonian fluid; at high levels of strain rate, momentum will diffuse faster through the Newtonian fluid

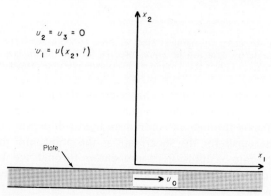

FIGURE 6 A flat plate lying adjacent to a semi-infinite fluid body. The plate is impulsively accelerated to the velocity u_0 in its own plane. In the fluid the velocity ($u_1 = u(x_2, t)$) lies parallel to the velocity of the plate and depends upon the coordinate (x_2) normal to the plate and upon time (t)

an accelerated plate, solved by Stokes[6] for the case of a Newtonian fluid. Figure 6 shows the geometry of this problem. The (x_1, x_2) coordinate axes are aligned (respectively) parallel and perpendicular to the plate, and the plate is impulsively accelerated to the velocity u_0 in the direction of the x_1 axis. The fluid undergoes shearing flow, so that the velocity components in the x_2 and x_3 directions vanish (the x_3 axis is normal to the plane of the figure and is not shown) and the velocity in the x_1 direction depends only on x_2 and time (t), as is expressed in the following equations:

The diffusion of boundary disturbances through a non-Newtonian mantle

$$u_2 = u_3 = 0; \qquad u_1 = u(x_2, t) \tag{2}$$

The deformation rates which derive from such a velocity field are given by*

$$\varepsilon_{ij} = (u_{i,j} + u_{j,i}) = \begin{pmatrix} 0 & u_{,2} & 0 \\ u_{,2} & 0 & 0 \\ 0 & 0 & 0 \end{pmatrix} \tag{3}$$

and the velocity field is to be determined so that it meets the following auxiliary data:

$$\begin{aligned} u(x_2, t) &= 0 & t &< 0 \\ u(0, t) &= u_0 & t &> 0 \\ u(\infty, t) &= 0 & t &> 0 \end{aligned} \tag{4}$$

Assuming that the mantle may be treated as an incompressible isotropic fluid, making use of the facts that the assumed flow field depends neither upon the coordinate normal to the plane of flow nor upon the coordinate parallel to the velocity of the plate, and, using the condition that the velocity field in the fluid vanishes at infinity, one finds (appendix) that the only equation of equilibrium which is not identically satisfied is that for equilibrium in the x_1 direction. Further, assuming that $u_{,2} > 0$, then the stress components acting in the (x_1, x_2) plane in the Andradean fluid described by (1) are

$$\sigma_{ij} = \begin{pmatrix} p & (\sigma_0/\varepsilon_0^{1/n})(u_{,2})^{1/n} \\ (\sigma_0/\varepsilon_0^{1/n})(u_{,2})^{1/n} & p \end{pmatrix}, \qquad (i, j) = (1, 2) \tag{5}$$

where p is a scalar function of x_2 and time (t). The equilibrium condition for the x_1 direction is

$$(u_{,2})^{(1-n)/n}(u_{,22}) = nu_{,T} \tag{6}$$

with

$$T = \frac{\sigma_0}{\varepsilon_0^{1/n}} \frac{t}{\rho} \tag{7}$$

This equation possesses a similarity solution in the parameter

$$\beta = x_2 \left(\frac{n+1}{n} \frac{T}{n} \right)^{-n/(n+1)} \tag{8}$$

and introducing the parameter β reduces (6) to

$$(u')^{(1-n)/n} u'' + \beta u' = 0 \tag{9}$$

where $( )' \equiv d/d\beta$.

The auxiliary data which must be met by this solution are

$$\begin{aligned} u(0) &= u_0 \\ u(\infty) &= 0 \end{aligned} \tag{10}$$

---

* Equation (1) describes the steady state creep of the material; neither elastic strains nor primary creep strains—both of which are usually very small—are represented by (1).

† A comma followed by an index indicates differentiation, i.e. $\phi_{,j} \equiv \partial\phi/\partial x_j$. Repeated indices in a product, adifferentiation, or a second-order tensor indicate summation over the range of the index, i.e. $\varepsilon_{kk} = \varepsilon_{11} + \varepsilon_{22} + \varepsilon_{33}$; the symbol δ_{ij} denotes the Kronecker delta, has the value 1 when $i=j$, and is zero otherwise.

One notes that the assumption that $u_{,2} > 0$ together with the condition (4) that u vanish as $x_2 \to \infty$ requires that u_0 be negative; thus one is concerned here with the case in which the plate of figure 6 is accelerated to the left. To obtain the fluid motion for plate acceleration to the right, one simply rotates the velocity field to be given below. The assumption $u_{,2} > 0$ was introduced to simplify calculations and carries no physical importance. Bearing this in mind the solution for equation (9) is

$$\frac{u(\beta)}{|u_0 \cdot|} + 1 = \int_0^\beta \left[\frac{n-1}{2n} \{\alpha^2 + \beta_0(n)^2\} \right]^{n/(1-n)} d\alpha \tag{11}$$

where $\beta_0(n)$ is determined by the condition that $u(\infty) = 0$, i.e.

$$\int_0^\infty \left(\frac{n-1}{2n} \right)^{n/(1-n)} \{\alpha^2 + \beta_0(n)^2\}^{n/(1-n)} d\alpha = 1 \tag{12}$$

Figure 7 shows the value of $\beta_0(n)$ as a function of n; the ratio of the fluid velocity at $\beta_0(n)$ to the plate velocity is given as a function of n in figure 8. One notes that $\beta_0(n)$ is of order unity and that at the distance $\beta_0(n)$ from the accelerated plate the fluid velocity is of the same order as the plate velocity. $\beta_0(n)$ then provides a measure of the distance from the plate over which the fluid has been 'captured' by the motion of the plate, and can therefore serve as a measure of the characteristic distance of diffusion of momentum through an Andradean fluid when the disturbing velocity at the boundary is fixed. If one were dealing with a case in which the boundary velocity were not constant but were to follow some prescribed program in time, the initial response of the fluid to the first step in boundary velocity would be described by the above calculation and the subsequent responses of the fluid to the subsequent boundary

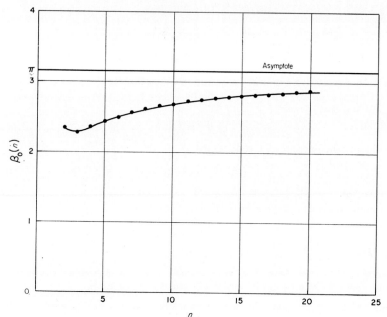

FIGURE 7 The behavior of $\beta_0(n)$ as a function of n. Note that $\beta_0(n)$ is always of order unity and approaches π as $n \to \infty$

velocity perturbations might be determined by treating the fluid as a nonhomogeneous viscous material, assigning a viscosity to each point in the material corresponding to the tangent of the Andradean flow curve (figure 4) associated with the point on the curve for which the shear–strain rate is the same as the local shear–strain rate.

The basic importance of the characteristic distance of diffusion is that it determines whether a given fluid system of typical size (say) L subjected to steady boundary disturbances for a given period of time (t) might be considered to be in steady state motion, or whether transient fluid motion (i.e. the diffusion of momentum across the fluid system) would predominate in the behavior of the system during the time of observation. If for the given time of observation (t) the characteristic length of diffusion $\delta(t)$ is large compared with the scale of the system (L), then the system may be considered to be in steady motion throughout most of the period of observation; otherwise transient motions are significant. It is emphasized that the geometry of flow (figure 6), to which the calculations above apply, is not assumed precisely to correspond to the geometry of convective flow under an oceanic ridge (except in the respect that the flow under the ridge has been assumed to be planar); the purpose of studying the flow of figure 6 is to determine the characteristic rate of diffusion of momentum through a non-Newtonian viscous mantle.

The specific problem under consideration here is the diffusion of momentum into the apparently quiescent mantle material adjacent to a convective rise, and the characteristic length of diffusion $\delta(t)$ serves to determine the period of time required after the onset of convective motion for the spread of deformation through, and the realization of steady flow in, the block of material adjacent to the convective rise.*

It is helpful to examine the limiting forms which the solution (11) takes on for various values of n. For $n = 1$, which is the case of the Newtonian fluid, the similarity parameter β becomes $x_2(2\nu t)^{1/2}$ and (9) becomes

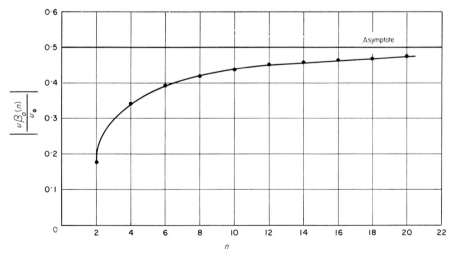

FIGURE 8 The ratio of the fluid velocity ($u(\beta_0(n))$) at a distance from the plate corresponding to $\beta_0(n)$, to the plate velocity u_0. This ratio is of order unity and so one may say that within the distance corresponding to $\beta_0(n)$ the fluid has been captured by the motion of the plate. $\beta_0(n)$ then serves to determine a characteristic distance of diffusion of momentum through the fluid

* The role of heat diffusion in determining the transient time is considered below.

$$u'' + \beta u' = 0 \qquad (13)$$

which, but for a choice of scale factor in the construction of the similarity parameter β, is identical with Stokes'[6] equation noted above. In the limit as $n \to \infty$ the material behavior approaches ideal nonhardening plasticity. For any fixed $x_2 > 0$, the parameter β becomes infinite as $n \to \infty$. Thus, at any station which is removed from the plate by a finite distance $(x_2 > 0)$, the flow field in the limit as $n \to \infty$ will be the same as if the station were infinitely far removed from the plate; that is, the velocity will vanish. The solutions of (11) therefore in the limit as $n \to \infty$ provide plastic slip flow on the face of the plate with no diffusion of momentum into the surrounding material. The velocity profile for the limiting case $n \to \infty$ is

$$u(\beta) = u_0 \left(1 + \frac{2}{\pi} \tan^{-1}\left(\frac{\beta u_0}{\pi} \right) \right) \qquad (14)$$

Figure 8 shows the velocity profiles as functions of the similarity parameter β for Andradean fluids having values of $n = 1, 4, 20$, and ∞. The arctangent velocity profile of the ideal nonhardening plastic material $(n \to \infty)$ appears to ascend much more rapidly than the velocity profiles of the Newtonian viscous $(n = 1)$ or Andradean $(n = 4, 20)$ fluids. However, in the limiting case of the ideal plastic material, the similarity parameter β at any finite distance from the boundary is infinite so that the velocity at the point vanishes. The velocity profiles shown in figure 9 are not plotted on the same physical scale of distance and one should not be misled by the apparent rapid ascent of the profiles.

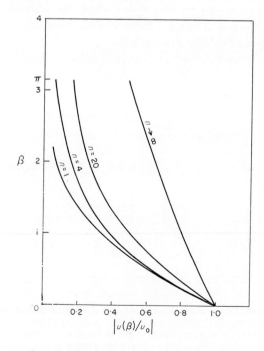

FIGURE 9 The nondimensional velocity profiles $(u(\beta)/u_0)$ in the fluid adjacent to the accelerated plate, plotted as functions of the nondimensional distance parameter β, for fluids having Andradean exponents $n = 1, 4, 20, \infty$

In passing from the Newtonian viscous case ($n = 1$) to the ideal plastic case ($n \to \infty$) the shape of the velocity profile changes only from that of the complementary error function to that of the arctangent. One could hardly distinguish such a small variation in shape from experimental measurements. For example, suppose that one had an experimental measurement of the velocity profile within a given distance from the plate, and suppose that one were to attempt to determine the nature of the fluid from the shape of the velocity profile. Figure 10 shows the velocity profiles for the Newtonian viscous fluid ($n = 1$) and the plastic material ($n \to \infty$) adjusted so that the fluid velocity in each case agrees at one point in the flow field. As the figure shows, the calculated points lying on the arctangent velocity profile of the ideal plastic case fall so close to the velocity profile of the Newtonian viscous case that the two profiles can hardly be distinguished.

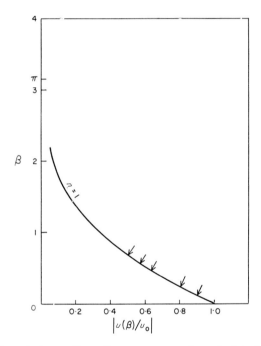

FIGURE 10 The velocity profiles adjacent to an accelerated plate for the case of a Newtonian fluid ($n = 1$) and the limiting Andradean fluid which represents ideal plastic behavior ($n \to \infty$) plotted in such a way that the velocities in the two fluids coincide at a given point, adjusted so that ($|u(\beta)/u_0| = 0.5$) occur at the same value of β. The full curve represents the Newtonian velocity profile and the arrows ($\downarrow$) show calculated points for the Andradean (plastic) fluid. The two profiles are indistinguishable, and from experimental measurements of the instantaneous shape of a velocity profile it would not be possible to determine whether the fluid were Newtonian or Andradean

Using the characteristic dimensionless parameter $\beta_0(n)$ to obtain the characteristic distance ($\delta(t)$) of diffusion of momentum during the time (t), one has

$$\delta = \left(\frac{n+1}{n} \frac{\sigma_0}{\varepsilon_0^{1/n}} \frac{t}{\rho n}\right)^{n/(n+1)} |u_0|^{(1-n)/(1+n)} \tag{15}$$

for the Andradean fluid ($n > 1$), and

$$\delta = (vt)^{1/2} \tag{16}$$

for the Newtonian fluid ($n = 1$).

From (15) one sees that for very large values of n (as the behavior of the fluid approaches plasticity) δ behaves like

$$\delta \sim \frac{\sigma_0 t \, | u_0 |}{\rho n}, \qquad n \gg 1 \tag{17}$$

Equation (17) indicates that δ becomes linear with respect to time so that the diffusion of momentum in the fluid becomes similar to wave propagation.

In order to determine the time required for steady flow to develop in a section of the mantle of the Earth which is subjected to steady boundary disturbances, assuming that the mantle undergoes permanent deformation via Andradean creep, one must know the typical scale of the distance (L) across which momentum must diffuse (i.e. the scale upon which the steady state velocity gradients must be measured), the magnitude of the boundary disturbance ($| u_0 |$), and one point (σ_0, ε_0) on the curve of stress against strain rate for the mantle material. If the mantle is indeed Newtonian or Andradean viscous, the typical scale on which velocity gradients must be measured in the steady state convective velocity field is the depth of the mantle, and accordingly it is assumed that L is approximately 3×10^3 km. The distance over which momentum should diffuse from the boundaries in order to bring the fluid flow into a steady state will be a few multiples of the size of the system; here we assume this distance to be approximately $5 \times L$ or in the case of the mantle 1.5×10^9 cm. The magnitude of the boundary disturbance $| u_0 |$ is assumed to be approximately one centimeter per year. To obtain a point on the stress–strain rate curve of the mantle material, the Fennoscandian data is used. As the map of figure 11 shows, the radius of the Fennoscandian uplift is approximately 550 km and the maximum velocity of uplift is approximately 1 cm/year; thus the characteristic strain rate in the Fennoscandian area is approximately 5.75×10^{-16} sec^{-1}. Previous estimates (e.g. Gutenberg[9], especially p. 28, assigns a Newtonian viscosity of approximately 10^{22} P to the mantle under Fennoscandia) place the characteristic driving stress in the Fennoscandian area at approximately 5.75×10^6 dyn/cm^2. Thus in the following calculations the values $\varepsilon_0 = 5.75 \times 10^{-16}$ and $\sigma_0 = 5.75 \times 10^6$ dyn/cm^2 will be used to represent a point on the stress–strain rate curve of the mantle material. The time t^* for a mantle sized block of Andradean fluid, with 'hardening' exponent n, to arrive at a condition of steady flow under fixed boundary disturbances is given in the table 1 for various values of n.

The extremely short times of table 1 ($t^* \sim 10^{-4}$ sec) may at first seem surprising. However, because of the extremely small magnitude of the disturbing velocity u_0 the mean level of strain rate being propagated away from the boundary is very small and so, locally, the apparent viscosity of the material (i.e. the slope of the stress–strain rate curve) is large. If the disturbing velocity were several orders of magnitude larger, the transient time (t^*) required for the establishment of steady flow in an Andradean ($n > 1$) fluid could be made correspondingly larger, as is shown by (15); for the Newtonian case ($n = 1$) the transient time is independent of the strength of the disturbance because of the linearity of the fluid. However, the fact is that with the

The diffusion of boundary disturbances through a non-Newtonian mantle

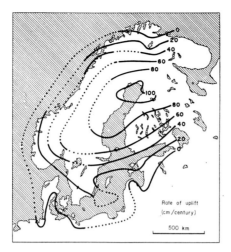

FIGURE 11 The Fennoscandian area showing contours of constant uplift. The radius of the Fennoscandian area is approximately 550 km and the maximum rate of uplift is approximately[9] 1 cm/year

TABLE 1 The time (t^*) required for attainment of steady flow in a block of the Earth's mantle 3×10^3 km deep and which behaves as an Andradean fluid having the hardening exponent n

n	t^* (sec)
1	6.7×10^{-4}
4	1.79×10^{-4}
5	2.38×10^{-5}
10	3.26×10^{-4}
20	5.26×10^{-4}

Fennoscandian data have been used to assign one point (σ_0, ε_0) to the stress–strain rate curve of the mantle material.

disturbing velocities ($u_0 \sim 1$ cm/year) strain rates and stresses (from Fennoscandian data) actually found in the Earth, if one assumes the mantle undergoes permanent deformation via Andradean (hot) creep, the transient times for establishment of steady flow under steady boundary disturbances is extremely short ($\sim 10^{-4}$ sec).

Transients in Thermal Convection

If thermal convection were occurring in the mantle, the driving force would, of course, derive from the excess of the vertical temperature gradient over the adiabatic temperature gradient for the mantle material. Thus, as the convective flow takes place the driving force will change as the hot material is transported by the flow. In order to estimate the time required for steady thermally driven convective flow to obtain in a fluid system, one must estimate the characteristic time for diffusion of heat across the system, as well as the characteristic time for diffusion of momentum (as has been done above). The system cannot approach steady state convective flow until sufficient time has elapsed to allow both heat and momentum to diffuse fully throughout the system. The characteristic distance (δ_{th}) over which heat will diffuse in a system having thermal conductivity κ and density ρ during the time t is

$$\delta_{th} = \left(\frac{\kappa t}{\rho}\right)^{1/2}$$

According to Lubimova[10] the thermal conductivity of the mantle lies between 4.7×10^{-2} and 3×10^{-2} cal/degC cm sec. Assuming the mean density of the mantle to be approximately[11] 3·5 g/cm^3 and choosing as a representative value of the thermal conductivity $k \sim 3.5 \times 10^{-3}$, which value is consistent with Lubimova's estimates of the greatest thermal conductivities in the mantle and will therefore provide somewhat of an underestimate of the time required for diffusion of heat through the mantle, one finds from (17) that the time required for diffusion of heat across a distance equal to five times the depth of the mantle (i.e. setting $\delta_{th} = 5 \times 3 \times 10^3$ km) is 0.72×10^{13} years. Thus, as is already well known, diffusion of heat in the mantle is an extremely slow process; indeed the entire history of continental drift seems to occupy only approximately 10^8 years which is only 10^{-5}—one thousandth of one per cent—of the time required for the establishment of steady state heat diffusion and steady state convective flow in a non-Newtonian fluid mantle.

There is an immense disparity in the typical transient diffusion times required for viscous diffusion of momentum ($\sim 10^{-4}$ sec) and diffusion of heat ($\sim 10^{13}$ years) in the mantle. If the mantle has Andradean viscous properties, the pattern of fluid flow in the mantle will adjust virtually instantaneously to changes in the forces driving the fluid flow. Since these forces arise from the thermal buoyancy, and since the typical relaxation time of the temperature field in the mantle is approximately 10^5 times as large as the time interval during which convective motion of the mantle is supposed to have been active, then if the mantle were an Andradean fluid the thermal convection patterns currently active in the mantle would be in their very early transient stages and would not have even approached steady state motion. How then could the apparently steady state convection patterns in the mantle have arisen within a few multiples of 10^8 years, as the observations of Wilson, Vine, and others indicate? It seems that the speed of diffusion of thermal disturbances dominates the analysis of the development of steady state convection in the mantle, and one is led to conclude that the distance over which heat must diffuse in order to establish steady state convection in the mantle must be much smaller than the dimension of the mantle itself. This rules out Newtonian or Andradean viscous flow in the mantle, for the distance over which heat must diffuse (in order to establish steady convection) is the typical scale of distance over which the velocity gradients in the steady state fluid motion are measured, and with an Andradean or Newtonian viscous mantle the scale of distance is just the depth of the mantle itself.

The Possibility of Plastic Deformation in the Mantle

The most natural mechanism of permanent deformation which can occur in polycrystalline materials and which can provide for concentrated zones of deformation, as opposed to the diffuse deformation which arises in Andradean or Newtonian viscous creep, is nonhardening plastic deformation. If the convection pattern in the mantle arises from nonhardening plastic deformation, then discontinuous slip across discrete surfaces (i.e. slip bands) may occur, and the material may deform so that whole blocks of material slide across each other as rigid bodies, while all the deformation is confined to the thin lamella of material between the two blocks. The scale of distance upon which the velocity gradients must be measured, in the case of nonhardening plastic deformation, is the width of the slip bands in the material, which need be only a few multiples of the typical grain diameter in the material.

Assuming that nonhardening plastic deformation takes place in the mantle, and that in order for steady state convective motion to obtain, the characteristic distance of diffusion of heat should be approximately five times the thickness of a typical slip band, and, once more using Lubimova's estimates of the thermal conductivity of the Earth one finds that, in order for the convective velocity field to attain steady state within approximately 10^8 years, the slip band thickness should be approximately 6 km. Estimates of the grain size in mantle material vary widely, ranging from micrometers to meters; however, a zone of concentrated plastic deformation of 6 km thickness would not be in conflict with even the largest estimates of the mantle grain size.

One question which may be raised in connection with the possibility of convection via nonhardening plastic deformation in the mantle is the possible presence of strain hardening in the mantle material. Experimental data on actual mantle material are not available; however, recent experiments on polycrystalline metals show that the strain hardening of the material disappears at large plastic strains. For example, Finnie[12] has deformed polycrystalline aluminium and other metals to plastic strains of 5·0 and has shown that between plastic strains of 1·5 to 2 the strain hardening of the material completely disappears; strain hardening appears to be significant when the plastic strain is of order unity or less. If a zone of highly concentrated shear begins to develop in a material, strain hardening will first tend to spread the zone, diffusing the slip band. However, as the strain in the zone proceeds the material will lose its capacity to harden and a true slip band will develop.

An objection has been raised to the possibility of convective motion in the mantle via plastic deformation; Vening-Meinesz[13] considers a convective pattern in a non-hardening plastic material in which a cylinder of plastic material rotates as a rigid body while the surrounding material stands still and a velocity discontinuity (i.e. a slip band) develops at the boundary of the cylinder. The driving force for the rotational motion of the cylinder is the moment of the buoyancy forces acting over the cylinder (see Vening-Meinesz[13], figure 3, p. 159). Vening-Meinesz points out that the motion of such a cylindrical convective pattern might be unstable, because as the displacement of the convection pattern increases the driving moment would also increase. However, this argument neglects horizontal heat transfer out of the cylindrical convection cell. Because the scale upon which the horizontal temperature gradients must be measured is just the thickness of the supposed shear band at the boundary of the cell (or at the boundary between any two blocks of plastic material which slide across each other as rigid bodies), the horizontal gradients of temperature can be very strong and can provide for rapid cooling of a rising hot spot and consequent stabilization of the type of convective motion considered by Vening-Meinesz. Since the mantle material carries with it its own radioactive heat source, it would seem that a steady state rotational convective pattern driven by the moment of the buoyancy forces and stabilized by the high horizontal rate of heat transfer into the surrounding mantle material might exist. Quantitative investigation of this possibility has as yet not been undertaken but would seem at this point to be a potentially fruitful avenue of exploration.

Conclusions

In order for steady convective flow to become established in a fluid system, sufficient time must elapse after the onset of convective motion to allow both heat and momentum to diffuse across a distance at least equal to several multiples of the characteristic scale of distance on which the velocity gradients of the steady convective flow should be measured. If the mantle deforms as an Andradean fluid, the typical scale of distance over which steady state velocity gradients should be measured is the depth of the mantle; the time required for diffusion of momentum across a distance equal to five times the mean depth of the mantle is very short ($\sim 10^{-4}$ sec) where as the time required for diffusion of heat across the same distance is very long ($\sim 10^{13}$ years). Geological evidence indicates that continental drift and mantle convection have been taking place for approximately 10^8 years, and that the process has apparently been in steady state for approximately the same period of time. If the principal mechanism of permanent deformation in the mantle were Andradean viscous (hot) creep, the very rapid diffusion of momentum which would take place would cause the flow pattern in the mantle to adjust instantaneously to the variations in the thermal driving forces caused by the convective transport of material, while the very slow diffusion of heat would strongly inhibit adjustments in the temperature of each material element of the mantle, which adjustments would be necessary in order to reduce the thermal driving force on a given element as it moves through the convective pattern and thus allow the convective motion to attain steady state. Given the rates of diffusion of heat and momentum in an Andradean viscous mantle, then, in the period over which continental drift and mantle convection is supposed to have taken place, the convective motion in the mantle would have barely passed into its initial transients. The history of the motion would show large accelerations, and nothing corresponding to a record of steady state motion would be found. Since geological evidence does indicate that convective motion has been in steady state for roughly 10^8 years, the distance over which heat and momentum must diffuse in order to allow the steady state motion to develop must be much less than the depth of the mantle. But the distance over which the diffusion must take place is the characteristic scale of distance on which the velocity gradients of the steady convective flow should be measured, and in an Andradean viscous mantle this distance is just the depth of the mantle. Thus one is led to conclude that the principal mechanism of permanent deformation in the mantle is most probably not Andradean viscous creep.

If the mantle is undergoing thermal convection, the mechanism of permanent deformation of the mantle material must provide a very small characteristic scale of length on which steady state velocity gradients should be measured, in order that, even with the very slow diffusion of heat in the mantle, the convective flow pattern could have come into the steady state within approximately 10^8 years, as current observations are interpreted to imply. Nonhardening plastic deformation of a polycrystalline mantle would provide a very short characteristic distance for measurement of steady state

The diffusion of boundary disturbances through a non-Newtonian mantle

velocity gradients and indeed seems to be the most natural choice for the mechanism of permanent deformation in the mantle. As matters stand at present it seems that continued investigation of mantle convection based upon the assumption that either Newtonian viscous flow or Andradean viscous creep is the principal mechanism of permanent mantle deformation is not likely to provide results which are useful in elucidating the quantitative aspects of the geophysics of mantle motion. Further investigation based upon the assumption of plastic deformation in the mantle may provide significant results relative to thermal convection in the mantle. Another possible line of attack for future investigation might be to discard the idea of thermal convection entirely and to seek possible alternative mechanisms which could have produced mantle motion and continental displacement. One tentative theory involving lamellar melting in the mantle, with consequent mantle motion and continental drift, has been advanced recently by Orowan[14] and will appear in another publication soon.

Acknowledgements

The author is pleased to acknowledge the helpful advice and criticism of Professor E. Orowan. In addition the efforts of Messrs. S. Stork and R. Sheldon, who numerically evaluated $\beta_0(n)$ of equation (12) and evaluated the dimensionless velocity profiles of equation (11) on an I.B.M. 360-65 computer at the Massachusetts Institute of Technology computation center, are gratefully acknowledged.

Appendix

In this appendix we show the reduction of the constitutive relations and equations of motion for the plane flow of an Andradean fluid to the forms given in equations (5) and (6) of the main body of this chapter. First, we recall that in the plane flow of an incompressible fluid in the (x_1, x_2) plane the velocity field

$$u_1 = u_1(x_1, x_2, t), \quad u_2 = u_2(x_1, x_2, t), \quad u_3 \equiv 0 \tag{A1}$$

provides the following strain rate field:

$$\varepsilon_{ij} = u_{i,j} + u_{j,i} = \begin{pmatrix} \varepsilon_{11} & \varepsilon_{12} & 0 \\ \varepsilon_{12} & \varepsilon_{22} = -\varepsilon_{11} & 0 \\ 0 & 0 & 0 \end{pmatrix} \tag{A2}$$

It is extremely important to note that because of the incompressibility of the fluid $\varepsilon_{11} = -\varepsilon_{22}$, i.e. that the velocity field is divergence free. Now the most general constitutive relation for an isotropic incompressible nonlinear viscous fluid connects the components of stress (σ_{ij}) to the components of strain rate (ε_{ij}) in the following way (e.g. Serrin[15]):

$$\sigma_{ij} = p'\delta_{ij} + \alpha(\theta_2, \theta_3)\varepsilon_{ij} + \beta(\theta_2, \theta_3)\varepsilon_{ik}\varepsilon_{kj} \tag{A3}$$

where p' is a scalar function of space and time and is to be determined so as to satisfy the requirements of the equilibrium relations and boundary conditions, θ_2 and θ_3 are, respectively, the second and third principal invariants of the strain rate tensor ε_{ij}, and α and β are scalar functions of θ_2 and θ_3. The third principal invariant (θ_3) of the ε_{ij} is the determinant of the strain rate components measured in a Cartesian frame. According to (A2), for the case of planar flow, θ_3 vanishes at every point in the flow field. In addition, if one expands the products $\varepsilon_{ik}\varepsilon_{kj}$ from (A2) and uses the incompressibility condition $\varepsilon_{11} = -\varepsilon_{22}$ explicitly in the result one obtains

$$\varepsilon_{ik}\varepsilon_{kj} = \begin{pmatrix} \varepsilon_{11}^2+\varepsilon_{12}^2 & 0 & 0 \\ 0 & \varepsilon_{11}^2+\varepsilon_{12}^2 & 0 \\ 0 & 0 & 0 \end{pmatrix} \quad (A4)$$

The equilibrium conditions which the stress components and velocity field must meet in the flow of any material through a space in which no body force acts are

$$\sigma_{ij,j} = \rho(u_{i,t}+u_k u_{i,k}) \quad (i,j,k = 1,2,3) \quad (A5)$$

Now in plane flow (A1) of an isotropic fluid having the constitutive relation given in (A3), the equilibrium relation for the direction (x_3) normal to the plane of flow is identically satisfied by requiring that p' be independent of x_3. Thus, in analyzing the plane flow of such a fluid one need be concerned only with the equilibrium equation for the directions (x_1, x_2) in the plane of flow and the stress components σ_{ij} $(i,j = 1, 2)$—i.e. $\sigma_{11}, \sigma_{12}, \sigma_{22}$—acting in the plane of flow. Once the x_1 and x_2 equilibrium relations have been satisfied by using a scalar pressure field (p' in (A3)) which is independent of x_3, then the condition of equilibrium for the x_3 direction will automatically be met. Recognizing that only $\sigma_{11}, \sigma_{12}, \sigma_{22}$—the stresses acting in the plane of flow—enter into the equilibrium relations for the axes (x_1, x_2) in the plane of flow, one may combine the scalar p' of (A3) with the product terms $\beta\varepsilon_{ik}\varepsilon_{kj}$, using (A4), as follows:

$$p' + \beta(\varepsilon_{11}^2+\varepsilon_{12}^2) = p \quad (A6)$$

which, when introduced into (A3), gives the following simplified representation of the stress components acting in the plane of flow:

$$\sigma_{ij} = p\,\delta_{ij} + \alpha(\theta_2, 0)\varepsilon_{ij}, \quad (i,j = 1, 2) \quad (A7)$$

It should be noted especially that the 'normal stress effects' represented by the final terms $\beta\varepsilon_{ik}\varepsilon_{kj}$ of (A3) have no influence upon a plane flow, as (A6) shows the 'normal stress effects' only add a symmetric hydrostatic component to the stresses acting in the plane of flow. The basic physical reason for the lack of influence of normal stress effects upon plane flow is that plane flow of an incompressible medium is 'pure shear' and the response of the material to pure shear is sufficient to describe its mechanical behavior on plane flow. This simple point greatly reduces the difficulty of analyzing plane flows of general nonlinear fluids, but seems to have been overlooked elsewhere.

The particular nonlinear fluid with which we are concerned here is one which in pure shear undergoes Andradean viscous flow, as does a polycrystal in the hot creep range. For such a 'fluid' the shear stress (σ) is related to the shear–strain rate (ε) by

$$\frac{\sigma}{\sigma_0} = \left(\frac{|\varepsilon|}{\varepsilon_0}\right)^{1/n} \operatorname{sgn} \varepsilon, \quad n > 1 \quad (A8)$$

where σ_0 and ε_0 are material constants, the 'hardening index' (n) is, in general, a function of temperature, sgn is the signum function, and $|\varepsilon|$ is the absolute value of ε. If the shear-strain rate (ε) is positive, (A8) reduces to the traditional form

$$\frac{\sigma}{\sigma_0} = \left(\frac{\varepsilon}{\varepsilon_0}\right)^{1/n} \tag{A9}$$

usually employed by metallurgists and engineers. In so far as interpreting the results of a pure shear test is concerned, the differences between (A8) and (A9) are trivial; however, in attempting to calculate the flow field in an Andradean fluid subject to given boundary loadings, it is imperative either to carry the constitutive relation in the form (A8) or otherwise to insure that the stress and strain rate have the same direction.

In the problem considered in the main body of this chapter, the restriction $u_{,2} = \varepsilon_{12} > 0$ was introduced in order to circumvent the difficulty of carrying the rather awkward form of the constitutive relation (A8) through the calculations; since the velocity field of the problem under study will, for obvious physical reasons, be monotonic with respect to the coordinate (x_2) normal to the accelerated plate, the restriction $u_{,2} > 0$ is permissible and stands only as a device for eliminating undue calculational complications rather than as having any special physical significance.

Having adopted the restriction $u_{,2} > 0$, the constitutive relation of the Andradean fluid in appropriate form for the study of the particular two-dimensional flow under consideration can be obtained by using equation (3) of the main body and (A9) in (A7) to give

$$\sigma_{ij} = \begin{pmatrix} p & (\sigma_0/\varepsilon_0^{1/n})(u_{,2})^{1/n} \\ (\sigma_0/\varepsilon_0^{1/n})(u_{,2})^{1/n} & p \end{pmatrix} \quad (i,j=1,2) \tag{A10}$$

which is equation (5) of the main body.

The general equilibrium equations (A5), when written for the plane flow of equation (4), are identically satisfied in the x_2 and x_3 directions by choosing p to be independent of x_1, x_2, and x_3. The remaining equilibrium equation (the equilibrium equation for the x_1 direction) reduces to

$$\frac{\partial \sigma}{\partial x_2} = \rho \frac{\partial u}{\partial t} \tag{A11}$$

and by introducing the expression for σ_{12} given in (A10) into (A11) one obtains equation (6) of the main body of the chapter.

In the arguments given above it has been assumed that no body forces act on the fluid. In the mantle the body force of the gravity field is present; however, provided that this body force may be regarded as being independent of time during the course of diffusion of momentum through the mantle ($\sim 10^{-4}$ sec) one can subtract the hydrostatic pressure field required to equilibrate the body force field from the stress field— without at all modifying the deformation field in the incompressible fluid under consideration—and proceed as above. The presence of a time-independent body force field which is exactly equilibrated by a hydrostatic pressure field produces no effect upon the rate of dynamic diffusion of momentum through an incompressible fluid.

References

1. E. Orowan, *Phil. Trans. Roy. Soc.* (*London*), *Ser. A*, **258,** 284–313 (1965).
2. J. T. Wilson, *Nature*, **197,** 536–8 (1963).
3. E. Orowan, *Science*, **154,** 413–16 (1966).
4. B. C. Heezen, 'The deep sea floor', *Continental Drift*, Academic Press, New York, 1962, pp. 235–88.
5. F. J. Vine, *Science*, **154,** 1405–15 (1966).
6. G. G. Stokes, *Camb. Phil. Trans.*, *Ser. IX*, **8** (1851).
7. E. N. Andrade, *Proc. Roy. Soc.* (*London*), *Ser. A*, **84,** 1 (1911).
8. F. H. Norton, *Creep of Steel at High Temperatures*, McGraw-Hill, New York, 1929.
9. B. Gutenberg, *Physics of the Earth's Interior*, Academic Press, New York, 1959.
10. H. A. Lubimova, 'Thermal history of the Earth with consideration of the variable thermal conductivity of its mantle', *Geophys. J.*, **1,** 115–34 (1958).
11. A. Holmes, *Principles of Physical Geology*, Ronald Press, New York, 1965.
12. I. Finnie, *Intern. Res. Prod. Engng.*, *Am. Soc. Mech. Engrs.*, 76–82 (1963).
13. F. A. Vening-Meinesz, *Continental Drift*, Academic Press, New York, 1962, pp. 148–76.
14. E. Orowan, private communication (1967).
15. J. Serrin, 'Mathematical principles of classical fluid mechanics', *Handbuch Phys.*, **8/1,** 125–263 (1959).

5

C. A. BERG
Department of Mechanical Engineering
Massachusetts Institute of Technology
Cambridge, Massachusetts, U.S.A.

The formation of oceanic trenches and the mechanism of permanent deformation in the mantle

Introduction

Previously[1] it was pointed out that the observations of Wilson[2] and Vine[3], indicating that mantle convection has been in progress for a few multiples of 10^8 years and that convection has also been in steady state for approximately 10^8 years, lead one to believe that the mechanism of permanent deformation in the mantle is probably not Newtonian or Andradean viscous flow, but is more likely to be plastic deformation. The reason for this conclusion is that with viscous creep occurring in the mantle the transient time for diffusion of momentum across the mantle is very small ($\sim 10^{-4}$ sec), whereas heat diffusion across the mantle requires a very long time ($\sim 10^{13}$ years). With the wide disparity of transient times for diffusion of momentum and heat, the motion of a viscous mantle within the first few multiples of 10^8 years after the onset of convective instability would have been very 'jerky' and would not have left a record even approximating steady state motion. If, on the other hand, the mantle deformed plastically, the heat would not have had to diffuse all the way across the mantle to bring a convection pattern into steady state, but only across the plastic slipbands (zones of concentrated plastic shear strain); the slipbands need be only a few multiples of the typical grain size in thickness, and so with plastic deformation in the mantle steady state convection could have been achieved in the time ($\sim 10^8$ years) indicated by the observations of Wilson and Vine.

Another observation bears on the identification of the mechanism of permanent deformation in the mantle. Since the thickness of the oceanic crust is small compared with the depth of the mantle, the large-scale surface features one finds on the ocean floor are indicative of flow and deformation patterns in the mantle itself. As Orowan[4] suggests, one may view the oceanic crust as a thin coating which simply rides along on, and conforms to, the flow field in the mantle itself. The oceanic trenches, which are found near continental boundaries (figure 1, from Heezen[5]) are attributed by Benioff[6] to 'down warping of the oceanic blocks of the mantle' in the vicinity of the descending convective flow which is assumed to occur near the edges of the continents. Near the descending convective flow the surface layer of the mantle would be pushed against the

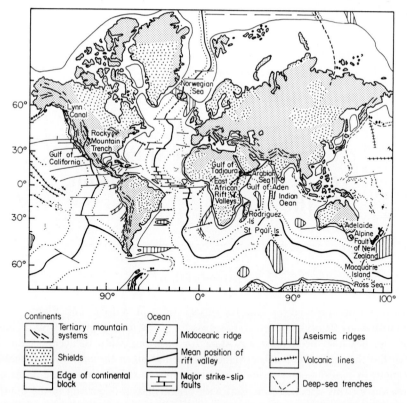

FIGURE 1 Heezen's[5] map of the ocean floor. At the edge of a continent a single oceanic trench is bound

continental edge by the drag of the convective flow beneath, and, therefore, the surface layer would be in compression, as indicated schematically in figure 2. One of the chief mechanisms contributing to the building of a trench, or any other type of irregularity, on the surface of the ocean floor in the vicinity of a continental boundary

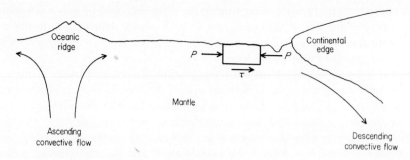

FIGURE 2 The descending convective flow which is assumed to occur in the mantle near a continental edge. The drag of the deep mantle material exerts a shear stress (τ) on the upper mantle, which pushes the upper mantle onto the continental block and causes a horizontal compressive loading (P) to be built up in the upper mantle in the vicinity of a continental edge

Formation of oceanic trenches and permanent deformation in the mantle

will be the amplification of existing surface perturbations ('ripples') by the compressive stress field (P' of figure 2) acting in the plane of the Earth's surface. A number of authors have considered the action of the convective drag in compressing the upper layer against the continental boundary (e.g. Chadwick[7], Elsasser, see section B III, chapter 1 of this book) and generally conclude that the compressive stress field extends approximately halfway across the convection cell (i.e. approximately one-quarter of the oceanic width away from the continental boundary, assuming that the ascending branch of the convective flow is situated at the middle of the ocean under the oceanic ridges). In considering the action of the compressive stress field in amplifying the 'ripples' in the surface layer of the mantle it would be reasonable to assume that the block of mantle material which is at present located under an oceanic trench has been subjected to compression during the history of its motion from a position directly over the center of the mantle convection cell to its present position, and that all of the surface layer of the mantle from the continental edge out to the halfway point between the oceanic ridge and the continental edge is currently subject to compression in the plane of the Earth's surface.

The Growth of 'Ripples' on a Viscous Surface

As best the writer can determine, the previous authors who have dealt with the effect of the compressive stress system, described in figure 2, on the surface features of the ocean floor or the continents, have limited their considerations to studies of buckling of a layer of the mantle or the continental edge. The type of surface disturbance to be considered here is quite different, and is therefore explained in some detail. If a block of viscous fluid having a slightly rippled free surface is placed under compression, as is shown in figure 3, two things happen simultaneously. First, the fluid body undergoes

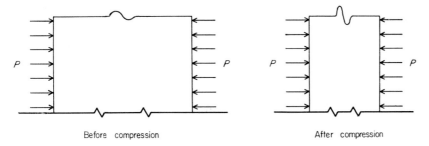

FIGURE 3 A viscous body having a slightly wrinkled surface and loaded by horizontal compression (P). As the body deforms (shortens) under the compressive loading, the height of the wrinkles increases. This is the basic mechanism of roughening of a viscous surface which undergoes compressive strain

overall compression due to the compressive loading; second, the depth of the surface ripple increases. The important thing to recognize about the behavior of the surface ripple is that the growth rate of its depth is as rapid as the rate of overall compression which the body as a whole undergoes; small ripples on the free surface of a viscous body are inherently unstable, in the sense that a very small amount of compression of the body in the plane of the free surface can cause an immense increase in the roughness of the surface. This mechanism of surface roughening can produce very large-scale surface features on the surface of a very deep body; buckling of a layer of the body is not at all required. If one supposes that the mantle is viscous (Newtonian or Andradean), then one must expect that one of the chief mechanisms for producing an oceanic trench—or any other type of large-scale roughness on the ocean floor—will be the amplification of surface irregularities by the compressive stress field acting in the upper layer of the mantle near the continental edge. In forming a quantitative appraisal of the mechanism, it would be extremely helpful to have at hand an analysis of the motion of a surface ripple on the free surface of an otherwise flat semi-infinite viscous body under compression acting in the plane of the free surface (as is shown in figure 3). Unfortunately such an analysis has not been carried out. However, the related and highly similar problem of determing the finite motion of the surface of a cylindrical hole of arbitrary cross section lying in an infinite body of incompressible Newtonian viscous fluid subject to arbitrary loading at infinity has been solved exactly[8] and provides sufficient information for examining the quantitative aspects of ripple growth on the surface of the semi-infinite body shown in figure 3.

The configuration of the problem solved earlier[8] is shown in figure 4. The motion of

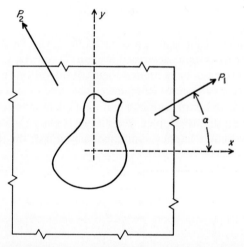

FIGURE 4 A cylindrical free surface of arbitrary cross-sectional shape lying in an infinite plane body of Newtonian viscous fluid loaded by the principal stresses P_1 and P_2 at infinity

each point on a cylindrical free surface of arbitrary initial cross-sectional configuration, lying in an infinite plane incompressible Newtonian viscous body, which undergoes plane creeping deformation due to the principal stresses P_1 and P_2 acting at infinity, with the axis of the principal stress P_1 inclined at the angle α to the x axis in the plane of deformation was determined. To relate the deformation hole of figure 4 to the deformation of a semi-infinite body having a slightly rippled surface (as is shown in figure 3) one may consider a very long crack with a slightly rippled surface (as is shown in figure 5); the deformation of the crack surface in the vicinity of the ripple will be essentially the same as if the ripple were located on an infinite plane surface. The motion of the free surfaces shown in figures 4 and 5 is most conveniently

Formation of oceanic trenches and permanent deformation in the mantle

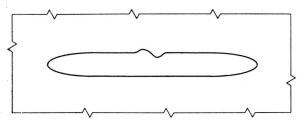

FIGURE 5 A long thin elliptical crack having a slight wrinkle in its surface

described via the conformal mapping of the contour of the surface onto a circle. If the mapping is constructed in such a way that the free surface in the $z = x + iy$ plane is mapped onto the unit circle centered on the origin of the ζ (image) plane, and in such a way that the points at infinity in the z and ζ planes correspond, the mapping which at any instant during the history of motion of the surface transforms the surface into the circle can be written in the form (e.g. Muskhelishvili[9])

$$x + iy \equiv z = R\zeta + \frac{Rm}{\zeta} + \sum_{n=2}^{\infty} \frac{RA_n}{\zeta^n} \tag{1}$$

The series representation of the mapping in (1) may be truncated to provide a polynomial approximation to any given mapping

$$z_{(M)} = R\left(\zeta + \frac{m}{\zeta} + \frac{A_2}{\zeta^2} + \ldots + \frac{A_M}{\zeta^M}\right) \tag{2}$$

and Sherman (see e.g. Muskhelishvili[9]) has demonstrated that the plane velocity fields in elastic or viscous bodies computed via the use of the polynomial approximation (2) to the exact mapping (1) of the surface of the body converge onto the true velocity field at least as fast as the approximate sequence of mappings ((2) for $M = 2, 3, \ldots$, etc.) converges onto the true mapping. With $|m| < 1$ in (1) the first two terms of (1) represent the mapping of an ellipse in the z (physical) plane onto the unit circle ($\zeta = e^{i\epsilon}$) in the ζ (image) plane; the remaining terms in (2) ($A_2, \ldots, A_M$) may be used to represent the 'ripple' on the surface of the thin nearly elliptical free surface shown in figure 5. Sherman's result assures one that the calculations of the velocity field based upon the polynomial approximations to the mapping which describes the 'slightly rippled' surface will provide a sequence of convergent approximations to the actual velocity field in the vicinity of the ripple. A few of the results of the calculation of free surface motion[8] are now quoted for later use.

(i) If a cylindrical free surface in an infinite plane viscous body is at any time mapped onto a unit circle by a polynomial, i.e. if the mapping ever has the form

$$z = R\left\{\zeta + \frac{m}{\zeta} + \frac{A_2}{\zeta^2} + \ldots + \frac{A_M}{\zeta^M}\right\} \tag{3}$$

the free surface will deform in such a way that it can always be mapped onto a circle by a polynomial mapping of *exactly* the same degree (i.e. the mapping will include terms up through $1/\zeta^M$, but will not 'pick up' terms of higher degree than $1/\zeta^M$)*;

---

* The only exception to this rule is the hole which is initially circular and for which the initial mapping is $z = R\zeta$; this hole becomes elliptical under shear loading at infinity and has a mapping of the form $z = R(\zeta + m/\zeta)$.

moreover, the mapping may be constructed in such a way that each material particle *on the free surface* is mapped onto the same point on the unit circle in the ζ (image) plane throughout the history of the motion.*

(ii) In view of result (i) above, one can construct a mapping of the form

$$z = R(t)\left\{\zeta + \frac{m(t)}{\zeta} + \frac{A_2(t)}{\zeta^2} + \ldots + \frac{A_M(t)}{\zeta^M}\right\} \tag{4}$$

in which the coefficients $R(t)$, $m(t)$, $A(t)$ depend upon the applied stress system at infinity (P_1, P_2 and α of figure 4), time (t), the viscosity (μ) of the fluid and the initial configuration of the free surface (specified by $R(0), m(0), A_j(0)$), which will continuously describe the configuration of the free surface as it deforms. It must be noted that the mapping of (4) will give the finite motion of the free surface, not just the incipient motion of particles on the free surface. The construction of the exact finite motion of cylindrical free surfaces via this technique has been described earlier[8, 10].

(iii) In the mapping

$$z = R(t)\left\{\zeta_1 + \frac{m(t)}{\zeta} + \frac{A_2(t)}{\zeta^2}\right\} \tag{5}$$

with $0 < 1 - |m| \ll 1$, the first two terms give the mapping of a thin elliptical contour onto the unit circle in the ζ (image) plane and, provided $A_2(t)$ is sufficiently small, the final term represents a small ripple distributed over the basic thin elliptical contour; the term involving $A_2(t)$ may be viewed as a first approximation to a surface ripple of the type shown in figure 5. When the mapping (5) represents the instantaneous configuration of a cylindrical free surface in a plane infinite viscous body, the coefficients $R(t)$, $m(t)$, and $A_2(t)$ behave as follows[8]:

$$R(t) = R(0)\exp\left\{(P_1 + P_2)\frac{t}{4}\right\} \tag{6}$$

$$m(t) = m(0) - \frac{P_1 - P_2}{P_1 + P_2}\exp(2i\alpha)\exp\left\{-(P_1 + P_2)\frac{t}{2\mu}\right\} + \frac{P_1 - P_2}{P_1 + P_2}\exp(2i\alpha) \tag{7}$$

$$A_2(t) = A_2(0)\exp\left\{-(P_1 + P_2)\frac{t}{2\mu}\right\} \tag{8}$$

where $R(0), m(0)$, and $A_2(0)$ are the mapping coefficients which describe the initial configuration of the free surface, μ is the viscosity of the material in which the free surface lies, P_1 and P_2 are the principal stresses applied at infinity, and α is the angle of inclination of the axis of P_1 relative to the x axis in the (physical) z plane as is shown in figures 4 and 5. Equation (7) gives the 'eccentricity' of the basically elliptical hole as it deforms, and is of minor concern in the present considerations (the interpretation of eccentricity has been given[11]). The quantity $R(t)$ describes the overall scale of size of the cavity and, as (6) indicates, the cavity grows like $\exp\{(P_1 + P_2)t/4\mu\}$. Now, the coefficient $A_2(t)$ gives the scale of the ripple on the deforming surface of the cavity,

---

* The points within the body are *not* mapped onto the same image point throughout the motion of the body; only the particles on the free surface have this property.

relative to the actual scale of size of the cavity, and, as (8) shows, the magnitude of the ripple grows like $\exp\{-(P_1+P_2)t/2\mu\}$. If the loading applied to the body (figure 5) is tensile $(P_1+P_2 > 0)$, then as the cavity deforms the scale of the cavity $(R(t))$ grows larger and the size of the small ripples on the cavity surface, represented by $A(t)$, decreases rapidly; the free surface of the hole becomes smoother as the tensile deformation of the body proceeds. This is why a relatively roughly cut slit in a sheet of viscous silicone putty loaded in tension will deform to become a nearly perfectly smooth circular hole[10]. If the body containing the cavity is loaded in compression $(P_1+P_2 < 0)$, then the relative amplitude of the ripples on the surface of the cavity, represented by $A_2(t)$, grows extremely rapidly. The surface in this case undergoes an immense increase in roughness; relatively smooth undulations in the original configuration of the surface are amplified by the compressive loading and rapidly become so large that they dominate the overall appearance of the surface. This explains why an apparently smooth circular hole in a large block of viscous silicone putty or plasticine subject to compression will rapidly develop a badly wrinkled surface.

There are two basic ways in which the roughness of a free surface in a body subject to compression may increase. First, the compressive deformation will tend to move the irregularities on the free surface closer together. If, as irregularities are squeezed closer together, the *absolute* amplitude of the irregularities stays fixed or does not decrease too rapidly, the surface will become rougher. Figure 6 shows a surface on

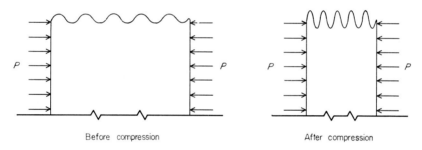

FIGURE 6 A body having a slightly wrinkled free surface and loaded by horizontal compression. As the body is compressed, the wrinkles on the free surface are brought closer together and thus the surface roughness increases even when the height of the wrinkles does not increase

which the absolute amplitude of surface irregularities remains fixed as the surface itself is compressed; the result is that the scale of the surface irregularities *relative* to the scale of the surface increases, and the surface roughness increases. On the other hand, the amplitude of the surface wrinkles themselves may increase as the body is compressed. This mechanism, which is illustrated in figure 7, causes very rapid surface roughening and absolute amplification of surface features. The type of surface roughness increase shown in figure 6 is not of great importance in the study of the building of an oceanic trench by amplification of initial surface irregularities. In order for an initial surface irregularity to be converted into a deep trench it is necessary that the absolute amplitude of the surface irregularity grow as the surface is compressed, as is illustrated in figure 7. Now, the absolute amplitude of the surface ripple represented by the final term A_2/ζ^2 in (5) is given by the product $R(t)A_2(t)$, and according to (6) and

(8) this product behaves as

$$R(t)A_2(t) = R(0)A(0) \exp\left\{-(P_1+P_2)\frac{t}{4\mu}\right\} \quad (9)$$

so that, with compressive loading ($P_1+P_2 < 0$) applied to the body, the absolute amplitude of the surface irregularity grows extremely rapidly; the type of growth of surface roughness which actually occurs in a viscous free surface is not just the growth of the amplitude of the irregularity relative to the scale of the surface, as is illustrated

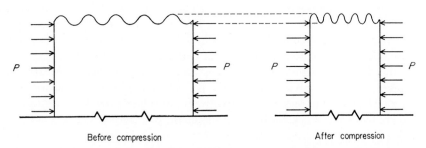

FIGURE 7 A body similar to the one shown in figure 6. However, in this case, as the body is compressed and the wrinkles are moved closer together, the height of the wrinkles grows. The type of free surface deformation illustrated here is what actually takes place on the surface of a viscous body, and can produce very large-scale surface features from small initial irregularities

in figure 6, but indeed the rapid growth of the amplitude of the irregularities themselves, as is illustrated in figure 7. Since $R(t)$ of (6) describes the overall scale of the surface itself, by comparing (9) and (6) one sees that with compressive loading applied to the body ($P_1+P_2 < 0$) the amplitude (9) of the surface irregularity grows as rapidly as the overall scale of the surface diminishes; thus a compressive loading which causes a fifty per cent diminution in the overall scale of the surface, i.e. $R(t)/R_0 = 0.5$ will cause the amplitude of the surface irregularity to double; $R(t)A(t)/R(0)A(0) = 2$. Recalling that a small ripple on a nearly flat free surface (figure 3) should behave approximately as would a small ripple on the surface of a thin nearly flat cavity (figure 5), then one would expect from the results quoted above that, if the mantle of the Earth were viscous, one of the chief mechanisms for the building of oceanic trenches and other large-scale features on the floors of the oceans would be the amplification of surface irregularities by the compressive stress field acting in the vicinity of the descending convective flow which presumably exists near the continental edges.

The results on free surface motion[8] quoted above leave two questions open. First, in the results (6), (7), and (8) only the first perturbation—represented by $A_2(t)/\zeta^2$ in (5)—on the shape of a basically flat elliptical free surface has been considered; one may wonder whether when higher-order terms—e.g. $A_3(t)/\zeta^3$, $A_4(t)/\zeta^4$, etc.—are included in (5) to represent more general shapes of surface irregularities the same general growth rates of the amplitudes of the surface irregularities are found. Second, the results quoted above derive from the study of free surfaces in Newtonian fluids, and one may wonder whether the basic mechanism of amplification of surface irregularities will be found in the case of a non-Newtonian fluid, such as the (hot polycrystalline) mantle of the Earth may be. As for the first question, Japiksie[12] studied the behavior of the higher-order perturbations (surface irregularities) in question. As it turns out, once one passes beyond terms of order ζ^{-3} in the mapping (4) the solutions for the coefficients $A_j(t)$ yield rather lengthy and complicated formulae which are not quoted here. However, Japiksie demonstrated that the asymptotic behavior of each coefficient at large values of time (i.e. $\lim_{t \to \infty} A_j(t)$) is as given in (8) for $A_2(t)$. Thus the answer to the first question is that the magnitudes of all free surface irregularities grow under compressive loading. As for the second question, the qualitative similarities between the physical and mathematical aspects of Newtonian and non-Newtonian flow (see e.g. Berg[1]) would suggest that the behavior of free surface irregularities on Newtonian and non-Newtonian fluids of the type commonly found in nature (e.g. Andradean fluids) would be quite similar; experimental evidence to support this conjecture will be given later in this writing.

Comparison of Surface Features on the Ocean Floor with the Appearance of Viscous Surfaces

Now we return to the central question of the present writing: knowing that in the vicinity of the continental edges, where the oceanic floors are presumably subject to horizontal compression, one finds a single oceanic trench parallel to the edge of a continent, and knowing that all irregularities on the free surface of a viscous body subject to compression grow rapidly in amplitude, can one justifiably assume that the mantle of the Earth is undergoing permanent deformation as a viscous fluid? Now, if the mantle were viscous, every surface ripple on the oceanic floor would grow under the horizontal compression found from the continental edge out to (approximately) one-quarter the width of the ocean; one would expect to find whole sheets of trenches on the oceanic floor, with the amplitude of the trenches diminishing as one proceeds from the continental edge where the horizontal compressive loading on the upper layer of the mantle is greatest, out to the point approximately one-quarter the way across the ocean where the horizontal compressive loading vanishes. Every minor surface disturbance on the ocean floor, within the zone where horizontal compressive loading is felt in the upper mantle, would provide a potential site for the formation of a large scale surface irregularity. The appearance of only one such irregularity—the oceanic trench bordering the continent—in the zone of horizontal compression of the upper mantle indicates that the main mechanism of permanent deformation in the mantle does not lead to amplification of all surface irregularities under horizontal compression. Instead we see, in the presence of the single oceanic trench lying near to and paralleling the continental boundary, evidence that the surface irregularities on the upper mantle are amplified only when the horizontal compressive loading attains some critically large value. This suggests that the deformation of the mantle takes place by plastic deformation rather than by viscous flow. In a plastic body, permanent deformation may take place only when the stress has reached a critically large value; the deformation may then become highly localized with concentrated zones of shear strain developing, while outside of these zones the body undergoes no deformation at all. Thus in a plastic mantle the production of a single trench in the zone of highest compressive loading near the continental edge is just what one would expect to occur.

Surface Instabilities in Non-Newtonian Viscous Materials

The arguments given above concerning growth rates of surface wrinkles on viscous fluids have been based upon exact calculations of the behavior of irregularities on surfaces of Newtonian fluids, and upon analogies between the physical and mathematical aspects of Newtonian and non-Newtonian flow. In order to determine whether, on the surface of a non-Newtonian fluid one might expect to find rapid proliferation and growth of large-scale surface irregularities, one may look at the surface of the La Brea tar lake. The tar which fills the lake is a non-Newtonian composite fluid consisting of asphaltic materials containing dispersions of remnants of ancient animal and plant life (bones, tree branches, etc.). Experiments[13] show that the rheological behavior of the La Brea tar approximates the behavior of an Andradean fluid with a

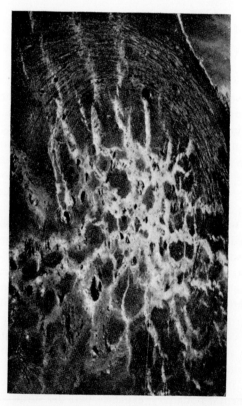

FIGURE 8 The surface of the La Brea tar lake (Holmes[14], p. 458) showing the top of a convection cell in the lake. The tar is (non-Newtonian) viscous. Note that at the boundary of the cell, where the descending convective flow has caused horizontal compressive loading in the surface layer of the tar, a very large number of large-scale surface irregularities, similar to mountains and trenches, has developed

hardening exponent between 3 and 5; therefore, the behavior of the tar lake should closely correspond to the behavior of the mantle, if the mantle is Andradean viscous. Large convection cells can be seen on the surface of the La Brea tar lake, and figure 8 shows an areal photograph of one of these convection cells. One may note that in the center of the cell the surface of the tar is very gently curved; no ridges, such as appear along the midspans of oceans, appear at the center of the convection cell. The inability of a viscous liquid to assume a sharply curved surface in the center of a convective upwelling has been used previously[1, 15] as evidence that the permanent deformation of the mantle under the oceanic ridges may not be treated as viscous flow. In addition to the gently curved surface of the center of the convection cell, one sees at the edge of the cell that a whole sheet of large-scale wrinkles has formed on the surface of the tar. At the edge of the cell the tar is just entering the descending branch of the convective flow pattern and the upper layer of the tar near the descending convective flow is subject to horizontal compression for the same reasons that horizontal compression acts on the upper mantle in the vicinity of the descending convective flow found at the continental edges. If the mantle were a (Newtonian or Andradean) viscous fluid, as is the tar in the La Brea lake, the surface of the ocean floor in the vicinity of the continental edge should resemble the surface of the tar lake near the edge of the convection cell shown in figure 8. A whole sheet of large-scale surface irregularities—i.e. trenches—should be found on the ocean floor near a continental boundary, and not just a single trench. In view of the appearance of only one major irregularity (trench) adjacent to a continental boundary, one is led to conclude that the mechanism of permanent deformation in the mantle is probably not viscous (Newtonian or Andradean) flow, but is much more likely to be plastic deformation.

Acknowledgements

The author is pleased to acknowledge the helpful advice and criticism of E. Orowan.

References

1. C. A. Berg, 'The character of plane deformation fields in isotropic incompressible non-linear viscous fluids', *J. Math. Mech.*, **46**, 448–58 (1967).
2. J. T. Wilson, *Nature*, **197**, 536–8 (1962).
3. F. J. Vine, *Science*, **154**, 1405–15 (1966).
4. E. Orowan, *Science*, **154**, 413–16 (1966).
5. B. E. Heezen, 'The deep sea floor', *Continental Drift*, Academic Press, New York, 1962, pp. 235–88.
6. H. Benioff, 'Seismic evidence for the fault origin of oceanic deeps', *Biol. Geol. Soc. Am.*, **60**, 1837–56 (1949).
7. P. Chadwick, 'Mountain building hypotheses', *Continental Drift*, Academic Press, New York, 1962, pp. 195–232.
8. C. A. Berg, 'The diffusion of boundary disturbances in a non-Newtonian mantle', *NATO Adv. Study Inst., Newcastle, 1964*, Wiley, New York, London, 1964.
9. N. I. Muskhelishvili, *Some Basic Problems of the Mathematical Theory of Elasticity*, Noordhoff, Grommgen, 1953.
10. C. A. Berg, 'The motion of cracks in plane viscous deformation', *Proc. U.S. Natl. Congr. Appl. Mech., 4th*, Vol. 2, 1962, pp. 885–92.
11. F. A. McClintock, S. M. Kaplan, and C. A. Berg, 'Ductile fracture by hole growth in shear bands', *Intern. J. Fracture Mech.*, December (1966).
12. B. Japiksie, M. S. Thesis, Department of Mechanical Engineering, Massachusetts Institute of Technology (1964).
13. F. Moavenzadeh, private communication.
14. A. Holmes, *Principles of Physical Geology*, Ronald Press, New York, 1965.
15. E. Orowan, *Phil. Trans. Roy. Soc. London, Ser. A*, **258**, 284–313 (1965).

C

HIGH-PRESURE PHYSICS AND THE EARTH'S INTERIOR

I. Chemical composition of the mantle

1

K. E. BULLEN

Department of Applied Mathematics
University of Sydney
Sydney, New South Wales, Australia

Seismic and related evidence on compressibility in the Earth

Introduction

For the purposes of this chapter, it is sufficient to treat the Earth as spherically symmetrical and to assume hydrostatic stress in the interior. Reference will be made (see Bullen[1, 2]) to internal concentric regions as follows: A, the crust; B, from the crust to 400 km depth; C, 400–1000 km; D, 1000–2900 km; D', 1000–2700 km; D'', 2700–2900 km; E' (the main outer core), 2900–4550 km; E'', 4550–4700 km; F, 4700–5150 km; G (the inner core proper), 5150–6370 km. This delineation, though known to need some amendments, serves adequately as a basis for the discussions to follow.

For convenience of exposition, a region of the Earth will be referred to as 'homogeneous' if the chemical composition is uniform and if there are no phase changes brought about by pressure. Because of the effects of the pressure p and temperature τ, the density ρ and incompressibility k will of course not be uniform in a homogeneous region. The parameters n_i will be used to represent chemical composition and phase.

In general, ρ and k will depend on p, τ, and n_i. When p exceeds 0.4×10^{12} dyn/cm² (the pressure at about 1000 km depth in the Earth), the dependence of k on the n_i is relatively slight (see later sections, also Bullen[3]). In some of the arguments to follow, it is for this reason sufficiently accurate to treat k as a function of p and τ alone. The symbol k will refer to the adiabatic incompressibility. Values of k and ρ are intimately connected through the relation

$$\frac{\partial \rho}{\partial p} = \frac{\rho}{k} \tag{1}$$

Inside the Earth, (1), along with $\mathrm{d}p = g\rho\,\mathrm{d}z$, gives the contribution to the density gradient arising from pure adiabatic compression; g denotes the gravitational intensity at depth z below the surface.

Standard seismological theory gives

$$\frac{k}{\rho} = \alpha^2 - \frac{4\beta^2}{3} = \phi, \quad \text{say} \tag{2}$$

where α and β are the velocities of P and S seismic bodily waves. Since values of α and β are fairly well known throughout much of the Earth's interior, (2) shows that the determinations of k and ρ inside the Earth are further intimately related. Both (1) and (2) play key roles in determining the distributions of k and ρ.

In practice, the most effective procedure is to determine values of ρ first and then to derive[1, 4–8] k using (2). In a somewhat (but not wholly) independent approach, Birch[9] has brought a simplified finite-strain theory to bear in estimating the variation of k in parts of the Earth, along with data from laboratory experiments on rocks and metals.

The present chapter will include an outline of the derivation of values of k in the Earth through extended forms of (1) and (2). In discussing the numerical detail, reference will be made to the comparatively simple behaviour of k and dk/dp at pressures beyond 0.4×10^{12} dyn/cm^2 and to Earth models which embody the main results. Attention will be given to the applications of compressibility theory to such questions as the estimation of density gradients in inhomogeneous regions of the Earth and the solidity of the inner core. The bearing of Birch's results on the present context will be briefly surveyed. The final section will refer to revisions of earlier work required by the recently changed estimate of the Earth's moment of inertia and current work on free Earth oscillation data.

Background Theory

For a point P of the Earth's interior, let z be the depth below the surface and r the distance from the centre. Let m be the mass inside the sphere of radius r, ϑ the excess of $d\tau/dz$ over the adiabatic temperature gradient at P, and G the constant of gravitation.

The Williamson–Adams equation (see Bullen[1], p. 229), namely

$$\frac{d\rho}{dz} = \frac{g\rho}{\phi} = \frac{Gm\rho}{r^2\phi} \tag{3}$$

gives an expression for the density gradient, where ϑ and changes in n_i can be neglected.

Williamson and Adams[10] used (3) and seismic data on ϕ to examine the extent to which pure compressibility can account for density variation in the Earth, and found that compressibility alone cannot account for the densities required in the deeper interior to meet the known value of the Earth's mass. By a compressibility argument, they were thus the first to show conclusively that the Earth's core is much denser than would be the case in a homogeneous Earth composed of the Earth's immediate subcrustal materials.

The relation (3) has since been much used to estimate $d\rho/dz$ in particular regions of the Earth. The relation is used appropriately where, as in the regions D' and E', there has been no strong evidence pointing to inhomogeneity and where ϑ may be assumed sufficiently small. It is also used to provide provisional estimates of $d\rho/dz$ even in regions where ϑ is thought to be significant.

The relation (3) has now been generalized to allow for effects[9, 11] of ϑ and continuous variation[1, 12] of n_i. A suitable form of the generalization[13] is

$$\frac{d\rho}{dz} = \frac{\eta g \rho}{\phi} \tag{4}$$

$$\eta = \frac{dk}{dp} - g^{-1}\frac{d\phi}{dz} \tag{5}$$

$$\eta = 1 - \frac{\gamma\phi\vartheta}{g} + \frac{\phi}{g\rho}\sum\frac{\partial\rho}{\partial n_i}\frac{dn_i}{dz} \tag{6}$$

The coefficient η is the ratio of the actual density gradient to the gradient given by (3), and γ is the coefficient of cubical expansion at constant pressure. In (5), dk/dp stands for $(dk/dz)/(dp/dz)$. By (5) and (6),

$$\frac{dk}{dp} = 1 + g^{-1}\frac{d\phi}{dz} - \frac{\gamma\phi\vartheta}{g} + \frac{\phi}{g\rho}\sum\frac{\partial\rho}{\partial n_i}\frac{dn_i}{dz} \tag{7}$$

The relation (7) focuses attention on the central place which dk/dp occupies in unravelling the mechanical properties of the Earth's interior. In regions where ϑ and dn_i/dz can be neglected, (7) gives

$$\frac{dk}{dp} = 1 + g^{-1}\frac{d\phi}{dz} \tag{8}$$

first derived and used in examining particular Earth models[14]. In such regions, (8) enables dk/dp to be estimated with good accuracy provided $d\phi/dz$ is sufficiently well determined from seismology. In regions where homogeneity cannot be assumed, estimates of dk/dp may be available independently of (7), and then (4) and (5) may enable η and $d\rho/dz$ to be estimated. Corrections for ϑ are usually carried out in a second approximation, to the extent that the evidence warrants it. The utility of the process lies in the fact that η as given by (5) depends only on dk/dp, $d\phi/dz$, and g: $d\phi/dz$ is observationally determined from seismology, and g can be estimated within fairly close bounds throughout most of the Earth's interior.

In addition to its prominence in the above equations, dk/dp proves to be a key physical property in the Earth below 1000 km depth.

It is of some incidental interest that the Legendre–Laplace density relation (see Bullen[1], p. 226)

$$\rho = Ar^{-1}\sin(Br) \tag{9}$$

much used in nineteenth century investigations of the interior of the Earth has a close connection with dk/dp. When ϑ and dn_i/dz are ignored, (9) corresponds to $dk/dp = 2$ and $k = C\rho^2$.

Earth Models Involving Compressibility

The first detailed assessments of the variation of k in the Earth were made when the seismic velocities α and β became sufficiently well determined to yield (in conjunction with other evidence) reliable values of ρ. The present section will discuss various aspects of the determinations of k as given in Earth models of A and B types[1], along with certain revisions[2] required in the lower core.

Models of A type

The A-type models[1,7] were based on the Jeffreys–Bullen seismological tables[15] and the values of α and β deduced by Jeffreys[16]. The process of construction included the assumption of near-homogeneity in regions where the seismically determined values of $d\phi/dz$ were 'normal'. (This assumption has since received a substantial measure of justification[3].) Thus (3) was originally applied in the regions B, D, and E'. Through the use of a moment-of-inertia criterion and certain other devices, a family of Earth models, with the central density ρ_c as parameter, was derived. The models consisted of values of ρ, k, and the rigidity μ as functions of z, and also gave values of p and g.

The models in which ρ_c had the values $17\cdot3$ g/cm^3, and the minimum possible value of $12\cdot26$ g/cm^3, were subsequently called models A and A', respectively.

Let k_1 and k_2 be the values of k at the bottom of the mantle and top of the core, N_1 and N_2 say, respectively. For $12\cdot26 \leq \rho_c \leq 17\cdot3$ g/cm^3, all the A-type models gave $k_2 < k_1$ and $|k_2 - k_1|/k_1 \leq 0\cdot05$. Such a small change in k at the mantle–core boundary, N say, was in marked contrast to the large changes in ρ and μ.

Application of (8) gave $dk/dp \simeq 3$ both at the bottom of the region D' and at N_2. (The values of $d\phi/dz$ imply inhomogeneity[14] in D'', so that dk/dp could not be immediately estimated in D''.)

The A-type calculations gave k ranging from $1\cdot2 \times 10^{12}$ dyn/cm^2 just below the crust, through about $6\cdot5 \times 10^{12}$ dyn/cm^2 at N, to between 12 and 13×10^{12} dyn/cm^2 at 5000 km depth.

The original model B

The near-continuity of k and dk/dp at the boundary N in the A-type models, along with such experimental evidence on k as was available at the time, led the writer (see Bullen[17] and later papers) to explore the hypothesis that k and dk/dp vary fairly smoothly with p throughout the whole Earth below 1000 km depth.

In a second type of Earth model—model B—full continuity of k and dk/dp were postulated[8] below 1000 km depth. The use of (3) was confined to the regions D' and E', where the evidence for near-homogeneity had now become moderately strong, and the compressibility postulate was used in conjunction with seismic data on ϕ to assist in determining ρ in D'' and the lower core. Values of k were deduced as before, using (2).

The procedure gave a fairly closely determined model in which the density gradient in D' had to be continued steadily upwards to within less than 100 km depth below the Earth's surface. Otherwise the values of the mass and moment of inertia of the Earth could not be fitted. The near-surface characteristics of model B were conventionally chosen to agree with the A-type models down to 80 km depth, to have a finite density jump (from $3\cdot36$ to $3\cdot87$ g/cm^3) at 80 km depth, and to have $d\rho/dz$ fairly steady for $80 < z < 2700$ km.

The following features of model B, in addition to the postulated continuity of k and dk/dp at N, have direct or indirect bearing on the properties of compressibility in the Earth.

(i) Since the application of (8) gives $dk/dp \simeq 3$ just above and just below D'', the compressibility postulate implies that $dk/dp \simeq 3$ also in D''. Taking $d\phi/dz = 0$ in D'' in accordance with the seismic data used, (4) and (5) then gave about three times the normal value of $d\rho/dz$ in D''.

(ii) The Jeffreys solution for α gave $d\alpha/dz$ strongly negative for a range of 140 km depth between the outer and inner cores proper. Assuming $d\phi/dz \simeq d\alpha^2/dz$, the compressibility postulate would then give η equal to at least[3] 30 units and a density increase of about 3 g/cm^3 through this range of depth (but see also p. 292).

(iii) Inside the inner core proper, all the available data gave $d\alpha/dz \geq 0$ but less than normal. Again assuming that $d\phi/dz$ is not greatly different from $d\alpha^2/dz$, η was found to be of the order of 4 or 5 units and the increase of ρ inside the inner core to be nearly another 3 g/cm^3. By this chain of argument, the compressibility theory led to the first observationally based[8] estimate (17·9 g/cm^3) of ρ_c.

(iv) On the compressibility postulate, the seismically observed jump in α from outer to inner core was attributed to an increase in rigidity, thus implying[17] a solid inner core.

None of the above inferences is likely to be significantly disturbed by allowances for ϑ, which is expected to be comparatively small below 1000 km depth (see Birch[9] for details).

Critique of model B

In view of the interesting formal consequences of the compressibility postulate, it became immediately important to seek evidence, additional to the empirical evidence from the A-type models, which would test the degree of reliability of the postulate, either directly or through its consequences (i) to (iv).

Evidence from laboratory experiments shows that differences in k and dk/dp for different materials do tend to diminish as the pressure increases beyond 10^5 atm (see also pp. 293–6). Theoretical evidence at pressures beyond 10^7 atm has been brought to bear by Elsasser[18] and adapted by Bullen[19], along with laboratory evidence, to show that, while the postulate is likely to hold to a good first approximation, small but significant changes of k with composition may occur in the Earth below 1000 km depth. The magnitudes of any sudden changes of k inside the core are, however, likley to be well short of what would be required to disturb the inference (iv) that the inner core is solid.

An analysis of evidence provided by Birch[9, 20], using finite-strain theory and related laboratory evidence, gives additional support to the postulate, at least as a useful first approximation. This evidence will be considered on pp. 293–6.

The overall evidence on k and dk/dp is now sufficiently strong to justify the parts of the arguments in (i), (ii), and (iii) which depend on the compressibility theory. For example, if the seismic evidence on $d\phi/dz$ is accepted, it is extremely improbable that dk/dp could be so small inside D″ as to invalidate the inference that the density gradient in D″ is significantly greater than normal.

In fact the principal uncertainties in the conclusions in (i), (ii), and (iii) now lie in uncertainties on the values of $d\phi/dz$ and in the assumption that $d\phi/dz$ is not greatly different from $d\alpha^2/dz$ throughout the core, rather than in uncertainties on the compressibility theory. In the following subsection, these conclusions will be looked into a little further.

A final comment on model B relates to its much higher density in the region B than in the A-type models. Some investigators have held this feature to be a serious objection to the original model B. On pp. 296–7, it will, however, be shown that, as a consequence of recent information from artificial satellite orbits, this difficulty no longer remains.

Revised evidence on the lower core

In seeking to accommodate seismological evidence[21, 22] later than 1940, Bolt[23] proposed a four-layer core with boundaries corresponding to the delineation made in the first section. Bolt proposed a P velocity distribution for the core which differed from the earlier Jeffreys solution, among other things, by having $d\alpha/dz$ nowhere negative. In each of three layers below 4560 km depth, Bolt found $d\alpha/dz$ to be not detectably different from zero. Other authors (e.g. Nguyen Hai[24], Adams and Randall[25]) have produced alternative solutions for α in the core, but the differences from Bolt's results are inconsequential for present purposes.

Using Bolt's velocities and taking $d\phi/dz \simeq d\alpha^2/dz \simeq 0$, the writer[26] used (4) and (5) to infer that

$$\frac{d\rho}{dz} \simeq \frac{g\rho}{\phi}\frac{dk}{dp} \tag{10}$$

inside each layer below 4560 km depth. Then, with the use of the compressibility postulate to set numerical bounds to dk/dp, it was shown that the earlier estimate of 17·9 g/cm³ for ρ_c could be reduced to about 15 g/cm³. This reduction was a direct consequence of abandoning the previously used evidence that $d\alpha/dz$ is strongly negative over a range of depth inside the core.

Meanwhile, Birch[27] had assembled evidence from shock-wave experiments at pressures exceeding 10^6 atm which led him to infer that $\rho_c \leq 13$ g/cm³. Implications of Birch's inference have been examined in some detail by the writer [2, 26, 28] in the light of the compressibility theory. Application of (4) shows that, if $\rho_c \leq 13$ g/cm³, then $\eta \simeq 1$ practically throughout the whole core. Evidence on dk/dp shows this to be incompatible with (5) so long as $d\phi/dz$ is taken as markedly less than normal for a sizeable range of depth. The seismological evidence does not appear to permit the raising of $d\alpha/dz$ to normal values in the lower core. If Birch's inference is correct, it therefore appears to follow that the assumption $d\phi/dz \simeq d\alpha^2/dz \simeq 0$ in the lower core must be abandoned. For most of the lower core, taking $d\alpha/dz \simeq 0$, (2) and (5) then give

$$\frac{4}{3}\frac{d\beta^2}{dz} \simeq 1 - \frac{dk}{dp} \tag{11}$$

Since dk/dp must here be equal to at least 3 units (see pp. 293–6), (11) requires $d\beta/dz$ to be negative. Numerical examination[2] shows further that $d\mu/dz$ must be negative, where μ denotes the rigidity.

The inference that $\rho_c \leq 13$ g/cm³, when examined in the light of the available evidence on seismic velocities and compressibility, thus entails, after the setting in of rigidity at some depth below 4560 km, a trend back towards fluidity with increasing depth; and, of course, the presence of a sustained rigidity gradient (positive or negative) is new evidence of the presence of significant rigidity in the lower core.

Numerical Estimates of k

The original model A and B calculations were carried out before artificial satellite observations[29] forced a rather drastic revision in the estimate of the Earth's moment of inertia I. Values of k which agree with the revised I and certain additional evidence are shown in table 1. Down to a depth of 4500 km, the values agree[30] within $0\cdot1 \times 10^{12}$ dyn/cm² with those in the original model A', and are probably correct within less than 5%.

TABLE 1 Values of p, ρ, k, μ, and g in model (A')

Depth (km)	ρ (g/cm³)	g (cm/sec²)	p (10^{12} dyn/cm²)	k (10^{12} dyn/cm²)	μ (10^{12} dyn/cm²)
33	3·32	984·7	0·009	1·14	0·62
200	3·36	990·7	0·064	1·33	0·71
400	3·41	999·4	0·132	1·59	0·82
600	4·01	1005·6	0·206	2·47	1·27
1000	4·66	1002·3	0·383	3·55	1·87
1400	4·90	996·7	0·57	4·14	2·13
1800	5·12	995·9	0·77	4·81	2·37
2200	5·33	1003·6	0·98	5·51	2·60
2600	5·54	1026	1·20	6·19	2·86
2700	5·59	1034	1·26	6·35	2·92
2883	5·68	1055	1·37	6·47	3·01
2883	9·79	1055	1·37	6·49	0
3000	9·97	1029	1·49	6·80	0
3500	10·65	907	1·99	8·53	0
4000	11·19	770	2·45	10·23	0
4500	11·60	620	2·84	11·66	0
4982	11·89	467	3·15		
6371	12·22	0	3·55		

Below 4500 km depth, the determination of k is somewhat more uncertain because of the complication of probable rigidity in the lower core and the absence of observational data on the seismic S velocity β. Seven lower-core models, each including a set of values of k, have been prepared[2] to cover the likely range of possibilities.

The models show the following conclusions (*inter alia*) to be reasonably well established. If $\rho_c \leq 13$ g/cm³, then the central value k_c of k must not be much greater than 15×10^{12} dyn/cm². If the core were entirely fluid, k_c could reach $18·5 \times 10^{12}$ dyn/cm²; but then ρ_c could not be much less than 15 g/cm³ and the variation of k with p would also depart improbably far from the compressibility postulate. The range of values of k_c in the seven models is $14·4–18·5 \times 10^{12}$ dyn/cm².

It may be noted that, over the pressure range $0·4 \times 10^{12} < p < 3·2 \times 10^{12}$ dyn/cm², the model B values of k agree with the quadratic relation

$$k = 2·25 + 2·86p + 0·16p^2 \qquad (12)$$

where the units are 10^{12} dyn/cm², within 2%. The relation (12), which is entirely empirical, is useful in representing for various test purposes the part of the Earth between 1000 and 5000 km depth.

Finite-strain Theory and Related Considerations

In the present section, a brief examination will be made of the bearing of work of Birch[9] on aspects of the compressibility theory under consideration. For simplicity of exposition, the small abnormalities connected with the region D″ will now be disregarded and D will be taken as running from 1000 km to the mantle–core boundary N.

Primes will be used to indicate isothermal values; and the subscript zero, values at zero pressure. The symbols f, c, and λ will denote the compression, specific heat at constant pressure, and Grüneisen's parameter (equal to $\gamma k/c\rho$), respectively.

Equations involving p and k

Birch developed a simplified finite-strain theory in which he assumed the form $af^2 + bf^3 + \ldots$ for strain energy, where $a, b, \ldots$ depend on τ alone, and used certain laboratory results in rejecting terms considered to be relatively unimportant. The following equations in the development apply to a homogeneous material:

$$p = 3k'_0 f(1+2f)^{5/2} \tag{13}$$

$$p = \frac{3k'_0}{2}\left\{\left(\frac{\rho}{\rho_0}\right)^{7/3} - \left(\frac{\rho}{\rho_0}\right)^{5/3}\right\} \tag{14}$$

$$k' = p\left\{\frac{7}{3} + (3f)^{-1}\right\} \tag{15}$$

$$\left(\frac{\partial k'}{\partial p}\right)' = \frac{12 + 49f}{3 + 21f} \tag{16}$$

Also for the particular case of a homogeneous material, Birch derived the following variant of (7):

$$\left(\frac{\partial k'}{\partial p}\right)' \simeq 1 + g^{-1}\frac{d\phi}{dz} + 5\tau\gamma\lambda + 2\gamma\phi\vartheta g^{-1} \tag{17}$$

His laboratory evidence and seismic data give $5\tau\gamma\lambda \simeq (0\cdot04, 0\cdot045, 0\cdot3) \times 10^{-3}\tau$ and $2\gamma\phi\vartheta g^{-1} \simeq (0\cdot2, 0\cdot1, 0\cdot2)\vartheta$ at the locations N_1, N_2, and a depth of 4500 km, respectively, where τ and ϑ are in °C, and degC/km, respectively. Since it is probable that τ and ϑ do not greatly exceed 4000 °C and 1 degC/km, (17) implies that $(\partial k'/\partial p)'$ exceeds $1 + g^{-1} d\phi/dz$ by less than 0·4 units in the whole region from the bottom of the mantle to the bottom of the outer core.

Combining (7) and (16) gives for a homogeneous region

$$\frac{dk}{dp} \simeq \left(\frac{\partial k'}{\partial p}\right)' - 5\tau\gamma\lambda - 3\gamma\phi\vartheta g^{-1} \tag{18}$$

The relation (18), which Birch does not give explicitly, shows that $(\partial k'/\partial p)'$ likewise exceeds dk/dp by less than 0·4 units in this part of the Earth.

The adiabatic and isothermal incompressibilities are known to be connected by

$$k = k'\left(1 + \frac{\tau k'\gamma^2}{\rho c}\right) \tag{19}$$

(see e.g. Bullen[1], section 4.6).

Considerations near the mantle–core boundary

The pressure at the mantle–core boundary N will be denoted as p_N and taken[1] as $1\cdot35 \times 10^{12}$ dyn/cm^2 when needed for numerical purposes.

Let the subscripts 1 and 2 refer to two different materials. It can be deduced from (13) and (15) that $|1 - k'_2/k'_1|$ diminishes steadily and indefinitely as p increases. The same applies to $|1-k_2/k_1|$, as found on substituting suitable numerical values into (19). Thus the finite-strain theory points in the same direction as the compressibility postulate earlier referred to. A numerical illustration of Birch taking $k_2/k_1 = 1.83$ at zero pressure gives $k_2/k_1 = 1.24$ at the pressure p_N.

Birch's theory also gives the same order of magnitude for k at the boundary N as holds for the A-type models. For the bottom of the mantle (N_1), Birch gives $f = 0.127$, whence (13) and (15) yield $k' = 6.7 \times 10^{12}$ dyn/cm². For the top of the core (N_2), an approximate estimate of f is 0.15, whence $k' = 6.2 \times 10^{12}$ dyn/cm².

Further, for these values of f, (16) yields $(\partial k'/\partial p)' = 3.21, 3.15$ at N_1, N_2, respectively. Thus the main indications of the model A results and Birch's theory are in fair agreement at N, in respect of both k and dk/dp.

Other estimates of dk/dp inside the Earth have been made by other authors, e.g. by Ramsey[31, 32], who gave $dk/dp = 3.7$ at N_1. But this value involved an *ad hoc* device, shown by the writer to be unnecessary, for securing agreement with a theory of planetary interiors. Most attempts to estimate k and dk/dp by methods independent of those described in earlier sections of this chapter appear to be inferior to Birch's approach.

Although it is well established that Birch's theory yields good order-of-magnitude results, and there is fairly good agreement with the model A calculations on the points mentioned, there is, however, not complete agreement.

A first difficulty arises in that Birch's theory requires $k_2/k_1 - 1$ to have the same sign for all p. In this respect, the theory is not in accord with the model A result that $k_2/k_1 < 1$ at N (taking k_1 and k_2 as relating now to the mantle and core materials, respectively), for it is improbable that $k_2/k_1 < 1$ at zero pressure. The model A result that $k_2/k_1 < 1$ at N may not apply in the real Earth, but the calculations carried out in constructing model B suggest that any excess of k_2/k_1 over unity must be very small in plausible Earth models—smaller in fact than the equations (13)–(16) would appear to allow when suitable numerical values are inserted.

A point of much interest, however, is that the assumption of near-continuity of k goes nearest to minimizing the differences between Birch's theory and the indications of the model A and B calculations.

A second difficulty will be mentioned in the following subsection.

Considerations inside the core

The A-type models have dk/dp increasing through the outer core E′ from about 3·1 to 3·8, this being a fairly direct consequence of assuming the Jeffreys P velocity distribution. In contrast, (13) and (16) give $(\partial k'/\partial p)'$ decreasing as p increases. (As previously shown, the difference between dk/dp and $(\partial k'/\partial p)'$ is less than 0·4). Birch sought to minimize this discrepancy by assuming an average value of 3 for dk/dp in the outer core, taking note of the seismic data of Gutenberg[33]. However, as will be shown in a later paper, his procedure is not altogether satisfactory and it appears that (16) is really significantly discordant with the seismological evidence in E′.

This and the difficulty previously mentioned suggest that, while Birch's theory gives good general support to the earlier findings and gives important evidence on the order of magnitude of dk/dp inside the core, it cannot be unreservedly relied on for the finer numerical detail.

A further point is that curves presented by Birch[27], in the course of his discussion on the value of ρ_c, implicitly give about 5 or 6 units for dk/dp for iron at the pressure reached at the Earth's centre. Griggs[34], quoting evidence from Verhoogen[35], has also questioned whether the form (16) is too simplified to meet requirements in the Earth's deeper interior to close accuracy.

The overall evidence thus appears to favour a fairly steady increase of dk/dp with depth inside the Earth's core, from about 3 units at the top to at least 4, and possibly 5 or 6 units, at the centre.

Current Work Bearing on Compressibility

In this section, account is taken of the revised estimate of the Earth's moment of inertia I, which has resulted in the lowering of I/Ma^2, where M and a are the Earth's mass and radius, from 0·3335 to 0·3309. The following discussion includes some details of work at present being carried out by Haddon and the writer[30], taking account also of data on free spheroidal and torsional Earth oscillations.

Because of the evidence that $\rho_c \leqq 13$ g/cm^3 (see earlier), the calculations in question have so far been mainly based on the earlier model A' which has the least value of ρ_c in A-type models. A series of new models has been constructed in which model A' has been corrected in various ways with a view to fitting the totality of evidence on I/Ma^2, seismic bodily waves, and the free oscillation data.

One of our best models to date, but by no means final, will for the purposes of this chapter be called model (A'). The model differs from model A' principally in having the core radius increased by 15 km to 3488 km and a density gradient in the region B significantly less than that given by the Williamson–Adams equation (3). (The latter property would imply a sizable super-adiabatic temperature gradient.) In broad terms, the change from model A' to model (A') involves little more than lowering the density for some distance on both sides of the B–C boundary, and transferring the mass thus removed to the outside of the core. There is remarkably little disturbance to the density at other depths.

The changes in k from model A' to model (A') are not very significant in themselves, but the values k_1 and k_2 of k at the locations N_1 and N_2 are interesting, giving $k_2/k_1 - 1$ now positive but less than 1%. Thus model (A') is less in disaccord with finite-strain theory and also gives near-continuity of k at the boundary N.

The lowered estimate of I/Ma^2 has an important repercussion on the original model B. By requiring a transfer of mass from the region B to lower down in the Earth, it removes the difficulty referred to on p. 291. An essential effect is a considerable reduction in the differences in the mantle earlier found between the models of A and B types. This effect in itself constitutes a measure of additional support for fairly smooth variation of k below 1000 km depth.

A slight complication arises, however, when the region D" (not allowed for in model (A')) is taken into account. Allowance for D" would affect the overall distributions of ρ and k only trivially, but there is a significant effect on $k_2/k_1 - 1$. As already seen, (5) shows that taking $d\phi/dz = 0$ in D" gives $\eta = dk/dp \simeq 3$. The main consequence on model (A') of taking $\eta \simeq 3$ in D" would be to increase ρ by 0·2 g/cm^3 at N_1, without noticeably altering ρ at N_2. Since k/ρ is being taken to be seismologically determined, $k_2/k_1 - 1$ would then be changed from +0·01 to −0·03. The value −0·03, though still quite small, is further away from agreement with the finite-strain theory, and is a little less close to continuity of k at N. It is possible that refinements yet to be made in model (A') in the upper mantle will restore the close continuity of k at N, also that the seismological evidence on $d\phi/dz$ may need some revision. Thus the apparent complication connected with D'' may prove to be unimportant.

The inductive inference from the whole evidence at present available is therefore that the compressibility postulate is likely to apply fairly closely in the Earth's interior where $p > 0.4 \times 10^6$ atm. Except where data on $d\phi/dz$ are in question, the main inferences based on the postulate are likely to be correspondingly reliable.

Table 1 gives values of p, ρ, k, μ, and g in model (A'). Values of k and μ are not included below 4500 km since allowance was not made in model (A') for the presence of rigidity in the lower core (for information on k and μ below 4500 km, see Bullen[2], Bullen and Haddon[36]).

References

1. K. E. Bullen, *Introduction to the Theory of Seismology*, 3rd ed. (reprinted with amendments), Cambridge University Press, Cambridge, 1965.
2. K. E. Bullen, *Geophys. J.*, **9**, 233 (1965).
3. K. E. Bullen, *Geophys. J.*, **9**, 195 (1965).
4. K. E. Bullen, *Monthly Notices Roy. Astron. Soc., Geophys. Suppl.*, **3**, 195 (1936).
5. K. E. Bullen, *Trans. Roy. Soc. New Zealand*, **70**, 137 (1940).
6. K. E. Bullen, *Trans. Roy. Soc. New Zealand*, **71**, 164 (1941).
7. K. E. Bullen, *Bull. Seismol. Soc. Am.*, **32**, 19 (1942).
8. K. E. Bullen, *Monthly Notices Roy. Astron. Soc., Geophys. Suppl.*, **6**, 50 (1950).
9. F. Birch, *J. Geophys. Res.*, **57**, 227 (1952).
10. E. D. Williamson and L. H. Adams, *J. Wash. Acad. Sci.*, **13**, 413 (1923).
11. K. E. Bullen, *Trans. Am. Geophys. Union*, **35**, 838 (1954).
12. K. E. Bullen, *Geophys. J.*, **7**, 584 (1963).
13. K. E. Bullen, *Geophys. J.*, **13**, 459 (1967).
14. K. E. Bullen, *Monthly Notices Roy. Astron. Soc., Geophys. Suppl.*, **5**, 355 (1949).
15. H. Jeffreys and K. E. Bullen, *Seismological Tables*, British Association, Gray–Milne Trust, London, 1940, 1958.
16. H. Jeffreys, *Monthly Notices Roy. Astron. Soc., Geophys. Suppl.*, **4**, 537, 548, 594 (1939).
17. K. E. Bullen, *Nature*, **157**, 405 (1946).
18. W. M. Elsasser, *Science*, **113**, 105 (1951).
19. K. E. Bullen, *Monthly Notices Roy. Astron. Soc., Geophys. Suppl.*, **6**, 383 (1952).
20. F. Birch, *Geophys. J.*, **4**, 295 (1961).
21. K. E. Bullen and T. N. Burke-Gaffney, *Geophys. J.*, **1**, 9 (1958).
22. B. Gutenberg, *Bull. Seismol. Soc. Am.*, **48**, 301 (1958).
23. B. A. Bolt, *Bull. Seismol. Soc. Am.*, **54**, 191 (1964).
24. Nguyen Hai, *Ann. Geophys.*, **19**, 285 (1963).
25. R. D. Adams and M. J. Randall, *Bull. Seismol. Soc. Am.*, **54**, 1299 (1964).
26. K. E. Bullen, *Proc. Natl. Acad. Sci. U.S.*, **52**, 38 (1964).
27. F. Birch, in *Solids under Pressure*, McGraw-Hill, New York, 1963, p. 137.
28. K. E. Bullen, *Nature*, **201**, 807 (1964).
29. A. H. Cook, *Space Sci. Rev.*, **2**, 355 (1963).
30. K. E. Bullen and R. A. W. Haddon, *Nature*, **213**, 574 (1967).
31. W. H. Ramsey, *Monthly Notices Roy. Astron. Soc., Geophys. Suppl.*, **5**, 409 (1949).
32. W. H. Ramsey, *Monthly Notices Roy. Astron. Soc., Geophys. Suppl.*, **6**, 42 (1950).
33. B. Gutenberg, *Trans. Am. Geophys. Union*, **32**, 373 (1951).
34. D. Griggs, *Trans. Am. Geophys. Union*, **35**, 93 (1954).
35. J. Verhoogen, *J. Geophys. Res.*, **58**, 337 (1953).
36. K. E. Bullen and R. A. W. Haddon, *Physics of Earth and Planetary Interiors*, **1**, 1 (1967).

2

F. BIRCH

Hoffman Laboratory
Harvard University
Cambridge, Massachusetts, U.S.A.

Composition and state of the Earth's interior in the light of recent experiments at high pressures

Abstract

Recent measurements employing shock-generated and static high pressures have largely removed the need for extrapolation in discussing density and compression in the Earth's interior. Theoretical equations of state remain of use chiefly for smoothing and interpolation. The major features of the mantle must be interpreted in terms of phase changes of light silicates, the difference between mantle and core in terms of chemical change. The role of temperature is minor except in the upper mantle.

3

T. HERCZEG

Hamburg Observatory
Hamburg, West Germany

Cosmogonical aspects in the theory of planetary interiors

Cosmogony may give us, at best, only 'second-class' information about the structure and composition of the planets. As is expected, information is mainly chanelled in the reverse direction: facts revealed by the physics and chemistry of the planets yield all the important starting points and basic criteria for cosmogonical theory. Nevertheless, planetary cosmogony is advancing very rapidly and seems to be already in a sufficiently developed stage to allow us at least an attempt to learn something about planetary interiors on this basis. In the following we try to tackle an important abundance problem in the region of the terrestrial planets in terms of the various hypotheses about their origin and their early evolution.

Planets of the solar system can be clearly divided into three distinct groups: the terrestrial planets, virtually lacking in hydrogen and helium; Jupiter and Saturn, giants having nearly solar composition; Uranus and Neptune, giants showing a strong deficit in hydrogen again, as indicated, for example, by their mean densities (1·6 and 2·2 g/cm^3 respectively).

The very existence of these divisions is a most important clue for cosmogonical theories: the group comprising Jupiter and Saturn is explicable without much difficulty on the assumption of gravitational retention of hydrogen and helium; in the case of Uranus and Neptune, the higher density suggests that evaporation of hydrogen (and perhaps helium) from the system took place during the formation of these planets. In both cases we obtain some important hints concerning temperature and density conditions—and also the time scale—in those remote phases of early planetary evolution.

We shall, however, restrict ourselves to the terrestrial planets and especially to questions of chemical composition. The problem can be stated as follows.

(i) First, there is the well-known concept—proposed by Ramsey—that the whole terrestrial group (including the Moon) forms a chemically homogeneous part of the system, showing essentially 'mantle composition'. Accordingly the cores in their various sizes consist of a high-density modification involving one of the chief constituents, probably the olivine. Promising as this hypothesis was, it, however, seems to be generally abandoned now, mainly because of the great difficulties in understanding the physics of this transition, especially the very considerable discontinuity required (about 1:2) in the density. At any rate, Mercury would be a very notable exceptional case even within the framework of this concept, the smallest of the planets showing a surprisingly high density which is comparable with that of the Earth.

(ii) Rejecting Ramsey's proposition means we must accept the traditional view of a chemically entirely different iron–nickel core. This composition is very strongly, in fact, necessarily suggested by the densities *and* the relatively high cosmic abundance of these elements, especially by the high iron content of the meteorites. But, in this case, we are confronted with the strange fact that the size of the core and, consequently, the overall iron content vary strongly among the objects considered, perhaps from 5 to 10% (Moon) to about 50% or more (Mercury). Moreover, there is a slight indication of a correlation between iron content and distance from the Sun: Mercury–Venus, Earth–Mars, and meteorites showing an order of a decreasing amount of iron and related elements. The Moon's position would be, from this point of view, exceptional.

(iii) The acceptance of chemically different cores involves a further, probably more serious, problem: an apparent overabundance of iron. In spite of the existence of the so-called iron peak, its cosmic abundance seems to be only about 0·02 by weight, in terms of *all* 'heavy elements' ($Z>2$). Taking the high volatility of some lighter elements into consideration, we still arrive at about the ratio 0·06–0·08, as compared with 0·20–0·30 in the terrestrial planets and the meteorites. To put it in another way: throughout this part of the solar system a Fe/Si ratio around 1 is indicated, whereas spectroscopic evidence gives about 0·2 or 0·3 for the stellar atmospheres, and notably for the Sun.

This discrepancy was first pointed out by Urey and also repeatedly discussed by him. His conclusion is that the difference between 'cosmic' and 'terrestrial' abundances may be real[1]. It would, indeed, be difficult for any astronomical spectroscopist to admit that recent stellar determinations can be in error by a factor of 4 or 5; the iron abundance, particularly, seems to be among the best-known ones. (It should, however, be remarked that this evidence is somewhat weakened by rather surprising indications of an Fe/Si ratio near 1 for the solar corona.)

Thus we can finally formulate our questions as follows. We have to find a fractionation mechanism (or several mechanisms?) which may have worked during the early phases of planetary evolution and which may have been capable of producing (*a*) the supposed enrichment of the iron group in the terrestrial region of the system, (*b*) especially the heavy core and high density of Mercury, and (*c*) possibly also the suggested tendency of containing less iron at a greater distance from the Sun.

It goes almost without saying that we can justifiably speak of an iron discrepancy alone. Any attempt to explain, for instance, the density of Mercury by relying on the heavy elements beyond the iron peak would be completely inadequate, implying deviations from the adopted cosmic abundance which are orders of magnitude wider than in the case of iron.

Before we consider these questions somewhat closer let us make a minor digression, enumerating those recent determinations which form the basis for the adopted high mean density of Mercury. To what extent are values like 5·5 or 5·6 g/cm^3 reliable?*

Concerning the mass of Mercury, early determinations led to the unusually wide range from about 1/9 000 000 to 1/3 000 000 (in units of the solar mass); the more reliable figures were in the vicinity of $m = 1/6\,000\,000$, a value which, following Newcomb's discussions, was in use for several decades. Recently, three determinations narrowed down this range considerably. Rabe found[2], from the perturbations of Eros

---

* This short review of data concerning Mercury's mass and size has been inserted here to represent a part of the discussion following the lecture.

by Mercury, $m = 1/6\,120\,000\,(1\pm0\cdot007)$ or $0\cdot0544$ in terms of the Earth's mass; Brouwer derived[3], from a provisional solution based on the secular perturbations of Mercury and the Earth, $1/6\,480\,000\,(1\pm0\cdot055)$; Duncombe rediscussed[4] the motion of Venus and from its perturbations by Mercury he obtained $1/5\,970\,000\,(1\pm0\cdot077)$. In spite of the high accuracy of Rabe's determination the question was reopened by the well-known discrepancy between astrometric and radar determinations of the astronomical unit. In an attempt to find a 'reconciling' set of data, Marsden[5] proposed three solutions. His solution A yields very nearly Rabe's figure for the mass of Mercury, whereas solutions B and C lower it to about $1/6\,400\,000$; the uncertainty is quoted as being around 3%. These determinations of Brouwer, Duncombe, and Marsden overlap—taking their probable errors literally—in a region roughly between $1/6\,250\,000$ and $1/6\,450\,000$, somewhat below Rabe's value.*

Early measurements of Mercury's 'particularly poorly known' diameter define, according to Dollfus[7], a mean of $6\cdot45''$, reduced to 1 A.U. (astronomical unit), but with a very modest accuracy of not higher than 5%. Thus corresponding linear values may lie between 4450 and 4950 km. A concerted effort of an improved determination was made on the occasion of Mercury's transit in front of the solar disk on November 7, 1960. Three series of measurements by double-image micrometer gave results around $6\cdot63''$; two photometric determinations—following a proposal by Hertzsprung—were about $0\cdot10''$ higher. As a final result, Dollfus[8] found $6\cdot67''\pm0\cdot05''$ (i.e. 4840 ± 35 km); Camichel, Hugon, and Rösch[9], however, being of the opinion that the photometric method is preferable, rediscussed these measurements and arrived at a value of $6\cdot83''\pm0\cdot03''$ (4960 ± 20 km). During the same transit, Gerharz[10] compared the solar chromospheric diameter with Mercury's diameter, on the basis of Hα pictures taken with the Climax coronograph; he obtained 4880 ± 50 km. Combined with Rabe's value for the mass, the resulting densities would be as follows:

$$D = 4500 \quad 4600 \quad 4700 \quad 4800 \quad 4900 \quad 5000 \text{ km}$$
$$\bar{\rho} = 6\cdot81 \quad 6\cdot38 \quad 5\cdot98 \quad 5\cdot61 \quad 5\cdot28 \quad 4\cdot97 \text{ g/cm}^3$$

Using Brouwer's or Duncombe's somewhat less accurate values, the mean density turns out to be about $0\cdot30$ g/cm^3 lower or $0\cdot15$ g/cm^3 higher, respectively. It is obvious that the high-density figures quoted in the literature cannot be merely results of erroneous diameter measurements: in fact, the best determinations seem to stabilize the 'low-density end' of the above scale. Taking, as a compromise, the values

$$m = 1/6\,200\,000 \quad \text{and} \quad D = 4900 \text{ km}$$

leads to $\bar{\rho} = 5\cdot21$ g/cm^3. This two-decimal accuracy is, of course, quite illusory and the most we can say at present is that the mean density probably lies between $5\cdot0$ and $5\cdot5$, possibly somewhat closer to the lower value.

Returning now to our cosmogonical question we give first a very short description of some of the main lines of approach to planetary cosmogony, the archetypes of cosmogonical hypotheses. Broadly speaking, they can be divided in two groups according to the basic assumption about the physical process of planetary origin and build-up, the birth and growth of a planet, so to speak.

---

* The system of planetary masses was also rediscussed by Clemence[6]; his result for Mercury amounts to $1/6\,110\,000\,(1\pm0\cdot007)$, in perfect agreement with Rabe's value.

(1) The first assumption is the 'condensation of gas spheres', i.e. the transition of the primordial nebula to small subunits, rapidly condensing into spherical, predominantly gaseous protoplanets. A 'formula' for this type of evolution would be as follows: gaseous medium → gas spheres → condensed spheres (= planets, satellites, asteroids, perhaps 'lunar-sized objects').

(2) The other basic assumption is 'dust accretion', gradual growth of the planets, satellites, etc., out of small condensed particles, according to the following formula: gas → dust → larger bodies.

This typology refers to the basic process and comprises, of course, also some arbitrary distinctions. The gas spheres of (1) may well have contained a dust component, too; on the other hand, the accretion process of (2) probably took place within a general gaseous substratum. There are, in particular, concepts based on a combination of these two classes, e.g. Urey's hypothesis of the two generations of objects, evolution of the primary one following essentially process (1), the secondary—after extensive collisional destruction—process (2).

Theories of type (1) were very common 30 or 40 years ago but are, in general, not favoured today. Concerning hypotheses of group (2), we meet two types of assumptions again, leading to very different sequences of events as a proposed evolutionary picture.

(2a) The primordial nebula was a relatively thin 'circumsolar' cloud of 10^{-9}–10^{-11} g/cm^3 density; the Sun itself was already in a late phase of its contraction. A representative of this concept is Hoyle[11].

(2b) The starting point of evolution was a dense *solar* nebula, with a density perhaps of 10^{-5}–10^{-6} g/cm^3; formation of the central mass (Sun) and accumulation processes in the outer regions (planets) were virtually simultaneous. Here we may mention as a representative Cameron's[12] elaborate theory.

It is important to mention here that all these types of early planetary evolution, at any rate both (2a) and (2b), involved a rather short time scale of 2–3×10^6 years, probably even less. This evidence is rather strong. In the case of (2b) this case is fixed by the short time scale of the pre-main-sequence evolution, but also by the rapid growth of solid particles in the relatively dense primordial nebula. In case (2a) similar reasoning is applicable, and considerable additional support is given by the indications of essential hydrogen losses from the edge of the nebula. The present author feels that it would be very difficult to accept a time scale extended to 50 or perhaps 100 million years. This circumstance—if confirmed definitely—may impose some limitations upon concepts including a complex history of the system, like Urey's proposal briefly mentioned above; this proposal is, on the other hand, the only significant representation relying on process (1), for the time being.

In the following discussion we briefly review some possible fractionation processes to find out whether one or other of them could be efficient enough to produce the enrichment of iron (taken as well established) and also, perhaps, its suggested correlation with the distance from the Sun. There is hardly sufficient room to give here more than the general lines of argument; only in the case of a possible radiation effect —the most important one, in the opinion of the writer—are we going to discuss it in more detail. We direct most attention to the dust accretion process (2) and especially to the possibility (2a). Though obviously not a rigorously proved theorem, we thereby follow a heuristic principle: theories based on models of type (2b) offer—from our point of view—a less advantageous position, because they require that early planetary evolution had taken place in a rather dense medium. Thus, from all heavier elements there existed, so to speak, a rich supply stored in the environment of the growing protoplanets; any fractionation between iron and silicon becomes increasingly difficult to understand. It seems, for our purpose, more promising to rely on hypotheses of a thin circumsolar cloud containing perhaps 0·005–0·010 solar mass. The amount of available mass was, therefore, not much higher than the total mass of the planetary system in its present form, and occasional losses from the growing objects might well have turned out irreparable. Besides Hoyle's concept already quoted, the entirely different theory due to Schmidt[13] is another important representation of this type (2a).

As possible causes of the fractionation we have to consider thermal, chemical, magnetic, and radiative processes. A mechanical factor, collision of particles (moving in highly elliptical orbits) with the Sun, may have played some minor role in the general thinning-out of matter in the immediate solar neighbourhood but, of course, cannot have produced any selective effect.

Thermal processes mean, basically, differential evaporation caused by intense heating. It would be difficult, and to a certain extent arbitrary, to distinguish them from chemical effects, the question being highly complicated by the presence of many possible compounds. Several thorough studies by Urey are devoted to this problem; perhaps the most important contributions are, in addition to his famous book[14], the papers listed under at the end of this chapter[15–17]. Though very great differences exist in the volatilities of the elements under the postulated circumstances of early planetary evolution, silicates seem to be, in general, somewhat more volatile than iron and its compounds. Thus we may expect, at least qualitatively, an effect in the observed sense; temperature requirements are, however, very difficult to meet.

The required temperatures lie, roughly, in the range 1500–2500 °C; they had to last for a considerable time interval of the order of several 10^5 years. In this respect, we cannot rely upon the early high-luminosity period of solar evolution: in all probability, the event of planet formation succeeded, and did not coincide with, the so-called Hayashi phase, as recently discussed, among others, by the present author[18]. (We wish to mention once again that, in the following, hypotheses mainly of type (2a) of the above classification scheme will be considered.) Later on, heating of all the accumulation products—from meteoric dimensions to growing planets—certainly took place but hardly to the required extent. Radioactivity played a major role; its influence, however, may have resulted in temperatures of the range 800–1200 °C, definitely too low for our present purpose. External heating by 'hot solar gases' or perhaps intense corpuscular radiation of the Sun—as sometimes proposed—turns out to be inefficient, especially for the interiors of the growing protoplanets; besides, it is rather doubtful whether the envisaged types of solar activity were at the required high level, at that time.

The most promising explanation of an early heating is still the conversion of gravitational into thermal energy, at least in the case of the two more massive objects, Venus and the Earth. The available energy is of the order of 10^{31}–10^{32} cal, enough to produce the required temperatures, though numerical estimates largely depend on assumptions concerning thermal losses, i.e. on the postulated cosmogonic model. In the case of Mercury and Mars, however, and *a fortiori* of lunar- or asteroid-sized bodies, gravitational energy is hopelessly insufficient as a source of heat. Thus we find, from all these considerations, temperatures which are too low by a margin of about 1000 degc.

These short summary-like remarks are, of course, only intended to give a mere outline of the argument; they certainly do no justice to the numerous, intricate details of the processes involved. In spite of the bewildering complexity of possible effects and phenomena, *chemical* evidence can be of extreme importance. Unfortunately, there is always a certain—evident or latent—interrelation with basic cosmogonical hypotheses accepted by the investigator; following the chemical clues means no purely 'inductive way' of approach, as indicated sometimes. In view of this crucial role chemistry may play in planetary cosmogony, astronomers are fortunate enough that recently two comprehensive articles did appear, by Suess[19] and Urey[20] respectively, summarizing the available chemical evidence bearing on the origin of the solar system.

Not being a chemist, we can hardly add new chemical points of view to these discussions. It is remarkable how many details, for instance minor differences in the Ni/Fe or Co/Fe ratio—the Sun compared with the meteorites—can be explained convincingly in terms of different chemical behaviour (cf. Suess[19], p. 220). For the more general, more marked iron–silicon shift, however, there seems to be no simple explanation found as yet. In fact, a proposal in Urey's paper indicates that he himself prefers a 'mechanical' explanation instead. During the collisonal destruction of the primary (lunar-sized) objects Urey[20] (p. 217) states as follows: 'The finely divided silicate materials could have been driven into space leaving larger fragments behind. These should have contained a larger proportion of metallic objects and hence the high densities of the terrestrial objects can be explained.' Acceptance of this proposal requires, obviously, the acceptance of the whole two-stage evolutionary concept briefly mentioned above. This is certainly still pending; there is even some reluctance among astronomers to agree to it; probably not because of the highly complicated succession of events postulated by Urey, rather because of the questionable point introduced by relying upon the large-scale collisional destruction.

There is a further, at first sight quite harmless, question: what happened to the lighter material which was removed? The question is far from being trivial. Either most of the lost matter existed already in the form of grains or it reached this stage rapidly by recondensation: this means, however, that both solar wind and radiation might have exerted only a limited influence. On the other hand, this silicon-rich component represented perhaps 25–30%, in any case a considerable fraction, of the total mass in the terrestrial region. To dispose of it quickly and completely, before light 'silicon planets' could have come into existence, requires a mechanism deserving closer study.

The very term 'iron anomaly' suggests the possibility of some *magnetic* effect involved. Let us suppose, in fact, that the protoplanets were sites of sufficiently strong magnetic fields; ferromagnetic particles in their vicinity could have 'slid down' into these centres of accumulation. The effect of this preferential accretion on the final composition would be just what we are looking for, but quantitative requirements are much too high. Even fields several orders of magnitude stronger than the Earth's present-day magnetic field could have influenced the motion of particles in the immediate neighbourhood of the developing planets only. Also, it seems that some abundance data yield further unfavourable evidence. As far as spectroscopic abundances for the Sun and the delicate analysis of meteoric samples are considered as giving reliable information (but, on turning to heavier and heavier atoms, reliability deteriorates rapidly), the iron–nickel excess is not confined to this group of elements. The most detailed study of this question is due to Urey; according to data given in his Harold Jeffreys Lecture[1], several other elements like copper, zinc, gadolinium, germanium, manganese, palladium, and perhaps also silver and antimony show a higher meteoric than photospheric abundance. If these differences can be substantiated by further studies, they would form a decisive argument in the case considered here.

An important *radiative* effect offers a more promising way of introducing a density separation among dust particles. This effect was first proposed by Poynting in 1903 but only a relativistic treatment by Robertson, more than 30 years later, elucidated it completely[21]. To put it briefly and very approximately: particles orbiting around the Sun, and with dimensions surpassing the wavelength of light, suffer a tangential drag constantly decreasing their angular momentum. (Particles absorb solar radiation coming from a radial or nearly radial direction but they are reemitting it isotropically; because of the Doppler effect higher frequency, carrying more momentum, is radiated away in the forward direction, thus exerting a type of recoil upon the orbiting particles.) Under plausible conditions, the time required for a grain of radius r (cm) and density ρ (g/cm^3) to spiral into the Sun, starting from an initial distance R (A.U.), is

$$7 \cdot 0 \times 10^6 r \rho R^2 \text{ years}$$

As this formula indicates, we may not only expect a separation according to the density of the particles but also a correlation with the distance from the Sun. Consideration of this effect as a selection factor was recommended by the author[18]; in fact, the Poynting–Robertson effect was already mentioned by Urey in 1952, though in a somewhat more restricted connection (in his book[14], p. 195). This effect would explain the disappearance of matter from the terrestrial environment in a natural way without involving momentum difficulties; some additional support could be found—if really proved—in the above-suggested overabundance of several heavy nuclides, too.

The general idea is, of course, that the growing dust particles consisted of a mixture composed of numerous elements (and possibly their compounds); the relative proportions of these components were subject to considerable fluctuations. Particles with a larger share of heavier elements could, by virtue of their higher density, resist this radiative braking longer, and had a better chance to survive. According to this concept, already during the earliest period of accretion a separation by specific weight took place, lighter elements becoming stronger depleted.

In order to estimate the influence of this effect on the final chemical composition of the terrestrial planets, we constructed a simple model of dust accumulation; it seems that the simplifying assumptions involved thereby do not introduce a significant shift in the numerical results. First, we defined two basic types of material, representing the 'silicates' and 'irons' (or mantle and core composition). As characteristic densities we have chosen 2·5 g/cm^3 and 8 g/cm^3, respectively. Proportions, by weight, for these two components are designated by $1-x$ for the silicates and x for irons. We assumed that the components in various growing particles show a random distribution, the same for all sizes of particles and all distances from the Sun. This we represented by a simple Maxwellian distribution

$$N(x) \sim \frac{x^2}{x_0^2} \exp\left(\frac{-x^2}{x_0^2}\right)$$

with $x_0 = 0\cdot 10$; this latter value (being probably somewhat on the low side) corresponds to a grain density of $3\cdot 05$ g/cm^3. Another assumption must relate to the growth of grains. We considered this as linear with time, which means that the resulting effect depends in a simple and direct way on the time scale of the early planetary evolution. Besides, grain formation was a stochastic process; some have grown rapidly, others slowly. The simplifying assumption we used here was the following: starting from a uniform radius of, say, 1 μm, after the time elapsed, linearly increasing grain radii became evenly distributed between 1 μm and 1 cm. During this period of growth, the Poynting–Robertson effect was constantly sweeping away part of the orbiting grains, preferentially the light ones. The size limits just mentioned are easy to explain: below 1 μm, the effect does not work: above 1 cm, particles are safe for 10^7 years or longer; the 'critical' interval of growth may be between a few micrometres and a few millimetres.

Even a cursory comparison between time intervals characteristic of the Poynting–Robertson effect (5–10 million years) and of early planetary evolution (1–3 million years) suggests that the radiative braking could not be very effective; dust grains have simply 'outgrown' the range of the Poynting–Robertson effect. More detailed calculations (made rather easy by our assumptions but omitted here) do show, indeed, that the *selective* effect is in the case of Venus, Earth, and Mars entirely negligible. The total available mass may have been reduced by some 10–20% only in these distances. In the case of Mercury ($R^2 \simeq 0\cdot 15$), however, the effect may have been significant. To be sure, taking $\Delta t = 10^6$ years, we obtain only about 50% loss and a very insufficient fractionation, changing the mean density by less than $0\cdot 1$ g/cm^3. But increasing Δt brings considerable gain; at $\Delta t = 3 \times 10^6$ years we have 96% loss of mass, whereas the mean values of x and ρ change, respectively, to 0·28 and 3·8 g/cm^3. This is still small enough; yet a minor additional increase in Δt give values at least comparable with the observations. Though time-scale requirements seem to be fairly high, it should perhaps be noted that losses amounting to 25–30 times the present mass of Mercury are in good agreement with the estimated density distribution in the primordial nebula (see e.g. Herczeg[18]).

These considerations are modified to some extent by the following circumstances. The preplanetary nebula we are dealing with must have been of a very high opacity especially during the earliest phase of accumulation; transparency was slowly increasing as a substantial part of grains reached meteoric dimensions. This means that at the very beginning of the dust accretion, at larger distances from the Sun, only extremely weakened, diffuse solar radiation was received and conditions for a radiative effect worth mentioning were not fulfilled. During this early period, the Poynting–Robertson braking was restricted to the inner edge of the nebular disk, steadily pushing its boundary outwards, as matter was continuously eaten away. It is obvious that this additional effect might further substantiate Mercury's exceptional position and bring our values closer to the actual mass and density data.

Summarizing the outcome of these studies we, perhaps, may say that understanding the so-called iron anomaly and accounting for the composition of massive iron cores, in terms of cosmogonical considerations, turned out to be unexpectedly difficult. We did not succeed in finding the relevant fractionation process; in particular, the Poynting–Robertson effect is hardly applicable as a selection mechanism beyond the orbit of Venus. The same effect can, however, explain the small mass, and also the high density, of Mercury.

If we are allowed to draw any conclusion or, at least, to obtain some hints—in the form of 'second-class evidence' mentioned at the beginning of this chapter—we must underline the following suggestions. The possibility that the 'cosmic', i.e. spectroscopic, abundance of iron has been underestimated (consequently, at this point solar system data would give more reliable information), is still not to be rejected completely. However, if the discussed iron anomaly really exists, our hope of an explanation is to be placed on the chemical facts, magnetic and radiative effects being certainly, and thermal segregation probably, not efficient enough.

In spite of these rather cautious conclusions, we hope that the composition and structure of Mercury are somewhat better understood now. If the explanation of its high density, as put forward in this chapter, can be further substantiated, this would again turn out to be a piece of evidence supporting the general concept of cold dust accumulation.

References

1. H. C. Urey, Harold Jeffreys Lecture, London (1966).
2. E. Rabe, *Astron. J.*, **55**, 112 (1950).
3. D. Brouwer, *Bull. Astron.*, **15**, 164 (1950).
4. R. L. Duncombe, *Astron. Papers, Am. Ephemeris, Wash.*, **16**, Parts 1–6 (1958).
5. B. G. Marsden, *Bull. Astron.*, **25**, 225 (1965).
6. G. M. Clemence, *Ann. Rev. Astron. Astrophys.*, **3**, 110 (1965).
7. A. Dollfus, *Handbuch Phys.*, **54**, 205 (1962).
8. A. Dollfus, *Icarus*, **2**, 219 (1963).
9. H. Camichel, M. Hugon, and J. Rösch, *Icarus*, **3**, 410 (1964).
10. R. Gerharz, *Mem. Soc. Astron. Ital.*, **35**, 167 (1964).
11. F. Hoyle, *Quart. J. Roy. Astron. Soc.*, **1**, 28 (1960).
12. A. G. W. Cameron, *Icarus*, **1**, 13 (1962).
13. O. J. Schmidt, *A Theory of the Origin of the Earth*, Wishart, London, 1959.
14. H. C. Urey, *The Planets, their Origin and Development*, Princeton University Press, Princeton, 1952.
15. H. C. Urey, *Astrophys. J., Suppl. Ser.*, **1**, No. 6 (1954).
16. H. C. Urey, *Phys. Chem. Earth*, **2** (1957).
17. H. C. Urey, '41st Guthrie Lecture', *Yearbook Phys. Soc.*, **1957**, 14.
18. T. Herczeg, Inaugural Dissertation, Hamburg (1966).
19. H. E. Suess, *Ann. Rev. Astron. Astrophys.*, **3**, 217 (1965).
20. H. C. Urey, *Monthly Notices Roy. Astron. Soc.*, **131**, 199 (1966).
21. H. P. Robertson, *Monthly Notices Roy. Astron. Soc.*, **97**, 423 (1937).

C

HIGH-PRESSURE PHYSICS AND THE EARTH'S INTERIOR

II. Equations of state

1

J. J. GILVARRY*

Theoretical Studies Branch
Space Sciences Division
Ames Research Center
National Aeronautics and Space Administration
Moffett Field, California, U.S.A.

Equations of state at high pressure from the Thomas–Fermi atom model

Introduction

The Thomas–Fermi approximation[1-3] was conceived originally as a semiclassical method of calculating the electron distribution in a heavy atom. In this connection, it possesses the salient advantage of requiring solution of total differential equations (in the spherically symmetric case) subject to initial and boundary values, rather than partial differential equations subject to boundary conditions implying eigenvalues of constants of the motion for many electrons, as in solution of the Schrödinger equation. Thus it constitutes essentially an approximate method of solution of the many-electron problem, in many respects analogous to the Debye–Hückel theory[4] of ionic solutions. The method has been applied to the approximate calculation of many atomic properties. The present review is concerned exclusively with its application to calculation of equations of state and thermodynamic functions at high pressure. For other applications, to atoms and to nuclei, the works of Gombas[5,6] can be consulted.

The principal interest here is in the problem of extracting the actual equation of state and the pertinent thermodynamic functions from the relevant differential equations of the statistical atom model. Thus, only a brief description of the theoretical basis will be given. The physical assumptions peculiar to the application of the statistical atom model at high pressure will be stated explicitly. Finally in this section, the field of application will be discussed.

Theoretical basis

The underlying idea of the Thomas–Fermi approximation is to treat the electrons classically except for the one respect that they satisfy the Pauli principle and therefore obey Fermi–Dirac statistics. The assumption of classical behavior makes it possible to calculate the electron density at any point within an atom self-consistently by means

---

* Senior Postdoctoral Resident Research Associate, National Academy of Sciences—National Research Council (Washington, D.C.), at Ames Research Center. Present address: The Rand Corporation, Santa Monica, California, U.S.A.

of the Poisson rather than the Schrödinger equation. To find the density of electrons which is the source at a point in the atom of the electric potential satisfying Poisson's equation, consider the atom divided into cells small enough so that the potential is essentially constant (over many wavelengths of an electron represented quantum mechanically by a plane wave) but large enough so that the electrons in a cell occupy a region of phase space large compared with h^3, where h is Planck's constant. For the absolute zero of temperature, the assumption of Fermi–Dirac statistics implies that the density of electrons in the (six-dimensional) phase space is $2/h^3$ (the factor 2 arises from the existence of two spin states) at all points corresponding to an energy less than some maximum value equal to the Fermi energy, and is zero at points corresponding to an energy greater than this maximum. At a given point of configuration space, those points of momentum space within a sphere of radius corresponding to the maximum momentum permitted by the Fermi energy are occupied by electrons, and those outside the sphere are unoccupied. The total number of electrons within the sphere in question fixes the density of electrons which act as the source of the electric potential in Poisson's equation, and solution of this equation fixes the potential and the electron distribution self-consistently.

In this physical picture, the electrons are represented quantum mechanically by plane waves satisfying periodic boundary conditions, and the density of electrons is fixed by the number of momentum states (corresponding to the different frequencies) allowed up to the maximum permitted by the value of the Fermi energy. To satisfy the requirements of the Pauli principle strictly, however, one must use antisymmetrized wave functions. The effect of this refinement was taken into account by Dirac[7]. A further error persists, since the entire wave function for all the electrons is assumed to be a simple product of one-electron wave functions, which neglects correlations between the positions of the electrons arising from the Coulomb repulsion. The energy difference involved is known as the correlation energy, so named by Wigner[8].

Rather than approach the Thomas–Fermi model from the statistical viewpoint outlined above, one can consider it from the standpoint of the semiclassical limit to quantum mechanics represented by the Wentzel–Kramers–Brillouin–Jeffreys approximation. This question has been discussed by Brillouin[9], by Fényes[10], and, most recently, by March and Plaskett[11]. The results show that the Thomas–Fermi equation can be regarded as the semiclassical limit (corresponding to the leading term in an expansion in powers of h) of the Hartree equations for the model of a self-consistent central field. The Thomas–Fermi–Dirac equation can be regarded as the similar limit of the Hartree–Fock equations. However, the statistical approach must be used to take account of an absolute temperature which is not vanishing, by specifying the average occupation number for an individual electron state entering the electron density by means of the Fermi factor for an elevated temperature.

Since the details of the electron shells of the atom, represented by the sublevels of angular momentum, are averaged over and thus do not enter the model, one expects it to describe reasonably only those properties of the atom which are relatively independent of the shell structure. At high pressure such structure must be largely inconsequential, so that the equation of state at the very highest compressions should certainly be given accurately (provided relativistic effects are ignored). On the other hand, properties that depend sensitively on features of the charge distribution at the edge of the atom or on the detailed structure of the outer electron shells are poorly given by the model, since the field of an electron itself at this point is of the same order as the total field.

In addition to the works of Gombas[5,6], a valuable review stressing the foundations of the model has been given by March[12].

Physical assumptions

The basic viewpoint in the present context is to consider an atom of a solid, a metal in the first instance, as a polyhedral cell which contains one nucleus and a number of electrons equal to the atomic number. At high pressure, this cell will exhibit a rather high degree of symmetry, and hence the electric field at points outside the cell arising from the charge within is small and decreases rapidly with increasing distance from the cell. It is therefore a good approximation to replace each polyhedral cell by a sphere of equal volume and to regard the electron distribution within the cell as spherically symmetric, exactly.

Thus, the approximation entailed rests on the same foundation as the Wigner–Seitz[13] calculation of the cohesive energy of a metal from the Schrödinger equation of quantum mechanics, as noted by Slater and Krutter[14]. The electric field within a cell is viewed as dependent only on the charge distribution within that particular cell. Hence, the same method of solving the basic differential equations, whether the Schrödinger or the Thomas–Fermi equation, for the isolated atom can be applied to a compressed atom, simply by proper adjustment of the boundary conditions. One infers immediately from Gauss's theorem that vanishing of the electric field at the surface of the atom is a necessary boundary condition in the statistical atom model, for neutrality of the atom as a whole.

These assumptions reduce the problem of calculating properties of the solid as a whole to that of determining the properties of a single atom. The volume of the solid is measured by the size of one of the spherical cells. Hence, the pressure is equal to the pressure of the electron cloud at the surface of a spherical cell, since this quantity is uniform throughout the solid as a whole. Further, the internal energy is simply the product of the energy per atom and the total number of atoms, since the assumptions made imply neglect of interatomic forces.

However, complete neglect of interatomic forces must be artificial. In the case of a metal at normal temperature and pressure, the Thomas–Fermi model on the physical assumptions noted would ascribe all the thermal properties to the effects of a degenerate Fermi–Dirac gas of free electrons. In such a case, the model would yield the very low heat capacity corresponding to a free-electron gas, rather than the value approaching the limit of Dulong and Petit appropriate to a harmonic lattice[15]. Correspondingly, the thermal expansion would not have the value predicted by Grüneisen's law from the model of a solid whose thermal behavior is specified by a set of characteristic lattice frequencies, but a far lower value[15].

It is clear that the concepts of the statistical atom model applied to an individual atom in the manner described should yield good approximations to the equation of state and the associated internal energy, where the contribution of the long-range order associated with the lattice is small. However, for certain thermodynamic functions, such as the heat capacity, the entropy, and the coefficient of thermal expansion, the contribution of the lattice can not be neglected realistically, as emphasized by Gilvarry[16]. However, it can be shown that it is possible to take account of the effect of the lattice by a perturbation procedure, provided the equation of state from the statistical atom model is known on the assumption that the atoms are independent. For this purpose, it is necessary to have available an approximation to the fusion curve of the solid, which can be obtained by application of the Lindemann law of melting, as discussed by Gilvarry[17]. For temperatures below the fusion value at a particular temperature, the effect of the long-range order on thermodynamic functions can be taken into account by means of the standard theory of lattice dynamics[16].

No mention of molecules has been made, since Teller[18] has proved that stable molecules do not exist within the framework of the statistical model (in the sense that the molecular state cannot yield a lower energy than the separate atoms). This result vitiates the conclusions of many prior studies on molecular properties by use of the statistical method.

Field of application

It is obvious that the equation of state inferred from the statistical atom model becomes more accurate the larger the atomic number Z and the higher the pressure P. The first condition follows because of the requirement that the number of electrons be large enough in a cell in the atom, for which the potential is relatively constant over an electron wavelength, so that the corresponding region occupied in phase space be large relative to h^3. The second condition is then a corollary, on the basis of the assumptions of the model, since it corresponds to the condition under which the detailed structure of the energy levels (corresponding to sublevels of angular momentum) can be ignored. However, a further physical factor enters in regard to the second condition. At low pressure, the effects of chemical binding and lattice structure of the solid are ignored in the statistical atom model. These effects account for binding in the solid state at normal pressure, and the procedure of viewing the effect of the lattice merely as a perturbation on properties inferred from the statistical model is valid only at high pressure.

It was emphasized by Feynman, Metropolis, and Teller[19] that equations of state inferred from the statistical atom model are likely to be valid only for pressures exceeding about 10 Mb. This figure is about the value required for the work corresponding to the integral of the pressure over the corresponding compression to approximate the usual energy of chemical binding at normal pressure, as pointed out by Gilvarry and McMillan[20]. However, the estimate on this basis probably is unduly optimistic. Elsasser[21] has suggested a few million megabars, beginning at somewhat higher pressures for the lighter elements and somewhat lower for the heavier elements. An extremely crude estimate by Ramsey[22] for the critical pressure P in question can be written in terms of atomic number Z as

$$P \simeq 10^{13} Z^{5/2} \, \text{dyn/cm}^2 \tag{1}$$

which implies a figure of 10^6 Mb for the heaviest elements. It should be noted, however, that the variation of equation (1) with Z is opposite in sense to that suggested by Elsasser[21], and is at variance with what one intuitively expects from the fundamental assumptions of the statistical atom model.

This question of this critical pressure can now be profitably examined on the basis of measurements on shock waves in metals produced by explosives. The original measurements by Walsh and Christian[23] for aluminum, copper, and zinc for shock pressures up to about 500 kb were extended by McQueen and Marsh[24] to many elements up to a pressure of 2 Mb. In subsequent work by Al'tschuler and his school[25], the upper limit of pressure has reached 5 Mb. The data for pressure versus density of Al'tshuler and associates and of McQueen and Marsh have independently been reduced to zero absolute temperature by Takeuchi and Kanamori[26] and compared with the predictions of the Thomas–Fermi–Dirac theory. They find that the Thomas–Fermi model yields densities that are too low at 10 Mb, but extrapolation of the experimental results indicates that the actual equation of state is well represented by the model at pressures slightly higher than 100 Mb. The conclusion applies to Ag, Au, Cd, Cu, Fe, Pb, and Zn, involving extrapolations in pressure by a factor 20 to 50, in general. On the basis of the results of Takeuchi and Kanamori, a figure for the critical pressure of about or slightly higher than 100 Mb, independent of atomic number, seems the most reasonable and will be adopted here.

Until the advent of the data from measurements on shock waves, laboratory determinations of equations of state could be carried out to pressures of only 0·1 Mb as an upper limit, by the techniques of Bridgman[27]. The procedure of interpolating the equation of state at intermediate pressures from the computations of the statistical model at high pressure and the measurements of Bridgman at low pressure has been used for geophysical purposes by Elasser[21], Bullen[28], Birch[29], and others, chiefly to infer the elemental composition of the core. Since data from shock-wave measurements now extend to 5 Mb, in excess of the pressure (3·6 Mb) at the Earth's center, the method is essentially of historical interest now. However, the method has its merits in spite of obvious deficiencies. The conclusion by Elasser that the core is composed of iron with a possible admixture of nickel and the deduction by Knopoff and MacDonald[30] that the iron of the core is probably alloyed with silicon, obtained with use of this interpolation procedure, have not been reversed by the direct experimental data obtained from the shock-wave measurements.

Only with use of such an interpolation scheme can the equation of state from the statistical atom model be applied with any validity to the planets of the solar system, since the maximum pressure in the interior (for Jupiter) is only about 30 Mb, less than the critical pressure 100 Mb adopted above. However, pressures considerably higher than this critical value occur in the interiors of white dwarf stars. Thus the equation of state from the statistical atom model can be applied in the determination of the mass–radius relation and the limiting maximum mass for white dwarfs. This problem has been studied extensively and summarized elegantly by Chandrasekhar[31] on the basis of the equation of state corresponding to a free-electron gas, which represents the limiting form corresponding to the Thomas–Fermi model when the pressure is so high that the effect of the electric field of the nucleus can be ignored. As pointed out by Marshak and Bethe[32], use of the Thomas–Fermi model yields a more realistic equation of state, and inclusion of the effect of nonvanishing temperature poses no problem. It should be noted, however, that, for the smallest and densest white dwarfs, the electrons move with such high velocities that it is necessary to take relativistic effects into account in determining the equation of state[31].

Among the astrophysical applications of the equation of state from the statistical atom model, one can mention the work of Brown[33] on the compositions and structures of the planets and the calculation of the mass–radius relation of the planets by

Scholte[34]. The latter used an empirical fit to the Thomas–Fermi equation of state calculated by Slater and Krutter[14] to show that the data for the planets fall on two curves of mass versus radius corresponding to the terrestrial and the major planets; for the former the average atomic number is close to that of silicon and for the latter close to that of hydrogen. He was able to determine a limiting maximum mass for the planets analogous to that for white dwarf stars.

As an example of the application of results from the statistical atom model to a question in planetary science, the problem of predicting the pressures and temperatures arising in explosive impact of large meteorites on the surface of the Moon and Earth can be discussed for astronomic meteorite velocities, as treated by Gilvarry and Hill[35]. On the basis of results presented in succeeding sections of this chapter, the problem is immediately tractable on the model of a one-dimensional approximation to the hydrodynamic flow. The purely theoretical conclusions confirm the explosive nature of the impact process inferred by empirical arguments.

It might be worth while to point out an application of the statistical atom model which has a peripheral connection with geophysics and planetary science. As is well known, equations of state from the statistical atom model are a basic tool in the design of atomic and thermonuclear bombs, for the engineering of the initial implosion process. The craters produced by these bombs when exploded on the surface of the Earth show all the characteristics of the meteoritic craters observed on the Earth and the Moon. This fact has played a role in the construction of theories of the origin of lunar surface features[36] and also of hypotheses on the origin of such gross terrestrial features as ocean basins and continents[37].

It is desirable to indicate schematically the domain of applicability of the statistical atom model in the field of the variables, temperature T (absolute) and pressure P. This is done in figure 1, adapted from a corresponding diagram of Wares[38] in terms of temperature and density, but with correction to the atomic number of iron in the nonrelativistic case and with addition of regions corresponding to condensed phases. The plane of the diagram is separated into two disjoint regions by the locus corresponding to the degeneracy temperature; for temperatures below this curve, Fermi–Dirac statistics must be used to characterize the electrons but, for temperatures above, the Fermi–Dirac statistics reduce to the usual Maxwell–Boltzmann form. The statistical atom model to be described in what follows applies to temperatures below the degeneracy locus, at pressures above the critical pressure of about 100 Mb discussed above but below the pressure at which relativistic effects become important. The degeneracy temperature appearing is approximately that for iron, and the corresponding domain of validity of the statistical atom model is shown cross-hatched. This region is cut into two parts by the fusion curve, separating the regions of existence of the liquid and solid phases. The melting line as drawn for the higher pressures corresponds roughly to iron[16]. For coordinate points below this curve, corrections to thermodynamic functions as inferred from the statistical atom model are necessary for the effect of the lattice[16]. One notes that the fusion curve does not fall far from the degeneracy locus; thus the domain of validity of the nonrelativistic statistical model to be described is roughly coextensive with the region of the solid phase. In actuality, the degeneracy transition from Fermi–Dirac to Maxwell–Boltzmann statistics is continuous (and not sharp) and this as well as the other boundaries of regions in figure 1 correspond to ranges of atomic number. Thus, all lines of demarcation in

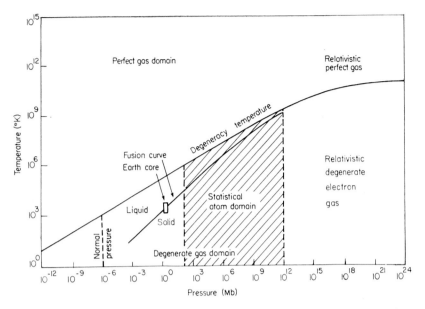

FIGURE 1 Schematic representation of the domain of applicability of the statistical atom model in the field of the variables, temperature T (absolute) and pressure P. The statistical atom model is approximately valid at temperatures below the degeneracy temperature, at pressures below that where relativistic effects become important, and at pressures above a limit of about 100 Mb necessary for the assumptions of the model to be valid; the corresponding region of the variables is shown cross-hatched. The lines for the degeneracy temperature and the fusion curve (at higher pressures) as shown correspond roughly to iron, and in general should be represented by bands for a range of atomic number. The rectangle indicates the coordinate point for the Earth's core. Adapted from Wares (for the relativistic case) with modifications and additions for the non-relativistic region

figure 1 should be drawn as bands (as done to some extent by Wares) but for clarity such refinements have been omitted. Hence, the corresponding numbers appearing on the coordinate axes are indicative of orders of magnitude only. It should be noted that the degeneracy criterion specified applies only to the electrons, while the heavy atomic particles (protons, neutrons, and nuclei) follow the classical Maxwell–Boltzmann statistics over essentially the entire field of the diagram.

The coordinate point corresponding to the Earth's core is shown in figure 1 by a rectangle whose height represents roughly the uncertainty in the temperature as inferred by Gilvarry[39–41]. Normal pressure is indicated by a vertical broken line. It should be noted that the free electrons in a metal are strongly degenerate under conditions of normal pressure and temperature.

Case of General Temperature

In this section, the Thomas–Fermi equation as generalized to arbitrary temperature will be presented. Its derivation from quantum statistical mechanics will be sketched[19,42]. Brachman[43] has derived expressions on this model for the Helmholtz function, the entropy, and the heat capacity at constant volume. It will be shown that

Brachman's results permit one to derive all the thermodynamic functions of the model in a form not requiring an integration over the volume of the atom[44]. A general perturbation method of solving the generalized Thomas–Fermi equation will be outlined, valid for any temperature below the degeneracy limit. This solution in terms of an asymptotic series in powers of the temperature yields corresponding series expansions of thermodynamic functions, and in particular the energy. Thus, it is possible to exhibit explicitly the thermodynamics of the generalized Thomas–Fermi model in any order of perturbation, asymptotically in the limit of low temperature[44, 45].

The effects of exchange and correlation will be neglected throughout, which makes the results valid only in the limit of high atomic number at sufficiently high compression.

Generalized Thomas–Fermi equation

The generalized Thomas–Fermi equation corresponds to Poisson's equation $\nabla^2 V = 4\pi e \rho$ or

$$r^{-1}\frac{\partial^2 (rV)}{\partial r^2} = 4\pi e \rho \tag{2}$$

for the self-consistent potential $V(r)$ at a point r in a spherically symmetric atom due to electrons of charge $-e$ with a number density $\rho(r)$ fixed by the Fermi distribution function for an absolute temperature T. The density is given by

$$\rho = \frac{4\pi}{h^3}(2m)^{3/2}(kT)^{3/2} I_{1/2}\left(\frac{eV}{kT}+\eta\right) \tag{3}$$

where m is the electron mass, $kT\eta$ is the chemical potential, and h and k are the Planck and Boltzmann constants, respectively. The function $I_{1/2}(\lambda)$ is defined by

$$I_k(\lambda) = \int_0^\infty \frac{y^k \, dy}{\exp(y-\lambda)+1} \tag{4}$$

where one notes that

$$\frac{dI_k}{d\lambda} = k I_{k-1} \tag{5}$$

Physically, the integrand (exclusive of the factor y^k) of the function I_k corresponds to the Fermi distribution function. Thus, the density ρ of equation (3) stems simply from the fundamental equation[46]

$$dn = g\left\{\exp\left(-\eta+\frac{H}{kT}\right)+1\right\}^{-1} h^{-m} dp_1 \ldots dp_m \, dq_1 \ldots dq_m \tag{6}$$

of quantum statistical mechanics for the differential number dn of fermions in an element of volume $dp_1 \ldots dp_m \, dq_1 \ldots dq_m$ in the phase space corresponding to the generalized coordinate q_i and its conjugate momentum p_i in m dimensions. Here, H is the Hamiltonian and g is the degeneracy factor (equal to 2 for electrons with two spin states). Equation (3) for ρ follows directly from

$$\rho = \int_p \frac{dn}{dq_1 \, dq_2 \, dq_3} \tag{7}$$

Thomas–Fermi atom model

by integrating over all possible momenta in three dimensions for a spherically symmetric atom.

The boundary condition on the potential V for a neutral atom is that the gradient vanish at the surface, as follows from Gauss's theorem, or

$$\left[\frac{dV}{dr}\right]_b = 0 \tag{8}$$

where the subscript b implies evaluation of a quantity on the boundary surface of the atom. Correspondingly, the total potential is zero at the atom boundary for a neutral atom. The initial condition to be imposed at the origin is that the potential V approach the nuclear potential corresponding to an atomic number Z at this point, or

$$V \to \frac{Ze}{r} \tag{9}$$

as the distance r from the nucleus vanishes. The modification in the boundary condition necessary for a positive ion is given by Gombas[5,6] (see p. 47 of Gombas[5] and p. 109 of Gombas[6]); it should be noted that no negative ions exist on the model. Actually, the boundary and initial conditions are easily generalized to take account of an arbitrary shape of the atom[47].

The development given above is due to Sakai[42] and to Feynman, Metropolis, and Teller[19].

Thermodynamic functions

The potential energy E_{pot} consists of two terms

$$E_{pot} = E_{ee} + E_{en} \tag{10}$$

where E_{ee} and E_{en} correspond to the electron–electron interaction and the electron–nucleus interaction, respectively. Of these two energies, the former equals the integral

$$E_{ee} = -\tfrac{1}{2}e \int_v \rho V_e \, d\tau \tag{11}$$

over the volume v of the atom (with volume element $d\tau = 4\pi r^2 \, dr$), and the latter is defined by

$$E_{en} = -e \int_v \rho V_n \, d\tau \tag{12}$$

and equals

$$E_{en} = ZeV_e(0) = Ze\left[\frac{d(Vr)}{dr}\right]_{r=0} \tag{13}$$

if $V_e(r)$ and $V_n(r) = Ze/r$ are the potentials due to the electrons and the nucleus, respectively. In the foregoing expressions, ρ is specified by equation (3) and V represents a solution of the generalized Thomas–Fermi equation.

The kinetic energy E_{kin} on this model is

$$E_{kin} = \int_v \varepsilon_{kin} \, d\tau \tag{14}$$

The kinetic energy density $\varepsilon_{kin}(r)$ is given by

$$\varepsilon_{kin} = \frac{4\pi}{h^3}(2m)^{3/2}(kT)^{5/2}I_{3/2}\left(\frac{eV}{kT}+\eta\right) \tag{15}$$

with $I_{3/2}(\lambda)$ fixed by the function $I_k(\lambda)$ of equation (4) and corresponds to the summation of the kinetic energies of the individual electrons, each weighted by the Fermi function. If one of the energies E_{kin}, E_{pot}, or the total energy E represented by the sum

$$E = E_{kin} + E_{pot} \tag{16}$$

is known, the other two energies can be determined directly from the virial theorem

$$3Pv = 2E_{kin} + E_{pot} \tag{17}$$

where P is the pressure. If r_b is the atom radius, the pressure P follows as

$$P = \tfrac{2}{3}\varepsilon_{kin}(r_b) \tag{18}$$

directly from the virial theorem (since the total potential vanishes at the atom radius, the potential energy of the electrons per unit volume is zero at this point). The virial theorem for the generalized Thomas–Fermi theory has been derived by Feynman, Metropolis, and Teller[19] by means of a similarity transformation. Gilvarry[47] has given a proof which has the virtue of being independent of the atom shape.

Brachman[43] has shown by an explicit temperature integration of the Gibbs–Helmholtz equation that the Helmholtz function F on the Thomas–Fermi model is given by

$$F = -\tfrac{2}{3}E_{kin} - E_{ee} + ZkT\eta \tag{19}$$

from which the entropy and heat capacity follow directly. By subtraction of the total energy E from F, the corresponding entropy S is given by

$$TS = \tfrac{5}{3}E_{kin} + 2E_{ee} + E_{en} - ZkT\eta \tag{20}$$

Brachman's expression[43] for the heat capacity C_v at constant volume can be written as[44]

$$TC_v = \tfrac{7}{4}E + \tfrac{3}{4}T^2\left(\frac{\partial}{\partial T}\right)_v\left(\frac{Pv + E_{en} + ZkT\eta}{T}\right) \tag{21}$$

by means of equation (13) and the virial theorem of equation (17).

The Gibbs free energy G in the generalized Thomas–Fermi theory has been given by Cowan and Ashkin[48] as

$$G = F + Pv + \tfrac{2}{3}E_{ee} - \tfrac{1}{3}E_{en} \tag{22}$$

It should be noticed that G is not simply the sum of F and Pv, since a Thomas–Fermi atom does not represent a homogeneous field-free system. The last two terms in G take into account the effect of the 'external' field produced by the nucleus, and the fact that, while the total volume is an adequate variable to describe the kinetic and exchange energies, it is insufficient to determine the changes under deformation in the electrostatic potential energies (which depend on the shape of the occupied volume or more accurately on individual interparticle distances). Equation (22) follows directly by taking into account possible changes in the charge of the nucleus and in the electron–nucleus and electron–electron distances[49,50]. Correspondingly, the enthalpy H is given by

$$H = E + Pv + \tfrac{2}{3}E_{ee} - \tfrac{1}{3}E_{en} \tag{23}$$

Cowan and Ashkin[48] have proved the relation

$$G = ZkT\eta \tag{24}$$

between the atomic number Z and the chemical potential $kT\eta$ per electron, which one expects to hold. Since V vanishes at the atom radius, the argument of $I_{3/2}(\lambda)$ in ε_{kin} entering equation (18) for P becomes simply η, and one has[51]

$$\left(\frac{\partial kT\eta}{\partial Z}\right)_{T,P} = 0 \tag{25}$$

Thus, Gilvarry[51] has shown directly from equation (24) that

$$\left(\frac{\partial G}{\partial Z}\right)_{T,P} = kT\eta \tag{26}$$

Hence, *both* basic properties of the Gibbs function, represented by equation (24) and equation (26), hold in the statistical atom model when G is suitably defined, in contrast with the discussion of Brachman[52] and of Reiss[53]. It should be noted that Z in these differential expressions designates the total number of electrons in a neutral atom and not the (numerically equal) nuclear charge.

The relations of Brachman have been obtained by March[54] from the Slater sum with a distribution function appropriate to a free-electron gas at the local electrostatic potential in the atom, and by Gilvarry and McMillan[55] from a model resembling the Debye–Hückel description of ionic solutions. These derivations as well as the one of Brachman require considerable mathematical manipulation, and in particular depend critically on the algorithm of equation (5), connecting the functions $I_{3/2}(\lambda)$ and $I_{1/2}(\lambda)$ entering the expressions for the kinetic energy and charge densities, respectively.

However, with the form of equation (22) for G, the Brachman relations follow directly by conjunction with the virial theorem, as Gilvarry[51] has pointed out. Substituting the value of Pv from equation (17) into equation (22) for G and replacing G by $ZkT\eta$ according to equation (24), one obtains precisely equation (19) for F.

It can be noticed that this derivation of the Brachman relations nowhere involves the assumption of spherical symmetry of the atom. Thus, the algorithm of equation (5) associated with the radial distribution functions is not used, and all thermodynamic quantities entering the argument are susceptible to definition independent of atom shape. Further, that the virial theorem holds independently of atom shape has been proved by Gilvarry[47]. For these reasons, he conjectured that the Brachman relations hold for an atom with an arbitrary shape of boundary surface[51]. That such a theorem is true in the restricted case with exchange represented by the approximate theory of Umeda and Tomishima has been shown by Gilvarry[47].

With use of the defining relation $C_v = (\partial E/\partial T)_v$, equation (21) of Brachman for C_v yields

$$T\left(\frac{\partial E}{\partial T}\right)_v - \tfrac{7}{4}E = \tfrac{3}{4}T^2\left(\frac{\partial}{\partial T}\right)_v \left(\frac{Pv + E_{en} + ZkT\eta}{T}\right) \tag{27}$$

One notes that P depends only on boundary parameters associated with a solution of the generalized Thomas–Fermi equation, as follows from equations (15) and (18); similarly, E_{en} depends only on initial parameters, from equations (13). Further, the chemical potential $kT\eta$ necessarily is a constant associated with a solution, fixed by the requirement that the total number of electrons is the atomic number Z. Hence, the right-hand side of equation (27) contains no terms depending on integrals over the volume.

Equation (27) is a first-order partial differential equation in T for the total energy E. Its solution yields the result of Gilvarry[44, 45, 56] for the total energy E of the atom as

$$E = \tfrac{3}{4}T^{7/4} \int^T T^{-3/4} \left(\frac{\partial}{\partial T}\right)_v \left(\frac{Pv + E_{en} + ZkT\eta}{T}\right) dT \qquad (28)$$

for arbitrary temperature, where the integration is carried out at constant volume v and the additive constant of integration is to be ignored. For $T = 0$, in which case the integrand appearing is singular, this equation gives correctly the result of Milne[57] and of Slater and Krutter[14] for the energy. The entropy S follows from the relation $T(\partial S/\partial T)_v = (\partial E/\partial T)_v$ by means of integration by parts as

$$S = \tfrac{3}{4}T^{3/4} \int^T T^{-7/4} \left(\frac{\partial}{\partial T}\right)_v \left(\frac{Pv + E_{en} + ZkT\eta}{T}\right) dT \qquad (29)$$

directly, where the additive constant of integration is to be neglected. This result is equivalent to the Brachman relation of equation (20) for S. The corresponding expression for the heat capacity C_v is

$$C_v = \tfrac{3}{4}T^{3/4} \int^T T^{-3/4} \left(\frac{\partial^2}{\partial T^2}\right)_v (Pv + E_{en} + ZkT\eta)\, dT \qquad (30)$$

The Helmholtz function F is given by

$$F = -\tfrac{3}{4}T^{7/4} \int^T T^{-11/4} (Pv + E_{en} + ZkT\eta)\, dT \qquad (31)$$

directly from the definition, where the additive constant of integration is to be taken as zero. This expression of Gilvarry[56] is entirely equivalent to Brachman's form of equation (19). It yields precisely the result of Milne and of Slater and Krutter for the energy of a compressed Thomas–Fermi atom at zero temperature.

These integral representations of Gilvarry[56] for thermodynamic functions have a special importance at elevated temperature, because they yield the entirety of thermodynamic functions without the necessity of performing an integration over the volume of the atom. Further, they must hold (at least in a limiting form) even at temperatures so high (approaching those appropriate for an ideal gas of electrons) that the Fermi–Dirac statistics reduce to the Maxwell–Boltzmann form. It is clear from the limits for zero temperature that equation (28) for E and equation (31) for F represent generalizations to arbitrary temperature of Milne's theorem expressing the energy of a Thomas–Fermi atom in terms of boundary and initial values of the Thomas–Fermi function. It is noteworthy that the rather involved sequence of repetitive uses of the differential equation with integration by parts entailed in the standard proof of Milne's theorem is unnecessary in the present derivation from the Brachman relations (which follow directly from the virial theorem, as shown).

Previously, it has been noted that the Brachman relations may apply to an atom of arbitrary shape, as conjectured by Gilvarry[51]. Since only the Brachman relations underlie the derivations of these integral representations of thermodynamic functions, he has conjectured further that they hold also for an atom with an arbitrary shape of boundary surface.

The set of linear dependences among thermodynamic functions represented by the Brachman relations is complemented by the further independent differential relationship of Gilvarry[58], which represents a consistency condition on them. From equation (31), the relation $P = -(\partial F/\partial v)_T$ yields

$$3\left(\frac{\partial E_i}{\partial v}\right)_T = 4T\left(\frac{\partial P}{\partial T}\right)_v - 3v\left(\frac{\partial P}{\partial v}\right)_T - 10P \qquad (32)$$

directly, where

$$E_i = E_{en} + ZkT\eta \qquad (33)$$

It will be shown that this differential constraint represents the necessary and sufficient condition that the basic thermodynamic functions E, H, F, and G be exact differentials in their arguments and that the Maxwell relations be satisfied.

Differentiating equation (29) for the entropy S with the operator $(\partial/\partial v)_T$ and using equation (32), one obtains

$$\left(\frac{\partial P}{\partial T}\right)_v = \left(\frac{\partial S}{\partial v}\right)_T \qquad (34)$$

which is the third Maxwell relation in the usual numeration[59]. On the assumption of this relation, conversely, equation (29) for S yields equation (32). The remaining Maxwell relations[59]

$$\left(\frac{\partial T}{\partial v}\right)_S = -\left(\frac{\partial P}{\partial S}\right)_v, \quad \left(\frac{\partial T}{\partial P}\right)_S = \left(\frac{\partial v}{\partial S}\right)_P, \quad \left(\frac{\partial v}{\partial T}\right)_P = -\left(\frac{\partial S}{\partial P}\right)_T \qquad (35)$$

now follow directly from equation (34) and conversely by the cyclic rule for differentiation of implicit functions, since the four Maxwell relations are not independent but follow one from another by use of this rule.

The Gibbs function G and the enthalpy H in the statistical atom model are given by equations (22) and (23), respectively, which differ from the usual expressions $G' = F + Pv$ and $H' = E + Pv$ for a homogeneous field-free system by terms linear in E_{ee} and E_{en}. However, these differences in definition do not affect the forms of the corresponding Maxwell relations, since E_{ee} and E_{en} are state functions. Hence, one concludes that equation (32) represents the necessary and sufficient condition for the basic thermodynamic functions E, H, F, and G to be exact differentials in their arguments and for the Maxwell relations to be satisfied.

It will be shown in the following section that E_i of equation (33) depends only on the initial slope of the generalized Thomas–Fermi function. Since the pressure P can always be expressed in terms of boundary parameters by equation (18), it follows that equation (32) yields a differential relation between initial and boundary parameters associated with the generalized Thomas–Fermi function at arbitrary temperatures. It represents a consistency condition on the thermodynamic functions and on the Brachman relations.

General perturbation method

It is convenient to define a Φ (generalizing in the manner of Marshak and Bethe[32] the usual Thomas–Fermi function ϕ corresponding to zero temperature) by

$$\frac{Ze^2\Phi}{\mu x} = eV + kT\eta \tag{36}$$

where x is defined by $r = \mu x$, and μ equals $a_0(9\pi^2/128Z)^{1/3}$ in terms of the radius a_0 of the first Bohr orbit for hydrogen. By means of an asymptotic expansion[60] for $I_{1/2}(\lambda)$ applied to ρ of equation (3), which is valid for $\lambda = Ze^2\Phi/\mu x kT > 1$, the asymptotic form of the generalized Thomas–Fermi equation for low temperatures becomes

$$\frac{\partial^2 \Phi}{\partial x^2} = \frac{\Phi^{3/2}}{x^{1/2}}\left\{1 + \sum_{n=1}^{N} \zeta_n \left(\frac{kTx}{\Phi}\right)^{2n}\right\} \tag{37}$$

in which $n \geq 1$,

$$\zeta_n = \left(\frac{\mu}{Ze^2}\right)^{2n} a_n \tag{38}$$

and the coefficients a_n are defined and tabulated by McDougall and Stoner[60]. Equation (37) reduces to the usual Thomas–Fermi equation for $T = 0$ (where $kT\eta$ approaches a constant η'). The function Φ is subject to a boundary condition at the radius x_b of the atom which makes it a function $\Phi(x, x_b)$ of the two variables x and x_b. The initial and boundary conditions on $\Phi(x, x_b)$ are, respectively,

$$\Phi(0, x_b) = 1 \tag{39a}$$

$$\Phi_b = x_b \left[\frac{\partial \Phi(x, x_b)}{\partial x}\right]_{x=x_b} \tag{39b}$$

where $\Phi_b = \Phi(x_b, x_b)$ is the boundary value of Φ. The distinction implied by the partial derivative notation in equations (37) and (39b) will be germane later. It is possible to show that Φ/x is a monotone-decreasing function of x for any neutral atom solution of the generalized Thomas–Fermi equation; hence the domain of validity $\lambda > 1$ of equation (37) becomes

$$kT < \frac{Ze^2}{\mu}\frac{\Phi_b}{x_b} \tag{40}$$

The asymptotic series on which equation (37) is based is divergent for N sufficiently large, so that an optimum N exists for a given accuracy when the inequality (40) is fulfilled for a given λ.

For this same domain of validity, the corresponding asymptotic expansion of $I_{3/2}(\lambda)$ yields for the pressure P, from equation (18),

$$P = \frac{Z^2 e^2}{10\pi\mu^4}\left(\frac{\Phi_b}{x_b}\right)^{5/2}\left\{1 + \sum_{n=1}^{N} \frac{5\zeta_n}{5-4n}\left(\frac{kTx_b}{\Phi_b}\right)^{2n}\right\} \tag{41}$$

where the chemical potential $kT\eta$ has been evaluated from equation (36) as

$$kT\eta = \frac{Ze^2}{\mu}\frac{\Phi_b}{x_b} \tag{42}$$

Thomas–Fermi atom model

(since $V = 0$ at the atom boundary). From equations (13) and (36), the energy E_{en} can be evaluated as

$$E_{en} = \frac{Z^2 e^2}{\mu}\left(\Phi'_i - \frac{\Phi_b}{x_b}\right) \tag{43}$$

where Φ'_i is the initial slope $[\partial \Phi / \partial x]_{x=0}$. The volume of the atom is determined by the boundary condition (39b) as $v = (4\pi/3)\mu^3 x_b^3$. These thermodynamic functions are the ones given directly by solution of the differential equation (37).

Evaluating the sum

$$E_i = E_{en} + ZkT\eta = \frac{Z^2 e^2}{\mu}\Phi'_i \tag{44}$$

from equations (42) and (43), one can write equation (28) for the total energy E as

$$E = \tfrac{3}{4}T^{7/4}\int^{T} T^{-3/4}\left(\frac{\partial}{\partial T}\right)_v \left\{T^{-1}\left(Pv + \frac{Z^2 e^2}{\mu}\Phi'_i\right)\right\} dT \tag{45}$$

where the integration is carried out at constant volume and the constant of integration is to be ignored. The corresponding form for the entropy S is

$$S = \tfrac{3}{4}T^{3/4}\int^{T} T^{-7/4}\left(\frac{\partial}{\partial T}\right)_v \left(Pv + \frac{Z^2 e^2}{\mu}\Phi'_i\right) dT \tag{46}$$

and for the heat capacity C_v is

$$C_v = \tfrac{3}{4}T^{3/4}\int^{T} T^{-3/4}\left(\frac{\partial^2}{\partial T^2}\right)_v \left(Pv + \frac{Z^2 e^2}{\mu}\Phi'_i\right) dT \tag{47}$$

From equation (31) the Helmholtz function F becomes

$$F = -\tfrac{3}{4}T^{7/4}\int^{T} T^{-11/4}\left(Pv + \frac{Z^2 e^2}{\mu}\Phi'_i\right) dT \tag{48}$$

Finally, the Gibbs function G is simply

$$G = \frac{Z^2 e^2}{\mu}\frac{\Phi_b}{x_b} \tag{49}$$

as implied by equation (24) and by equation (42). It should be noted that G is given directly by the boundary value of the generalized Thomas–Fermi function.

The integrands in these integral representations of thermodynamic functions depend only on the variables P and Φ'_i, apart from the temperature. However, P is expressed by equation (41) as an asymptotic series in T^2 whose coefficients depend on Φ only through boundary values. Hence, the integrals on the temperature in all these thermodynamic functions can be evaluated if Φ'_i can be obtained as a function of temperature. In such a case, the entirety of thermodynamic functions at low temperature become available from the expressions obtained.

The parameter Φ'_i can be obtained as a function of T^2 by expanding Φ as an asymptotic series

$$\Phi = \phi + \sum_{n=1}^{N} \zeta_n \chi_n (kT)^{2n} \tag{50}$$

in T^2, where ϕ is the Thomas–Fermi function corresponding to $T=0$, and the coefficients χ_n are functions only of x and x_b, determined by solution of the asymptotic Thomas–Fermi equation (37). Determination of the χ_n requires the solution of N separate differential equations, whose analytical form can be found by substituting equation (50) into equation (37). The differential equation for χ_n is an inhomogeneous linear equation involving ϕ and all perturbations of lower order. With the functions χ_n available, Φ'_i becomes

$$\Phi'_i = \phi'_i + \sum_{n=1}^{N} \zeta_n \chi'_{n,i}(kT)^{2n} \qquad (51)$$

where ϕ'_i and $\chi'_{n,i}$ are the initial slopes corresponding to ϕ and χ_n, respectively. Since the integrands in equations (45)–(48) then depend on Φ only through boundary and initial parameters, these equations yield asymptotic series for thermodynamic functions analogous to the corresponding expansions in the case of a degenerate Fermi–Dirac gas[46]. For the energy E and the Helmholtz function F, the expansions are in powers of T^2 up to T^{2N}; for the entropy S and the heat capacity C_v, the expansions are in odd powers of T from T to T^{2N-1}.

In the literature, two methods of solving the differential perturbation equations (for the case $N=1$) have been used. In the method of Gilvarry[44, 45], the radius of the atom is kept at its unperturbed value x_b, and the boundary condition (39b) is met by requiring

$$\chi_{n,b} = x_b \left[\frac{\partial \chi_n(x, x_b)}{\partial x} \right]_{x=x_b} \qquad (52)$$

where $\chi_{n,b}$ is the boundary value of χ_n at the atom boundary χ_b. In this case, the initial slope of the perturbation must be selected (by trial) so that equation (52) is met at the boundary. In the method of Feynman, Metropolis, and Teller[19], the initial slope is set equal to zero, and a perturbed radius x_b^* of the atom is determined at which the boundary condition (39b) is met. The method of Feynman, Metropolis, and Teller is computationally far more convenient, but the method of Gilvarry avoids the complication of introducing a perturbation in the volume. In any event, solutions of the differential perturbation equation of any order by the two methods can differ only by a solution of the corresponding homogeneous equation. Thus it is possible to express the pertinent parameters on the method of Gilvarry in terms of corresponding parameters derived by the computationally more convenient method of Feynman, Metropolis, and Teller; on pp. 375–83 this process is carried out explicitly for the case $N=1$.

The preceding development is due to Gilvarry[44, 45, 51].

Thermodynamic parameters

For the subsequent development of the thermodynamics, it is convenient to modify the total energy E by subtraction of the energy of a standard state to obtain a total energy U which is always positive. The standard state is taken as a neutral atom of infinite radius at zero temperature, for which the energy is $(3/7)(Z^2 e^2/\mu)\phi'_{i,\infty}$, where $\phi'_{i,\infty}$ is the corresponding initial slope of ϕ. Accordingly, one has

$$U = E - \frac{3}{7} \frac{Z^2 e^2}{\mu} \phi'_{i,\infty} \qquad (53)$$

Thomas–Fermi atom model

In terms of U, a parameter γ can be defined at this point by

$$U = \frac{Pv}{\gamma-1} \qquad (54)$$

Further, it is convenient to introduce two differential parameters:

$$\varepsilon_S = -\left(\frac{\partial \ln P}{\partial \ln v}\right)_S, \qquad \varepsilon_T = -\left(\frac{\partial \ln P}{\partial \ln v}\right)_T \qquad (55)$$

which are the (negative) slopes of the pressure–volume curves in log–log coordinates for constant entropy and for constant temperature, respectively. These parameters are connected by the relation[59]

$$\frac{\varepsilon_S}{\varepsilon_T} = \frac{C_P}{C_v} \qquad (56)$$

where C_P is the heat capacity at constant pressure. The energy equation of thermodynamics[59] provides a connection

$$\varepsilon_S = \gamma - \left[\frac{\partial \ln(\gamma-1)}{\partial \ln v}\right]_S \qquad (57)$$

between ε_S and γ. An integral relation between ε_S and γ can be derived directly from the first law of thermodynamics; it is

$$\gamma = U^{-1} \int_0^U \varepsilon_S \, dU = \langle \varepsilon_S \rangle_U \qquad (58)$$

which one can show by direct mathematical processes to be an integral of equation (57). An integration by parts on the integral in equation (58) shows that $\gamma \geq \varepsilon_S$.

Case of Zero Temperature

The Thomas–Fermi equation for zero temperature will be discussed in this section. This equation is nonlinear and hence resort must be made to numerical methods of solution, in general. However, limiting cases and asymptotic solutions corresponding to the two limits of an infinitesimal and an infinite atom will be considered. The thermodynamic functions for the case of zero temperature can be obtained directly from the results of the preceding section. The corresponding thermodynamic functions can be made directly accessible by fitting semitheoretically as a function of atom radius the pertinent parameters derived from solution of the Thomas–Fermi equation.

For convenience in reference, results obtained by Gilvarry[44] for this case will be referred to by the abbreviation GI. For similar reasons of brevity in reference, the results of Slater and Krutter[14] will be referred to as SK, and those of Feynman, Metropolis, and Teller[19] will be designated by FMT.

Thomas–Fermi equation

For the case $T = 0$, the Thomas–Fermi equation becomes

$$\frac{\partial^2 \phi}{\partial x^2} = \frac{\phi^{3/2}}{x^{1/2}} \qquad (59)$$

from equation (37). The initial and boundary conditions on $\phi(x, x_b)$ are

$$\phi(0, x_b) = 1 \tag{60a}$$

$$\phi_b = x_b \left[\frac{\partial \phi(x, x_b)}{\partial x}\right]_{x=x_b} \tag{60b}$$

respectively, where $\phi_b = \phi(x_b, x_b)$ is the boundary value of ϕ. These conditions are the direct analogs, of course, of those of equations (39). The initial slope of ϕ at the origin will be designated by ϕ_i'.

The distinction implied by the partial derivative notation will be important in the treatment of the next section of the effect of a temperature pertubation, as will appear.

The basic method of solution of the Thomas–Fermi equation is to select an initial slope ϕ_i', integrate the equation numerically from the origin outward, and determine the atom boundary x_b through equation (60b). The corresponding atom volume thus becomes

$$v = \frac{4\pi}{3}(\mu x_b)^3 \tag{61}$$

and hence is not known *a priori*, but only after the fact. It is usual to start the solution at the origin by making use of the semiconvergent expansion of Baker[61]

$$\phi = 1 + \phi_i' x + \frac{4}{3}x^{3/2} + \frac{2}{5}\phi_i' x^{5/2} + \frac{1}{3}x^3 + \frac{3}{70}(\phi_i')^2 x^{7/2} + \frac{2}{15}\phi_i' x^4 + \ldots \tag{62}$$

It should be noted that ϕ does not possess a Taylor expansion at the origin, since $\partial^2 \phi/\partial x^2 \to \infty$ at this point.

By selecting different values of ϕ_i' and determining the corresponding values of x_b and ϕ_b in this manner, a functional relationship of ϕ_i' and ϕ_b dependent on the parameter x_b can be constructed, represented by a table. The march of the solutions as the

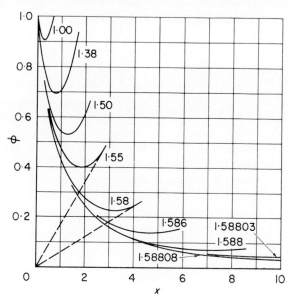

FIGURE 2 The Thomas–Fermi function ϕ as a function of x for different initial slopes [after SK]

Thomas–Fermi atom model

independent variable x_b is changed is indicated in figure 2, constructed from results of SK; the numbers associated with the different curves represent the corresponding absolute values of the initial slopes. One notes that the atom boundary is fixed by the graphical requirement that a solution of equation (59) be tangent at this point to a straight line passing through the origin (shown broken in figure 2), as demanded by equation (60b). For a particular and unique value of the initial slope $\phi'_i = \phi'_{i,\infty}$, the atom boundary x_b approaches infinity. Evaluations of $\phi'_{i,\infty}$ by different investigators yield results which vary somewhat; the figure indicated in absolute value by the largest number in figure 2 represents the estimate of SK.

One notes that equation (59) is written in terms of dimensionless variables which are independent of the atomic number Z. Hence, a particular solution of the Thomas–Fermi equation yields a one-parameter family of values for a particular thermodynamic function, which scales with Z. For a fixed value of x_b, for example, equation (61) implies that the volume v of the atom varies inversely as Z, when one notes the definition of μ.

Thermodynamic quantities

In this section, the thermodynamic functions corresponding to the case $T = 0$ will be developed. For reasons that will become clear in the next section, thermodynamic variables corresponding to this case will be distinguished by a special notation. Since the entropy is zero, the isothermal and isentropic equations of state are identical in this case; the common value of ε_S and ε_T fixed by equations (55) will be denoted by ε_0, and the value of γ from equation (54) by γ_0. For the other thermodynamic functions, the values corresponding to the case $T = 0$ will be distinguished by use of lower case letters.

The basic thermodynamic function is the total energy u (corresponding to U of equation (53)), evaluated by Milne[57] and SK as

$$u = \frac{Z^2 e^2}{\mu} \left\{ \frac{3}{7}(\phi'_i - \phi'_{i,\infty}) + \frac{2}{35} x_b^{1/2} \phi_b^{5/2} \right\} \tag{63}$$

The pressure p is

$$p = \frac{Z^2 e^2}{10\pi\mu^4} \left(\frac{\phi_b}{x_b} \right)^{5/2} \tag{64}$$

from equation (41). Equation (24) implies that the enthalpy h is given at zero temperature by

$$h = Z \lim_{T \to 0} kT\eta = Z\eta' \tag{65}$$

where $\eta'(x_b)$ depends on x_b. From equation (49), one has

$$h = \frac{Z^2 e^2}{\mu} \frac{\phi_b}{x_b} \tag{66}$$

The enthalpy $h' = u + pv$ as usually defined is

$$h' = \frac{Z^2 e^2}{\mu} \left\{ \frac{3}{7}(\phi'_i - \phi'_{i,\infty}) + \frac{4}{21} x_b^{1/2} \phi_b^{5/2} \right\} \tag{67}$$

The Helmholtz function is identical, of course, with the total energy, and the Gibbs function is identical with the enthalpy. The connection between the three variables x_b, ϕ_b, and ϕ'_i which enter the functions is provided by $p = -du/dv$, which yields the differential relation

$$x_b^{1/2} \frac{d\phi'_i}{dx_b} = -\phi_b^{5/2} - \tfrac{1}{3} x_b \phi_b^{3/2} \frac{d\phi_b}{dx_b} \tag{68}$$

It should be noted that this relation represents the limit of equation (32) corresponding to $T \to 0$.

The parameter ε_0 can be evaluated directly from equation (60) as

$$\varepsilon_0 = \frac{5}{6}\left(1 - \frac{d \ln \phi_b}{d \ln x_b}\right) \tag{69}$$

and the parameter γ_0 can be evaluated from equations (63) and (64) as

$$(\gamma_0 - 1)^{-1} = \frac{3}{7}\left\{1 + \frac{15}{2} \frac{\phi'_i - \phi'_{i,\infty}}{x_b^{1/2} \phi_b^{5/2}}\right\} \tag{70}$$

For the case $T = 0$, the partial derivative in equation (57) becomes a total derivative, so that ε_0 and γ_0 are equal whenever either is constant. A lower bound on γ_0 can be written immediately from the virial theorem in the form of equation (17); since E_{pot} is necessarily negative, it follows that $\gamma_0 \geq 5/3$ at $T = 0$.

It follows from equation (63) that the energy u scales as $Z^{7/3}$ and from equation (64) that the pressure p scales as $Z^{10/3}$, for a fixed value of x_b and a corresponding solution of the Thomas–Fermi equation. Actually, these scaling rules are not peculiar to the limit of zero temperature, but hold also for the Thomas–Fermi equation for general temperature discussed in the preceding section.

Limiting cases and fitted functions

The limiting values of the thermodynamic functions for high and low pressures can be evaluated if the corresponding asymptotic forms of ϕ_b and ϕ'_i as a function of x_b can be determined for $x_b \to 0$ and $x_b \to \infty$. For sufficiently high pressures, the Thomas–Fermi atomic model must pass over into the degenerate Fermi–Dirac gas, since the kinetic energy (varying as $1/r^2$ in the limit of small volume) dominates the potential energy (varying as $1/r$). Accordingly, the potential inside the atom for this limiting case is

$$V = \frac{Ze}{\mu}(x^{-1} + \tfrac{1}{2}x_b^{-3}x^2 - \tfrac{3}{2}x_b^{-1}) \tag{71}$$

which corresponds to the sum of the nuclear potential and the potential due to a uniform distribution of Z electrons within the atomic volume. The corresponding value of ϕ is thus

$$\phi = 1 + \frac{1}{2}\left(\frac{x}{x_b}\right)^3 - \frac{3}{2}\left(\frac{x}{x_b}\right) + \frac{3^{2/3} x}{x_b^2} \tag{72}$$

where the last term arises by evaluating the chemical potential

$$\eta' = \lim_{T \to 0} kT\eta$$

Thomas–Fermi atom model

per electron for the Thomas–Fermi atom from the corresponding expression[46] $\eta' = (h^2/8m)(3Z/\pi v)^{2/3}$ for the chemical potential of a Fermi–Dirac gas of Z electrons in the atomic volume v, as

$$\eta' = \frac{3^{2/3} Z e^2}{\mu x_b^2} \tag{73}$$

in terms of x_b. Since the last term of equation (72) dominates in ϕ for x_b small, one obtains

$$\phi_b = \frac{3^{2/3}}{x_b}, \qquad \phi'_i = \frac{3^{2/3}}{x_b^2} \tag{74}$$

in the limit $x_b \to 0$. With these asymptotic forms for x_b small, all the thermodynamic functions derived reduce to the corresponding functions for a Fermi–Dirac gas[46], as one can verify. The common limiting value of ε_0 and γ_0 in this case is 5/3.

The corresponding forms for the limit of low pressure (infinite atom) can be obtained from an asymptotic formula due to Sauvenier[62] (derived from a result of Sommerfeld[63]), which represents an approximate solution of the Thomas–Fermi equation which is perturbed slightly from the solution for an infinite atom. The solution is

$$\phi = (1+z)^{-\lambda_1/2} \left\{ 1 + \frac{4z_b + 1}{(\lambda_1 - 4)z_b - 1} \left(\frac{1+z}{1+z_b} \right)^{\lambda_1/\lambda_2} \right\} \tag{75}$$

where

$$z = \left(\frac{x}{12^{2/3}} \right)^{\lambda_2}, \qquad z_b = \left(\frac{x_b}{12^{2/3}} \right)^{\lambda_2} \tag{76}$$

and

$$\lambda_1 = \tfrac{1}{2}(73^{1/2} + 7), \qquad \lambda_2 = \tfrac{1}{2}(73^{1/2} - 7) \tag{77}$$

This expression yields

$$\phi_b = 16(3 + 2\lambda_1) x_b^{-3} \left\{ 1 - \frac{(3 + 8\lambda_1) 12^{2\lambda_2/3}}{18 x_b^{\lambda_2}} \right\} \tag{78}$$

as the asymptotic form of ϕ_b for $x_b \to \infty$. The corresponding asymptotic form of ϕ'_i follows from Sauvenier's result, by means of equation (68) as

$$\phi'_i = \phi'_{i,\infty} + \frac{256}{9}(2 + 3\lambda_2)(3 + 2\lambda_1)^{5/2} 12^{2\lambda_2/3} x_b^{-\lambda_1} \tag{79}$$

and is determined only by the second term of equation (78). The common limiting value of ε_0 and γ_0 is 10/3 from these asymptotic forms. It is possible to show that ϕ_b and $d\phi_b/dx_b$ are monotonic functions of x_b, which shows that ε_0 and γ_0 between the limits 5/3 and 10/3 are monotonic functions of x_b.

To determine thermodynamic functions for a given volume v, one needs values of ϕ_b and ϕ'_i (and their derivatives) corresponding to $x_b = (3v/4\pi)^{1/3}/\mu$. Fourteen values of ϕ_b and $\phi'_i - \phi'_{i,\infty}$ corresponding to values of x_b are available from numerical results of FMT and SK, derived from six solutions of FMT (table III of their paper), and from eight solutions of SK, as tabulated in table V of Gombas[5] (pp. 53, 357).

The value of $\phi'_{i,\infty}$ (-1.58875) given by FMT was associated with their data, and the value (-1.58808) given by SK was associated with theirs, in obtaining the difference $\phi'_i - \phi'_{i,\infty}$. In the region of x_b where they overlap, the results of FMT and SK are discrepant; the discrepancy is more significant in $\phi'_i - \phi'_{i,\infty}$ than in ϕ_b. The data were smoothed in this region with major weight given to the data of FMT. For this reason, and because insufficient significant figures apparently appear in the difference $\phi'_i - \phi'_{i,\infty}$, results of a seventh and eighth solution of SK corresponding to the largest volumes, as given by Gombas, have been ignored. The discrepancy in question has been noted by Umeda[64].

To reduce the labor of interpolation, and in the interest of direct accessibility of the data, it is convenient to fit these results by empirical formulae chosen to have the correct asymptotic forms in the two limits $x_b \to 0$ and $x_b \to \infty$. A fitting function for ϕ_b yielding the correct asymptotic forms can be written as

$$\phi_b = \left(\sum A_n x_b^{n/2}\right)^{-1} \tag{80}$$

if n ranges over the sequence $n = 2, 3, \ldots, 6$ and if the coefficients A_2 and A_6 are chosen to agree with the corresponding coefficients in the asymptotic forms (74) and (78), respectively. It has been found possible to choose the intermediate coefficients A_n to represent the smoothed data of FMT and SK within about 1.5%. The coefficients A_n are tabulated in table 1; since they are all positive, the fitted function is monotone,

TABLE 1 Coefficients[a] of fitted functions, zero-temperature case [from GI]

n	A_n	B_n
2	4.8075×10^{-1}	4.8075×10^{-1}
3	7.009×10^{-3}	4.3462×10^{-1}
4	7.003×10^{-2}	6.9203×10^{-2}
5	8.901×10^{-3}	5.9472×10^{-2}
6	3.3704×10^{-3}	-4.9688×10^{-3}
7		4.3386×10^{-4}
$\lambda_1 = 7.77200$		1.5311×10^{-6}

[a] The coefficients are given to four or five figures to minimize round-off error and to yield smoothness in computational work.

as is ϕ_b. The fitted function ϕ_b is shown in figure 3 for comparison with the data points.

The corresponding fitted function for $\phi'_i - \phi'_{i,\infty}$ could be determined, in principle, from equation (68) but it is simpler to fit the data for $\phi'_i = \phi'_{i,\infty}$ directly. A reasonable fitting function is

$$\phi'_i - \phi'_{i,\infty} = \left(\sum B_n x_b^n\right)^{-1} \tag{81}$$

in which n ranges over the sequence $n = 2, 3, \ldots, 7, \lambda_1$, and the coefficients B_2 and B_{λ_1} are chosen to agree with the corresponding coefficients in the asymptotic forms (74) and (79), respectively. The smoothed data of FMT and SK for the difference $\phi'_i - \phi'_{i,\infty}$ can be reproduced within about 1.5% by means of the coefficients B_n

Thomas–Fermi atom model

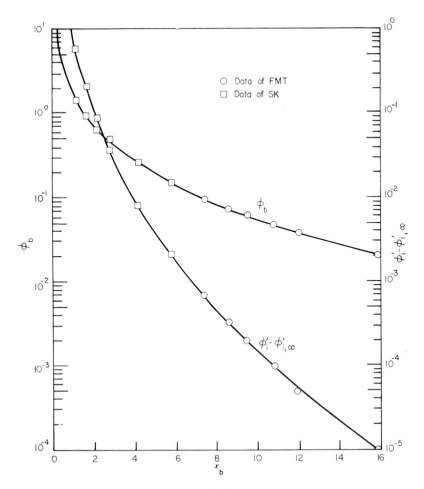

FIGURE 3 Fitted functions for ϕ_b and $\phi'_i - \phi'_{i,\infty}$ against boundary radius x_b in zero-temperature case [from GI]

tabulated in table 1. It is possible to show that ϕ'_i is a monotonic function of x_b; the fitted function has this property likewise. The fitted function and the data points are shown in figure 3.

With these fitted functions for ϕ_b and ϕ'_i, all the thermodynamic functions discussed previously can be determined by direct algebraic or differential processes. In the range covered by the data of FMT and SK, thermodynamic functions can be obtained with an accuracy of the order of a few per cent, on the basis of the data employed. Outside this range the fitted functions yield results which are correct in both asymptotic limits, and their monotonic character gives some assurance of approximate validity, at least, outside of the regions fitted directly. The fitted function for ϕ_b yields the equation of state

$$p^{2/5} \sum_{n=2}^{6} A_n \left(\frac{3v}{4\pi\mu^3} \right)^{(n+2)/6} = \left(\frac{Z^2 e^2}{10\pi\mu^4} \right)^{2/5} \tag{82}$$

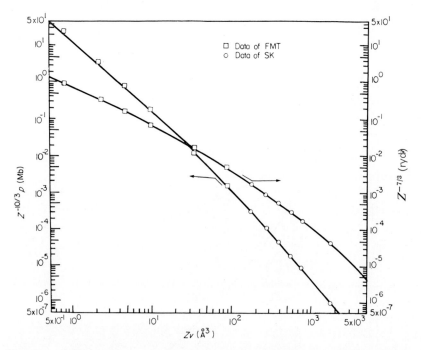

FIGURE 4 Scaled pressure and energy from fitted functions for ϕ_b and $\phi'_i - \phi'_{i,\infty}$, against scaled volume in cubic ångströms; zero-temperature case [from GI]

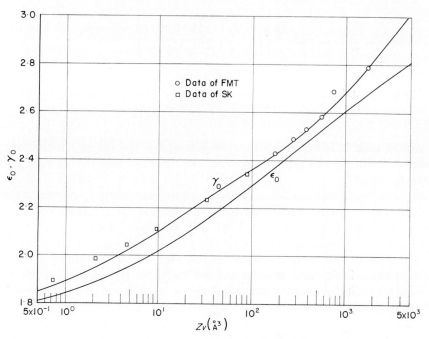

FIGURE 5 Parameters ε_0 and γ_0 from fitted functions for ϕ_b and $\phi'_i - \phi'_{i,\infty}$, against scaled volume in cubic ångströms; zero-temperature case [from GI]

Thomas–Fermi atom model

Pressures from this equation are shown in figure 4 for comparison with the corresponding quantities as calculated directly from the data of FMT and SK. The energy u from equation (63) is shown likewise; the unit of energy is the rydberg, $R = e^2/2a_0$. A similar comparison with directly calculated values is shown in figure 5 for γ_0 (directly calculated values are not given for ε_0 since it is a differential parameter). It should be noted that pressure and energy scale as $Z^{10/3}$ and $Z^{7/3}$, respectively, while the volume scales as Z^{-1}.

The preceding discussion is based on the results of GI. The fitted functions given have been superseded by the more accurate expressions derived on p. 363. However, the functions presented here are the ones actually used in connection with the effect of the lattice on thermodynamic functions[16], and used in the treatment of the explosive impact of meteorites[35]. For this reason, and because they illustrate more clearly the principles used in the construction of the more accurate fitted functions, it was felt desirable to treat them briefly here.

Asymptotic solution for small atom radius

In the previous subsection, asymptotic solutions of the Thomas–Fermi equation were developed for the two limits of an infinitesimal and an infinite atom, as given in GI. The results were used in the construction of analytic expressions, fitting the available numerical data on parameters derived from the Thomas–Fermi equation, and having the correct asymptotic forms for the two limits in question. This discussion of the limiting forms of boundary and initial parameters for small atom radius was based on a physical argument, that the results of the Thomas–Fermi method must go over into those appropriate to a Fermi–Dirac gas of completely free electrons when the pressure is sufficiently high. Subsequently, March[65] showed that a direct analytic solution of the Thomas–Fermi equation in the limit of small atom radius could be obtained by means of a Taylor expansion of the Thomas–Fermi function about a point corresponding to the atom radius. In this manner, he verified the results of GI and, furthermore, obtained terms of higher order. March's method of solution and his results for boundary and initial parameters will be discussed in this subsection.

The corresponding results of March will be referred to by the abbreviation M.

Expanding the solution $\phi(x)$ of the Thomas–Fermi equation in a Taylor series about the boundary point $x = x_b$, one obtains

$$\phi - \phi_b = \sum_{n=1}^{\infty} \frac{(-1)^n}{n!} \phi^n(x_b)(x_b - x)^n = \sum_{n=1}^{\infty} t_n(x_b - x)^n \tag{83}$$

where $\phi^n(x)$ is the nth derivative of $\phi(x)$ and

$$t_n = \left\{ \frac{(-1)^n}{n!} \right\} \phi^n(x_b) \tag{84}$$

It is convenient at this stage to introduce a parameter a defined by

$$\frac{\phi_b}{x_b} = \left[\frac{\partial \phi}{\partial x} \right]_{x=x_b} = a^2 \tag{85}$$

and then to express the coefficients t_n in terms of a and x_b. The results are shown in table 2, where coefficients up to and including t_{12} are given.

TABLE 2 Coefficients t_n in the Taylor series solution of the Thomas–Fermi equation [from M]

n	t_n	n	t_n
1	$-a^2$	7	$\dfrac{a^5}{3360}$
2	$\dfrac{a^3 x_b}{2}$	8	$\dfrac{a^5 x_b^{-1}}{1344} + \dfrac{a^6 x_b}{2240}$
3	$\dfrac{-a^3}{6}$	6	$\dfrac{a^5 x_b^{-2}}{1728} + \dfrac{a^6}{20160}$
4	$\dfrac{a^4 x_b}{16}$	10	$\dfrac{a^5 x_b^{-3}}{2160} + \dfrac{43 a^6 x_b^{-1}}{1209600} + \dfrac{3 a^7 x_b}{89600}$
5	$-\dfrac{a^4}{80}$	11	$\dfrac{a^5 x_b^{-4}}{2640} + \dfrac{a^6 x_b^{-2}}{172800} + \dfrac{47 a^7}{2956800}$
6	$\dfrac{a^5 x_b}{160}$	12	$\dfrac{a^5 x_b^{-5}}{3168} - \dfrac{a^6 x_b^{-3}}{71280} + \dfrac{481 a^7 x_b^{-1}}{26611200} + \dfrac{13 a^8 x_b}{7884800}$

The solution of the Thomas–Fermi equation specified by equation (83) satisfies of course the required boundary condition at $x = x_b$, by virtue of equation (85) However, it is necessary to impose the initial condition $\phi(0) = 1$. It seems from investigation of the series corresponding to equation (83) that it diverges for the radius x_b actually coincident with the origin. Nevertheless, the series is still useful, from a practical point of view, as a semiconvergent series and it can be applied first of all to determine the limiting form of ϕ_b as $x_b \to 0$. One proceeds by writing down $\phi(0)$ as far as terms in a^3 (the terms in a^2 cancel one another). In this manner, one obtains

$$\phi(0) = \tfrac{1}{3} a^3 x_b^3 = 1 \tag{86}$$

and hence

$$a = \frac{3^{1/3}}{x_b} \tag{87}$$

Thus, ϕ_b has the limiting form

$$\phi_b = \frac{3^{2/3}}{x_b} \tag{88}$$

which is just the result of GI.

However, the method of M permits the computation of further terms in ϕ_b. First, one notes the consistency of the approximation made in writing equation (86), since it can be seen that the first neglected term in ϕ_b is of order x_b. Including the terms in a^4, one obtains

$$\phi(0) = \tfrac{1}{3} a^3 x_b^3 + \tfrac{1}{20} a^4 x_b^5 = 1 \tag{89}$$

Hence, by substituting for a in the correction term from equation (87), one finds

$$a = \frac{3^{1/3}}{x_b}\left(1 - \frac{3^{1/3}}{20}x_b\right) \qquad (90)$$

and thus

$$\phi_b = \frac{3^{2/3}}{x_b}\left(1 - \frac{3^{1/3}}{10}x_b\right) \qquad (91)$$

Unfortunately, a difficulty appears in the computational scheme when one attempts to go farther than corresponds to the approximation of equation (91). As can be seen from table 2, every coefficient t_n of order higher than corresponds to t_5 contains a term in a^5. Even though the term in a^5 in the nth coefficient can be written down and the terms easily summed, the method of M does not yield a procedure which can be justified to calculate the next term in equation (91). This deficiency has been remedied by Rijnierse[66], who recently has developed a method of obtaining systematically any desired number of terms in the solution of the Thomas–Fermi equation for small atom radius. Nevertheless, equation (91) is very useful, as will appear.

Before going on, it is interesting to compare the expression of M for $\phi(x)$ as far as terms in a^3 with that given in GI (equation (72)). The result of M may be written as

$$\phi = \tfrac{1}{3}a^3 x_b^3 + (a - \tfrac{1}{2}a^3 x_b^2)x + \tfrac{1}{6}a^3 x^3 \qquad (92)$$

corresponding to the complete expression of equation (91) for ϕ_b. If one substitutes in this equation for ϕ the one-term approximation of equation (87) for a, one finds simply the result of GI, which was determined from the potential due to a point charge Ze and a uniform distribution of Z electrons within the atomic volume, combined with a term calculated from the chemical potential of a Fermi–Dirac gas. It is clear from the treatment of M that this procedure will give correctly only the leading term (of order x_b^{-2}) in the coefficient of x.

The other parameter entering the expressions for thermodynamic functions, namely the initial slope ϕ_i' of the Thomas–Fermi function, can also be obtained from the Taylor expansion to the same approximation as ϕ_b given by equation (91). The result is

$$\phi_i' = \frac{3^{2/3}}{x_b^2}\left(1 - 3\frac{3^{1/3}}{5}x_b\right) \qquad (93)$$

where the leading term agrees with that of GI. As a check on the correctness of the results, it should be noted that ϕ_b and ϕ_i' as given by equations (91) and (93), respectively, satisfy the relation of Gilvarry (equation (68)) connecting ϕ_b and ϕ_i' which any exact solution of the Thomas–Fermi equation must obey.

It is of interest to compare the findings of M from numerical data on the boundary and initial parameters of the Thomas–Fermi equation (in part computed by himself for small values of x_b) with results of GI. The form used in the latter work to fit the available numerical data for the boundary value is

$$\phi_b = \left(\sum A_n x_b^{n/2}\right)^{-1} \qquad (94)$$

where n ranges from 2 to 6, and the coefficients are given in table 1. The coefficients

A_2 and A_6 were chosen so that ϕ_b has the correct limiting forms for $x_b \to 0$ and $x_b \to \infty$, respectively. Equation (94) predicts that

$$\phi_b = \frac{3^{2/3}}{x_b}(1 - 3^{2/3} A_3 x_b^{1/2}) \tag{95}$$

for small x_b, whereas the correct form is given by equation (91). However, it is noteworthy that the coefficient A_3 found in GI is smaller than A_4 by a factor 10. If one puts $A_3 = 0$, in order to obtain agreement with equation (91), then the expression of GI gives

$$\phi_b = \frac{3^{2/3}}{x_b}(1 - 0.1457 x_b) \tag{96}$$

for small x_b, whereas the form of M yields

$$\phi_b = \frac{3^{2/3}}{x_b}(1 - 0.1442 x_b) \tag{97}$$

with remarkable agreement between the numerical coefficients.

These considerations point to the fact that the expression of GI for ϕ_b will give good results for small values of x_b, outside the range of directly computed data. However, they suggest further that good agreement with the numerical data can still be obtained when the expression is placed on a rather better theoretical basis by choosing $A_3 = 0$ and changing A_4 slightly to $3^{-1/3}/10$. In this case, one finds by comparison with existing numerical data that a fit to better than 1% can be obtained by changing the remaining adjustable coefficient A_5 somewhat, to 0.0097. The agreement is quite remarkable when it is remembered that A_5 is the one parameter fitted to the data. Finally, it is worth noting that the simple formula of equation (91) gives a quite reasonable fit of the numerical values even for a value of x_b as large as 3, where it is in error by 9%.

With regard to the initial slope ϕ'_i, the data were fitted in GI with the function

$$\phi'_i - \phi'_{i,\infty} = (\sum B_n x_b^n)^{-1} \tag{98}$$

where n ranges over the sequence $n = 2, 3, \ldots, 7, \lambda_1$, in which $\lambda_1 = 7.77200$. The coefficients B_2 and B_{λ_1} were chosen to give the correct asymptotic forms. Examining again the connection with the theoretical results for small x_b, one finds that this time the form of GI is in agreement with that of M as regards the powers of x_b appearing, and from table 1 the numerical coefficient of the term proportional to x_b^{-1} is -1.88, whereas the value from equation (93) is -1.80. Again, agreement is quite good.

Finally, one can note that, if the coefficients A_3 and A_4 are assigned the suggested values 0 and $3^{-1/3}/10$, respectively, in the fitted function of equation (94), then the corresponding equation of state reduces for very high pressures to the correct result

$$p = \frac{h^2}{5m}\left(\frac{3}{8\pi}\right)^{2/3}\left(\frac{Z}{v}\right)^{5/3}\left\{1 - \frac{2\pi m e^2}{h^2}(4Zv)^{1/3}\right\} \tag{99}$$

where e and m are the charge and mass of the electron, respectively. Omitting the second term in this equation, one has simply the expression for the pressure exerted by Z free electrons enclosed in a volume v, and this term is of course independent of

Thomas–Fermi atom model

the electronic charge. On the other hand, the second term in equation (99) shows the way in which the equation of state is modified at the highest pressures by including electron–nuclear and electron–electron interactions. It should be noted that this term is independent of Planck's constant h, since it has a purely classical origin.

Asymptotic solution for large atom radius

In the limit of large atom radius, the leading coefficients for this case required for the fitted functions discussed on p. 334 were obtained in GI from an approximate solution of the Thomas–Fermi equation given by Sauvenier[62] (modified from results of Sommerfeld[63] for the positive ion). This solution is not exact, since the result corresponds merely to the leading term in a slowly convergent expansion. Further, the problem of solution of the Thomas–Fermi equation in this limit is apparently not tractable to machine solution, for a reasonable expenditure of coding and computing time. In this subsection, the rigorous asymptotic solution of the Thomas–Fermi equation in the limit of large atom radius given by Gilvarry and March[67] will be discussed. This solution makes possible relatively accurate specification of the boundary and initial parameters needed to construct fitted functions of the atom radius for the boundary value and initial slope of the Thomas–Fermi function.

For brevity in reference, the results of Gilvarry and March will be referred to under the designation GM.

Asymptotic solutions. Sommerfeld[68] has given the asymptotic expression

$$\phi_0 = \frac{144}{x^3} \tag{100}$$

for the solution of the Thomas–Fermi equation in the limit $x, x_b \to \infty$, corresponding to the isolated atom. One notes that this form contains no constants of integration. Further, it fails to meet the initial condition of equation (60a).

To remedy this deficiency of equation (100) as regards the initial condition, Sommerfeld resorted to a mathematical artifice. He noticed that a more general solution of the Thomas–Fermi equation in the same limit is

$$\phi = \frac{144}{x^3}\left(1 - \frac{c}{x^{\lambda_2}}\right) \tag{101}$$

with λ_2 given by equations (77). In this form, the correction term in parentheses represents the first term of a slowly convergent expansion. He then proved that, for z as given by equations (76), the solution

$$\phi = (1+z)^{-\lambda_1/2} \tag{102}$$

reduces asymptotically to equation (101) for x large and furthermore satisfies the required initial condition at $x = 0$. However, it fails to yield the correct initial slope $\phi'_{i,\infty}$ for an infinite atom. Sommerfeld[63] has used equation (101) as the basis for an analytical treatment of the positive ion on the Thomas–Fermi model. His results have been adapted by Sauvenier[62] to obtain the Thomas–Fermi function for a compressed atom which is perturbed only slightly from the infinite atom.

Rather than resorting to analytical approximations, GM considered the problem of obtaining the rigorous solution of the Thomas–Fermi equation for large atom radius in series form. In this connection, it should be noted that c of equation (101) corresponds on the basis of equation (102) to $c = -(\lambda_1/2)12^{2\lambda_2/3}$. In fact, this value of c corresponds to an analytic approximation to a constant of integration E_{1-} entering the rigorous series solution. Its value has been obtained numerically with high accuracy by Kobayashi and others[69], by the requirement that the asymptotic solution of the Thomas–Fermi equation for large atom radius must be continuous, when extended inwards, in value and slope with the corresponding solution near the origin provided by Baker's series (p. 330) with the proper initial slope $\phi'_{i,\infty}$. Their value of E_{1-} was used by GM in a rigorous procedure free of analytical approximations.

To obtain the general solution of equation (59) in the limit $x, x_b \to \infty$, the Sommerfeld solution of equation (100) for this limit was taken as a point of departure. The solution was expressed as a perturbation on ϕ_0 by writing

$$\phi = \phi_0(1+\chi) \tag{103}$$

where χ is a perturbation function. The basic assumption is thus the same as made by Sommerfeld in his determination of the (approximate) asymptotic solution of equation (102) corresponding to ϕ_0 which meets the initial condition and in his treatment of the positive ion. It is also the assumption made by Coulson and March[70, 71] in obtaining the asymptotic solution

$$\phi = \phi_0\left\{1 + \sum_{n=1}^{\infty} (-1)^n E_{n-} x^{-n\lambda_2}\right\} \tag{104}$$

of equation (59) which tends to zero at infinity. In this expression, λ_2 is a constant which will be defined later, and the successive coefficients E_{n-} for $n \geq 2$ are fixed by the differential equation in terms of E_{1-}. The estimate of Coulson and March has been superseded by the very accurate numerical value

$$E_{1-} = 13 \cdot 270974 \tag{105}$$

given by Kobayashi and others[69] for the constant E_{1-} required to make the solution (5) meet the initial condition when extended inwards, for the case of the isolated atom.

For the purposes of the following paragraphs on boundary and initial parameters, it is convenient to make the substitution

$$\rho = \frac{x}{x_b} \tag{106}$$

which transforms the boundary of the atom into the point $\rho = 1$. The perturbation function χ then satisfies the differential equation

$$\frac{d^2[\rho^{-3}(1+\chi)]}{d\rho^2} = 18\rho^{-5}(1+\chi)^{3/2} \tag{107}$$

It should be noted that the slope of a solution, and in particular its initial slope, transforms according to

$$\frac{d\phi}{d\rho} = x_b \frac{d\phi}{dx} \tag{108}$$

under the substitution (106).

Thomas–Fermi atom model

Since equation (107) is nonlinear, one does not necessarily expect a solution as a power series in the independent variable ρ, as is the case for a linear equation when its singularities are suitably restricted. However, any part of the general solution which can be expressed as a power series in a variable ρ^λ can be isolated by writing

$$\chi = \sum_{n=1}^{\infty} F(n, \lambda) \rho^{n\lambda} \tag{109}$$

where λ and $F(n, \lambda)$ are subject to determination. One finds that solutions of this form exist, provided that the condition

$$(\lambda^2 - 7\lambda - 6) F(1, \lambda) = 0 \tag{110}$$

holds, which is required to make the initial term in ρ^λ meet equation (107). For $F(1, \lambda)$ nonvanishing, this equation has two roots λ_+ and λ_- of opposite sign, given by

$$\lambda_+ = \lambda_1, \qquad \lambda_- = -\lambda_2 \tag{111}$$

where λ_1 and λ_2 are the positive quantities

$$\lambda_1 = \tfrac{1}{2}(73^{1/2} + 7), \qquad \lambda_2 = \tfrac{1}{2}(73^{1/2} - 7) \tag{112}$$

introduced by Sommerfeld[68]. For either choice of λ, the corresponding coefficient $F(1, \lambda)$ is clearly arbitrary, so that two constants of integration appear. When the solution corresponding to the other value of λ is ignored for the moment, the coefficients $F(n, \lambda)$ for one value of λ can be computed recursively from the differential equation and obtained in terms of the corresponding $F(1, \lambda)$ and λ for $n \geq 2$. The two sets of coefficients $F(n, \lambda)$ obtained in this manner will be designated as F_{n+} and F_{n-} corresponding to λ_+ and λ_-, respectively. For either choice of λ_r, a solution χ_r of equation (107) for the case $x, x_b \to \infty$ can be written as

$$\chi_r = \sum_{n=1}^{\infty} F_{nr} \rho^{n\lambda_r} \tag{113}$$

where the range of r is the set $+$ and $-$ of indices. For the former selection of r the series is in ascending and for the latter in descending powers of ρ, and two disposable coefficients F_{1+} and F_{1-} appear.

The result of the recursive solution for F_{nr} can be written for $n \geq 2$ as

$$F_{nr} = \frac{12}{\gamma_{nr}} \sum_{i=2}^{n} \binom{\tfrac{3}{2}}{i} \begin{bmatrix} i \\ nr \end{bmatrix} \tag{114}$$

where

$$\gamma_{nr} = (n-1)\{6(n+1) + 7n\lambda_r\} \tag{115}$$

and

$$\binom{\tfrac{3}{2}}{i}, \qquad \begin{bmatrix} i \\ nr \end{bmatrix} \tag{116}$$

are respectively a binomial coefficient, and the coefficient of $\rho^{n\lambda_r}$ in the expansion of χ^i in powers of ρ^{λ_r}. The latter coefficient (square-bracket symbol) for $i \geq 2$ cannot depend on F_{nr} but involves all such coefficients of lower order down to $n = 1$; the general term exclusive of coefficient is of form $F_{pr}^j F_{qr}^k \ldots$ with the indices r all the same, and with

$$jp + kq + \ldots = n \qquad (117)$$

for $j, k, \ldots \geq 1$. Thus, the right-hand side of equation (114) does not involve F_{nr} since the summation appearing starts out at $i = 2$. It should be noted that the square-bracket symbol represents a definite function which can be defined independently of the differential equation (107); aside from dependence on powers of F_{nr}, it is simply a multinomial coefficient. In table 3 values of the coefficients F_{nr} computed from equation (114) are shown for n up to 4.

TABLE 3   Coefficients $F_{n\pm}$ for the perturbation χ
[from GM]

n	F_{n+}	F_{n-}
1	F_{1+}	F_{1-}
2	$\dfrac{9}{4}\left(\dfrac{F_{1+}^2}{9+7\lambda_1}\right)$	$\dfrac{9}{4}\left(\dfrac{F_{1-}^2}{9-7\lambda_2}\right)$
3	$-\dfrac{F_{1+}^2}{48}\left(\dfrac{7\lambda_1 - 18}{61 + 77\lambda_1}\right)$	$\dfrac{F_{1-}^3}{48}\left(\dfrac{18 + 7\lambda_2}{61 - 77\lambda_2}\right)$
4	$\dfrac{F_{1+}^4}{1152}\left(\dfrac{16230 + 20489\lambda_1}{51973 + 67319\lambda_1}\right)$	$\dfrac{F_{1-}^4}{1152}\left(\dfrac{16230 - 20489\lambda_2}{51973 - 67319\lambda_2}\right)$

Some general properties of the coefficients $F_{n\pm}$ can be shown easily. At any point in the recursive solution for $F_{n\pm}$ in terms of $F_{1\pm}$ one must solve an algebraic equation involving $\lambda_\pm$ and $F_{n\pm}$ of all lower order (upper or lower signs go together in the ambiguous signs, here and in what follows). However, the coefficients of $\lambda_\pm$ and $F_{n\pm}$ in the equation in question are fixed by equation (107), and hence are independent of the particular choice λ_+ or λ_- of λ. Thus, one concludes that

$$F_{n+} = F(n, \lambda_+, F_{1+}), \qquad F_{n-} = F(n, \lambda_-, F_{1-}) \qquad (118)$$

for $n \geq 2$, where the dependence on the initial member $F_{1\pm}$ of the set of coefficients $F(n, \lambda)$ of equation (113) has been introduced explicitly into the notation. The function F appearing in the two cases of equations (118) must be the same function of its three arguments; in other words, F_{n+} goes into F_{n-} and vice versa under the reciprocal transformation

$$\lambda_1 \rightleftarrows -\lambda_2, \qquad F_{1+} \rightleftarrows F_{1-} \qquad (119)$$

This rule is consistent with the results of table 3. Further, one notes from table 3 that $F_{2\pm}$, as directly computed, differs from $F_{1\pm}^2$ only by a constant of proportionality. With use of equation (117), this fact is sufficient to prove by induction that $F_{n\pm}$ differs from $F_{1\pm}^n$ only by a constant of proportionality. For each value of n in table 3, the recursion relations for the coefficients of lower order have been used to reduce $F_{n\pm}$ to a function of $F_{1\pm}$ (and $\lambda_\pm$) only.

The descending power series χ_- of equation (113) can be identified immediately as corresponding to the Coulson–March[70] solution (104). Comparing coefficients, one obtains

$$F_{n-} = (-1)^n E_{n-} x_b^{-n\lambda_2} \tag{120}$$

for F_{n-} in terms of the Coulson–March coefficients E_{n-}. In contrast with E_{n-}, which approaches a constant as $x_b \to \infty$, one sees that F_{n-} varies with x_b in this limit. This behavior arises from the change in initial slope of ϕ, given by equation (108), which is induced by the substitution (106), since E_{1-} for a neutral atom is fixed as the value of equation (105) by the requirement of continuity in value and slope with the result of Baker's series[61] for small x. The ascending power series χ_+ of equation (113) yields an asymptotic solution analogous to the Coulson–March form but not vanishing at infinity, expressible in terms of the variable x as

$$\phi = \phi_0 \left\{ E_{1+} x^{\lambda_1} + \sum_{n=2}^{\infty} (-1)^n E_{n+} x^{n\lambda_1} \right\} \tag{121}$$

where $E_{1+} = F_{1+}$ and $E_{n+} = (-1)^n F_{n+} x_b^{-n\lambda_1}$ for $n \geq 2$.

The recursion relations for the coefficients E_{n-} are available for n up to 10 from unpublished results of Coulson and March. These relations have been used to obtain F_{n+} numerically for values of n beyond those given in table 3 by means of equation (120) and the transformation (119). The results can be expressed most conveniently in terms of coefficients c_{n+}, defined by

$$c_{n\pm} = \frac{F_{n\pm}}{F_{1\pm}^n} \tag{122}$$

and which thus are pure numbers. Numerical values of c_{n+} are given in table 4 for n up to 10; Kobayashi and others[69] tabulate values of c_{n-} for n up to 17. The signs appearing for c_{n+}, and the fact (shown on p. 351) that F_{1+} is positive, imply that the coefficients E_{n+} of equation (121) are all positive.

TABLE 4 Coefficients c_{n+} for the perturbation χ [from GM]

n	c_{n+}	n	c_{n+}
1	1	6	0·0000309
2	0·0354867	7	−0·0000138
3	−0·0011501	8	0·0000069
4	0·0002648	9	−0·0000038
5	−0·0000803	10	0·0000022

The general solution of equation (107) in the limit $x, x_b \to \infty$ is not simply the sum $\chi_+ + \chi_-$ of the two solutions of equation (113), because one set of the coefficients $F_{n\pm}$ was disregarded in order to obtain the other set. However, this sum contains two disposable constants, and hence one can obtain the general asymptotic solution by adding further series to it, such that substitution of the sum of the series in equation (107) satisfies the differential equation to every power of x appearing. It is clear that the additional series can entail no further disposable constants; in addition, any coefficients appearing which depend on $F_{n\pm}$ can involve only cross-terms between

F_{n+} and F_{n-} (or ultimately F_{1+} and F_{1-}), since all terms whose coefficients are proportional to powers of F_{1+} or F_{1-} alone are included in one or the other of the series of equation (113). By inspection, one concludes that the additional series are of the form

$$\sum_{n=1}^{\infty} G_{n0} \rho^{7n} + \sum_{n=1}^{\infty} \sum_{m=1}^{\infty} G_{nm+} \rho^{7n+m\lambda_+} + \sum_{n=1}^{\infty} \sum_{m=1}^{\infty} G_{nm-} \rho^{7n+m\lambda_-} \qquad (123)$$

where one can note that $\lambda_+ + \lambda_- = 7$. The general asymptotic solution for the limit $x, x_b \to \infty$ can then be written succinctly as

$$\chi = \sum_{n=1}^{\infty} \sum_{r} F_{nr} \rho^{n\lambda_r} + \sum_{n=1}^{\infty} \sum_{m=0}^{\infty} \sum_{r} G_{nmr} \rho^{7n+m\lambda_r} \qquad (124)$$

where the range of r is the set $+$ and $-$ of indices; the index r on G_{nmr} is superfluous when $m = 0$.

The recursive solution for G_{nmr} can be written for $n \geq 1$, $m \geq 0$ as

$$G_{nmr} = \frac{12}{\delta_{nmr}} \sum_{i=2}^{2n+m} \binom{3/2}{i} \left[\begin{array}{c} i \\ nmr \end{array} \right] \qquad (125)$$

where

$$\delta_{nmr} = 49n(n-1) + 6(m^2 - 1) + 7m(2n + m - 1)\lambda_r \qquad (126)$$

and the square-bracket symbol represents the coefficient of $\rho^{7n+m\lambda_r}$ in the expansion of χ^i in powers of ρ^{λ_+} and ρ^{λ_-} jointly. This coefficient for $i \geq 2$ cannot depend on G_{nmr}; the general term aside from coefficient is $F_{pr}^j F_{qr}^k G_{str}^l \ldots$ with the indices r *not* all the same if G coefficients are absent, and with

$$jp + kq + l(2s + t) + \ldots = 2n + m \qquad (127)$$

for $j, k, l, \ldots \geq 1$. Thus, G_{nmr} does not appear in the right-hand side of equation (125). In table 5, the coefficients G_{nm+} actually needed for the purposes of obtaining numerical results have been tabulated, as determined from equation (125).

TABLE 5 Coefficients G_{nm+} for the perturbation χ [from GM]

n	m	G_{nm+}
1	0	$-\frac{3}{2}F_{1-}F_{1+}$
1	1	$-(3\lambda_2/112)F_{1-}(7F_{1+}^2 - 4F_{2+})$
1	2	$\frac{3}{2}(3+7\lambda_1)^{-1}\{G_{11+}F_{1+} + G_{10}F_{2+} + F_{1-}F_{3+}$ $- \frac{1}{4}(G_{10}F_{1+}^2 + 2F_{1-}F_{1+}F_{2+}) + \frac{1}{8}F_{1-}F_{1+}^3\}$
1	3	$\frac{3}{4}(4+7\lambda_1)^{-1}\{G_{12+}F_{1+} + G_{11+}F_{2+} + G_{10}F_{3+} + F_{1-}F_{4+}$ $- \frac{1}{4}(G_{11+}F_{1+}^2 + 2G_{10}F_{1+}F_{2+} + F_{1-}F_{2+}^2 + 2F_{1-}F_{1+}F_{3+})$ $+ \frac{1}{8}(G_{10}F_{1+}^3 + 3F_{1-}F_{1+}^2F_{2+}) - (5/64)F_{1-}F_{1+}^4\}$

By arguments similar to those used in connection with $F_{n\pm}$, one finds that

$$G_{nm+} \rightleftarrows G_{nm-} \qquad (128)$$

under the transformation (119). In particular, G_{n0} transforms into itself under the transformation in question and thus must be symmetric in λ_+ and λ_-; this fact can

Thomas–Fermi atom model

be checked from the entry for G_{n0} in table 5. The transformation (128) makes immediately available the values of G_{nm-} corresponding to the coefficients G_{nm+} tabulated. Further, one can show by means of equation (127) that $G_{nm\pm}$ differs from $(F_{1+}F_{1-})^n F_{1\pm}^m$ only by a constant of proportionality, as is clear in the cases of the first two coefficients of table 5, and can be verified for the others. Hence, coefficients $d_{nm\pm}$ can be defined by

$$d_{nm\pm} = G_{nm\pm}(F_{1+}F_{1-})^{-n}F_{1\pm}^{-m} \tag{129}$$

which are pure numbers. In table 6, numerical values of the coefficients d_{1m+} needed for the purposes of obtaining numerical results have been tabulated.

TABLE 6 Coefficients d_{nm+} for the perturbation χ [from GM]

n	m	d_{nm+}
1	0	$-1\cdot 5000000$
1	1	$-0\cdot 1418151$
1	2	$0\cdot 0074749$
1	3	$-0\cdot 0023828$

One notes that the Sommerfeld form (100) represents the only term in the general solution for ϕ which is independent of the disposable constants. The results obtained yield the general form of that part of the asymptotic solution of the Thomas–Fermi equation, for an atom of large radius, which depends on constants of integration. The solution corresponding to the series (124) thus generalizes the Coulson–March expression (104).

Boundary and initial parameters. In this section, the problem of specifying the disposable coefficients F_{1+} and F_{1-} to meet the boundary and initial conditions will be discussed.

The general boundary condition in the Thomas–Fermi atom model is

$$\left[\frac{d\phi}{dx}\right]_{x=x_b} = \frac{\phi_b}{x_b} \tag{130}$$

in terms of the boundary value ϕ_b of ϕ at the atom radius x_b. For the solution to correspond to the isolated atom when the atom radius becomes infinite, the condition

$$[\phi_b]_{x_b \to \infty} \to 0 \tag{131}$$

must be met in addition. In terms of the variable ρ, equation (130) becomes

$$\left[\frac{d\phi}{d\rho}\right]_{\rho=1} = \phi_b \tag{132}$$

which, applied to equation (103), yields

$$\left[\frac{d\chi}{d\rho}\right]_{\rho=1} = 4(1+\chi_b) \tag{133}$$

as the boundary condition on the perturbation χ, in terms of its boundary value $\chi_b = \chi(1)$.

Setting $\rho = 1$ in equation (124), one obtains

$$\chi_b = \sum_{n=1}^{\infty} (F_{n+} + F_{n-} + G_{n0}) + \sum_{n=1}^{\infty} \sum_{m=1}^{\infty} (G_{nm+} + G_{nm-}) \tag{134}$$

as the boundary value of χ. One notes from equation (120) that the coefficients F_{n-} depend on $x_b^{-n\lambda_2}$ as x_b tends to infinity, and, further, $G_{nm\pm}$ always involves F_{n-}. Hence, all terms of equation (134) in F_{n-} and $G_{nm\pm}$ (with $m = 0$ included) contribute terms to χ_b which vary as inverse powers of $x_b^{\lambda_2}$.

To find the complete dependence of χ_b on atom radius, it remains to determine the dependence of the two disposable coefficients F_{1-} and F_{1+} on x_b. It is necessary to note that the value (105) of Kobayashi and others for E_{1-} represents only the limit for $x_b \to \infty$ of a function $E_{1-}(x_b)$ of the atom radius. This fact follows from the relation

$$\frac{d\phi_i'}{dx_b} = -\left(\frac{\phi_b^{5/2}}{x_b^{1/2}}\right)\left(1 + \frac{1}{3}\frac{d \ln \phi_b}{d \ln x_b}\right) \tag{135}$$

given in GI (see p. 332) for the variation with atom radius of the initial slope ϕ_i' of the Thomas–Fermi function. Such a variation perturbs ϕ_i' from its value for the isolated atom, and $E_{1-}(x_b)$ must change correspondingly to maintain continuity in value and slope with the result obtained from Baker's series for small x. It will be assumed that $E_{1-}(x_b)$ can be expanded as a power series

$$E_{1-}(x_b) = e_{0-} + \sum_{n=1}^{\infty} e_{n-} x_b^{-n\lambda_2} \tag{136}$$

in the variable $x_b^{-\lambda_2}$, where the first coefficient e_{0-} is fixed by $e_{0-} = E_{1-}$ as given by Kobayashi and others. Correspondingly, equation (120) implies that the disposable coefficient F_{1-} is a function $F_{1-}(x_b)$ expressible as

$$F_{1-}(x_b) = \sum_{n=1}^{\infty} f_{n-} x_b^{-n\lambda_2} \tag{137}$$

where $f_{n-} = -e_{(n-1)-}$, and the first coefficient is given by

$$f_{1-} = -e_{0-} = -E_{1-} \tag{138}$$

in terms of E_{1-} as evaluated by Kobayashi and others. The possibility of the expansion (136) and thus of (137) will be shown by exhibiting a method of calculating the coefficients f_{n-} for $n \geq 2$, when use is made of numerical results from solutions of the Thomas–Fermi equation for different atom radii (as will be discussed in the final paragraph of this subsubsection). One can note that the coefficients F_{n-} of order higher than one in equation (124) must be specified by the Coulson–March recursion relations in terms of $F_{1-}(x_b)$ of equation (137) when x_b is not infinite, in order to satisfy equation (107).

When the series (137) for F_{1-} is substituted into the series (124) for χ, and the boundary condition (133) is imposed to evaluate F_{1+}, one finds that all terms in F_{n-} or $G_{nm\pm}$ can be ignored to first order, because of their dependence on powers of $x_b^{-\lambda_2}$. Hence the leading term f_{0+} in F_{1+} is independent of the parameters f_{n-} and is a

Thomas–Fermi atom model

constant determined only by the coefficients F_{n+} in χ. By a similar argument, the next term of F_{1+} must be of the form $f_{1+}x_b^{-\lambda_2}$, where f_{1+} is a constant fixed by all the terms in χ which contain F_{1-} to the first power, or those with coefficients F_{1-}, G_{10}, and G_{1m+} with $m \geq 1$ arbitrary. Thus, f_{1+} must depend on f_{1-}. Generalizing, one infers that imposition of the boundary condition on χ yields the power series

$$F_{1+}(x_b) = f_{0+} + \sum_{n=1}^{\infty} f_{n+} x_b^{-n\lambda_2} \tag{139}$$

in $x_b^{-\lambda_2}$ for F_{1+}, where f_{n+} depends on all the terms in χ containing F_{1-}^n, or that with coefficient F_{n-}, those with coefficients of form G_{pq-} with $p+q=n$ subject to the inequalities $1 \leq p \leq n$ and $0 \leq q \leq n-1$, and, finally, those with coefficients G_{nm+} with $m \geq 1$ arbitrary. The value of f_{n+} for $n \geq 1$ thus depends recursively on the coefficients $f_{1-}, f_{2-}, \ldots, f_{n-}$ as well as $f_{0+}, f_{1+}, \ldots, f_{(n-1)+}$; it should be noted that the coefficient f_{n-} is included in the former set.

When the values of $F_{n\pm}$ and $G_{nm\pm}$ are determined from equations (137) and (139) by means of the recursion relations of the previous subsection, and are substituted into equation (134) for χ_b, it is clear that all terms appearing depend on inverse powers of $x_b^{\lambda_2}$, exclusive of those involving f_{0+} alone. If terms of the same order are then collected in χ_b, the value of the product $x_b^3 \phi_b$ can be represented by the power series

$$x_b^3 \phi_b = a_0 + \sum_{n=1}^{\infty} a_n x_b^{-n\lambda_2} \tag{140}$$

in $x_b^{-\lambda_2}$, where the first coefficient is given by

$$a_0 = 144\left(1 + \sum_{n=1}^{\infty} c_{n+} f_{0+}^n\right) \tag{141}$$

and the further coefficients a_n depend on $f_{0+}, f_{1+}, \ldots, f_{n+}; f_{1-}, f_{2-}, \ldots, f_{n-}$. Equation (135) then yields the power series

$$x_b^7(\phi_i' - \phi_{i,\infty}') = \sum_{n=1}^{\infty} b_n x_b^{-n\lambda_2} \tag{142}$$

determining the general asymptotic form for the initial slope in the limit $x_b \to \infty$, where $\phi_{i,\infty}'$ is the initial slope for the isolated atom. The first coefficient b_1 is given by

$$b_1 = -\frac{1}{3}\frac{\lambda_2}{\lambda_1} a_0^{5/2} a_1 \tag{143}$$

and hence depends on the coefficients f_{0+}, f_{1+}, and f_{1-}; the further coefficients b_n are functions of $a_0, a_1, \ldots, a_n$ and thus of $f_{0+}, f_{1+}, \ldots, f_{n+}; f_{1-}, f_{2-}, \ldots, f_{n-}$. Equation (142) gives

$$x_b^8 \frac{d\phi_i'}{dx_b} = -\sum_{n=1}^{\infty} (7+n\lambda_2) b_n x_b^{-n\lambda_2} \tag{144}$$

as the asymptotic form specifying the derivative of the initial slope with respect to atom radius. One can note that the requirement of the limit (131) is satisfied by equation (140). Further, one observes that equation (139) permits definition of a function $E_{1+}(x_b)$ analogous to $E_{1-}(x_b)$, expressible as a power series in $x_b^{-\lambda_2}$ corresponding to equation (136), but with coefficients e_{n+}.

The general method of imposing the boundary condition on χ consistently to obtain the f coefficients in F_{1+} and F_{1-} as numerical values can now be outlined. If powers of x_b up to $x_b^{-n\lambda_2}$ are retained in F_{1+} and F_{1-} in imposing the boundary condition, the necessity of satisfying it for each power $x_b^{-i\lambda_2}$ appearing yields $n+1$ relations of form

$$a_i = a_i(f_{0+}, f_{1+}, \ldots, f_{i+}; f_{1-}, f_{2-}, \ldots, f_{i-}) \tag{145}$$

for $i = 0, 1, \ldots, n$, where the left-hand side is a number and the right-hand side is a function corresponding to equation (141) for the case $i = 0$. The coefficients b_i for $i = 1, 2, \ldots, n$ must be evaluated by numerical methods from solutions of the Thomas–Fermi equation for different atom radii, by means of either equation (142) or equation (144). Accordingly, in the n relations

$$b_i = b_i(f_{0+}, f_{1+}, \ldots, f_{i+}; f_{1-}, f_{2-}, \ldots, f_{i-}) \tag{146}$$

for $i = 1, 2, \ldots, n$, the left-hand side is available as a number and the right-hand side is a function corresponding for the case $i = 1$ to equation (143) (when a_0 and a_1 are replaced by their values in terms of f_{0+}, f_{1+}, and f_{1-}). Thus, one has $2n+1$ equations of form (145) and (146) to determine the $2n+1$ unknown f coefficients, and, by induction, the problem is a determinate one if it is determinate (as will appear) for the case $n = 0$. Retention of powers of x_b through order $x_b^{-n\lambda_2}$ requires an infinite number of terms in the right-hand sides of equations (145) and (146) at each stage of the process, and one in principle must pass to the limit $n \to \infty$, so that the procedure can be carried out in practice only by a method of successive approximations.

One can note that use of $x_b^{-n\lambda_2}$ as the expansion variable in equation (136) for $E_{1-}(x_b)$ leads to the simple forms of equations (140) and (142) for $x_b^3 \phi_b$ and $x_b^7(\phi_i' - \phi_{i,\infty}')$, respectively; for any other choice of expansion variable, the results would be considerably more complicated.

Numerical results. The foregoing results will be used in this subsection to obtain actual values of boundary and initial parameters.

Evaluation of the first approximation f_{0+} to F_{1+} involves imposing the boundary condition (133) on the infinite series χ_+ of equation (113) in ascending powers of ρ. Since each coefficient F_{n+} is proportional to F_{1+}^n, the process must be carried out iteratively. Retaining at first only the term $F_{1+} \rho^{\lambda_1}$ in the series, one finds immediately that

$$f_{0+} = \frac{2}{9}(\lambda_1 - 3) \tag{147}$$

The value of a_0 to the same order of approximation is

$$a_0 = 16(3 + 2\lambda_1) \tag{148}$$

from equation (141), when the infinite series appearing is truncated after the first term. The value of equation (148) for a_0 is precisely that obtained in GI from the approximate solution of Sauvenier[62], and hence the value of equation (147) for f_{0+}

Thomas–Fermi atom model   351

corresponds also to this solution. The numerical values of f_{0+} and a_0 on this approximation are shown in the second column of table 7. It is clear from the present argument that these results do not represent the true limiting values. Table 8 shows the effect

TABLE 7 Values of computed coefficients in boundary and initial parameters [from GM] compared with values from the Sauvenier solution

Coefficient	From Sauvenier solution	From GM
f_{0+}	1·0604	0·9637
$10^{-2}a_0$	2·967^a	2·874
$-10^{-1}f_{1+}$	—	3·07 ± 0·01
$-10^{-3}a_1$	3·86^a	3·47 ± 0·01
$10^{-5}b_1$	6·53^a	5·60 ± 0·02

^{a} From GI.

TABLE 8 Successive approximations to the leading terms f_{0+} and a_0 in F_{1+} and $x_b^3 \phi_b$, respectively, on inclusion of increasing order n of F_{n+} [from GM]

n	f_{0+}	a_0
1	1·0604	296·7
2	0·9603	287·0
3	0·9647	287·5
4	0·9633	287·3
⋮	⋮	⋮
10	0·9637	287·4

on f_{0+} and a_0 of including terms $F_{n+}\rho^{n\lambda_1}$ through various orders n in imposing the boundary condition on χ; final values are shown in the last column of table 7. One sees from the tables that the boundary value a_0 changes relatively little from the value given by equation (148), although the coefficient f_{0+} changes rather more significantly. A value[72] for a_0, stated to be indicated by some approximations, is in good agreement with the final value in table 7.

To obtain the coefficient f_{1+} in the first variable term of F_{1+}, coefficients c_{n+} to $n = 9$ and coefficients d_{1m+} to $m = 3$ were taken into account by a method of successive approximations in imposing the boundary condition, with use of the value (105) of Kobayashi and others for E_{1-} in equation (138) for f_{1-}. Successive convergents are shown in table 9 to demonstrate that good convergence has been achieved. The final result is tabulated in the last column of table 7. The value of the coefficient a_1 in the first variable term of $x_b^3 \phi_b$ is fixed in terms of f_{0+}, f_{1+}, and f_{1-} by infinite processes which are not easily specified explicitly. Successive convergents corresponding to those in f_{1+} are shown in table 10, and the final result is given in the last column of table 7. The coefficient b_1 of the leading term in the series for $x_b^7(\phi_i' - \phi_{i,\infty}')$ is fixed algebraically by equation (143) in terms of a_0 and a_1; its value is shown as the final entry in the last column of table 7, corresponding to the entries for a_0 and a_1 above it.

TABLE 9 Successive approximations to the coefficient $-10^{-1}f_{1+}$ in F_{1+}, on inclusion of c_{n+} of order n and d_{1m+} of order m [from GM]

n \ m	0	1	2	3
1	3·20	3·70	3·66	3·68
2	2·65	3·06	3·03	3·04
3	2·69	3·10	3·07	3·08
4	2·67	3·09	3·05	3·07
⋮	⋮	⋮	⋮	⋮
9	2·68	3·09	3·06	3·07

TABLE 10 Successive approximations to the coefficient $-10^{-3}a_1$ in $x_b^3 \phi_b$, on inclusion of c_{n+} of order n and d_{1m+} of order m [from GM]

n \ m	0	1	2	3
1	3·76	4·22	4·18	4·20
2	3·09	3·45	3·42	3·43
3	3·15	3·51	3·48	3·49
4	3·12	3·49	3·45	3·47
⋮	⋮	⋮	⋮	⋮
9	3·14	3·49	3·46	3·47

For comparison purposes, the numerical value is shown in the second column of table 7 for the approximation

$$a_1 = \frac{8}{9}(105 + 142\lambda_1)12^{2\lambda_2/3} \qquad (149)$$

obtained in GI for a_1 from the approximate solution of Sauvenier. In view of the approximate nature of the Sauvenier solution, the agreement with a_1 of this work is good; the excellent agreement shown by the value from equation (149) with the first convergent (for $n = 1$ and $m = 0$) of table 10 for a_1 can be noted. Shown also in the second column is the numerical value of the approximation

$$b_1 = \frac{256}{9}(2 + 3\lambda_2)(3 + 2\lambda_1)^{5/2} 12^{2\lambda_2/3} \qquad (150)$$

of GI for b_1 from the Sauvenier solution; one sees that agreement with b_1 of this work is reasonable. Thus, one conclusion from the present work is that the approximation in question is remarkably good in spite of its simplicity.

Discussion. Umeda[73] has pointed out that the assumption of the Sommerfeld solution ϕ_0 of equation (100) for the isolated atom appears to be in conflict with the Sommerfeld–Sauvenier limit corresponding to a_0 of equation (148). As Umeda

Thomas–Fermi atom model

notes, the correct solution in this limit for the boundary value ϕ_b of $\phi(x, x_b)$ must satisfy the requirement

$$\phi_b = \phi(x_b, x_b) \tag{151}$$

with respect to which equation (100) and (130) seem inconsistent. He attempted to resolve the difficulty by obtaining a solution for a compressed atom which was based on an approximation to the solution for a free atom as given by Kerner[74]. The result satisfies equation (151) but leads to poorer agreement with the computed data for large x_b than does the original Sommerfeld–Sauvenier approximation.

From a mathematical point of view, no paradoxical results appear in the treatment given here. The values of ϕ_b and $\phi(x, x_b)$ implied by equations (134) and (124), respectively, clearly satisfy equation (151) of Umeda. Further, the general solution for $\phi(x, x_b)$ meets the boundary condition (130) and satisfies the limit (131) demanded for an isolated neutral atom. The slope of this solution is alway positive for x and x_b large, and thus is consistent with the presence of the vertical asymptotes which the Thomas–Fermi equation possesses, as Brillouin[9] has shown. Finally, the general solution corresponding to equation (124) reduces to the Sommerfeld solution ϕ_0 when x_b is large but $x \ll x_b$, as one can see by writing it in the form

$$\phi(x, x_b) = \frac{144}{x^3}\left\{1 + \sum_{n=1}^{\infty}\sum_r F_{nr}\left(\frac{x}{x_b}\right)^{n\lambda_r} + \sum_{n=1}^{\infty}\sum_{m=0}^{\infty}\sum_r G_{nmr}\left(\frac{x}{x_b}\right)^{7n+m\lambda_r}\right\} \tag{152}$$

Thus, no internal inconsistency appears in the treatment given.

Since Umeda's paradox depends on the assumption that ϕ_0 of Sommerfeld be valid in an infinite range of x, a viewpoint can be suggested on which it disappears. Equation (152) for $\phi(x, x_b)$ yields $\phi_0 = \phi(x, \infty)$ for $x_b \to \infty$ when x is *kept* finite. However, the demand $x_b \to \infty$ for *all* x finite is impossible, since the range of x extends up to x_b, and values of x close to x_b clearly violate the assumed condition. Thus, mathematical consistency in this procedure requires exclusion of values of the argument x of $\phi_0(x) = \phi(x, \infty)$ for which $x \to x_b (\to \infty)$, as enter in taking $\phi_0(x_b)$ as the boundary value of ϕ, unless one can show consistency with the result of the direct order of the limits in equation (152) (as is not the case). From this standpoint, Sommerfeld's solution ϕ_0 is valid only for $x \ll x_b$ when x_b is very large, and cannot be used to infer a boundary value, which is correctly given by equation (140). On this view, the inconsistency of the negative slope of ϕ_0 with the presence of vertical asymptotes has no significance. It can be noted that the boundary condition (130) demands

$$\left[\frac{d \ln \phi}{d \ln x}\right]_{x=x_b} = 1 \tag{153}$$

with which the general solution of equation (152) is consistent; Sommerfeld's solution yields the incorrect value -3 for the derivative as $x \to x_b \to \infty$, since the boundary condition is satisfied only in the sense that both sides of equation (130) vanish for the point at infinity.

Since the condition $x \ll x_b$ for x_b very large still permits a large finite value of x and a correspondingly small value of ϕ, it seems definite that no differences between predicted physical properties of free atoms should appear, whether one determines ϕ as the limit for $x_b \to \infty$ of solutions for compressed atoms, or on the other hand from the intuitive and usual procedure of requiring $\phi \to \phi_0$ for $x \to \infty$. In any event it is very unlikely that much physical significance can be attached to the predictions of the Thomas–Fermi model for both x and x_b large, since the electron density in the corresponding region of the atom becomes so low that the basic assumptions of the model tend to break down. Further, a main purpose of the work of GM was to examine the Sommerfeld limit as a means of developing accurate interpolation formulas for finite values of x_b, and there is no doubt that their solution yields the correct limit of the solutions for compressed atoms as $x_b \to \infty$.

By extension of the method used by GM to evaluate a_0 and a_1, it is possible to calculate the coefficients $a_2, a_3, \ldots$ of higher order in the expansion (140) of ϕ_b. One can note, however, that computation of a_2 requires knowledge of the coefficient e_{1-} in the expansion (136) of $E_{1-}(x_b)$, and this coefficient is not available numerically as yet; hence the numerical results of GM for the coefficients a_n represent what is possible at the present time.

The coefficients a_0, a_1, and b_1 determined in the preceding subsection were used by GM in the construction of an accurate fitted function for the boundary value of the Thomas–Fermi function, superseding that given on p. 334 from GI. Although this fitted function is used in the following subsection in the determination of accurate values of the differences in initial slopes of the Thomas–Fermi function entering the expression for the energy, it was thought desirable to defer actual discussion of the fitted function of GM to p. 363.

Initial slopes of the Thomas–Fermi function

The energies U of the subsection on p. 328 and u of the foregoing sections represent a total energy E modified by subtraction of the energy of a standard state corresponding to a neutral atom of infinite radius at zero temperature. This modification of the energy is convenient in applications, where one generally is interested only in changes in energy, and not in absolute values. Such is the case in hydrodynamic applications of the theory, as in the theoretical treatment of the explosive impact of meteorites at high velocity[35]. However, it has the consequence that the initial slope of the Thomas–Fermi function enters the expression for the energy only in a difference from the corresponding value for an infinite atom. Accordingly, the difference in question can be accurately known only if a large number of significant figures are retained in the individual slopes, in general.

As an example, one can note that the data of SK show a vanishing value of the difference in question for $x_b \geq 8\cdot6$. An effort has been made to overcome this problem in a machine calculation[75], by evaluating energies with a precision of one part in 10^7 or 10^8. However, in this particular formulation of the problem, dimensionless variables independent of atomic constants were not used. Since the basic atomic constants e, m, and h are known only to six significant figures at best, it is a moot question what a difference of two significant figures between quantities calculated to eight figures according to this procedure actually represent. In any event, later work[72] (designated by L in what follows) by this method fails to show any significant figures in the difference of initial slopes for $x_b \geq 16$.

In the present subsection, the procedure of Gilvarry for obtaining accurate values of the differences in initial slopes at zero temperature will be described, which is based essentially on integration of the pressure with respect to volume[76]. For convenience in reference, results by this method will be referred to by the designation GII, whereas

Thomas–Fermi atom model

the earlier results from a fitted function (given on p. 334) will be referred to under the designation GI. For similar reasons of brevity in reference, the numerical results of Slater and Krutter[14] will be referred to as SK, those of Feynman, Metropolis, and Teller[19] will be designated by FMT, and M will be used to refer to the results of March[65]. Correspondingly, the work of Gilvarry and March will be referred to by GM.

Theory. The total energy E of the atom can be written as

$$E = \frac{3}{7}\left(\frac{Z^2 e^2}{\mu}\phi'_i + pv\right) \tag{154}$$

with use of p of equation (64). This total energy of the atom is indefinite, in general, to the extent of an additive constant; the value of equation (154) corresponds to the choice of a total electrostatic potential which vanishes at infinity. As noted, what is important in most applications is the difference of the total energy from that corresponding to a standard state. To make this difference u always positive, the standard state is generally taken as corresponding to a neutral atom of infinite radius at zero temperature, for which the energy E_∞ is given by

$$E_\infty = \frac{3}{7}\frac{Z^2 e^2}{\mu}\phi'_{i,\infty} \tag{155}$$

in terms of the corresponding initial slope $\phi'_{i,\infty}$ (from equation (154) with $pv = 0$). The energy difference u defined by

$$u = E - E_\infty \tag{156}$$

becomes

$$u = u' + \frac{3}{7}pv \tag{157}$$

accordingly, where u' fixed by

$$u' = \frac{3}{7}\frac{Z^2 e^2}{\mu}(\phi'_i - \phi'_{i,\infty}) \tag{158}$$

depends only on initial slopes.

The standard method of solution of the Thomas–Fermi equation is to assume an initial slope ϕ'_i, to determine ϕ for x small by means of Baker's series, and to continue the solution numerically to find the point x_b and boundary value ϕ_b meeting equation (60b). For small values of x_b, the initial slope ϕ'_i differs sufficiently from $\phi'_{i,\infty}$ that this procedure can yield an accurate value of the difference $\phi'_i - \phi'_{i,\infty}$ appearing in equation (158) for the energy u'. For large values of x_b, and for most of the intermediate range of this variable, ϕ'_i is so close numerically to $\phi'_{i,\infty}$ that most of the significant figures are lost in the difference. Over the range $x_b \simeq 4$ to $x_b \to \infty$, the first three figures of ϕ'_i coincide with those for $\phi'_{i,\infty}$, and hence computations carried to five figures can yield only two significant figures in the part u' of the energy depending on initial slopes.

If E_{pot} is the total potential energy corresponding to the choice of a total electrostatic potential which vanishes at infinity, one can define a potential energy u_{pot} analogous to u of equation (156) by

$$u_{pot} = E_{pot} - E_{pot,\infty} \tag{159}$$

where

$$E_{pot,\infty} = \frac{6}{7}\frac{Z^2 e^2}{\mu}\phi'_{i,\infty} \tag{160}$$

represents the total potential energy corresponding to an infinite atom. Similarly, a kinetic energy u_{kin} can be defined by

$$u_{kin} = E_{kin} - E_{kin,\infty} \tag{161}$$

in which

$$E_{kin,\infty} = -\frac{3}{7}\frac{Z^2 e^2}{\mu}\phi'_{i,\infty} \tag{162}$$

corresponds to the infinite atom. By means of the virial theorem, one can show that

$$u' = \frac{5}{7}u + \frac{1}{7}u_{pot} = \frac{6}{7}u - \frac{1}{7}u_{kin} \tag{163}$$

and hence u' can be evaluated if any two of the three energies u, u_{pot}, and u_{kin} are known.

To assess the effect on u of the uncertainty in u', the ratio u'/u can be expressed as

$$\frac{u'}{u} = \frac{3}{7}\left(\frac{10}{3} - \gamma_0\right) \tag{164}$$

where the parameter γ_0 is defined by

$$u = \frac{pv}{\gamma_0 - 1} \tag{165}$$

Since $\gamma_0 \to 10/3$ for $x_b \to \infty$, the term in the initial slopes makes a vanishing contribution to the energy in this case. The contribution is a maximum for $x_b \to 0$, in which case $\gamma_0 \to 5/3$ and $u'/u = 5/7$. For $x_b \simeq 4$, however, γ_0 is about 2, and u' is of the order of one-half of the total energy. Accordingly, a range of intermediate values of x_b exists over which the accuracy in the energy is limited by the uncertainty in $\phi'_i - \phi'_{i,\infty}$. In the present context, the range of intermediate values of x_b will be defined as $5 \lesssim x_b \lesssim 50$, roughly; a value of x_b satisfying $x_b \lesssim 5$ will be referred to as small, and a value such that $x_b \gtrsim 50$ will be viewed as large.

It should be emphasized that the effect of u' in contributing to the energy u becomes small when x_b becomes sufficiently large to approach infinity, and the significant effect of u' occurs for intermediate values of x_b. To see this result, consider the expansions

$$\phi_b = x_b^{-3}\left(a_0 + \sum_{n=1}^{\infty} a_n x_b^{-n\lambda_2}\right) \tag{166}$$

$$\phi'_i - \phi'_{i,\infty} = x_b^{-7} \sum_{n=1}^{\infty} b_n x_b^{-n\lambda_2} \tag{167}$$

Thomas–Fermi atom model

of ϕ_b and $\phi'_i - \phi'_{i,\infty}$, respectively, given by GM (see p. 349) for x_b large. The parameter λ_2 is fixed by

$$\lambda_2 = \tfrac{1}{2}(73^{1/2} - 7) \tag{168}$$

and relatively accurate values of the leading coefficients a_0, a_1, and b_1 are given in table 7 from results of GM. From these expansions, one calculates

$$\frac{u'}{u} = \frac{15}{2}\frac{b_1}{a_0^{5/2}} x_b^{-\lambda_2} \tag{169}$$

for x_b large. Hence, one has $u'/u \to 0$ for $x_b \to \infty$, in agreement with equation (164). However, λ_2 is 0·772 numerically and hence equation (169) corresponds to a relatively slow decrease of u' with increasing x_b.

Values of the difference $\phi'_i - \phi'_{i,\infty}$ comparable in accuracy with values of ϕ_b can be obtained by integrating the relation

$$\frac{d\phi'_i}{dx_b} = -\frac{\phi_b^{5/2}}{x_b^{1/2}}\left(1 + \frac{1}{3}\frac{d\ln\phi_b}{d\ln x_b}\right) \tag{170}$$

in GI (see p. 332). Accurate integration of equation (170) requires an interpolation procedure, since the relatively accurate values of ϕ_b available in L fall at uneven intervals of x_b. In the work of GM, these values of ϕ_b were fitted by a function φ_b defined by

$$\varphi_b = \left(\sum_n A_n x_b^n\right)^{-1} \tag{171}$$

where n ranges over the sequence

$$n = 1, 2, n', 3 - \lambda_2, 3 \tag{172}$$

The form (171) shows the proper limiting behavior of ϕ_b through the two leading terms in both limits $x_b \to 0$ and $x_b \to \infty$, and only one term (that in $x_b^{n'}$) was not obtained theoretically. The term $A_{n'} x_b^{n'}$ was evaluated semiempirically by a least-squares fit to the values of ϕ_b over the intermediate range of x_b, from data of L. Values of the coefficients and exponents of GM are given on p. 363.

The numerical data of L for ϕ_b are given to five significant figures. These data are reproduced by the fitted function of GM (see p. 363) within an accuracy at least 0·1% in general. This accuracy clearly is not enough to achieve an accuracy in $\phi'_i - \phi'_{i,\infty}$ comparable with that in ϕ_b as tabulated. Hence, an interpolating formula for ϕ_b was written in the form

$$\phi_b = \varphi_b\{1 - \delta(x_b)\} \tag{173}$$

where φ_b has the analytical form (171) and δ represents a small correction. For a reason explained later, the semiempirical term $A_{n'} x_b^{n'}$ appearing in equation (171) was changed slightly from the result of GM; the exponent n' and coefficient $A_{n'}$ used in GII are tabulated in table 11.

TABLE 11 Values of the constants appearing in the functions of equations (171), (174), and (175) [from GII]

Equation	Constant	Value	Constant	Value
(171)	n'	2·09290	$A_{n'}$	−0·0333550
(174)	v	2·07295	α	0·518461
(174)	D	0·626200		
(175)	X	16·4778	β	0·136575

The analytical form of the function δ in equation (173) was chosen as

$$\delta = Dz \exp\{-\alpha(z^2)^{1+v}\} \quad (174)$$

in which

$$z = \ln\left(\frac{x_b}{X}\right)\left\{1 - \beta \ln\left(\frac{x_b}{X}\right)\right\} \quad (175)$$

and D, α, v, X, and β represent disposable constants. The values of the parameters selected are given in table 11. On the basis of these tabulated values, the fitted function of equation (173) reproduces the values of L for ϕ_b within five significant figures over a range of x_b from about 5 to 50, as figure 6 indicates. Over the range $x_b \simeq 1$ to

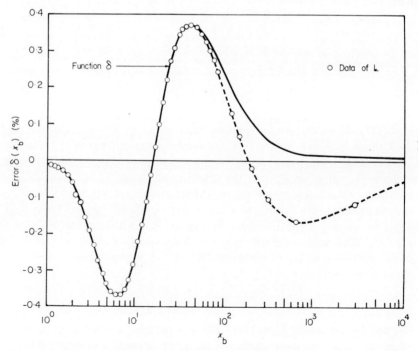

FIGURE 6 The correction factor $\delta(x)_b$ versus x_b applied to the function for ϕ_b of GM (as modified) in the integration to obtain $\phi'_i - \phi'_{i,\infty}$ as a function of atom radius x_b [from GII]

Thomas–Fermi atom model 359

$x_b \simeq 5$, discrepancies of a few units in the fifth significant figure appear in the agreement for some cases, but these are explainable partly by a lack of smoothness in the results of L in this range. The function δ of equation (174) is the product of an odd by an even function of z, and thus is symmetric about the origin in the (z, δ) plane, where δ vanishes when x_b has a value equal to the parameter X. Thus, the minimum of δ is the negative of its maximum, occurring for $z^2 = \alpha^{-1/(1+\nu)}$. To make the data for $\delta = 1 - \phi_b/\varphi_b$ show the same behavior, the parameters n' and $A_{n'}$ in the function φ_b of equation (171) were assigned the values in GII shown in the first line of table 11, rather than those of p. 363 from GM. Otherwise, the constants of equation (171) were taken as tabulated by GM. The resulting symmetry of δ as a function of z is obvious from figure 6.

For the change $\Delta_n \phi_i'$ in ϕ_i' over the interval $x_{b,n}$ to $x_{b,n+1}$ of x_b, one obtains

$$\Delta_n \phi_i' = \frac{14}{15} \int_{x_{b,n}}^{x_{b,n+1}} \frac{\phi_b^{5/2}}{x_b^{1/2}} \, dx_b - \frac{2}{15} \Delta_n(\phi_b^{5/2} x_b^{1/2}) \tag{176}$$

from equation (170), where the operator Δ_n implies the difference between values of the operand at $x_{b,n+1}$ and $x_{b,n}$. For purposes of the numerical integration, it is convenient to apply the transformation

$$y = x_b^{-7} \tag{177}$$

Equation (176) becomes

$$\Delta_n \phi_i' = \frac{2}{15} \{I_n - \Delta_n(\phi_b^{5/2} x_b^{1/2})\} \tag{178}$$

for the change $\Delta_n \phi_i'$ in the interval y_n to y_{n+1}, where

$$I_n = \int_{y_n}^{y_{n+1}} \frac{\phi_b^{5/2}}{y^{15/14}} \, dy \tag{179}$$

Using the series of equation (166), one finds that the integrand of I_n approaches a constant value $a_0^{5/2}$ as $x_b \to \infty$, or $y \to 0$. For purposes of integration by Simpson's rule, the integrand of I_n therefore has the desirable feature of tending asymptotically to a constant in the region of intermediate and large values of x_b where one wishes to attain the highest accuracy.

With values of $\Delta_n \phi_i'$ available for contiguous intervals extending from x_b to infinity, the corresponding value of $\phi_i' - \phi_{i,\infty}'$ is given by the cumulation

$$\phi_i' - \phi_{i,\infty}' = \sum_n \Delta_n \phi_i' \tag{180}$$

from infinity to the value of x_b in question, or from $y = 0$ to $y = x_b^{-7}$.

The fitted function represented by equations (173), (171), and (174) should have a wider field of application than the use made here as an interpolating means in integration, whenever it is required to reproduce values of the Thomas–Fermi function ϕ_b to five significant figures for values of x_b (up to about 50) not covered by available tabulations.

Initial slopes and parameters. Previous determinations of the differences $\phi_i' - \phi_{i,\infty}'$ of initial slopes from results of L, SK, FMT, and M are available for 24 values of x_b, as given in the first column of table 12, spanning a range of x_b from about unity to

TABLE 12 Comparison of results from GII for the initial slope $\phi'_i - \phi'_{i,\infty}$ with those of L, SK, FMT, and M, as a function of atom radius x_b

x_b	GII	L	SK	FMT	M
1·190	0·5853$_6$	0·5855	0·59		
1·690	0·2066$_4$	0·2066	0·21		
2·200	0·08814$_4$	0·0882	0·09		
2·800	0·03807$_7$	0·0381	0·04		0·04
3·043	0·02808$_9$	0·0280			0·03
3·285	0·02109$_0$	0·0210			0·021
3·704	0·01326$_9$	0·0135			0·0133
4·230	0·0^{2}7790$_2$	0·0^{2}78	0·01		
4·330	0·0^{2}7077$_6$	0·0^{2}72			0·0^{2}7
5·229	0·0^{2}3182$_0$	0·0^{2}32			0·0^{2}32
5·401	0·0^{2}2761$_8$	0·0^{2}28			0·0^{2}28
5·850	0·0^{2}1936	0·0^{2}19	0·0^{2}2		
6·206	0·0^{2}1481$_1$	0·0^{2}15			0·0^{2}15
7·385	0·0^{3}6557$_4$	0·0^{3}7		0·0^{3}69	
7·790	0·0^{3}5067$_1$	0·0^{3}5			0·0^{3}51
8·015	0·0^{3}4409$_6$	0·0^{3}5			0·0^{3}43
8·588	0·0^{3}3134$_6$	0·0^{3}5		0·0^{3}33	
8·590	0·0^{3}3130$_9$	0·0^{3}5	0·0^3		
9·565	0·0^{3}1818$_4$	0·0^{3}2		0·0^{3}19	
10·804	0·0^{4}9655$_8$	0·0^{3}1		0·0^{3}10	
11·300	0·0^{4}7612$_2$	0·0^{3}1	0·0^{4}5		
11·963	0·0^{4}5607$_6$	0·0^{3}1		0·0^{4}5	
15·870	0·0^{4}1165$_3$	0·0^4		0·0^{4}1	
16·000	0·0^{4}1112$_1$	0·0^4			

$x_b = 16$. To facilitate direct comparison of the results with prior ones, the basic intervals of integration were selected in GII as the 23 corresponding to these values of x_b up to $x_b = 16$. Within each such interval, values of I_n of equation (179) were obtained by repeated use of Simpson's rule from values of ϕ_b computed from the fitted function of equation (173). In the successive integrations, the subinterval between each set of two consecutive points in the prior integration was halved until the value of I_n became stationary in the sixth significant figure. The computation required from three to seven repetitions in individual cases, with an average number of five, corresponding to a grid on each interval containing 8 to 128 points, as needed. This procedure ensured that maximum advantage be taken of the accuracy of the fitted function.

The results for $\phi'_i - \phi'_{i,\infty}$ from GII for the 24 values of x_b in question are shown in the second column of table 12. It is felt that these values almost certainly are accurate when rounded to four figures from the five shown. For comparison, values from L are given in the third column. One sees that less than four significant figures appear for $x_b > 2$, and only one (or none) appears for the larger values of x_b. For the region of x_b between 3 and 4, discrepancies appear in the third figure, when referred to rounded values from the present work. Similar discrepancies appear when the values from L contain only two or one significant figure. Results of SK, FMT, and M are shown also, for comparison. The results of M were obtained by taking $\phi'_{i,\infty}$ to correspond

Thomas–Fermi atom model

to the largest value of x_b appearing in his work; in other cases, values of $\phi'_{i,\infty}$ given by the authors in question were used in deducing values of $\phi'_i - \phi'_{i,\infty}$. For the cases at larger x_b where results of L are discrepant from those given here, one notes that values of SK, FMT, or M frequently are in better agreement with the present ones.

In addition to the values of the differences of initial slopes shown in table 12, 11 further values were obtained for higher values of x_b extending up to $x_b = 50$. In these cases, the subinterval of integration was halved until the difference of initial slopes became stationary in the fourth figure. Thus, accuracy can be guaranteed only for three figures. A tabulation is not shown, since the values to three figures are represented adequately by the fitted function for $\phi'_i - \phi'_{i,\infty}$, to be discussed.

In the construction of a fitted function for $\phi'_i - \phi'_{i,\infty}$, accuracy at large values of x_b obviously is enhanced by evaluating theoretically as many terms in this limit as is possible. As noted on p. 354, the method of GM of direct calculation from the Thomas–Fermi equation of the coefficients a_n and b_n in the asymptotic series of equations (166) and (167), respectively, can yield only the values of a_0, a_1, and b_1. In the work of GII, approximate values of b_2 and a_2 were obtained by a less elegant but quite direct method, requiring only knowledge of the coefficients a_0, a_1, and b_1 of lower order and independent of requirements on boundary parameters fixed by the behavior of the Thomas–Fermi function at the origin.

Define a function f of x_b by

$$f(x_b) = x_b^{\lambda_2}\{x_b^{\lambda_1}(\phi'_i - \phi'_{i,\infty}) - b_1\} \tag{181}$$

where

$$\lambda_1 = \tfrac{1}{2}(73^{1/2} + 7) \tag{182}$$

With use of equation (167), one finds $f \to b_2$ as $x_b \to \infty$. Now, b_1 in table 7 as given by GM is positive and is known to an accuracy of ± 2 in the third significant figure. Since b_1 enters a difference in the curly brackets of equation (181), no gain is obtained by computing values of $\phi'_i - \phi'_{i,\infty}$ to better than this accuracy in obtaining b_2 from the expression. However, the fitted function of equation (171) is accurate to within five figures out to $x_b \simeq 50$, and from figure 6 one infers that it should be reliable within about three figures out to $x_b \simeq 200$. Beyond this point, the error can be estimated with sufficient accuracy from figure 6 and taken into account as needed in the numerical integration of equation (179). From values of $\phi'_i - \phi'_{i,\infty}$ computed in this manner out to $x_b \simeq 1000$, results for $-f$ are shown as a function of x_b in figure 7 (on linear scales in the inset). One notes that it approaches a constant for $x_b \geq 400$, which yields the entry for b_2 given in table 13. The corresponding value of a_2 in the expansion of equation (16) for ϕ_b can be obtained by means of equation (170) as

$$a_2 = -\tfrac{3}{2}a_0^{-3/2}\left\{\left(1 + \frac{\lambda_1}{\lambda_2}\right)b_2 + \tfrac{1}{2}a_0^{1/2}a_1^2\right\} \tag{183}$$

Using the values of GM in table 7 for a_0 and a_1, one obtains the figure for a_2 given in table 13.

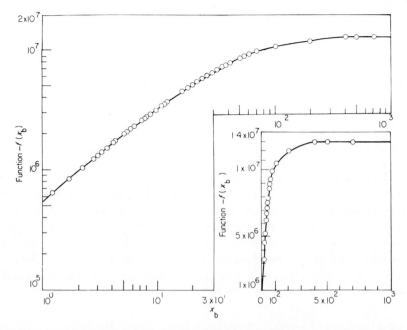

FIGURE 7   The function $-f(x_b)$ versus x_b which becomes equal to $-b_2$ asymptotically for large atom radius x_b. Scales are logarithmic in the main diagram and linear in the inset. [From GII]

TABLE 13   Values of computed coefficients in initial and boundary parameters from GII, compared with values from the Sauvenier solution

Coefficient	From Sauvenier solution	From GII
b_2	$-1 \cdot 14 \times 10^7$	$-1 \cdot 3 \times 10^7$
a_2	$3 \cdot 27 \times 10^4$	$1 \cdot 3 \times 10^4$

An asymptotic solution of the Thomas–Fermi equation in the limit $x, x_b \to \infty$ for an infinite atom has been given by Sauvenier[62] as equation (75). In the work of GM, it was found that this analytic solution yielded remarkably good results for a_0, a_1, and b_1, in spite of its simplicity. The value of a_2 implied by this asymptotic form can be found by expansion of the corresponding ϕ_b in powers of $x_b^{\lambda_2}$ and comparison with equation (166) as

$$a_2 = \frac{2}{81}(9132 + 12019\lambda_1)12^{4\lambda_2/3} \tag{184}$$

and the corresponding value of b_2 follows as

$$b_2 = -\frac{128}{243}(754944 + 977299\lambda_1)(3 + 2\lambda_1)^{5/2}12^{4\lambda_2/3} \tag{185}$$

Thomas–Fermi atom model

from equation (170) or (183). Numerical values from these analytic results are shown in table 13 for comparison with those calculated directly. One sees that agreement between values of b_2 is excellent. The somewhat poorer agreement in the case of the coefficient a_2 may arise because a_2 from the Sauvenier solution is fixed by equation (183) as a difference of order 10^4 between two positive terms of order 10^7; in such a case, however, the excellent agreement between values of b_2 should be partly fortuitous, in view of the method of derivation. In any event, the conclusion stands that the Sauvenier form provides a reasonable approximation, even for the coefficients of relatively high order under discussion here.

Accurate fitted functions

In this subsection, the theoretical development of GM (as described on p. 341 ff.) for large atom radius will be conjoined with that of GI (p. 332 ff.) and of M (p. 337 ff.) to construct a highly accurate fitted function for the boundary value of the Thomas–Fermi function. Similarly, the accurate numerical results of GII (p. 360) for the differences in initial slopes will be used to construct a fitted function of correspondingly high accuracy for this parameter. The results given here supersede the ones given briefly on p. 334.

Boundary values. The values obtained for the parameters a_0 and a_1 by GM (pp. 351–2) were used by them in the construction of an improved fitted function for the Thomas–Fermi boundary value, based on relatively accurate numerical values[75] of ϕ_b which have become available since the original work of GI and M. The form selected for the fitted function is

$$\phi_b = \left(\sum_n A_n x_b^n\right)^{-1} \tag{186}$$

where n ranges over the sequence

$$n = 1, 2, n', 3-\lambda_2, 3 \tag{187}$$

and the disposable exponent n' is required to satisfy

$$2 < n' < 3 - \lambda_2 \tag{188}$$

It follows from equation (140) that the correct asymptotic form of ϕ_b through the two leading terms in x_b as $x_b \to \infty$ is given by equation (186) if one takes

$$A_3 = a_0^{-1}, \qquad A_{3-\lambda_2} = -a_0^{-2} a_1 \tag{189}$$

From the results of GI and of M, equation (186) yields also the correct asymptotic form of ϕ_b through the two leading powers of x_b for $x_b \to 0$, provided that

$$A_1 = 3^{-2/3}, \qquad A_2 = \frac{3^{-1/3}}{10} \tag{190}$$

as specified by equation (91).

That inclusion in equation (186) of one empirical term $A_{n'} x_b^{n'}$ is adequate to fit the numerical data accurately can be seen from the logarithmic plot of figure 8. The data points shown represent values from the numerical data[75] of the function

$$-g(x_b) = \{\phi(x_b)\}^{-1} - {\sum_n}' A_n x_b^n \tag{191}$$

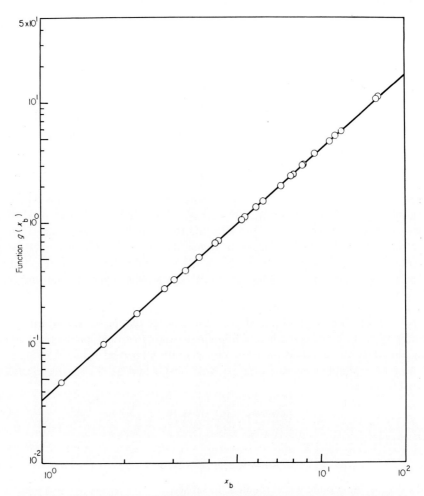

FIGURE 8 The function $g(x_b)$ against x_b, as evaluated from the numerical data. The straight line represents a least-squares fit to the data points shown. Scales are logarithmic. [From GM]

where the prime on the summation sign indicates that n runs through the sequence (187) exclusive of n'. The straight line $g(x_b) = A_{n'} x_b^{n'}$ of figure 8 corresponds to a least-squares fit of the data to evaluate n' and $A_{n'}$. The value of $A_{n'}$ is tabulated in table 14 against n', which satisfies inequality (188); values of the other coefficients from equations (189) and (190) are given also. With these coefficients to the number of significant figures shown, the fitted function of equation (186) reproduces the numerical data to an accuracy of at least 0·1%, in general. This figure can be compared with an accuracy of better than one per cent (relative to the data used) achieved by M with a form containing one adjustable coefficient corresponding to an exponent n selected as $\frac{5}{2}$; for the form in GI, the figure is about 1·5% in general (relative to the data fitted). These functions contain three and two theoretically determined terms, respectively, with the coefficient of x_b^3 fixed by the value of equation (148) from the Sauvenier solution. With respect to the more accurate data on which equation (186) is based,

TABLE 14 Coefficients[a] A_n in the fitted function for ϕ_b [from GM]

n	A_n
1	$4{\cdot}807 \times 10^{-1}$
2	$6{\cdot}934 \times 10^{-2}$
$n' = 2{\cdot}093$	$-3{\cdot}293 \times 10^{-2}$
$3 - \lambda_2 = 2{\cdot}228$	$4{\cdot}20_0 \times 10^{-2}$[a]
3	$3{\cdot}480 \times 10^{-3}$

[a] The coefficients are given to four figures to minimize round-off error.

the fitted functions of both M and GI show an error which is at most 1·3% and is about 0·5% on the average.

While it has proved possible here to achieve an accuracy of 0·1% in a fitted function for numerical values of the Thomas–Fermi boundary value ϕ_b, the error corresponding to the function given originally in GI is not large. Hence, the conclusions drawn by use of the latter function on thermodynamic functions of the lattice and the course of the fusion curve at high pressure by Gilvarry[16] and on the nature of meteorite impacts by Gilvarry and Hill[35] retain qualitative and semiquantitative validity, within the limitations of the physical models employed in these treatments.

The fitted function of equation (186) yields the equation of state

$$p^{2/5} \sum_n A_n \left(\frac{3v}{4\pi\mu^3}\right)^{(n+1)/3} = \left(\frac{Z^2 e^2}{10\pi\mu^4}\right)^{2/5} \quad (192)$$

for the pressure p in terms of the atom volume $v = (4\pi/3)\mu^3 x_b^3$. The scaled pressure from this equation is shown as a function of the scaled volume in figure 9, for comparison with results as evaluated directly from the numerical data. The directly calculated pressures are reproduced within an accuracy of 0·3%, in general. Hence, figure 9 does not show the small deviations between direct and fitted values which appear in the corresponding figure 4 from GI.

Because of the higher accuracy of fit to the value of ϕ_b, one expects equation (186) to yield a more accurate value for $d\phi_b/dx_b$ than do the previous fitted functions. This derivative appears in the expression

$$\varepsilon_0 = \frac{5}{6}\left(1 - \frac{d \ln \phi_b}{d \ln x_b}\right) \quad (193)$$

for the differential parameter

$$\varepsilon_0 = -\frac{d \ln p}{d \ln v} \quad (194)$$

associated with the equation of state. The value of ε_0 from equation (186) is compared in figure 10 with the value from the function of M for ϕ_b. It is seen that a significant difference exists between the two, particularly for x_b large (the difference is actually somewhat less for all values of x_b in the case of the function given in GI). One can note that values of $d\phi_b/dx_b$ and thus of ε_0 are not immediately available from the numerical data except through some such means as a fitted function.

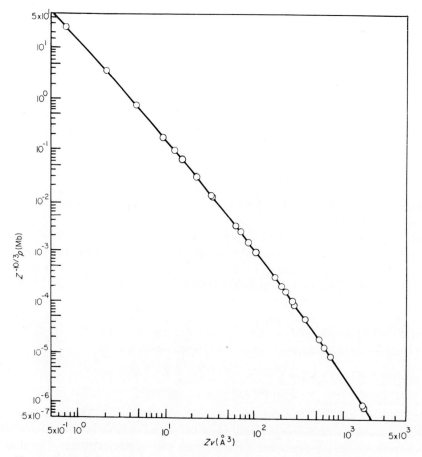

FIGURE 9 Scaled pressure from fitted function for ϕ_b, against scaled volume in cubic ångströms, for comparison with data points as computed directly [from GM]

It is rather remarkable that the numerical values of the Thomas–Fermi boundary value ϕ_b can be reproduced so accurately by an expression in which only one term is not obtained theoretically.

Initial slopes. The values available for b_n from the results of GII as presented on p. 361 and from the results of GM on p. 349, in conjunction with values of the corresponding coefficients from GI and M were used in GII in the construction of a fitted function for $\phi'_i - \phi'_{i,\infty}$ based on the data of table 12. The form selected for the fitted function is

$$\phi'_i - \phi'_{i,\infty} = (\sum_n B_n x_b^n)^{-1} \tag{195}$$

where n ranges over the sequence

$$n = 2, 3, 4, n', 5, 6, 7, \lambda_1 \tag{196}$$

with the exponent n' and the coefficient $B_{n'}$ disposable. It follows from equation (167)

Thomas–Fermi atom model

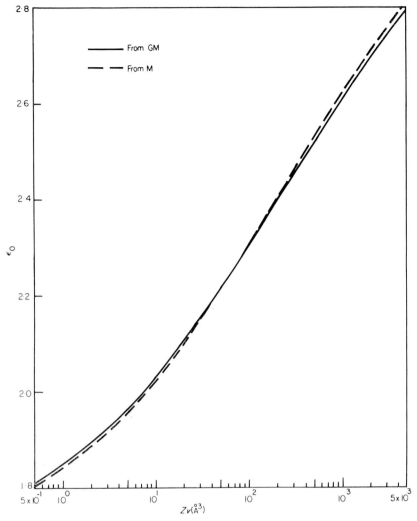

FIGURE 10 Parameter ε_0 from fitted function for ϕ_b against scaled volume in cubic ångströms, for comparison with corresponding result from fitted function of M [from GM]

that the correct asymptotic form of $\phi'_i - \phi'_{i,\infty}$ through the two leading terms in x_b as $x_b \to \infty$ is given by equation (195) if one takes

$$B_{\lambda_1} = b_1^{-1}, \qquad B_7 = -\frac{b_2}{b_1^2} \tag{197}$$

in terms of coefficients available from results of GM and of GII, as gven in tables 7 and 13. Equation (195) yields also the correct asymptotic form of $\phi'_i - \phi'_{i,\infty}$ through the two leading powers of x_b for $x_b \to 0$, provided that

$$B_2 = 3^{-2/3}, \qquad B_3 = \frac{1}{5} 3^{2/3} \tag{198}$$

from results of GI and M, respectively. This procedure fixes four of the eight terms in the parenthetic sum of equation (195) theoretically, and leaves four to be evaluated semiempirically from numerical data. To determine the latter, n' was provisionally assigned the exact value 4·5, to which preliminary analysis indicated it was closely equal, and the coefficients B_4, B_5, and B_6 were obtained by a least-squares fit to the data on this basis. With these coefficients fixed in this manner, a function $h(x_b)$ was constructed, defined by

$$h(x_b) = (\phi'_i - \phi'_{i,\infty})^{-1} - {\sum_n}' B_n x_b^n \qquad (199)$$

where the prime on the summation sign indicates that n runs through the sequence (196) exclusive of n'. The straight line $h(x_b) = B_{n'} x_b^{n'}$ of figure 11 corresponds to a least-squares fit of the data to evaluate n' and $B_{n'}$ in this manner from the data of

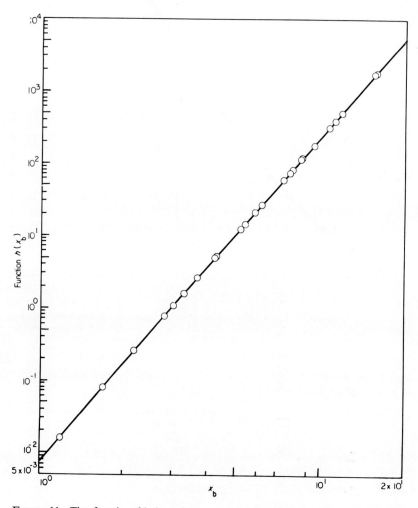

FIGURE 11 The function $h(x_b)$ against x_b, as evaluated from the numerical data. The straight line represents a least-squares fit to the data points shown. Scales are logarithmic. [From GII]

table 12. The value of $B_{n'}$ is tabulated in table 15 against n' with the corresponding values of B_4, B_5, and B_6; values of the theoretically available coefficients from equations (197) and (198) are given also.

TABLE 15   Coefficients[a] B_n in the fitted function for $\phi'_i - \phi'_{i,\infty}$ [from GII]

n	B_n
2	4.8075×10^{-1}
3	4.1602×10^{-1}
4	1.1052×10^{-1}
$n' = 4.5058$	3.2589×10^{-2}
5	1.0578×10^{-2}
6	2.7271×10^{-3}
7	$4.1_{500} \times 10^{-5}$
$\lambda_1 = 7.7720$	$1.78_{00} \times 10^{-6}$

[a] The coefficients are given to five figures to minimize round-off error.

With the coefficients of table 15 to the number of significant figures shown, the fitted function of equation (195) reproduces the numerical results of the second column of table 12 within an accuracy of 0.2% in all cases, with an average accuracy of 0.1%. This figure can be compared with an accuracy of about 1.5%, relative to the data of SK and FMT, obtained previously in the fitted function of GI (p. 334). This function contained two theoretically evaluated terms, with the coefficient B_{λ_1} determined from the value of b_1 obtained from Sauvenier's asymptotic solution of the Thomas–Fermi equation. The two fitted functions are compared in figure 12 with the 35 values of $\phi'_i - \phi'_{i,\infty}$ available from the data of GII; one notes that the prior result shows discrepancies in the intermediate region of x_b from $x_b \simeq 5$ onwards. The discrepancy tends to become serious as one approaches the boundary $x_b \simeq 50$ of the intermediate range of x_b, in conformity with the discussion on p. 354. On the other hand, the small error in the fitted function of the present work decreases as x_b increases.

The fitted function for $\phi'_i - \phi'_{i,\infty}$ constructed in GII yields the explicit equation

$$u' \sum_n B_n \left(\frac{3v}{4\pi\mu^3} \right)^{n/3} = \frac{3}{7} \frac{Z^2 e^2}{\mu} \qquad (200)$$

with n ranging over the sequence (196), for the energy u' of equation (157) in terms of the atom volume v. This equation provides values of u' agreeing within 0.2% in all cases with numerical results from GII, with an average accuracy of 0.1%. The fitted function for ϕ_b from GII yields the corresponding equation of state as

$$p^{2/5} \sum_n A_n \left(\frac{3v}{4\pi\mu^3} \right)^{(n+1)/3} = \frac{Z^2 e^2}{10\pi\mu^4}(1-\delta) \qquad (201)$$

where n ranges over the sequence (172) with n' (and $A_{n'}$) given in table 11. When the parameter δ is included as shown, the error in p can be kept far below a level of 0.1% in the range of small and intermediate values of x_b; for this purpose, values of δ

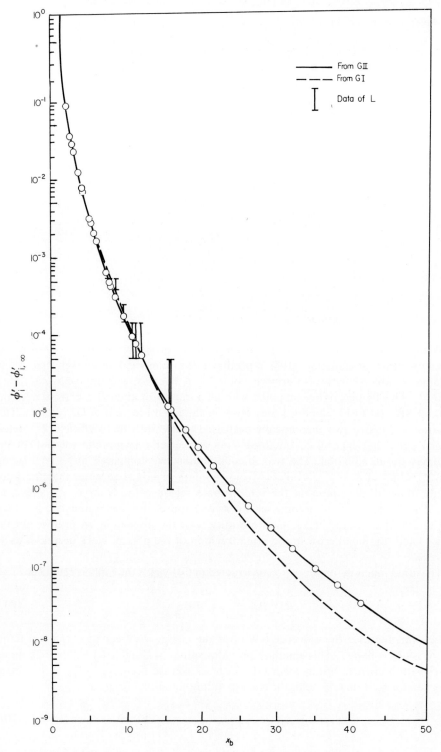

FIGURE 12 Fitted function for $\phi'_i - \phi'_{i,\infty}$ from GII against x_b, for comparison with data points and with fitted function (broken) given in GI. Vertical bars indicate error in values from L. [From GII]

can simply be read from the graph of figure 6. It follows that the total energy u of equation (157), computed with use of equations (200) and (201) for u' and p respectively, agrees with directly computed values within an average accuracy of 0·1% for values of x_b up to 50. The error tends to be less than this average for small x_b, for values of x_b in the higher region of the intermediate range, and for x_b large, in all cases susceptible to check. In the regime of large x_b not subject to direct check, it is felt that the accuracy is at least 1% throughout the range to infinity. If the fitted function of GM for ϕ_b is used to compute p, accuracy in u can be guaranteed only within a limit of 0·3%, in general.

The ratio u'/u computed by means of equations (200) and (201) is shown in figure 13 as a function of the scaled atom volume Zv. Data points obtained with use of the 35 values of $\phi'_i - \phi'_{i,\infty}$ from GII and of ϕ_b from L appear also. The results of using the

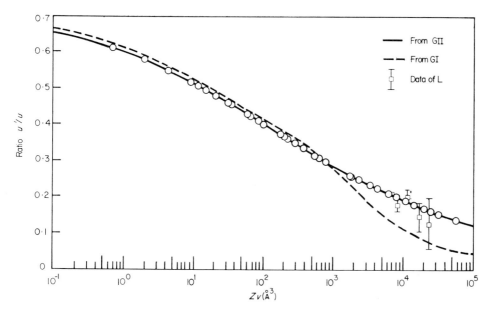

FIGURE 13 The ratio u'/u against scaled volume in cubic ångströms, from the fitted functions of GII for $\phi'_i - \phi'_{i,\infty}$ and of GM for ϕ_b, for comparison with the data points and with results (broken) from the fitted functions of GI. Vertical bars indicate errors in values from L. [From GII]

original fitted functions of GI for $\phi'_i - \phi'_{i,\infty}$ and ϕ_b are shown broken, for comparison; one sees that significant error sets in for the larger values of Zv, precisely as the theoretical discussion on p. 355 would lead one to expect. In analogous fashion figure 14 shows the scaled total energy $Z^{-7/3}u$ as a function of scaled atom volume. Results from the original fitted functions of GI show better agreement with the present ones in the case of u than is true for the ratio u'/u. The reason is the decreased pejorative effect on u of inaccuracy in u' for the same value of Zv, because of the presence of the term in pv in equation (157).

The problem of accuracy in the energy raised on p. 355 was attacked in L in a fashion different from that used here. Instead of bypassing the problem for the inter-

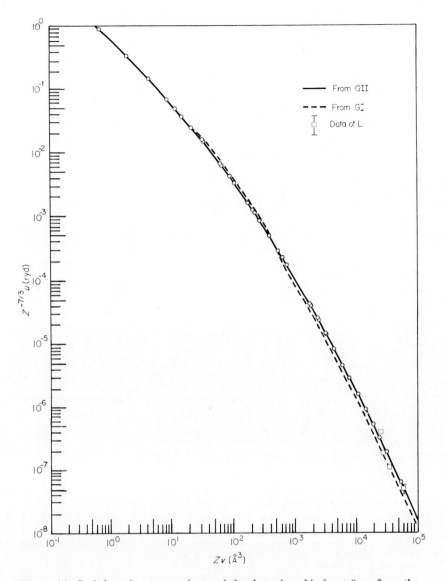

FIGURE 14 Scaled total energy against scaled volume in cubic ångströms, from the fitted functions of GII for $\phi'_i - \phi'_{i,\infty}$ and of GM for ϕ_b, for comparison with the data points and with results (broken) from the fitted functions of GI. Vertical bars indicate errors in values from L. [From GII]

mediate range of x_b by obtaining accurate values of the energy by integration of the pressure, individual integrals involved in the potential energy E_{pot} and kinetic energy E_{kin} were evaluated directly to seven or eight significant figures, and then u_{pot} and u_{kin} as defined by equations (159) and (161) were computed. The actual quantities tabulated in L are u_{pot} and u_{kin}. However, u is simply their sum and the energy u' can be obtained by use of equations (163).

However, it follows from the discussion on p. 356 that all accuracy in the values of

Thomas–Fermi atom model

u' and u computed by the procedure of L eventually must be lost as the atom radius x_b or scaled volume Zv increases. To determine the corresponding limiting values, table 16 shows values of u' and u from results of L for comparison with values implied by equations (200) and (201). One sees that all significant figures in u' disappear at $x_b \simeq 40$, and in the case of u vanish at a value of x_b only slightly exceeding 50 (and thus at a point lying effectively in the range of intermediate values of x_b). However, when sufficient figures exist in the results of L to permit rounding to three figures, agreement with values from the present work is essentially exact, as the first line of table 16 indicates. Limits of error corresponding to results of L are shown in figures 13 and 14 for u'/u and u, respectively, when they are significant on the scale used; the large limits arise from lack of significant figures.

TABLE 16 Comparison of the energies u' and u from GII with those of L

x_b	Zv ($\times 10^3$ Å^3)	$Z^{-7/3}u'$ (L) ($\times 10^{-16}$ erg)	$Z^{-7/3}u'$ (GII) ($\times 10^{-16}$ erg)	$Z^{-7/3}u$ (L) ($\times 10^{-16}$ erg)	$Z^{-7/3}u$ (GII) ($\times 10^{-16}$ erg)
17·457	2·2916	1·42	1·42	5·75	5·76
19·416	3·1525	0·76	0·763	3·27	3·27
21·456	4·2546	0·42	0·421	1·90	1·90
24·123	6·0462	0·21	0·207	1·00	0·999
27·224	8·6907	0·09	0·0983	0·50	0·507
29·898	11·511	0·06	0·0547	0·30	0·297
34·035	16·982	0·02	0·0240	0·14	0·141
37·666	23·018	0·01	0·0125	0·08	0·0776
40·391	28·384	0·00	0·00793	0·05	0·0513
45·028	39·324	0·00	0·00388	0·03	0·0267
52·491	62·299	0·00	0·00139	0·01	0·0105
59·338	89·994	0·00	0·000608	0·00	0·00500

The thermodynamic parameter γ_0 of equation (70) is important for hydrodynamic and other applications of the model. Its value as a function of scaled atom volume is shown in figure 15, as determined from equations (200) and (201), and as fixed directly by the values of $\phi'_i - \phi'_{i,\infty}$ available from the present work and the values of L for ϕ_b. The points corresponding to use of u' and u of L from table 16 indicate a large error in γ_0 for the larger values of Zv. This result is in agreement with the data of figure 13, since u' and γ_0 are related linearly by virtue of equation (164); however, this expression involves a difference, and hence γ_0 is more sensitive to error in u' than is the ratio u'/u. The broken curve determined from the original fitted functions of GI also shows relatively large error as Zv increases.

Shown also in figure 15 is the differential parameter ε_0 associated with the equation of state, defined by equation (194) and calculated from equation (193) by means of the fitted function of GM for ϕ_b. One notes that ε_0 and γ_0 approach each other monotonically in the two extreme limits of Zv large and small. This result is demanded by the theory on p. 332, and the fact that it is satisfied by results from the fitted function of the work of GII for $\phi'_i - \phi'_{i,\infty}$ and the prior one of GM for ϕ_b represents a stringent check on the accuracy and consistency of both results.

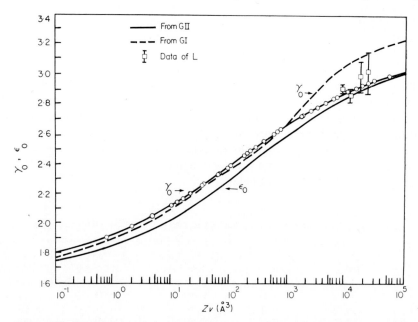

FIGURE 15 Parameters ε_0 and γ_0 against scaled volume in cubic ångströms, from the fitted functions of GII for $\phi'_i - \phi'_{i,\infty}$ and of GM for ϕ_b, for comparison with the data points and with results (broken) from the fitted functions of GI. Vertical bars indicate errors in values from L. [From GII]

Case of Low Temperature

In this section, the general perturbation method of solution of the Thomas–Fermi equation discussed in the second section will be treated in detail for the case of a temperature perturbation of first order. Two methods of solution of the first-order temperature-perturbed equation as used in practice will be described, that of Gilvarry and that of Feynman, Metropolis, and Teller, which differ in the respect that the radius of the atom is maintained unperturbed in the former case but is perturbed in the latter. A third method, that of Mayer and Gilvarry, will be discussed briefly. From the general expressions for thermodynamic functions given on p. 327, it is possible to exhibit explicitly the thermodynamics of the atom model for the case of a first-order temperature perturbation, on the basis of any of these methods of solution of the differential equation. An analytical solution of the temperature-perturbed equation of general order in terms of quadratures will be given. This result yields a general solution of the Thomas–Fermi equation as generalized to arbitrary temperature as an infinite series with coefficients in analytic form for the case of positive chemical potential (for the temperature below the degeneracy limit). As in the case of zero temperature, the thermodynamic functions can be made directly accessible by means of fitted functions constructed from solutions of the temperature-perturbed Thomas–Fermi equation.

For convenience of reference, the pertinent work of Gilvarry[44] again will be referred to as GI, that of Feynman, Metropolis, and Teller[19] as FMT, and that of Slater and Krutter[14] as SK.

Thomas–Fermi atom model

Temperature-perturbed Thomas–Fermi equation

In what follows, the thermodynamic functions derived for the case $T = 0$ will be generalized to include terms in T^2. The procedure corresponds to retaining terms through T^2 in the expansion of the right-hand side of the asymptotic Thomas–Fermi equation (37). Setting $\chi_1 = \chi$ and $\zeta_1 = \zeta$ in equation (50), one can write

$$\Phi = \phi + \chi\zeta(kT)^2 \tag{202}$$

as a solution of the resulting equation, where ϕ is an unperturbed solution corresponding to $T = 0$, and the perturbation χ satisfies

$$\frac{\partial^2 \chi}{\partial x^2} = \frac{3}{2}\left(\frac{\phi}{x}\right)^{1/2}\chi + \frac{x^{3/2}}{\phi^{1/2}} \tag{203}$$

under the initial condition $\chi(0, x_b) = 0$. The coefficient ζ can be evaluated (with $a_1 = \pi^2/8$) as $\pi^2\mu^2/8Z^2 e^4$, or as $(2{\cdot}034 Z^{4/3} R)^{-2}$ in terms of the rydberg. The limitation (40) on the temperature can be written

$$Z^{-4/3}\frac{kT}{R} \ll 8\left(\frac{2}{9\pi^2}\right)^{1/3}\frac{\phi_b}{x_b} \tag{204}$$

A particular perturbation χ always corresponds to a definite $\phi(x, x_b)$; the atom corresponding to this ϕ will be referred to as the unperturbed atom to which the perturbation χ corresponds.

Methods of solution

In the following subsection, the thermodynamic functions will be developed in terms of parameters corresponding to the method of Gilvarry, given in GI. These parameters will be expressed on p. 381 in terms of parameters on the method of Feynman, Metropolis, and Teller and on p. 382 on the method of Mayer and Gilvarry, in order to yield the thermodynamic functions in these cases. In a final subsection, numerical results will be presented.

Method of Gilvarry. In this case, the perturbation leaves the volume v unchanged. The boundary value $\chi_b = \chi_{1,b}$ satisfies the boundary condition (52) at the radius x_b of the unperturbed atom. The initial slope $\chi'_i = \chi'_{1,i}$ must be selected by trial so that this condition is met.

It is convenient to introduce two parameters,

$$\sigma = \frac{\chi_b}{\phi_b} \qquad \omega = \frac{\chi'_i}{x_b^{1/2}\phi_b^{5/2}} \tag{205}$$

in terms of which the boundary value Φ_b and initial slope Φ'_i of Φ can be written

$$\Phi_b = \phi_b\{1 + \sigma\zeta(kT)^2\} \tag{206a}$$

$$\Phi'_i = \phi'_i + x_b^{1/2}\phi_b^{5/2}\omega\zeta(kT)^2 \tag{206b}$$

respectively. With use of equation (206a), the pressure P of the perturbed atom follows directly from equation (41), by retention only of terms through T^2, as

$$P = p\left\{1 + \frac{5}{2}(\sigma + 2\tau)\zeta(kT)^2\right\} \tag{207}$$

where p is the pressure of equation (64) corresponding to the unperturbed atom, and

$$\tau = \left(\frac{x_b}{\phi_b}\right)^2 \tag{208}$$

The parameters σ, τ, and ω are functions of x_b, and thus of v.

Equations (207) and (206b) provide expansions through terms in T^2 of P and Φ'_i, respectively, so that the total energy can be evaluated from equation (45). The energy U is

$$U = u + \frac{15}{2} pv(\sigma + 2\tau + 3\omega)\zeta(kT)^2 \tag{209}$$

where u is the energy of the unperturbed atom, given by equation (63). The entropy S is

$$S = 15pv(\sigma + 2\tau + 3\omega)\zeta k^2 T \tag{210}$$

from the relation $T(\partial S/\partial T)_v = (\partial E/\partial T)_v$. The Helmholtz function F is

$$F = u - \frac{15}{2}(\sigma + 2\tau + 3\omega)\zeta(kT)^2 \tag{211}$$

One can verify (with the aid of the virial theorem) that F satisfies the Brachman relation (19).

From equation (49), the Gibbs function G is

$$G = h\{1 + \sigma\zeta(kT)^2\} \tag{212}$$

where h is the enthalpy at zero temperature, given by equation (66). Accordingly, the enthalpy $H = G + TS$ is fixed by

$$H = h + \{pv(\sigma + 2\tau + 3\omega) + h\sigma\}\zeta(kT)^2 \tag{213}$$

The Gibbs function $G' = F + Pv$ as usually defined is

$$G = h' - \frac{5}{2}pv(\sigma + 2\tau + 3\omega)\zeta(kT)^2 \tag{214}$$

where h' is fixed by equation (67).

The parameter γ is determined by

$$\gamma = \gamma_0 - \frac{5}{2}(\gamma_0 - 1)\{(3\gamma_0 - 4)(\sigma + 2\tau) + 9(\gamma_0 - 1)\omega\}\zeta(kT)^2 \tag{215}$$

where γ_0 corresponds to the unperturbed atom. The differential parameter ε_T can be obtained from the definition as

$$\varepsilon_T = \varepsilon_0 - \left(\frac{5}{6}\frac{d\sigma}{d\ln x_b} + 4\varepsilon_0 \tau\right)\zeta(kT)^2 \tag{216}$$

where use has been made of the relation

$$\frac{d\tau}{d\ln x_b} = \frac{12}{5}\varepsilon_0 \tau \tag{217}$$

and ε_0 corresponds to the unperturbed atom. To obtain ε_S, it should be noted that P of equation (207) can be expressed in terms of S rather than T by means of equation (210). In this manner one obtains

$$\varepsilon_S = \varepsilon_T + 5\gamma_e(\sigma+2\tau)\zeta(kT)^2 \tag{218}$$

where the parameter

$$\gamma_e = \frac{1}{3}\frac{\sigma+2\tau}{\sigma+2\tau+3\omega} \tag{219}$$

is the electronic Grüneisen parameter[16]. It should be noted that the ratio $\varepsilon_S/\varepsilon_T$ yields the ratio C_P/C_v of the heat capacities directly from equation (56). From this equation, one has

$$C_P = C_v = S \tag{220}$$

in this approximation, where the heat capacity C_v is equal to S.

The parameters σ, τ, and ω are subject to an equation of connection fixed by the identity $(\partial P/\partial T)_v = (\partial S/\partial v)_T$, which yields the differential relation

$$\frac{d(\sigma+3\omega)}{d\ln x_b} = (3\varepsilon_0 - 2)\sigma + \left(\frac{6}{5}\varepsilon_0 - 4\right)\tau + 9(\varepsilon_0 - 1)\omega \tag{221}$$

As one can verify, this relation is necessary and sufficient for the differentials of U, H, F, and G, as derived above, to be exact differentials of their arguments, and for the Maxwell relations to be satisfied. Equation (221) implies a differential relation between χ_b, χ_i', and ϕ_b which is the analog of equation (68) corresponding to the zero-temperature case and can be obtained directly from equation (32). Thus, the present results are in agreement with the general conclusions of p. 325, regarding the conditions for the Maxwell relations to be satisfied.

In the limit $x_b \to 0$ of vanishing atomic volume, the thermodynamic functions derived must reduce to the corresponding functions for a degenerate Fermi–Dirac gas through terms of order T^2. Substitution in the differential equation (203) for χ of the dominant terms of equation (72) for ϕ in the Fermi–Dirac limit, solution of the resulting equation, and imposition of the boundary condition (52) yields

$$\chi_i' = -\frac{2}{3^{5/3}}x_b^2 \tag{222}$$

for the initial slope. The corresponding solution for χ in the Fermi–Dirac limit is

$$\chi = -\frac{2}{3^{5/3}}x_b^2 x \tag{223}$$

which is a straight line in x. The boundary value of χ is thus

$$\chi_b = -\frac{2}{3^{5/3}}x_b^3 \tag{224}$$

and the parameters σ and ω become

$$\sigma = -\frac{2}{3^{7/3}}x_b^4 \qquad \omega = -\frac{2}{3^{10/3}}x_b^4 \tag{225}$$

with use of the first of equations (74). The parameter τ is

$$\tau = \frac{1}{3^{4/3}} x_b^4 \qquad (226)$$

One can verify that, with these asymptotic forms, all the thermodynamic functions derived reduce to the corresponding functions for a degenerate Fermi–Dirac gas[46] through terms of order T^2. The inequality (204) in this case requires that $Z^{-4/3}kT/R$ be less than a quantity proportional to x_b^{-2}, so that large temperatures are permitted.

In the opposite limit of an infinite atom, for $x_b \to \infty$, the corresponding asymptotic forms can be obtained from the differential equation (203) by substitution of the approximation for ϕ in this case

$$\phi = \frac{144}{x^3}\left(1 - \frac{\lambda_1\, 12^{2\lambda_2/3}}{2x^{\lambda_2}}\right) \qquad (227)$$

which corresponds, in equation (75), to replacing the curly bracket by unity and expanding the factor $(1+z)^{-\lambda_1/2}$. The solution for χ in this limit is

$$\chi = Ax^{\lambda_1-3}\left(1 + \frac{3\lambda_1\, 12^{2\lambda_2/3}}{4x^{\lambda_2}}\right) + \frac{x^5}{24}\left\{1 + \frac{(4+5\lambda_1)12^{2\lambda_2/3}}{6x^{\lambda_2}}\right\} \qquad (228)$$

where the constant A is fixed by the boundary condition (52) as

$$A = -\frac{\lambda_1-3}{108} x_b^{8-\lambda_1}\left\{1 + \frac{(\lambda_1-1)12^{2\lambda_2/3}}{6x_b^{\lambda_2}}\right\} \qquad (229)$$

This solution yields

$$\chi_b = -\frac{2\lambda_2-1}{216} x_b^5\left(1 + \frac{\lambda_1\, 12^{2\lambda_2/3}}{2x_b^{\lambda_2}}\right) \qquad (230)$$

for the boundary value of χ. The value of σ follows as

$$\sigma = -\frac{4\lambda_2-3}{2^7 3^4} x_b^8\left\{1 + \frac{(3+17\lambda_1)12^{2\lambda_2/3}}{6x_b^{\lambda_2}}\right\} \qquad (231)$$

with use of equation (78), and the value of τ is

$$\tau = \frac{33-40\lambda_2}{2^8 3^6} x_b^8\left\{1 + \frac{(3+8\lambda_1)12^{2\lambda_2/3}}{9x_b^{\lambda_2}}\right\} \qquad (232)$$

The parameter ω is determined by the differential identity (221), and is independent of the first-order terms of σ and τ in x_b^8. With inclusion of a second-order term in ε_0 by

$$\varepsilon_0 = \frac{10}{3}\left\{1 - \frac{(\lambda_1+9)12^{2\lambda_2/3}}{24x_b^{\lambda_2}}\right\} \qquad (233)$$

integration of equation (221) yields

$$\omega = -\frac{66+169\lambda_2}{2^9 3^7} 12^{2\lambda_2/3} x_b^{8-\lambda_2} \qquad (234)$$

Thomas–Fermi atom model

The corresponding value of the initial slope χ_i' is

$$\chi_i' = -\frac{2}{3^7}(66+169\lambda_2)(3+2\lambda_1)^{5/2}12^{2\lambda_2/3}x_b^{1-\lambda_2} \tag{235}$$

The inequality (204) limiting the temperature requires that $Z^{-4/3}kT/R$ be much less than a quantity proportional to x_b^{-4} in this case, and thus demands temperatures approaching zero as $x_b \to \infty$.

The values of ω and of χ_i' given by equations (234) and (235), respectively, are corrected from those given in GI. Calculation of the parameter χ_i' from equation (221) is facilitated through use of the integral

$$\chi_i' = \int_0^{x_b} \frac{\phi_b^{5/2}}{x_b^{1/2}} \left\{ \left(\varepsilon_0 - \frac{2}{3}\right)\sigma + \left(\frac{2}{5}\varepsilon_0 - \frac{4}{3}\right)\tau - \frac{x_b}{3}\frac{d\sigma}{dx_b} \right\} dx_b \tag{236}$$

directly.

The preceding development is drawn from GI.

Method of Feynman, Metropolis, and Teller. In this case, the perturbed volume of the atom is allowed to differ from the unperturbed volume. The perturbation function χ^* satisfies equation (203), with an initial slope equal to zero. The radius of the atom boundary is fixed by the condition (39b).

For a definite zero-temperature solution ϕ and corresponding perturbation χ^*, the radius x_b^* can be determined approximately from the boundary condition in terms of the radius x_b corresponding to the unperturbed atom, if ϕ, χ^*, and their derivatives are expanded in Taylor series about the point $x = x_b$. Making use of the differential equations (203) and (59) and the boundary condition on ϕ, one obtains in first order

$$x_b^* = x_b\{1 + v\zeta(kT)^2\} \tag{237}$$

where the parameter v is defined by

$$v = \frac{\chi_b^* - x_b \chi_b^{*\prime}}{(x_b \phi_b)^{3/2}} \tag{238}$$

in which ϕ_b is the value of ϕ at the boundary $x = x_b$ of the unperturbed atom, and χ_b^* and $\chi_b^{*\prime}$ are the values of χ^* and $\partial\chi^*/\partial x$, respectively, evaluated at the same point $x = x_b$. The volume v^* of the perturbed atom is accordingly

$$v^* = v\{1 + 3v\zeta(kT)^2\} \tag{239}$$

in terms of the volume v of the corresponding unperturbed atom.

To relate χ^* with χ for the same volume v^*, where χ is the perturbation function computed by the method of Gilvarry, one notes that χ must satisfy equation (203) with a $\phi = \phi^*$ chosen to correspond to the perturbed atom radius x_b^*. The value of ϕ^* is, accordingly,

$$\phi^* = \phi(x, x_b) + \frac{\partial \phi(x, x_b)}{\partial x_b}(x_b^* - x_b) \tag{240}$$

where $\phi(x, x_b)$ is the unperturbed solution to which χ^* corresponds. Substitution of

ϕ^* into equation (203) to obtain the differential equation for χ, and comparison of the result with the differential equation for χ^*, yields

$$\chi^* = \chi + v x_b \frac{\partial \phi}{\partial x_b} \tag{241}$$

with use of equation (237).

The function $\partial \phi / \partial x_b$ is a solution of the homogeneous equation

$$\frac{\partial^2}{\partial x^2} \frac{\partial \phi}{\partial x_b} = \frac{3}{2}\left(\frac{\phi}{x}\right)^{1/2} \frac{\partial \phi}{\partial x_b} \tag{242}$$

corresponding to the inhomogeneous perturbation equation, as one can verify by differentiating the zero-temperature equation (59) with respect to x_b. At the origin, the function vanishes and its slope satisfies the initial condition,

$$\left[\frac{\partial}{\partial x} \frac{\partial \phi}{\partial x_b}\right]_{x=0} = \frac{d\phi'_i}{dx_b} \tag{243}$$

where ϕ'_i is the initial slope of the zero-temperature solution $\phi(x, x_b)$. The function ϕ^* then satisfies the boundary condition (60b) at x_b^* automatically.

Evaluation at $x = x_b$ of equation (241) for χ^*, with use of

$$\frac{d\phi_b}{dx_b} = \frac{\phi_b}{x_b} + \left[\frac{\partial \phi(x, x_b)}{\partial x_b}\right]_{x=x_b} \tag{244}$$

yields the relation

$$\chi_b = \chi_b^* + \frac{6}{5} \varepsilon_0 v \phi_b \tag{245}$$

between boundary values computed under the two methods. Differentiation of the same equation with respect to x and evaluation at $x = 0$ yields the further relation

$$x_b^{1/2} \phi_b^{3/2} \chi_b + 3\chi'_i = x_b^{1/2} \phi_b^{3/2} \chi_b^* + 4v x_b^{1/2} \phi_b^{5/2} \tag{246}$$

when equations (68) and (243) are employed. The last two equations give

$$\chi'_i = \left(\frac{4}{3} - \frac{2}{5}\varepsilon_0\right) v x_b^{1/2} \phi_b^{5/2} \tag{247}$$

Equations (245) and (247) fix completely the significant parameters χ_b and χ'_i on the method of Gilvarry in terms of quantities computed by the method of Feynman, Metropolis, and Teller. In terms of the parameters σ and ω, one obtains, from equation (245),

$$\sigma = \sigma^* + \frac{6}{5}\varepsilon_0 v \tag{248}$$

where $\sigma^* = \chi_b^*/\phi_b$, and one obtains

$$\omega = \left(\frac{4}{3} - \frac{2}{5}\varepsilon_0\right) v \tag{249}$$

from equation (247).

Thermodynamic functions for the volume v are available from preceding results in terms of the parameters σ and ω corresponding to the method of Gilvarry. To determine a thermodynamic function for the volume v^* in terms of the parameters σ^* and v corresponding to the method of Feynman, Metropolis, and Teller, it is necessary merely to take account of the perturbation in the temperature-independent term due to the volume change, $v^* - v$, and to transform σ and ω into σ^* and v by the preceding equations. Thus, if P^* is the pressure corresponding to the volume v^*, one has

$$P^* = P + \frac{dp}{dv}(v^* - v) \tag{250}$$

if P corresponds to v. With use of equation (239) for $v^* - v$ and the defining equation for ε_0, one obtains

$$P^* = p\left\{1 + \frac{5}{2}(\sigma^* + 2\tau)\zeta(kT)^2\right\} \tag{251}$$

In a similar manner, it can be shown that the energy U^* corresponding to v^* is

$$U^* = u + \tfrac{3}{2}pv(5\sigma^* + 10\tau + 18v)\zeta(kT)^2 \tag{252}$$

and that the entropy S^* is

$$S^* = 15pv(\sigma^* + 2\tau + 4v)\zeta k^2 T \tag{253}$$

Method of Mayer and Gilvarry. The connection between χ^* and χ given above was derived in GI, essentially by noting that these functions can differ only by a solution of the corresponding homogeneous perturbation equation. However, Hartle and Gilvarry[77] have obtained the relation in a more perspicuous manner by a direct physical argument, taking account in the final results of the use of an arbitrary initial slope (rather than zero) in the solution of the perturbation differential equation. Such a procedure represents an obvious generalization of the method of Feynman, Metropolis, and Teller, and was first discussed by Mayer and Gilvarry[78].

The density ρ of the electrons is a functional $\rho[\Phi, x]$ of the dimensionless variables Φ and x of the model. Poisson's equation in terms of Φ and x must be solved with this electron density subject to the initial and boundary conditions corresponding to equations (39). The method of Gilvarry for solution of the perturbation differential equation presumes that the atom radius x_b is maintained unperturbed at its value for zero temperature, which implies a proper selection (by trial or iteration) of the initial slope χ'_i, while the method of Feynman, Metropolis, and Teller assumes $\chi'_i = 0$. A simpler computational procedure than either, generalizing the latter method in which the initial slope is equated to zero, consists in determining a function χ^* by solution of the differential equation for an initial value $\chi^*_i = 0$ with an initial slope $\chi^{*'}_i$ set equal to an arbitrary value. The corresponding complete Thomas–Fermi function Φ^* defined by

$$\Phi^* = \phi + \chi^*\zeta(kT)^2 \tag{254}$$

must satisfy the boundary condition of equation (39b) at a perturbed radius x_b^* of the atom. The value of x_b^* is fixed by equation (237) in terms of the perturbation parameter v.

In the generalized method of Feynman, Metropolis, and Teller or the Mayer–Gilvarry case, consider an atom with Thomas–Fermi function $\phi(x, x_b)$ and radius x_b at zero temperature, where the latter becomes x_b^* for nonvanishing temperature. On the method of Gilvarry, the Thomas–Fermi function for zero temperature must be taken as a function $\phi^*(x, x_b^*)$ for a radius x_b^* which remains the same for this atom under a temperature perturbation. These two atoms of the same radius x_b^* at an arbitrary low temperature can be identical if and only if the electron density $\rho^*[\Phi^*, x]$ in the Mayer–Gilvarry case equals the density $\rho[\Phi, x]$ on the method of Gilvarry at every interior point x, which implies

$$\Phi^*(x, x_b) = \Phi(x, x_b^*) \tag{255}$$

or equivalently

$$\phi(x, x_b) + \chi^*(x, x_b)\zeta(kT)^2 = \phi^*(x, x_b^*) + \chi(x, x_b^*)\zeta(kT)^2 \tag{256}$$

from equation (202). Taylor expansion of $\phi^*(x, x_b^*)$ about the point x_b, substitution of this expression into equation (256), and use of equation (237) yields the relation of equation (241). Thus, this connection between χ^* and χ obviously holds independently of the choice of initial slope for χ^*.

With introduction of $\omega^* = \chi_i^{*\prime}/(x_b^{1/2}\phi_b^{5/2})$ depending on the initial slope in the method of Mayer and Gilvarry, the connection with perturbation parameters on the method of Gilvarry is given by

$$\omega = \omega^* + \left(\frac{4}{3} - \frac{2}{5}\varepsilon_0\right)v \tag{257}$$

with the relation between σ and σ^* unchanged. The corresponding entropy S^* becomes

$$S^* = 15pv(\sigma^* + 2\tau + 3\omega^* + 4v)\zeta k^2 T \tag{258}$$

in the Mayer–Gilvarry method. The energy U^* is

$$U^* = u + \tfrac{3}{2}pv(5\sigma^* + 10\tau + 15\omega^* + 18v)\zeta(kT)^2 \tag{259}$$

The pressure P^* remains as given by equation (251).

The development above connecting Thomas–Fermi perturbation functions and thermodynamic quantities for the cases where the temperature perturbation leaves the volume unchanged and where it perturbs the volume was carried out for the method of Feynman, Metropolis, and Teller involving zero initial slope in GI. The expressions for thermodynamic functions involving use of an arbitrary initial slope in the generalized method were first given by Mayer and Gilvarry[78].

Numerical results. Three solutions by their method of the first-order temperature-perturbation equation have been tabulated by FMT. The corresponding values of χ_b^* and $\chi_b^{*\prime}$ have been determined in GI by quartic interpolation from their data and are tabulated in table 17 against the unperturbed radius x_b to which they correspond.

Thomas–Fermi atom model

TABLE 17 Boundary and initial parameters, temperature-perturbed case [from GI]

x_b	χ_b^*	$\chi_b^{*'}$	χ_b	χ_i'
9·5651	1747·3	834·82	−812	−13·4
10·8038	3068·2	1315·7	−1320	−15·7
15·8698	18731	6689·1	−6420	−24·8

The corresponding values of χ_b and χ_i' on the method of Gilvarry have been determined in GI also by use of equations (245) and (247), respectively, and are tabulated likewise.

Analytic solution in quadratures

The analytic solution in quadratures given by Gilvarry in GIII will be described under three headings. First, the general solution of the prototype of the perturbed differential equation will be given. This result will then be applied to the case of a temperature perturbation of first order. Finally, it will be applied to a perturbation of general order.

General solution. The Thomas–Fermi function Φ is connected with the electrostatic potential V at a point r in an atom of atomic number Z by

$$\frac{Ze^2\Phi}{\mu x} = eV + kT\eta \tag{260}$$

where e is the electronic charge, k is Boltzmann's constant, T is the temperature, $kT\eta$ is the chemical potential, and x is defined by $r = \mu x$, where $\mu = a_0(9\pi^2/128Z)^{1/3}$ in terms of the radius a_0 of the first Bohr orbit for hydrogen. The asymptotic form for low temperature of the generalized Thomas–Fermi equation which Φ satisfies is

$$\frac{d^2\Phi}{dx^2} = \frac{\Phi^{3/2}}{x^{1/2}}\left\{1 + \sum_n \zeta_n\left(\frac{kTx}{\Phi}\right)^{2n}\right\} \tag{261}$$

in which $n \geq 1$,

$$\zeta_n = \left(\frac{\mu}{Ze^2}\right)^{2n} a_n, \tag{262}$$

and the coefficients a_n are defined and tabulated by McDougall and Stoner[60].

The function Φ can be expanded as an asymptotic series:

$$\Phi = \phi + \sum_n \chi_n \zeta_n (kT)^{2n} \tag{263}$$

where ϕ is the Thomas–Fermi function corresponding to $T = 0$, and the perturbation χ_n of order n is a function of x. The function ϕ satisfies the differential equation

$$\frac{d^2\phi}{dx^2} = \frac{\phi^{3/2}}{x^{1/2}} \tag{264}$$

and each perturbation χ_n satisfies an inhomogeneous linear differential equation of the form

$$\frac{d^2\chi_n}{dx^2} = \frac{3}{2}\left(\frac{\phi}{x}\right)^{1/2}\chi_n + f_n \tag{265}$$

where f_n is a function of x, ϕ, and the perturbations $\chi_1, \chi_2, \ldots, \chi_{n-1}$. The form of f_n can be found by substituting the series (263) into the differential equation (261). The initial and boundary conditions on $\chi_n(x)$ are

$$\chi_n(0) = 0 \tag{266a}$$

$$\chi_{n,b} = x_b\left[\frac{d\chi_n}{dx}\right]_{x=x_b} \tag{266b}$$

where $\chi_{n,b}$ is the boundary value of χ_n at the boundary x_b of the unperturbed atom. The atomic volume corresponding to the boundary condition (266b) is

$$v = \left(\frac{4\pi}{3}\right)(\mu x_b)^3$$

and thus the volume is unperturbed.

Fermi[79] has shown that one solution of the homogeneous equation corresponding to equation (265) is $\xi_1 = \phi + \tfrac{1}{3}x\,d\phi/dx$ if ϕ satisfies equation (264). It follows that a second solution of the homogeneous equation is $\xi_2 = \xi_1 \int \xi_1^{-2}\,dx$. By the method of variation of parameters, one can then obtain the general solution of equation (265) in terms of integrals on ϕ and f_n. It is convenient to define four integrals, of which the first

$$J_1(x) = \int_0^x \left(\phi + \tfrac{1}{3}x'\frac{d\phi}{dx'}\right)^{-2} dx' \tag{267}$$

is a function only of x, and the three

$$J_2(f, x) = \int_0^x f(x')\left(\phi + \tfrac{1}{3}x'\frac{d\phi}{dx'}\right) dx' \tag{268a}$$

$$J_{12}(f, x) = \int_0^x J_1(x')f(x')\left(\phi + \tfrac{1}{3}x'\frac{d\phi}{dx'}\right) dx' \tag{268b}$$

$$J_{21}(f, x) = \int_0^x J_2(f, x')\left(\phi + \tfrac{1}{3}x'\frac{d\phi}{dx'}\right)^{-2} dx' \tag{268c}$$

are functionals of a function f of x. These four integrals satisfy the identity

$$J_{12}(f, x) + J_{21}(f, x) = J_1(x)J_2(f, x) \tag{269}$$

The general solution of equation (265) is then

$$\chi_n = \{c_1 + c_2 J_1(x) + J_{21}(f_n, x)\}\left(\phi + \tfrac{1}{3}x\frac{d\phi}{dx}\right) \tag{270}$$

where c_1 and c_2 are constants.

Thomas–Fermi atom model

First-order temperature perturbation. The first-order differential perturbation equation corresponds to equation (265) with

$$f_1 = \frac{x^{3/2}}{\phi^{1/2}} \tag{271}$$

The solution satisfying the initial condition $\chi_1(0) = 0$ is, from equation (270),

$$\chi_1 = \{\chi'_{1,i} J_1(x) + J_{21}(\ _1, x)\}\left(\phi + \tfrac{1}{3}x\frac{d\phi}{dx}\right) \tag{272}$$

where $\chi'_{1,i}$ is the initial slope of the solution. Substitution of $x = x_b$ in this solution yields one equation connecting the boundary value $\chi_{1,b}$ with the initial slope $\chi'_{1,i}$, and the boundary condition (266b) provides another. The solution of these two equations yields, with use of the identity (269),

$$\chi_{1,b} = -12\phi_b J_{12}(f_1, x_b)\{9 + 4x_b^{1/2}\phi_b^{5/2} J_1(x_b)\}^{-1} \tag{273a}$$

$$\chi'_{1,i} = -\frac{9J_2(f_1, x_b) + 4x_b^{1/2}\phi_b^{5/2} J_{21}(f_1, x_b)}{9 + 4x_b^{1/2}\phi_b^{5/2} J_1(x_b)} \tag{273b}$$

where ϕ_b is the boundary value of ϕ. The boundary and initial parameters ($\chi_{1,b}$ and $\chi'_{1,i}$, respectively) of the solution, necessary for the thermodynamics, are thus obtained directly in terms of quadratures on ϕ.

In GI, thermodynamic functions for the first-order temperature-perturbed case were given in a form in which the coefficient of T^2 or T was expressed in terms of the parameters σ, τ, and ω, of which

$$\tau = \left(\frac{x_b}{\phi_b}\right)^2 \tag{274}$$

depends on the solution of the zero-temperature equation, and

$$\sigma = \frac{\chi_{1,b}}{\phi_b} \qquad \omega = \frac{\chi'_{1,i}}{x_b^{1/2}\phi_b^{5/2}} \tag{275}$$

depend on the solution of the corresponding differential perturbation equation (see p. 375). Thus, the pressure P from GI is

$$P = p\left\{1 + \tfrac{5}{2}(\sigma + 2\tau)\zeta_1(kT)^2\right\} \tag{276}$$

where p is the pressure

$$p = \frac{Z^2 e^2}{10\pi\mu^4}\left(\frac{\phi_b}{x_b}\right)^{5/2} \tag{277}$$

corresponding to the unperturbed atom of the same volume. By means of equations (273), the parameters σ and ω can be written directly in terms of quadratures on ϕ. Hence, one obtains Mayer's result[80]: the coefficient of T^2 or T in the first-order perturbation of every thermodynamic function can be expressed in terms of quadratures on ϕ. This statement does not imply that the coefficient of T^2 or T may not contain a derivative with respect to x_b (or volume), as is necessarily the case, for example, in the differential parameter ε_T defined by

$$\varepsilon_T = -\left(\frac{\partial \ln P}{\partial \ln v}\right)_T \tag{278}$$

and evaluated in GI as

$$\varepsilon_T = \varepsilon_0 - \left(\frac{5}{6}\frac{d\sigma}{d\ln x_b} + 4\varepsilon_0\tau\right)\zeta_1(kT)^2 \tag{279}$$

where $\varepsilon_0 = (5/6)(1 - d\ln\phi_b/d\ln x_b)$ corresponds to zero temperature.

The basic thermodynamic functions do not involve σ, τ, or ω differentially. In such a case, the coefficient of T^2 or T in the first-order perturbation can be evaluated algebraically in terms of quadratures on ϕ. Thus, the pressure P from equation (276) becomes

$$P = p\left[1 + 5\left\{\frac{x_b^2}{\phi_b^2} - \frac{6J_{12}(f_1, x_b)}{9 + 4x_b^{1/2}\phi_b^{5/2}J_1(x_b)}\right\}\zeta_1(kT)^2\right] \tag{280}$$

Computation of the entropy S in this manner yields the result of Mayer[80]

$$S = 4\frac{Z^2e^2}{\mu}\zeta_1 k^2 T \int_0^{x_b} x^{3/2}\phi^{1/2}\,dx \tag{281}$$

where the special result

$$J_2(f_1, x) = \frac{2}{3}\left(x^{5/2}\phi^{1/2} - \int_0^x x^{3/2}\phi^{1/2}\,dx\right) \tag{282}$$

for a first-order perturbation has been used. The energy U is

$$U = u + 2\frac{Z^2e^2}{\mu}\zeta_1(kT)^2 \int_0^{x_b} x^{3/2}\phi^{1/2}\,dx \tag{283}$$

where u is the energy at zero temperature corresponding to the unperturbed atom of the same volume, given by

$$u = \frac{Z^2e^2}{\mu}\left\{\frac{3}{7}(\phi_i' - \phi_{i,\infty}') + \frac{2}{35}x_b^{1/2}\phi_b^{5/2}\right\} \tag{284}$$

in terms of the initial slope ϕ_i' and boundary value ϕ_b of ϕ ($\phi_{i,\infty}'$ is the initial slope corresponding to an infinite atom). The quadrature expression of the Helmholtz function follows directly. The method of Mayer[80] does not yield the pressure in the form of an algebraic function of quadratures on ϕ, since the pressure must be determined from equation (281) by means of the relation $(\partial P/\partial T)_v = (\partial S/\partial v)_T$.

Mayer derived the result that the first-order perturbation in every thermodynamic function can be expressed in terms of quadratures on ϕ independently of (and prior to) the considerations of GIII. It follows directly from the quadrature expression (281) for the entropy, since every thermodynamic function can be derived from the entropy. There is a slight difference between the result obtained by the method of Mayer and the result by the method of GIII: in the former case, thermodynamic functions are

Thomas–Fermi atom model

expressed in terms of quadratures on ϕ, while the method of GIII expresses them in terms of quadratures on ϕ and $d\phi/dx$ (since ϕ is given in tabular form for a solution, $d\phi/dx$ is also directly available). The pressure obtained by Mayer's method contains an integral on the function $\partial\phi/\partial x_b$, which is not obtained usually in a solution.

The statement made without proof in GI that $\chi_{1,b}$ and $\chi'_{1,i}$ are monotonic functions of the radius x_b follows directly from equations (273).

General temperature perturbation. The functions f_n of equation (265) are

$$f_2 = (x\phi)^{-1/2}\left\{\frac{x^4}{\phi^2} - \frac{a_1^2}{a_2}\left(\frac{1}{2}\frac{x^2}{\phi}\chi_1 - \tfrac{3}{8}\chi_1^2\right)\right\} \tag{285}$$

and

$$f_3 = (x\phi)^{-1/2}\left\{\frac{x^6}{\phi^4} + \frac{a_1^3}{a_3}\left(\frac{3}{8}\frac{x^2}{\phi^2}\chi_1^2 - \frac{1}{16\phi}\chi_1^3\right) - \frac{a_1 a_2}{a_3}\left(\frac{5}{2}\frac{x^4}{\phi^3}\chi_1 + \frac{1}{2}\frac{x^2}{\phi}\chi_2 - \tfrac{3}{4}\chi_1\chi_2\right)\right\} \tag{286}$$

for $n = 2, 3$. A general (but rather complex) formula can be given for arbitrary n. The general solution of the differential perturbation equation of nth order is, from equation (270),

$$\chi_n = \{\chi'_{n,i} J_1(x) + J_{21}(f_n, x)\}\left(\phi + \frac{1}{3}x\frac{d\phi}{dx}\right) \tag{287}$$

where $\chi'_{n,i}$ is the initial slope of the solution. It has been shown in the preceding that χ_1 can be expressed in terms of quadratures on ϕ. It follows by induction that f_n (which in the first instance is a function of x, ϕ and $\chi_1, \chi_2, \ldots, \chi_{n-1}$) and thus χ_n can be so expressed. The boundary value $\chi_{n,b}$ and initial slope $\chi'_{n,i}$ of χ_n are then obtainable from the equations derived by replacing f_1 with f_n in equations (273a) and (273b), respectively. Thus the solution of the perturbation equation of nth order and the associated boundary and initial parameters can all be expressed in terms of quadratures on ϕ.

The thermodynamic functions that are available directly from a solution of the Thomas–Fermi equation are essentially the pressure and the volume. The pressure P is given by the asymptotic series

$$P = \frac{Z^2 e^2}{10\pi\mu^4}\left(\frac{\Phi_b}{x_b}\right)^{5/2}\left\{1 + \sum_n \frac{5\zeta_n}{5-4n}\left(\frac{kTx_b}{\Phi_b}\right)^{2n}\right\} \tag{288}$$

where Φ_b, given by

$$\Phi_b = \phi_b + \sum_n \chi_{n,b}\zeta_n(kT)^{2n} \tag{289}$$

is the boundary value of Φ. Thus, the pressure can be expressed as an asymptotic series in T^2, with coefficients which can be evaluated in terms of quadratures on ϕ. From GI (see p. 327) the energy E is given by

$$E = \tfrac{3}{4}T^{7/4}\int^T T^{-3/4}\left(\frac{\partial}{\partial T}\right)_v\left\{T^{-1}\left(Pv + \frac{Z^2 e^2}{\mu}\Phi'_i\right)\right\}dT \tag{290}$$

where the integration is carried out at constant volume, and Φ'_i, defined by

$$\Phi'_i = \phi'_i + \sum_n \chi'_{n,i}\zeta_n(kT)^{2n} \tag{291}$$

is the initial slope of Φ. If this series for Φ'_i and the series (288) for P are substituted into equation (290), E can be expressed as an asymptotic series in T^2:

$$E = E_0 + \sum_n E_n \zeta_n (kT)^{2n} \qquad (292)$$

where the term E_0 corresponds to zero temperature, and the temperature-independent coefficients E_n are functions in the first instance of $x_b, \phi_b; \chi_{1,b}, \ldots, \chi_{n,b}; \chi'_{1,i}, \ldots, \chi'_{n,i}$. Thus, the energy likewise can be expressed as an asymptotic series in T^2, with coefficients in terms of quadratures on ϕ. This expansion is analogous to the corresponding result[46] in the case of a degenerate Fermi–Dirac gas. From the energy, one can obtain every thermodynamic function by algebraic, differential, or integral processes. Hence the result of GIII follows: every thermodynamic function at low temperature can be written as an asymptotic series, either in the even or the odd powers of the temperature, with coefficients in terms of quadratures on ϕ. This statement is the generalization to a perturbation of nth order of Mayer's result for the case $n = 1$ cited previously. Alternatively, this conclusion can be obtained from the expressions for thermodynamic functions in terms of quadratures given on p. 327.

The coefficients of the asymptotic series corresponding to the basic thermodynamic functions can be written in algebraic form in terms of quadratures on ϕ. This statement is obvious in the case of the pressure and the energy, from equations (288) and (292), respectively. The entropy S, from equation (292) and the relation $T(\partial S/\partial T)_v = (\partial E/\partial T)_v$ is

$$S = \sum_n \frac{2n}{2n-1} E_n \zeta_n k^{2n} T^{2n-1} \qquad (293)$$

from which the statement follows for the corresponding coefficients. The result for the Helmholtz and Gibbs functions follows directly from the results on p. 327.

The method given in GIII for solving the temperature-perturbed Thomas–Fermi equation can be used likewise to obtain a solution of the Thomas–Fermi–Dirac equation[7] in powers of the parameter $\varepsilon = (3/32\pi^2)^{1/3} Z^{-2/3}$. The corresponding approximation improves for higher Z and for higher compressions. To a degree depending on the powers of ε retained, it should be valid except at the immediate boundary of the limiting atom (atom of largest radius); the type of difficulty that enters at this boundary is the same as appears in Sommerfeld's perturbation treatment of the positive ion[81]. This approximation yields a method of determining the exchange correction to the binding energy of an atom, alternative to that of Scott[82] or March[83]. Difficulties associated with the boundary of the atom should be unimportant in this application, since the boundary region of the limiting atom contributes little to the energy.

Numerical results. The quadrature solutions of p. 386 yield the boundary and initial parameters of a solution, which enter the thermodynamic functions, as explicit integrals on the zero-temperature function to which the solution refers. From these results, accurate values of the requisite boundary and initial parameters can be obtained by numerical integration from the zero-temperature solutions. The available data consist of six neutral-atom zero-temperature solutions from the work of FMT and seven such solutions obtained by SK, as tabulated by Gombas[5] (p. 357). The solutions of SK refer to considerably smaller atomic volumes than do those of FMT;

Thomas–Fermi atom model

the smallest volumes approach those corresponding to a degenerate Fermi–Dirac gas. The subscript 1 will be deleted from the symbols corresponding to the first-order temperature-perturbation function χ and to its boundary value χ_b and initial slope χ'_i.

The values of χ_b and χ'_i obtained in this manner in GIII from the solutions of FMT are tabulated in table 18 against the radius of the unperturbed atom to which they

TABLE 18 Boundary and initial parameters, first-order temperature perturbation [from GIII]

x_b	$-\chi_b$ (GIII)	$-\chi'_i$ (GIII)	$-\chi_b$ (GI)	$-\chi'_i$ (GI)
7·3851	297·$_9$	9·41$_5$		
8·5880	532·$_7$	11·6$_2$		
9·5651	810·$_6$	13·4$_2$	81$_2$·	13·$_4$
10·8038	131$_7$·	15·7$_4$	13$_{20}$·	15·$_7$
11·9634	198$_5$·	17·9$_1$		
15·8698	634$_7$·	25·1$_5$	64$_{20}$·	24·$_8$

correspond. Shown also for comparison are the values with much lower accuracy shown in table 17 from the three solutions of the first-order temperature-perturbed equation given by FMT. These results involve use of the relations of equations (245) and (247) to convert parameters on the method of Feynman, Metropolis, and Teller, into corresponding parameters on the method of Gilvarry. One sees that agreement is good.

Gilvarry and Peebles[84] extended the results in GIII by making use of the numerical solutions of SK for the case of zero temperature. To obtain accurate values of the integrals entering χ_b and χ'_i, the tabulated zero-temperature solutions of SK (given in all but one case to four figures at the atom boundary) were used to obtain improved

TABLE 19 Boundary and initial parameters from data of SK; zero-temperature case [from Gilvarry and Peebles]

$-\phi'_i$ (from SK)	ϕ_b	x_b	x_b (from SK)
1·00	1·447	1·194$_2$	1·19
1·38	0·946$_9$	1·696$_0$	1·69
1·50	0·6650	2·20$_4$	2·20
1·55	0·477$_7$	2·80$_4$	2·80
1·58	0·2556	4·24$_4$	4·23
1·586	(0·150$_2$)[a]	(5·86$_4$)[a]	5·85
1·588	(0·073)[b]	(8·5)[b]	8·59
1·58808	0		∞

[a] The values of SK for ϕ in this case have a discontinuity of jump 0·0037 at the point $x = 1·46$. It was concluded that the error probably lay in the region $x < 1·46$, since subtraction for $x > 1·46$ of the saltus 0·0037 to make ϕ and its derivatives continuous yielded a value of ϕ_b (and of χ_b and χ'_i) not smooth with respect to the other values of the table. The tabulated boundary values were derived from unchanged data.

[b] In this case, only two figures are given in the data at the boundary, so that this point is of low weight. The point is bracketed by two points from the data of FMT, which yield by interpolation (for the tabulated x_b) a value of ϕ_b differing by about 4% from the tabulated ϕ_b. This case was ignored in obtaining the fitted functions of GI for ϕ_b and $\phi'_i - \phi'_{i,\infty}$.

values of the boundary radius x_b (given to only three figures by SK), and of the boundary value ϕ_b (not given by SK). The computed values of ϕ_b and x_b are tabulated in table 19 against the initial slope ϕ_i' to which they correspond, with the values of x_b from SK for comparison. Parenthetic values in the table involve minor discrepancies, as noted. With the exception of the second parenthetic case (in the region of which smoothing was used to give major weight to the data of FMT), the values of ϕ_b given in this table are essentially the ones used to obtain the fitted function of GI for ϕ_b (see p. 334). As is evident from the table, the number of significant figures given by SK for the initial slope ϕ_i' (or, more important, for the difference $\phi_i' - \phi_{i,\infty}'$, where

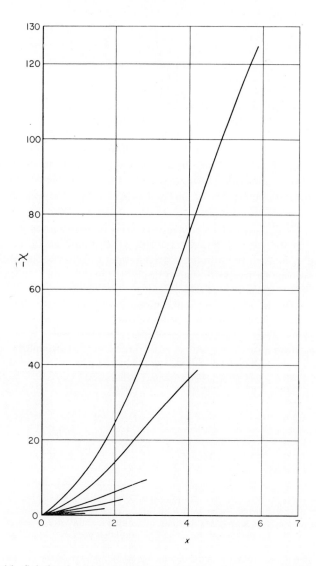

FIGURE 16 Solution for the temperature-perturbation function as a function of radial distance x in the atom [from data of SK]. The solution corresponding to the second parenthetic case in table 19 has been omitted [from Gilvarry and Peebles]

Thomas–Fermi atom model

$\phi'_{i,\infty}$ is the initial slope corresponding to an atom of infinite radius) is not commensurate in general with the number possible in the other parameters, on the basis of the solutions as tabulated.

With these boundary parameters for the unperturbed case, the values of χ_b and χ'_i corresponding to a temperature perturbation were determined from equations (273) by numerical integration on the zero-temperature solutions of SK. The results are tabulated in table 20 against the unperturbed radius to which they correspond. The

TABLE 20 Boundary and initial parameters from data of SK, temperature-perturbed case [from Gilvarry and Peebles]

x_b	$-\chi_b$	$-\chi'_i$
$1 \cdot 194_2$	$0 \cdot 587_5$	$0 \cdot 429_5$
$1 \cdot 696_0$	$1 \cdot 78_5$	$0 \cdot 826_6$
$2 \cdot 20_4$	$4 \cdot 18_7$	$1 \cdot 33_2$
$2 \cdot 80_4$	$9 \cdot 32$	$2 \cdot 03_4$
$4 \cdot 24_4$	$38 \cdot 8_5$	$4 \cdot 07_2$
$5 \cdot 86_4$	$(124 \cdot _7)^a$	$(6 \cdot 73_5)^a$
$8 \cdot 5$	$(51_1 \cdot)^b$	$(11 \cdot _4)^b$

[a] It was found that the values of χ_b and χ'_i depend very little on the values of ϕ to the left of the discontinuity noted in table 19 for this case, and hence the saltus in ϕ was removed by fairing the data into the value unity at $x = 0$. Different but reasonable fairings yielded changes in χ_b and χ'_i less than their corresponding possible errors. The tabulated values for this case seem smooth with respect to the other values of the table.

[b] The values of χ_b and χ'_i are smooth, within less than 0·5%, with respect to the bracketing values from the data of FMT, in spite of the difficulty noted in table 19.

values of χ'_i yield, from equation (272) the solutions plotted in figure 16 for the temperature-perturbation function χ as a function of radial distance x in the atom. The solutions corresponding to the smaller values of x_b do not differ much from the straight lines characteristic (as noted on p. 377) of the Fermi–Dirac limit.

Solution for small atom radius

March has extended his solution of the Thomas–Fermi equation for zero temperature in the limit of small atom radius to the corresponding case of the first-order temperature-perturbed equation[85]. His treatment of the problem will be discussed here.

If one expands the solution $\chi(x)$ of the first-order differential perturbation equation about the boundary point $x = x_b$, one obtains

$$\chi - \chi_b = \sum_{n=1}^{\infty} \frac{(-1)^n}{n!} \chi^n(x_b)(x_b - x)^n = \sum_{n=1}^{\infty} s_n(x_b - x)^n \qquad (294)$$

where $\chi^n(x)$ is the nth derivative of $\chi(x)$ and

$$s_n = \frac{(-1)^n}{n!} \chi^n(x_b) \qquad (295)$$

It is convenient to express the results in terms of two further parameters a and c. The first of these was introduced previously in connection with the case of zero temperature (p. 337) and was defined by

$$\frac{\phi_b}{x_b} = \left[\frac{d\phi}{dx}\right]_{x=x_b} = a^2 \tag{296}$$

and the second will be chosen as

$$\frac{\chi_b}{x_b} = \left[\frac{d\chi}{dx}\right]_{x=x_b} = -c^2 \tag{297}$$

Differentiating the first-order differential perturbation equation repeatedly, one finds the coefficients s_n in terms of x_b, a, and c as tabulated in table 21 for $n = 1$ to $n = 5$.

TABLE 21 Coefficients s_n in the Taylor series solution of the temperature-perturbed Thomas–Fermi equation of first order [from March]

n	s_n	n	s_n
1	c^2		
2	$\frac{1}{2}\left(\frac{1}{a} - \frac{3}{2}ac^2\right)x_b$	4	$\frac{1}{24}\left(1 - 3a^2c^2\right)x_b$
3	$-\frac{1}{6}\left(\frac{1}{a} - \frac{3}{2}ac^2\right)$	5	$\frac{1}{120}\left(1 - 3a^2c^2\right)$

The solution thus obtained satisfies the boundary condition on χ, but it is necessary to find the parameter c (since a is known from the results for the case of zero temperature given on p. 337) by imposing the initial condition $\chi(0) = 0$. One begins by neglecting the coefficients s_n in equation (294) of order higher than corresponds to s_3, and finds

$$\chi(0) = \frac{1}{3}\left(a^{-1} - \frac{3}{2}ac^2\right)x_b^3 = 0 \tag{298}$$

which yields

$$a^2 c^2 = \tfrac{2}{3} \tag{299}$$

Putting in the limiting form of a^2 from the results for zero temperature (p. 338), namely

$$a^2 = \frac{3^{2/3}}{x_b^2} \tag{300}$$

one obtains

$$c^2 = \frac{2}{3^{5/3}} x_b^2 \tag{301}$$

or

$$\chi_b = -\frac{2}{3^{5/3}} x_b^3 \tag{302}$$

which is simply the result of GI in equation (224) obtained by another method.

Thomas–Fermi atom model

However, the treatment of March permits examination of the next term in the expansion of χ_b for small x_b. Retaining all coefficients up to and including s_5 in equation (294), one finds

$$\chi(0) = \frac{1}{3}\left(a^{-1} - \frac{3}{2}ac^2\right)x_b^3 + \frac{1}{30}(1 - 3a^2c^2)x_b^5 = 0 \tag{303}$$

By substitution in the correction term of the values from equations (300) and (301) for a^2 and c^2, respectively, and in the leading term the expression

$$a^2 = \frac{3^{2/3}}{x_b^2}\left(1 - \frac{3^{1/3}}{10}x_b\right) \tag{304}$$

for a^2 to second order from the results (p. 338) for zero temperature, it is easy to show that the coefficient of the term of χ_b in x_b^4 is identically zero. Thus, the argument demonstrates that equation (302) of GI is in fact accurate to better than $O(x_b^4)$.

In a similar manner, the initial slope χ'_i of χ at the origin can be obtained to the same order of accuracy from the Taylor expansion and the result is

$$\chi'_i = -\frac{2}{3^{5/3}}x_b^2 \tag{305}$$

Again, the coefficient of the second term (in x_b^3) is identically zero, and equation (305) is simply the result of GI.

Since it has now been established that equations (302) and (305) are more accurate than was evident from the derivation used in GI, it is possible to calculate the parameters entering the temperature perturbations of thermodynamic functions for small x_b more accurately than was possible in that treatment. The parameters entering these functions which depend on solution of the differential perturbation equation are σ and ω. From the definitions on p. 375 and using the result

$$\phi_b = \frac{3^{2/3}}{x_b}\left(1 - \frac{3^{1/3}}{10}x_b\right) \tag{306}$$

of March (equation (91)) for zero temperature, one obtains

$$\sigma = -\frac{2}{3^{7/3}}x_b^4\left(1 + \frac{3^{1/3}}{10}x_b\right) \tag{307}$$

and

$$\omega = -\frac{2}{3^{10/3}}x_b^4\left(1 + \frac{3^{1/3}}{4}x_b\right) \tag{308}$$

which are correct through the two terms given. For completeness, the parameters τ and ε_0, which depend only on the solution of the Thomas–Fermi equation for zero temperature, will be given also. Their values are

$$\tau = \frac{1}{3^{4/3}}x_b^4\left(1 + \frac{3^{1/3}}{5}x_b\right) \tag{309}$$

and

$$\varepsilon_0 = \frac{5}{3}\left(1 + \frac{3^{1/3}}{20}x_b\right) \tag{310}$$

In all four cases, the first terms in these parameters agree with the results of GI. Further proof of the correctness of the results embodied in equations (307) to (310) is given by the fact that the parameters as expressed by these equations satisfy the differential relation of GI (equation (221)) connecting σ, τ, ω, and ε_0 which must be obeyed by any exact solution.

From these results for σ, τ, ω, and ε_0 for small x_b, the corresponding expressions for thermodynamic functions follow directly from the discussion of p. 376. It should be noted that the pressure P depends only on the parameters σ and τ and can be written as

$$P = p\left\{1 + \frac{10}{3^{7/3}} x_b^4 \left(1 + \frac{3^{1/3}}{4} x_b\right) \zeta(kT)^2\right\} \tag{311}$$

for x_b small, where p is the corresponding pressure at zero temperature.

Fitted functions

As in the zero-temperature case, it is desirable to represent the pertinent quantities, which the differential equation yields, by fitted functions having the proper asymptotic behavior in the two limits $x_b \to 0$ and $x_b \to \infty$. It is convenient, of course, to use parameters corresponding to the method of Gilvarry, to avoid the complication of a perturbation in the volume. The available results are represented by the numbers of table 18 from GIII, corresponding to the data of FMT, and the numbers of table 20, obtained by Gilvarry and Peebles from the data of SK.

The parameter χ_b can be represented by

$$\chi_b = \sum_n C_n x_b^n \tag{312}$$

if n ranges over the sequence

$$n = 3, n', 5 \tag{313}$$

with n' and the corresponding coefficient $C_{n'}$ disposable. The coefficients C_3 and C_5 are chosen to agree with the corresponding coefficients in the asymptotic forms of χ_b for $x_b \to 0$ and $x_b \to \infty$, respectively, as given from GI on p. 380. The values of C_3 corresponding to the powers x_b^3 and x_b^5, respectively, in equation (312) are thus

$$C_3 = -\frac{2}{3^{5/3}}, \qquad C_5 = -\frac{2\lambda_2 - 1}{216} \tag{314}$$

However, it has been shown by March (see the preceding subsection) that the result of GI for χ_b in the limit $x_b \to 0$ is actually accurate to better than $O(x_b^4)$, since the coefficient of x_b^4 in χ_b vanishes. For the fitted function of equation (312) to display this property, the disposable exponent n' must satisfy the constraint

$$4 < n' < 5 \tag{315}$$

Similarly, the parameter χ_i' can be represented by

$$\chi_i' = \left(\sum_m D_m x_b^{-m}\right)^{-1} \tag{316}$$

if m ranges over the sequence

$$m = 1 - \lambda_2, \; m', \; 2 \tag{317}$$

Thomas–Fermi atom model

with m' and the corresponding coefficient $D_{m'}$ disposable. As in the case of χ_b, the coefficients D_2 and $D_{1-\lambda_2}$ are chosen to yield the proper asymptotic behavior of the fitted function in the two limits $x_b \to 0$ and $x_b \to \infty$, respectively. The condition

$$1 - \lambda_2 < m' < 1 \tag{318}$$

on m' is necessary if equation (315) is to reflect the fact that the coefficient of the second term (in x_b^3) in χ_i' is identically zero as shown by March (see preceding subsection). The coefficients D_2 and $D_{1-\lambda_2}$ are given by

$$D_2 = -\tfrac{1}{2} 3^{5/3} \tag{319}$$

and

$$D_{1-\lambda_2} = \tfrac{1}{2} 3^{-7} \{(66 + 169\lambda_2)(3 + 2\lambda_1)^{5/2} \, 12^{2\lambda_2/3 - 1}\}^{-1} \tag{320}$$

respectively.

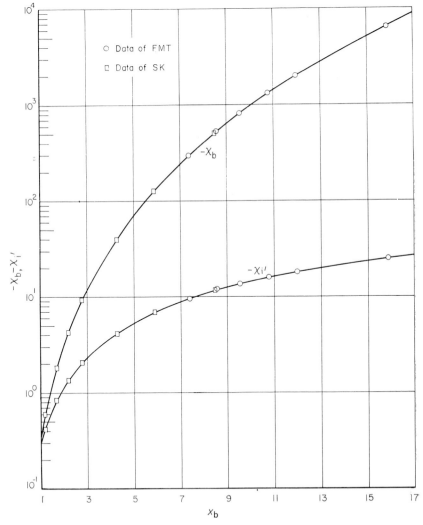

FIGURE 17 Fitted functions for χ_b and χ_i' against boundary radius x_b of atom. [From Gilvarry and Peebles]

The values of the coefficients and exponents in these fitted functions for χ_b and χ_i' are tabulated in table 22. The entry for $D_{1-\lambda_2}$ has been corrected from that given

TABLE 22 Coefficients[a] of fitted functions from data of SK and FMT; temperature-perturbed case [from Gilvarry and Peebles, with values of D_m and m' corrected]

n	C_n	m	D_m
3	$-3 \cdot 025 \times 10^{-1}$	$1-\lambda_2 = 0 \cdot 2280$	$-1 \cdot 046 \times 10^{-3}$
$n' = 4 \cdot 215$	$-2 \cdot 331 \times 10^{-2}$	$m' = 0 \cdot 6358$	$-1 \cdot 715 \times 10^{-1}$
5	$-2 \cdot 519 \times 10^{-3}$	2	$-3 \cdot 120$

[a] The coefficients are given to four figures to minimize round-off error.

in GI and GIII and by Gilvarry and Peebles[84], to take into account the correction from the value of GI in the coefficient of χ_i' in the limit $x_b \to \infty$, as represented by

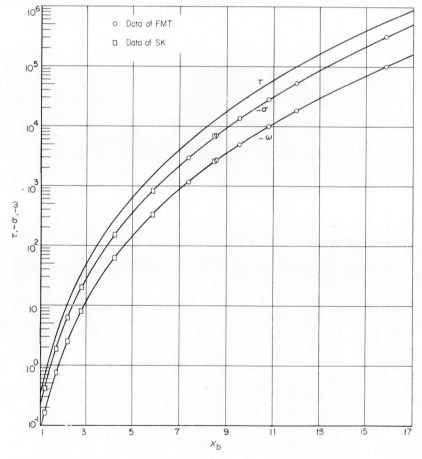

FIGURE 18 Perturbation parameters σ, τ, and ω from fitted functions, against boundary radius x_b of atom [from Gilvarry and Peebles]

Thomas–Fermi atom model

equation (235). Correspondingly the disposable exponent m' and coefficient $D_{m'}$ have been redetermined to fit the data. One notes that the disposable exponents n' and m' satisfy the requirements of inequalities (315) and (318), respectively. The corresponding fitted functions reproduce the data of tables 18 and 20 within 2·3% in χ_b 1·5% in χ_i'. The fitted functions are shown in figure 17, with data points from tables 18 and 20 for comparison.

Values of the perturbation parameters σ and ω from equations (205), computed by means of these fitted functions and the fitted function of GI for ϕ_b given on p. 334, are shown in figure 18 with directly computed values from table 18 of GIII and table 20 of Gilvarry and Peebles for comparison. The perturbation parameter τ is shown likewise from results of GI (p. 333). In figure 19, the effect of the first-order

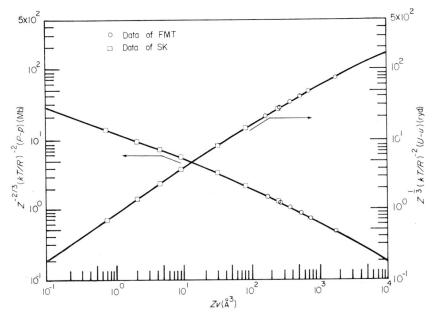

FIGURE 19 Scaled pressure and energy perturbations from fitted functions, against scaled volume in cubic ångströms [from Gilvarry and Peebles]

temperature perturbation on the equation of state is shown graphically by plotting $Z^{-2/3}(kT/R)^{-2}(P-p)$, where R is the rydberg, against the scaled volume Zv; the ordinate is independent of temperature. The energy perturbation $U-u$ in units of R is shown similarly; the value of $Z^{1/3}(kT/R)^{-1}S$ in $R/\text{deg}K$, where S is the entropy, differs from the plotted quantity by the numerical factor $1·27 \times 10^{-5}$. In figure 20, the dimensionless parameters $\varepsilon_T - \varepsilon_0$ and $\gamma - \gamma_0$ associated with the equation of state are shown in similar manner; data points corresponding to $\varepsilon_T - \varepsilon_0$ (for which fitted functions must be used to obtain the necessary derivatives) are omitted. The dimensionless quantity $8(2/9\pi^2)^{1/3}\phi_b/x_b$ shown in figure 20 represents a limit relative to which the scaled temperature $Z^{-4/3}kT/R$ must be small, for validity of the perturbation method.

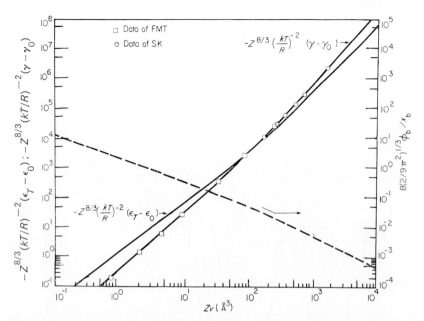

FIGURE 20 Scaled perturbations in parameters ε_T and γ from fitted functions, against scaled volume in cubic ångströms. The scaled temperature $Z^{-4/3}kT/R$ must be small relative to the quantity shown by the broken curve. [From Gilvarry and Peebles]

Discussion and Conclusions

The preceding development has yielded fitted functions over the entire span of atom radius from zero to infinity for the pertinent parameters derived from the zero-temperature Thomas–Fermi equation and the first-order temperature-perturbed equation, corresponding to accuracies of the order of some tenths of a per cent and a few per cent, respectively. The accuracy in the second case is consistent with that of the first, in the situation discussed where the effect of nonvanishing temperature represents merely a perturbation. The accuracies attained are desirable in the case at hand, where the corresponding physical terms represent dominant contributions to the equation of state and thermodynamic functions and thus should be calculated with the precision appropriate to unperturbed leading terms. Other physical factors represent perturbations, and hence can be computed with reduced accuracy, in general, relative to the dominant terms considered in the development given.

However, it should be emphasized that the physical terms corresponding to a Thomas–Fermi computation for zero-temperature and first-order temperature perturbation represent merely the first two terms in an asymptotic series expansion. For physical validity, one must include at least the effect of exchange, correlation, and the influence of the lattice. Any mathematical accuracy attained in the Thomas–Fermi case (except at the very highest pressures) is illusory unless the corresponding corrections are included. In particular, the asymptotic validity of the equation of state above a pressure of 100 Mb implied by the discussion of Takeuchi and Kanamori[26] (see p.317) is valid only if the effect of exchange corresponding to the Thomas–Fermi–

Dirac equation is included in the theoretical results for zero temperature. Further, it can be noted that at the lower pressures the effect of exchange is not represented merely by a perturbation, but becomes comparable (and with opposite sign) with the magnitude of the effect given by the presumably dominant Thomas–Fermi term.

The original model of Thomas and Fermi ignores the quantum-mechanical interaction of electrons except for the mean screening effect of the electron cloud as manifested in the electrostatic energies. The first step in improving the basic approximation was taken by Dirac[7] who included the exchange energy of a free-electron gas. This procedure takes into account the quantum-mechanical tendency for electrons of like spin to stay apart from one another by virtue of the exclusion principle, through the use of a wave function for the electrons which is an antisymmetrized product of one-electron wave functions. The effect of this procedure evidently is to reduce the energy of the system, through its effect on the potential energy corresponding to the electron–electron interaction.

In general, the physical properties of the atom sensitive to the electron density at the atom boundary are poorly given by the simple Thomas–Fermi model. Thus, the model does not permit any cohesion, since the atom under zero pressure displays an infinite radius. Further, the pressure for a compressed atom is too high in general. With introduction of exchange, a limiting radius for an atom under zero pressure appears, as shown by Jensen[86]. Concomitantly, the pressure of a compressed atom is decreased relative to the Thomas–Fermi value. Both improvements result from the more accurate description of conditions near the atom boundary by the Thomas–Fermi–Dirac equation, which implies a lower electron density at this point.

Numerical solutions of the Thomas–Fermi–Dirac equation have been given by Feynman, Metropolis, and Teller[19], by Metropolis and Reitz[87], and by Thomas[88]. The results of the latter for zero pressure encompass essentially the whole atomic table.

Physical properties specified by the Thomas–Fermi–Dirac equation do not scale directly with atomic number, as is true in the Thomas–Fermi case. Thus, it has been shown that a range of solutions of the Thomas–Fermi equation spanning values of the atom radius from zero to infinity is sufficient to specify the thermodynamic functions of *every* atom through a fitted function; in the Thomas–Fermi–Dirac case, the process would have to be carried out separately for *each* atomic number, and the vector of coefficients A_n in equation (80), for example, would become a matrix. The asymptotic relations between the Thomas–Fermi–Dirac and Thomas–Fermi atom models considered by Gilvarry[89], by Gilvarry, March, and Hartle[90], and by Hartle and Gilvarry[91] serve to bypass this problem. As first suggested by Gilvarry[89], attention was directed in this work not to solutions of the Thomas–Fermi–Dirac equation, as such, but to the problem of obtaining asymptotic relations between properties of a Thomas–Fermi–Dirac atom and a Thomas–Fermi atom. At least in one case, that of low atomic number, where the program has been carried out, this procedure permits one to express rigorously thermodynamic functions of a Thomas–Fermi–Dirac atom directly in terms of corresponding functions of the Thomas–Fermi atom of the same radius. Results in other cases are only approximate, but are nevertheless useful.

The effect of exchange at low temperature has been treated by Umeda and Tomishima[92] and by Gilvarry[47] on the basis of an approximate theory. Subsequent to this work, the rigorous theory of the effect of exchange at arbitrary temperature was given by Cowan and Ashkin[93]. Limitation of the results for thermodynamic functions to the case of a first-order temperature perturbation is not a serious one, since the effects of exchange become relatively unimportant for elevated temperature[19]. A far more important limitation on the physical validity of the results obtained arises from neglect of correlation.

In addition to exchange, it is necessary to take into account the tendency of electrons, independently of spin orientation, to stay apart because of their Coulomb repulsion. The corresponding effect clearly is to lower the potential energy corresponding to the electron–electron interaction, as does the effect of exchange. This lowering of the total energy is ignored in any computational scheme based on one-electron wave functions. Thus, the correlation energy represents the true total energy corresponding to exact wave functions for the many-electron problem less the energy including the effect of exchange computed on the basis of one-electron wave functions.

As compared with the case with exchange, the theory of the statistical atom model including the effect of correlation is in a very incomplete state. Wigner[8, 94] has pointed out that the total correlation energy can be estimated simply for the case of large atom radius, where the kinetic energy of the free electrons is negligible. He has given also an approximate calculation of the correlation energy for the case of high electron density. Further, he has suggested an approximate interpolation formula yielding asymptotically the results derived for the two limits of low and high electron density.

The results of Wigner for high density have been superseded by the exact calculations of Gell-Mann and Brueckner[95] for this limit. These investigators considered the idealized problem of the ground-state energy of a gas of electrons in the presence of a uniform background of positive charge. Their computational method is based on summing the most highly divergent terms of a perturbation series under the integral sign to give a convergent result, by a technique similar to Feynman's method in field theory. Some of their results were anticipated by or could have been inferred from previous work of Wigner[8, 94] and Pines[96]. The work of Pines made use of the theory of collective electron oscillations.

On the basis of the results of Gell-Mann and Brueckner, Lewis[97] has given a semi-theoretical modification of the Thomas–Fermi–Dirac equation to include the effect of correlation. The scheme is approximate in that correlation is incorporated in the analog of the Thomas–Fermi–Dirac equation in a manner which is exact at small atom radius on the basis of the results of Gell-Mann and Brueckner and is also exact at large atom radius on the basis of Wigner's theory. Since the method yields a reasonable interpolation between these two extreme limits, it probably yields results not far from the truth at any intermediate value of atom radius.

It should be mentioned that various suggestions have been made (for example, by Landsberg[98]) for the use of a screened Coulomb potential as an artifice to estimate the effect of correlation. In this connection, it can be noted that a basic error of the Thomas–Fermi model, which is not unrelated to ignoration of the correlation energy, has its origin in the physical circumstance that electrons do not interact with themselves. A crude effort to take this fact into account was made by Fermi and Amaldi[99], simply by multiplying the charge distribution each electron sees by the factor $(Z-1)/Z$, where Z is the atomic number.

For low temperature, Gell-Mann[100] has treated the case of excited states of a degenerate gas of free electrons at high density by the same method used by

Gell-Mann and Brueckner to determine the effect of correlation for the ground state. He has used the results to calculate the heat capacity of such a gas at low temperature.

While an approximate treatment of the effect of exchange on the properties of a statistical atom at low temperature is possible through use of the Umeda–Tomishima equation or the corresponding perturbation equation of Cowan and Ashkin, no corresponding equation incorporating the effect of correlation is available. Thus, no method is known at present to take into account the effect of nonvanishing temperature when *both* exchange and correlation are included, except in the limit of high density where Gell-Mann's results become applicable. Since the two effects tend individually to be profound for low temperature and show generally opposite signs, this circumstance has the important consequence that the temperature-perturbed Thomas–Fermi theory described in this chapter yields numerical results which are more realistic physically than can be obtained with inclusion of exchange but neglect of correlation. This point has been emphasized by Gilvarry[47].

The question of evaluating the effect of the lattice on thermodynamic functions at high pressure has been discussed by Gilvarry[16]. The Grüneisen constant of the solid is fundamental in determining the contribution of the lattice to the pressure under conditions of high compression. Since the theoretical results are strictly valid only for a body in the solid phase, it is necessary to determine the course of the fusion curve at high pressure. This can be done on the basis of a reformulation of the Lindemann law of melting given by Gilvarry[17]. This formulation of the law yields an immediate connection with the Grüneisen law determining the thermal expansion of a solid. Thus, the results permit valid computation of the thermal expansion, heat capacity, and entropy of a solid at high pressure, which are specified only incompletely by the statistical atom model alone.

References

1. L. H. Thomas, *Proc. Cambridge Phil. Soc.*, **23**, 542 (1927).
2. E. Fermi, *Atti. Accad. Nazl. Lincei, Rend., Classe Sci. Fis., Mat. Nat.*, **6**, 602 (1927).
3. E. Fermi, *Z. Physik*, **48**, 73 (1928).
4. P. Debye and E. Hückel, *Physik. Z.*, **24**, 185 (1923).
5. P. Gombas, *Die statistische Theorie des Atoms und ihre Anwendungen*, Springer Verlag, Vienna, 1949.
6. P. Gombas, *Handbuch Phys.*, **36** (1956).
7. P. A. M. Dirac, *Proc. Cambridge Phil. Soc.*, **26**, 376 (1930).
8. E. Wigner, *Phys. Rev.*, **46**, 1002 (1934).
9. L. Brillouin, *Actualites Sci. Ind.*, **160**, 1 (1934).
10. I. Fényes, *Z. Physik*, **125**, 336 (1949).
11. N. H. March and J. S. Plaskett, *Proc. Roy. Soc. (London), Ser. A*, **235**, 419 (1956).
12. N. H. March, *Advan. Phys.*, **6**, 1 (1957).
13. E. Wigner and F. Seitz, *Phys. Rev.*, **43**, 804 (1933).
14. J. C. Slater and H. M. Krutter, *Phys. Rev.*, **47**, 559 (1935).
15. J. C. Slater, *Introduction to Chemical Physics*, McGraw-Hill, New York, 1939.
16. J. J. Gilvarry, *Phys. Rev.*, **102**, 317 (1956).
17. J. J. Gilvarry, *Phys. Rev.*, **102**, 308 (1956).
18. E. Teller, *Rev. Mod. Phys.*, **34**, 627 (1962).
19. R. P. Feynman, N. Metropolis, and E. Teller, *Phys. Rev.*, **75**, 1561 (1949).
20. J. J. Gilvarry and W. G. McMillan, unpublished.
21. W. M. Elsasser, *Science*, **113**, 105 (1951).

22. W. H. Ramsey, *Monthly Notices Roy. Astron. Soc.*, **110**, 444 (1950).
23. J. M. Walsh and R. H. Christian, *Phys. Rev.*, **97**, 1544 (1955).
24. R. G. McQueen and S. P. Marsh, *J. Appl. Phys.*, **31**, 1253 (1960).
25. L. V. Al'tshculer, S. B. Kormer, M. J. Brazhnik, L. A. Vladimorov, M. P. Speranskaya, and A. I. Funtikov, *Soviet Phys.—JETP (English Transl.)*, **11**, 766 (1960).
26. H. Takeuchi and H. Kanamori, *J. Geophys. Res.*, **71**, 3985 (1966).
27. P. W. Bridgman, *The Physics of High Pressure*, Bell, London, 1949.
28. K. E. Bullen, *Monthly Notices Roy. Astron. Soc., Geophys. Suppl.*, **6**, 383 (1952).
29. F. Birch, *J. Geophys. Res.*, **57**, 227 (1952).
30. L. Knopoff and G. J. F. MacDonald, *Geophys. J.*, **3**, 68 (1960).
31. S. Chandrasekhar, *An Introduction to the Study of Stellar Structure*, University of Chicago Press, Chicago, 1939, Chap. XI.
32. R. E. Marshak and H. A. Bethe, *Astrophys. J.*, **91**, 239 (1940).
33. H. Brown, *Astrophys. J.*, **111**, 641 (1950).
34. J. G. Scholte, *Monthly Notices Roy. Astron. Soc.*, **107**, 237 (1947).
35. J. J. Gilvarry and J. E. Hill, *Astrophys. J.*, **124**, 610 (1956).
36. J. J. Gilvarry, *Nature*, **188**, 886 (1960).
37. J. J. Gilvarry, *Nature*, **190**, 1048 (1961).
38. G. Wares, *Astrophys. J.*, **100**, 159 (1944).
39. J. J. Gilvarry, *Nature*, **178**, 1249 (1956).
40. J. J. Gilvarry, *J. Atmos. Terrest. Phys.*, **10**, 84 (1957).
41. J. J. Gilvarry, *Phys. Rev. Letters*, **16**, 1089 (1966).
42. T. Sakai, *Proc. Phys. Math. Soc. Japan*, **24**, 259 (1942).
43. M. K. Brachman, *Phys. Rev.*, **84**, 1263 (1951).
44. J. J. Gilvarry, *Phys. Rev.*, **96**, 934 (1954).
45. J. J. Gilvarry, *Phys. Rev.*, **96**, 944 (1954).
46. J. E. Mayer and M. G. Mayer, *Statistical Mechanics*, Wiley, New York, 1940, p. 374.
47. J. J. Gilvarry, *Phys. Rev.*, **107**, 33 (1957).
48. R. D. Cowan and J. Ashkin, *Phys. Rev.*, **105**, 144 (1957).
49. R. Fowler and E. A. Guggenheim, *Statistical Thermodynamics*, Cambridge University Press, Cambridge, 1956, pp. 58ff.
50. J. L. Finck, *Thermodynamics from the Classic and Generalized Standpoints*, Bookman Associates, New York, 1955, pp. 67ff., 108ff.
51. J. J. Gilvarry, *J. Chem. Phys.*, in press (1968).
52. M. K. Brachman, *J. Chem. Phys.*, **22**, 1152 (1954).
53. H. Reiss, *J. Chem. Phys.*, **21**, 1209 (1953).
54. N. H. March, *Phil. Mag.*, **44**, 346 (1953).
55. J.J. Gilvarry and W. G. McMillan, *Phys. Rev.*, **105**, 579 (1957).
56. J. J. Gilvarry, *J. Chem. Phys.*, in press (1968).
57. E. A. Milne, *Proc. Cambridge Phil. Soc.*, **23**, 794 (1927).
58. J. J. Gilvarry, *J. Chem. Phys.*, in press (1968).
59. M. W. Zemansky, *Heat and Thermodynamics*, McGraw-Hill, New York, 1937.
60. J. McDougall and E. C. Stoner, *Phil. Trans. Roy. Soc. London, Ser. A*, **237**, 67 (1938).
61. E. Baker, *Phys. Rev.*, **36**, 630 (1930).
62. H. Sauvenier, *Bull. Soc. Roy. Sci. Liege*, **8**, 313 (1939).
63. A. Sommerfeld, *Z. Physik*, **78**, 19 (1932).
64. K. Umeda, *Phys. Rev.*, **83**, 651 (1951).
65. N. H. March, *Proc. Phys. Soc. (London), Ser. A*, **68**, 726 (1955).
66. P. J. Rijnierse, *Proc. Roy. Soc. (London), Ser. A*, **292**, 288 (1966).
67. J. J. Gilvarry and N. H. March, *Phys. Rev.*, **112**, 140 (1958).
68. A. Sommerfeld, *Rend. Reale Accad. Nazl. Lincei*, **15**, 788 (1932).
69. S. Kobayashi, T. Matsukuma, S. Nagai, and K. Umeda, *J. Phys. Soc. Japan*, **10**, 759 (1955).
70. C. A. Coulson and N. H. March, *Proc. Phys. Soc. (London), Ser. A*, **63**, 367 (1950).
71. N. H. March, *Proc. Cambridge Phil. Soc.*, **48**, 665 (1952).
72. R. Latter, *J. Chem. Phys.*, **24**, 280 (1956).

73. K. Umeda, *J. Phys. Soc. Japan*, **9**, 291 (1954).
74. E. H. Kerner, *Phys. Rev.*, **83**, 71 (1951).
75. R. Latter, *Phys. Rev.*, **99**, 1854 (1955).
76. J. J. Gilvarry, *J. Chem. Phys.*, in press (1968).
77. R. E. Hartle and J. J. Gilvarry, *J. Chem. Phys.*, **45**, 4367 (1966).
78. M. G. Mayer and J. J. Gilvarry, to be published.
79. E. Fermi, *Mem. Reale Accad. Ital., Classe Sci. Fis., Mat. Nat.*, **1**, 1 (1930).
80. M. G. Mayer, unpublished.
81. A. Sommerfeld, *Z. Physik*, **78**, 283 (1932).
82. J. M. C. Scott, *Phil. Mag.*, **43**, 859 (1952).
83. N. H. March, *Phil. Mag.*, **44**, 1193 (1952).
84. J. J. Gilvarry and G. H. Peebles, *Phys. Rev.*, **99**, 550 (1955).
85. N. H. March, *Proc. Phys. Soc. (London), Ser. A*, **68**, 1145 (1955).
86. H. Jensen, *Z. Physik*, **93**, 232 (1935).
87. N. Metropolis and J. R. Reitz, *J. Chem. Phys.*, **19**, 555 (1951).
88. L. H. Thomas, *J. Chem. Phys.*, **22**, 1758 (1954).
89. J. J. Gilvarry, *J. Chem. Phys.*, **27**, 150 (1957).
90. J. J. Gilvarry, R. E. Hartle, and N. H. March, *J. Chem. Phys.*, **47**, 1029 (1967).
91. R. E. Hartle and J. J. Gilvarry, *J. Chem. Phys.*, in press (1968).
92. K. Umeda and Y. Tomishima, *J. Phys. Soc. Japan*, **8**, 360 (1953).
93. R. D. Cowan and J. Ashkin, *Phys. Rev.*, **105**, 144 (1957).
94. E. Wigner, *Trans. Faraday Soc.*, **34**, 678 (1938).
95. M. Gell-Mann and K. Brueckner, *Phys. Rev.*, **106**, 364 (1957).
96. D. Pines, *Phys. Rev.*, **92**, 626 (1953).
97. H. W. Lewis, *Phys. Rev.*, **111**, 1554 (1958).
98. P. T. Landsberg, *Proc. Phys. Soc. (London), Ser. A*, **62**, 806 (1949).
99. E. Fermi and E. Amaldi, *Mem. Accad. Sci. ist. Bologna*, **6**, 117 (1935).
100. M. Gell-Mann, *Phys. Rev.*, **106**, 369 (1957).

2

N. H. MARCH

Department of Physics
The University
Sheffield, England

High-pressure behaviour of solids

Introduction

The purpose of this chapter is to consider the behaviour of solids, and in particular of metals, under very high pressure. It is to be expected that, under sufficient compression, all solids will become metallic, and therefore it is quite appropriate to focus attention on the metallic state.

So far, theoretical treatments have relied heavily on the statistical atom model discussed in the previous chapter by Gilvarry (see section C II, chapter 1 of this book). While this model has the merit that a self-consistent field is established for each lattice spacing, it has the defect that it does not expose the shell structure characteristic of atoms and is therefore only valid when the shell structure is broken down. Unfortunately the statistical atom model gives us no obvious way of telling when such pressure ionization occurs. In this chapter we shall make a start on the problem of determining how this critical pressure varies with atomic number by estimating for the light elements when the K shell is broken open. We do this, on pp. 407–10, by considering the formation of bound states in appropriately screened ionic potentials. Then the problem of the conductivity of metals under pressure is dealt with on pp. 410–17, and, in particular, elementary theories are given for scattering cross sections of electrons off individual ions and for the Debye temperature. This work complements important low-pressure studies of Young and his collaborators which are also briefly discussed here. Finally, a few remarks are made about interionic forces in metals at exceedingly high pressures and possible stable crystal structures which might occur in this regime.

Equations of State

Gilvarry, in the previous chapter, has discussed how equations of state may be calculated from the Thomas–Fermi statistical model. This, as we discuss further below, is a band theory, but without energy gaps.

In principle, we know how to apply the full apparatus of band theory (i.e. avoiding the Thomas–Fermi approximation, which assumes that the potential acting on an electron varies only slowly over a distance of the order of a de Broglie wavelength of an electron at the Fermi surface). Restricting ourselves to $T = 0$, the total energy (electronic) of the metal may be expressed as the sum of four contributions:

$$E_{\text{total}} = E_0 + E_{\text{Fermi}} + E_{\text{exchange}} + E_{\text{correlation}} \tag{1}$$

E_0 is the energy of the lowest state in the band, E_{Fermi} describes the energy associated with the filling up of the band, while the last two terms arise from the electron interactions.

To carry out such calculations self-consistently in band theory, as a function of lattice spacing, is a formidable task. However, results have been obtained on the high-pressures forms of helium by Simcox and March[1] and for lithium by Johnson and March[2]. The equation of state which was obtained for helium may be written in the form

$$10^{-8} p = \frac{1 \cdot 620}{r_W^5} - \frac{1 \cdot 098}{r_W^4} - \frac{0 \cdot 0144}{r_W^3} + \frac{0 \cdot 0316}{r_W^2} + \ldots \tag{2}$$

where r_W is the radius of the Wigner–Seitz equal volume sphere. Here the pressure p is in atmospheres and r_W is measured in units of the Bohr radius a_0. To establish some

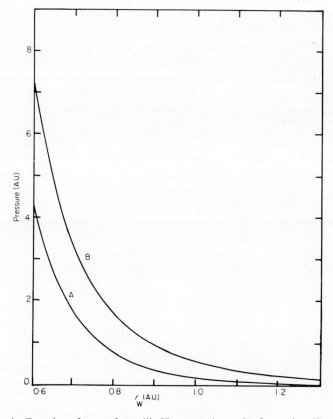

FIGURE 1 Equation of state of metallic He: curve A, result of equation (2); curve B, result for non-interacting Fermi gas. N.B. 1 atomic unit of pressure = $2 \cdot 90 \times 10^8$ atm (1 A.U. = 1 atomic unit)

magnitudes, let us take r_w actually equal to a_0. While it is not quite clear that there is adequate convergence, it will be seen that the first two terms contribute about 0.52×10^8 to the pressure, whereas the last two in equation (2) give a contribution of only 0.02×18^8 atm. In figure 1 we plot this equation of state, and it will be seen that the non-interacting Fermi gas theory appreciably overestimates the pressure in the range of densities shown.

The equation of state (2) is valid when the first and second energy bands overlap, and we shall discuss the pressure at which this occurs in the next section. However, the significance of the numbers above is that, at $r_w = a_0$ (which we shall see, in the next section, represents roughly the point where the K shell merges into the conduction band!), the first two terms are adequate. In fact, these agree with the equation of state given earlier, from Thomas–Fermi theory, for a general atomic number Z by March[3]. The result may be written

$$p = \frac{h^2}{5m}\left(\frac{3}{8\pi}\right)^{2/3} \frac{Z^{5/3}}{v^{5/3}} \left\{1 - \frac{2\pi me^2}{h^2}(4Zv)^{1/3} - \frac{10\pi me^2}{3^{2/3}h^2}(4Zv)^{1/3}\frac{6^{1/3}}{4(\pi Z)^{2/3}} + \ldots\right\} \quad (3)$$

where v is the atomic volume. We shall not record here the explicit results obtained by Johnson and March for the high-pressure equation of state of trivalent lithium, for we expect the Thomas–Fermi result (3) to work well at pressures just higher than are required to make Li go over from a monovalent to a trivalent metal. Instead, we turn immediately to discuss such electronic transitions.

Electronic Transitions and Critical Pressures

Simcox and March[1], following earlier work by Ten Seldam[4], studied by band theory the overlap of the first two energy bands in face-centred cubic helium. The Brillouin zone for this lattice is shown in figure 2. The energy levels W_s and $L_{p'}$, which lie respectively in the lowest and next higher band, will become equal for $r_w = 0.91 a_0$, where a_0 is the Bohr radius, the corresponding pressure being 1.0×10^8 atm. The calculation for X_s verifies that W_s is higher than X_s.

This type of calculation is lengthy and tedious and we therefore sketch below a cruder method of estimating the transition pressure at which bands overlap. Instead of starting from a periodic potential, say that given by the Thomas–Fermi theory, which would involve us again in a full band structure calculation, we adopt the point of view that each ion is screened by the Fermi gas. In calculating the screening, we must remember that all Z electrons in each atom are to be included in the regime of pressure ionization. Then, dealing with this screening by the linearized Thomas–Fermi theory, the potential around a single ion is given by

$$V = -\frac{Z}{r}\exp(-qr); \qquad q^2 = \frac{4k_F}{\pi a_0} \quad (4)$$

where k_F is the Fermi wave number. The periodic potential at each lattice spacing is then thought of as built up by superposing the localized potentials (4) centred on every lattice site. Not only is this picture useful for estimating critical pressures but, as we shall see in the next section, it is also valuable in calculating the electrical conductivity of metals, and for discussing interatomic forces (see p. 417).

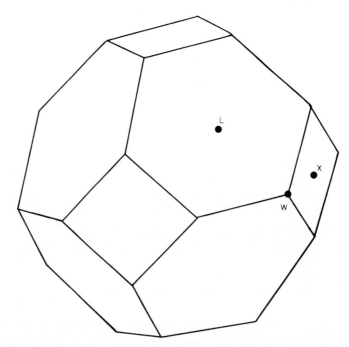

FIGURE 2 Brillouin zone for face-centred cubic lattice, showing high-symmetry points W, X, and L

We now consider the formation of bound states around screened point ions, as the density is reduced from the high-pressure regime in which the atoms are completely ionized. It is known that, for $Z = 1$, the critical value of the reciprocal screening radius q for bound-state formation round the potential (4) is $q = 1 \cdot 25/a_0$ (see, for example, Isenberg[5]). However, since the dependence of the screening radius on atomic number is required, we shall be content with a variational estimate using the hydrogenic $1s$ wave function

$$\phi = \left(\frac{\alpha}{\pi}\right)^{3/2} \exp(-\alpha r) \tag{5}$$

as the trial function, with α varied to minimize the energy. Then the critical value of q is given by

$$q_{(\text{bound})} = \frac{Z}{a_0} \tag{6}$$

which underestimates q by 25% for $Z = 1$. Hence, from (4),

$$k_{F\,(\text{bound})} = \frac{\pi Z^2}{4a_0} \tag{7}$$

This estimate of the density at which a band gap develops is too crude and seriously overestimates the transition pressure. This is because overlap of the wave functions[5] occurs, which broadens the bound state. In the appendix (see section on the broadening of bound states by the tight-binding method) we show how this broadening can be

High-pressure behaviour of solids

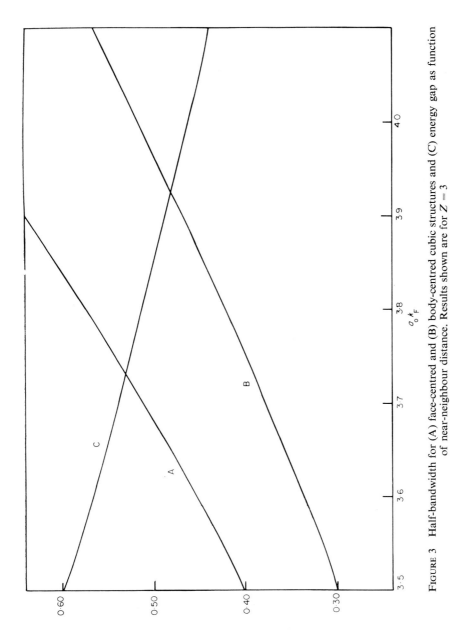

FIGURE 3 Half-bandwidth for (A) face-centred and (B) body-centred cubic structures and (C) energy gap as function of near-neighbour distance. Results shown are for $Z = 3$

estimated from the wave function overlap using the tight-binding approximation. The criterion used to define the electronic transition is then that the energy gap equals half the bandwidth. Results are shown in figure 3 for the face-centred cubic and body-centred cubic structures, for $Z = 3$. We have obtained similar results for the bandwidth of the 'bound state' and the energy gap, as a function of k_F, for $Z = 1, 2$, and 4, and the result (7) suggests that we represent the critical value of k_F as a power law. Doing so, we find

$$k_{F(\text{critical})} \simeq \frac{0 \cdot 6 Z^{5/3}}{a_0} \tag{8}$$

where, we stress, the exponent and the coefficient are only approximate. It turns out that at, such high densities, the equations of state are largely dominated by the Fermi gas term in (3) and we find for the transition pressure

$$p_{(\text{critical})} \propto k_{F(\text{critical})}^5 \propto Z^{25/3} \tag{9}$$

showing a rapid rise with atomic number. The increase in the transition pressure would have been like Z^{10} if we used the criterion for bound-state formation given by (7). It should be emphasized that our calculations were carried out only up to $Z = 4$, and some caution is therefore required in applying (8) to very large Z.

Two other points are worth noting. First, the coefficient of $Z^{5/3}$ in (8) is relatively insensitive to structure, differing by only 5% between face-centred and body-centred cubic structures. Secondly, the critical k_F for $Z = 2$ according to equation (8) is $1 \cdot 90/a_0$ whereas the band overlap calculation of Simcox and March gave a value of $2 \cdot 66/a_0$. This suggests that a more accurate estimate of the coefficient of $Z^{5/3}/a_0$ will be 0·7 or 0·8. Using the value 0·8, we find for Li from equation (3) that the K shell breaks open at a pressure of 5×10^9 atm, and, as seen from equation (9), this value will increase very rapidly with Z.

Thus, the conclusion is somewhat discouraging: namely that the Thomas–Fermi results, which do not account for discrete energy levels, are probably only valid at pressures in excess of $p_{(\text{critical})}$ calculated above. To calculate accurate equations of state for a solid composed of heavy atoms, reflecting the breaking open of the various shells, is a formidable task.

Electrical Conductivity of Metals

We expect the changing number of electrons per atom in the conduction band, as we apply exceedingly high pressures, to reflect itself in the electrical properties, particularly the conductivity and the Hall effect. We shall therefore outline below a theory of the conductivity of pressure-ionized solid (and liquid) metals. The theory is crude but may be a useful guide for future work.

Theory for liquid metals

It will be convenient to present the theory first in a form suitable for dealing with the liquid state. Let ρ_0 be the number of ions per unit volume, related to the number of electrons per unit volume, n say, by

$$n = Z\rho_0 \tag{10}$$

High-pressure behaviour of solids

Also n is related to k_F by the usual Fermi gas result

$$n = \frac{k_F^3}{3\pi^2} \tag{11}$$

The ions scatter electrons and we denote the scattering potential energy for a single ion by $U(K)$, where we have chosen to work with the Fourier transform of the potential energy $U(r)$. $U(K)$ is chosen so that

$$U(0) = -\tfrac{2}{3}E_F \tag{12}$$

where E_F is the Fermi energy; this follows only in the linear theory we are using. We shall later identify $U(r)$ with $V(r)$ of equation (4) in the pressure-ionized regime but *not* otherwise.

The structure of the liquid is also needed. This gives us the instantaneous picture of the distribution of the other ions around an ion we choose to 'sit' on. This is generally described by the radial distribution function $g(r)$, such that $g(r) \to 1$ as $r \to \infty$. The structure factor $S(K)$ is the Fourier transform of $g(r) - 1$. For liquids under normal conditions, $S(K)$ can be measured by X-ray or neutron experiments. Such measurements are, of course, not feasible at the high pressures of interest to us. Fortunately, an approximate theory can be given in terms of $U(K)$, as we shall see below.

Then the basic formula given by Ziman[6] for the resistivity ρ_l of the liquid, in the Born approximation, is

$$\rho_l = \frac{3\pi}{\hbar e^2 v_F^2 \rho_0} \int_0^1 S(K) |U(K)|^2 4\left(\frac{K}{2k_F}\right)^3 d\left(\frac{K}{2k_F}\right) \tag{13}$$

where $v_F = \hbar k_F/m$ is the Fermi velocity, with m the electron mass. For a given temperature the problem is characterized then by the Fermi wave number k_F and the valency Z, ρ_0 being then known from equations (10) and (11).

It will be convenient to write $S(K)$ and $U(K)$ in terms of the dielectric function $\varepsilon(K)$ of the Fermi gas, in the extreme high-pressure regime. $\varepsilon(K)$ gives us the linear response of the Fermi gas to a charged ion, and in terms of it we have

$$U(K) = -\tfrac{2}{3}E_F \frac{q^2}{K^2\varepsilon(K)} \tag{14}$$

$$S(K) = \frac{1}{1 - ZU(K)/k_B T} \tag{15}$$

Two expressions for $\varepsilon(K)$ are available.
(i) Using semiclassical theory

$$K^2\varepsilon(K) = K^2 + q^2 \tag{16}$$

This follows immediately by taking the Fourier transform of equation (4) and comparing the result with equation (14).
(ii) Using wave theory

$$K^2\varepsilon(K) = K^2 + \frac{k_F}{\pi a_0} g\left(\frac{K}{2k_F}\right) \tag{17}$$

where

$$g(x) = 2 + \frac{x^2-1}{x}\ln\left|\frac{1-x}{1+x}\right|$$

The wave theory includes the de Broglie wavelength $2\pi/k_F$ for electrons at the Fermi surface in an essential way (actually π/k_F is the precise wavelength which dominates the problem).

Results for liquid metals are recorded in the appendix (see section on conductivity of solid and liquid metals in the regime of complete pressure ionization) where we also discuss a primitive theory of the melting temperature. We turn immediately then to discuss the application to solids, which represent our main interest here.

Application to solid metals

If we were dealing with independent ions, the structure factor $S(K)$ in equation (13) is simply unity. In the solid, for temperatures high compared with the Debye temperature θ, it turns out to be a useful, if rough, approximation to replace $S(K)$ by its long-wavelength limit $S(0)$. This in turn is given by the thermodynamic result

$$S(0) = \rho_0 k_B T K_T \tag{18}$$

where K_T is the isothermal compressibility. Introducing the velocity of sound v_s, we shall write (18) to sufficient accuracy for our purposes as

$$S(0) \simeq \frac{k_B T}{M v_s^2} \tag{19}$$

where M is the ionic mass. Recalling that elementary Debye theory gives

$$\theta = \frac{h}{k_B} \left(\frac{3}{4\pi}\right)^{1/3} \frac{v_s}{v^{1/3}} \tag{20}$$

where v is the atomic volume, we can write (13) in the approximate form

$$\rho_s = \frac{K_i k_B T}{M \theta^2} \tag{21}$$

where K_i clearly depends on the scattering from a single ion, and from (13) is proportional to

$$\int_0^1 |U(K)|^2 \left(\frac{K}{2k_F}\right) d\left(\frac{K}{2k_F}\right)$$

Consistent with the model of dielectric screening of ions, we can also construct a first-order theory for the Debye temperature θ. This was done in essence by Bohm and Staver[7], who pointed out that the velocity of sound is given by

$$v_s = \left(\frac{Zm}{3M}\right)^{1/2} v_F \tag{22}$$

where m is the electronic mass. Combining (20) and (22) we obtain

$$\theta = \frac{3^{1/6}}{2^{5/3} \pi^{2/3}} \frac{h^2}{m k_B} \left(\frac{m}{M}\right)^{1/2} \frac{Z^{5/6}}{v^{2/3}} \tag{23}$$

We have evaluated ρ_s using both equations (16) and (17) for the dielectric function $\varepsilon(K)$ and some details are recorded in the appendix. The semiclassical results can then be expressed in universal form, by plotting $C\rho_s k_F^3/T$ against $k_F a_0$, where $C = 4\hbar e^2/27\pi^3 m k_B$.

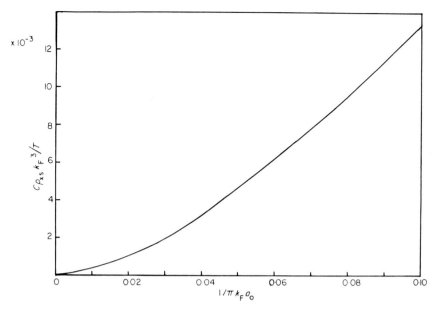

FIGURE 4 Restivity ρ_s of solid metals in the regime of complete pressure ionization. ρ_s is proportional to temperature T in this approximation when T is much higher than the Debye temperature θ. Constant $C = 4\hbar c^2/27\pi^3 m k_B$

Results are shown in figure 4 and we have verified that for large k_F the results are affected only in a minor way by using the wave theory equation (17). The essential point we wish to stress is that the dependence of ρ_s on volume v, in the extreme high-density limit $v \to 0$ is given by

$$\rho_s \propto v^{5/3} \ln v \qquad (24)$$

In terms of the pressure p, if we use simply the free Fermi gas equation of state given by the first term in equation (3), we find

$$\rho_s \propto p^{-1} \ln p \qquad (25)$$

The presence of the logarithmic term in (24), while weak compared with the $v^{5/3}$ term, is interesting theoretically, in that this transport property *cannot* therefore be expanded as a Taylor series in the lattice parameter. The result (24) indicates that the resistivity should decrease with increasing pressure, and, while we expect this to become true at sufficiently high pressures, if we examine measurements on the alkalis under pressure, then out to about 100 000 atm the behaviour is quite different. This problem has been examined by Dickey, Meyer, and Young[8] and we shall briefly summarize their results here.

Alkali metals with bound electrons in cores

Evidently, the presence of bound electrons in cores must be crucial, since otherwise equation (25) should apply at least qualitatively. Dickey, Meyer, and Young[8] thus construct a single-centre potential, which transcends the point ion form (4) in the following way.

(i) They take the Hartree–Fock–Slater fields for $Li^+ \ldots Cs^+$.

(ii) They screen these ions, not by the dielectric functions (16) or (17) but with a shell of electrons at a distance s from the ion, with s chosen to satisfy the Friedel sum rule

$$\frac{2}{\pi} \sum_{l=0}^{\infty} (2l+1)\eta_l(k_F) = Z \qquad (26)$$

with $Z = 1$ for the alkali metals. Here, the η_l's are the phase shifts of the partial waves, evaluated at the Fermi surface. s is determined for each value of the lattice parameter, and hence again the change in the self-consistent field with compression is at least crudely simulated.

It will be clear from the above discussion that the scattering off a single ion is calculated beyond the Born approximation, and hence, in the formula (21), K_i is obtained from the phase shifts directly. In addition, Dickey, Meyer, and Young estimate the variation of the Debye temperature with volume semiempirically, rather than via the crude formula (23). Thus a comparison of the quantity K in (21) with experiment can be made. Scaling the results in terms of K_0, the value of K_i at atmospheric pressure, the 'experimental' results at room temperature, based on Bridgman's work and taken from

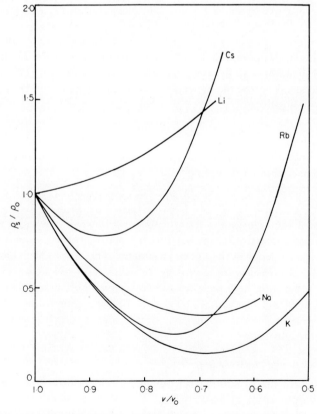

FIGURE 5 Measured resistivities ρs of alkali metals as function of atomic volume v, in units of atmospheric pressure values ρ_0 and v_0 respectively

High-pressure behaviour of solids

a paper by Dugdale[9] (for a more recent discussion see Dugdale and Phillips[10]), are shown in figures 5 and 6, and the theory for K_i/K_0 in figure 7. While the agreement with experiment is not fully quantitative (neither are the absolute values of ρ_s very good), the main features of the experiments are clearly revealed. Obviously, then, electrons in bound states dominate the problem in this pressure regime. Some qualitative difficulties remain, apart from the quantitative discrepancies referred to above. We mention only the following.

(i) The resistivity of Li drops suddenly, by a factor of about 4, at about 100 000 atm according to Drickamer[11]. If this is a crystal structure change, it is hard to understand it within the above framework. It must mean that the treatment of the phonons will have to be refined, along lines discussed by Darby and March[12] and Greene and Kohn[13].

(ii) The resistivity of K increases by a factor of the order of 100 in the pressure range 10^5–5×10^5 atm. It is tempting to ascribe this to the breaking open of the M shell. Young[14] has made qualitative estimates which suggest that a large increase (though substantially less than 100) in ρ_s could result from such a transition. The method outlined in the last section can, with small modifications, be used to estimate the pressure at which the M shell is expected to merge into the conduction band, and this matter is now being examined in this Department.

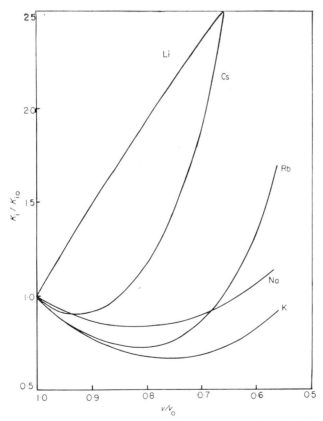

FIGURE 6 K_i/K_{i0} for alkali metals under pressure, as determined 'experimentally'. K_i is defined by equation (21), while K_{i0} is the value of K_i at atmospheric pressure

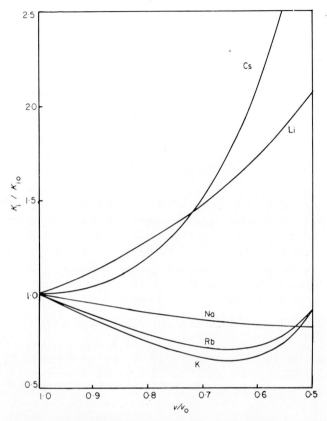

FIGURE 7 Theoretical values of K_i/K_{i0} for alkali metals under pressure

Pressure dependence of Hall effect

In connection with the resistivity results of Dickey, Meyer, and Young[8], Professor W. Paul has recently drawn my attention to measurements of the pressure dependence of the Hall constant at room temperature by Deutsch, Paul, and Brooks[15]. The Hall constat R is written as $1/nex$, where, as usual, n is the number of conduction electrons per cubic centimetre and x expresses the deviation from the free-electron value of the Hall constant.

For spherical energy surfaces, but with an anisotropic relaxation time τ defined by

$$\tau = \tau_0(1 + CK_4 + C_1K_6 + \ldots) \tag{27}$$

where K_4 and K_6 are cubic harmonics corresponding to $l = 4$ and $l = 6$, it has been shown by Davis[16] that

$$x = 1 - \frac{4}{21}C^2 - \frac{8}{13}C_1^2 \tag{28}$$

Over this range of pressure, x decreases for Li, Na, K, and Rb, but increases for Cs beyond about 11 000 or 12 000 kg/cm². It follows for Cs that we must have also non-spherical energy surfaces, since x begins to exceed unity when the pressure is above about 11 000 kg/cm².

Even though there appears some similarity between the pressure variation of the conductivity σ and x, we can say that, to order of K_4, there is no C^2 term in the conductivity. Thus, it appears valid to calculate the conductivity with a spherical approximation to the relaxation time, even though the free-electron Hall constant is not found. This argument is less well-founded for Cs, where we should consider the departure of the energy surfaces from the spherical symmetry more carefully.

In summary, we can say that very complex phenomena are to be expected in the variation of ρ_s with pressure, before the asymptotic form of equation (25) will be remotely relevant. An examination of the way the $3d$ electrons in Fe, say, both in the solid and liquid states, influence the electrical transport under pressure would be of considerable interest, both in geophysics and in current methods of producing very high magnetic fields using high explosive techniques. Further, even in a relatively simple metal like K, it would be very valuable if Hall measurements could be made eventually up to 5×10^5 atm to see whether rather direct experimental evidence can be found for the breaking open of the M shell.

Interionic Forces and Metallic Structures at Very High Pressures

From the model of dielectric screening of point ions in the pressure-ionized regime, the interionic forces can be calculated, as described by Corless and March[17] and Johnson and March[18]. In spite of the presence of the electron gas, a pair potential $\phi(r)$ can be defined (quite precisely when the Born approximation applies) and its asymptotic form for large r is given by

$$\phi(r) \sim \frac{\cos 2k_F r}{r^3} \tag{29}$$

These oscillations arise from the 'kink' at $2k_F$ in the wave theory form (17) for the dielectric function $\varepsilon(K)$. The presence of the subsidiary minima in the potential curve, and the facts that (a) the first minimum is not simply related to the near-neighbour distance, and (b) the first minimum is not always the deepest, lead one to speculate that the conventional close-packed structures may not be the stable ones in the extreme high-pressure regime. Structures with a lower coordination number, and chosen such that the subsidiary minima are carefully exploited, may well turn out to be of interest, and further work might be worth while on this aspect of the problem.

Acknowledgements

I am much indebted to Dr. S. Baranovsky who generously made the numerical calculations relating to the electronic transitions, and to Mr. P. C. Gibbs who has helped me greatly with the conductivity calculations. Dr. M. D. Johnson also contributed considerably to aspects of the work at an early stage.

I am most grateful to Dr. W. H. Young for allowing me to discuss the work of Dickey, Meyer, and Young before publication, and to him and to Dr. J. E. Enderby for valuable discussions. Finally, it is a pleasure to acknowledge that Professor G. Lehner first stimulated my interests in the problem of the conductivity of metals under pressure, in connection with the production of megagauss fields.

Appendix

Broadening of bound states by tight-binding method

We wish to obtain an approximation to the states of the crystal defined by the Hamiltonian

$$H = -\frac{\hbar^2}{2m}\nabla^2 + V_p \tag{A1}$$

where V_p is the periodic lattice potential. Let us consider the electron moving in the field of an isolated pseudoatom, with potential $V(r)$. Let the Schrödinger equation be

$$\nabla^2\phi + \frac{2m}{\hbar^2}(E-V)\phi = 0 \tag{A2}$$

We assume that V_p is built up as a sum of the V's centred on the sites $\mathbf{R}_n$:

$$V_p - V = \sum_{\mathbf{R}_n \neq 0} V(\mathbf{r} - \mathbf{R}_n) \tag{A3}$$

Then as discussed, for example, by Mott and Jones[19] we find the following results.

(a) *For the body-centred cubic lattice*

$$E = E_0 - \alpha - 8\gamma \cos\tfrac{1}{2}k_x a \, \cos\tfrac{1}{2}k_y a \, \cos\tfrac{1}{2}k_z a \tag{A4}$$

(b) *For the face-centred cubic lattice*

$$E = E_0 - \alpha - 4\gamma(\cos\tfrac{1}{2}k_y a \, \cos\tfrac{1}{2}k_z a + \cos\tfrac{1}{2}k_z a \, \cos\tfrac{1}{2}k_x a + \cos\tfrac{1}{2}k_x a \, \cos\tfrac{1}{2}k_y a) \tag{A5}$$

Here α and γ are defined by

$$\int \phi^*(\mathbf{r})\{V_p - V(\mathbf{r})\}\phi(\mathbf{r})\,d\mathbf{r} = -\alpha \tag{A6}$$

and

$$\int \phi^*(\mathbf{r}-\mathbf{R})\{V_p(\mathbf{r}) - V(\mathbf{r})\}\phi(\mathbf{r})\,d\mathbf{r} = -\gamma \tag{A7}$$

For the body-centred cubic case, the shortest lattice vectors $\mathbf{R}$ are given by

$$\mathbf{R} = (\pm\tfrac{1}{2}a, \; \pm\tfrac{1}{2}a, \; \pm\tfrac{1}{2}a) \tag{A8}$$

while for the face-centred cubic structure

$$\mathbf{R} = (0, \pm a, \pm a), \quad (\pm a, 0, \pm a) \quad \text{and} \quad (\pm a, \pm a, 0) \tag{A9}$$

The bandwidth for the body-centred cubic case is given by 16γ and for the face-centred cubic lattice 24γ. It is therefore necessary to calculate γ defined by equation (A7), taking the orbital $\phi(\mathbf{r})$ to be given by equation (5) and the local potential $V(r)$ to be defined by equation (4). Then we find

$$\gamma = \frac{\alpha^3 Z}{\pi} \int e^{-\alpha r} \frac{e^{-qr}}{r} e^{-\alpha|\mathbf{r}-\mathbf{R}|} \, d\mathbf{r} \tag{A10}$$

and performing the integration over angles equation (A10) yields

$$\gamma = \frac{2\alpha^3 Z}{R} \int_0^\infty dr\, e^{-(\alpha+q)r} \left[e^{-\alpha|r-R|}(\alpha|r-R|+1) - \frac{e^{-\alpha(r+R)}}{\alpha^2}\{\alpha(r+R)+1\} \right] \quad (A11)$$

The radial integration can now be completed, and the final result is

$$\gamma = \frac{2\alpha^3 Z}{R(2\alpha+q)^2} \left\{ \frac{(3\alpha+q)e^{-(\alpha+q)R}}{\alpha^2} - \frac{(\alpha+q)e^{-(3\alpha+q)R}}{\alpha^2} - \frac{2e^{-\alpha R}}{\alpha} \right\} \quad (A12)$$

From equation (A12), with the best variational values of α for each k_F, and for the corresponding near-neighbour distance R, the bandwidth shown in figure 3 was constructed as a function of k_F. Results are displayed for $Z = 3$ and for both body-centred cubic and face-centred cubic lattices.

Conductivity of solid and liquid metals in regime of complete pressure ionization

Solid in extreme high-pressure limit. As remarked earlier, a rough theory of the solid resistivity ρ_s for temperatures high compared with the Debye temperature θ is given by putting $S(K)$ in equation (13) equal to $S(0)$ given by equation (19). Using the Bohm–Staver result (22) for the velocity of sound, we find almost immediately

$$\rho_s = \frac{9\pi m k_B T}{4\hbar e^2 \rho_0 Z} \int_0^1 \frac{|U(K)|^2}{E_F^2} 4\left(\frac{K}{2k_F}\right)^3 d\left(\frac{K}{2k_F}\right) \quad (A13)$$

We make a first approximation to this by evaluating it with

$$U(K) = -\tfrac{2}{3} E_F \frac{q^2}{K^2+q^2} \quad (A14)$$

corresponding to the dielectric response function (22). We then obtain the result

$$\rho_s = \frac{6\pi m k_B T}{k_F^5 \hbar e^2 a_0^2} \left\{ \ln\left(1+\frac{1}{y}\right) - \frac{1}{1+y} \right\}, \qquad y = \frac{q^2}{4k_F^2} \quad (A15)$$

Noting that $q^2 \propto k_F$ from equation (4), we see that $y \propto k_F^{-1}$ which becomes smaller as the metal is compressed. Thus the curly brackets in equation (A15) tends to $-\ln y$ as $y \to 0$. But $y \propto k_F^{-1} \propto v^{1/3}$, and hence in the limit we have

$$\rho_s \to \frac{6\pi m k_B T}{k_F^5 \hbar e^2 a_0^2} \ln(k_F a_0) \quad (A16)$$

the result quoted in equation (24) when expressed in terms of v.

More detailed results for the variation of ρ_s with volume are given in figure 4. We should stress once again that the calculation neglects the details of the phonon spectrum and is, at best, a rough guide for temperatures higher than θ. It is worth noting that, if we write the conductivity σ in the form

$$\sigma = \frac{n e^2 \tau}{m} = \frac{n e^2 \lambda}{m v_F} \quad (A17)$$

where τ is the relaxation time and λ the electronic mean free path, then we find as $k_F \to \infty$

$$\lambda \to \frac{v_F}{4} \frac{(k_F a_0)^2}{\ln(k_F a_0)} \frac{h}{k_B T} \qquad (A18)$$

The results using the wave theory $\varepsilon(K)$ of equation (17) are not significantly different from the semiclassical theory for large k_F.

Resistivity of liquid metal. We have also calculated the resistivity of the liquid metals in the extreme high-pressure regime, with the structure factor (15) inserted in equation (13). The temperature dependence is less simple than in the approximation employed above for the high-temperature solid, because the structure factor (15) is now involved.

However, the resistivity can again be obtained in closed form if we use the dielectric constant $\varepsilon(K)$ of the semiclassical theory, and we shall briefly summarize the results. By analogy with the solid, it will be convenient to work with the quantity $C\rho_l k_F^3/T$, which from (13), (15) and (16) may be written as

$$\frac{Ck_F^3 \rho_l}{T} = \frac{q^4}{9k_F^4} \frac{2ZE_F}{3k_B T} \int_0^1 \frac{x^3 \, dx}{(x^2+z)(x^2+y)} \qquad (A19)$$

where $z = y\{1 + \tfrac{2}{3}(ZE_F/k_B T)\}$. The integral is readily evaluated and we find

$$\frac{Ck_F^3 \rho_l}{T} = \frac{8}{9} \frac{2ZE_F}{3k_B T} \frac{y^2}{z-y} \left\{ z \ln\left(1+\frac{1}{z}\right) - y \ln\left(1+\frac{1}{y}\right) \right\} \qquad (A20)$$

Hence again we have the resistivity explicitly as a function of k_F, Z, and the temperature T.

In order to ensure that we use the results in the physically relevant region, we need an estimate of the melting temperature T_m under pressure. Fortunately a rough theory leading to such an estimate has recently been given by March[20] and Enderby and March[21]. The roughest estimate (probably generally overestimating the melting point) is

$$k_B T_m \sim \frac{ZE_F}{30} \qquad (A21)$$

The relevance of this formula to the melting points for simple metals at atmospheric pressure is shown in figure 8; it is correct only as $Z \to 0$. Whether (A21) works well at high pressures depends on the validity of the Born approximation. We expect this to become valid in the extreme high-pressure limit. At $T = T_m$ as given by (A21) and as k_F tends to infinity, we can simplify (A20) and we find

$$\frac{C\rho_l k_F^3}{T_m} \to -\frac{8}{9} y^2 \ln y \quad 20 \qquad (A22)$$

This implies a very substantial increase in the resistance on melting, by a factor of order 20. While some increase is to be anticipated, this factor is probably considerably too large, and we believe the present estimate of the resistivity in the liquid to be more reliable than the calculation in the solid, owing presumably to the crude way the

High-pressure behaviour of solids

phonons are treated in that case. However, the calculations agree with regard to the form of the pressure dependence of ρ at the melting point, for it again follows from (A22) that

$$\rho_1 \propto p^{-1} \ln p \tag{A23}$$

in the extreme high-pressure regime. The temperature dependence in the liquid region is readily calculated from (A20), but we shall not go into further details here, except to remark finally that the use of the wave theory form of $\varepsilon(K)$ again appears to introduce only minor corrections for large k_F.

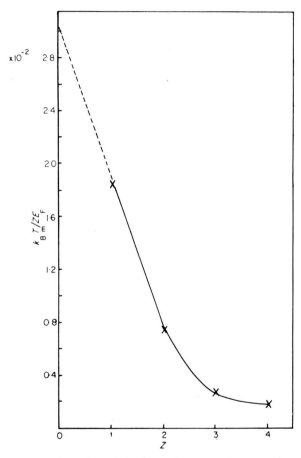

FIGURE 8 Melting points of simple metals at normal pressure (see appendix)

References

1. L. N. Simcox and N. H. March, *Proc. Phys. Soc. (London)*, **80**, 830 (1962).
2. M. D. Johnson and N. H. March, unpublished.
3. N. H. March, *Proc. Phys. Soc. (London)*, Ser. A, **68**, 726 (1955).
4. C. A. Ten Seldam, *Proc. Phys. Soc. (London)*, Ser. A, **70**, 97, 529 (1957).
5. I. Isenberg, *Phys. Rev.*, **79**, 737 (1950).

6. J. M. Ziman, *Phil. Mag.*, **6,** 1013 (1961).
7. D. Bohm and T. Staver, *Phys. Rev.*, **84,** 836 (1952).
8. M. Dickey, A. Meyer, and W. H. Young, *Proc. Phys. Soc.* (*London*), **92,** 460 (1967).
9. J. S. Dugdale, *Science*, **134,** 77 (1961).
10. J. S. Dugdale and D. Phillips, *Proc. Roy. Soc.* (*London*), *Ser. A*, **287,** 381 (1965).
11. H. G. Drickamer, *Solid State Phys.*, **17,** 1 (1965).
12. J. K. Darby and N. H. March, *Proc. Phys. Soc.* (*London*), **84,** 591 (1964).
13. M. P. Greene and W. Kohn, *Phys. Rev.*, **137,** 513 (1964).
14. W. H. Young, private communication.
15. T. Deutsch, W. Paul, and H. Brooks, *Phys. Rev.*, **124,** 753 (1961).
16. L. Davis, *Phys. Rev.*, **56,** 93 (1939).
17. G. K. Corless and N. H. March, *Phil. Mag.*, **6,** 1285 (1961).
18. M. D. Johnson and N. H. March, *Phys. Letters*, **3,** 313 (1963).
19. N. F. Mott and H. Jones, *The Theory of the Properties of Metals and Alloys*, Clarendon Press, Oxford, 1936, p. 65.
20. N. H. March, *Phys. Letters*, **20,** 231 (1966).
21. J. E. Enderby and N. H. March, *Proc. Phys. Soc.* (*London*), **88,** 717 (1966).

C

HIGH-PRESSURE PHYSICS AND THE EARTH'S INTERIOR

III. Laboratory data

O. L. ANDERSON

and

R. C. LIEBERMANN

Lamont Geological Observatory of Columbia University
Palisades, New York, U.S.A.

Elastic constants of oxide compounds used to estimate the properties of the Earth's interior*

Introduction

The difficulties inherent in attempting to infer the composition of the lower crust and the upper mantle on the basis of the observed seismic velocities are well known[1,2]. The approach adopted in the Mineral Physics Laboratory at the Lamont Geological Observatory has been to investigate thoroughly the physical properties of the materials which serve as basic building blocks in the structure of Earth-forming rocks and minerals. In figure 1, we have constructed a tetrahedron with the following oxides at

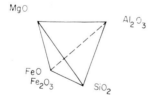

FIGURE 1   Composition tetrahedron within which are represented most of the minerals which are thought to be of interest in the mantle

its corner: Al_2O_3, MgO, SiO_2, FeO, or Fe_2O_3. Within this tetrahedron we can include the chemical composition of most of the minerals which are thought to be of interest in the mantle. The elastic constants and their pressure and temperature derivatives are well established for corundum, periclase, α-quartz, forsterite, spinel, and garnet. These data are presented in summary form in table 1. The properties of the iron oxides are currently under investigation in our laboratory. We have ignored the effect of CaO, since it is not usually considered to be an important constituent in the Earth's mantle.

It should be noted that only compositional and not structural relationships are

* Lamont–Doherty Geological Observatory Contribution No. 1296.

TABLE 1 Physical properties

Property (1 bar; 298 °K)	Corundum α-Al$_2$O$_3$ Polycrystalline	Periclase MgO Polycrystalline[a]	Periclase MgO Single crystal	Spinel MgO.2.6Al$_2$O$_3$ Single crystal	Quartz α-SiO$_2$ Single crystal	Garnet Single crystal	Forsterite Mg$_2$SiO$_4$ Single crystal	Forsterite Mg$_2$SiO$_4$ Polycrystalline
ρ, theoretical	3·987[54]	3·584[54]	3·584[54]	3·6193[29]	2·6483[54]	4·1602[30]		3·223[54]
ρ, bulk	3·972[3]	3·5800[5]	3·5833[28]		2·6485[11]			3·021[6]
M/p	20·39	20·16	20·16	20·36[29]	20·03	23·79[30]		20·10
$\alpha \times 10^6$	16·3[55]	31·8[53]	31·5[53]	22·5[58]	35·4[11]	21·6[30]		24·0[52]
$C_P \times 10^{-6}$	7·83[56]	9·41[57]	9·25[57]	8·03[56]	7·27[60]	7·61[30]		8·38[56]
v_p	10·845[3]	9·766[5]	9·692	9·914	6·047[51]	8·531[30]		7·586[6]
v_s	6·373[3]	5·964[5]	6·041	5·645	4·092[51]	4·762[30]		4·359[6]
$K_S = K_0$	2521[3]	1717[5]	1622[28]	2020[29]	377[51]	1770[30]		974[6]
$k_T = k_0$	2505[3]	1690[5]	1598[28]	1999[29]	374	1757[30]		968[6]
μ	1613[3]	1273[5]	1308[28]	1153[29]	443	943[30]		574
σ_s	0·236[3]	0·203[5]	0·182	0·260[29]	0·078	0·274[30]		0·254[6]
Pressure derivatives (per kb)								
$(\partial v_p/\partial P)_T \times 10^3$	5·18[3]	7·71[5]	8·29	4·91[29]	13·7[51]	7·84[30]		10·3[6]
$(\partial v_s/\partial P)_T \times 10^3$	2·21[3]	4·35[5]	3·97	0·43[29]	−3·4[51]	2·17[30]		2·45[6,61]
$(\partial K_S/\partial P)_T = K_0'$	3·98[3]	3·92[5]	4·49[28]	4·18[29]	6·42[51]	5·43[30]		4·8[6]
$(\partial k_T/\partial P)_T = k_0'$	3·99[3]	3·94[5]	4·52[28]	4·19[29]	(6·4)	5·45[30]		(4·8)
$(\partial \mu/\partial P)_T$	1·76	2·60[5]	2·56	0·75	0·45	1·40[30]		1·2
$(\partial \sigma_s/\partial P)_T \times 10^4$	1·02[3]	0·56[5]	2·04	3·00	50	3·03[30]		5·8[6]

Elastic constants of oxide compounds

Temperature derivatives (per degK)						
$(\partial v_p/\partial T)_P \times 10^4$	-3.7^3	-5.0^8	-4.9	3.25^{59}	-2.75^{51}	-3.93^{30}	-4.1^9
$(\partial v_s/\partial T)_P \times 10^4$	-3.1^3	-4.8^8	-4.0	2.85^{59}	$+0.05^{51}$	-2.18^{30}	-2.9^9
$(\partial K_S/\partial T)_P$	-0.15	-0.13^8	-0.16^{28}	0.13	-0.096^{51}	-0.20^{30}	-0.11
$(\partial k_T/\partial T)_P$	-0.20	-0.22	-0.24	-0.14	-0.105	-0.25	-0.12
$(\partial \mu/\partial T)_P$	-0.18	-0.25	-0.23	-0.11	-0.02	-0.11^{30}	-0.09
$(\partial \sigma_s/\partial T)_P \times 10^5$	1.2	2.8	1.3	0.54	-7.1	-0.023^{30}	0.92
Critical thermal gradients						
$(\partial T/\partial P)_{v_p}$ degK/kb	14.0	15.4^8	16.9	15.3	50.7	19.9^{30}	25.1
$(\partial T/\partial P)_{v_s}$ degK/kb	7.1	9.1^8	10.0	1.5	680	9.95	8.4
$(\partial T/\partial Z)_{v_p}$ degK/km	5.5	5.5	6.0	5.5	13.3	8.2	7.5
$(\partial T/\partial Z)_{v_s}$ degK/km	2.8	3.2	3.5	0.6	178	4.1	2.5
Grüneisen parameters						
γ	1.32^3	1.62	1.55^2	1.56	0.69^{51}	1.21	0.92
δ_s	3.4	2.4	3.1	2.9	7.4^{51}	5.3^{30}	4.7
Debye temperature θ	1029	931	941	887	572	745^{30}	647

[a] Data at 307·6 °K.
Symbols and units are as defined in the appendix.
Numerals have been calculated from the other data in the column; quantities in parentheses represent estimated values based on the other data.

implied in figure 1: one cannot obtain the properties of spinel simply by combining periclase and corundum in equal stoichiometric parts, since spinel has a different packing structure. However, by determining precisely the elastic properties of the Earth's constituent oxides, it may be possible to delimit the behavior of materials in the interior of the real Earth.

The elastic constants of oxide materials and their pressure and temperature derivatives exhibit patterns which are believed to hold for all of the major constituents of the Earth's mantle. The independent parameters are the ambient density and the mean atomic weight. Using the measured pressure and temperature derivatives of the elastic-wave velocities to estimate the thermal gradient necessary to produce a low-velocity layer, it is found that this critical gradient is lower for shear waves than for compressional waves for all of the materials studied. The law of corresponding states and the other observed patterns relate the parameters in Murnaghan's equation of state in such a way that chemical composition has little effect upon either the density–pressure trajectories for the lower mantle or the limiting density of the mantle near the core–mantle boundary.

Elastic Constants and their Dependence upon Pressure and Temperature

The adiabatic bulk modulus or incompressibility (K_S) is defined as

$$K_S = -V\left(\frac{\partial P}{\partial V}\right)_S = \rho\left(\frac{\partial P}{\partial \rho}\right)_S \qquad (1)$$

For an isotropic substance, we may express the elastic-wave velocities in terms of the bulk modulus (K_S), the shear modulus (μ), and density (ρ).

Compressional-wave velocity: $\quad v_p = \left(\dfrac{K_S + \tfrac{4}{3}\mu}{\rho}\right)^{1/2} \qquad (2)$

Shear-wave velocity: $\quad v_s = \left(\dfrac{\mu}{\rho}\right)^{1/2} \qquad (3)$

Solving for K_S, we obtain

$$(K_S = v_p^2 - \tfrac{4}{3}v_s^2) \qquad (4)$$

Now v_p and v_s are directly measurable in the laboratory by refined ultrasonic techniques[3,4]. The individual velocity measurements using high-quality specimens have a precision ranging from 0.03%[5] to 0.25%[6]; this precision permits us to determine the pressure derivatives of the velocities with an error of less than 5%[6]. An example of the data for v_p and v_s as a function of pressure is given in figure 2, using the measurements of Schreiber and Anderson[3] on polycrystalline Al_2O_3.

If we express K_S as a function of pressure and temperature by a Maclaurin series expansion about $P=0$ and $T=T_0$, we find

Elastic constants of oxide compounds

$$K_S(P,T) = K_S(0,T_0) + P\left(\frac{\partial K_S}{\partial P}\right)_{P=0,T} + \left(\frac{\partial K_S}{\partial T}\right)_{P,T_0}(T-T_0) + \left(\frac{\partial^2 K_S}{\partial P^2}\right)_{P=0,T}\frac{P^2}{2!}$$
$$+ \left(\frac{\partial^2 K_S}{\partial T \partial P}\right)_{T_0,P=0}(T-T_0)P + \left(\frac{\partial^2 K_S}{\partial T^2}\right)_{P,T_0}\frac{(T-T_0)^2}{2!} \cdots \qquad (5)$$

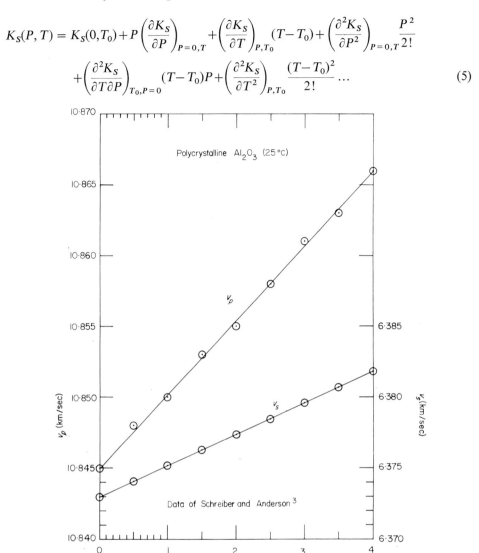

FIGURE 2 Compressional- and shear-wave velocities in polycrystalline Al_2O_3 as a function of pressure from 1 b to 4 kb [from data of Schreiber and Anderson[3]]

As an initial approximation justified in a previous paper by Schreiber and Anderson[3] we let

$$\left(\frac{\partial^n K_S}{\partial P^n}\right)_{P=0,T} = 0 \quad \text{for} \quad n \geq 2 \qquad (6)$$

In figure 3 we have plotted a theoretical relationship between K_S and T given by the following equation (from Soga and Anderson[7])

$$K_S(T) = K_S(0) - \frac{\gamma\delta}{V_0}\{H(T) - H(0)\} \qquad (7)$$

where δ is an important physical constant (virtually independent of T) given by

$$\delta = -\frac{1}{\alpha}\left(\frac{\partial \ln K_S}{\partial T}\right) \qquad (8)$$

Several examples of comparisons of equation (7) with experimental data for MgO, Al_2O_3, and Mg_2SiO_4 may be found in previous papers[7-10]. In figure 3 we see that $(\partial^2 K_S/\partial T^2)$ is very nearly zero for $T > \theta/2$ (θ = Debye temperature). By induction we infer that the higher temperature derivatives may also be neglected. In a strict sense, this approximation is valid only for $T > \theta/2$. For convenience in this chapter, however, we have expanded about $T_0 = 298$ °K.

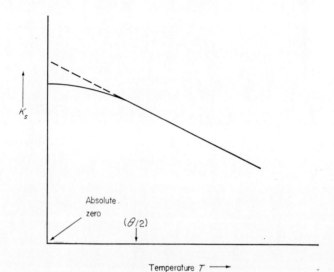

FIGURE 3 Theoretical relationship between K_S and T given by [from Soga and Anderson[7]]
$$K_S(T) = K_S(0) + (\gamma\delta/V_0)\{H(T) - H(0)\}$$

For MgO, Schreiber and Anderson[8] have found that $(\partial K_S/\partial P)_T$ changes only by 0·0008 per degree K between 194·5 °K and 307·6 °K. Similar results were found for α-quartz by McSkimin, Andreatch, and Thurston[11]. Such results indicate that, in many cases, the mixed partial derivative

$$\left(\frac{\partial^2 K_S}{\partial T \partial P}\right)_{T_0, P=0}$$

may be sufficiently small to be neglected in the high-pressure high-temperature regime. With this final approximation, our expansion reduces to

$$K_S(P, T) = K_S(0, T_0) + \left(\frac{\partial K_S}{\partial P}\right)_{P=0, T} P + \left(\frac{\partial K_S}{\partial T}\right)_{P, T_0} (T - T_0) \qquad (9)$$

It will be shown in a later section that the pressure effect tends to overshadow the effect of temperature upon K_S for Earth-forming materials and for real Earth conditions. Our expression for K_S then may be written

$$K_S = K_0 + K'_0 P \qquad (10)$$

where
$$K_0 = K_S(0, T_0)$$
$$K'_0 = \left(\frac{\partial K_S}{\partial P}\right)_{P=0,T}$$

It is this equation (10) which formed the basis of Murnaghan's[12] law of compressibility[13] and also the compressibility–pressure hypothesis of Bullen[14]. K_0 may be calculated directly from the measured velocity data using equation (4). K'_0 may be determined graphically[3] from $K_S(P)$ or analytically[8] from

$$\left(\frac{\partial K_S}{\partial P}\right)_T = \left(\frac{\partial \rho}{\partial P}\right)_T (v_p^2 - \tfrac{4}{3}v_s^2) + 2\rho\left\{v_p\left(\frac{\partial v_p}{\partial P}\right)_T - \tfrac{4}{3}v_s\left(\frac{\partial v_s}{\partial P}\right)_T\right\} \quad (11)$$

where

$$\left(\frac{\partial \rho}{\partial P}\right)_T = \frac{\rho}{k_T} = \frac{\rho}{K_S}(1 + T\alpha_v \gamma) \quad (12)$$

It should be noted that from dimensional arguments (see equation (4)), the velocity and the bulk modulus cannot both exhibit a strictly linear dependence upon pressure. We note that (see figure 2)

$$v_p = a + bP \quad (13)$$

where a and b are constants determined empirically. As we shall show later (p. 439) by combining (10) and (12), the density can be represented by

$$\rho = \rho_0\left(1 + k'_0\frac{P}{k_0}\right)^{1/k'_0} \quad (14)$$

At the pressures for which (13) is strictly true ($P < 4$ kb), $P/k_0 \ll 1$ and we may expand (14) as follows:

$$\rho = \rho_0\left\{1 + \frac{P}{k_0} + \frac{1/k'_0 - 1}{2!}\left(\frac{P}{k_0}\right)^2 + \cdots\right\}$$

Now since
$$K_S \sim \rho v_p^2$$
we may write

$$\begin{aligned}K_S &\sim \rho_0\left\{1 + \frac{P}{k_0} + \frac{1/k'_0 - 1}{2!}\left(\frac{P}{k_0}\right)^2\right\}(a^2 + 2abP + b^2P^2) \\ K_S &\sim \rho_0\left[a^2 + \left(\frac{a^2}{k_0} + 2ab\right)P + \left\{\frac{2ab}{k_0} + b^2 + \frac{a^2}{2k_0^2}\left(\frac{1}{k'_0} - 1\right)\right\}P^2\cdots\right]\end{aligned} \quad (15)$$

From table 1 we note that

$$O[b] = O\left[\frac{1}{k_0}\right] = 10^{-3}$$

and we may then neglect the P^2 term in (15). Within the experimental precision of our data, we may then have both v_p and K_S appear as linear functions of the pressure up to $P = 4$ kb. This is illustrated in figures 2 and 4, in which we have plotted the data of

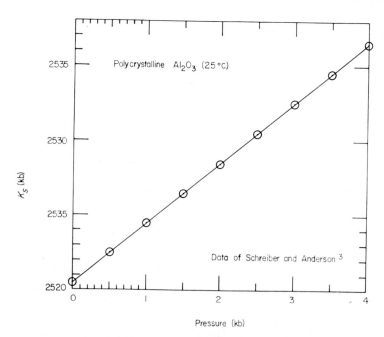

FIGURE 4 K_S for polycrystalline Al_2O_3 as a function of pressure from 1 b to 4 kb. K_S calculated directly from v_p and v_s data in figure 2 by Schreiber and Anderson[3]

Schreiber and Anderson[3] for v_p and K_S. At higher pressures, the justification for (10) comes from the agreement of the extrapolated equation of state (14) with the experimental shock-wave data[13].

Patterns Relating Equation of State Parameters

One of our principal goals in this chapter is to point out a pattern in the values of the elastic constants of the oxides we have studied. The pattern is a systematic dependence of the elastic data upon two parameters, the ambient (zero-pressure) density (ρ_0) and the mean atomic weight (M/p). We would now like to describe this pattern using primarily the data of table 1.

Anderson and Nafe[15] showed that the bulk modulus–ambient density (ρ_0) relationship for oxide compounds differs in a remarkable degree from that found for other materials such as alkali halides, fluorides, selenides, sulfides, and covalent compounds. In figure 5, we have plotted the (K_0, ρ_0) relationship for various oxide compounds whose mean atomic weight (M/p) is about 20. This (log, log) plot indicates a power function dependence of K_0 upon ρ_0. We find that for these materials

$$K_0 \sim \rho_0^4 \qquad (16)$$

which has been called the law of corresponding states[16].

From equation (2) we can see that (16) implies that

$$\left. \begin{array}{l} v_{po} \sim \rho_0^{1.5} \\ v_{so} \sim \rho_0^{1.5} \end{array} \right\} \qquad (17)$$

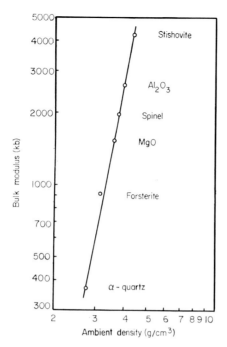

FIGURE 5 K_0 (bulk modulus) versus ρ_0 (ambient density) relationship for oxide compounds with $M/p \simeq 20$. Full line represents the law of corresponding states
$$K_0 \sim \rho_0^4$$

if we assume that μ obeys a relationship similar to (16). Since these relationships (17) stem from the original derivation that the bulk modulus is a second derivative of the free energy, they are soundly based on thermodynamics. However, in the density range $\rho_0 = 2\cdot5$–$5\cdot0$, this power law (17) is almost indistinguishable from a simple linear law of the form

$$v_p = a_0 + a_1 \rho_0 \qquad (18)$$

where a_0, a_1 are arbitrary constants established by a least-squares fit to the data. This is demonstrated in figure 6.

A linear dependence of velocity upon density was first suggested on empirical evidence by Birch[17]. Subsequent work [18–20] has confirmed this evidence and extended the applicability of 'Birch's law' to shear velocities.

Figure 7 is a plot of v_p versus ρ_0 for various materials which have been the object of special attention in our laboratory. The broken lines represent Birch's law for constant mean atomic weights (from Birch[17], figure 3). For $M/p \simeq 20$, Birch's law appears to hold for spinel, periclase, corundum, forsterite, and α-quartz.

The key point at this stage is that, in terms of physical properties, we are sorting the Earth-forming materials on the basis of two parameters: ρ_0, the ambient density, and M/ρ, the mean atomic weight.

From our results, it appears that the two parameters, ρ_0 and M/ρ, are also sufficient to sort out two other measured properties, K_0' and $(\partial K_S/\partial T)_P$.

In figure 8, we have plotted K_0' as a function of ρ_0 for a number of materials. With the exception of α-quartz, the oxides with $M/p \simeq 20$ exhibit values of K_0' between 4 and 5. But note that K_0' decreases with density at constant M/p and increases with M/p at constant density.

Values of $(\partial K_S/\partial T)_P$ for these materials are seen in figure 9. Here the trend is that $(\partial K_S/\partial T)_P$ increases with density at constant M/p and decreases with M/p at constant density.

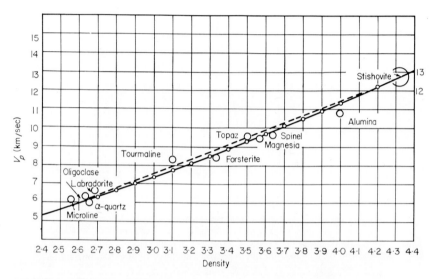

FIGURE 6 v_p versus ρ_0 relationships for oxide compounds with $M/p \simeq 20$. Full line is computed from $v_p = 1\cdot42\rho^{1\cdot5}$ km/sec. Broken line is arbitrary straight line $v_p = a_0 + a_1\rho$

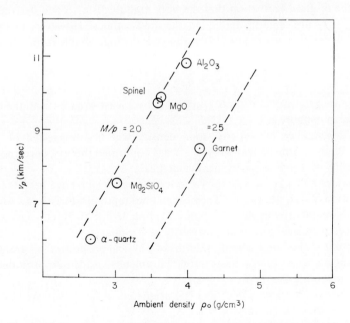

FIGURE 7 v_p versus ρ_0 relationships for materials as a function of M/p

Elastic constants of oxide compounds

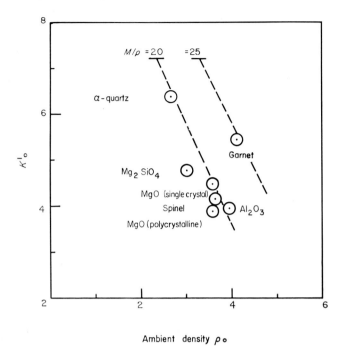

FIGURE 8 $K_0' = (\partial K_S/\partial P)_{P \to 0, T}$ versus ρ_0 relationships for materials as a function of M/p

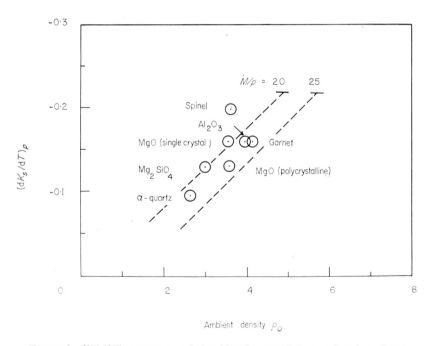

FIGURE 9 $(\partial K_S/\partial T)_P$ versus ρ_0 relationships for materials as a function of M/p

From figures 5, 8, and 9, we can illustrate the behavior of K_S as a function of pressure and temperature, using density and M/p as parameters. For mantle materials the value of M/p is largely controlled by the iron content.

For Al_2O_3 with *high density* and *low iron* content, we have

$$K_S = 2521 + 3·98P - 0·15(T - T_0) \tag{19}$$

Note that since P (in kilobars) and T (in degrees K) are of the same order of magnitude at the mantle–core boundary and their coefficients are in the ratio

$$\frac{K'_0}{(\partial K_S/\partial T)_P} \simeq \frac{25}{1}$$

the pressure term will dominate as we approach this discontinuity.

For garnet with *high density* and *high iron* content

$$K_S = 1770 + 5·43P - 0·20(T - T_0) \tag{20}$$

For forsterite with *low density* and *low iron* content

$$K_S = 974 + 4·8P - 0·11(T - T_0) \tag{21}$$

Values of $K'_0 < 4$ have figured prominently in the theoretical core models of Ramsey[21] ($K'_0 = 3·7$) and Lyttleton[22] ($K_0 = 3·5$). The present ultrasonic data do not exhibit such small values of K'_0. However, in the Earth's mantle it may be possible to represent the combined effects of pressure and temperature upon K_S by an effective value of $K'_0 = 3·5$–$4·0$.

Thermal Gradients and the Low-velocity Layer

The existence of a low-velocity layer at depths of the order of 100–150 km in the Earth was first suggested by Gutenberg[23] (see also Gutenberg and Richter[24]) on the basis of seismic amplitude data. While the existence of a zone of negative velocity gradient with depth is not accepted by all seismologists, especially for compressional waves, it is of interest to investigate possible causes for such an occurrence. One of the most plausible explanations is that the thermal gradient becomes sufficiently large to cause the effect of a velocity increase with pressure. This critical thermal gradient is determined by parameters which can be derived from the ultrasonic laboratory data.

If a low-velocity layer (LVL) does exist, then by definition we must have

$$\frac{dv}{dZ} < 0 \tag{22}$$

where v is v_p or v_s and Z is depth. Now since $v = v(P, T)$, the total derivative may be written as

$$\frac{dv}{dZ} = \left(\frac{\partial v}{\partial P}\right)_T \frac{dP}{dZ} + \left(\frac{\partial v}{\partial T}\right)_P \frac{dT}{dZ} \tag{23}$$

Eliminating dv/dZ from (22) and (23) and solving for dT/dZ, we find

$$\frac{dT}{dZ} > \frac{dP}{dZ} \left\{ \frac{-(\partial v/\partial P)_T}{(\partial v/\partial T)_P} \right\} = \left(\frac{dT}{dZ}\right)_{\text{critical, LVL}} \tag{24}$$

Equation (24) represents the critical thermal gradient necessary to produce a negative velocity gradient in the Earth[8].

The necessary condition for the existence of a ray with its lowest point at a given distance from the center of the Earth (r) is (see Bullen[25], p. 112)

$$\frac{r}{v}\frac{dv}{dr} \leq 1 \qquad (25)$$

If $(r/v)(dv/dr)$ exceeds 1, then at the bottom of its path the downward curvature of the ray will be greater than the curvature of the level surface through r; the ray will be unable to reach the surface of the Earth and a 'shadow zone' will be created. A negative velocity gradient dv/dz, corresponding to a positive dv/dr, may exist without the presence of a shadow zone if the condition (25) is satisfied.

If, however, we wish to determine the maximum thermal gradient allowable without a shadow zone in a radially symmetrical sphere of radius r, then (23) should be substituted in the equality form of (25), noting that $dv/dr = -dv/dZ$ (see Birch and Bancroft[26], see also Birch[27]),

$$\left(\frac{\partial v}{\partial P}\right)_T \frac{dP}{dZ} + \left(\frac{\partial v}{\partial T}\right)_P \frac{dT}{dZ} = -\frac{v}{r}$$

which can be rewritten as

$$\left(\frac{dT}{dZ}\right)_{\text{critical, shadow zone}} = \frac{-v/r - (dP/dZ)(\partial v/\partial P)_T}{(\partial v/\partial T)_P} = \left(\frac{dT}{dZ}\right)_{\text{critical, LVL}} + \left\{\frac{-v/r}{(\partial v/\partial T)_P}\right\} \qquad (26)$$

where $(dT/dZ)_{\text{critical, LVL}}$ is given by (24). If the actual temperature gradient were larger than this critical value (26), then a shadow zone would exist.

Since we are primarily interested in the implications of a low-velocity layer on the thermal gradients, equation (24) is the formula to use. The values of these critical gradients are given in table 2 for various materials. We have assumed in all cases that the hydrostatic law is obeyed

$$\frac{dP}{dZ} = \rho g \qquad (27)$$

TABLE 2 Critical thermal gradients for a low-velocity layer calculated from laboratory data

Material	ρ_0 (g/cm^3)	$(dT/dZ)_{vp}$ (degK/km)	$(dT/dZ)_{vs}$ (degK/km)
Al_2O_3[a]	3·972	5·5	2·8
MgO[b]	3·5803	5·5	3·2
MgO[c]	3·5833	6·0	3·5
Mg_2SiO_4[d]	3·021	7·5	2·5
spinel[e]	3·6193	5·5	0·6
garnet[f]	4·1602	8·2	4·1

References for raw data: [a] Schreiber and Anderson[3,8]; [b] Anderson and Schreiber[5]; [c] Anderson and Andreatch[28]; [d] Schreiber and Anderson[6]; [e] Schreiber[29]; [f] Soga[30].

and we have used ρ_0 for density and $g=990$ cm/sec^2 (in the Earth, this value is correct within 1% to a depth of 2400 km (see Bullen[25], p. 234).

The data of table 2, when plotted in figure 10, tend to fall into two levels—one for

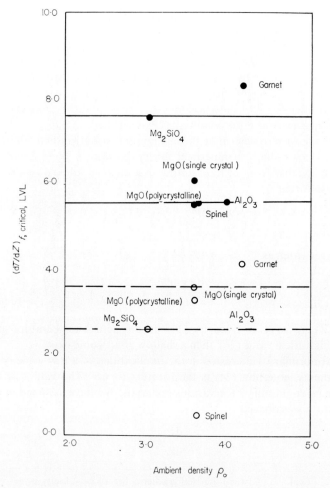

FIGURE 10   Critical thermal gradients as a function of ρ_0 for compressional- and shear-wave low-velocity layers. ● $f = v_p$; ○ $f = v_s$

v_p and one for v_s. There are several exceptions to this grouping: the gradients for garnet are high, probably due to a higher density and a high iron content; the gradient at constant v_s for spinel is anomalously small owing to a very low $(\partial v_s/\partial P)_T = 0.43$ km/sec kb, which in turn results from a low value[29] of $(\partial C_{44}/\partial P)_T$.

On the basis of the data in figure 10, we may offer a few comments concerning the temperature gradient in the neighborhood of a low-velocity zone: if

$$(dT/dZ)_{\text{Earth}} > 3.5 \text{ degc/km}$$

then v_s will decrease with depth; if $(dT/dZ)_{\text{Earth}} > 7.5$ degc/km, then v_p will also decrease with depth; if, however, 3.5 degc/km $< (dT/dZ)_{\text{Earth}} \leq 5.5$ degc/km, then there

Elastic constants of oxide compounds

will be a low-velocity zone for shear waves but not for compressional waves. These limits would correspond to temperatures of 575 °C and 875 °C, respectively, at a depth of 150 km if these thermal gradients were held constant.

We are not suggesting that the upper mantle is composed entirely of corundum, periclase, forsterite, spinel, or garnet, or any specific combination of these materials. Nor are we implying that the values of $(dT/dZ)_{\text{critical, LVL}}$ are precise estimates of the thermal gradient in the upper mantle. However, all of these materials or their isomorphs have been suggested in various theories of the Earth's interior[31-39], and their elastic properties have been intensively studied using ultrasonic techniques. The thermal gradients inferred from these elastic properties are intended to provide only order-of-magnitude estimates of possible gradients in the neighbourhood of a low velocity zone. Perhaps the most significant result of the data in figure 10 is that $(dT/dZ)_{v_s} < (dT/dZ)_{v_p}$ for all of the materials studied. Thus, in a model based entirely on our laboratory data, it is possible to find a thermal gradient sufficiently large to cause v_s to decrease with depth and small enough to permit v_p to be a continuously increasing function of depth.

Density–pressure Trajectories for the Upper and Lower Mantle

We now wish to turn our attention to the density profile of the Earth's mantle. This profile can be constructed from the measured elastic constants and their pressure and temperature derivatives if a suitable equation of state is employed

$$\rho = \rho(P, T) \tag{28}$$

Neglecting for the moment the temperature effect upon density, let us consider only the pressure dependence. Using the proposed linearity of the bulk modulus as a function of pressure, we may write[12-14]

$$K_S = K_0 + K_0' P \tag{10}$$

or

$$k_T = k_0 + k_0' P \tag{29}$$

where k_T is the isothermal bulk modulus, $k_0 = k_T(P=0)$, $k_0' = (\partial k_T/\partial P)_{P=0,T}$. Anderson[13] has presented evidence to support the applicability of (10) at high pressures. Murnaghan[12] integrated (29) directly using (12) to obtain

$$P = \frac{k_0}{k_0'} \left\{ \left(\frac{V}{V_0}\right)^{k_0'} - 1 \right\} \tag{30}$$

which can be rewritten as

$$\rho = \rho_0 \left(1 + k_0' \frac{P}{k_0}\right)^{1/k_0'} \tag{14}$$

For materials with $k_0' \sim 4$, we note that, as P/k_0 becomes large with respect to unity, we have

$$\rho \sim \rho_0 \left(4 \frac{P}{k_0}\right)^{1/4} \tag{31}$$

From equation (16) we recall that

$$K_0 \sim \rho_0^4 \tag{13}$$

and since

$$k_0 \sim K_0 \sim \rho_0^4 \tag{32}$$

we see that

$$\rho \sim \rho_0(\rho_0^{-4})^{1/4} = \text{constant} \tag{33}$$

Thus the effects of ρ_0 and k_0 in Murnaghan's equation tend to offset each other at large P/k_0, so that for all materials for which $k_0' \sim 4$, we see that the ρ versus P curves will approach the same limiting value. Thus we come to the rather surprising conclusion that the density at the mantle–core boundary is largely independent of composition. This recalls Bullen's[14] earlier ideas.

At low pressures we can estimate the slope $(\partial \rho/\partial P)_T$ of the curve by differentiating Murnaghan's equation or recalling the definition of k_T

$$\rho = \rho_0 \left(1 + k_0' \frac{P}{k_0}\right)^{1/k_0'} \tag{14}$$

$$\left(\frac{\partial \rho}{\partial P}\right)_T = \rho_0 \frac{1}{k_0'} \left(1 + k_0' \frac{P}{k_0}\right)^{1/k_0' - 1} \frac{k_0'}{k_0} \tag{34}$$

$$\left(\frac{\partial \rho}{\partial P}\right)_T = \frac{\rho_0}{k_0} \frac{\{1 + k_0'(P/k_0)\}^{1/k_0'}}{\{1 + k_0'(P/k_0)\}} \tag{35}$$

$$\left(\frac{\partial \rho}{\partial P}\right)_T = \frac{\rho}{k_0 + k_0' P} \tag{36}$$

$$\left(\frac{\partial \rho}{\partial P}\right)_T = \frac{\rho}{k_T} \tag{12}$$

which is just the definition of k_T. If we evaluate (35) as $P \to 0$, we find that

$$\left(\frac{\partial \rho}{\partial P}\right)_{P \to 0, T} \sim \frac{\rho_0}{k_0}$$

or using (16) again

$$\left(\frac{\partial \rho}{\partial P}\right)_{T, P \to 0} \sim \rho_0^{-3}$$

Thus the lower ρ_0, the greater the slope $(\partial \rho/\partial P)_{P \to 0, T}$. This effect can be seen in figures 11 and 12. Materials with a smaller initial density change density faster as a function of increasing pressure.

In figure 11, we have plotted the ρ versus P trajectory predicted by Murnaghan's equation for α-quartz and for stishovite, the high-pressure polymorph of α-quartz. The parameters for α-quartz (ρ_0, k_0, k_0') were calculated from the single-crystal data of McSkimin, Andreatch, and Thurston[11] and are given in table 3. These parameters represent the Voigt–Reuss–Hill (VRH) values[40] for an isotropic solid. For stishovite, we used[41] $\rho_0 = 4.28$ and the bulk sound velocity $(K_S/\rho)^{1/2} = 10$ km/sec (see McQueen,

Elastic constants of oxide compounds

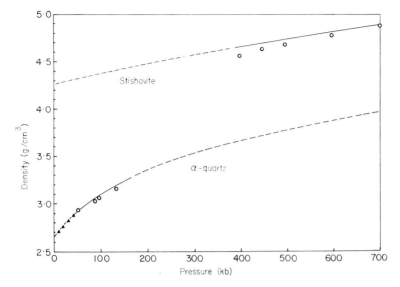

FIGURE 11 Murnaghan's ρ versus P trajectories for SiO_2. ▲ Data by Bridgman[44]; ○ data by Wackerle[43]

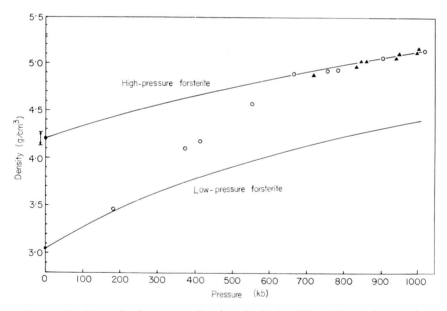

FIGURE 12 Murnaghan's ρ versus P trajectories for Mg_2SiO_4. ○ Forsterite ceramic; ▲ dunite (Twin Sisters Mt., Washington) (McQueen and Marsh[45])

Fritz, and Marsh[42] from data of Wackerle[43]) to obtain $K_S = 4280 \sim k_T \sim k_0$; we let $k_0' = 4.0$ arbitrarily. These values are also given in table 3. The open circles in figure 11 represent the experimental shock-wave data of Wackerle[43]. These data fall on the α-quartz Murnaghan curve for $P < 150$ kb and on the stishovite curve for $P > 400$ kb, according to the interpretation of Wackerle's data by McQueen, Fritz, and Marsh[42]. The closed triangles represent the static compression data of Bridgman[44].

TABLE 3 Parameters for Murnaghan's equation of state[a]

Material	ρ_0 (g/cm^3)	k_0 (kb)	k'_0
α-quartz	2·65	374	6·4
stishovite	4·28	4280	4·0
low-pressure forsterite	3·02	968	4·8
high-pressure forsterite	4·20	3300	4·0
low-pressure spinel	3·62	1999	4·2
high-pressure spinel	4·25	3800	4·0
Al$_2$O$_3$	3·97	2505	4·0
MgO (s.c.)	3·58	1598	4·5
MgO (poly.)	3·58	1690	3·9
garnet	4·16	1757	5·4

[a] The sources for these data are given in table 1 or are discussed in the text.

The Murnaghan trajectories for low-pressure and high-pressure forsterite are plotted in figure 12, using the parameters given in table 3. The low-pressure forsterite parameters are from Schreiber and Anderson[6]. We obtained the zero-pressure density for high-pressure forsterite from an extrapolation of the shock-wave data of McQueen and Marsh[45] for ceramic forsterite above 600 kb. Using this $\rho_0 = 4\cdot2$, we then invoked the law of corresponding states to estimate $k_0 = 3300$, and chose $k'_0 = 4\cdot0$ arbitrarily. The agreement in figure 12 of the Murnaghan trajectories with the shock-wave data of McQueen and Marsh[45] for ceramic forsterite and Twin Sisters' dunite is very encouraging.

The Murnaghan parameters for low-pressure spinel[29], Al$_2$O$_3$[3], polycrystalline[5] MgO, and garnet[30] are also listed in table 3. These tabulated parameters are derived from ultrasonic data on the elastic properties of these materials and are not merely best-fit values for compression data. The parameters (ρ_0, k_0) for the high-pressure spinel were derived from the data of McQueen and Marsh[45] and the law of corresponding states in the same manner as that described for high-pressure forsterite; again, we took $k'_0 = 4\cdot0$ arbitrarily.

The Murnaghan trajectories for these materials are shown in figure 13 as functions of depth (Z) in the Earth. As equation (32) implies, for all materials with $k'_0 \sim 4$, these trajectories tend to approach the limit of $\rho = 5\cdot3$ g/cm^3 at the mantle–core boundary ($Z = 2900$ km). Note that this is true even though the values of k_0 differ by as much as 60%. The trajectory for garnet ($k'_0 = 5\cdot4$) leads to a somewhat higher limit ($\rho = 5\cdot6$ at $Z = 2900$ km) which agrees with the behavior predicted by equation (34).

For reference in figure 13, we have indicated the density solutions of D. Anderson[46] (see Kovach and Anderson[47]) and Bullard[48]. The sigmoid character of D. Anderson's[46] distribution strongly resembles the behavior observed in materials such as α-quartz and forsterite when they undergo a phase change to the high-pressure form (see figures 11 and 12). Bullard's density value at 1000 km appears to fall near the base of the transition zone to the high-pressure phase.

In figure 14, we have simplified the density–depth picture for the Earth. The Murnaghan trajectory for high-pressure forsterite ($M/p = 20$) exhibits a behavior in the lower mantle very similar to that of D. Anderson's[46] solution, but falls slightly

Elastic constants of oxide compounds

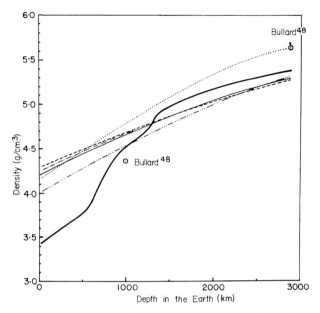

FIGURE 13 Murnaghan trajectories for various materials based on laboratory data and compared with the density solutions for the Earth of D. Anderson[46] and Bullard[48]. — · · · — Corundum; —— high-pressure forsterite; — · — · high-pressure spinel; – – – – stishovite; · · · · · garnet; —— D. Anderson's solution

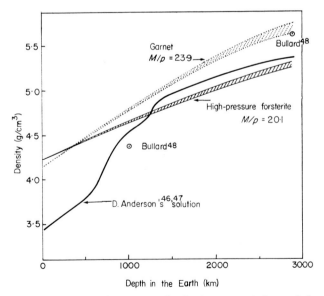

FIGURE 14 Comparison of laboratory and seismic representations of the density–depth relationship in the Earth

lower in absolute value. Bullard's density at the mantle–core boundary agrees very well with the lower limit of the garnet ($M/p = 23 \cdot 9$) trajectory. Since the higher M/p for garnet is due to its iron content, we might infer that high-pressure forsterite, slightly enriched by iron to give $M/p = 22$, would be in better agreement with the density distributions for the lower mantle deduced from seismological data (for earlier discussions of this point, see Birch[49] and McQueen, Marsh, and Fritz[50]).

The important point in figure 14 is that many very different assumptions upon chemical composition will yield density trajectories which closely approximate the density values computed for the lower mantle from seismological data. This arises because, as shown by equation (13), the value of ρ_0 in a (ρ, P) plot is not independent of the value of the initial slope, $(\partial \rho / \partial P)_{P \to 0, T}$. These two parameters are related in such a fashion that a change in one requires a compensating change in the other.

Widely different assumptions about temperature conditions do not significantly change the computed density trajectories. The value of the computed density can be altered somewhat by variation in the iron content. The effect of iron content on the computed density at the mantle–core boundary is of the order of the differences computed from different seismological models.

Conclusions

The elastic constants of oxide materials and the derivatives of these constants as functions of pressure and temperature are measurable in the laboratory using refined ultrasonic techniques. The values of these elastic constants and their pressure and temperature derivatives appear to be a function of the ambient density (ρ_0) and the mean atomic weight (M/p).

The determination of the pressure and temperature derivatives of the elastic-wave velocities enables us to estimate the thermal gradients necessary to produce a low-velocity zone. For the materials studied, it is possible to deduce a thermal gradient which will produce a velocity inversion for shear waves but not for compressional waves.

Elastic constants of oxide materials measured in the laboratory have an important bearing on equations of state for the Earth's interior. Using the simplified version of the pressure–temperature expansion, which is known as Murnaghan's[12] equation of state, it is possible to construct density–pressure trajectories which approximate the density models derived for the Earth's mantle from seismological data. These trajectories may be constructed without information on the specific chemical composition of the mantle materials if there is information on the two critical parameters, ambient density and mean atomic weight.

The effect of pressure on the elastic constants dominates over the temperature effect in the lower mantle so that the previous conclusion is independent of the temperature model.

The value of the ambient bulk modulus (K_0) is controlled by the value of the ambient density (ρ_0) according to the law of corresponding states in such a way that the density–pressure trajectories for oxide compounds (with no iron) approach the same limiting value of density as pressures corresponding to the mantle–core boundary. The amount of iron has a slight modifying effect on this density limit.

Acknowledgements

This research was supported by the National Science Foundation, Grant NSF GP-4437. The experimental work in large part was performed by Drs. E. Schreiber and N. Soga, and has been published elsewhere with the support of the Advanced Research Projects Agency monitored by the Air Force Office of Scientific Research under contract AF-49(638)–1355. Support for the authors' attendance at the Institute was provided by AFOSR (for O. L. Anderson) and NATO (for R. C. Liebermann).

Appendix

C_P	erg/g degK	specific heat at constant pressure
H	cal/mole	enthalpy
K_S	kb	adiabatic bulk modulus
K_0	kb	K_S evaluated at $P=0$
K'_0		$(\partial K_S/\partial P)_T$ evaluated at $P=0$
k_T	kb	isothermal bulk modulus
k_0	kb	k_T evaluated at $P=0$
k'_0		$(\partial k_T/\partial P)_T$ evaluated at $P=0$
M	g	molecular weight
P	kb	pressure
p		number of atoms in the formula
M/p	g	mean atomic weight
r	km	radius of the Earth
S		entropy
T	°K	temperature (Kelvin)
T_0	°K	298 °K
V	cm^3	volume
V_0	cm^3/g-atom	specific volume per 'average' atom at room temperature ($V_0 = M/p\rho$)
v_p	km/sec	compressional wave velocity
v_s	km/sec	shear wave velocity
Z	km	depth within the Earth
α	degK^{-1}	coefficient of volume thermal expansion
γ		Grüneisen parameter
δ		$\left\{-\dfrac{1}{\alpha}\left(\dfrac{\partial \ln K_S}{\partial T}\right)_P\right\}$
ρ	g/cm^3	density
ρ_0	g/cm^3	ambient density (at $P=0$)
θ	°K	Debye temperature
μ	kb	shear modulus
σ_s		Poisson's ratio, adiabatic

References

1. F. Birch, 'Elasticity and constitution of the Earth's interior', *J. Geophys. Res.*, **57**, 227–86 (1952).
2. M. A. Tuve, H. E. Tatel, and P. J. Hart, 'Crustal structure from seismic exploration', *J. Geophys. Res.*, **59**, 415–22 (1954).
3. E. Schreiber and O. L. Anderson, 'Pressure derivatives of the sound velocities of polycrystalline alumina', *J. Am. Ceram. Soc.*, **49**, 184–90 (1966).
4. H. J. McSkimin, 'Pulse superposition method for measuring ultrasonic wave velocities in solids', *J. Acoust. Soc. Am.*, **33**, 12–16 (1961).
5. O. L. Anderson and E. Schreiber, 'The pressure derivatives of the sound velocities of polycrystalline magnesia', *J. Geophys, Res.*, **70**, 5241–8 (1965).
6. E. Schreiber and O. L. Anderson, 'Pressure derivatives of the sound velocities of polycrystalline forsterite, with 6 % porosity', *J. Geophys. Res.*, **72**, 762–4 (1967).
7. N. Soga and O. L. Anderson, 'High-temperature elastic properties of polycrystalline MgO and Al_2O_3', *J. Am. Ceram. Soc.*, **49**, 355–9 (1966).
8. E. Schreiber and O. L. Anderson, 'Temperature dependence of the velocity derivatives of periclase', *J. Geophys. Res.*, **71**, 3007–12 (1966).
9. N. Soga, E. Schreiber, and O. L. Anderson, 'Estimation of bulk modulus and sound velocities of oxides at very high temperatures', *J. Geophys. Res.*, **71**, 5315–20 (1966).
10. O. L. Anderson, 'A derivation of Wachtman's equation for the temperature dependence of elastic moduli of oxide compounds', *Phys. Rev.*, **144**, 553–7 (1966).
11. H. J. McSkimin, P. Andreatch, and R. N. Thurston, 'Elastic moduli of quartz versus hydrostatic pressure at 25° and -195.8 °C', *J. Appl. Phys.*, **36**, 1624–32 (1965).
12. F. D. Murnaghan, 'The compressibility of media under extreme pressures', *Proc. Natl. Acad. Sci. U.S.*, **30**, 244–7 (1944).
13. O. L. Anderson, 'The use of ultrasonic measurements under modest pressure to estimate compression at high pressure', *J. Phys. Chem. Solids*, **27**, 547–65 (1966).
14. K. E. Bullen, 'A hypothesis on compressibility at pressures of the order of a million atmospheres', *Nature*, **157**, 405–6 (1946).
15. O. L. Anderson and J. E. Nafe, 'The bulk modulus–volume relationship for oxide compounds and related geophysical problems', *J. Geophys. Res.*, **70**, 3951–63 (1965).
16. O. L. Anderson, 'A proposed law of corresponding states for oxide compounds', *J. Geophys. Res.*, **71**, 4963–71 (1966).
17. F. Birch, 'The velocity of compressional waves in rocks to 10 kilobars, Part 2', *J. Geophys. Res.*, **66**, 2199–224 (1961).
18. G. Simmons, 'Velocity of compressional waves in various minerals at pressures up to 10 kilobars', *J. Geophys. Res.*, **69**, 1117–21 (1964).
19. G. Simmons, 'Velocity of shear waves in rocks to 10 kilobars, I', *J. Geophys. Res.*, **69**, 1123–30 (1964).
20. N. Soga, 'New measurements on the sound velocity of CaO, and its relation to Birch's law', *J. Geophys. Res.*, **72**, 5157–9 (1967).
21. W. H. Ramsey, 'On the nature of the Earth's core', *Monthly Notices Roy. Astron. Soc., Geophys. Suppl.*, **6**, 42–9 (1950).
22. R. A. Lyttleton, 'On the phase-change hypothesis of the structure of the Earth', *Proc. Roy Soc. (London), Ser. A*, **287**, 471–93 (1965).
23. B. Gutenberg, 'Über Gruppengeschwindigkeit bei Erdbebenwellen, *Z. Physik*, **27**, 111–14 (1926).
24. B. Gutenberg and C. F. Richter, 'New evidence for a change in physical conditions at depths near 100 kilometers', *Bull. Seismol. Soc. Am.*, **29**, 531–7 (1939).
25. K. E. Bullen, *An Introduction to the Theory of Seismology*, Cambridge University Press, Cambridge, 1963.
26. F. Birch and D. Bancroft, 'The effect of pressure on the rigidity of rocks', *J. Geol.*, **46**, 113–41 (1938).
27. F. Birch, 'Interpretation of the seismic structures of the crust in the light of experimental

studies of wave velocities in rocks', in *Contributions to Geophysics in Honor of Beno Gutenberg*, Pergamon Press, Oxford, 1958, Chap. 12, pp. 158–70.
28. O. L. Anderson and P. Andreatch, Jr., 'Pressure derivatives of elastic constants of single-crystal MgO at 23° and −195·8 °C', *J. Am. Ceram. Soc.*, **49**, 404–9 (1966).
29. E. Schreiber, 'Elastic moduli of single-crystal spinel at 25 °C and to 2 kbar', *J. Appl. Phys.*, **38**, 2508–11 (1967).
30. N. Soga, 'The elastic constants of garnet under pressure and temperature', *J. Geophys. Res.*, **72**, 4227–34 (1967).
31. J. D. Bernal, *Observatory*, **59**, 268 (1936).
32. C. S. Ross, M. D. Foster, and A. T. Myers, 'Origin of dunites and of olivine-rich inclusions in basaltic rocks', *Am. Mineralogist*, **39**, 693–737 (1954).
33. J. F. Lovering, 'The nature of the Mohorovicic discontinuity', *Trans. Am. Geophys. Union*, **39**, 947–58 (1958).
34. A. E. Ringwood, 'The constitution of the mantle—I: Thermodynamics of the olivine–spinel transition', *Geochim. Cosmochim. Acta*, **13**, 303–21 (1957).
35. A. E. Ringwood, 'The constitution of the mantle—II: Further data on the olivine–spinel transition', *Geochim. Cosmochim. Acta*, **15**, 18–29 (1958).
36. J. P. Eaton, 'Crustal structure and volcanism in Hawaii', *Am. Geophys. Union Monograph*, No. 6 (1962).
37. C. B. Sclar, L. C. Carrison, and C. M. Schwartz, 'High-pressure reaction of clinoenstatite to forsterite plus stishovite', *J. Geophys. Res.*, **69**, 325–30 (1964).
38. S. P. Clark, Jr., and A. E. Ringwood, 'Density distribution and constitution of the mantle', *Rev. Geophys.*, **2**, 35–88 (1964).
39. T. J. Ahrens and Y. Syono, 'Calculated mineral reactions in the Earth's mantle', *J. Geophys. Res.*, **72**, 4181–8 (1967).
40. O. L. Anderson, 'Determination and some uses of isotropic elastic constants of polycrystalline aggregates using single crystal data', in *Physical Acoustics* (Ed. W. P. Mason) Vol. IIIB, Academic Press, New York, 1965, pp. 43–95.
41. E. C. T. Chao, J. J. Fahey, Janet Littler, and D. J. Milton, 'Stishovite, SiO_2, a very high pressure new mineral from Meteor Crater, Arizona', *J. Geophys. Res.*, **67**, 419–21 (1962).
42. R. G. McQueen, J. N. Fritz, and S. P. Marsh, 'On the equation of state of stishovite', *J. Geophys. Res.*, **68**, 2319–22 (1963).
43. J. Wackerle, 'Shock-wave compression of quartz', *J. Appl. Phys.*, **33**, 922–37 (1962).
44. P. W. Bridgman, 'Rough compressions of 177 substances to 40 000 kg/cm^2', *Proc. Am. Acad. Arts Sci.*, **76**, 71–87 (1948).
45. R. G. McQueen and S. P. Marsh, 'Hugoniot data for various rocks and minerals', quoted by F. Birch, 'Compressibility; elastic constants', in *Handbook of Physical Constants* (Ed. S. P. Clark, Jr.) (*Geol. Soc. Am., Mem.*, **97**, Section 7, 97–173 (1966)).
46. D. L. Anderson, 'Densities of the mantle and the core', *Trans. Am. Geophys. Union*, **45**, 101 (1964).
47. R. L. Kovach and D. L. Anderson, 'The interiors of terrestrial planets', *J. Geophys. Res.*, **70**, 2873–82 (1965).
48. E. C. Bullard, 'The density within the Earth', *Verhandel, Ned., Geol.-Mijnbouwk, Genoot.*, **18**, 23–41 (1957).
49. F. Birch, 'Composition of the Earth's mantle', *Geophys. J.*, **4**, 295–311 (1961).
50. R. G. McQueen, S. P. Marsh, and J. N. Fritz, 'The Hugoniot equation of state of twelve rocks', *J. Geophys. Res.*, **72**, 4999–5036 (1967).
51. N. Soga, 'Temperature and pressure derivatives of isotropic sound velocities of α-quartz', *J. Geophys. Res.*, **73**, 827–9 (1968).
52. N. Soga and O. L. Anderson, 'High-temperature elasticity and expansivity of forsterite and steatite', *J. Am. Ceram. Soc.*, **50**, 239–42 (1967).
53. G. K. White and O. L. Anderson, 'Grüneisen parameter of magnesium oxide', *J. Appl. Phys.*, **37**, 430–2 (1966).
54. R. A. Robie, P. M. Bethke, M. S. Toulmin, and J. L. Edwards, 'X-ray crystallographic data, densities, and molar volumes of minerals', in *Handbook of Physical Constants* (Ed. S. P. Clark, Jr.) (*Geol. Soc. Am., Mem.*, **97**, Section 5, 27–53 (1966)).

55. J. B. Wachtman, Jr., T. G. Scuderi, and G. W. Cleek, 'Linear thermal expansion of aluminum oxide and thorium oxide from 100° to 1100 °K', *J. Am. Ceram. Soc.*, **45**, 319–23 (1962).
56. O. Kubaschewski and E. Ll. Evans, *Metallurgical Thermochemistry*, 3rd ed., Pergamon Press, New York, 1958.
57. A. C. Victor and T. B. Douglas ,'Thermodynamic properties of magnesium oxide and beryllium oxide from 298° to 1200 °K, *J. Res. Natl. Bur. Std., Ser. A*, **67**, 325–9 (1963).
58. B. J. Skinner, 'Thermal expansion', in *Handbook of Physical Constants* (Ed. S. P. Clark, Jr.) (*Geol. Soc. Am., Mem.*, **97**, Section 6, 75–96 (1966)).
59. E. Schreiber, unpublished.
60. R. J. Corruccini and J. J. Gniewek, 'Specific heats and enthalpies of technical solids at low temperatures', *Natl. Bur. Std. (U.S.), Monograph*, No. 21, 17 (1960).
61. E. Schreiber and O. L. Anderson, "Correction to paper by E. Schreiber and O. L. Anderson 'Pressure derivatives of sound velocities of polycrystalline forsterite with 6% porosity' ", *J. Geophys. Res.*, **72**, 3781 (1967).

2

T. J. AHRENS

Division of Geological Sciences
California Institute of Technology
California, U.S.A.

and

C. F. PETERSEN

Stanford Research Institute
Menlo Park, California, U.S.A.

Shock wave data and the study of the Earth*

Introduction

The shock wave studies on solids which have been reported since[1] 1955 have resulted in a body of data on the behavior of a host of elements and compounds at pressures ranging from tens of kilobars to several megabars. Equation of state data, primarily for metals, have been reviewed by Rice, McQueen, and Walsh[2] and more recently for a wider class of materials by Al'tschuler[3]. Reviews emphasizing thermodynamic concepts and experimental techniques have been presented by Duvall and Fowles[4] and Doran[5], respectively. Measurements of the physical properties of solids during shock compression are discussed by Alder[6] and both during and after compression are discussed by Doran and Linde[7]. Shock wave research was first extended to the study of minerals and rocks by Hughes and McQueen[8] in 1958. Later studies by Wackerle[9], McQueen, Marsh, and Fritz[10,11], McQueen and Marsh[12,13], Ahrens and Gregson[14], and Ahrens[15], have made equation of state data available for many of the rock-forming minerals and for common rock types.

In general the shock wave research on various solid materials, including that on rocks and minerals, has been directed toward answering one of the following questions.

(i) How do the characteristics of the material affect the propagation of shock waves?

(ii) What is the net effect of the passage of a shock wave disturbance on the material?

Knowledge of the constitutive relation is required to answer question (i). A constitutive relation is the locus of stress–volume–energy states achievable by the various thermodynamic processes such as shock compression and isothermal and isentropic compressions and expansions. In regimes where the compressional stresses are high compared with the strength of the material, the term pressure is used interchangeably with stress. At relatively low shock stresses, the material strength effects must be taken into account. This can, in principle, be done if the three-dimensional time-dependent

---

* Contribution No. 1501.

yield criterion is known and a formulation of the rheological behavior of the yielded material is available.

Although strength effects present difficulties in interpreting shock wave data at low stress levels, the pressure–volume–energy relations obtained from shock experiments at high pressures have been employed with a high degree of success in correlations with Earth density models inferred from seismology[16–18]. The occurrence of phase changes in silicates and in iron which are observed in shock experiments are of particular significance in application to the study of the Earth's interior. Resistivity is another property which is of interest to the study of the Earth and is measurable in the high-pressure shock-induced state. Some high-pressure resistance data for iron and MgO have been reported by Fuller and Price[19] and Ahrens[20] respectively.

Shock wave and other equation of state data may also be applied to calculation of the Gibbs' free energy as a function of pressure and temperature and, hence, the prediction of the pressures and temperatures required to induce certain phase changes within the Earth's interior[17, 21].

Studies of post-shock effects in solids (relating to question (ii)) have led to important results in the area of shock-induced high-pressure polymorphs and the production of macro- and microdefects in solids upon the passage of shock waves. Although most of the research on minerals in this area has been concerned with impact effects on planetary surfaces, the pressures required for shock-induced formations of such high-pressure polymorphs as coesite, stishovite[22, 23], and diamond[24] are of interest because of their probable presence in the Earth's interior.

The basic concepts underlying equation of state measurements on Earth materials using shock wave techniques are discussed below. Some recent results for olivine and their implications as to the constitution of the Earth's mantle are also discussed.

Basic Concepts

The collision of two bodies travelling at high relative initial velocity, or the sudden application of intense pressure from a detonating explosive, causes a mechanical shock wave to propagate into a normal solid (figure 1). The (P, V, E) (pressure, volume, and

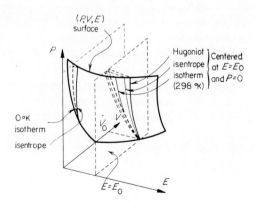

FIGURE 1   Pressure–volume–energy surface of a solid

internal energy) state of material encompassed by the shock wave is uniquely determined by a Hugoniot curve for the solid. The Hugoniot curve represents the locus of states that may be achieved by shocking a series of specimens from some initial state (which is usually at atmospheric, essentially zero, pressure and room temperature). A Hugoniot curve may be centered at an initial state which is at a high pressure; this state could be the result of a previous shock compression.

The Hugoniot curve, centered at atmospheric pressure at room temperature is called the principal Hugoniot, and is represented in figure 2 as lying on the general (P, V, E)

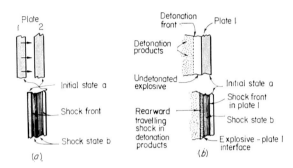

FIGURE 2 Generation of plane shock waves: (a) plate 1 impacts plate 2 and, as a result, shock fronts propagate into both plates; (b) upon interaction of detonation front at the explosive–plate 1 interface, forward shock propagates into plate 1 and rearward shock propagates back into detonation products

surface of a solid. The position on the (P, V, E) surface of the curves representing the locus of states achieved by isothermal and isentropic compression from the 298 °K, and $P=0$, initial state are also indicated in figure 2. The usual representation of the pressure–volume plane is indicated by the projection of the (P, V, E) curves in the plane $E=E_0$. Here E_0 represents the internal energy of the solid at $P=0$ and $V=V_0$.

The Rankine–Hugoniot conservation relations

$$\frac{V_b}{V_a} = \frac{U_b - u_b}{U_b - u_a} \tag{1}$$

$$P_b - P_a = \frac{(U_b - u_a)(u_b - u_a)}{V_a} \tag{2}$$

$$E_b - E_a = \frac{(P_a + P_b)(V_a - V_b)}{2} \tag{3}$$

express the conservation of mass (1), momentum (2), and energy (3) across the shock front. Here U and u are shock and particle velocity (in laboratory coordinates).

The subscripts a and b refer to the states in front of, and behind, the shock front, respectively. The Rankine–Hugoniot relations were derived originally so as to apply to fluids, but they are also valid for solids if the pressure is taken to be the stress component normal to the shock front. For sufficiently strong shocks, the mean stress and the stress normal to the shock front are assumed to be equivalent. At lower shock pressures, where the material's yield strength is taken into account, the stresses acting parallel to the shock front must generally be inferred.

In order to establish the initially unknown Hugoniot curve of a solid, a series of shock states are generated in the material. A measurement of any of two of the five variables E_b, U_b, u_b, P_b, or V_b is required in order to determine the shock state completely, using equations (1), (2), and (3). The initial state (P_a, V_a) is assumed to be known. Usually u_b and U_b are experimentally measured. Reviews of techniques used to measure U_b and u_b are given by Duvall and Fowles[4] and Doran[5].

Hugoniot data for a wide variety of rocks and minerals have been tabulated in terms of the quantities measured, shock and particle velocity. Data reported to 1963 are tabulated by Rinehart[25] and a more recent tabulation which includes recent Soviet and Los Alamos results is given in the *Compendium of Shock Wave Data*[26]. A useful tabulation of Hugoniot data for basic rock-forming minerals is given by Birch[27] and extensive series of measurements of various rocks has recently been reported by McQueen, Marsh, and Fritz[11]. Hugoniot data for porous rocks and carbonates are given by Adadurov, Balashov, and Dremin[28], Ahrens and Gregson[14], Flanagan[29], and Bass[30]. The studies of iron–silicon and iron–nickel alloys by Balchan and Cowan[31] and McQueen and Marsh[13] have considerably narrowed the range of compositions of the Earth's core which are compatible with seismic data.

It has been observed in most laboratory experiments performed with crystalline solids that the shock wave disturbance forms two (and in some cases more) shock fronts (in at least one stress range). For example, when a shock is driven to state B in a material for which the projection of the principal Hugoniot into the plane $E=E_0$ is given as ABCDEF (figure 3), an intermediate shock front forms. The material first

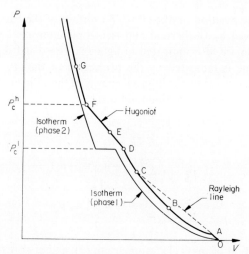

FIGURE 3 Generalized Hugoniot curve for a silicate. Transition from phase 1 to phase 2 is indicated at pressure P_c^l along isotherm. As a result a mixed phase region exists along Hugoniot from state D to state F

undergoes a shock transition from state 0 to the intermediate shock state A, which represents what is often referred to as the Hugoniot elastic limit shock state. The amplitude of the shock transforming material from 0 to A is believed to represent the maximum stress that the solid may withstand under rapid one-dimensional compression without rearrangement through plastic flow or fracture at the shock front. The shock velocity associated with the transition to state A is approximately equal to the longitudinal elastic-wave velocity. The Hugoniot curve between 0 and A is referred to as the elastic portion because material shocked to these states is able to support the resulting large stress differences. Finite-strain elasticity theory should describe the states of stress and strain on the elastic Hugoniot.

Final shock states, such as B, lying at the pressures greater than the Hugoniot elastic limit, are presumably achieved when internal rearrangement (deformation) occurs at the shock front. A deformational shock state will be achieved via one or more intermediate shock fronts if it lies above a region of the Hugoniot (such as above A) for which the curvature of the Hugoniot is anomalous (i.e. $d^2P/dV^2 \leq 0$). In certain cases intermediate shock fronts also occur when a phase transition takes place in the material and gives rise to an anomalous curvature along the Hugoniot as indicated at point D in figure 3.

When an intermediate shock front forms, the material is driven from an initial shock state (P_0, V_0) to a final shock state (P_2, V_2) via two shocks of amplitude $P_1 - P_0$ and $P_2 - P_1$. The specific volumes associated with each of these shock fronts are V_1 and V_2, respectively. The second shock, bringing the material from state (P_1, V_1) to (P_2, V_2) must travel at a slower velocity $U_2 - u_1$ with respect to the material behind the first shock. Thus the condition for the formation of two shocks is

$$U_2 - u_1 < U_1 - u_1$$

or from equations (1) and (2) this is equivalent to

$$\frac{P_2 - P_1}{V_1 - V_2} < \frac{P_1 - P_0}{V_0 - V_1} \qquad (4)$$

It follows from (4) that a single shock from an initial arbitrary state (P_0, V_0) to (P_2, V_2) will result only if the straight (Rayleigh) line in the (P, V) plane between the two states lies everywhere above the Hugoniot curve. Thus for states lying between points A and C (such as B) two shocks form, while states above C are achieved via a single shock front.

Some Hugoniot Data for Olivine

The Hugoniot of magnesium-rich olivine $(Mg, Fe)_2SiO_4$ is of interest because material of similar composition is believed to be a major constituent of the Earth's mantle and because this mineral is an important constituent of common rocks such as gabbro and basalt. The olivine Hugoniot is especially interesting since it exhibits the phenomena of double shock fronts as well as at least one shock-induced phase transition.

The Hugoniot data listed in table 1 were obtained using Twin Sisters olivine which has an index of refraction in the α direction which varies from 1·657 to 1·659. This, and a positive optical sign, indicates that the olivine had a composition of $\sim 92\%$ (by mass) Mg_2SiO_4 (forsterite). The measured density of the present samples were 3·29–3·32 g/cm^3 as compared with 3·214 g/cm^3 for pure Mg_2SiO_4. Except for one sample this material probably had a slightly lower iron content than the material extensively studied by McQueen, Marsh, and Fritz[11]. The present data were measured using the inclined mirror method described in Ahrens and Gregson[14]. These data and the extensive Los Alamos results on olivine and some Soviet data are plotted in figure 4.

TABLE 1 Hugoniot data, olivine

	Initial state		Elastic shock state				Deformational shock state				
Shot No.	Density ρ_0 (g/cm^3)	Specimen thickness d (mm)	Shock velocity U_1 (mm/μsec)	Particle velocity u_1 (mm/μsec)	Hugoniot elastic limit P_1 (kb)	Compression V_1/V_0	Shock velocity U_2 (mm/μsec)	Free surface velocity u_{fs2} (mm/μsec)	Particle velocity u_2 (mm/μsec)	Shock pressure P_2 (kb)	Compression V_2/V_0
11897	3·289	8·90	8·45	0·323	89·6	0·962	6·58	1·24	0·665	160	0·907
12026	3·315	4·81	8·72	0·272	75·6	0·968	8·21	3·37	1·59	437	0·809
11899	3·289	4·82	—	0·323^a	93·2	0·963	8·39	4·22	2·19	604	0·739
11898	3·289	4·83	—	—	—	—	8·49	5·41	2·80	777	0·670

^{a} Value obtained from shot 11897.

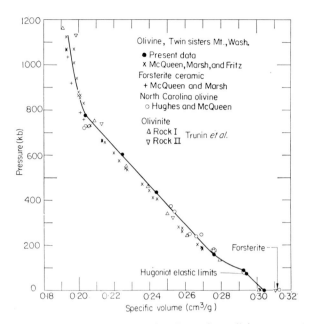

FIGURE 4 Hugoniot data for various olivines

An initial shock which appears to be an elastic precursor of 76 to 93 kb amplitude was detected in all the experiments except shot 11 898. The precursor velocity observed is slightly higher than the mean longitudinal velocities observed between 4 and 10 kb of 8·32 to 8·42 mm/μsec which are reported by Birch[32] for a nominally similar material. This interpretation of the data implies that the elastic precursor would be just overdriven by a deformational shock of approximately 700 kb amplitude. Hence the precursor would not be observed in the 777 kb experiment. It appears that the flash gaps employed in the McQueen, Marsh, and Fritz[11] experiments were not excited by this elastic shock. The elastic shock state persists to this unusually high stress level because of the shape of the olivine deformational Hugoniot in the (P, V) plane is nearly linear over a wide pressure range from approximately 200 to 780 kb. This implies that the apparent compressibility increases with pressure as approximately V^{-1}. The significance of this result is discussed below.

The Olivine Hugoniot and its Relation to the Mantle

The Hugoniot data for different compositions appears to depend strongly on initial density and hence iron content and to a lesser extent initial porosity. The data of Trunin and coworkers[33] for rock II which has a density of 3·21 g/cm^3 is fortuitously close to that of pure forsterite. McQueen, Fritz, and Marsh's extensive data for Twin Sisters dunite (only representative points are plotted) appear to contain slightly more iron than our samples.

The breakdown of the silicates $(Mg, Fe)_2SiO_4$ and $(Mg, Fe)SiO_3$ has long been thought to occur in the Earth's mantle in the 200 to 800 km depth range on the basis of thermodynamic calculations and the marked increase in K_0/ρ_0 (bulk modulus/density) in this region of the Earth[34]. A recent analysis by Anderson[18] has correlated the increases in the observed compressional velocity, starting at depths of ~ 350 and ~ 650 km in the Earth with the pressure and temperature required for two transitions in material having a composition in the range $(Mg_{0.6-0.8}, Fe_{0.2-0.4})_2SiO_4$. These velocity increases are consistent with the olivine to spinel transition and the expected post-spinel transition respectively. The pressure–temperature phase diagrams for various olivines transforming to spinel structure have been studied under static high pressure by Akimoto and Fujisawa[35], Akimoto and Ida[36], and Ringwood and Major[37,38]. The pressure–temperature required for the breakdown of various olivines to spinel and spinel or olivine going to post-spinel phases have been calculated by Anderson[17,18], Ahrens and Syono[21], and Mao[39]. For the forsterite end member these reactions may be represented by

$$Mg_2SiO_4 \text{ (olivine)} \rightarrow Mg_2SiO_4 \text{ (spinel)}$$
$$Mg_2SiO_4 \text{ (spinel)} \rightarrow 2Mg_2O + SiO_2 \text{ (stishovite)}$$
$$Mg_2SiO_4 \text{ (olivine)} \rightarrow 2MgO + SiO_2 \text{ (stishovite)}$$

The phase equilibrium studies, theoretical calculations, and low-pressure elastic constant data[27] suggest that at low pressures (below ~ 200 kb and perhaps as low as ~ 100 kb) the Hugoniot represents the behavior of normal olivine. The regime of anomalous compression extending to ~ 780 kb then probably corresponds to direct post-spinel transition of the olivine by one or more of the above reactions. In order to evaluate the Hugoniot data in the region of anomalous compressibility we have assumed that for shock states above ~ 200 kb the transformation of forsterite to oxides takes place directly in the shock wave. The ~ 200 kb transition pressure was chosen so as to be consistent with the thermochemical calculations for a direct transition to the oxides. In the calculations given below, however, only the change in energy for the transition, and not the transition pressure, is required in the calculaton.

Reduction of Olivine Data in High-pressure Regime

In order to calculate the position in (P, V, E) space of the 25 °C isotherm of olivine in the high-pressure state, we assume it to be a mixture of $2MgO + SiO_2$ (stishovite). The increase in internal energy for any state along the Hugoniot curve is obtained from the Rankine–Hugoniot conservation relation equation (3) (path a, figure 5) which is rewritten as

$$\Delta E = \frac{P_H^h(V_0^l - V_H^h)}{2} \qquad (5)$$

where P_H^h is the Hugoniot pressure, and V_0^l and V_H^h are the values of the low-pressure phase at zero pressure and the high-pressure phase at the Hugoniot state. This internal energy increase is to be equated with the increase in internal energy achieved by isothermally compressing the high-pressure phase from V_0^h to V_H^h (path c, figure 5) and then heating it at constant volume from T_0 (taken to be 298 °K) to T_H (path d, figure 5) *plus* the difference in internal energy $\Delta E_{l \rightarrow h}$ between the low- and high-pressure phase

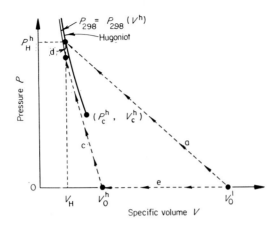

FIGURE 5 Relation of thermodynamic paths to Hugoniot and 25 °C isotherm

(path e, figure 5). Here T_H is the temperature at any point on the Hugoniot. The internal energy increase associated with paths c, d, and e is then

$$\Delta E = \int_{V_0^h}^{V_H} (T b^h C_V^h - P^h)_{T_0} \, dV + \int_{T_0}^{T_H} C_V^h \, dT + \Delta E_{l \to h} \qquad (6)$$

where b^h is the ratio $\gamma^h/V^h = $ constant and γ^h is the Grüneisen ratio. The dependence of the specific heat at constant volume, C_V^h, on temperature and compression is assumed to be given by the Debye formulae

$$C_V^h = 3k \left[12 \left(\frac{T}{\theta^h} \right)^3 \int_0^{\theta^h/T} \frac{y^3 \, dy}{\exp(y) - 1} - \frac{3 \theta^h}{T \{ \exp(\theta^h/T) - 1 \}} \right] \qquad (7)$$

and

$$\theta^h = \theta_0^h \exp \left\{ \frac{(V_0^h - V^h) \gamma^h}{V^h} \right\} \qquad (8)$$

where θ_0^h is the Debye temperature estimated from the empirical correlation by Anderson[40] as 987 °K at standard conditions.

At a given volume, the Hugoniot pressure P_H^h exceeds the pressure P_T^h along the T_0 isotherm by

$$P_H^h - P_T^h = b^h \int_{T_0}^{T_H} C_V^h \, dT \qquad (9)$$

where $b^h = 5 \cdot 37$ g/cm^3 is assumed constant. Then eliminating

$$\int_{T_0}^{T_H} C_V^h \, dT$$

between equations (9) and (6) and substituting ΔE from equation (5) into equation (6) yields

$$\frac{P_T^h}{b^h} + \int_{V_0^h}^{V_H} P_{T_0}^h \, dV = \int_{V_0^h}^{V_H} (T_0 \, b^h C_V^h)_{T_0} \, dV + \Delta E_{l \to h} + P_H^h \, b^h - \frac{P_H^h (V_0^l - V_H^h)}{2} \qquad (10)$$

Equation (7), (8), and (10) are simultaneously solved at a series of volumes for P_T^h, θ^h, and C_V^h. Then, at each interval of volume, equation (9) is solved for T_H along the Hugoniot. The results given in table 2 were calculated using the data above 780 kb from McQueen and Marsh[12, 13].

TABLE 2 Reduced pressure–volume data for high-pressure phase of olivine

Specific volume V (cm³/g)	Hugoniot pressure P_H (kb)	25 °C isotherm pressure P_T^h (kb)	Specific heat along Hugoniot C_V (cal/mole degK)	Hugoniot temperature T_H (°K)
0·202	823	742	5·78	1730
0·200	873	781	5·81	1910
0·198	928	823	5·83	2100
0·196	989	870	5·85	2330
0·194	1060	920	5·87	2590
0·192	1130	976	5·88	2890
0·190	1210	1040	5·90	3230

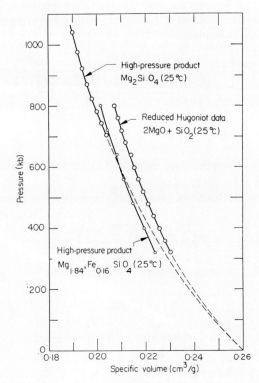

FIGURE 6 Comparison of 25 °C isotherms reduced from Hugoniot data for Mg₂SiO₄ and 2MgO + SiO₂ (stishovite), uncorrected, and corrected for 8% FeO content

Discussion

In order to compare the 25 °C isotherm obtained from the high-pressure olivine data to that expected for a mixture of $2MgO + SiO_2$ (stishovite), the Hugoniot data above 300 kb for SiO_2 of Wackerle[9] (assumed to represent stishovite) and that of McQueen and Marsh[12, 13] for MgO were treated similarly as those for olivine. The sum of the 25 °C volumes are plotted in figure 6. As can be seen the reduced olivine data predict volumes that are approximately 1% lower than given by the $2MgO + SiO_2$ curve at 300 kb. If the reduced olivine data are corrected at high pressures for iron content, assuming 8% FeO, using the extrapolated pressure volume isothermal data of Clendenen and Drickamer[41], the agreement as to the volume, but not the slope of the (P, V) curve above 300 kb is improved. This result and similar results obtained by Al'tschuler[42] and Mao[39] using somewhat different data provide direct evidence for the breakdown of olivine to a yet unknown structure having the density of the oxides; this presumably occurs in the Earth at depths of about 650 km. Furthermore, it appears that this type of reaction can occur in the submicrosecond time scale of a shock experiment.

Acknowledgements

The experimental data reported were obtained at Stanford Research Institute under contracts NASr-49 (24) and DA-49-46-XC-277. The explosive experiments were carried out by J. T. Rosenberg. Discussions of these results with D. L. Anderson were especially helpful and contributed materially to this work.

References

1. J. M. Walsh and R. H. Christian, 'Equations of state of metals from shock wave measurements', *Phys. Rev.*, **97**, 1544–56 (1955).
2. M. H. Rice, R. G. McQueen, and J. M. Walsh, 'Compression of solids by strong shock waves', *Solid State Phys.*, **6**, 1–63 (1958).
3. L. V. Al'tschuler, 'Use of shock waves in high-pressure physics', *Soviet Phys.—Uspekhi (English Transl.)*, **85**, No. 2 (1965).
4. G. E. Duvall and G. R. Fowles, 'Shock waves', in *High Pressure Physics and Chemistry* (Ed. R. S. Bradley), Vol. 2, Academic Press, New York, 1963, Chap. 9.
5. D. G. Doran, 'Measurement of shock pressures in solids', in *High Pressure Measurement* (Eds. A. A. Giardini and E. C. Lloyd), Butterworths, Washington, D.C., 1963.
6. B. Alder, 'Physics experiments with strong pressure pulses', in *Physics Under Pressure* (Eds. W. Paul and D. M. Warschauer), McGraw-Hill, New York, 1963, Chap. 13.
7. D. G. Doran and R. K. Linde, 'Shock effects in solids', *Solid State Phys.*, **19**, 229–90 (1967).
8. D. S. Hughes and R. G. McQueen, 'Density of basic rocks at very high pressures', *Trans. Am. Geophys. Union*, **39**, 959–965 (1958).
9. J. Wackerle, 'Shock-wave compression of quartz', *J. Appl. Phys.*, **33**, 922-37 (1962).
10. R. G. McQueen, S. P. Marsh, and J. N. Fritz, 'On the composition of the Earth's interior', *J. Geophys. Res.*, **69**, 2047–965 (1964).
11. R. G. McQueen, S. P. Marsh, and J. N. Fritz, 'The Hugoniot equation of state of twelve rocks', *J. Geophys. Res.*, **72**, 4999–5036 (1967).
12. R. G. McQueen and S. P. Marsh, in *Handbook of Physical Constants* (Ed. S. P. Clark, Jr.) (*Geol. Soc. Am., Mem.*, **97** (1966)).

13. R. G. McQueen and S. P. Marsh, 'Shock wave compression of iron–nickel alloys and the Earth's core', *J. Geophys. Res.*, **71**, 1751–6 (1966).
14. T. J. Ahrens and V. G. Gregson, Jr., 'Shock compression of crustal rocks: Data for quartz, calcite, and plagioclase rocks', *J. Geophys. Res.*, **69**, 4839–74 (1964).
15. T. J. Ahrens, 'Shock metamorphism: experiments on quartz and plagioclase', *Proc. Conf. on Shock Metamorphism of Natural Materials*, in press (1967).
16. F. Birch, 'Density and composition of mantle and core', *J. Geophys. Res.*, **69**, 4377–88 (1964).
17. D. L. Anderson, 'A seismic equation of state', *Geophys. J.*, **13**, 9–30 (1967).
18. D. L. Anderson, 'Phase changes in the upper mantle', *Science*, **157**, 1165–73 (1967).
19. P. J. A. Fuller and J. H. Price, 'Electrical conductivity of manganin and iron at high pressures', *Nature*, **193**, 262–3 (1962).
20. T. J. Ahrens, 'High-pressure electrical behavior and equation of state of magnesium oxide from shock wave measurements', *J. Appl. Phys.*, **37**, 2532–41 (1966).
21. T. J. Ahrens and Y. Syono, 'Calculated mineral reactions in the Earth's mantle', *J. Geophys. Res.*, **72**, 4181–8 (1967).
22. P. S. DeCarli and D. J. Milton, 'Stishovite: synthesis by shock wave', *Science*, **147**, 144–5 (1965).
23. A. A. Deribas, N. L. Dobretson, V. M. Kudinov, and N. I. Zyuzin, 'Shock compression of SiO_2 powders', *Soviet Phys.—Doklady (English Transl.)*, **188**, 127–30 (1966).
24. P. S. DeCarli and J. C. Jamieson, 'Formation of diamond by explosive shock', *Science*, **133**, 1821–1922 (1961).
25. J. S. Rinehart, 'Compilation of dynamic equation of state data for solids and liquids', *U.S. Naval Ordnance Test Station Rept.*, No. NOTS TP-3738 (1965).
26. Lawrence Radiation Laboratory, *Compendium of Shock Wave Data* (Ed. M. van Thiel), *Univ. Calif., Lawrence Radiation Rept.*, No. UCRL-50108, Vols. I and II (1966).
27. F. Birch, 'Compressibility; elastic constants', in *Handbook of Physical Constants* (Ed. S. P. Clark, Jr.) (*Geol. Soc. Am., Mem.*, **97**, 97–174 (1966)).
28. G. A. Adadurov, D. B. Balashov, and A. N. Dremin, 'A study of the volumetric compressibility of marble at high pressures', *Bull. Acad. Sci. USSR, Geophys. Ser. (English Transl.)*, **1964**, 463–6.
29. T. J. Flanagan, 'The Hugoniot equation of state of materials for the Ferris wheel program', *Sandia Corp. Tech. Rept.*, No. SC-M-66-451 (1966).
30. R. C. Bass, 'Additional Hugoniot data for geologic materials', *Sandia Corp. Res. Rept.*, No. SC-RR-66-548 (1966).
31. A. S. Balchan and G. R. Cowan, 'Shock compression of two iron–silicon alloys to 2·7 megabars', *J. Geophys. Res.*, **71**, 3577–88 (1966).
32. F. Birch, 'The velocity of compressional waves in rocks to 10 kilobars', *J. Geophys. Res.*, **65**, 1083–102 (1960).
33. R. F. Trunin, V. I. Gon'shakova, G. V. Simakov, and N. E. Galdin, 'A study of rocks under the high pressures and temperatures created by shock compression', *Izv. Earth Phys. Ser.*, No. 9, 1–12 (1965).
34. F. Birch, 'Elasticity and constitution of the Earth's interior', *J. Geophys. Res.*, **57**, 227–286 (1952).
35. S. Akimoto and H. Fujisawa, 'Olivine–spinel transition in the system Mg_2SiO_4–Fe_2SiO_4 at 800 °C', *Earth Planetary Sci. Letters*, **1**, 237–40 (1966).
36. S. Akimoto and Y. Ida, 'High-pressure synthesis of Mg_2SiO_4 spinel', *Earth Planetary Sci. Letters*, **1**, 358–9 (1966).
37. A. E. Ringwood and A. Major, 'High-pressure transformation of $FeSiO_3$ pyroxene to spinel plus stishovite', *Earth Planetary Sci. Letters*, **1**, 135–6 (1966).
38. A. E. Ringwood and A. Major, 'Synthesis of Mg_2SiO_4–Fe_2SiO_4 spinel solid solutions, *Earth Planetary Sci. Letters*, **1**, 241–5 (1966).
39. H. Mao, *The Pressure Dependence of the Lattice Parameters and Volume of Ferromagnesion Spinels and its Application to the Earth's Mantle*, Ph.D. Thesis, University of Rochester (1967).

40. O. L. Anderson, 'Determination and some uses of isotropic elastic constants of polycrystalline aggregates using single-crystal data', in *Physical Acoustics* (Ed. W. P. Mason), Academic Press, New York, 1965, Chap. 2, pp. 43–97.
41. R. L. Clendenen and H. G. Drickamer, 'Lattice parameters of nine oxides and sulphides as a function of pressure', *J. Chem. Phys.*, **44**, 4223–8 (1966).
42. L. V. Al'tschuler, 'Shock-wave compression of periclase and quartz and the composition of the Earth's lower mantle', *Izv. Earth Phys. Ser.*, No. 10, 1–6 (1965).

D

DEVELOPMENTS IN TECHNIQUES

I. Application of techniques of modern physics to rock magnetism

1

D. L. ANDERSON
and
C. G. SAMMIS
Seismological Laboratory
California Institute of Technology
Pasadena, California, U.S.A.

R. A. PHINNEY
Department of Geological and Geophysical Sciences
Princeton University
Princeton, New Jersey, U.S.A.

Brillouin scattering—a new geophysical tool[*]

Introduction

The complete interpretation of seismic data in terms of the physics and chemistry of the Earth's interior involves an understanding of the effects of composition, crystal structure, pressure, and temperature on the elastic wave velocities, attenuation, and density. Conventional ultrasonic methods of determining the elastic constants of materials important in geophysics have been summarized in recent review articles by Simmons[1] and Anderson and Lieberman[2]. These methods use transducers that are bonded to the specimen and require that the sample be accurately shaped. The frequency of vibration of the sample, or the transit times of an elastic pulse, combined with an accurate measurement of length yield the elastic wave velocities. The change of frequency, or transit time, with a change in temperature or pressure depends on both the change of elastic velocity and change in length. In high-purity, high-Q specimens, the accuracy in velocity measurements is controlled by the accuracy with which lengths and changes in length can be measured. The definition of 'length' is also critical in experiments in which the transducer size and bond thickness are not negligible compared with the sample size. In natural rock samples the accuracy is limited by the ability to pick the beginning of a pulse which has lost the high-frequency components.

There is an alternate way to obtain elastic wave velocities which is completely nondestructive, which requires no physical contact with the sample, and which requires no absolute length measurements. The method of 'Brillouin scattering' uses thermally generated sound waves, or phonons, to scatter light waves. The spectrum of the scattered light provides the velocity of thermally excited sound waves[3]. This method became particularly useful with the advent of optical masers which are ideal Brillouin sources because of their spectral purity, directivity, and power[4, 5]. The method has been mainly used in the study of liquids[5–7] but enough measurements have been performed on solids[4, 8–12] to indicate that this method provides an important new tool for the geophysicist.

[*] Contribution No. 1482, Division of Geological Sciences, California Institute of Technology, Pasadena, California, U.S.A.

The method looks particularly promising for measuring the temperature derivatives of the elastic constants, particularly on small crystals.

The effect of temperature on the elastic properties of rocks and rock-forming minerals is still very poorly known. Present estimates of the temperature derivative of the wave velocity at elevated temperatures are based on theoretical extrapolation of low-temperature data, and vary by as much as an order of magnitude. Accurate temperature derivatives are required in both empirical and theoretical treatments of the composition and temperature of the Earth's mantle. The intrinsic effect of temperature on the elastic velocities and density is difficult to measure on natural rock samples because of the effects of porosity and differential thermal expansion. Intrinsic properties of rocks pertinent to discussions of the Earth's mantle must be measured at high pressure in order to suppress these effects. An alternate approach is to measure the elastic properties and their temperature derivatives on single crystals and then to compute the appropriate properties for rocks from knowledge of the properties of their constituents.

Theory of Brillouin Scattering

The phenomenon of Brillouin scattering may be described as either a classical Bragg reflection from the wave fronts of thermally generated plane waves, or as a quantum-mechanical photon–phonon interaction. Since both descriptions are equivalent except at very low temperatures, $\sim 4\,°K$, where quantum effects become important[8], we shall discuss only the more lucid classical development.

According to Debye theory, the thermal energy of a body may be identified with the excitation of elastic sound waves. These thermal sound waves may be pictured as plane waves of all frequencies up to the Debye cutoff which propagate in all directions throughout the crystal[13].

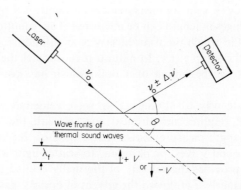

FIGURE 1 Schematic diagram showing Brillouin scattering as a Doppler-shifted Bragg reflection from the fronts of thermal elastic waves

If a beam of monochromatic light of wavelength λ_0 is directed on such a thermally excited crystal of refractive index n, and the scattered light is detected at an angle θ

away from the forward direction, the reflections from all wave fronts interfere destructively except those of wavelength λ_f for which the Bragg relation holds (see figure 1):

$$\frac{\lambda_0}{n} = 2\lambda_f \sin\frac{\theta}{2} \tag{1}$$

Because these thermal sound waves move with the elastic wave velocity V, the incident light of frequency v_0 is Doppler shifted upon reflection by an amount $\pm\Delta v$ given by

$$\frac{\Delta v}{v_0} = 2\frac{V}{C} n \sin\frac{\theta}{2} \tag{2}$$

This shift is positive for waves moving toward the incident light beam and negative for those moving away. The spectrum of the scattered light beam consists of two peaks separated in frequency by $2\Delta v$. Besides this Brillouin doublet, the spectrum also contains an unshifted component centered at the incident frequency v_0. This is caused by reflection from static entropy fluctuations and other static lattice imperfections.

A measurement of the frequency shift Δv and the scattering angle θ allow a direct calculation of the elastic wave velocity V according to equation (2). The wavelength of the elastic waves which are doing the scattering is given by equation (1). It should be noted that by simply changing the scattering angle one can measure phase velocity as a function of wavelength and thus delineate a segment of the dispersion curve.

The elementary analysis just presented does not provide the amplitude or shape of the spectrum of the scattered light. The more rigorous classical theory of scattering necessary for such calculations is developed by Benedek and Greytak[6] for liquids and by Benedek and Fritsch[8] for solids.

Brillouin scattering is a nonlinear process in which the incident light generates high-frequency sound waves within the medium. If the incident laser beam is sufficiently powerful, these secondary sound waves may give rise to a Brillouin spectrum which is orders of magnitude more intense than that resulting from ordinary thermal Brillouin scattering. This phenomenon, known as stimulated Brillouin scattering, should be useful for those crystals in which ordinary Brillouin scattering is weak. The effect has been observed by Chiao, Townes, and Stoicheff[14], Garmire and Townes[15], Brewer and Rieckhoff[16], and Brewer[17,18].

Advantages over Transducer Techniques

The Brillouin scattering method of measuring elastic constants offers several advantages over standard transducer techniques.

(i) *Temperature limit not controlled by transducer properties*: Piezoelectric transducers are temperature sensitive. Their resonant frequency changes with temperature until they completely lose piezoelectric properties at high temperature.

(ii) *Smaller samples*: The Brillouin scattering technique is basically a much higher-frequency technique than standard ultrasonic methods and, in principle therefore, can be applied to much smaller samples. In fact, different interior portions of a crystal may be measured. The size of sample required is limited only by the power required in the scattered beam for adequate detection. Brillouin measurements should be possible on crystal only a few cubic millimeters in volume.

(iii) *No absolute measurement of lengths is required*: The accuracy of standard ultrasonic 'echo' techniques is a function of sample length since the accuracy of direct velocity measurements is proportional to path length. This is an important consideration since it is difficult to obtain certain mineral crystals in large size.

(iv) *Greater accuracy*: Since sample dimension does not enter into the Brillouin method, the accuracy of measured velocities is limited only by the precision to which the Doppler shift in the scattered beam can be measured. Using the optical heterodyning technique outlined in a later section, this frequency shift may be measured directly with a precision approaching that of the electronic counter (1 part in 10^8).

(v) *Sample purity assured*: Only that small part of the sample illuminated by the focused laser beam contributes to the scattering. Crystal imperfections in this region are visible through their reflected light and may be avoided by repositioning the sample.

(vi) *Non-destructive*: Samples can be borrowed from mineral collections since the technique is totally nondestructive, unlike transducer techniques which require the grinding of flat parallel surfaces and bonding.

(vii) *Convenience*: Once a sample is mounted, velocity can be determined along the various crystallographic axes more conveniently than is possible with conventional techniques. By varying the detection angle it is also possible to obtain a limited amount of information regarding the phonon spectrum.

Experimental Methods

The experimental systems used to detect and record Brillouin spectra may be divided into three general categories according to the spectrometer system employed: (i) Fabry–Pérot etalon (*a*) photographically recorded, (*b*) pressure-scanned and electronically recorded; (ii) diffraction grating spectrometer; (iii) optical superheterodyne spectrometer.

The Fabry–Pérot pressure-scanned interferometer has been the most popular detection system. Chiao and Stoicheff[5], Rank and others[7], Chiao and Fleury[19],

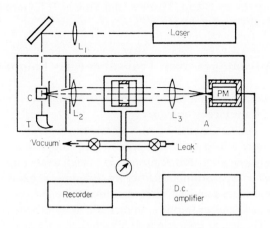

FIGURE 2 Typical Fabry–Pérot pressure-scanned system: PM, photomultiplier; L_1, collecting lens; L_2, L_3, focusing lenses; A, aperture; C, crystal; T, transverse [after Cummins and Gammon[20]]

Benedek and Greytak[6], Cummins and Gammon[20], and O'Connor and Schlupf[21] have used this system to measure the Brillouin spectrum of liquids, while Gammon and Cummins[9] and Shapiro, Gammon, and Cummins[10] have applied it to solids. The experimental apparatus is usually arranged as shown in figure 2. A typical spectral display is shown in figure 3. Scattered light emerging from the illuminated section

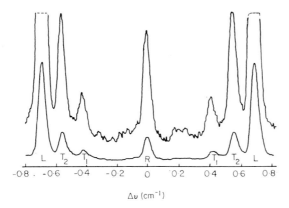

FIGURE 3   Typical spectral display of pressure scanned Fabry–Pérot system. The unshifted Rayleigh peak (R), the longitudinal (L) and two transverse (T_1, T_2) Brillouin peaks of crystalline quartz are shown at two different gains. [After Shapiro, Gammon, and Cummins[10]]

of the sample is made parallel by the collecting lens L_1. The range of scattering angles, and thus the width of the Brillouin components, is controlled by the aperture. After passing through the spectrometer the concentric ring interference pattern is focused on the pinhole by lens L_2. This pinhole is made smaller than the central fringe and becomes alternatively light or dark as the interferometer pattern is scanned.

The pressure-scanned Fabry–Pérot is a simple etalon of the type described in Jenkins and White[22] which has been mounted in a vacuum chamber. The frequency separation between successive interference orders is called the free spectral range:

$$f = \frac{c}{2nd}$$

where c is the speed of light in vacuum, n is the index of refraction between the plates, and d is their separation. As a rough rule, the spacing d is chosen such that the free spectral range is three times the expected frequency difference between the Rayleigh and transverse Brillouin component. For liquids this spacing is of the order of one centimeter, but for solids of geophysical interest it must be smaller than 2 mm if the Brillouin components of successive orders are not to overlap.

The pattern is scanned by allowing nitrogen gas to leak linearly into the previously evacuated interferometer housing, thus scanning the index of refraction from $n = 1.0000$ to $n = 1.0003$. For a 1-cm spacing this corresponds to approximately ten free spectral ranges. The nitrogen is forced through a supersonic orifice or other flow control device which assures constant leak rate independent of the back-pressure in the interferometer. The spectra may thus be calibrated by linear interpolation over the known free spectral range between corresponding peaks of successive orders.

The output of the phototube is either fed directly into a d.c. amplifier and plotted on a strip recorder or it is first processed by a pulse-discriminating network before being amplified and plotted. The pulse discriminator increases the signal-to-noise ratio by rejecting spontaneous tube emissions.

The advantages of this interferometer system are the ease with which it is aligned and calibrated and the relatively low light intensities at which it will operate. Its disadvantages are limited resolution and the fact that calibration is dependent upon a measurement of the spacer length. However, this is still preferable to a measurement of the sample length which changes under temperature and pressure.

In some applications the concentric ring diffraction pattern of the Fabry–Pérot is photographed directly, but the pressure-scanned system described above is capable of greater resolution.

The diffraction grating spectrograph has been used to resolve successfully the Brillouin spectrum by Benedek and others[4], and by Benedek and Fritsch[8]. In this method the diffraction pattern is sampled through a moving slit and recorded using a photomultiplier as in figure 4. A typical spectrum is shown in figure 5. This spectrometer system has the advantage of a very small light requirement but the disadvantage

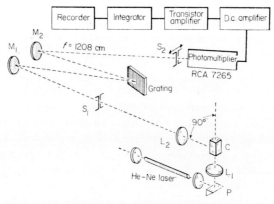

FIGURE 4 Typical diffraction grating spectrometer system: S_1, entrance slit; S_2, exit slit; M_1, M_2, mirrors; L_1, focusing lens; L_2, collecting lens; P, prism; C, crystal [after Benedek and Fritsch[8]]

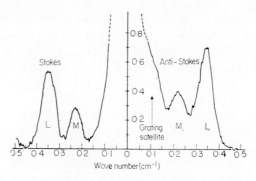

FIGURE 5 Typical spectral display of diffraction grating spectrometer system showing the longitudinal (L) and mixed (M) modes of light scattered in the [110] plane of RbCl [after Benedek and Fritsch[8]]

that the grating introduces Rowland ghosts and satellite lines which must be laboriously subtracted from the recorded spectrum.

Optical heterodyning has been successfully employed by Jennings and Takuma[23] and by Lastovka and Benedek[24]. In this method the scattered beam is mixed with an unscattered beam on the surface of a photodiode. The Brillouin frequency shift Δv is measured directly as a beat frequency between scattered and unscattered beams and is thus independent of laser stability. Because of the inherent accuracy of frequency measurements, this method is capable of resolving powers which are[24] 'several orders of magnitude greater than available using the best optical spectrometers'. The basic experimental design is sketched in figure 6.

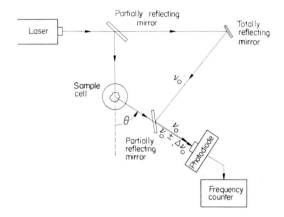

FIGURE 6   Typical optical heterodyne spectrometer system

Although this is the most elegant and accurate way to perform the Brillouin experiment, it is also the most difficult. Alignment is critical. Unless the wave fronts of the scattered and unscattered beams are parallel to $\lambda/4$ on the photodiode surface, destructive interference occurs and the output is decreased. All heterodyne measurements to date have been performed in liquids and at very small scattering angles ($\theta < 0.5°$). This is because Brillouin scattering in liquids is typically thirty times more intense than in solids. The small scattering angles reflect the fact that the frequency of the beat note, Δv, must be kept low enough to be measurable with available photodiodes. These are only temporary technological restrictions which will be removed by more powerful gas lasers and faster, more efficient photodiodes.

Experimental Results of Brillouin Scattering in Solids

The first observation of the Brillouin components of scattered light was reported by Gross[25] for crystalline quartz but photographic evidence was not presented until 1938. Raman and Venkateswaran[26] photographed the interference pattern of a Lummer–Gehrke plate identifying the Brillouin components of gypsum. Sibaiya[27] presented similar evidence for Rochelle salt. Krishnan[28] recorded the Brillouin spectrum of

TABLE 1 Summary of Brillouin scattering measurements in solids for longitudinal modes (L) and transverse modes (T)

Sample	Reference No.	Date	Method	Temperature (°C)	Wavelength (Å) of scattering phonon	Quantity measured: velocity (10^5 cm/sec) or elastic constant (10^{11} dyn/cm^2)		Brillouin determination	Ultrasonic determination	Ultrasonic frequency (10^6 Hz)
KCl	8	1966	He–Ne laser and diffraction grating spectrometer	22·8	3007	elastic constant	C_{11}	4·06 ±0·01	4·078 ±0·008	?
							C_{44}	0·63 ±0·01	0·633 ±0·006	?
							C_{12}	0·69 ±0·04	0·69 ±0·014	?
						$C_{44}+C_{12}$		1·95 ±0·02	1·96 ±0·03	?
KI				22·4	2695		C_{11}	2·73 ±0·01	2·76 ±0·006	?
							C_{44}	0·375 ±0·015	0·37 ±0·004	?
							C_{12}	0·40 ±0·03	0·45 ±0·01	?
						$C_{44}+C_{12}$		1·15 ±0·02	1·19 ±0·02	?
RbCl				22·6	3000		C_{11}	3·74 ±0·01	3·72 ±0·06	10 and 20
							C_{44}	0·535 ±0·02	0·503 ±0·01	10 and 20
							C_{12}	0·72 ±0·04	—	
						$C_{44}+C_{12}$		1·79 ±0·02		
KI	4	1964	He–Ne laser and diffraction grating spectrometer	22·0	3000	velocity in the [110]L direction		2·472 ±0·015	2·508	10
CsI				22·4		velocity in the [111]L direction		2·132 ±0·005	2·154	
TiO$_2$ (rutile)				22·0		velocity in the [101]L and T direction		4·872 ±0·022	74·20	
crystalline quartz	10	1966	He–Ne laser and Fabry–Pérot interferometer	room amb.	2872	longitudinal velocity for waves 45° to x and y axes		5·86	5·85	10
					2878	transverse velocities for waves 45° to x and y axes	T_1	4·72	4·72	
							T_2	3·57	3·56	
fused quartz					3062	longitudinal velocity		5·94	5·97	
						transverse velocity		3·74	3·72	
Pyrex glass					3033	longitudinal velocity		5·52	5·64	
						transverse velocity		3·37	3·28	
LiF	11	1955	Hg arc and 3-m quartz spectrograph	room amb.	~1200	velocity in the [110]L direction		6·90	6·87	
NaCl					~1200	velocity in the [110]L direction		4·71	4·46	
KCl					~1200	velocity in the [110]L direction		4·10	3·84	
diamond	11	1955	Hg arc and 3-m quartz spectrograph	room amb.	~1200	velocity in direction	[111]T	10·12	9·69	10
							[4̄1,10]L	17·85	17·28	
							[4̄1,10]T	11·44	10·49	
							[1̄44]L	17·53	17·76	

Material	Ref	Year	Apparatus	Conditions	Pressure	Quantity	Direction/Constant	Value 1	Value 2	Value 3
α-quartz	11	1955				velocity in direction	[811]L	16.6	16.6	
							[811]T	12.0	11.00	
							[341]L	17.45	17.77	
							[341]T	10.12	10.30	
							[011]L	6.39	6.09	
							[011]T	6.96	6.98	
							[101]L	6.51	6.64	
							[110]L	5.88	5.84	
							[100]L	5.61	5.68	
							[010]L	5.28	5.93	
							[001]L	5.96	6.32	
calcite	11	1955	Hg arc and 3-m quartz spectrograph	room amb.	~1200	velocity in direction	[011]L	6.85	7.12	10
							[011]L	6.71	7.12	
							[011]T	2.47	3.08	
							[011]T	3.27	3.08	
alumina barite						elastic constant	C_{33}	21.4	56.3	
							C_{11}	10.84	8.83	
							C_{22}	9.74	7.81	
							C_{33}	10.23	10.38	
diamond	12	1950	Hg arc and 3-m quartz spectrometer	room amb.	~1200	Brillouin shift (cm^{-1}) in direction	[144]L	8.56	8.66	
							[144]T	5.63	5.39	
							[811]L	8.09	8.11	
							[811]T	5.88	5.36	
							[341]L	8.50	8.68	
							[341]T	5.58	5.02	
							[4̄1, 10]L	5.58	8.43	
							[4̄1, 10]L	5.59	5.12	
							[111]L	5.59	12.47	
							[111]T	7.0	6.69	
diamond	29	1947		room amb.	~1200	velocity	L	18.08	17.80	
							T	11.00	10.30	
							L	16.4	16.5	
							T	11.5	11.0	
							L	17.5	17.3	
							T	11.2	10.5	
quartz	31	1949	Hg arc and 3-m quartz spectrograph	room amb.	~1200	Brillouin shift (cm^{-1}) wave normal to	elastic [011]	1.90	1.8	10
							[01̄1]	2.07	2.07	
							[101]	1.96	1.97	
							[1̄10]	1.75	1.73	
							[100]	2.36	2.38	
							[01̄0]	2.22	2.49	
							[001]	2.51	2.64	

quartz by photographing the interference pattern of a Fabry–Pérot etalon. However, even with a 1-mm plate separation, the components of the violet Hg line of successive orders overlapped and no analysis was attempted. Using a quartz spectrograph Krishnan[29] measured the Brillouin frequency shift for selected propagation directions in diamond. The velocities thus determined were in satisfactory agreement with the ultrasonic determinations. Chandrasekharan[12] refined these measurements using a 3-m quartz spectrometer bringing them in better agreement with ultrasonics. Krishnan[11] used this 3-m spectrometer to measure the Brillouin shift in the alkali halides LiF, NaCl, and KCl and the birefringent crystals α-quartz, calcite, alumina, and borite, as well as diamond. With the exception of alumina, all measurements were consistent with the ultrasonic velocities and it was concluded that measurement of the Brillouin component offered a new and accurate method of determining elastic constants.

Development of the continuous gas laser stimulated renewed interest in Brillouin scattering. Although most of the early work was done in liquids, Benedek and others[4] measured longitudinal sound velocities for a single direction in CsI and KI and a mixed mode in TiO_2 using a 10-mw He–Ne laser and a grating spectrometer. The Brillouin spectrum of crystalline quartz, fused quartz, and glass were measured by Shapiro, Gammon, and Cummins[10] using a 50-mw He–Ne laser and a pressure-scanned Fabry–Pérot etalon. Benedek and Fritsch[8] reported the first actual values of elastic constants determined by Brillouin scattering. They used a He–Ne laser and a diffraction grating spectrometer to measure the Brillouin shift as a function of propagation direction to the [110] plane of the cubic crystals KCl, KI, and RbCl. This completely determined the elastic constants C_{11}, C_{12}, and C_{44}. Working with a He–Ne laser and pressure-scanned Fabry–Pérot interferometer Gammon[30] has determined the elastic constants of the monoclinic crystal triglycine sulfate, thus demonstrating the applicability of the Brillouin method to lower symmetry crystals. The results of Brillouin scattering in solids to date have been compiled in table 1.

Application of Brillouin Scattering to Geophysical Problems

Knowledge of the temperature and pressure dependence of elastic constants of mantle-candidate minerals is essential if the temperature, pressure, composition, and phase of the mantle are to be inferred from seismic velocity profiles. Brillouin scattering offers a convenient new way to make these measurements, the only requirement being a sample cell with three small optical windows.

To measure the temperature derivative of the velocity, the sample cell may be used directly as a thermal bath. A circulating liquid serves the dual purpose of controlling the sample temperature and matching the index of refraction to minimize reflected light from its surfaces. The sample can be rotated to allow measurements in enough crystallographic directions to determine the elastic constants without loosing the Brillouin components in the tails of the reflected component. This technique should work up to those temperatures at which such index-matching liquids decompose. At higher temperatures it may be necessary to shape the sample so the light can enter and leave the crystal normal to its surface. Measurements would then be made in a furnace provided with three optical ports.

Using the simple heterodyne system shown in figure 6, it should be possible to measure the velocity change accompanying a 0·1 degC temperature change to 0·5% as is shown in the following brief design calculation.

Frequency counters are available which are stable to 1 part in 10^8 over a three-day period. Thus a 5-gigacycles beat note could be measured to

$$\Delta v = \pm \frac{5 \times 10^9}{10^8} = 50 \text{ cycles/sec}$$

Using equation (2) this corresponds to a velocity uncertainty of

$$\Delta V = \frac{C\Delta v}{2v_0 n \sin(\theta/2)} = \frac{(3 \times 10^5)(\Delta v)}{2(4\cdot7 \times 10^{14})(1\cdot86)(8\cdot7 \times 10^{-2})} = 1\cdot97 \times 10^{-9} \Delta v$$

$$= 9\cdot85 \times 10^{-8} \text{ km/sec}$$

Assuming the temperature coefficient[32] is of the order 2×10^{-4} km/sec degC, we would be able to resolve a 0·1 degC temperature change to

$$\frac{10^{-7}}{2 \times 10^{-5}} = 0\cdot5 \times 10^{-2} = 0\cdot5\%$$

In reality the accuracy will be somewhat less owing to a Doppler broadening of the laser modes. It is important to notice that, while the temperature derivative can be measured with this excellent accuracy, the absolute measurement of velocity is dependent upon the measurement of θ as shown by equation (2). The absolute value of the lower-temperature measurements should thus be checked with ultrasonic data. It is also important to notice that the refractive index appears in equation (2) and must be measured as a function of temperature before the observed frequency shifts may be used to compute velocities as a function of temperature.

To obtain pressure derivatives of the elastic constants, it may be possible either to scatter in a pressure cell having NaCl windows or to scatter from a sample compressed between diamond anvils as outlined in the review article by Whatley and Van Valkenburg[33]. X-ray diffraction could be used to simultaneously measure the lattice spacing.

There are other optical techniques which can also be used to determine the elastic constants of crystals. The bulk modulus of diatomic solids may be determined from a measurement of the infrared reflection spectrum[34]. Knowledge of the index of refraction, dielectric constant, and reststrahlen wavelength is required to interpret the spectrum. Like Brillouin scattering this method is nondestructive and requires no contact with the sample and should therefore be useful on small samples at high temperature and pressure. Unfortunately this method is not applicable to Fe_2O_3, SiO_2, quartz, and other nondiatomic solids.

Brillouin scattering broadens the X-ray diffraction lines. Therefore the X-ray data itself can be used to determine the elastic constants and dispersion curves (see Quate, Wilkinson, and Winslow[35] for references).

Conclusion

Brillouin scattering offers a new, convenient, and accurate technique for measuring the elastic properties of materials of geophysical interest. Because it is nondestructive, requires no physical contact with the sample, and can be used on small crystals, the Brillouin technique will be particularly useful at elevated temperatures and pressures.

Acknowledgements

We are grateful for helpful conversations with J. Hopfield. We would also like to acknowledge the advice, encouragement, and courtesy of H. Z. Cummins, R. W. Gammon, S. M. Shapiro, and D. O'Shea at the John Hopkins University and J. Litster at the Massachusetts Institute of Technology who have allowed us to inspect their facilities.

References

1. G. Simmons, 'Ultrasonics in Geology', *Proc. Inst. Elec. Electron. Engrs.*, **53**, 1337–45 (1965).
2. O. L. Anderson and R. L. Lieberman, 'Sound velocity in rocks and minerals', *VESIAC State of the Art Rept.*, in press (1967).
3. L. Brillouin, 'Diffusion de la lumière et des rayons X par un corps transparent homogène', *Ann. Phys. (Paris), 9th Ser.*, **17**, 88 (1922).
4. G. B. Benedek, J. B. Lastovka, K. Fritsch, and T. Greytak, 'Brillouin scattering in liquids and solids using low-power lasers', *J. Opt. Soc. Am.*, **54**, 1284–5 (1964).
5. R. Y. Chiao and B. P. Stoicheff, 'Brillouin scattering in liquids excited by the He–Ne maser', *J. Opt. Soc. Am.*, **54**, 1286–7 (1964).
6. G. B. Benedek and T. Greytak, 'Brillouin scattering in liquids', *Proc. Inst. Elec. Electron. Engrs.*, **53**, 1623–9 (1965).
7. D. H. Rank, F. M. Kriess, V. Fink, and T. A. Wiggens, 'Brillouin spectra of liquids using He–Ne laser', *J. Opt. Am.*, **55**, 925–7 (1965).
8. G. B. Benedek and K. Fritsch, 'Brillouin scattering in cubic crystals', *Phys. Rev.*, **149**, 647–62 (1966).
9. R. W. Gammon and H. Z. Cummins, 'Brillouin scattering dispersion in ferroelectric triglycine sulfate', *Phys. Rev. Letters*, **17**, 193–5 (1966).
10. S. M. Shapiro, R. W. Gammon, and H. Z. Cummins, 'Brillouin scattering spectra of crystalline quartz, fused quartz, and glass', *Appl. Phys. Letters*, **9**, 157–9 (1966).
11. R. S. Krishnan, 'Elastic constants of crystals from light scattering measurements', *Proc. Indian Acad. Sci., Sect. A*, 91–7 (1955).
12. V. Chandrasekharen, 'Thermal scattering of light in crystals II', *Proc. Indian Acad. Sci., Sect. A*, **32**, 379–85 (1950).
13. P. M. Morse and K. U. Ingard, *Theoretical Acoustics*, McGraw-Hill, New York, 1968.
14. R. Y. Chiao, C. H. Townes, and B. P. Stoicheff, 'Stimulated Brillouin, scattering and coherent generation of intense hypersonic waves, *Phys. Rev. Letters*, **13**, 334–6 (1964).
15. E. Garmire and C. H. Townes, 'Stimulated Brillouin scattering in liquids', *Appl. Phys. Letters*, **5**, 84–5 (1964).
16. R. G. Brewer and K. E. Rieckhoff, 'Stimulated Brillouin scattering in liquids', *Phys. Rev. Letters*, **13**, 334–6 (1964).
17. R. G. Brewer, 'The ruby laser as a Brillouin light amplifier', *Appl. Phys. Letters*, **5**, 127–8 (1964).
18. R. G. Brewer, 'Growth of optical plane waves in stimulated Brillouin scattering', *Phys. Rev.*, **140**, 800–5 (1965).
19. R. Y. Chiao and P. A. Fleury, 'Brillouin scattering and the dispersion of hypersonic waves', in *Physics of Quantum Electronics* (Eds. P. L. Kelly, D. Lox, and P. E. Tannenwald), McGraw-Hill, New York, 1966.
20. H. Z. Cummins and R. W. Gammon, 'Rayleigh and Brillouin scattering in liquids—the Landau–Placzek ratio', *J. Chem. Phys.*, **44**, 2785–96 (1966).

21. C. L. O'Connor and J. P. Schlupf, 'Brillouin scattering and thermal relaxation in benzene', *J. Acoust. Soc. Am.*, **40**, 633–66 (1966).
22. F. A. Jenkins and H. E. White, *Fundamentals of Optics*, McGraw-Hill, New York, 1957.
23. D. A. Jennings and H. Takuma, 'Optical heterodyne detection of the forward-stimulated Brillouin scattering', *Appl. Phys. Letters*, **5**, 241–2 (1964).
24. J. B. Lastovka and G. B. Benedek, 'Light-beating techniques for the study of the Rayleigh–Brillouin spectrum', in *Physics of Quantum Electronics* (Eds. P. L. Kelly, D. Lox, and P. E. Tannenwald), McGraw-Hill, New York, 1966.
25. E. P. Gross, 'Change of wavelength of light due to elastic heat waves at scattering in liquids', *Nature*, **126**, 201–2 (1930).
26. C. V. Raman and C. S. Venkateswaran, 'Optical observation of the Debye heat waves in crystals', *Nature*, **142**, 250 (1938).
27. L. Sibaiya, 'Scattering of light in a Rochelle salt crystal', *Proc. Indian Acad. Sci., Sect. A*, 393–7, (1938).
28. R. S. Krishnan, 'Raman spectra of the second order in crystals III, *Proc. Indian Acad. Sci., Sect. A*, **22**, 329–40 (1945).
29. R. S. Krishnan, 'The scattering of light in a diamond and its Raman spectra', *Proc. Indian Acad. Sci., Sect. A*, **26**, 399–418 (1947).
30. R. W. Gammon, Ph.D. Thesis, The John Hopkins University, Baltimore, Maryland (1967).
31. R. S. Krishnan and V. Chandrasekharan, 'Thermal scattering of light in crystals', *Proc. Indian Acad. Sci., Sect. A*, **31**, 427–34 (1950).
32. N. Soga, E. Schrieber, and O. L. Anderson, 'Estimation of the bulk modulus and sound velocities of oxides at very high temperatures', *J. Geophys. Res.*, **71**, 5315–20 (1966).
33. L. S. Whatley and A. Van Valkenberg, 'High pressure optics', *Advan. High Pressure Res.*, **1**, Chap. 6, 327–71 (1966).
34. O. L. Anderson and P. Glynn, 'Measurement of the compressibility in polycrystalline MgO using the reflectivity method', *J. Phys. Chem. Solids*, **26**, 1961–7 (1965).
35. C. F. Quate, C. D. W. Wilkinson, and D. K. Winslow, 'Interaction of light and microwave sound', *Proc. Inst. Elec. Electron. Engrs.*, **53**, 1604–20 (1965).

2

W. O'REILLY

Department of Geophysics and Planetary Physics
School of Physics
University of Newcastle upon Tyne
England

Application of neutron diffraction and Mössbauer effect to rock magnetism

Introduction

Palaeomagnetic problems, such as the origin of magnetization of red sandstones or the possibility of self-reversal in rocks, have resulted in the field of study known as rock magnetism. Since rocks are aggregates of minerals, this field of study is really a branch of mineral magnetism which itself forms part of the wider discipline known as materials science. It is not surprising, therefore, that techniques of materials science such as neutron diffraction and Mössbauer spectra are now being applied to minerals found in rocks of palaeomagnetic interest, and the results of such experiments occasionally have a direct bearing on geophysical problems.

The purpose of this chapter is to give one example of the use of each of the above techniques in rock magnetic problems.

Neutron Diffraction

The diffraction of neutrons by solids[1] offers a number of advantages over the more familiar technique of X-ray diffraction. Firstly, because neutrons are scattered by nuclei, the scattering factor is isotropic. In contrast, X-rays are scattered by electrons and the scattering factor falls off as the scattering angle increases and intensities of high-angle lines are low. Secondly, the scattering factors of the different transition metals for X-rays are very similar to one another, but are quite different for neutrons thus allowing the accurate determination of cation distribution in ferrites. Thirdly, and most important, in the case of magnetic atoms an additional scattering occurs owing to the interaction between the neutron moment and the magnetic moment of the atom. This produces coherent diffraction peaks for ferromagnetic and antiferromagnetic materials. It is this last property which particularly interests us here.

Because of their usefulness in palaeomagnetism, the origin of magnetization of red beds is of great interest. According to van Houten[2], goethite (α-FeOOH) is a major constituent of the pigment of some red beds, laterites, and brown soils, so that it is

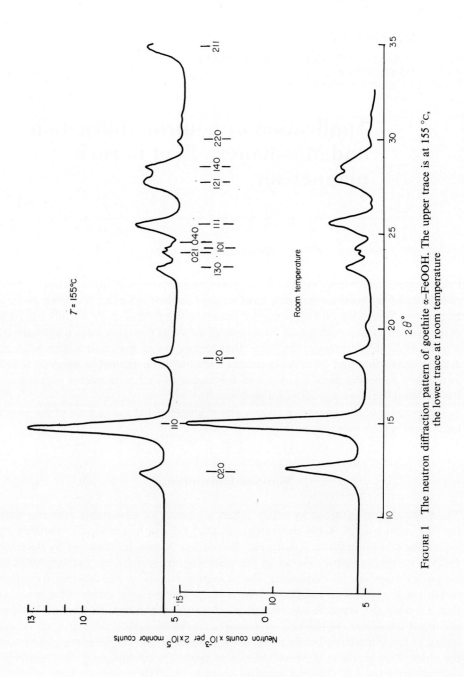

FIGURE 1 The neutron diffraction pattern of goethite α-FeOOH. The upper trace is at 155 °C, the lower trace at room temperature

important to study the magnetic structure of goethite. A neutron diffraction study has been made by I. G. Hedley of the University of Newcastle upon Tyne to determine the spin structure of goethite. The neutron diffraction pattern is shown in figure 1. The upper trace shows the diffraction pattern at 155 °C, above the antiferromagnetic Néel temperature, at which the scattering should be purely nuclear. The lower trace, at room temperature, shows that some peaks, notably the (110) and (020) are considerably enhanced by magnetic scattering. The temperature dependence of nuclear scattering is corrected for by considering those lines with no magnetic component such as (120) and (140). The magnetic pattern may then be compared with calculated patterns for different spin structures. Figure 2 shows the result of Hedley's calculation. The

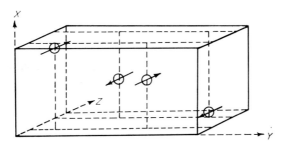

FIGURE 2   Proposed spin structure for goethite in the unit cell of von Hoppe. Hydrogen and oxygen atoms are omitted for simplicity

hydrogen and oxygen ions are omitted for simplicity. The spin configuration may be described as $+ - + -$ and the spins lie in the [001] direction. Some samples of goethite show ferromagnetism, and it is possible that this is the result of spin canting like that found in haematite α-Fe_2O_3. More accurate determinations may reveal this. One experimental difficulty is that short-range order may persist at temperatures not very far above the Néel point, but that heating to still higher temperatures is not possible because of dehydration to α-Fe_2O_3.

Mössbauer Effect

Briefly, Mössbauer effect spectroscopy is the resonant absorption of γ-rays by nuclei (e.g. ^{57}Fe) contained in the material under investigation. A comprehensive account may be found in Wertheim[3]. The γ-rays are emitted from the source without recoil of the parent atoms and with an extremely narrow energy bandwidth. The energy of the γ-rays is modulated by imparting a relative motion to the source and absorber with a velocity usually in the range of about ± 1 cm/sec. The spectrum is displayed by plotting the transmission of the γ-rays through the sample as a function of the relative velocity. In general the centre of the absorption pattern does not occur at zero velocity. This is because the radiative transition between the ground and excited states of the nucleus involves a change in diameter of the nucleus. The work done in the transition is therefore a function of the electrostatic interaction between the nuclear and electronic charges and different isomer shifts (the shift of the centre of the absorption pattern from zero velocity) are found for, for example, ferrous and ferric ions in the case of ^{57}Fe resonance.

The excited state of ^{57}Fe, with spin $\frac{3}{2}$, having an asymmetric nuclear charge distribution and therefore an electric quadrupole moment, interacts with any electric field gradient which may originate from the atom's own electrons or from surrounding ions in a non-cubic configuration. The ground state of spin $\frac{1}{2}$ having zero quadrupole moment remains unsplit and the result is the characteristic two-finger absorption spectrum.

In the case of paramagnetic ions, for example, iron ions, the magnetic field due to the ion's electrons splits the excited state in four hyperfine levels of spin $\pm\frac{3}{2}$, $\pm\frac{1}{2}$ (for ^{57}Fe) and the ground state into two levels $\pm\frac{1}{2}$. Six radiative transitions are permitted resulting in the well-known six-line Zeeman pattern which is usually observed when the material is in the ferrimagnetic or antiferromagnetic state.

Such a pattern is shown in figure 3. This is the Mössbauer spectrum for γ-Fe$_2$O$_3$,

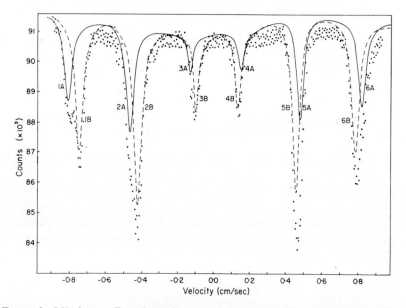

FIGURE 3 Mössbauer effect absorption pattern for maghemite in the presence of an external field of 17 kOe. Pattern A is for tetrahedral sites and B for octahedral sites

maghemite, obtained by Armstrong, Morrish, and Sawatzky[4]. Maghemite has the spinel crystal structure with ferric ions on two types of lattice site, one of tetrahedral and the other of octahedral symmetry. The magnetic moments of ions on the two sites are coupled antiparallel and so the magnetic field at the nuclei of ions on the two sites are also antiparallel. By applying an external field of 17 kOe the pattern was resolved into two parts, one being due to ferric ions on tetrahedral sites and the other to ferric ions on octahedral sites. In this way it was shown that both the isomer shift and nuclear field for tetrahedral sites are less than those for octahedral sites (488 kOe, 0·036 cm/sec and 499 kOe, 0·05 cm/sec respectively). As the valency of the iron is the same in both cases this implies that the bonding for tetrahedral site cations is covalent. The bonding in octahedral sites is probably almost purely ionic. This is important when one is considering the mechanism of oxidation of minerals of spinel structure such as the titanomagnetites Fe$_{3-x}$Ti$_x$O$_4$ ($0 < x < 1$).

If one assumes, as a result of this covalent bonding, that tetrahedral cations are much less free to diffuse through the lattice and therefore do not take part in the oxidation process, the degree of oxidation which is possible for oxidation of octahedral sites only can be calculated. This is shown on the FeO-Fe_2O_3-TiO_2 ternary diagram in figure 4. The details of the calculation may be found elsewhere[5]. Region 1 in the

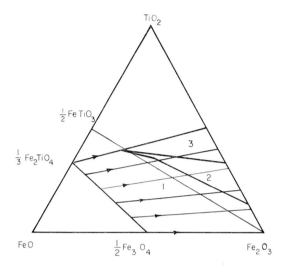

FIGURE 4 Oxidation of titanomagnetites shown on the FeO-Fe_2O_3-TiO_2 ternary diagram. Region 1 represents oxidation of octahedral sites only, regions 2 and 3 oxidation of octahedral sites. In region 3 the magnetic moment is reversed with respect to the original titanomagnetities

figure represents chemical compositions which can be obtained by oxidation of octahedral sites only. The arrows indicate the change in composition as oxidation proceeds.

It is interesting to note that the spontaneous magnetization of titanomagnetites can be reversed on oxidation. Originally, for the Fe_3O_4-Fe_2TiO_4 series, the magnetic moment of the octahedral (B) sublattice predominates over the tetrahedral (A) sublattice moment. As oxidation proceeds the A and B sublattice moments become equal at the boundary between regions 2 and 3 of figure 4. In region 3, the A sublattice moment predominates. To obtain this A sublattice predominance it is necessary to oxidize tetrahedral site ferrous ions and so it may not be probable that such a mechanism could produce self-reversal of titanomagnetites in rocks. However, whether or not self-reversal is an important phenomenon, because of the ubiquity of oxidized titanomagnetites in rocks, the study of the mechanism of oxidation of titanomagnetites is important to rock magnetism.

Mössbauer effect measurements on the titanomagnetites are complicated by line broadening effects[6, 7].

Acknowledgements

Thanks are due to I. G. Hedley for making his neutron diffraction data available prior to publication.

References

1. G. E. Bacon, *Neutron Diffraction*, 2nd ed., Oxford University Press, London, 1962.
2. F. B. van Houten, 'Origin of red beds—some unsolved problems', in *Problems in Palaeoclimatology* (Ed. A. E. M. Nairn), Interscience, New York, London, Sydney, 1964, pp. 652–3.
3. G. K. Wertheim, *Mössbauer Effect: Principles and Applications*, Academic Press, New York, London, 1964.
4. R. J. Armstrong, A. H. Morrish, and G. A. Sawatzky, *Phys. Letters*, **23**, 414 (1966).
5. W. O'Reilly and S. K. Banerjee, *Nature*, **211**, 26 (1966).
6. S. K. Banerjee, W. O'Reilly, T. C. Gibb, and N. N. Greenwood, *Phys. Letters*, **20**, 455 (1966).
7. S. K. Banerjee, W. O'Reilly, and C. E. Johnson, *J. Appl. Phys.*, **38**, 1289 (1967).

3

C. E. JOHNSON

Atomic Energy Research Establishment
Harwell, Berkshire, England

Mössbauer effect studies of magnetic minerals

Introduction

The Mössbauer effect makes possible the observation of the resonant absorption of γ-rays by nuclei in solids without any line broadening by thermal vibrations. Thus it provides a method for detecting the spectrum of the γ-rays with such a precision that the small changes in the energy of the nuclei produced by their electronic environment may be measured. These energy changes take the form of shifts and splittings of the γ-ray line and result from the interactions between the nuclei and electrons which may be conveniently described by the effective fields produced at the nuclei by the electrons. The Mössbauer effect enables these fields to be measured and used to determine the state of the atoms in the crystal, and has had a great impact on solid-state physics in recent years.

Nuclear resonance fluorescence, i.e. the excitation of nuclei by the absorption of γ-rays, has been for a long time a subject of great interest to nuclear physicists, but its realization presented very great difficulties to the experimentalist. Although it was first attempted in the 1920's, many years elapsed and much effort was expended before it was first successfully observed in 1951. The main difficulty arises from the effects of the nuclear recoil. Usually the nucleus recoils with an energy $E_R = p^2/2M$, where p is the momentum imparted to it which must be of equal magnitude and opposite in sign to the momentum of the γ-ray, and M is the mass of the nucleus. If the energy of the excited state of the nucleus is E_0, then $p = E_0/c$ and so

$$E_R = \frac{E_0^2}{2Mc^2} \qquad (1)$$

and the energy of the emitted γ-radiation is $E_0 - E_R$. For resonance to occur the energy of the incident radiation must be equal to $E_0 + E_R$, since a further energy loss occurs in absorption, and so the γ-ray is usually out of resonance by $2E_R$. This is large compared with the thermally produced linewidth $\Delta = (E_R \bar{\varepsilon})^{1/2}$, where $\bar{\varepsilon}$ is the average thermal energy of the nuclei, and makes nuclear resonance fluorescence difficult to observe. The same problem is negligible for optical resonance because of the much smaller transition energy. Several methods have been used successfully to supply the energy difference $2E_R$ by means of the Doppler effect. In one the source of γ-rays was moved on a high-speed rotor to give linear velocities of the order of 10^4 cm/sec and a nuclear resonance of width 2Δ was observed. (The factor 2 comes in because the observed line is the convolution of the line shapes of the source and absorber.) In another both the source and absorber were heated so that the widths Δ of the emission and absorption lines was increased; consequently, their overlap, and hence the resonant absorption, increased and the transmitted counting rate decreased.

Mössbauer's discovery[1,2] enabled the problems to be bypassed for many nuclei. He performed the second kind of experiment just described with the 129 kev γ-ray of ^{191}Ir, but he found that when he cooled the source and absorber the absorption increased, whereas it would have been expected to decrease. Mössbauer showed that this was a result of the binding of the nucleus in the solid, and arises from the quantum nature of the lattice vibrations (phonons) in the crystal. If the recoil energy E_R is small compared with the energy required to excite a phonon, there is a finite probability that the γ-rays will be emitted without either the shift in energy due to the recoiling nucleus or the line broadening due to thermal vibrations. The probability is largest for low-energy γ-rays, and for tight crystal binding (i.e. for solids with a high Debye θ) and increases with decreasing temperature. For a Debye solid it is given by the Debye–Waller factor

$$f = \exp\left[-\frac{3}{2}\frac{E_R}{k\theta}\left\{1+4\left(\frac{T}{\theta}\right)^2\int_0^{\theta/T}\frac{x\,dx}{e^x-1}\right\}\right] \quad (2)$$

This fraction of the γ-rays may be resonantly absorbed without using elaborate apparatus. Mössbauer traced out the spectrum of the ^{191}Ir γ-ray by Doppler shifting the energy

$$\frac{\Delta E}{E_0} = \frac{v}{c} \quad (3)$$

with relatively modest velocities v of the order of 1 cm/sec, and obtained a line with a width 2Γ, where Γ is the natural linewidth equal to h/τ, where τ is the mean lifetime of the excited nuclear state. This is narrower by a factor of about 10^4 than the thermal linewidth Δ, and the resonance was extremely sharp with a width of $9\cdot2 \times 10^{-6}$ ev which is an energy resolution of better than one part in 10^{10}.

Following this work, resonance has been observed in many other nuclei with low-energy γ-rays and an enormous number of applications of the effect, frequently to measurements which had previously been thought to be quite impossible, has made it of great value to physicists working in many fields. Among the applications have been such diverse topics as a terrestrial experiment to test the general theory of relativity, the study of lattice vibrations in crystals, and measurements on alloys, ferromagnetic and antiferromagnetic crystals, inorganic complexes and haemoglobin, and other biological molecules[3].

The Mössbauer Effect as a Research Tool

Since Mössbauer's original work, much thought has been given to the design of the spectrometer for Doppler shifting the energy of the γ-rays, and commercially produced apparatus is now available from several firms. In most designs the source is mounted on a vibrator which is driven through a series of velocities by a waveform generator. The most convenient waveform is probably one with constant acceleration (parabolic in displacement) so that the time spent at each velocity is the same. The γ-rays transmitted by the absorber are counted and stored in a multichannel analyser operating in the 'time' mode where the channels are advanced in step with the increasing velocity. The display then gives a plot of counting rate against velocity, i.e. the spectrum of the λ-rays.

The main features of this spectrum are as follows.

(a) *The chemical shift* (*often called the isomer shift*): this is related to the electronic charge density at the nucleus. This can provide a measure of the transfer of electrons to or from the surrounding ions by the chemical bonding, and leads to information on the identification of the chemical state of the atom.

(b) *The quadrupole splitting*: this is produced by the electric field gradient at the nucleus. It is caused by the distortion of the crystal lattice or the molecule from cubic symmetry, and so leads to information on the crystallography or stereochemistry in the neighbourhood of the Mössbauer nucleus.

(c) *The hyperfine field*: this arises from the interaction between the nucleus and the magnetic moment of the atom when there are unpaired electrons present. This may produce an effective magnetic field at the nucleus of a few hundred kilogauss up to 10 MG, and causes a Zeeman splitting of the Mössbauer lines.

Measurement of the Mössbauer spectrum thus enables chemical, crystallographic, and magnetic information to be derived simultaneously. Its power as a technique for investigating the solid state stems from many factors. It provides a resonant spectroscopic method which operates equally well for diamagnetic and paramagnetic materials, i.e. it is not a *magnetic* resonance technique. It requires no special change in the experimental set-up to measure magnetically ordered solids (ferromagnets, ferrimagnets, and antiferromagnets) unlike, for example, nuclear magnetic resonance where the technique is changed from a fixed frequency, variable field method to variable frequency, zero field on going below the transition temperature. There are no problems connected with the skin depth in a conductor, so it may be applied to metals and alloys as well as to dielectric crystals.

As it measures a spectrum rather than obtains a value for some average property of a solid (e.g. susceptibility) it enables very detailed studies to be made. It may be used as its own chemical analyser, and small amounts of impurities may be detected and allowed for without affecting the accuracy of the measurement. Impurities not containing the Mössbauer isotope do not contribute to the spectrum. It is of great advantage to have a method for analysing different chemical states in the form in which they occur in the solid, as conventional methods of chemical analysis usually only measure the total amount of each element present without regard for how they are combined in the sample. Also of course the oxidation state may be altered during the process of chemical analysis. Thus the Mössbauer effect is a very powerful non-destructive method which may be applied to the study of quite small quantities of material, which makes it especially valuable in mineralogy.

The best-known Mössbauer nucleus is ^{57}Fe which has a γ-ray of energy of 14·4 kev. f values of 70% at room temperature are common, so the effect is easy to observe. The γ-ray is emitted by ^{57}Co which has a half-life of 270 days, and measurements on a large number of iron-containing absorbers have been made. Since iron forms a large number of minerals, magnetic alloys and compounds, inorganic complexes, and biological materials, the scope of possible applications is very wide. Natural iron contains 2% of ^{57}Fe and specimens containing 10 mg of iron are adequate for making accurate measurements. Very much smaller quantities than this may be detected by the Mössbauer effect, however, especially if they can be enriched in ^{57}Fe. As well as being able to distinguish between different chemical species, the spectrum may allow two or more different ions in different sites in the same crystal to be studied, e.g. the Fe^{2+} and Fe^{3+} ions in the spinels, which we shall describe as an example later.

The minerals to be used as absorbers may be either in the form of a single crystal or a powdered specimen. The sample thickness must be small enough for the attenuation of the γ-rays caused by electronic absorption not to be too severe, and in practice it is usually between one and twenty thousandths of an inch, depending upon what other elements are present.

The Mössbauer Spectra of Iron Compounds

The ^{57}Fe Mössbauer effect spectra of several iron compounds are shown in figure 1 in order to illustrate how the chemical state and magnetic properties influence the γ-ray energies. The data are for (a) ferric chloride $FeCl_3$, (b) ferrous chloride tetrahydrate $FeCl_2 \cdot 4H_2O$, (c) pure iron, and (d) ferric oxide Fe_2O_3 (haematite). The measurements were made at room temperature and the zero of the velocity scale is the centre of the pure iron spectrum.

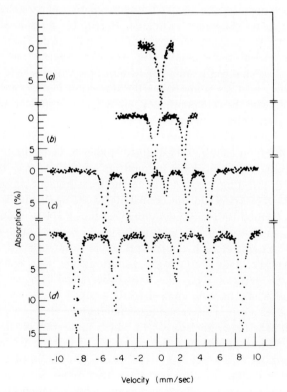

FIGURE 1 Mössbauer spectra at room temperature of (a) ferric chloride $FeCl_3$, (b) ferrous chloride tetrahydrate $FeCl_2 \cdot 4H_2O$, (c) pure iron, and (d) ferric oxide Fe_2O_3

(a) and (b) are non-magnetic at room temperature. (c) is ferromagnetic and (d) antiferromagnetic, both showing a Zeeman splitting into six lines by hyperfine fields of 330 and 510 kG respectively. The ferric compounds (a) and (d) have the chemical shift of about 0·5 mm/sec characteristic of the Fe^{3+} ion, and show at most only a small quadrupole splitting since the ion is almost spherical. The ferrous salt (b) has a chemical shift of 1·5 mm/sec and a quadrupole splitting of 3·0 mm/sec characteristic of the highly anisotropic Fe^{2+} ion.

As well as enabling the energies of the γ-ray transitions to be measured, the Mössbauer effect makes possible measurements on other properties of γ-rays which are analogous to those of optical radiation. For example, their linear and circular polarization have been measured and used to determine the electric and magnetic axes in crystals.

We shall describe the application of the Mössbauer effect to the study of magnetic minerals with reference to two examples: (i) the determination of the valence states of iron in Fe_3O_4 (magnetite) and other crystals with the spinel structure, and the measurement of the distribution of the Fe^{2+} and Fe^{3+} ions over the tetrahedral and octahedral sites as titanium impurities are introduced into the crystal, and (ii) the determination of the axis of magnetic alignment in antiferromagnetic goethite α-FeOOH.

Application to the Study of Iron Spinels

The spinels have the general formula XY_2O_4, where one-third of the cations are in sites of tetrahedral symmetry (the A sites) and two-thirds have octahedral symmetry (the B sites). In 'normal' spinels each site contains only one kind of cation, e.g. zinc ferrite $Zn^{2+}Fe_2^{3+}O_4^{2-}$. In 'inverse' spinels they are mixed, e.g. magnetite $Fe^{2+}[Fe^{2+}Fe^{3+}]O_4^{2-}$. The crystal structure is shown in figure 2. A large number of crystals with this structure occur naturally as minerals, the commonest being magnetite Fe_3O_4.

Normal spinels are non-magnetic, and the Mössbauer effect has been used[4-6] to determine the oxidation state of the iron at each site. In $ZnFe_2O_4$ the chemical shift showed the iron to be in the Fe^{3+} state, and the large (for Fe^{3+}) quadrupole splitting of 0·6 mm/sec showed it to be on the octahedral sites which have a trigonal distortion.

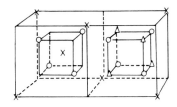

Symbol	Normal structure		Inverse structure	Coordination number
	$MgAl_2O_4$	MFe_2O_4	MFe_2O_4	
X	1 Mg^{2+}	1 M^{2+}	1 Fe^{3+}	4 (tetrahedral)
△	2 Al^{3+}	2 Fe^{3+}	1 M^{2+}, 1 Fe^{3+}	6 ('octahedral)
○	4 O^{2-}	4 O^{2-}	4 O^{2-}	

FIGURE 2 Crystal structure of the spinels

In Fe_2GeO_4 the iron is Fe^{2+} in octahedral sites, showing a 3 mm/sec quadrupole splitting. Fe^{2+} in tetrahedral sites shows a smaller shift than the more usually encountered octahedral symmetry, and has a value of about 0·9 mm/sec in $FeAl_2O_4$, $FeCr_2O_4$, and FeV_2O_4.

Inverse spinels are magnetic. Magnetite Fe_3O_4 has a tetrahedral Fe^{3+} and octahedral Fe^{2+} and Fe^{3+} ions. Ulvöspinel (Fe_2TiO_4) has a tetrahedral Fe^{2+} ion and mixed Fe^{2+} and Ti^{4+} on the octahedral sites. A series of compounds with the same structure may be formed from these oxides with the formula $Fe_{3-x}Ti_xO_4$ ($0 \leq x \leq 1$), the substitution of titanium atoms occurring according to the process

$$2\ Fe^{3+} \rightarrow Fe^{2+} + Ti^{4+}$$

so that charge is conserved. An important problem in the magnetism of the ferrospinels is which of the ferric ions is replaced and to which sites are the ferrous and titanium ions distributed. The study of this system may be of technical importance, since these compounds have high coercivities, and it is also important for rock magnetism, since titanium in all proportions is the most common impurity in naturally occurring magnetite.

Recent measurements[7] of the magnetic moment of the $Fe_{3-x}Ti_xO_4$ system shows that the situation is more complicated than had been originally thought, and that the mechanism for the substitution of the Ti^{4+} ions depends upon the concentration x. Figure 3 shows the variation of the magnetic moment with concentration, and it is

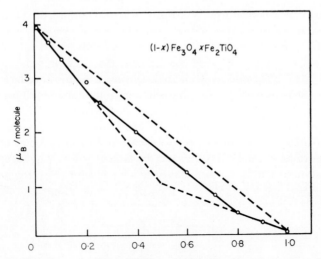

FIGURE 3 Variation of the magnetic moment for the titanomagnetite system $Fe_{3-x}Ti_xO_4$ with Ti content x

seen that there are two discontinuities in the curve, at $x = 0·2$ and $x = 0·8$. The model[7] proposed to account for these data is

$$0 < x < 0·2 \qquad Fe^{3+}[Fe^{2+}_{1-x}Fe^{3+}_{1-2x}Ti^{4+}_x]O_4^{2-}$$

$$0·2 < x < 0·8 \qquad Fe^{2+}_{x-0·2}Fe^{3+}_{1·2-x}[Fe^{2+}_{1·2}Fe^{3+}_{0·8-x}Ti^{4+}_x]O_4^{2-}$$

$$0·8 < x < 1·0 \qquad Fe^{2+}_{2x-1}Fe^{3+}_{2-2x}[Fe^{2+}_{2-x}Ti^{4+}_x]O_4^{2-}$$

i.e. initially pairs of octahedral Fe^{3+} ions are replaced by Fe^{2+} and Ti^{4+} ions going to the octahedral sites. After $x = 0.2$ an octahedral and a tetrahedral Fe^{3+} are replaced by a tetrahedral Fe^{2+} and an octahedral Ti^{4+} until at $x = 0.8$ there are no octahedral Fe^{3+} ions left. Then pairs of tetrahedral Fe^{3+} ions are replaced by a tetrahedral Fe^{2+} and an octahedral Ti^{4+} while simultaneously a Fe^{2+} migrates from an octahedral to a tetrahedral site.

The electrical resistivity of Fe_3O_4 shows a sharp increase when the temperature is raised above 118 °K. This has been explained as being due to the occurrence of electron hopping between the Fe^{3+} and Fe^{2+} ions on the octahedral sites which takes place above this temperature. The Mössbauer spectrum[8-10] provides a striking confirmation of this phenomenon. Figure 4(a) shows the spectrum of Fe_3O_4 at 300 °K and it may be

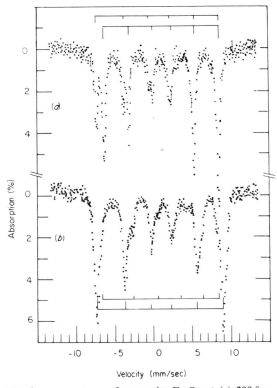

FIGURE 4   Mössbauer spectrum of magnetite Fe_3O_4 at (a) 290 °K and (b) 77 °K

analysed into two six-line hyperfine patterns, one arising from the tetrahedral sites (Fe^{3+}) and one from the octahedral sites ($Fe^{2+}-Fe^{3+}$). The latter is the averaged spectrum of a Fe^{2+} and a Fe^{3+} ion produced by the hopping which is rapid compared with the nuclear precession frequency, and has a smaller hyperfine field and twice the intensity of the former, since there are twice as many octahedral ions. Below 118 °K the hopping ceases, and the spectrum changes as shown in figure 4(b) which was measured at 77 °K. The ions on the octahedral sites now give their individual spectra separately instead of their average. The Fe^{3+} ion gives a spectrum the same as that of the tetrahedral Fe^{3+} ion, so that the larger hyperfine field now has the greater intensity. The data obtained from these spectra are summarized in table 1.

TABLE 1   Mössbauer effect data on Fe_3O_4

T (°K)	Site	Ion	Hyperfine field (kG)
290	tetrahedral	Fe^{3+}	491
	octahedral	Fe^{2+}–Fe^{3+}	453
77	tetrahedral octahedral	Fe^{3+}	503
	octahedral	Fe^{2+}	480

In Fe_2TiO_4 the Mössbauer spectrum[10] at room temperature shows a quadrupole splitting into two rather broad lines. This may be analysed into contributions from the two Fe^{2+} sites, that from the tetrahedral site presumably having the smaller shift and quadrupole splitting. The compound becomes antiferromagnetic below its Néel temperature of about 120 °K, and the Mössbauer spectrum shows hyperfine splitting but with lines very much broader than in most magnetic materials (figure 5). This

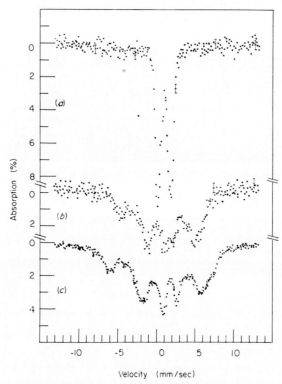

FIGURE 5   Mössbauer spectra of ulvöspinel Fe_2TiO_4 at (a) 290 °K, (b) 77 °K, and (c) 4·2 °K

arises from the randomness of the distribution of the Fe^{2+} and Ti^{4+} ions on the octahedral sites which causes the hyperfine field to vary from site to site. The resultant spectrum, which is the superposition of contribution from all the inequivalent sites, shows very broad lines in a way similar to that found in disordered alloys.

The Mössbauer spectra of the $Fe_{3-x}Ti_xO_4$ series[10, 11] show features associated both with the hopping process and with the line broadening due to the local random disorder. Figure 6 shows some of the spectra which have been measured and which throw

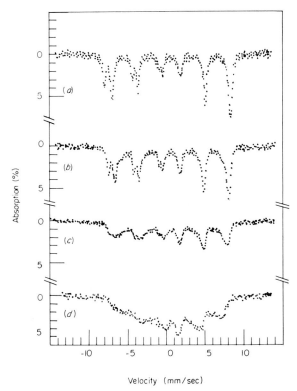

FIGURE 6 Mössbauer spectra at room temperature for (a) Fe_3O_4, (b) $Fe_{2.8}Ti_{0.2}O_4$, (c) $Fe_{2.6}Ti_{0.4}O_4$, and (d) $Fe_{2.4}Ti_{0.6}O_4$

light on the mechanism of titanium substitution. These may prove useful for analysing minerals of this series.

Measurement of the Axis of Magnetization in Goethite

The mineral goethite α-FeOOH is formed from the ageing of precipitated ferric hydroxide and it is an antiferromagnet at ordinary temperatures. By measuring the absorption of polarized γ-rays by a single crystal using the Mössbauer effect the axis of magnetization may be determined. In a magnetic material the hyperfine components of the γ-rays will be polarized in directions which depend upon the direction of magnetization[12], just as the Zeeman components in optical spectra are polarized. Since we now have a method for detecting the hyperfine components of γ-rays, we may use it to measure their polarization.

The γ-rays emitted by a source of ^{57}Co diffused in a ferromagnetic iron foil which is unmagnetized consist of six lines with relative intensities 3:2:1:2:3 and are similar to the absorption spectrum of iron (figure 1(c)). If such a source is moved with respect to an iron absorber, the Mössbauer spectrum shows rather a large number of lines, as seen in figure 7(a). At zero velocity all the components are at resonance, so there is a

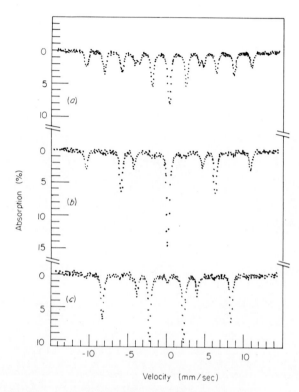

FIGURE 7 Mössbauer spectra for a source of ^{57}Co in iron and an iron absorber (a) unmagnetized, (b) source and absorber magnetized parallel to each other, and (c) magnetized perpendicular to each other

deep central line. When the iron source is magnetized in its own plane, which requires a field of a hundred gauss or so, the intensities of the γ-rays emitted perpendicular to the foil change to $3:4:1:1:4:3$ and these components become linearly polarized in directions $\parallel : \perp : \parallel : \parallel : \perp : \parallel$ with respect to the magnetic field. If the iron absorber is also magnetized, absorption can occur at each of the resonant energies only if the radiation has the right polarization. Hence, if the magnetic fields on the source and absorber are parallel, all the γ-rays will still be at resonance at zero velocity and there will be a large absorption dip (figure 7(b)). If the fields are perpendicular, however, there will be no central line in the spectrum, as seen in figure 7(c). Similarly the intensities of all the other lines in the spectrum will depend upon the relative directions of the magnetizations of the source and absorber. This is analogous to the variation of intensity of polarized light transmitted by a Nicol prism, or other polarization-sensitive detector. The source acts as a polarizer and the absorber acts as an analyser.

In an antiferromagnetic crystal the Mössbauer absorption will also be orientation dependent, but the direction of magnetization is an internal property of the structure of the crystal. This direction may be determined by a measurement analogous to that just described, but with the unknown axis of the crystal replacing the known direction of the magnetic field on the iron absorber. Figure 8 shows the spectra obtained for a

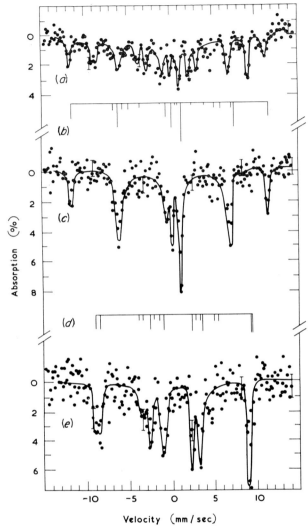

FIGURE 8 Mössbauer spectra of single-crystal absorber of α-FeOOH with a source of ^{57}Co (a) unmagnetized, (c) magnetized parallel to the crystal c axis, and (e) magnetized perpendicular to the c axis. Also shown are the predicted spectra for the source magnetization (b) parallel to and (d) perpendicular to the sublattice magnetization of the crystal. The γ-rays are incident along the b axis

single crystal of goethite with (a) an unmagnetized source, and (c) a source magnetized parallel to, and (e) perpendicular to, the c axis of the crystal. From the complete extinction of the lines as the magnetization is rotated it is concluded that the magnetic moments are collinear, and further analysis shows that they are all parallel to the c axis of the crystal.

It is not possible to tell by this method how the magnetic moments are arranged in detail, e.g. whether they are like ↑↓↑↓ or ↑↑↓↓. Such information may be obtained from neutron diffraction measurements with polarized neutrons. Measurements done on this crystal[13] confirm that the c axis is the direction of magnetic alignment and have established the spin structure in the antiferromagnetic state.

Conclusion

An outline has been given of some of the possible applications of the Mössbauer effect to the study of magnetic minerals. The speed and relative simplicity of the measurements, and the detail and variety of information which they can yield on the state of iron atoms make the technique very powerful. Work is being carried out in many laboratories to further the study of solids of geophysical interest, e.g. by Sprenkel-Segal and Hanna[14], and the volume of useful data obtained is likely to expand enormously in the next few years.

References

1. R. L. Mössbauer, *Z. Physik*, **151**, 124 (1958).
2. R. L. Mössbauer, *Naturwissenschaften*, **45**, 538 (1958).
3. A. H. Muir, Jr., K. J. Ando, and H. M. Coogan, *Mössbauer Effect Data Index, North American Science Center*, Wiley, New York, 1966.
4. T. Mizoguchi and M. Tanaka, *J. Phys. Soc. Japan*, **18**, 1301 (1963).
5. M. Tanaka, T. Tokudo, and Y. Aiyama, *J. Phys. Soc. Japan*, **21**, 262 (1966).
6. M. J. Rossiter, *Phys. Chem. Solids*, **26**, 775 (1965).
7. S. K. Banerjee and W. O'Reilly, *Trans. Inst. Elec. Electron. Engrs.*, No. MAG-2, 463 (1966).
8. R. Bauminger, S. G. Cohen, A. Marinov, S. Ofer, and E. Segal, *Phys. Rev.*, **122**, 1447 (1961).
9. K. Ono, Y. Ishikawa, A. Ito, and E. Hirahara, *J. Phys. Soc. Japan, Suppl. B-I*, **17**, 125 (1962).
10. S. K. Banerjee, W. O'Reilly, and C. E. Johnson, *J. Appl. Phys.*, **38**, 1289 (1967).
11. C. E. Johnson and S. K. Banerjee, to be published.
12. S. S. Hanna, J. Heberle, C. Littlejohn, G. J. Perlow, R. S. Preston, and D. H. Vincent, *Phys. Rev. Letters*, **4**, 177, (1961).
13. J. B. Forsyth, I. G. Hedley, and C. E. Johnson, *J. Phys. C (Proc. Phys. Soc. (London))*, Ser. 2, **1**, 179 (1968).
14. E. L. Sprenkel-Segal and S. S. Hanna, *Geochim. Cosmochim. Acta*, **28**, 1913 (1964).

D

DEVELOPMENTS IN TECHNIQUES

II. New applications of atomic and nuclear physics to geophysical measurements

1

R. L. FLEISCHER

and

P. B. PRICE

General Electric Research and Development Center
Schenectady, New York, U.S.A.

R. M. WALKER

Washington University
St. Louis, Missouri, U.S.A.

Fission track dating and processes in the Earth's interior

Introduction

Among the existing dating methods that utilize radioactive decay, fission track dating[1,2] presents extensive and largely untapped opportunities for the dating of thermal, magnetic, and mechanical events. The unique simplicity of the technique, conceptually and experimentally, makes it an attractive tool for probing the past.

The technique depends on the fact that the fragments emitted in spontaneous fission of uranium-238 can create and store narrow, but continuous, damage trails in most minerals (including natural glasses). These sites can be displayed by the use of preferential chemical attack to etch out tube-like or conical holes to a size where they can be viewed and counted in an ordinary optical microscope. Since their number depends only on the uranium content and on the time since track storage began, this age can be measured provided the uranium content can be determined. This latter step is in fact taken by merely giving the sample a known dose of thermal neutrons in a nuclear reactor, and subsequently counting the new fissions (of uranium-235) induced by the reactor irradiation.

It is fortunate that for terrestrial samples there are no known alternative radiations which would directly produce tracks in nature, and it is only in the most unusual circumstances (such as fissions induced by the neutron burst from an atomic bomb) that induced tracks have been found in natural samples.

The permanence of tracks depends on the specific material and *upon its thermal history*, so that, once the response of a particular material to thermal treatments has been calibrated, it can be used to learn about thermal events in the past. To calibrate a sample containing fission tracks, one merely divides it into smaller pieces and gives each one a different thermal treatment. In most of the cases examined so far [2-4] we have found that the time (t)–temperature (T) relation for the beginning of track fading is described by equations of the form

$$t\,e^{-a/T} = b \tag{1}$$

where a and b are different constants for each given material. Extrapolations of relations such as equation (1) are useful for assessing the likely stability of tracks over times which are much longer than are accessible in the laboratory. Although the question of the validity of such extrapolations obviously must be approached with caution, they are our sole access to the information we seek. So far we have encountered no discrepancies with such extrapolations on either terrestrial or meteoritic samples. Those which we expect should store tracks for geological times do (for example, diopside[5, 6], hornblende[7], olivine[5], zircon[8], and most natural glasses[3, 9]); those which we expect should not did not (for example, meteoritic feldspar glass[2]); in one case to be mentioned later[6] extrapolations of three such relations over 4.5×10^9 years gave quantitative agreement with results from another entirely independent procedure.

Applications

Thermal dating

Because each different mineral type possesses a different temperature below which tracks are retained, a mineral assemblage records the thermal history of the rock through the range of track retention temperatures. In cases of slow cooling, a cooling rate may be obtainable[6]; in the case of stepped cooling or (more likely) reheating into the annealing range, the time and temperature of the second event and the time of the original cooling should be accessible[3].

As an example, let us consider the track annealing temperature given in table 1. If samples of diopside, zircon, and hornblende from a given rock have ages of 450, 400,

TABLE 1   Track annealing properties—track fading temperatures for various materials (1 hour heating)

Material	Temperature (°C)
diopside	825
bytownite	725
zircon	700
silica glass	650
sphene	600
hornblende	575
tektite glass	525
pigeonite	525
hypersthene	475
enstatite	475
apatite	450
muscovite	450
phlogopite	375

and 350 million years, we infer that the rock cooled at a rate of 2·5 degC/million years through the temperature range where tracks begin to be retained in these minerals. Here we are assuming that the same temperature differences apply for much longer cooling times, an assumption that must be tested by measuring the relations such as equation (1) for the three minerals involved.

As a second example, if diopside and zircon gave an age of 450 million years while sphene, hornblende, and apatite (all from the same rock) yielded a 7 million year age, we infer a two-stage process: original solidification and cooling occurred at 450 million years followed much later by a reheating to a temperature between the zircon and sphene annealing points 7 million years ago. Since ages can be measured with a sensitivity of $\sim 5\%$, we can place lower limits on the cooling rates at the 450 million years and 7 million years events of ~ 5 degK/million years and ~ 400 degK/million years respectively, the detectable cooling rate increasing as the age of the event decreases.

In one case a cooling rate has actually been measured[6]. The dating in that special case involved the use of tracks of ^{244}Pu, which are not found in terrestrial material, but which have been identified in meteorites. The idea in this case is that ^{244}Pu, along with various other isotopes, must have been produced in the stars and have been mixed into the planetary material from which the solar system formed. Once the material condenses to form planetary bodies, there is no means of further injection of ^{244}Pu into the system; hence, the ^{244}Pu begins to die out in place and to store fission tracks as soon as the material is sufficiently cool. Of the known isotopes ^{244}Pu is of primary interest because it has a moderately long half-life (82 million years) and a high fission branching ratio of greater than 10^{-3} compared with less than 10^{-6} for ^{238}U. As a result, ^{244}Pu tracks will be much more numerous than ^{238}U tracks, provided that matter cools within a few ^{244}Pu half-lives of the beginning of the formation of the solar system.

During cooling, the more retentive minerals will reach their track retention temperatures first, will begin to store tracks earlier, and will therefore contain a larger fraction of the fission tracks that are due to ^{244}Pu than will the minerals which start measuring time later, when lower temperatures are reached. Thus from the fission tracks we can calculate the time of cooling to a given temperature[10]; from an extrapolation of laboratory track annealing data we can infer what that temperature is for geological times; hence we can construct a curve with as many points on it as we have different minerals containing fission tracks.

For the iron meteorite Toluca, results[6] on three minerals gave a cooling rate of 1 degK/million years through the temperature range 925–500 °K, and this result agreed within estimated errors with that inferred by Wood[11] and Goldstein and Short[12] on the same meteorite. The slowness of this cooling indicates that we are sampling material from a considerable depth within a massive body, probably of asteroidal size. Further details, as well as discussion of this point, are presented elsewhere[6].

Paleomagnetic dating

The relation of fission track ages to paleomagnetic results can be made clear by comparing table 1, which presents track annealing temperatures of potential track storing minerals, with table 2, which lists temperatures at which ferromagnetic or ferrimagnetic minerals begin to store magnetization. For each magnetic mineral there are a number of minerals which potentially are appropriate for dating the last heating to the temperature below which magnetization is frozen in. It follows that track dating of minerals from a rock fragment that also contains magnetic minerals measures the time of recording of the magnetic information. This possibility for magnetic dating has not to our knowledge been made use of as yet.

TABLE 2 Magnetic annealing properties—Curie or Néel
temperatures of magnetic minerals

Material	Temperature (°C)
iron	770
magnemite	675
hematite	670
magnetite	578
(magnetite) 0·85 : (ulvöspinel) 0·15	500
(magnetite) 0·7 : (ulvöspinel) 0·3	400
pyrrhotite	325
(hematite) 0·40 : (ilmenite) 0·60	230

Dating of mechanical processes

Results that relate mechanical effects to track storage are limited, and hence so too is our knowledge of what mechanical processes can be dated by fission tracks. We do know that hydrostatic stresses of up to 80 kb have negligible effects on crystalline materials[4] and that pressures of greater than 10 kb are needed to affect the one natural glass that has been examined[4]. In both cases the magnitude of the effect is so small relative to thermal effects that no information about deep Earth pressures is obtainable without precise knowledge of the temperature distribution within the Earth. In fact minerals pushed up from deep within the Earth are potentially a means of observing the temperature distribution at depth. Shoemaker[13] has suggested that kimberlite pipes would be worth examining with such a possibility in mind.

In contrast with their lack of response to static elastic deformation, tracks are likely indicators of plastic deformation. In mica, shear strains have been seen to cut and fragment tracks[4], so that it is clear that large strains will so disperse the regions of damage that recognizable fission tracks are no longer present. In such a case the age obtained will be the time since the plastic deformation occurred. It seems clear that tracks can be effectively erased in the manner just described for any crystal that deforms by dislocation motion. Preliminary experiments have also shown that shock loading can erase tracks in crystals and glass[14]. A decision as to whether the erasure is due to the associated plastic deformation or to other effects awaits further experimental work.

The conclusion with respect to processes within the Earth is that, although static stresses are probably not detectable, plastic strains should be.

Conclusions

Fission track dating with its wide accessible time span and many temperature indicators presents multiple opportunities for learning about the chronology of thermal, magnetic, and mechanical processes in planetary bodies. Some measurements involving thermal effects in nature have been made[6, 15, 16], but the paleomagnetic and paleomechanical opportunities are as yet untapped.

References

1. R. L. Fleischer, P. B. Price, and R. M. Walker, *Science*, **149**, 383 (1965).
2. R. L. Fleischer, P. B. Price, and R. M. Walker, *Ann. Rev. Nucl. Sci.*, **15**, 1 (1965).
3. R. L. Fleischer and P. B. Price, *J. Geophys. Res.*, **69**, 331 (1964).
4. R. L. Fleischer, P. B. Price, and R. M. Walker, *J. Geophys. Res.*, **70**, 1497 (1965).
5. R. L. Fleischer, P. B. Price, R. M. Walker, M. Maurette, and G. Morgan, *J. Geophys. Res.*, **72**, 355 (1967).
6. R. L. Fleischer, P. B. Price, and R. M. Walker, '^{244}Pu fission tracks and the cooling curve for the parent body of the Toluca meteorite', in press (1967).
7. R. L. Fleischer and P. B. Price, *Geochim. Cosmochim. Acta*, **28**, 1705 (1964).
8. R. L. Fleischer, P. B. Price, and R. M. Walker, *J. Geophys. Res.*, **69**, 4885 (1964).
9. R. L. Fleischer and P. B. Price, *Geochim. Cosmochim. Acta*, **28**, 755 (1964).
10. R. L. Fleischer, P. B. Price, and R. M. Walker, *J. Geophys. Res.*, **70**, 2703 (1965).
11. J. A. Wood, *Icarus*, **3**, 429 (1964).
12. J. I. Goldstein and J. M. Short, *Geochim. Cosmochim. Acta*, **31**, 1733 (1967).
13. E. Shoemaker, private communication.
14. R. L. Fleischer, P. B. Price, R. M. Walker, and M. Maurette, *J. Geophys. Res.*, **72**, 331 (1967).
15. R. L. Fleischer, C. W. Naeser, P. B. Price, R. M. Walker, and M. Maurette, *J. Geophys. Res.*, **70**, 1491 (1965).
16. R. L. Fleischer, P. B. Price, and R. M. Walker, *Geochim. Cosmochim. Acta*, **29**, 161 (1965).

2

A. H. COOK

Division of Quantum Metrology
National Physical Laboratory
Teddington, Middlesex, England

Gas lasers and the measurement of long period and secular changes of strain

Requirements of a Geophysical Strain Gauge

Strain gauges at present in use for geophysical observations are sensitive to changes of strain of about 1 part in 10^9, but are not suitable for measurements over long periods because the zero drifts. A strain gauge which would record reliably changes of strain over periods from a few days up to a year or so is very desirable. It would enable observations to be made of the fortnightly tides, for example, and would enable secular changes to be detected. Thus, not only would it be possible to measure movements across faults but also perhaps the postulated permanent sets due to distant earthquakes might be detected. An optical interferometer using a laser as a source may provide such a strain gauge, and a number of groups are working on developing the possibilities of this scheme. The author explains how an interferometer with a gas laser source may be used to measure strain and considers the possible accuracy that may be obtained and the difficulties still to be overcome in making a reliable strain gauge, concentrating on the problems associated with measurements over a long period of time, say from a few days to a year or more. In another chapter (see section D II, chapter 3 of this book) Davies and King discuss the use of a laser interferometer for measurements of free periods of the Earth and other short-period phenomena.

The Interferometer

A diagram of a simple Michelson interferometer suitable for Earth strain measurements is shown in figure 1. Light incident on a semireflecting surface is divided into two beams, each of which is reflected from a plane mirror back to the semireflecting surface where they are combined. If we consider light travelling in one direction through the interferometer, the intensity of the combined beam after the semireflecting surface is of the form $1 + a\cos(\theta + \phi)$, where $\theta = 2\pi\, 2D/\lambda$, $2D$ being the optical path difference between the beams reflected from the two mirrors, and ϕ is a phase angle introduced at the beam divider. If the light from the beam divider is incident on a photodetector, the current from the photodetector will vary with the angle θ and the number of times

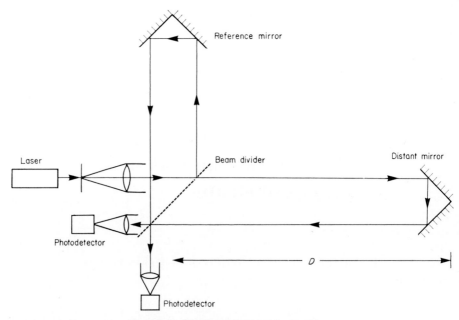

FIGURE 1 Simple Michelson interferometer

that this current reaches a maximum, that is to say, the number of fringes of the interference pattern passing across the cathode of the detector in a given time can be counted electronically. The two beams emerging from the beam divider have different values of ϕ and it is possible to arrange for the difference to be about 90°. Then the outputs from photodetectors on which the two beams fall can be applied to a counter that adds or subtracts counts according to whether the path difference is increasing or decreasing. It is a fairly straightforward matter to record automatically the accumulated count as a function of time and to use it to determine how the strain of the ground on which the interferometer is set up changes as a function of time.

If the wavelength of the laser light is about 0·6 μm, one fringe count corresponds to a change of about 0·3 μm in D. To make measurements to 1 part in 10^9, the path difference should be approximately 5×10^8 wavelengths or 250 m. Bearing in mind that it is possible to detect changes of less than one fringe, for example it is easy to arrange for a bidirectional counter to count quarter fringes, a path difference of about 100 m is quite suitable for measurements to 1 part in 10^9.

The number of fringes passing the detector depends of course on the wavelength, which should be stable to 1 part in 10^9 over the period for which the measurements are to be conducted. The wavelength is the effective wavelength within the medium between the beam divider and the mirrors. The refractive index of air can neither be kept constant enough, nor can it be estimated accurately enough, for measurements to 1 part in 10^9, and so the interferometer should be in a vacuum. The measured path difference is the separation of the mirrors and careful thought must be given to the way in which the mirrors are mounted on the ground if the change in the distance of the mirrors is indeed to be the same as a change in strain of the ground surface. Apart from the requirement that the wavelength should be very stable, the source used for the interferometer must satisfy other stringent requirements. It is desirable that it should be an intense source because then the electronic problems of the detector are simplified, and it is essential that the spread of energy with respect to wavelength should be very restricted. Ordinary gas discharge sources have spectra with relative linewidths of about 1 in 10^6 and as a result the fringes in an interferometer become indistinct at distances appreciably less than 1 m. Very much narrower lines are necessary if interference fringes are to be seen and recorded over distances of 100 m or so. Lasers provide such narrow lines. Lastly the angular spread of the beam should be very small. If the angular spread of the nominally parallel beam between the beam divider and the mirrors is appreciable, light is lost from the mirrors and the visibility of the fringes formed on recombination is reduced. The beam spread can be reduced by isolating a very narrow range of angle by means of a pinhole placed in front of the detector at the focus of the telescope lens as indicated in figure 1, but the smaller this pinhole the less light is passed. Because it is coherent in space, the beam from a gas laser has an extremely small angular spread and it is therefore possible to illuminate a mirror at a distance of 100 m or so without using unduly large optics or losing a significant amount of light. The gas laser is a source which meets all the requirements just set out, subject to the proviso that it is not yet known how stable the wavelength can be maintained over a period of some months.

The Gas Laser

In a gas laser, and in particular in the helium–neon laser, a gas is excited by an electrical discharge in such a way that the population of the upper energy level involved in a transition is greater than that of the lower energy level. The energy level system for the helium–neon laser is shown in figure 2. In these conditions stimulated emission can

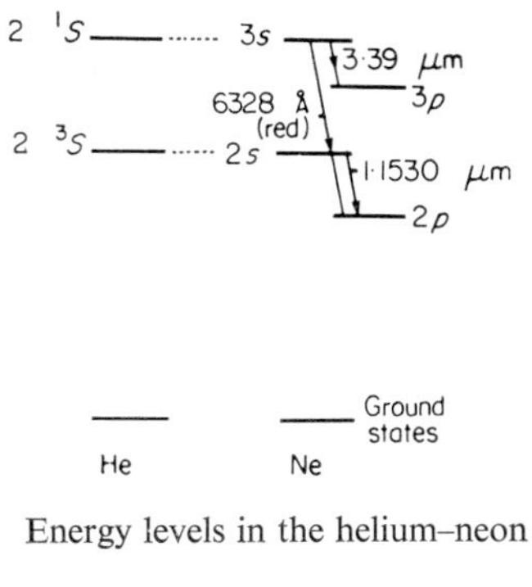

FIGURE 2 Energy levels in the helium–neon system

occur and a beam of light having a wavelength corresponding to the difference of energy between the two levels is amplified as it passes through the gas. The gas is contained between two highly reflecting mirrors, and oscillation will occur for those wavelengths such that the distance between the mirrors is an integral number of half wavelengths, for then light reflected twice between the mirrors returns to a given point in the gas with the same phase. If, as is common, concave spherical mirrors are used, the light is emitted in a conical beam a few millimetres in diameter, with a Gaussian radial intensity distribution and an almost perfectly spherical wavefront. The beam can be changed by a lens into an almost parallel one with very small divergence, such that a 10 mm diameter beam increases to only 13 mm in 100 metres.

The amplification of a beam of light passing through the gas is proportional to the spontaneous emission intensity from the gas which varies with wavelength as shown in figure 3. The half-width of the gain curve is about 2 parts in 10^6. Within that

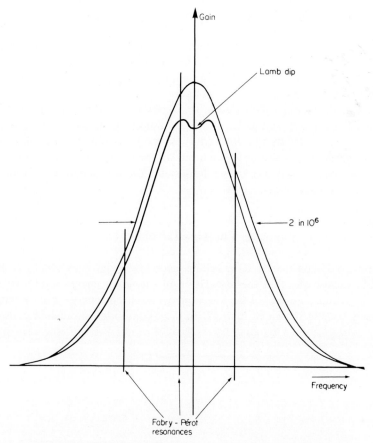

FIGURE 3   Gain curves of the laser and wavelength of oscillation

range there may be a number of wavelengths at which oscillation can occur, depending on the separation d of the mirrors; the spacing of these wavelengths is given by the formula

$$\delta\lambda = \frac{\lambda^2}{2d}$$

For a laser working at 6000 Å, and with mirrors 50 cm apart, the relative spacing of the wavelengths at which oscillation occurs is about 6×10^{-7}. Thus, as shown in figure 3, oscillation can occur at two or three wavelengths within the range of the gain curve. By reducing the power applied to the gas it is possible to reduce the gain so that oscillation occurs at just one frequency close to the peak of the gain curve. A laser operated in these conditions, in one mode, as it is termed, will give a wavelength

Gas lasers

known to about 1 part in 10^7, the exact value depending on the mechanical separation of the mirrors. The width of the line emitted by the laser is very much narrower than this; indeed it may amount to no more than some 100 c/s in 5×10^{14} c/s. The ultimate limit set by the properties of the laser itself is not known because there is strong evidence that the narrowest lines so far obtained are still broadened by the mechanical movements of the mirrors which of course cause a spread in the radiated wavelength. Provided care is taken in the mounting of the laser to ensure that it is mechanically stable and isolated from air vibration, it is possible to get lines which have widths of much less than 1 in 10^9, so that the gas laser provides a source which enables interference fringes to be seen quite clearly at path differences of 100 m or more, and there is every reason to suppose that, if the purely mechanical and optical difficulties of setting up an interferometer with a path difference of 1 km can be overcome, fringes at that distance will also be sharp.

In order to measure a distance to 1 in 10^9 it is necessary that the wavelength of the laser, in terms of which the distance is measured, should be constant to better than 1 in 10^9.

Various schemes for stabilizing the wavelength of a laser and making it independent of the arbitrary separation of the mirrors have been devised. The simplest works with a laser having a natural mixture of neon isotopes and excited by a radiofrequency discharge. In that case the gain of the laser varies with wavelength in just the same way as the spontaneous emission, and the wavelength can be stabilized to that corresponding to the peak of the curve by using a servo system to make small adjustments to the separation of the mirrors so as to keep the intensity a maximum. Such a system works very satisfactorily although it may be thought that it is limited by the rather large width of the spontaneous emission curve. However, stabilities of about 1 part in 10^9 have been achieved[1]. Rather higher precisions can be obtained by more sophisticated schemes. If pure neon-20 is used in the laser with a d.c. discharge, the intensity of the light shows a small dip at the peak, the Lamb dip (figure 3), and a somewhat higher accuracy is obtained by stabilizing to the dip rather than to the peak of the curve for the natural mixture. Practically, it may be a somewhat less satisfactory scheme because the servo mechanism can go out of control more readily than with a simple one.

In a typical servo system one of the mirrors is made to oscillate backwards and forwards a few hundred times a second so that the power output also oscillates at this frequency if the wavelength does not correspond to maximum intensity or to the Lamb dip, whereas it oscillates at double the frequency if the wavelength does correspond to maximum intensity or the Lamb dip. The detector of the servo system compares the frequency, phase, and amplitude of the laser output with that of the signal driving the mirror, and from this comparison a signal is derived which displaces the mean position of the mirror. It will be evident that in addition to causing a variation of power the oscillation of the mirror causes a variation of frequency of the laser output, the range of the variation being perhaps a few Mc/s. In consequence the fringes seen at large path differences are blurred out because the phase is changing by perhaps a radian or so a few hundred times per second. A simple slow detector system would see fringes of very poor visibility or perhaps none at all; there are two ways of overcoming this difficulty. One method is to use a second laser with a steady frequency, the output of which is mixed with that from the first laser with the oscillating frequency, the beats from the mixed output being used to control the frequency of the second laser. In this way the frequency of the second oscillator can be made equal to the mean wavelength of the first but has no frequency modulation imposed on it. A second method is to use a fast detector in the interferometer and to make use in the detector system of the fact that the light is frequency modulated.

By these methods it is possible to stabilize the frequency of the laser to 1 part in 10^9 or so over perhaps one day and to make measurements over path differences of up to 100 m with precisions of the same order. A strain gauge consisting of a Michelson interferometer with a gas laser so stabilized would therefore be suitable for observing the changes of strain produced by the free vibrations of the Earth. If, however, one considers the stability of the laser over a longer time, the change of wavelength of the peak of the gain curve must be taken into account. The position of the peak depends on the pressures of the gases in the discharge tube, and these pressures may change with time so that the wavelength to which the laser is stabilized may also change. At present very little is known about the changes over a long period of time, nor indeed about how long a single laser tube will last at one gas filling. The helium–neon system is probably not the most suitable for work requiring long-term operation and stability of the wavelength for a period of more than a month, because the behaviour does depend rather critically on the relative proportions of the helium and neon in the gas. It would almost certainly be better to use a laser working in a single gas. One possibility is to use infrared lines in krypton in which laser action occurs at wavelengths of about 1·2 μm. The desirable pressure of krypton is about 30 μb, a convenient pressure because it is the vapour pressure of krypton at a temperature close to the triple point of nitrogen. The laser tube could be attached to a reservoir maintained at that temperature and containing an excess of krypton so that the pressure, and hence the wavelength, was stabilized over a long time. This system for controlling the pressure of krypton is used in the lamp by which the standard orange line (λ 6058 Å) which provides the wavelength standard of length is emitted and gives wavelengths which are apparently stable to 1 part in 10^9. It may be expected therefore that laser wavelengths from a krypton source with the pressure stabilized in the same way will also be stable to about 1 part in 10^9. The short-term relative stability of two lasers operating at the same wavelength can be determined readily by mixing the outputs photoelectrically and observing the radiofrequency beats between them, but the only way at present of determining the long-term stability of a laser source is to compare its wavelength with the orange standard line from krypton-86 using the methods of optical interference which, working with the orange line, are limited to precisions that are not much better than 1 part in 10^8. Optical methods will similarly have to be used for comparisons between lasers emitting wavelengths that are appreciably different, say by more than a few hundred Mc/s, but a higher precision than 1 in 10^9 should be possible since the widths of laser lines are less than those of the krypton standard line.

To summarize, the gas laser to a large degree meets the requirements for a source for an interferometric strain gauge. The angular spread of the beam is very small and the intensity is high, thus simplifying the optical design of the interferometer. The spread of energy in the source, that is the linewidth, is very small, so that interference fringes can be obtained over long distances and lastly, over a short time, a high degree of stability of the laser wavelength can be achieved. The system of a Michelson interferometer working with a laser source is therefore quite suitable for short-term measurements of strain to 1 part in 10^9 or may be better, but its behaviour over the

Gas lasers

longer times necessary to detect and estimate either fortnightly tides or secular changes of strain has still to be evaluated and it may well be that lasers other than the helium–neon laser are more suitable for such long-term measurements. The possible applications of a system stabilized over a long period have already been indicated; they are measurements of the fortnightly tides and detection of secular changes of strain. Secular changes may be those due to movements of faults or those due to movements produced by distant earthquakes. England is a rather stable country from both points of view and it is to be expected that delicate measurements will be necessary to detect such changes. At the same time it would be very valuable to be able to compare measurements in this country with those in regions which are far more seismically disturbed such as, for example, the west coast of North America. Plans so far have concentrated on horizontal strain measurements but there is no reason, provided suitable sites can be obtained, why there should be this restriction and there would be considerable value in operating a laser strain gauge for vertical observations in association with one making horizontal measurements.

Measurements of the Differential Strain Tensor

The previous remarks have concerned the use of optical methods for determining change of strain together with a laser source emitting a stabilized wavelength. It is, however, possible to record relative changes of strain between different laser sources by observing the beats between them. Thus, instead of maintaining the wavelength of the laser constant, the mirrors of the laser may be fixed to the ground and the change of wavelength consequent upon changes in the separation of the mirrors may be determined from beats between the one laser and a comparison one. A system for determining the components of the differential strain tensor based on this idea has been suggested[2]. Corresponding to the six independent components of the strain tensor, six lasers could be set up with their mirrors defining six different directions fixed to the rock. Beats between the outputs of these lasers and a seventh stabilized laser provide measurements of the changes of strain in the six directions. The change of frequency is given by

$$\delta v = -v \frac{\delta d}{d}$$

and thus for a change of strain in 1 part in 10^9 the change of frequency would be about 500 kc/s. The spacing of the mirrors must be kept small enough that the changes of strain that may occur do not cause the oscillations to change from one longitudinal mode to another, and this seems to limit the separation of the mirrors to about 1 m. It will be seen that in a sense the method of measuring frequency change of a laser is more flexible than that of the optical interferometric method, since the change of frequency for a given strain is independent of the separation of the mirrors. Since the lasers which detect the changes of strain are not stabilized, the effects of gas pressure variation and other sources of change in the gain curve will be less important than in a system using a stabilized laser, but there may well be trouble due to variations of the refractive index of the plasma.

The problem of mounting the laser mirrors may be more serious than with the long-distance optical interferometer. In the latter device very small changes of position of the mirrors due to changes in the mounting system may not be too important provided they correspond to less than 1 part in 10^9 of the overall distance, that is to say, to less than about one-tenth of a wavelength of light. In a system with the laser mirrors separated by 1 m, however, the stability of the mounting must be some hundred times better and it is by no means sure that such high stabilities can be attained. It is indeed not merely a question of stability but of knowing that the movements of the mirrors do correspond to the changes of strain that are representative for a large volume of rock. This is the crucial problem in the application of laser strain gauges to seismometry and will be discussed more thoroughly by Davies and King (see section D II, chapter 3 of this book).

Acknowledgements

The development and application of stabilized lasers form part of the research programme of the National Physical Laboratory.

References

1. G. Birnbaum, *URSI Rept.*, No. 15 (1966).
2. A. H. Cook, A. Marussi, and W. R. C. Rowley, *Geophys. J.*, **9**, 281 (1965).

3

G. C. P. KING

and

D. DAVIES

Department of Geodesy and Geophysics
University of Cambridge
England

A laser seismometer

'Deep in unfathomable mines . . .
He treasures up his bright designs . . . '—William Cowper (1731–1800)

Introduction

This chapter describes some of the problems and techniques involved in the development and installation of a seismometer using a frequency-stabilized laser of the type described by Cook in the preceding chapter (see section D II, chapter 2 of this book). For this reason no discussion of the workings of the laser will be given—instead, the emphasis is on the emplacement, operation, and capabilities of such a device.

It is of interest to note that similar developments are proceeding in other laboratories: Boeing at Seattle (see Vali, Krogstad, and Moss[1]) have had a Fabry–Pérot type interferometer in operation using a non-frequency-stabilized laser for about two years —Lamont Geological Observatory (see van Veen, Savino, and Alsop[2]), Scripps Institution of Oceanography at La Jolla, and the University of Trieste are among other institutions currently involved in the development of laser strain gauges. Within a year or two it should be possible to make a fairly comprehensive evaluation of the merits and demerits of various systems and sites.

Purpose of the Instrument

The instrument was originally devised as being a strain seismometer using the Michelson interferometer principle and having a 'long arm' of about 100 m. Although not initially envisaged as necessary, it has since become apparent that 'semiportability' is a desirable feature—that is, that the instrument should not be uniquely associated with any particular site but should be capable of movement (in perhaps a week) to another site. This addition to the specification (which has not proved too severe in terms of instrumental development) has arisen through the many questions that have emerged about the 'representativeness' of any particular site and the spatial coherence of geophysical data.

It may reasonably be asked why we are concerned to build a laser strain gauge when an ultra-long-period device (the Benioff quartz rod) is already well established in many sites throughout the world. The answers to this question are as follows.

(i) Although at present the Benioff quartz rod is more sensitive by an order of magnitude than our laser seismometer will be, it is a device that requires, for its maximum sensitivity, an environment so restrictive that there are few places in the World where it may be installed satisfactorily and it is doubtful whether such a site exists in the United Kingdom.

(ii) Quite apart from the relative cheapness of even a frequency-stabilized laser plus its associated electronics (the initial budget for the whole instrument and site preparations is less than £8000), the light from one laser may be used for more than one experiment simultaneously; for instance, two orthogonal arms can be installed, or paths along the same direction, but of different length, may be considered. Thus the investment in one laser opens up the possibility of several experiments.

(iii) There seems to be no major difficulty in removing the laser to a new environment fairly rapidly (our requirement of semiportability).

Theory of a Near-surface Strain Instrument

Benioff[3] and Major and others[4] very thoroughly document the instrumental capabilities of a strain seismometer, and, for response curves for incident P and S waves, reference may be made to their papers.

At a free surface three of the six independent components of the stress tensor p_{ij} are necessarily zero and thus, under assumptions of isotropy and known Lamé constants λ and μ, we may write (where z is the vertical axis)

$$p_{xx} = \lambda\Delta + 2\mu e_{xx} \tag{1}$$

$$p_{yy} = \lambda\Delta + 2\mu e_{yy} \tag{2}$$

$$p_{xy} = p_{yx} = 2\mu e_{xy} \tag{3}$$

$$p_{zz} = \lambda\Delta + 2\mu e_{zz} = 0 \tag{4}$$

where

$$\Delta = e_{xx} + e_{yy} + e_{zz}$$

There are thus only three independent components of the strain tensor, say e_{xx}, e_{yy}, and e_{xy}. It is possible to devise a system of three strain gauges to extract these components. In practice, it would also be of interest to have e_{zz} but this requires a borehole of adequate depth in the vicinity of two orthogonal strain gauges. For the moment we are concerned to get *one* device into operation.

The rotational terms (ω_x, ω_y, ω_z) are likewise not within our capabilities, but of course two of the three components may be dealt with by tiltmeters.

The Instrument

The laser injects light into a Michelson interferometer arrangement (figure 1). Apart from the semireflecting surfaces, the near end of the instrument also comprises the

A laser seismometer

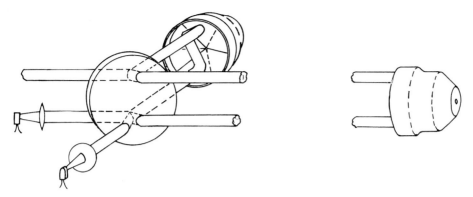

FIGURE 1 The optical components and light paths for the interferometer

'short arm' which is defined by an Invar base. The interferometer sits on a table attached to the geology. At a distance (initially) of 100 m a corner-cube reflector is attached to the geology. The whole optical system is under a vacuum of 1/100 torr and by means of push–pull bellows no stress can be transmitted between the optical components via the enclosing pipe. The bellows are shown in figure 2.

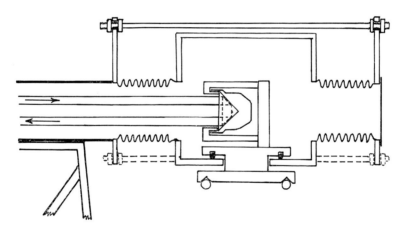

FIGURE 2 The 'far end' corner cube seated on a mounting decoupled from the pipe by a push–pull bellows

Fringes formed by the combination of light (of wavelength 6328 Å) from the two optical paths will be detected by photocells, and by an appropriate treatment of the optical surfaces it is possible to observe two sets of fringes, one about 90° out of phase with the other. In this way it will be possible to detect not only *movement* of fringes but also the *direction* in which they have travelled. A reversible counter and digital-to-analogue conversion equipment complete the system initially, and results will be recorded on an analogue paper recorder but it is anticipated that, at an early stage, advantage will be taken of the essentially digital nature of the system to record in digital form, either on paper tape or on an incremental digital tape recorder. The capabilities of the counting system are the tracking of frequencies up to 500 kc/s—a figure far in excess of the number of fringes we normally anticipate to pass in one

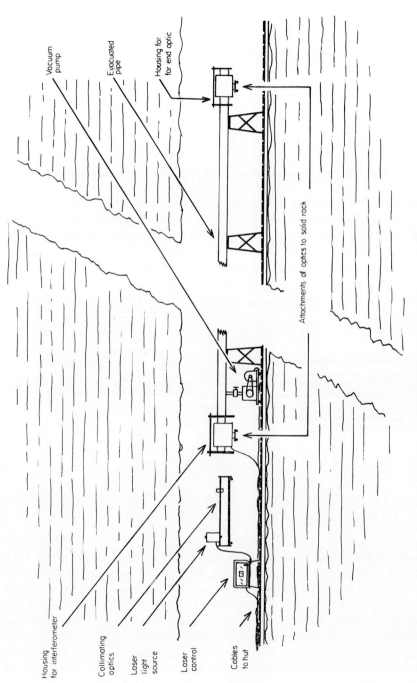

FIGURE 3 The complete layout of the system in a mining gallery

second, but giving a wide safety margin against losing a fringe during a large local disturbance of high seismic frequency.

The laser and electronics are powered by accumulators (24 v, 3 A). Trickle charging of the accumulators, and their emplacement (together with as much as possible of the electronics) at a distance of at least 100 m from the near end of the strain gauge, will minimize the need for visits to the proximity of the gauge itself and will greatly assist efforts to maintain long-term stability.

The present counting capability is the passage of one fringe, corresponding to the detection of strain changes of about 3 parts in 10^9. Currently the long-term laser stability is quoted at about 1 part in 10^9, so we are approaching what is at the moment the limit in useful sensitivity. However, with an eye to the future, it is clear that fringe subdivision (even to 1/100 of a fringe) or path multiplexing (causing the light to travel the length of the pipe several times before entering the interferometer) can significantly increase the sensitivity without increasing appreciably the physical dimensions of the system.

In figure 3 is shown a representation of the laser seismometer. It anticipates the contents of the next section in showing the instrument in a mining gallery.

Site

The establishment of a strain seismometer has been as much as anything a problem in the selection of an adequate site. It was early established that a mine or tunnel was most desirable. A buried pipe would be too near the surface in such a varied climate as ours to guarantee that the instrument was not merely recording the effect of temperature and pressure fluctuations on the surface. With the requirement that the instrument should be decoupled as far as possible from surface effects went the requirement that there should be good coupling to the solid interior of the Earth. This we have interpreted as meaning that there should be an underlay of 'solid' rocks beneath the instrument, using solid in the sense of excluding clays and loosely compacted sedimentary rocks. This has been an exacting criterion to meet since it has required us to place the instrument about 200 miles away from Cambridge, but, in the absence of any precise knowledge of the behaviour of soft sediments and particularly boundaries between sediments and hard rocks at ultra long periods, it has satisfied us that the results that we hope to obtain will be more representative than would be the case if the instrument sat on a sedimentary basin.

The site selected for the first experiment (brought to our attention by Professor K. C. Dunham) is a disused mining level at Stanhope, Co. Durham. The location of the site, together with the solid geology of the area and the position of one or two other places of importance, is given in figure 4. The subsurface geology of this area is very well documented owing to the drilling of a borehole nearby at Rookhope for geological investigations[5]. The whole region is known as the Alston Block and is dominantly Carboniferous. The complexities of figure 4 merely reflect the effect of topography on the boundaries between geological series[6]. At Rookhope, granite was encountered at −222 ft ordnance datum and was still present when drilling ceased at −1600 ft ordnance datum. The sequence down to the granite is Lower Carboniferous and mainly limestones and sandstones.

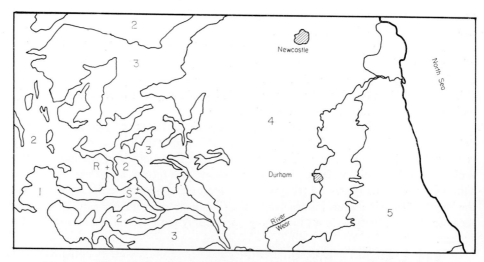

FIGURE 4 The solid geology[6] of the region. S, Stanhope; R, Rookhope. The geological series are as follows: 1, Lower Carboniferous limestone; 2, Upper Carboniferous limestone; 3, Millstone Grit; 4, coal measures; 5, Magnesian limestone. The boundaries of the map are 54° 40′ N, 55° N and 1° 10′ W, 2° 20′ W

The mine at Stanhope (which was worked for galena and fluorspar) has practically the same negative Bouguer gravity value as that at Rookhope and it seems an entirely reasonable assumption that the so-called 'Weardale granite' is about 1000 ft below the surface and that the geological succession above the granite will not be fundamentally different. Unless the granite rests on something quite extraordinary, we have considerable confidence that the strain gauge rests on a very firm substructure.

Figure 5 is a cross section of the mine and it is characteristic of many mines in the area in being long, straight, and passing through Lower Carboniferous limestone. Indeed there is, in the tunnel, the possibility of extending the 'long arm' to at least 800 m. The presence of other similar mines in the vicinity is, as we have indicated

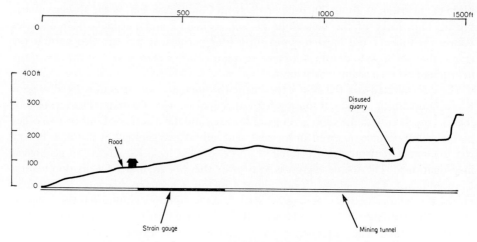

FIGURE 5 A cross section of the Stanhope level (no vertical exaggeration)

earlier, of value in that future work on spatial coherence is possible. The mine is a fortunate blend of being within reach of power lines and yet about half a mile from the nearest major source of noise—a main road. Figure 6 shows the main gallery along

FIGURE 6 The gallery in which the instrument is being sited. Attachments of the optical components will be made to the walls

which the instrument is being sited, and also shows the water flow along the floor of the tunnel. A steady drainage flow runs along the floor and is about three inches deep. Original intentions were to dam this flow beyond the far end of the pipe and channel it all, by means of a conduit, past the instrument, thereby creating a relatively dry underground laboratory (there are a few places where a little water falls from the ceiling). However, this would have proved relatively expensive and, as the following discussion will show, would have removed a potentially useful feature.

Signals from undesirable sources may be considered under several different categories, and we have attempted in figure 7 to illustrate diagrammatically these sources. It is of importance at least to get an order-of-magnitude feel for the contributions to spurious signal, and so each potential source will be considered briefly if only to show its relative unimportance.

Direct loading on the ground overhead

This (denoted by A in figure 7) produces two effects, one of which is recognizable, small, and removable, the other of which is not necessarily recognizable but is small in most cases. The former is the signal generated by a moving load, for instance on the minor road which runs over the mine. Vehicles on this road, or wind, generate microseismic disturbances; these, however, will both be confined largely to surface waves and will have sufficiently high frequencies (2–10 c/s) that an appropriate low-pass filter incorporated in the recording system will remove them.

The other noise source will arise from the d.c. or nearly d.c. effects of slowly moving

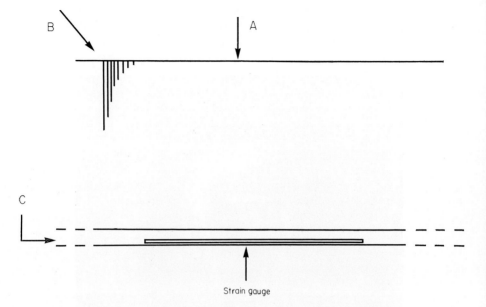

FIGURE 7 Sources of spurious signal. The letters A, B, and C are explained in the text

loads or large movements of surface water. It is fairly easy to calculate what effect such loads will have on strain distributions in the system—the appropriate formulae may, for instance, be found in Love[7]. To give an idea of the values involved, if a ten-ton lorry were parked immediately over the centre of the gauge, with 50 m of rock intravening, a 100 m long instrument would record a strain change of about 1 in 10^{10}.

Diurnal and annual temperature fluctuations at the surface

It is well known that the amplitude of such fluctuations (denoted by B in figure 7) falls off exponentially with depth, and is down to less than 1% of the surface value at 1 m for the diurnal variation and at 20 m for the annual variation. Further, if the medium were homogeneous and isotropic, vertical strains only would be generated by this mechanism. If the assumption of homogeneity is dropped or a variable surface temperature distribution is permitted, horizontal strains are introduced—the magnitude of these is not obviously trivial, and theoretical studies are being pursued at the moment to put a figure to conceivable strains. The actual problem of this area is much more complex in practice as there is a steady flow of ground water which may well completely alter the temperature pattern. For the purpose of gaining an impression of thermal effects (also those discussed in the next section), temperature recorders are being installed at strategic points along the mine.

Temperature and pressure fluctuations entering the mine

Although, as we have already noted, there is a steady water flow along the tunnel floor which, in addition to the minuteness of the effects expected in the last subsection, should act as a thermal 'shunt' to ensure that the rocks are at an almost constant temperature, the air in the mine will completely reflect the external pressure and will slightly reflect the external temperature. The fluctuations in air temperature and pressure (denoted by C in figure 7) may be expected to have effects on both the laser and the optical system. The frequency-stabilized laser will resist attempts to change its frequency (by cavity distortions) by self-alterations in cavity length to stay on the emission peak. Effects on the optical system (which may be regarded as changes in the length standard via variations in refractive index) may be estimated in the following way. Let us consider a system at atmospheric pressure. The coefficients for the change in refractive index, or wavelength, with temperature and pressure are 1 in 10^6 $degc^{-1}$ and 3 in 10^7 mb^{-1} approximately. Extraction of all but 1 in 10^5 of the molecules in the pipe will reduce these coefficients by a factor of 10^5 and these figures will be for variations in temperature and pressure within the pipe. Atmospheric fluctuations in temperature will be transmitted through the pipe but are unlikely to cause more than annual variations in wavelength of 1 in 10^9. Pressure fluctuations in the tunnel will not be transmitted to the pipe and the only possible factor entering via the pressure coefficient is the efficiency of the vacuum pump, which is unlikely to disturb the wavelength by more than 1 in 10^{10}.

By making the 'short arm' an Invar mount no trouble is anticipated from thermal expansion and corresponding fringe movements in the system.

Once the interferometer is established, visits to the vicinity of the laser will be kept to an absolute minimum, not because there is any permanent strain change to be expected from such visits, but because the optical alignment at the near end is crucial and could be upset with a complete loss of fringe count by a large differential tilt between the laser and the optical components.

One further source of spurious signal which cannot be predicted, and which may be unavoidable in this region, is that caused by slips in mines and quarries in the vicinity of the tunnel. It is the nature of things that mines should only be excavated in solid rock when there is extensive veining and mineralization. This implies weaknesses in the rock system and movements along faults and slips in quarries are a not uncommon feature of the area. However, if such slippages do occur, they will manifest themselves at the strain gauge almost simultaneously both as microearthquakes with a clear signature (which can be monitored on a high-pass filtered record run in parallel with the strain gauge's main output) and also as permanent changes in strain. Thus it seems possible that we may be able to recover slow strain changes in the presence of a large permanent change provided we have a record of simultaneous high-frequency behaviour.

Results from the Strain Gauge

The account so far has been concerned purely with the nature of the instrument and the problems of satisfactory installation. We conclude with a very brief résumé of what we may expect to be able to obtain from this instrument. The instrument, because of its d.c. character, should provide data of use to a wide variety of elasticity problems. Its capabilities as a short-period seismometer will not be used extensively as there is a comprehensive short-period seismic station already in operation at Rookhope. For periods of a minute up to an hour it will act as a long-period seismometer with the capabilities of observing surface waves and (when the sensitivity can be raised by a factor of about 10) as an instrument for the recording of the free oscillations of the Earth. Beyond the hour period it will record Earth tides and thus will be a useful addition to the Earth Tide Observatory currently being designed by the University of Liverpool. At periods greater than several days much depends on the ability of the laser to maintain its frequency constant (the composition of the He–Ne mixture in the tube varies slightly as the tube ages). If accuracies of a part or two in 10^9 can be maintained over periods of months, secular rates of change of strain in an aseismic region should be determinable. Certainly permanent strain changes associated with distant large earthquakes (such as Press[8] has shown) will be accessible to the instrument.

Acknowledgements

This work is financed by the Natural Environment Research Council and has been the result of fruitful collaboration with the Department of Quantum Metrology, National Physical Laboratory. We are particularly grateful to Dr. A. H. Cook, to Dr. W. R. C. Rowley, and to Mr. D. Wilson of that Establishment for their co-operation and assistance. Dr. D. L. McKeown, currently at the Department of Geodesy and Geophysics, Cambridge, has played a large part in the conversion of the instrument from concept to hardware and we are greatly indebted to him.

References

1. V. Vali, R. S. Krogstad, and R. W. Moss, *Rev. Sci. Instr.*, **36,** 1352 (1965).
2. H. J. van Veen, J. Savino, and L. E. Alsop, *J. Geophys. Res.*, **71,** 5478 (1966).
3. H. Benioff, *Bull. Seismol. Soc. Am.*, **25,** 283 (1935).
4. M. W. Major, G. H. Sutton, J. Oliver, and R. Metsger, *Bull. Seismol. Soc. Am.*, **54,** 295 (1964).
5. K. C. Dunham, A. C. Dunham, B. L. Hodge, and G. A. L. Johnson, *Quart. J. Geol. Soc. London*, **121,** 383 (1965).
6. ¼ in. to mile Geological Map, Geol. Surv. (1952).
7. A. E. H. Love, *A Treatise on the Mathematical Theory of Elasticity*, 4th ed., Cambridge University Press, London, 1927, Section 135.
8. F. Press, *J. Geophys. Res.*, **70,** 2395 (1965).

4

C. O. ALLEY
Department of Physics and Astronomy
University of Maryland
College Park, Maryland, U.S.A.

P. L. BENDER
Joint Institute for Laboratory Astrophysics
National Bureau of Standards
and
University of Colorado
Boulder, Colorado, U.S.A.

D. G. CURRIE
Department of Physics and Astronomy
University of Maryland
College Park, Maryland, U.S.A.

R. H. DICKE
Department of Physics
Princeton University
Princeton, New Jersey, U.S.A.

and

J. E. FALLER
Department of Physics
Wesleyan University
Middletown, Connecticut, U.S.A.

Some implications for physics and geophysics of laser range measurements from Earth to a lunar retro-reflector

Introduction

Recent developments in modern physics and in modern engineering—quantum electronics as embodied in the atomic clock and the laser, and the ability to land scientific instruments on the Moon—make possible for the first time the precise measurement of distances from points on the Earth to points on the Moon. The general technique and implications for several areas of science were discussed briefly in an earlier publication[1]. A more detailed discussion of the method and how it could improve our knowledge of geocentric longitude is to be published in the proceedings of the *IAU/IUGG Symposium on Continental Drift, Secular Motion of the Pole, and Rotation of the Earth*, held at Stresa, Italy, in March 1965[2].

Although the talk actually given at the NATO Advanced Study Institute in March 1967 was devoted mainly to the topics presented at Stresa (at the invitation of Professor Runcorn who was present at that earlier symposium), we would refer the interested reader to that publication and wish to amplify here some other geophysical and physical implications mentioned only briefly during the talk. These are (i) the possibility of increased understanding of the Chandler wobble through better measurement of the tipping of the Earth with respect to its axis of rotation, and (ii) a definitive test of the conjectured secular decrease of the gravitational constant.

Technique

Retro-reflector

The gain in the retro-direction of diffraction-limited corner reflectors is sufficiently large to permit a package (described below) of about 8 kg mass (including support and pointing mechanisms) to serve as an effective point bench mark on the Moon's surface. For a laser-illuminated spot of about 10 km diameter, the return from the corner reflector package will be much larger than that from the lunar surface and, in addition, will not suffer the time smearing of several microseconds produced by the curvature of the lunar surface.

The problems which arise in placing corners on the Moon stem from the fact that one must meet and simultaneously satisfy many different and sometimes conflicting requirements. In an ideal environment, the choice is relatively simple since, for a given geometry and allowed weight (payload), one maximizes the return signal by making a single diffraction-limited retro-reflector as large as the weight restrictions and fabrication techniques will permit.

Two aspects of the practical problem which vitiate the above conclusion are (i) a sideways displacement of the returned laser beam because of the relative velocity between the Moon and the laser transmitter (velocity aberration), and (ii) the wide lunar temperature variation from full Moon to new Moon as well as the exposure of the reflector to an energy input from direct sunlight of 2 cal/min cm^2 during essentially half the time.

The velocity aberration (which displaces the centre of the returned diffraction pattern between 1·5 and 2 km) limits the diameter of a diffraction-limited retro-reflector to about 12 cm (4·5 in.) unless one employs two telescopes spatially separated from one another, one for transmitting and another for receiving. One of these must be movable along an arc of about 1·6 km to allow for the varying direction of the relative velocity vector, and both must have high precision of pointing and tracking. For a wide range of corner sizes, however, the loss in efficiency for a given total payload weight, which results from using a larger number of small diameter corners, is almost exactly compensated by the increased diffraction spreading which locates a combination transmitting–receiving instrument site further up on the side of the returned diffraction pattern. This results in essentially the same optical efficiency for a given total payload weight for corner sizes ranging from 1·5 in. to 4 or 5 in. Below the 1·5 in. size one experiences an overall loss in efficiency because the diffraction spreading is so large that the combined site is essentially at the centre of the returned diffraction pattern.

It is therefore possible to minimize the thermal gradients which would distort the diffraction pattern by choosing a size of 1·5 in., and using a sufficient number in an array to permit observation of a reflected signal. Actual tests in a simulated lunar environment on a single solid fused silica corner reflector have confirmed the conclusions of theoretical calculations which indicate that the proposed array consisting of some 90 corner cubes 1·5 in. in size will remain essentially diffraction-limited throughout the lunar day and night. Some sacrifice in returned signal is accepted by using total internal reflection, rather than aluminizing the back surfaces, in order to increase the lifetime and to prevent additional thermal distortion. An array of this type is being constructed for earliest possible emplacement under the NASA *Apollo* lunar surface experiments programme[3].

Laser ranging system

The return to a laser radar transceiver from the corner reflector described above will be adequate for a round trip transit time measurement using range gating and single photoelectron detection techniques with currently available Q-switched ruby lasers. By processing the data from about 100 returns it will be possible to measure the time to about 10^{-9} sec which gives a relative time uncertainty of 15 cm/c. (The absolute uncertainty involves the value of the speed of light, c, which is known with a probable error of about three parts in 10^7. Assuming the constancy of the velocity of light, the relative range is all that is needed.) Typical laser system parameters are

pulse duration	10^{-8} sec
energy	10 J
beam divergence (angular radius from 2-cm diameter aperture)	10^{-3} rad
transceiver aperture	150 cm
bandpass filter width at 6943 Å	3 Å
detector quantum efficiency (multi-pass)	10%
pulse repetition rate	1 pulse/3 sec

Continued progress in laser technology is expected to lead soon to field operational systems having near diffraction-limited performance and pulse lengths less than 10^{-9} sec. This would allow much smaller transmitting apertures and less elaborate signal processing.

Method of measurement and accuracy

The basic uncertainty in the range will be introduced by uncertainty in the total atmospheric delay. However, this can be predicted to about 6 cm for a zenith distance of 70° from a knowledge of local temperature and pressure. The technique is independent of angular errors introduced by fluctuations in the atmospheric index of refraction.

When accurate lunar range measurements have been made, the residuals from the best available ephemeris will be used in a least-squares analysis to determine improved values for the various lunar and geophysical parameters which affect the range. However, a simplified model can be used for estimating the expected accuracy with which the parameters can be determined. The model assumes that one range measurement D_1 from a given ground station is made at time t_1 about 4 hours before the time t_m of minimum distance. A second measurement D_m is made at about t_m, and a third D_2 at a time t_2 about 4 hours after t_m.

Uncertainties in the distance R of the observing station from the Earth's axis of rotation and in the time t_m at which the distance to the target is minimum will give mainly range residuals with a period of one day[4]. Since the only important residuals due to errors in the lunar parameters will have periods of 14 days or longer, it is easy to divide the problem. Measurements of the minimum distance D_m made over an extended period can give improved values for a number of the lunar parameters and for the distance of the station from the equatorial plane of the Earth. The difference $D_1 - D_2$ plus the times t_1 and t_2 will give t_m, while the quantity $\frac{1}{2}(D_1 + D_2) - D_m$ can be used to find R.

The accuracy with which t_m can be determined has been discussed elsewhere[2]. It is easily shown that

$$\delta t_m = \frac{\delta(D_1 - D_2)}{2\omega R \sin \phi}$$

where ω is the Earth's rotation rate and $\phi = (\omega/2)(t_2 - t_1)$. For $\delta(D_1 - D_2) = 15$ cm, $R = 5 \times 10^6$ m, and $\phi = 60°$, we find that δt_m is about 0·25 msec. Similar measurements at two stations would give the difference in geocentric longitude to the equivalent of 15 cm in distance. Since atmospheric, timing, and Earth tide effects would be expected to give little systematic error when averaged over periods of a month or more[2], somewhat higher accuracy may be achievable for the average difference in geocentric longitudes. This is, of course, important in connection with the hypotheses of ocean floor spreading and continental drift. The accuracy with which the lunar motion can be determined from the measurements made near the time of minimum range is high enough so that the contribution to the uncertainty in the difference in geocentric longitude between two stations will be negligible.

Motion of the Pole and the Chandler Wobble

In this chapter we wish to concentrate mainly on the problem of the tipping of the Earth with respect to its axis of rotation (motion of the pole)[5-7]. For the past 60 years the International Latitude Service has coordinated and analysed observations of astronomical latitude carried out at approximately five stations located around the Earth at 39° 8′ N latitude. From these observations and from latitude and longitude determinations at other observatories, it is known that the position where the Earth's axis of rotation intersects the surface traces out a roughly elliptical path with a mean radius of about 5 m. However, the motion contains both an annual term and one with a period of about 14 months corresponding to a free nutation of the elastic Earth. As these terms get in and out of phase, the amplitude of the motion changes considerably. The annual term can be driven by various perturbations which have the proper frequency, but the means for exciting the 14-month 'Chandler wobble' term is not known. In addition there are both shorter period motions of the pole, including fortnightly, monthly, and semiannual tides, and slower motions consisting of an apparent secular motion of the pole and possibly a 24-year libration.

The analyses of the Chandler wobble which have been carried out would be consistent with a Q for the Earth in the range of 10 to 50[3]. However, the problem of noise in the observations is a severe one. It is normally assumed that the uncertainty in the monthly mean position of the pole is about a metre. The results can also be affected by systematic errors due to slow changes in the horizontal gradient of the atmospheric index of refraction or due to variations in the local vertical. Because of the noise it is not yet possible to say whether the Chandler wobble can be characterized by its power spectrum alone, as would be the case if it resulted from excitation of a damped resonance by many small disturbances with a broad range of frequencies. The other alternative would be occasional strong excitation followed by exponential decay. In addition, we do not yet know whether the damping takes place in (or due to interactions with) the atmosphere, the oceans, the mantle, or the core. The conclusion of Munk and MacDonald's discussion of damping[5] still seems valid: 'The situation is appallingly uncertain.'

We shall assume that observations of the lunar range are made from three ground stations located at latitude $B = 35°$ N and at longitudes L_1, L_2, and L_3 which differ by $120°$. If the position of the pole moves a distance μ towards longitude λ through an angle μ, then

$$\Delta(L_2 - L_1) = \sqrt{3} \tan B \ \{\mu \cos(\lambda - L_{12})\}$$

$$\Delta(2L_3 - L_1 - L_2) = 3 \tan B \ \{\mu \sin(\lambda - L_{12})\}$$

where $L_{12} = (L_1 + L_2)/2$. Assuming that the uncertainties in determining the times t_m for the three stations are independent and each is equal to 0·25 msec, we find that the components of the polar motion can be found with an uncertainty of

$$\delta\{\mu \cos(\lambda - L_{12})\} = \delta\{\mu \sin(\lambda - L_{12})\} = \sqrt{\frac{2}{3 \tan B}} \, \omega \, \delta t_m \sim 15 \text{ cm}$$

This value is the proper one to use in discussing fortnightly or monthly tides. However, somewhat higher accuracy may be achievable for slower terms in the polar motion due to the averaging out of systematic atmospheric or other errors which have a monthly period. It thus appears that a regular, long-term programme of lunar range measurements may be capable of improving our knowledge of the Chandler wobble by a substantial amount.

In the above discussion it was assumed that the rotation rate of the Earth is known. Actually, observations of t_m at the same station on successive days will give the single-day rotation rate to 5 parts in 10^9. This will permit a check with improved accuracy on short-term variations in the rotation rate, but it seems unlikely that variations will be found with periods short enough to increase significantly the uncertainty in the polar motion. Although the rotation rate is measured with respect to the retro-reflector package on the Moon, the laser range measurements will give the lunar motion and librations with sufficient accuracy so that the rotation rate with respect to the Sun can also be obtained.

Test for a Secular Decrease of the Gravitational Constant

The existence of one or more corner reflectors on the Moon would permit a new check of general relativity. It has frequently been suggested that the gravitational 'constant' may not be constant, that gravitation may be steadily weakening in comparison with the strength of electromagnetism and other interactions[8-10]. These theories are frequently misunderstood, it being thought that they are non-relativistic. Within the framework of standard relativistic field theory, all that is required for the validity of the Brans–Dicke cosmology is the existence of still another elementary particle, a particle with neither spin, charge, nor mass. The massless and chargeless particles known or presumed to exist consist of the photon, electron, and muon neutrinos, and the graviton, particles with spin 1, 1/2, 1/2 and 2, respectively. The Brans–Dicke cosmology requires in addition the scalaron, a particle with zero spin. Actually, it is not the existence of the scalar particle which is critical, the quantum effects not yet being important, but rather the associated zero mass scalar field, treated classically.

It has been shown[11] that within the framework of the formalism of Einstein's general relativity the inclusion of the zero mass scalar field, plus assumptions of initial conditions and coupling, leads directly to the Brans–Dicke formalism. While this theory is general relativistic it is not general relativity, for part of the gravitational force is due to the scalar field and the whole of the gravitational effect is not geometrical as Einstein assumed.

A peculiar feature of the coupling of a particle (such as a proton or electron) to a scalar field is the resulting variability of the particle's mass, being a function of the scalar field variable. This results in the dimensionless gravitational coupling constant $Gm^2/\hbar c$ being a function of the scalar. When this result is expressed in ordinary centimetre, second, gram units of length, time, and mass, G becomes a function of the scalar. Hence it changes with the change in the scalar that results from the expansion of the Universe.

It has been shown that the existence of a zero mass scalar field would not affect any non-gravitational physics. It has also been shown[12] that the mass regulation effect mentioned above serves to adjust the strengths of the scalar and gravitational interactions automatically to make them the same order of magnitude.

An ideal way to test for the presence or absence of a secular decrease in 'G' would be to follow the Moon's motion with great precision for a period of years[13–15]. The secular increase in the Moon's period to be expected within the framework of the Brans–Dicke cosmology is in the range of $(2 \text{ to } 6) \times 10^{-11}$ per year, if one takes the value of the parameter ω contained in the theory to have the value of about 6, as suggested by measurements of the solar oblateness and by other information[16]. Based upon the telescope observations of the past 250 years, it is known that the tidal-induced increase in the Moon's period is 12.9×10^{-11} per year[17], which is known to 0.6×10^{-11} per year. This results from comparisons of the motion of the Moon with that of the Sun and of Mercury. The time scale is thus provided by planetary time. Inasmuch as it is believed that the tidal couple on the Moon is reasonably constant over a period of a few hundred years (this belief is strengthened by telescopic observations) the next decade should witness the same tidal-induced increase in the lunar period. However, a further increase of 2 to 6 parts in 10^{11} per year should occur *not* on a planetary but on an atomic time scale, if the secular decrease in the gravitational 'constant' should exist.

An accurate method for determining the rate of change of the lunar motion is to observe the phase of the $\cos 2D$ term in the lunar range. Here D is the mean longitude of the Moon with respect to the Sun. The amplitude of this term[18] is about 3×10^6 m. An error of 2.5×10^{-8} rad in D would give a residual in the observed range of the form $\sin 2D$ with amplitude 15 cm. Such a term should be easily observable for the following reasons: (i) uncertainties in the other lunar parameters will not give this frequency, (ii) no systematic errors with this frequency are expected, and (iii) this term is well resolved in frequency from other terms expected in the residuals.

If the difference between n, the mean motion of the Moon, and n', the mean motion of the Sun, is given by $n-n' = (n-n')_0(1-\beta t)$ with $\beta = 4 \times 10^{-11}$ year^{-1}, then

$$D = D_0 + (n-n')_0 t - \tfrac{1}{2}\beta(n-n')_0 t^2$$

From observations of the phase of the $\cos 2D$ term at times near $t = 0$, T, and $2T$, we can find β. Its uncertainty is given by

$$(n-n')_0 T^2(\delta\beta) = 2\delta\{D(T)\} - \delta\{D(2T)\} - \delta\{D(0)\}$$

Assuming that the uncertainties in D are independent and that each is $2\cdot 5 \times 10^{-8}$ rad, we find for $T = 5$ years that $\delta\beta \sim 3 \times 10^{-11}$. It thus appears that the lunar deceleration predicted by the Brans–Dicke theory could be observed in about 10 years. The rotational rate of the Earth with respect to the Moon obtained by laser range measurements could also be used if the rotation rate of the Earth with respect to the fixed stars could be determined with sufficient accuracy[19].

Acknowledgements

It is a pleasure to thank our colleagues in the design and scientific analysis of the lunar laser ranging retro-reflector experiment, W. Kaula, G. MacDonald, H. Plotkin, S. Poultney, H. Richard, and D. Wilkinson for many valuable discussions. We would also like to acknowledge the assistance given in the early phases of planning by Professors D. Brouwer and U. Van Wijk before their untimely deaths.

References

1. C. O. Alley, P. L. Bender, R. H. Dicke, J. E. Faller, P. A. Franken, H. H. Plotkin, and D. T. Wilkinson, 'Optical radar using a corner reflector on the Moon', *J. Geophys. Res.*, **70**, 2267–9 (1965).
2. C. O. Alley and P. L. Bender, 'Information obtainable from laser range measurements to a lunar corner reflector', *IAU/IUGG Symposium on Continental Drift, Secular Motion of the Pole, and Rotation of the Earth, Intern. Astron. Union Symp., 32nd, Stresa, Italy, 1967*.
3. H. E. Newell and L. Jaffe, 'Impact of space research on science and technology', *Science*, **157**, 29–39 (1967).
4. D. W. Trask and C. J. Vegos, 'Intercontinental longitude differences of tracking stations as determined from radio tracking data', *IAU/IUGG Symposium on Continental Drift, Secular Motion of the Pole, and Rotation of the Earth, Intern. Astron. Union Symp., 32nd, Stresa, Italy, 1967*.
5. W. H. Munk and G. J. F. MacDonald, *The Rotation of the Earth*, Cambridge University Press, Cambridge, 1960.
6. W. Markowitz, 'Latitude and longitude and the secular motion of the pole', in *Methods and Techniques in Geophysics* (Ed. S. K. Runcorn), Interscience, New York, 1960, pp. 325–61.
7. Numerous articles in *IAU/IUGG Symposium on Continental Drift, Secular Motion of the Pole, and Rotation of the Earth, Intern. Astron. Union Symp., 32nd, Stresa, Italy, 1967*.
8. P. A. M. Dirac, 'A new basis for cosmology', *Proc. Roy. Soc. (London), Ser. A*, **165**, 199–208 (1938).
9. P. Jordan, *Schwerkraft und Weltall*, Vieweg, Braunschweig, 1955.
10. C. Brans and R. H. Dicke, 'Mach's principle and a relativistic theory of gravitation', *Phys. Rev.*, **124**, 925–935 (1961).
11. R. H. Dicke, 'Mach's principle and invariance under transformation of units', *Phys. Rev.*, **125**, 2163–7 (1962).

12. R. H. Dicke, 'Long-range scalar interaction', *Phys. Rev.*, **126,** 1875–7 (1962).
13. R. H. Dicke, 'The significance for the solar system of time-varying gravitation', in *Gravitation and Relativity* (Eds. H. Y. Chiu and W. F. Hoffmann), Benjamin, New York, 1964, pp. 142–74.
14. R. H. Dicke, 'The secular acceleration of the Earth's rotation rate', in *The Earth–Moon System* (Eds. B. G. Marsden and A. G. W. Cameron), Plenum Press, New York, 1966, pp. 98–164.
15. R. H. Dicke and P. J. Peebles, 'Gravitation and space science', *Space Sci. Rev.*, **4,** 419–61 (1965).
16. R. H. Dicke and H. M. Goldenberg, 'Solar oblateness and general relativity', *Phys. Rev. Letters*, **18,** 313–16 (1967).
17. H. S. Jones, *Monthly Notices Roy. Astron. Soc.*, **99,** 541 (1939).
18. E. W. Brown, 'Theory of the motion of the Moon—Part IV', *Mem. Roy. Astron. Soc.*, **57,** 51–145 (1908).
19. G. J. F. MacDonald, 'Implications for geophysics of the precise measurement of the Earth's rotation', *Science*, **157,** 204–5 (1967).

D

DEVELOPMENTS IN TECHNIQUES

III. Space technology

1

V. R. WILMARTH

NASA Headquarters
Washington D.C., U.S.A.

Apollo lunar sample analysis program—a progress report

One of the most exciting scientific programs of the 20th century is the landing of the first U.S. astronauts on the Moon and their subsequent return to Earth with samples of surface materials. This spectacular event will provide the scientific world with its first opportunity to study materials from another celestrial body of the solar system. With the return of lunar materials scheduled by 1970, a comprehensive analytical program that will no doubt continue for many years will be initiated. The results of lunar sample analysis will provide first-hand information on the composition and structure of the lunar surface materials. These data are necessary for understanding the origin of the Moon, its history, the processing that formed its present configuration, and for planning future scientific missions.

For this progress report some of the events leading up to the manned lunar landing, the scientific activities of the astronaut, particularly those related to sample collection, and subsequent return of the astronauts to Earth are briefly described. The investigations planned for the returned lunar material are discussed in considerable detail.

The first astronauts will land in the rectangular *Apollo* zone outlined in black on figure 1. From Earth-based telescopic observations, approximately twenty-five sites

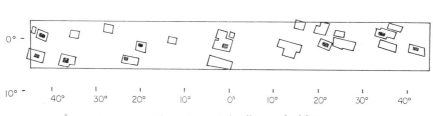

FIGURE 1   First astronauts landing on the Moon

were selected for further study and photographing during *Lunar Orbiter* missions. On the first, second, and third missions, the areas shown as open rectangles were photographed at 1 and 30 m resolution. From these data, the eight tentative sites indicated in black on figure 1 were chosen for the first manned landing. Further geologic and related studies of these sites using high-resolution *Lunar Orbiter* photographs are underway. Figure 2 is a photograph of a potential landing area in the Mare Tranquillitates in the eastern part of the *Apollo* landing zone. The area photographed

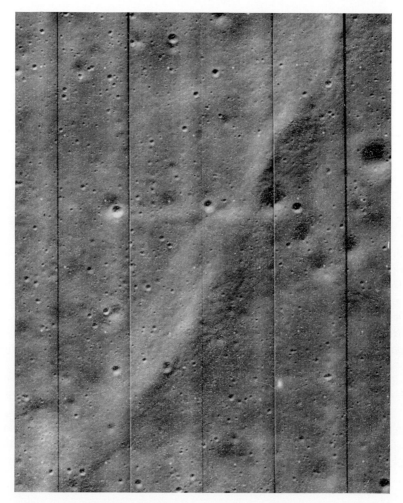

FIGURE 2 Photographs of potential landing site in maria Mare Tranquillitatis, *Apollo* landing zone. Craters 1 m across are visible

is about one mile on the side and craters as small as one meter in diameter are visible. From these photographs and data obtained by landed spacecraft, the final site for the manned landing will be selected. Once this site has been chosen, several maneuvers by the manned spacecraft are planned. These consist of a brief Earth parking orbit, before injections into a translunar trajectory for a transit period of 3 days, and, finally, injection into a lunar orbit before the actual landing. The estimated total time to complete the first mission is 7 to 10 days including no more than two days stay on the lunar surface.

Once the astronauts are on the surface their first scientific activity is to gather, from the immediate area about the lunar module, samples of lunar material referred to as 'grab samples', in the event that their stay is unexpectedly curtailed. Their next activity is to deploy the *Apollo* lunar scientific experiment package (ALSEP). Finally, they will traverse a short distance making geologic observations, photographing lunar features, and carefully collecting and documenting representative samples of the rock types. The scientific data obtained from the geologic investigations and from ALSEP will provide the background information for interpreting the data that results from the analysis of returned samples.

The ALSEP is designed to operate for a year and contains the scientific experiments shown in table 1. The first ALSEP is planned to carry a three-axis surface magnetometer, a passive seismometer, and instruments to determine the nature of the solar

TABLE 1 Objectives and instruments for the *Apollo* lunar surface experiment package (ALSEP)

	Objective	Instrument
interior	state of interior	passive seismometer
		heat flow detectors
	diffusivity	lunar surface magnetometer
	magnetic field	lunar surface magnetometer
	source of internal heat	heat flow detectors
environment	solar wind	solar wind spectrometer
	ambient charged particles	charged particle detector
	magnetic wake of Earth	lunar surface magnetometer
		charged particle detector
	atmospheric pressure	cold cathode gauge
	thermalized positive ions	suprathermal ion detector

wind and to make atmospheric pressure measurements. Active seismic and heat flow experiments will be carried on later flights.

Field geologic investigations are the last scientific activities of the astronauts. The objective of these investigations are similar to terrestrial geologic studies, that is, by a traverse conduct systematic examination of the surface to determine the lithologic features, structure, and thickness of the surficial and underlying materials. An integral part of the field study is to collect representative samples of the rocks or lithologic units for use in later investigations. Because the geologic significance of samples in part depends on the field relations, each sample will be carefully photographed in place and accurate descriptions of the field relations will be made.

Because of limitations of the astronauts' life support systems, and potential difficulties in the conduct of geologic investigations, special tools are being developed to assist the astronaut in these activities. In particular, considerable effort has been expended to provide the astronaut with tools to aid him in collecting and packaging of samples. Figure 3 is an artist's concept of the astronauts traversing the lunar surface with their sample boxes and geologic tools. Figure 4 is a view of the tools arranged on the deployed carrier and a close-up view of the aseptic sampler and coring tubes. Figure 5 is a close-up view of other sampling tools and a spring scale to weigh the samples. Once the samples have been collected they will be stowed in two sample containers for the return trip to Earth. The containers are 8 in. wide, 19 in. long, $11\frac{1}{2}$ in. high and are to be constructed of aluminium with a sealing mechanism which will be activated on the lunar surface (figure 6). This procedure is planned to prevent possible contamination of the Earth's environments, but more important it may

FIGURE 3 Astronauts conducting lunar geologic investigations

preserve for later analysis any gases that might escape from the samples during Earth return. The container consists of removable compartments for easy storage of samples of various sizes.

Development and construction of the sampling tools and sample containers was planned with these objectives. First, we must prevent, however remote, the possibility of the return and escape of microbial life forms or organic substances that would be injurious to Earth life and, secondly, we must preserve, in as near a pristine state as possible, the lunar material so that any chemical compound or organism detected on the sample can positively be identified as of lunar origin. To meet these objectives, care was taken to construct the astronaut's tools and the sample containers from materials that would not compromise any of the experiments planned for the lunar sample analysis program.

Additionally, the astronauts will be trained in the use of special tools such as the aseptic sampler for collecting samples from below the lunar surface and thereby avoid as much as possible the contaminants from the descent engines of the lunar module as well as from the astronaut. As a further precaution against contamination of the lunar samples, all equipment will be sterilized prior to launch.

The total allowance for scientific materials returned to Earth is 36·2 kg of which 23 kg are samples. Three general types of samples will be collected. The first are representative samples from the geologic traverse for use in the overall analytical program.

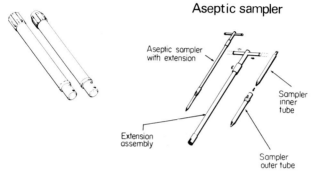

FIGURE 4 View of tools arranged on the deployed carrier and a close-up view of the aseptic sampler and coring tubes

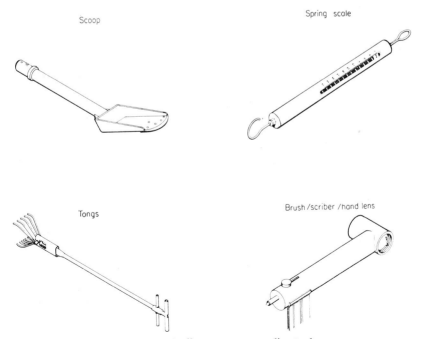

FIGURE 5 *Apollo* astronauts sampling tools

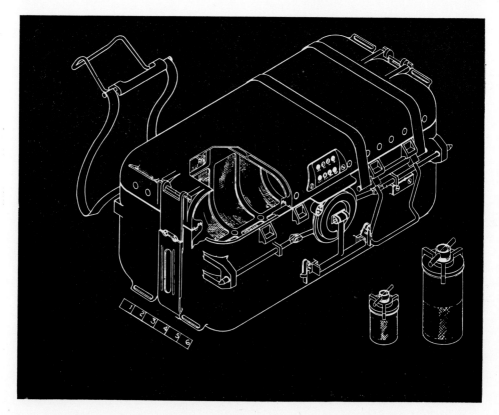

FIGURE 6 *Apollo* lunar sample return container with special containers for gas analysis

These will be placed in small bags that are not hermetically sealed. The second are special purpose samples including a lunar environment sample and a sample for gas analysis. These samples will be returned in special containers designed to maintain as near as possible the actual lunar conditions under which the sample was collected. A third is the aseptic samples for use in biological investigations.

It is worthwhile at this point to consider briefly the type of lunar material expected to be returned to Earth. As previously indicated, the site of the first manned landing will be in the *Apollo* landing zone. Because of navigational and guidance constraints, such sites must be smooth and the approach path be free of large changes in elevation. From study of *Lunar Orbiter* photographs supplemented by geologic and other data obtained from Earth-based telescopic observations, sites having these characteristics are in the maria areas of the *Apollo* zone of interest. Typical of the lunar landscape the first astronauts will observe is indicated in this photograph taken by *Surveyor I* cameras (figure 7). The surface is characterized by depressions and dotted by boulders or protuberances as much as several feet across embedded in what appears to be fine-grained material. Using photographic data obtained by *Lunar Orbiters*, *Surveyors* and *Luna 9* and *13* as examples of what the astronauts will encounter, a prediction as to the type of material to be returned can be made with some confidence. Several types of lunar material may occur in the maria of the *Apollo* site. These include fine-grained particulate somewhat cohesive materials observed in this close-up photograph of the

Apollo lunar sample analysis program

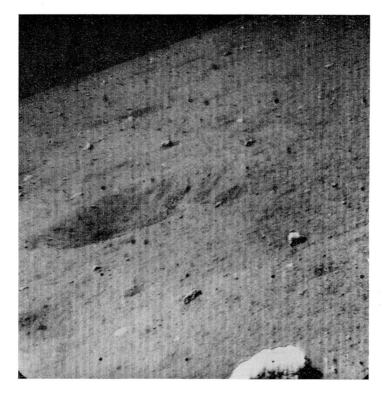

FIGURE 7 Lunar surface taken by *Surveyor I* cameras

lunar surface taken by the *Surveyor I* cameras (figure 8). Similar type material was found at the surface and to a depth of a few inches at the *Surveyor III* site.

Rocks ranging from pebbles to boulders occur on the surface and if possible will be sampled. Similar to terrestrial geologic investigations, the astronauts will obtain samples from as many different types of rocks and particulate materials as possible.

Preparatory to return of lunar samples, construction of a lunar receiving laboratory (LRL) was started in 1966. This facility, now nearing completion, contains facilities to provide for quarantine of the astronauts, spacecraft, and samples, to perform time-dependent analyses of the lunar materials, and to serve as a distribution and storage center for the samples and related data. The laboratory has a two-way quarantine system designed not only to prevent escape of lunar material that might contaminate the terrestrial environment, but also to prevent contamination of the samples by terrestrial matter.

Figure 9 is a cut-away view of the laboratory showing the cabinets for biologic preparation and preliminary mineralogic–petrologic examination in the foreground. A part of the containment system is the vacuum system shown on the first and second floors. Most of the operational handling and examination of the sample will be done under a vacuum of 1×10^{-6} torr; however, as it is desirable to keep some samples as near a lunar environment as possible, part of the system will be operated at 1×10^{-11} torr.

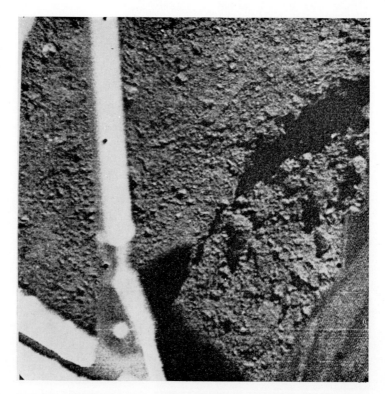

FIGURE 8 Photograph of fine-grained lunar material at *Surveyor I* Site, in Oceanus Procellarium

An important function at the LRL is to determine the type and amounts of short-lived radionuclides and gases in the lunar materials. The low-level counting laboratory as shown in figure 10 is about 15 m underground and consists of a counting room surrounded on all sides by 91 cm of dunite supported by steel plate liners. The counting room is equipped with a special ventilation system, a main counter shield, and an anticoincidence shield enclosing two 23 cm by 13 cm sodium iodide detectors. The estimated radioactive background is 0·1 to 0·2 counts per minute per cubic centimeter of detector in the 0·1 to 2·0 MeV range.

The gas analysis laboratory in the LRL will contain four mass spectrometers for organic and inorganic, including rare-gas, analysis. The gas present in the sample containers, individual sample bags, as well as the gases that evolve during the preliminary atmospheric reaction testing and sample splitting will be analyzed. A special sample is to be collected to obtain a complete gas composition–temperature profile for lunar materials.

A vital scientific function at the LRL is the preliminary examination of lunar samples. Included in this examination will be several types of tests for quarantine release, atmosphere reactions testing, and cursory mineralogic, petrologic, and chemical testing. These results will provide a basis for planning the sequence and types of analyses that will be carried out later in other laboratories throughout the World. Such plans will be prepared during the quarantine period estimated to be about 30 days.

Once the spacecraft, astronauts, and lunar samples have returned to Earth, the

Apollo lunar sample analysis program

FIGURE 9 Cutaway view of part of the lunar receiving laboratory, manned spacecraft center, showing location of gas analysis equipment (third floor), vacuum system (first and second floor), and biological physical–chemical testing facilities in foreground

samples will be transported as quickly as possible to the LRL where the quarantine investigations, time critical analysis, and sample processing can be undertaken (figure 11). From the LRL, samples will be shipped to the principal investigators' laboratories where the physical properties, mineral phases, and textures will be studied, the chemical–isotopic composition determined, and bioscience investigations carried out (figure 11).

Many scientists in the United States and other countries have been selected to participate in the lunar sample analysis program. In response to the May 3, 1966, announcement of this program, 130 proposals involving more than 475 scientists were received from institutions in the United States and other countries. From these proposals, 113 principal investigators were selected. Their distribution by country and area is shown in table 2.

Within each of these areas of investigation many individual analyses were proposed. The types of mineralogic and petrographic studies proposed for lunar samples are shown in table 3. Determinative mineralogic and petrographic studies using many standard laboratory techniques will provide data on the minerals and mineral phases that are present. In addition, study of specific minerals such as the pyroxenes, diamonds, plagioclase, and mineral groups such as sublimates, if present, are included. Of particular significance are the studies proposed to determine what changes have been brought about by the impact of meteorites or other bodies. Terrestrial studies have shown that definitive changes in minerals occur in and around impact structures and

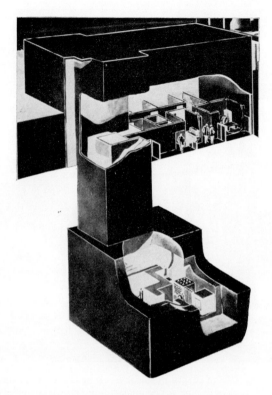

FIGURE 10   Cutaway view of low-level counting facility at lunar receiving laboratory, manned spacecraft center, Houston, Texas

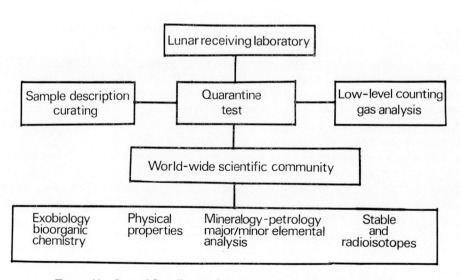

FIGURE 11   General flow diagram for processing of returned lunar samples

Apollo lunar sample analysis program

TABLE 2 Principal investigators for the lunar sample analysis program

Country	Mineralogy/ Petrology	Chemical/ isotope	Physical properties	Biochemical organic analysis
Canada	1	2	2	
Finland		1		
Great Britain	9	2	2	1
Germany	2	3		
Japan	1		2	
Switzerland		1		
United States	25	37	10	12
Total	38	46	16	13

TABLE 3 Mineralogic and petrologic analysis

Investigation	Method
determinative mineralogy, phase studies	many techniques, optical
radioactive minerals and radiation effect	autoradiography, X-ray, optical, microprobe, fossil track studies
opaque minerals	optical, X-ray, microprobe
specific mineral studies, diamonds, plagioclase, pyroxenes, sublimates	optical, X-ray, microprobe
petrographic studies	optical
shock effects	optical, electron microscopy, X-ray diffraction
crystal structure	Mössbauer, nuclear magnetic resonance, electron spin resonance, X-ray
luminescence, polarization, fluorescence	spectrographs, polarimeter

underground nuclear explosions. Similar effects will be sought in lunar material. The valence state of iron and other elements, and the energy state within the crystal lattice will be investigated using electron spin and nuclear magnetic resonance as well as Mössbauer techniques. By means of autoradiographs and microprobe analysis, discrete minerals that are radioactive may be located and their composition determined. In addition, a special study of the fossil fission tracks resulting from cosmic-ray bombardment and spontaneous fission of instrinsic radioactive materials are planned to aid in calculating the age of the rocks and events that have occurred on the Moon. Investigations of the luminescence spectra and polarization properties of the lunar material may provide information for use in the interpretation of Earth-based telescopic observational data.

The chemical and isotopic analysis shown in table 4 will provide information on the elemental and isotopic composition of the lunar material and thereby an insight into the geologic evolution of the Moon and the processes and events that have occurred. Ideally, the abundances of all elements, their isotopes, and their distribution should be

TABLE 4 Chemical and isotopic analysis

Investigation	Method
major elements	wet chemical, flame photometry X-ray fluorescence, neutron activation
minor and trace elements	emission spectroscopy
major, minor, rare-earth elements	neutron activation
alkali, alkaline earths, rare earths and appropriate isotopes	mass spectrometry
heavy elements and isotopes	mass spectrometry
rare gases and isotopes	mass spectrometry
light stable isotopes	mass spectrometry
cosmic-ray induced radionuclides	mass spectrometry

determined. The age of the lunar material can be calculated from information on the relative abundances of elements and isotopes such as uranium, lead, K/Ar, and Sr/Rb, or I/Xe isotope ratios. The abundance and composition of the rare gases in the lunar samples will be determined in part at the LRL and at the principal investigators' laboratories. From these data it may be possible to further our understanding of the Moon's solar wind and nuclear bombardment histories.

Distribution of trace elements, in particular the rare-earth and alkali metals, and the stable isotopes of oxygen, sulphur, strontium, hydrogen, and nitrogen are expected to provide information on the geochemical fractionation that may have occurred during lunar history. The presence of volatile elements such as mercury, cadmium, lead, and bismuth may indicate something about the temperatures during formation of the Moon. From the analysis for uranium, thorium, and potassium, further models of the Moon's thermal regime can be calculated.

As in terrestrial studies, physical property measurements of lunar materials will

TABLE 5 Physical property measurements

Property	Measurement
magnetic	remanent, orientation, susceptibility, Curie point, shock effects on magnetic fields, thermal degradation of magnetic fields, monopoles
thermal	reflectivity, conductivity, diffusivity, inertia, heat capacity
electrical	resistivity, emissivity, dielectric constant, velocity, attentuation
elastic–mechanical	velocity (P and S waves), pressure, temperature effects, compressibility, mechanical strength
electromagnetic	optical, radiations, signal backscatter
miscellaneous	density, porosity, permeability

yield data for use in the interpretation of lunar field data such as obtained from ALSEP or unmanned landers. The properties shown in table 5 are listed in priority, that is, the measurement of magnetic and thermal properties are considered to have a high priority. The standard magnetic properties, remanent and susceptibility, are important in understanding the Moon's magnetic and cooling history. One of the most interesting magnetic experiments proposed is the search for magnetic monopoles. Though this experiment is considered a 'long shot' because monopoles have not been found on Earth, lunar samples may have the advantage of having retained any monopoles formed during primary cosmic-ray interactions.

Earth-based infrared and microwave radiation measurements have shown there is a wide range in lunar surface temperatures. To aid in the interpretation of these data many thermal properties of lunar materials will be measured (table 5). For several properties, their variation with pressure will be determined. The thermal conductivity measurement is especially pertinent because the data will be valuable in the conduct of the heat flow experiment planned for a later ALSEP. Measurement of the P- and S-wave velocity will be made under varying temperatures and pressures. This information will aid in the interpretation of the data obtained from the *Apollo* active seismic experiment.

The electrical property most desired is the dielectric constant for comparison with radar data obtained from Earth-based observations. The effect of radiation in various wavelengths, including gamma and X-ray, and the bombardment by hydrogen and helium ions will be measured. Hopefully, these data may shed some light on the darkening of the lunar surface materials observed in *Surveyor I* and *III* photographs.

The biochemical and organic investigations planned for the lunar samples are shown in table 6. These analyses, using primarily mass spectrometry, gas chromatography,

TABLE 6   Biochemical and organic analysis

Investigation	Method
lipids, amino acids, polmero-type organic matter	mass spectrometry
organic compound identification	gas chromatography, mass spectrometry, nuclear magnetic resonance
isotopes of C, H, O, and S	mass spectrometry
carbonaceous and organogenic matter	gas chromatography, mass spectrometry
form, structure organic matter	electron microscopy
viable organism studies	cultivation techniques

and optical–microprobe techniques, are aimed at determining whether molecules of biological or prebiological origin were ever present or could develop on the Moon. Search for amino acids or structures, that is, the exact arrangement of molecules within an organic compound will be carried out and the data compared with that from Earth. Determination of the presence of viable organisms are part of the biological studies and will be done at the LRL to meet the quarantine requirements and at several investigators' laboratories.

The comprehensive analysis program planned for the returned lunar materials may provide answers to some of the major questions, shown in table 7, that have been

TABLE 7 Major scientific questions about the Moon

Origin of surface features
Chemical evolution of the Moon
Age of the Moon
Nature of gravitational and magnetic fields
Role of internal/external processes in shaping the Moon
Nature of thermal regime

asked about the Moon. Hopefully, partial answers will be obtained on the age of the Moon and its chemical evolution through analysis for major and minor elements. However, the Moon may be unique; in which case, there will be many more questions than answers.

E

MAGNETO-HYDRODYNAMICS

I. Geomagnetic reversals

1

N. D. OPDYKE

Lamont Geological Observatory
Columbia University
Palisades, New York, U.S.A.

The Jaramillo event as detected in oceanic cores*

The changes of polarity history of the Earth's magnetic field during the last 4 million years has been delineated in a series of papers by Cox, Doell, and Dalrymple[1, 2], Doell, Dalrymple, and Cox[3] of the U.S. Geological Survey and by McDougall, Allsopp, and Chamalaun[4], McDougall and Tarling[5], McDougall, and Wensink[6] at the Australian National University. This magnetic polarity history has been divided into magnetic epochs with durations of from 700 000 to 1 800 000 years and magnetic events which are of shorter duration, probably 200 000 years or less. This magnetic history which was derived from lavas has been confirmed by work on oceanic sediments[7, 8].

The Jaramillo normal polarity event which occurs in the upper part of the Matuyama reversed epoch was established and named from a site in New Mexico by Doell and Dalrymple[9] and is the most recent of the magnetic events. Evidence for its existence has also been obtained from oceanic sediments[10].

Reliable data from a good many cores are now available at Lamont as a result of the general study of the paleomagnetism of oceanic sediments. Fifteen cores have been selected for this study which pass through the Jaramillo event. The purpose of this study is to define better the length of time during the event and to try and determine the time of its lower and upper boundaries.

Cox and Dalrymple[11] in a recent paper have given the time of the last reversal of the Earth's field which defines the boundary between the Brunhes normal polarity epoch and the Matuyama reversed polarity epoch as occurring at 700 000 years with an average estimated standard deviation of 41 000 years. This datum is easily seen in all the cores used in this study. An example of the inclination change in one of the cores is shown in figure 1. Figure 2 shows a plot in which the depth of the last reversal in the core is plotted against the time of this last reversal which is taken to be 700 000 years. The slope of the resulting line is then a function of the sedimentation rate. If the sedimentation rate is constant, then the age of the boundaries and the duration of the Jaramillo event can be determined.

It can be seen from an inspection of figure 2 that the Jaramillo event in all these cores falls between 786 000 and 1 125 000 years. The highest variability in the time of occurrence of the event is seen in cores with slow rates of sedimentation. In cores with rates of sedimentation above 4 mm per 1000 years the Jaramillo event is more closely confined in time to the interval between 786 000 and 1 000 000 years. In order to try

* Lamont Geological Observatory Contribution No. 1147.

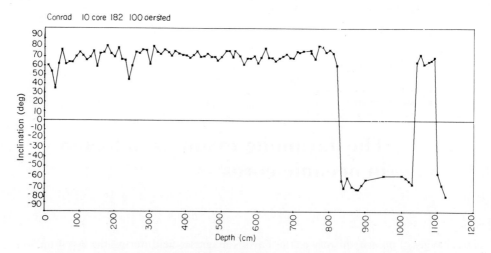

FIGURE 1 Plot of inclination against depth in core Conrad 10-182 (44° 05' N, 176° 50' E). The change from the Brunhes normal series to the Matuyama reversed series occurs at 825 cm. The Jaramillo normal event occupies the interval between 1035 and 1095 cm

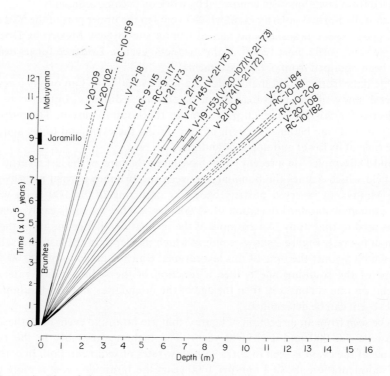

FIGURE 2 Plot of the depth of the last reversal against the time of its occurrence (700 000 years). This allows an estimate of the time of the beginning and end of the event and of its duration

and determine the most probable interval of time represented by the Jaramillo event and the most probable time for its beginning and end, averages were computed using cores in which the rate of sedimentation exceeded 5 mm per 1000 years. The cores with higher sedimentation rates are preferred because the interval of core representing the Jaramillo event can be more accurately determined.

Results

The time for the end of the Jaramillo event ranges from 786×10^3 years to 932×10^3 years. The average is $870 \cdot 5 \times 10^3$ years with a standard deviation of $32 \cdot 56 \times 10^3$ years. The average age obtained for the beginning of the Jaramillo is $926 \cdot 5 \times 10^3$ years.

The length of time included in the Jaramillo event ranges from $34 \cdot 6 \times 10^3$ years to $93 \cdot 4 \times 10^3$ years with the average duration being $56 \cdot 05 \times 10^3$ years with a standard deviation of $14 \cdot 6 \times 10^3$ years.

TABLE 1 Data for Jaramillo event

Core	Sedimentation rate (mm/1000 years)	End of Jaramillo event ($\times 10^3$ years)	Beginning of Jaramillo event ($\times 10^3$ years)	Length (cm)	Duration of Jaramillo event ($\times 10^3$ years)
R–C10–182	11·8	877	928	60	50·8
V–20–108	11·3	889	947	65	57·5
RC–10–206	11·1	898	953	61	54·9
RC–10–181	10·9	876	913	40	36·7
V–20–184	10·35	932	990	60	57·9
V–20–104	8·65	884	942	50	57·8
V–21–74	8·10	846	907	50	61·7
V–21–172	8·10	870	932	50	61·7
V–21–73	7·50	847	913	50	66·6
V–20–107	7·50	900	950	38	50·6
V–19–153	7·50	849	884	26	34·6
V–21–145	6·93	859	902	30	43·3
V–21–175	6·93	786	830	30	43·3
V–21–75	6·43	895	964	45	70·0
V–21–173	5·35	850	943	50	93·4
Average		879·5	926		56·05
Standard deviation		32·56			14·60

The values obtained from this study can be compared with those derived from K–Ar ages on lava flows. Cox and Dalrymple[11] estimate the mean age of the Jaramillo event as being 0·91 million years which is in substantial agreement with the present study since the value would fall within the event as delineated in this chapter. They estimate the length of time involved in the Jaramillo event as 120 000 years. This estimate would seem to be a factor of about 2 too long. However, given the uncertainties inherent in both methods, it would seem that the agreement is reasonable.

McDougall and Chamalaun[12] conclude in a recent review of the polarity time scale that the lower boundary of the Jaramillo event lies close to 1×10^6 years boundary polarity ±0·04. Doell and Dalrymple[9] have dated a transition lava from New Mexico at 0·88 million years which most probably represents the upper boundary. Cox and Dalrymple[11] have estimated the probable dating errors in the 0–1 million years range for ^{40}K–^{40}Ar to be about 18%. In view of the inherent errors which are involved in dating and the errors involved in extrapolating sedimentation rates it is clear that the results given by the two methods are in good agreement.

Acknowledgements

The author would like to thank Mr. J. Foster and Dr. D. Ninkovich who provided the data from some of the cores. This work was supported by the National Science Foundation under Grant NSF GA-824. This support is gratefully acknowledged.

References

1. A. Cox, R. R. Doell, and G. B. Dalrymple, *Nature*, **198**, 1049–51 (1963).
2. A. Cox, R. R. Doell, and G. B. Dalrymple, *Science*, **142**, 382–5 (1963).
3. R. R. Doell, G. B. Dalrymple, and A. Cox, *J. Geophys. Res.*, **71**, 531–41 (1966).
4. I. McDougall, H. L. Allsopp, and F. H. Chamalaun, *J. Geophys. Res.*, **71**, 24 (1966).
5. I. McDougall and D. H. Tarling, *Nature*, **200**, 54–6 (1963).
6. I. McDougall and H. Wensink, *Earth Planetary Sci. Letters*, **1**, 232–6 (1966).
7. C. G. A. Harrison and B. M. Funnell, *Nature*, **204**, 566 (1964).
8. N. D. Opdyke, B. Glass, J. D. Hays, and J. H. Foster, *Science*, **154**, 349–57 (1966).
9. R. R. Doell and G. B. Dalrymple, *Science*, **152**, 1060–1 (1966).
10. D. Ninkovich, N. D. Opdyke, B. C. Heezen, and J. H. Foster, *Earth Planetary Sci. Letters*, **1**, 476–92 (1966).
11. A. Cox and G. B. Dalrymple, *Abstr., U.S. Geol. Surv.* (1966).
12. I. McDougall and F. H. Chamalaun, *Nature*, **212**, 1415–18 (1966).

E

MAGNETO-HYDRODYNAMICS

II. The dynamo problem

1

J. G. TOUGH

and

R. D. GIBSON

School of Mathematics
University of Newcastle upon Tyne
England

The Braginskiĭ dynamo

Introduction

The kinematic dynamo problem is the search for non-zero solenoidal solutions to the vector equation

$$\frac{1}{R}\frac{\partial \mathbf{B}}{\partial t} = \mathrm{curl}\,(\mathbf{u}\times\mathbf{B}) + \frac{1}{R}\nabla^2\mathbf{B} \tag{1}$$

subject to the condition $\mathbf{B} = O(1/r^3)$ as $r \to \infty$, and which do not vanish as the time, t, tends to infinity. Here $\mathbf{B}$ is the magnetic field, and $\mathbf{u}$ is the velocity, supposed given, of the conducting fluid which, it is imagined, lies in a finite simply connected volume V. The exterior of V is an insulator. Equation (1) has been written in dimensionless form using L, a characteristic dimension of the flow, as unit of length and L^2/η as unit of time, where η is the magnetic diffusivity. The non-dimensional parameter $R = UL/\eta$, where U is a characteristic flow speed, is the magnetic Reynolds number. The condition on $\mathbf{B}$ as distance, r, tends to infinity excludes the presence of external sources.

Braginskiĭ[1] initiated the study of homogeneous dynamos in a cylindrically symmetric volume in which the conducting fluid moves almost symmetrically, and in which the magnetic Reynolds number of the symmetric components is large. He showed that, in a first approximation, the effects of the small non-axisymmetric components of flow could be simply represented in the equation for the symmetric part of the magnetic field by a generating term, provided new 'effective' components of symmetric flow and field were defined.

The First Approximation

First, let us consider the case when the magnetic field and velocity are axisymmetric. Then, by introducing a vector potential $\mathbf{A}$ for the magnetic field, we can write

$$\mathbf{B} = \mathrm{curl}\,(A_\phi \mathbf{1}_\phi) + B_\phi \mathbf{1}_\phi$$

where cylindrical polar coordinates $(\tilde{\omega}, \phi, z)$ are used along with a triad of unit vectors $\mathbf{1}_{\tilde{\omega}}, \mathbf{1}_\phi,$ and $\mathbf{1}_z$.

Consequently we are able to express the magnetic field by two equations; namely the ϕ component of the induction equation and the ϕ component of its vector potential form

$$\frac{1}{R}\frac{\partial \mathbf{A}}{\partial t} + \operatorname{grad} \Phi = \mathbf{u} \times \mathbf{B} - \frac{1}{R}\operatorname{curl} \mathbf{B} \tag{2}$$

where Φ is the electric potential. These then give

$$\frac{1}{R}\frac{\partial A_\phi}{\partial t} + \tilde{\omega}^{-1}\mathbf{u}_p \cdot \boldsymbol{\nabla}(\tilde{\omega} A_\phi) = \frac{1}{R}\Delta_1 A_\phi \tag{3}$$

$$\frac{1}{R}\frac{\partial B_\phi}{\partial t} + \tilde{\omega}\mathbf{u}_p \cdot \boldsymbol{\nabla}(\tilde{\omega}^{-1} B_\phi) = \frac{1}{R}\Delta_1 B_\phi + \left\{ \boldsymbol{\nabla}\left(\frac{u_\phi}{\tilde{\omega}}\right) \times \boldsymbol{\nabla}(\tilde{\omega} A_\phi) \right\}_\phi \tag{4}$$

where

$$\Delta_1 = \boldsymbol{\nabla}^2 - \frac{1}{\tilde{\omega}^2}$$

and the subscript p denotes the meridional component of a vector.

Outside the volume of conducting fluid we have that curl $\mathbf{B} = 0$, which implies that $\Delta_1 A_\phi = B_\phi = 0$, and across the boundary of the fluid $\mathbf{B}$ is continuous. Then using these facts together with equations (3) and (4) we obtain the following results:

$$\frac{1}{2R}\frac{d}{dt}\int_V (\tilde{\omega} A_\phi)^2 \, dV = -\frac{1}{R}\int_{V+\hat{V}} \{\boldsymbol{\nabla}(\tilde{\omega} A_\phi)\}^2 \tag{5}$$

$$\frac{1}{2R}\frac{d}{dt}\int_V (\tilde{\omega}^{-1} B_\phi)^2 \, dV = -\frac{1}{R}\int_{V+\hat{V}} \{\boldsymbol{\nabla}(\tilde{\omega}^{-1} B_\phi)\}^2 \, dV + \int_V \tilde{\omega}^{-2} B_\phi \left\{ \boldsymbol{\nabla}\left(\frac{u_\phi}{\tilde{\omega}}\right) \times \boldsymbol{\nabla}(\tilde{\omega} A_\phi) \right\}_\phi dV \tag{6}$$

From equation (5) we see that A_ϕ tends to zero with increasing time, i.e. the meridional component of the magnetic field decays with increasing time. Then, substituting $A_\phi = 0$ in equation (6) we see that B_ϕ likewise tends to zero with increasing time; hence all components of the magnetic field decay after a sufficient length of time. This result, due to Braginskiĭ, furnishes a simple and elegant proof of Cowling's theorem, which states that an axisymmetric magnetic field cannot be maintained by dynamo action.

If a mechanism can be found which will maintain the meridional field, i.e. so that A_ϕ does not decay to zero with increasing time, then the azimuthal field, B_ϕ, can be maintained by the last term in equation (4). This is, of course, provided that the lines of constant angular velocity, $u_\phi/\tilde{\omega} = \text{constant}$, do not coincide with the magnetic lines of force, $\tilde{\omega} A = \text{constant}$. In an attempt to obtain such a generating mechanism, Braginskiĭ examined the effect on dynamo action of small deviations from axisymmetry.

For simplicity in writing the necessary equations, let us introduce the notation used by Braginskiĭ in his analysis. Firstly there is an averaging process over ϕ, defined for scalars, F, and vectors, $\mathbf{u}$, by

$$\langle F \rangle = \frac{1}{2\pi}\int_0^{2\pi} F(\tilde{\omega}, \phi, z) \, d\phi$$

$$\langle \mathbf{u} \rangle = \langle u_{\tilde{\omega}} \rangle \mathbf{1}_{\tilde{\omega}} + \langle u_\phi \rangle \mathbf{1}_\phi + \langle u_z \rangle \mathbf{1}_z$$

The Braginskiĭ dynamo

Then any quantity can be expressed as the sum of two parts, one axisymmetric and the other purely variable in ϕ, for example

$$\mathbf{u} = \langle \mathbf{u} \rangle + \mathbf{u}'$$

where

$$\langle \mathbf{u}' \rangle = 0$$

Further simplicity can be obtained by omitting the angled brackets and writing, for example,

$$\langle \mathbf{u} \rangle = \mathbf{u}, \qquad \langle \mathbf{B} \rangle = \mathbf{B}$$

Secondly, we shall use $\partial_1/\partial\phi$ to denote differentiation with respect to ϕ holding $\mathbf{1}_{\tilde{\omega}}$ and $\mathbf{1}_\phi$ constant. Finally, we introduce a notation for integration with respect to ϕ holding $\mathbf{1}_{\tilde{\omega}}$ and $\mathbf{1}_\phi$ constant; we denote by $\hat{\mathbf{F}}'$ a quantity which satisfies $\partial_1 \hat{\mathbf{F}}'/\partial\phi = \mathbf{F}'$.

If we now average equations (1) and (2) over ϕ and consider the scalar product of these with $\mathbf{1}_\phi$, remembering that $\mathbf{B}$ and $\mathbf{u}$ are no longer axisymmetric, we obtain

$$\frac{1}{R}\frac{\partial A_\phi}{\partial t} + \tilde{\omega}^{-1}\mathbf{u}_p \cdot \nabla(\tilde{\omega} A_\phi) = \frac{1}{R}\Delta_1 A_\phi + S_\phi \tag{7}$$

$$\frac{1}{R}\frac{\partial B_\phi}{\partial t} + \tilde{\omega}\mathbf{u}_p \cdot \nabla(\tilde{\omega}^{-1} B_\phi) = \frac{1}{R}\Delta_1 B_\phi + \left\{\nabla\left(\frac{u_\phi}{\tilde{\omega}}\right) \times \nabla(\tilde{\omega} A_\phi)\right\}_\phi + (\operatorname{curl} \mathbf{S})_\phi \tag{8}$$

where

$$\mathbf{S} = \langle \mathbf{u}' \times \mathbf{B}' \rangle \tag{9}$$

Braginskiĭ was the first to suggest that $\mathbf{S}$ might be evaluated asymptotically for $R \to \infty$. It will be recognized that, in this limit, (1) is a singular perturbation problem and that its solution can be thought of as a mainstream (in which $\partial/\partial t = O(1)$ and $\nabla = O(1)$) together with boundary layers (in which $\mathbf{n} \cdot \nabla \gg 1$, where $\mathbf{n}$ is the unit normal to the boundary). Braginskiĭ confined his analysis to flows of the type

$$\mathbf{u} = \mathbf{u}_0 + \mathbf{u}'_1 + \mathbf{u}_2$$

where $u_n = O(R^{-n/2})$, $\mathbf{u}_2$ being the velocity used in defining R. Furthermore, $\mathbf{u}_0$ was azimuthal and axisymmetric, $\mathbf{u}_0$ was meridional and axisymmetric, and $\mathbf{u}'_1$ was non-axisymmetric.

Then, by equation (1), the asymptotic form of the mainstream solution for $\mathbf{B}$ is

$$\mathbf{B} = \sum_{n=0}^{\infty} \mathbf{B}_n$$

where

$$B_n = O(B_0 R^{-n/2})$$

By equating like powers of $R^{-1/2}$ in equation (1) an infinite sequence of equations for $\mathbf{B}_0, \mathbf{B}_1, \ldots$ is obtained. This then allows us to write $\mathbf{S}$ in terms of a power series, namely

$$\mathbf{S} = \sum_{n=2}^{\infty} \mathbf{S}^n = \left\langle \mathbf{u}'_1 \times \left(\sum_{r=1}^{\infty} \mathbf{B}_r \right) \right\rangle$$

where

$$S^n = O(B_0 R^{-n/2})$$

In order to evaluate the highest-order terms of S_ϕ and $(\text{curl } \mathbf{S})_\phi$, which are $O(B_0 R^{-2})$ and $O(B_0 R^{-1})$ respectively, it is sufficient to find $\mathbf{B}_0$, $\mathbf{B}_1$, and $\mathbf{B}_2$. On performing this reduction, a most remarkable simplification of equations (7) and (8) was discovered by Braginskiĭ. He proved that by introducing 'effective' quantities for A_ϕ and $\mathbf{u}_2$ by

$$A_e = A_\phi + w B_\phi \tag{10}$$

$$\mathbf{u}_{2e} = \mathbf{u}_2 + \text{curl}(w \mathbf{u}_0) \tag{11}$$

where

$$w = \frac{\tilde{\omega}}{2 u_0^2} \langle (\mathbf{u}_p' \times \hat{\mathbf{u}}_p')_\phi \rangle \tag{12}$$

(7) and (8) could be written as

$$\frac{1}{R}\frac{\partial A_e}{\partial t} + \tilde{\omega}^{-1} \mathbf{u}_{2e} \cdot \nabla(\tilde{\omega} A_e) = \frac{1}{R}\Delta_1 A_e + \frac{1}{R} B_\phi \Gamma \tag{13}$$

$$\frac{1}{R}\frac{\partial B_\phi}{\partial t} + \tilde{\omega} \mathbf{u}_{2e} \cdot \nabla(\tilde{\omega}^{-1} B_\phi) = \frac{1}{R}\Delta_1 B_\phi + \left\{ \nabla\left(\frac{u_\phi}{\tilde{\omega}}\right) \times \nabla(\tilde{\omega} A_e) \right\}_\phi \tag{14}$$

where Γ is a generating term given by

$$\Gamma = \frac{1}{\tilde{\omega} u_0^2}\left\langle (\mathbf{u}_p' \times \hat{\mathbf{u}}_p')_\phi + \left(\mathbf{u}_p' \times \frac{\partial_1 \mathbf{u}_p'}{\partial \phi}\right)_\phi \right\rangle + 2\left\langle \nabla_p(r u_r') \cdot \nabla_p\left(\frac{\hat{u}_z'}{u_0}\right) \right\rangle \tag{15}$$

and

$$r u_r' = \frac{\tilde{\omega} u_{\tilde{\omega}}' + z u_z'}{u_0}$$

This powerful result was first anticipated in qualitative arguments by Parker[2].

If we Fourier analyse $\mathbf{u}_p'$, i.e. write

$$\mathbf{u}_p' = \sum_{m=1}^{\infty} (\mathbf{u}_p^{mc} \cos m\phi + \mathbf{u}_p^{ms} \sin m\phi)$$

then Γ can be expressed in the form

$$\Gamma = \sum_{m=1}^{\infty} m^{-1}\left[\frac{1}{\tilde{\omega} u_0^2}\{\mathbf{u}_p^{ms} \times \mathbf{u}_p^{mc}\}_\phi (1-m^2) + \nabla(r u_r^{ms}) \cdot \nabla\left(\frac{u_z^{mc}}{u_0}\right) - \nabla(r u_r^{mc}) \cdot \nabla\left(\frac{u_z^{ms}}{u_0}\right)\right] \tag{16}$$

If $\Gamma = 0$, equations (13) and (14) reduce to the same form as the axisymmetric equations, (3) and (4), when, as shown above, A_e and B_ϕ decay with increasing time. If $\Gamma \neq 0$, the last term in equation (13) may act as a source term and prevent A_e from decaying. Then the last term in equation (14) may, as we have already seen, act as a source for B_ϕ, thereby allowing the magnetic field to be maintained indefinitely.

Dynamo in a Cylinder

Having developed the generation equations, which have been reformulated in the second section, Braginskii continued by solving them in a special case. In one study[3],

The Braginskii dynamo

he considered the maintenance of a field by fluid motions in an infinite plane layer of fluid. By appropriate choice of the velocity field he could show that the generation equations have simple analytic solutions, and that dynamo action was possible. In fact, Braginskii's work provided the first analytic proof of dynamo action by *stationary fluid* velocities.

In this section, we show that it is possible to solve (13) and (14) in the case where the conducting fluid is confined to lie within the infinite cylinder $\tilde{\omega} = 1$. The exterior $\tilde{\omega} > 1$ is insulating. (It should be noted that we have chosen the radius of the cylinder as a dimension characteristic of the flow.) We choose, as velocity field,

$$\mathbf{u}_0 = \tilde{\omega} V(\tilde{\omega}) \mathbf{1}_\phi \tag{17}$$

$$\mathbf{u}_1' = \left(p(\tilde{\omega})V(\tilde{\omega}) \sin\phi, \frac{d}{d\tilde{\omega}}\{p(\tilde{\omega})V(\tilde{\omega})\}\cos\phi, \tilde{\omega}q(\tilde{\omega})V(\tilde{\omega})\cos\phi \right) \tag{18}$$

and, defining w as in (12), we choose

$$\mathbf{u}_2 = -\operatorname{curl}(w\mathbf{u}_0) \tag{19}$$

The purpose of this very special choice is to reduce (11) simply to

$$\mathbf{u}_{2e} = 0 \tag{20}$$

Also (16), (17), and (18) give

$$\Gamma = \frac{dp}{d\tilde{\omega}}\frac{dq}{d\tilde{\omega}}$$

so that the generation equations (13) and (14) become

$$\left(\frac{\partial}{\partial t} - \Delta_1\right) A_e = \frac{dp}{d\tilde{\omega}}\frac{dq}{d\tilde{\omega}} B_\phi \tag{21}$$

$$\left(\frac{\partial}{\partial t} - \Delta_1\right) B_\phi = -\tilde{\omega} R \frac{dV}{d\tilde{\omega}}\frac{\partial A_e}{\partial z} \tag{22}$$

The pair of linear homogeneous equations (21) and (22) have coefficients which are independent of z and t. In fact, we may continue by seeking normal modes; that is, we write

$$A_e, B_\phi \propto \exp(\sigma t + iaz) \tag{23}$$

where σ is the time constant and is, in general, complex; a is the wave number. It then follows that we may replace (21) and (22) by

$$\left(D^2 + \frac{1}{\tilde{\omega}}D - a^2 - \frac{1}{\tilde{\omega}^2} - \sigma\right) A_e = -(Dp)(Dq)B_\phi \tag{24}$$

$$\left(D^2 + \frac{1}{\tilde{\omega}}D - a^2 - \frac{1}{\tilde{\omega}^2} - \sigma\right) B_\phi = iaR\tilde{\omega}(DV)A_e \tag{25}$$

where $D = d/d\tilde{\omega}$. We solve (24) and (25) in the case when

$$p = P(1 - \tilde{\omega}) \tag{26}$$

$$q = Q\tilde{\omega} \tag{27}$$

$$V = \tfrac{1}{3}U\tilde{\omega}^3 \tag{28}$$

where P, Q, and U are constants. It may be noted that Γ ($= -PQ$) is a homogeneous source of field.

If we set
$$B_\phi^* = PQB_\phi \tag{29}$$
and introduce the parameter
$$\lambda = PQRU \tag{30}$$
then (24) and (25) become
$$\left(D^2 + \frac{1}{\tilde{\omega}}D - a^2 - \frac{1}{\tilde{\omega}^2} - \sigma\right)A_e = B_\phi \tag{31}$$

$$\left(D^2 + \frac{1}{\tilde{\omega}}D - a^2 - \frac{1}{\tilde{\omega}^2} - \sigma\right)B_\phi = ia\lambda\tilde{\omega}^3 A_e \tag{32}$$

where we have dropped the asterisks. The boundary condition on B_ϕ at $\tilde{\omega} = 1$ is
$$B_\phi = 0 \tag{33}$$
To obtain the boundary condition on A_e, we note that the field equation in the exterior insulator is
$$\Delta_1 A_e = 0$$
and the solution of this which tends to zero as $\tilde{\omega} \to \infty$ is
$$A_e = lK_1(a\tilde{\omega})\exp(\sigma t + iaz)$$
where l is a constant and K is a modified Bessel function of the second kind. By the continuity of A_e and $\partial A_e/\partial \tilde{\omega}$ across the conductor–insulator boundary, we have
$$\frac{DA_e}{A_e} = \frac{aK_1'(a)}{K_1(a)} \tag{34}$$
at $\tilde{\omega} = 1$. Equations (33) and (34), together with the fact that A_e and B_ϕ must be bounded for $\tilde{\omega} < 1$, provide a complete set of conditions for the fourth-order system posed by (31) and (32). Table 1 shows the results derived from the numerical integration of this system, using a Chebyshev collocation method[4]. We have considered only

TABLE 1 Results using a Chebyshev collocation method

a	$-\lambda$	$\mathscr{I}(\sigma)$
2·0	14707·01	−49·94
2·2	14071·89	−51·49
2·4	13613·43	−53·18
2·6	13294·28	−55·04
2·8	13087·77	−57·05
3·0	12973·73	−59·22
3·2	12936·74	−61·55
3·4	12964·41	−64·05
3·6	13046·52	−66·70

The Braginskiĭ dynamo

solutions in which the field is purely oscillatory; that is $\mathscr{R}(\sigma) = 0$. The values of λ and $\mathscr{I}(\sigma)$ are shown as functions of a in figures 1 and 2 respectively. It is possible to minimize λ over all the wave numbers, and hence to determine the mode which is most easily excited.

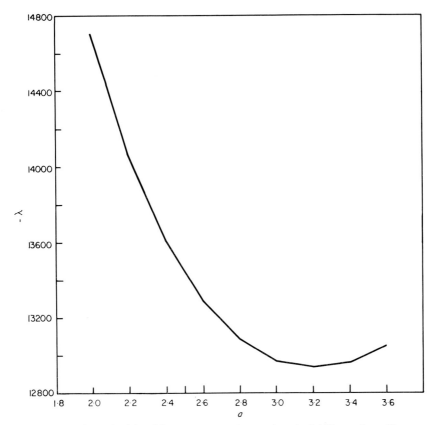

FIGURE 1   Values of $-\lambda$ for different wave numbers a when the field is purely oscillatory

In figures 3 and 4, we show the modulus and phase of the functions A_e and B_ϕ for the particular case $a = 3\cdot2$, and $\lambda = -\,12\,936\cdot74$.

The Second Approximation

We now leave the simple application of Braginskiĭ's theory and consider what form the theory itself would take in a higher approximation. In doing this we shall also show how the results of Braginskii, quoted in the second section, are derived; i.e. we shall obtain the first two terms in each of the series for $(\operatorname{curl} \mathbf{S})_\phi$ and S_ϕ, where

$$\mathbf{S} = \left\langle \mathbf{u}' \times \left(\sum_{n=1}^{\infty} \mathbf{B}'_n \right) \right\rangle \tag{35}$$

It will be found convenient to restate some equations used by Braginskiĭ in his paper.

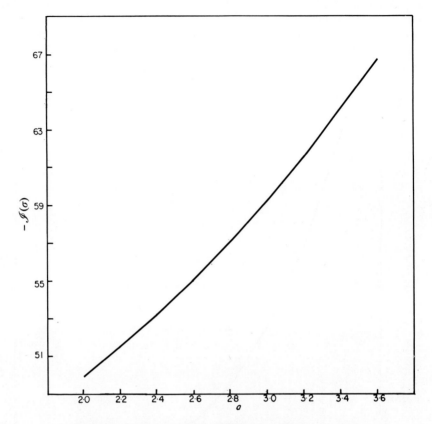

FIGURE 2 Frequency of the oscillation for different values of a

Braginskiĭ showed that, if equation (1) is averaged over ϕ and then subtracted from its original form, and the meridional component of this equation is taken, then (see Braginskiĭ[1], p. 729, equation (3.7)))

$$\mathbf{B}'_p = \left(\sum_{n=1}^{\infty}\mathbf{B}'_n\right)_p = \frac{B_0}{u_0}\mathbf{u}'_p + \frac{\tilde{\omega}}{u_0}\left[\operatorname{curl}(\hat{\mathbf{G}}' + \hat{\mathbf{u}}' \times \mathbf{B}_p)\}_p + \frac{1}{R}(\nabla^2\hat{\mathbf{B}}')_p + \hat{\mathbf{B}}'_p \cdot \nabla\mathbf{u}_2 - \frac{1}{R}\frac{d}{dt}\hat{\mathbf{B}}'_p\right] \quad (36)$$

where

$$\frac{1}{R}\frac{d}{dt} = \frac{1}{R}\frac{\partial}{\partial t} + \mathbf{u}_2 \cdot \nabla$$

$$\mathbf{G}' = \sum_{n=2}^{\infty}\mathbf{G}^n = \mathbf{u}' \times \left(\sum_{n=1}^{\infty}\mathbf{B}'_n\right) - \left\langle \mathbf{u}' \times \left(\sum_{n=1}^{\infty}\mathbf{B}'_n\right)\right\rangle$$

and (see Braginskiĭ[1], p. 729, equation (3.9))

$$(\operatorname{curl}\hat{\mathbf{G}}')_p = \tilde{\omega}^{-1}\mathbf{1}_\phi \times (\mathbf{G}'_p - \nabla(\tilde{\omega}\hat{G}'_\phi)) \quad (37)$$

Further, if equation (36) is substituted into equation (35), then (see Braginskiĭ[1], p. 730, equation (3.17))

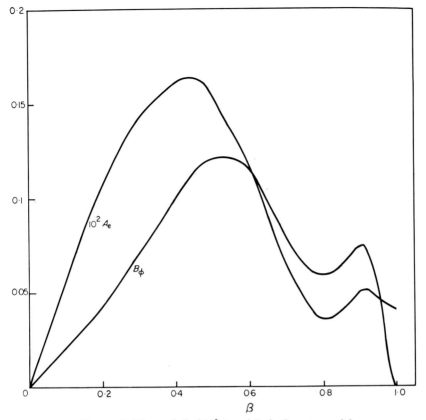

FIGURE 3 The moduli of $10^2 A_e$ and B_ϕ in the case $a = 3\cdot 2$

$$S_\phi = \langle (\mathbf{u}'_p \times \mathbf{B}'_p)_\phi \rangle$$

$$= \frac{1}{u_0} \operatorname{div} \langle \tilde{\omega}(\mathbf{u}'_p \times \mathbf{B}'_p)_\phi \hat{\mathbf{u}}'_p \rangle + \frac{1}{R}\left\langle \left\{ \tilde{\omega}\frac{\mathbf{u}'_p}{u_0} \times (\nabla^2 \hat{\mathbf{B}})_p \right\}_\phi \right\rangle$$

$$+ \frac{1}{u_0}\operatorname{div}\langle \tilde{\omega}(\mathbf{u}'_p \times \mathbf{B}_p)_\phi \hat{\mathbf{u}}'_p \rangle$$

$$+ \left\langle \tilde{\omega}\left[\frac{\mathbf{u}'_p}{u_0} \times \left\{(\hat{\mathbf{B}}' \cdot \nabla)\mathbf{u}_2 - \frac{1}{R}\frac{d\hat{\mathbf{B}}'_p}{dt}\right\}\right] \right\rangle \tag{38}$$

From equation (36) by equating like powers of $R^{-1/2}$ we have

$$\mathbf{B}'_{1p} = \frac{B_0}{u_0}\mathbf{u}'_p \tag{39}$$

and

$$\mathbf{B}'_{2p} = \frac{\tilde{\omega}}{u_0}(\operatorname{curl} \hat{\mathbf{G}}^2)_p$$

$$= \frac{1}{u_0}\mathbf{1}_\phi \times \{\mathbf{G}_p^2 - \nabla(\tilde{\omega}\hat{G}_\phi^2)\} \tag{40}$$

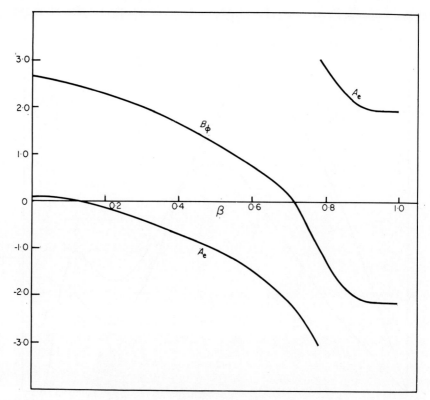

FIGURE 4 The phases of A_e and B_ϕ in the case $a = 3 \cdot 2$

$(\mathbf{B}'_n)_\phi$ can then be obtained by using the relationship

$$(\mathbf{B}'_n)_\phi = -\tilde{\omega}\operatorname{div}\hat{\mathbf{B}}_{np} \tag{41}$$

In order to evaluate the terms in the series $(\operatorname{curl}\mathbf{S})_\phi$ it is necessary to find the terms of the series $\mathbf{S}_p$; the first of these, which is of order $O(B_0 R^{-1})$, is given by

$$\mathbf{S}_p^2 = \langle(\mathbf{u}'_1 \times \mathbf{B}'_1)_p\rangle$$

Then, using equation (39) in conjunction with (41), it can be shown that (see Braginskiĭ[1], p. 730, equation (3.13))

$$\mathbf{S}_p^2 = w u_0^2 \nabla\left(\frac{B_\phi}{u_0}\right)$$

where

$$w = \frac{\tilde{\omega}}{2u_0^2}\langle(\mathbf{u}'_p \times \hat{\mathbf{u}}'_p)_\phi\rangle$$

Hence

$$(\operatorname{curl}\mathbf{S}_p^2)_\phi = -\tilde{\omega}\operatorname{curl}(w\mathbf{u}_0)\cdot\nabla(\tilde{\omega}^{-1}B_\phi) + \left\{\nabla\left(\frac{u_\phi}{\tilde{\omega}}\right)\times\nabla(\tilde{\omega}wB_\phi)\right\}_\phi$$

The Braginskiĭ dynamo

If we now define 'effective' quantities for A_ϕ and $\mathbf{u}_2$, as in the second section, by

$$A_e = A_\phi + wB_\phi \tag{42}$$

and

$$\mathbf{u}_{2e} = \mathbf{u}_2 + \operatorname{curl}(w\mathbf{u}_0) \tag{43}$$

equation (8) reduces to the form (14).

The first two terms in the series S_ϕ given by

$$S_\phi^2 = \langle (\mathbf{u}_p' \times \mathbf{B}_{1p}')_\phi \rangle$$

and

$$S_\phi^3 = \langle (\mathbf{u}_p' \times \mathbf{B}_{2p}')_\phi \rangle$$

are, by equations (39) and (40) respectively, zero. In evaluating the next term S_ϕ^4 it is easier to use equation (38), since the terms on the right-hand side of (38) of the same order as S_ϕ^4, i.e. $O(B_0 R^{-2})$, can be obtained by substituting $\mathbf{B}_{1p}'$ and $\mathbf{B}_{2p}'$ in the first term and $\mathbf{B}_{1p}'$ in the others. The first term in equation (38) then becomes

$$-\frac{wu_0^2}{\tilde\omega}\frac{\partial \lambda}{\partial z} - \frac{1}{u_0}\frac{\partial \lambda}{\partial z}\frac{\partial}{\partial \tilde\omega}(w^2 u_0^3) + \frac{1}{u_0}\frac{\partial \lambda}{\partial \tilde\omega}\frac{\partial}{\partial z}(w^2 u_0^3) \tag{44}$$

where

$$\lambda = \frac{B_\phi}{u_0}$$

and the third and fourth terms are, respectively,

$$-\frac{\partial A_\phi}{\partial z}\left(u_0 \frac{\partial w}{\partial \tilde\omega} + 2w \frac{\partial u_0}{\partial \tilde\omega}\right) + \frac{1}{\tilde\omega}\frac{\partial}{\partial \tilde\omega}(\tilde\omega A_\phi)\left(u_0 \frac{\partial w}{\partial z} + 2w \frac{\partial u_0}{\partial z}\right) \tag{45}$$

and

$$-\frac{1}{R}\frac{\partial}{\partial t}(wB_\phi) - \mathbf{u}_2 \cdot \nabla(wB_\phi) - w\frac{1}{R}\frac{d}{dt}B_\phi \tag{46}$$

From equation (14) we are able to replace dB_ϕ/dt by

$$\frac{1}{R}\frac{dB_\phi}{dt} = \frac{1}{R}\Delta_1 B_\phi + \left[\nabla\left(\frac{u_\phi}{\tilde\omega}\right) \times \nabla\{\tilde\omega(A_\phi + wB_\phi)\}\right]_\phi$$

$$+ \frac{B_\phi}{\tilde\omega}\left\{u_{2\tilde\omega} - \frac{\partial}{\partial z}(wu_0)\right\} - \operatorname{curl}(w\mathbf{u}_0) \cdot \nabla B_\phi \tag{47}$$

Then from equations (44), (45), (46), and (47), the terms involving $\mathbf{u}_2$ give

$$-\frac{1}{\tilde\omega}\mathbf{u}_2 \cdot \nabla(\tilde\omega w B_\phi) \tag{48}$$

those involving A_ϕ give

$$-\frac{1}{\tilde\omega}\operatorname{curl}(w\mathbf{u}_0) \cdot \nabla(\tilde\omega A_\phi) \tag{49}$$

and the remaining terms in w give

$$-\frac{1}{R}\frac{\partial}{\partial t}(wB_\phi) - \frac{1}{\tilde{\omega}}\mathrm{curl}\,(w\mathbf{u}_0)\cdot\nabla(\tilde{\omega}wB_\phi) \tag{50}$$

When we now substitute $\mathbf{B}'_{1p}$ in the second term of equation (38), and combine it with the first term on the right in (47) we obtain

$$\frac{1}{R}B_\phi\Gamma^2 = \frac{1}{R}B_\phi\left\{\frac{1}{\tilde{\omega}u_0^2}\left\langle(\mathbf{u}'_p\times\hat{\mathbf{u}}'_p)_\phi + \left(\mathbf{u}'_p\times\frac{\partial_1\mathbf{u}'_p}{\partial\phi}\right)_\phi\right\rangle + 2\left\langle\nabla_p(r\hat{u}'_r)\cdot\nabla_p\left(\frac{\hat{u}'_z}{u_0}\right)\right\rangle\right\} \tag{51}$$

Then substituting for S_ϕ, i.e. equation (48), (49), (50), and (51), in (7), and using the 'effective' quantities A_e and $\mathbf{u}_{2e}$, equations (42) and (43), we obtain an equation of the same form as (13).

In evaluating the next approximation[5], we follow the analysis through in the same way as Braginskiĭ in his first approximation. The next term in the series S_p, of order $O(B_0 R^{-3/2})$, is given by

$$S_p^3 = \langle(\mathbf{u}'_1\times\mathbf{B}'_2)_p\rangle$$

After straightforward manipulation this becomes

$$S_p^3 = \mathbf{1}_{\tilde{\omega}}\left\{-W_{\tilde{\omega}}u_0^3\frac{\partial}{\partial\tilde{\omega}}\left(\frac{1}{u_0}\frac{\partial\lambda}{\partial\tilde{\omega}}\right) - W_z u_0^3\frac{\partial}{\partial\tilde{\omega}}\left(\frac{1}{u_0}\frac{\partial\lambda}{\partial z}\right) + \frac{\partial\lambda}{\partial\tilde{\omega}}\frac{\partial}{\partial z}(u_0^2 W_z) - \frac{1}{\tilde{\omega}}\frac{\partial\lambda}{\partial z}\frac{\partial}{\partial\tilde{\omega}}(\tilde{\omega}u_0^2 W_z)\right\}$$

$$+\mathbf{1}_z\left\{-W_{\tilde{\omega}}u_0^3\frac{\partial}{\partial z}\left(\frac{1}{u_0}\frac{\partial\lambda}{\partial\tilde{\omega}}\right) - W_z u_0^3\frac{\partial}{\partial z}\left(\frac{1}{u_0}\frac{\partial\lambda}{\partial z}\right) - \frac{\partial\lambda}{\partial\tilde{\omega}}\frac{\partial}{\partial z}(u_0^2 W_{\tilde{\omega}}) + \frac{1}{\tilde{\omega}}\frac{\partial\lambda}{\partial z}\frac{\partial}{\partial\tilde{\omega}}(\tilde{\omega}u_0^2 W_{\tilde{\omega}})\right\} \tag{52}$$

where

$$\mathbf{W} = \frac{\tilde{\omega}^3}{3u_0^3}\langle\hat{\mathbf{u}}'_p(\mathbf{u}'_p\times\hat{\mathbf{u}}'_p)_\phi\rangle$$

Then, by taking the curl of equation (52), the resulting terms can be arranged to give

$$(\mathrm{curl}\,\mathbf{S}_p^3)_\phi = -\tilde{\omega}\,\mathrm{curl}\left\{\frac{\mathbf{u}_0}{u_0^2}\nabla\cdot(u_0^2\mathbf{W})\right\}\cdot\nabla(\tilde{\omega}^{-1}B_\phi)$$

$$-\left[\nabla\left(\frac{u_\phi}{\tilde{\omega}}\right)\times\nabla\left(\tilde{\omega}\left\{u_0\mathbf{W}\cdot\nabla\left(\frac{B_\phi}{u_0}\right) - \frac{B_\phi}{u_0^2}\nabla\cdot(u_0^2\mathbf{W})\right\}\right)\right]_\phi$$

Hence, by redefining the 'effective' quantities A_e and $\mathbf{u}_{2e}$ by

$$A_e = A_\phi + wB_\phi + \frac{B_\phi}{u_0^2}\nabla\cdot(u_0^2\mathbf{W}) - u_0\mathbf{W}\cdot\nabla\left(\frac{B_\phi}{u_0}\right) \tag{53}$$

and

$$\mathbf{u}_{2e} = \mathbf{u}_2 + \mathrm{curl}\,(w\mathbf{u}_0) + \mathrm{curl}\left\{\frac{\mathbf{u}_0}{u_0^2}\nabla\cdot(u_0^2\mathbf{W})\right\} \tag{54}$$

equation (8) can again be reduced to the same form as equation (14).

In order to evaluate the next term in S_ϕ, i.e. S_ϕ^5, it is sufficient to substitute $\mathbf{B}'_{3p}$ into

The Braginskiĭ dynamo

the first term of equation (38) and $\mathbf{B}'_{2p}$ into the other terms. We can rearrange equation (36) to give $\mathbf{B}'_{3p}$ in the form

$$\mathbf{B}'_{3p} = \frac{\tilde{\omega}}{u_0}\left\{(\mathbf{B}_p\cdot\nabla\hat{\mathbf{u}}')_p - (\hat{\mathbf{u}}'\cdot\nabla\mathbf{B}_p)_p + \frac{B_\phi}{u_0}\hat{\mathbf{u}}'_p\cdot\nabla\mathbf{u}_2 - \mathbf{u}_2\cdot\nabla\left(\frac{B_\phi}{u_0}\hat{\mathbf{u}}'_p\right)\right.$$

$$\left. + (\operatorname{curl}\hat{\mathbf{G}}^3)_p + \frac{1}{R}(\nabla^2\hat{\mathbf{B}}'_1)_p - \frac{1}{R}\frac{\partial}{\partial t}\left(\frac{B_\phi}{u_0}\hat{\mathbf{u}}'_p\right)\right\} \quad (55)$$

When the first five terms and the last in equation (55) are substituted into (38) the following are obtained:

$$\frac{1}{u_0}\operatorname{div}\left[\mathbf{1}_{\tilde{\omega}}\left\{\frac{1}{u_0}\mathbf{B}_p\cdot\nabla(W_{\tilde{\omega}}u_0^3) - 2\frac{W_{\tilde{\omega}}u_0^2}{\tilde{\omega}}\mathbf{B}_{\tilde{\omega}}\right\} + \mathbf{1}_z\left\{\frac{1}{u_0}\mathbf{B}_p\cdot\nabla(W_z u_0^3) - 2\frac{W_z u_0^2}{\tilde{\omega}}\mathbf{B}_{\tilde{\omega}}\right\}\right] \quad (56)$$

$$-\frac{1}{u_0}\operatorname{div}\left(u_0^2\mathbf{W}\cdot\nabla\mathbf{B}_p - \frac{u_0^2 B_{\tilde{\omega}}}{\tilde{\omega}}\mathbf{W}\right) \quad (57)$$

$$\frac{1}{u_0}\operatorname{div}\left\{B_\phi u_0\mathbf{W}\cdot\nabla(u_0\mathbf{u}_2) - \frac{B_\phi u_0^2 u_{2\tilde{\omega}}}{\tilde{\omega}}\mathbf{W}\right\} \quad (58)$$

$$-\frac{1}{u_0}\operatorname{div}\left\{\frac{B_\phi}{u_0}\mathbf{u}_2\cdot\nabla(Wu_0^3) - \frac{2B_\phi u_0^2 u_{2\tilde{\omega}}}{\tilde{\omega}}\mathbf{W} + 3u_0^3\mathbf{W}\mathbf{u}_2\cdot\nabla\left(\frac{B_\phi}{u_0}\right)\right\} \quad (59)$$

$$\frac{1}{u_0}\nabla\left(\frac{B_\phi}{u_0}\right)\cdot\operatorname{curl}\left[\mathbf{1}_\phi\{-2wu_0\nabla\cdot(u_0^2\mathbf{W}) - 2u_0^2\mathbf{W}\cdot\nabla(wu_0) + u_0^3\mathbf{W}\cdot\nabla w\}\right] \quad (60)$$

$$-\frac{1}{\tilde{\omega}u_0}\frac{\partial}{\partial\tilde{\omega}}\left\{\tilde{\omega}\left(3W_{\tilde{\omega}}u_0\frac{\partial B_\phi}{\partial t} + B_\phi u_0\frac{\partial W_{\tilde{\omega}}}{\partial t}\right)\right\} - \frac{1}{u_0}\frac{\partial}{\partial z}\left\{3W_z u_0\frac{\partial B_\phi}{\partial t} + B_\phi u_0\frac{\partial W_z}{\partial t}\right\} \quad (61)$$

The last term in (38) gives

$$\frac{\partial\lambda}{\partial\tilde{\omega}}\left\{\frac{W_{\tilde{\omega}}u_0^2}{\tilde{\omega}}u_{2\tilde{\omega}} - W_{\tilde{\omega}}u_0\frac{\partial}{\partial z}(u_0 u_{2z}) + W_z u_0\frac{\partial}{\partial z}(u_0 u_{2\tilde{\omega}})\right\}$$

$$+\frac{\partial\lambda}{\partial z}\left\{\frac{W_z u_0^2}{\tilde{\omega}}u_{2\tilde{\omega}} - W_z u_0\frac{\partial}{\partial\tilde{\omega}}(u_0 u_{2\tilde{\omega}}) + W_{\tilde{\omega}}u_0\frac{\partial}{\partial\tilde{\omega}}(u_0 u_{2z})\right\}$$

$$+2\frac{\partial}{\partial t}(u_0\mathbf{W}\cdot\nabla\lambda) + \mathbf{W}\cdot\left\{\frac{\partial}{\partial t}(u_0\nabla\lambda)\right\} + 2\mathbf{u}_2\cdot\nabla(u_0\mathbf{W}\cdot\nabla\lambda)$$

$$+W_{\tilde{\omega}}u_0\mathbf{u}_2\cdot\nabla\left(u_0\frac{\partial\lambda}{\partial\tilde{\omega}}\right) + W_z u_0\mathbf{u}_2\cdot\nabla\left(u_0\frac{\partial\lambda}{\partial z}\right) - \frac{u_0^2 u_{2\tilde{\omega}}}{\tilde{\omega}}\mathbf{W}\cdot\nabla\lambda \quad (62)$$

We also obtain an extra term from the last term in (38) when we substitute $\mathbf{B}'_{1p}$; this comes about as a result of the new form for dB_ϕ/dt given by equation (14); it should be remembered though that we need only take terms of order $O(B_0 R^{-1})$ in (14) to obtain the order to which we are working, viz. $O(B_0 R^{-5/2})$. This extra term is

$$-w\left[\nabla\left(\frac{u_\phi}{\tilde{\omega}}\right)\times\nabla\left\{\tilde{\omega}\left(\frac{\lambda}{u_0}\nabla\cdot(u_0^2\mathbf{W}) - u_0\mathbf{W}\cdot\nabla\lambda\right)\right\} - \tilde{\omega}\operatorname{curl}\left\{\frac{\mathbf{u}_0}{u_0^2}\nabla\cdot(u_0^2\mathbf{W})\right\}\cdot\nabla(\tilde{\omega}^{-1}B_\phi)\right] \quad (63)$$

If as before we now go through equations (56) to (63) collecting first terms involving $\mathbf{u}_2$, then A_ϕ, and lastly w, we get, respectively,

$$-\frac{1}{\tilde{\omega}}\mathbf{u}_2 \cdot \nabla(\tilde{\omega}A^3) \tag{64}$$

$$-\frac{1}{\tilde{\omega}}\mathbf{u}_p^3 \cdot \nabla(\tilde{\omega}A_\phi) \tag{65}$$

$$-\frac{1}{\tilde{\omega}}\operatorname{curl}(w\mathbf{u}_0) \cdot \nabla(\tilde{\omega}A^3) - \frac{1}{\tilde{\omega}}\mathbf{u}_p^3 \cdot \nabla(\tilde{\omega}wB_\phi) \tag{66}$$

where

$$A^3 = A_e - A_\phi - wB_\phi$$

$$\mathbf{u}_p^3 = \mathbf{u}_{2e} - \mathbf{u}_2 - \operatorname{curl}(w\mathbf{u}_0)$$

When we now substitute $\mathbf{B}'_{2p}$ in the second term of (38) and add to it the sixth term in (55) along with the remaining terms in equations (56) to (63), we obtain

$$\frac{1}{R}B_\phi \Gamma^3 = \frac{1}{R}B_\phi \left[\nabla^2\left(\frac{1}{u_0}\mathbf{W}\cdot\nabla u_0\right) - \mathbf{W}\cdot\nabla^2\left(\frac{1}{u_0}\nabla u_0\right) - \frac{1}{u_0}\nabla u_0 \cdot \nabla^2 \mathbf{W} \right.$$
$$-\frac{6\tilde{\omega}}{u_0}\frac{\partial}{\partial \tilde{\omega}}\left(\frac{\mathbf{W}}{\tilde{\omega}^2}\right)\cdot \nabla u_0 - \frac{2\tilde{\omega}}{u_0}\frac{\partial u_0}{\partial \tilde{\omega}}\nabla\cdot\left(\frac{\mathbf{W}}{\tilde{\omega}^2}\right) - 4\nabla\cdot\left\{\tilde{\omega}\frac{\partial}{\partial \tilde{\omega}}\left(\frac{\mathbf{W}}{\tilde{\omega}^2}\right)\right\}$$
$$\left. +\frac{2}{u_0 \tilde{\omega}^3}\frac{\partial}{\partial \tilde{\omega}}(\tilde{\omega}u_0 W_{\tilde{\omega}}) + \frac{2}{u_0^2}\nabla\cdot(\tilde{\omega}^2 u_0^2 \mathbf{C}) \right] \tag{67}$$

where

$$\mathbf{C} = \mathbf{1}_{\tilde{\omega}}\left\langle u'_{\tilde{\omega}}\nabla \hat{u}'_{\tilde{\omega}} \cdot \nabla \hat{u}'_z + \hat{u}'_{\tilde{\omega}}\nabla u'_{\tilde{\omega}} \cdot \nabla \hat{u}'_z + \hat{u}'_z \nabla u'_{\tilde{\omega}} \cdot \nabla \hat{u}'_{\tilde{\omega}} + \frac{1}{\tilde{\omega}}u'_z \hat{u}'_{\tilde{\omega}}\frac{\partial \hat{u}'_{\tilde{\omega}}}{\partial \tilde{\omega}} + \frac{1}{\tilde{\omega}}u'_z \hat{u}'_{\tilde{\omega}}\frac{\partial \hat{u}'_z}{\partial z}\right\rangle$$

$$-\mathbf{1}_z\left\langle \hat{u}'_{\tilde{\omega}}\nabla u'_z \cdot \nabla \hat{u}'_z + \hat{u}'_z \nabla \hat{u}'_{\tilde{\omega}} \cdot \nabla u'_z + u'_z \nabla \hat{u}'_{\tilde{\omega}} \cdot \nabla \hat{u}'_z - \frac{1}{\tilde{\omega}}u'_z \hat{u}'_z \frac{\partial \hat{u}'_{\tilde{\omega}}}{\partial \tilde{\omega}} - \frac{1}{\tilde{\omega}}u'_z \hat{u}'_z \frac{\partial \hat{u}'_z}{\partial z}\right\rangle$$

If we now combine (64), (65), (66), and (67) and substitute them into equation (8), then, using the new definitions for 'effective' quantities A_e and $\mathbf{u}_{2e}$, equation (53) and (54), an equation of the same form as (14) is arrived at, where now

$$\Gamma = \Gamma^2 + \Gamma^3$$

Thus we have that equations (13) and (14) are valid to orders $O(B_0 R^{-5/2})$ and $O(B_0 R^{-3/2})$ respectively, provided we use the effective quantities (53) and (54).

It seems likely, from these results, that Braginskiĭ's discovery is a foretaste of a stronger result in which equations (13) and (14) are true to arbitrary order, provided A_e, $\mathbf{u}_{2e}$, and Γ are defined correctly as power series in $R^{-1/2}$. We have, however, not as yet succeeded in proving this generalisation, which, if true, would, we feel, be of considerable significance in dynamo theory.

Acknowledgements

It is with great pleasure that we record our gratitude to Professor P. H. Roberts for helping and encouraging us to develop the ideas presented here.

References

1. S. I. Braginskiĭ, *Zh. Eksperim. i Teor. Fiz.*, **47,** 1084 (1964) (*Soviet Phys.—JETP* (*English Transl.*), **20,** 726 (1965)).
2. E. N. Parker, *Astrophys. J.*, **122,** 293 (1955).
3. S. I. Braginskiĭ, *Zh. Eksperim. i Teor. Fiz.*, **47,** 2178 (1964) (*Soviet Phys.—JETP* (*English Transl.*), **20,** 1462 (1965)).
4. K. Wright, *Comput. J.*, **6,** 358 (1964).
5. J. G. Tough, *Monthly Notices Roy. Astron. Soc., Geophys. Suppl.*, in press (1967).

2

R. D. GIBSON

School of Mathematics
University of Newcastle upon Tyne
England

The Herzenberg dynamo

Introduction

No series of chapters devoted to the geomagnetic dynamo problem would be complete without some reference to the existence proofs, which are of fundamental importance to the subject. To this end we shall discuss here the model due to Herzenberg[1] and describe some extensions of his model. In fact the present chapter summarizes two recent studies[2,3], and extends these, on pp. 573–6, to the case of fluid rotors. We do not repeat any of the mathematics which may be found in the above-mentioned references; rather we describe those models which appear to be of some interest. In the section on pp. 573–6, which does include some algebra, we draw freely on the notation and terminology of the papers by Gibson[2,3].

The Basic Model

In the first instance it is perhaps best to think of the Herzenberg dynamo as consisting of the arrangement shown in figure 1. Two spherical rotors, L_1 and L_2, of equal radii, a, and conductivity σ, are embedded at distance R apart in a rigid conductor V, of

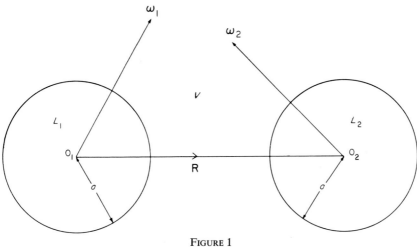

FIGURE 1

infinite spatial extent, also of conductivity σ. The rotors, L_1 and L_2, turn with angular velocities ω_1 and ω_2 respectively, where $|\omega_1| = |\omega_2| = \omega$. It is a conceptual advantage to divide the field **B** into two parts: $\mathbf{B}_1$, the field created by induction in L_1, $\mathbf{B}_2$, that created by L_2. In other words the rotation of L_2 in $\mathbf{B}_1$ creates $\mathbf{B}_2$, and the problem is to arrange matters so that $\mathbf{B}_2$ maintains, through the rotation of L_1, the very field $\mathbf{B}_1$ by which the process began.

We now observe that the high harmonics of $\mathbf{B}_2$ (or $\mathbf{B}_1$) die away with distance from L_2 (or L_1) more rapidly than do the low harmonics. Indeed by making the motion of L_1 and L_2 sufficiently vigorous and R sufficiently great the effect of the high harmonics can be majorized.

More precisely, the conditions of large R and ω are

$$\frac{a}{R} \ll 1 \quad \text{and} \quad \frac{a^2\omega}{\eta} \gg 1 \tag{1}$$

where η is the diffusivity ($= 1/\mu\sigma$). At the same time we must insist that

$$R_m \equiv \frac{a^2\omega}{\eta}\left(\frac{a}{R}\right)^3 = O(1) \tag{2}$$

where R_m is a magnetic Reynolds number. As a result of (1) and (2) it can be shown that the condition for dynamo action is of the form

$$\lambda^2 = O\left(\frac{a^2}{R^2}\right) \tag{3}$$

where λ depends on the orientation of ω_1 and ω_2 and upon R_m. The right-hand side of (3) arises from the neglected harmonics. Unfortunately there is the possibility that the right-hand side of (3) is negative definite; λ and therefore R_m would then be imaginary, an impossible state of affairs.

Two possible ways out of this dilemma are dealt with elsewhere[2]. First, as in section 5 of that article, we may evaluate the second-order term on the right of (3), and examine its sign explicitly. Unfortunately we again find that the condition for dynamo action may be expressed as the vanishing of a perfect square. That is we still cannot say whether or not the dynamo will function. The second possibility, which was adopted by Herzenberg himself, is to modify the model by limiting V to the interior of a large sphere in the exterior, ε, of which there is an insulator (see section 4 of Gibson[2]). Currents reflected from the interface between these regions introduce new terms on the right-hand side of (3) which, under certain circumstances, dominate those already appearing and are positive. Dynamo action has then been conclusively proved.

Extensions: the Calculus of Rigid Rotors

As we have noted in the previous section, it is no small matter to prove that the two-rotor model will function as a dynamo when V is unbounded, and, indeed, we have not succeeded in doing so. Let us, however, retain the unbounded conductor and increase the number of rotors to three. In this case the condition for dynamo action cannot, in general, be expressed to leading order as the vanishing of a perfect square (see

The Herzenberg dynamo

section 2 of Gibson[3]). This proves dynamo action irrespective of the $O(a^2/R^2)$ terms omitted in the analysis.

A further model considered in the paper by Gibson[3] consists of two rotors of conductivity different to that of the infinite medium in which they are embedded. Again dynamo action may be proved without reference to the $(a/R)^2$ terms neglected.

In section 4 of Gibson[3] two attempts to construct a single-rotor model are presented in which there is a feedback on the rotor due to currents reflected from some conductor–insulator interface. Cases considered are those in which the conductor boundaries are plane, and then use can be made of a simple image theorem. However, in the cases considered, which are those of a conducting corner and a conducting wedge, the condition for dynamo action is of the form of equation (3). And as in the case of the two-rotor model we may not conclude that dynamo action is possible, though such may well be the case if the conducting region is again made bounded by an insulating cut-off, ε, far from the rotor.

Extensions: Laminated Rotors

In this section we study the effect of non-rigid body rotations on the dynamo behaviour of the Herzenberg model. By a non-rigid body rotation we shall mean that the angular velocity is the same on shells spherical about the centre of each rotor. Incidentally the model we construct has 'fluid' moving parts as distinct from the solid moving parts of the previous two sections. In particular, it provides a means of eliminating the infinite shear gradients of Herzenberg's model which are objectionable on hydrodynamic grounds.

As a preliminary to the analysis we must find the field induced by the motion of a fluid, spherical rotor L, of radius a and centre O, in a given field. Inside L, we consider the induction equation

$$\eta \, \text{curl}^2 \mathbf{B} = \text{curl}\,(\mathbf{u} \times \mathbf{B}) \qquad (4)$$

where $\mathbf{B}$ is the field and $\mathbf{u}$ is the fluid velocity. In the conducting exterior, V, which is of infinite spatial extent and has the same electrical properties as L, we have

$$\text{curl}^2 \mathbf{B} = 0 \qquad (5)$$

By the solenoidal condition on $\mathbf{B}$ we may write

$$\mathbf{B} = \text{curl}\,T\mathbf{r} + \text{curl}^2 S\mathbf{r} \qquad (6)$$

where $\mathbf{r}$ is the radius vector from O; T and S are the torodial and poloidal defining scalars being functions of (r, θ, ϕ), the usual spherical polar coordinates. We treat the case in which

$$\mathbf{u} = \omega(r)\, r \sin\theta \, \mathbf{1}_\phi \qquad (7)$$

$\omega(r)$ being a non-uniform rotation and $\mathbf{1}_\phi$ being a unit vector in the azimuthal direction.

It then follows, from (4), (5), and (6), that the field equations in V are

$$\nabla^2 S = 0 \qquad (8)$$

$$\nabla^2 T = 0 \qquad (9)$$

and inside L are

$$\nabla^2 S = 0 \tag{10}$$

$$\nabla^2 T = -\frac{\omega'}{\eta} \sin\theta \frac{\partial S}{\partial \theta} \tag{11}$$

where $\omega' = d\omega/dr$ and we are considering the axisymmetric case $T = T(r, \theta)$ and $S = S(r, \theta)$. Poloidal inducing fields are determined by

$$S = A_n \left(\frac{r}{a}\right)^n P_n(\cos\theta) \tag{12}$$

which is a solution, finite at O, of (8) and (10); here P_n is the Legendre polynomial. Toroidal fields, being azimuthal, induce no further fields at all and hence may be neglected. The complementary function associated with (11) is

$$G_m \left(\frac{r}{a}\right)^m P_m(\cos\theta) \tag{13}$$

the particular integral being determined by the non-homogeneous equation

$$\nabla^2 T = -A_n \left(\frac{r}{a}\right)^n \frac{\omega'}{\eta} \sin\theta \frac{dP_n(\cos\theta)}{d\theta} = \frac{n(n+1)}{2n+1} \left(\frac{r}{a}\right)^n \frac{\omega'}{\eta} A_n \{P_{n-1}(\cos\theta) - P_{n+1}(\cos\theta)\} \tag{14}$$

This may be readily evaluated by the method of variation of parameters, and we obtain, after some manipulation,

$$T = \frac{n(n+1)}{2n+1} \frac{A_n}{a^n \eta} \left[\frac{P_{n-1}(\cos\theta)}{2n-1} \left\{ \frac{2n+1}{r^n} \int_0^r r^{2n} \omega(r) \, dr \right. \right.$$

$$\left. - 2r^{n-1} \int_0^r r\omega(r) \, dr \right\} - \frac{P_{n+1}(\cos\theta)}{r^{n+2}} \int_0^r r^{2n+2} \omega(r) \, dr \right] \tag{15}$$

Combining (12), (13), and (15) we show that the solution of the induction problem is given by

$$S = A_n \left(\frac{r}{a}\right)^n P_n(\cos\theta) \tag{16}$$

$$T = C_n \left(\frac{r}{a}\right)^{n-1} P_{n-1}(\cos\theta) + D_n \left(\frac{r}{a}\right)^{n+1} P_{n+1}(\cos\theta)$$

$$+ \frac{n(n+1)}{2n+1} \frac{A_n}{a^n \eta} \left[\frac{P_{n-1}(\cos\theta)}{2n-1} \left\{ \frac{2n+1}{r^n} \int_0^r r^{2n} \omega(r) \, dr \right. \right.$$

$$\left. - 2r^{n-1} \int_0^r r\omega(r) \, dr \right\} - \frac{P_{n+1}(\cos\theta)}{r^{n+2}} \int_0^r r^{2n+2} \omega(r) \, dr \right] \tag{17}$$

inside L, and

$$S = A_n \left(\frac{r}{a}\right)^n P_n(\cos\theta) \tag{18}$$

$$T = E_n \left(\frac{a}{r}\right)^n P_{n-1}(\cos\theta) + F_n \left(\frac{a}{r}\right)^{n+2} P_{n+1}(\cos\theta) \tag{19}$$

outside L, where C_n, D_n, E_n, and F_n are constants. The boundary conditions to be satisfied are that all components of the field and the tangential components of the electric field should be continuous. In the case when the fluid velocity satisfies the no-slip condition* ($\omega(a) = 0$) these conditions reduce to the continuity of T and $\partial T/\partial r$ across $r = a$. And, furthermore, they are sufficient to determine the constants C_n, D_n, E_n, and F_n:

$$C_n = \frac{2n(n+1)}{(2n-1)(2n+1)} \frac{A_n}{a\eta} \int_0^a r\omega(r)\,dr \tag{20}$$

$$D_n = 0 \tag{21}$$

$$E_n = \frac{n(n+1)}{2n-1} \frac{A_n}{a^{2n}\eta} \int_0^a r^{2n}\omega(r)\,dr \tag{22}$$

$$F_n = -\frac{n(n+1)}{2n+1} \frac{A_n}{a^{2n+2}\eta} \int_0^a r^{2n+2}\omega(r)\,dr \tag{23}$$

Having solved the induction problem we now proceed, as in Gibson[2,3], to consider dynamo action. The situation of interest is that of two fluid rotors, L_1 and L_2, embedded in a rigid conductor, V, of infinite spatial extent and having the same electrical properties as L_1 and L_2; their angular velocities are $\omega_1(r_1)$ and $\omega_2(r_2)$, where $\mathbf{r}_1$ and $\mathbf{r}_2$ are the radius vectors from the centres, O_1 and O_2, of L_1 and L_2 respectively. Arguments similar to those given in section 3 of Gibson[2] show that we need only retain that part of $\mathbf{B}_2$, the inducing field at rotor L_1, which is axisymmetric with respect to $\boldsymbol{\omega}_1$. Indeed, if we write the inducing field at L_1 as

$$\mathbf{B}_2(\mathbf{r}_1) = a \,\mathrm{grad}\, \sum_{n=1}^{\infty} A_n^{(2)} \left(\frac{r_1}{a}\right)^n P_n(\cos\theta_1) \tag{24}$$

and ignore all other contributions to $\mathbf{B}_2$, then, to an adequate approximation, the induced field is

$$\mathbf{B}_1(\mathbf{r}_1) = \left\{ -\frac{A_1^{(2)}}{a^3\eta} \int_0^a r^4\omega_1(r)\,dr \left(\frac{a}{r_1}\right)^3 \sin\theta_1 \cos\theta_1 \right.$$

$$\left. + \frac{2A_2^{(2)}}{3\,a^3\eta} \int_0^a r^4\omega_1(r)\,dr \left(\frac{a}{r_1}\right)^2 \sin\theta_1 \right\} \mathbf{1}_{\phi_1} \tag{25}$$

where θ_1 is the polar angle relating to $\boldsymbol{\omega}_1$ and $\mathbf{1}_{\phi_1}$ is the unit vector in the azimuthal

---

* We consider this case for simplicity. If, however, we assume that there is slip on the boundary we find that (20), (21), (22), and (23) below are unaltered.

direction. Equation (25) is to be compared with (26) of Gibson[2]. Indeed the comparison shows that if we replace the fluid rotor L_1 by a rigid rotor having the same axis of rotation, and angular speed Ω_1, given by

$$\Omega_1 = \frac{5}{a^5} \int_0^a r^4 \omega_1(r) \, dr \qquad (26)$$

then the very field $\mathbf{B}_1$, given by (25), would be induced. In other words, the analysis of fluid rotor dynamos can be reduced to studying rigid rotor dynamos. Equation (26) has significance in terms of the kinemetics of these two velocity fields: it implies that a fluid rotor and its rigid rotor counterpart have the same moment of momentum about their common axis of rotation.

The condition for dynamo action relating to a rigid rotor system having different angular speeds may be found in Gibson and Roberts[4]. It is

$$\{1 + \tfrac{1}{3}\Lambda_1 \Lambda_2 (\sin \Theta_1 \sin \Theta_2 \cos \Phi - \cos \Theta_1 \cos \Theta_2)\}^2 = 0 \qquad (27)$$

where

$$\Lambda_i = \frac{a^2 \Omega_i}{\eta} \left(\frac{a}{R}\right)^3 \frac{\mathbf{R} \cdot (\mathbf{\Omega}_1 \times \mathbf{\Omega}_2)}{5R\Omega_1 \Omega_2}, \qquad i = 1, 2 \qquad (28)$$

Θ_i is the angle between $\mathbf{\Omega}_i$ and $\mathbf{R}$ ($i = 1, 2$), and Φ is the angle between the planes defined by $\mathbf{\Omega}_1$ and $\mathbf{R}$ and $\mathbf{\Omega}_2$ and $\mathbf{R}$. All comments in the text which have been associated with (3) may now be similarly associated with (27).

Acknowledgements

Professor P. H. Roberts has provided much help and encouragement during the progress of this work. The author also wishes to thank Dr. S. Childress for a helpful discussion.

References

1. A. Herzenberg, *Phil. Trans. Roy. Soc. London, Ser. A*, **250**, 43 (1958).
2. R. D. Gibson, *Quart. J. Mech. Appl. Math.*, in press (1968).
3. R. D. Gibson, *Quart. J. Mech. Appl. Math.*, in press (1968).
4. R. D. Gibson and P. H. Roberts, in *Magnetism and the Cosmos* (Eds. W. R. Hindmarsh, F. J. Lowes, P. H. Roberts, and S. K. Runcorn), Oliver and Boyd, Edinburgh, 1966, p. 108.

3

R. D. GIBSON
and
P. H. ROBERTS

School of Mathematics
University of Newcastle upon Tyne
England

with an appendix by

S. SCOTT

Department of Mathematics
Rutherford College of Technology
Newcastle upon Tyne, England

The Bullard–Gellman dynamo

Introduction

The dynamo problem, stated in a form most relevant to planetary interiors, is that of determining whether a simply connected body V of conducting fluid (such as a sphere) can, by the inductive effect of motions in its interior, maintain a magnetic field, even though its exterior $\hat{V}$, a vacuum (or stationary conductor), contains no sources of field. And, if it can do so, what form the motions and fields must take. That answers to these questions are required is made clear by the case of the Earth: the electromagnetic decay time of the geomagnetic dipole (assuming the Earth's core is stationary and of conductivity 3×10^5 mho/m) is only of the order of 10^4 years. The age of the geomagnetic field is, however, far greater than this, as palaeomagnetic studies have established[1]. In other words a mechanism must exist which replenishes the field against ohmic losses. And, *a priori*, what could be more likely than that the Earth's core should generate fields and currents in the same way that the familiar commercial self-exciting dynamo does? There are, however, important distinctions to be made: whereas the man-made machine is multiply connected and highly asymmetric, the Earth's core is simply connected, and is almost spherically symmetric and homogeneous in structure. These differences are so crucial that one cannot conclude, from the obvious success of the commercial dynamo, that a geomagnetic dynamo could function; indeed, for many years it was not known theoretically whether homogeneous dynamos could exist at all. Celebrated existence proofs, provided independently by Herzenberg and Backus in 1958, have now made it clear that they can.

Formally, if we denote by **B**, the magnetic field, by **u** the fluid velocity, and by η the magnetic diffusivity ($= 1/\mu\sigma$; $\mu =$ permeability; $\sigma =$ electrical conductivity; m.k.s. units), we seek solutions of the induction equations

$$\frac{\partial \mathbf{B}}{\partial t} = \operatorname{curl}(\mathbf{u} \times \mathbf{B}) + \eta \nabla^2 \mathbf{B} \qquad \text{in } V \qquad (1)$$

$$\operatorname{div} \mathbf{B} = 0 \qquad (2)$$

with the added requirement that the solution selected must be continuous, in all components, with the field in the (insulating) exterior $\hat{V}$ of V. This external field $\hat{\mathbf{B}}$ may be derived from a single-valued potential $\hat{\Omega}$ satisfying Laplace's equation

$$\hat{\mathbf{B}} = -\operatorname{grad}\hat{\Omega}, \qquad \nabla^2\hat{\Omega} = 0 \qquad \text{in } \hat{V} \qquad (3)$$

and vanishing with distance r from V

$$\hat{\Omega} = O(r^{-2}), \qquad r \to \infty \qquad (4)$$

Condition (4) reflects the presumed absence of sources in $\hat{V}$. In the absence of fluid motions, it is known that $\mathbf{B}$ decays aperiodically to zero in a characteristic time of L^2/η, where L is a typical dimension of V. This 'free decay time' provided the estimate of 10^4 years for the Earth given above. To account for the Earth's field it is necessary to find a flow $\mathbf{u}$ such that $\mathbf{B}$ does not vanish as $t \to \infty$. The main emphasis in theoretical work until the present time has been the search for steady dynamos. Then, in dimensionless units based on a typical dimension L of V, and a typical velocity U within it, equation (1) becomes simply

$$0 = R\operatorname{curl}(\mathbf{u} \times \mathbf{B}) + \nabla^2\mathbf{B} \qquad \text{in } V \qquad (5)$$

where R is the magnetic Reynolds number:

$$R \equiv \frac{UL}{\eta} \equiv \mu\sigma UL \qquad (6)$$

Given $\mathbf{u}$, equations (2) to (5) pose an eigenvalue problem for R (see Roberts[2] or our review article[3]).

The mathematical problem formulated so far may be described as 'given $\mathbf{u}$, find $\mathbf{B}$'. Such mathematical models are referred to as *kinematic dynamos*. Of more relevance are the *hydromagnetic dynamos*, which we may describe as 'given $\mathbf{f}$ (a body force driving the flow), find $\mathbf{u}$ and $\mathbf{B}$'. The hydromagnetic dynamo does not suffer from the obvious disadvantage of the kinematic dynamo, viz. that, if we find a flow $\mathbf{u}$ which maintains a field $\mathbf{B}$, according to (1) to (4), we should, by increasing its overall speed slightly, obtain a field which grows without bound. In practice, of course, the field would only increase until its Lorentz force became so potent that it modified and reduced (Lenz's law) the flow, and therefore diminished its electromotive effects. And the details of this process falls within the province of hydromagnetic dynamo theory. Another significant distinction between the models arises from the difference in the nature of the boundary layer at the surface Σ of V. In the hydromagnetic dynamo this layer is essentially the union of a Hartmann layer (present at a fluid–insulator interface in a non-rotating flow at large Hartmann numbers) to an Ekman layer (present at the boundary of a non-magnetic flow at large rotation rates). Both these types of boundary layer decisively *control* the flow deep within[2] V. In the kinematic dynamo, on the other hand, the boundary layer is essentially electromagnetic (i.e. it is akin to the familiar 'skin effect' on a solid conduction in an oscillating field). This is a *passive* layer, i.e. the flow deep in V can be determined without reference to it, and the boundary layer can be subsequently matched to it without affecting it, in the first approximation.

Although the hydromagnetic dynamo is far more realistic than the kinematic dynamo, it does pose, for a given $\mathbf{f}$, a *non*-linear problem for $\mathbf{u}$ and $\mathbf{B}$, the non-linearities arising from the $\mathbf{u} \times \mathbf{B}$ term of the induction equation and the Lorentz force $\mathbf{j} \times \mathbf{B}$ of the equation of motion. Moreover, the kinematic dynamo problem, which merely poses a linear problem for **B** given **u**, is far from trivial. It is not surprising, then, that the study of hydromagnetic dynamos is in its infancy.

The difficulties of the kinematic dynamo arise from the fact that the values of R required for self-excitation are $O(1)$. Thus the convergence of a perturbation expansion of the solution of (5) in powers of R will be poor or non-existent. In fact, so far all successful approaches to the dynamo problem have depended on the (often artificial) introduction of a small parameter in terms of which a successful perturbation expansion can be developed. The difficulties of doing otherwise are, we feel, well illustrated by the present chapter, in which we continue a search, initiated by Bullard and Gellman[4], for a dynamo in which **u** is a simple steady field whose *only* length scale is that of V. The field **B** is developed as a series expansion the higher terms being derived from the lower by interaction with **u** via the first term on the right of equation (5). The series is truncated arbitrarily after, say, N terms and the corresponding approximation (R_N) to the eigenvalue R of equation (5) is found. It is hoped that, as N is increased, R_N will give a convincing appearance of converging to a definite limit R. Bullard and Gellman's[4] results did not give any indication of such a convergence. We have studied Bullard and Gellman's levels of truncation in more detail, and have increased N beyond their values, and have also found no evidence that R_N converges; indeed it appears to increase remorselessly with N. To establish convergence, if it does exist, we would have to resort to machines beyond the capacity of those available to us at the present time, or to much more elaborate methods.

Spherical Dynamos

In this section we describe briefly the method used by Bullard and Gellman[4] to systematize spherical homogeneous dynamos. By equation (2), we may represent **B** by the sum of toroidal and poloidal vectors:

$$\mathbf{B} = \operatorname{curl} T\mathbf{r} + \operatorname{curl}^2 S\mathbf{r} \tag{7}$$

where T and S are, respectively, the toroidal and poloidal defining scalars, and **r** is the radius vector from the centre of the sphere O. Furthermore, we divide T and S into their spherical harmonic components:

$$rT = \sum_{n=1}^{\infty} \sum_{m=0}^{n} \{T_n^{mc}(r) \cos m\phi + T_n^{ms}(r) \sin m\phi\} P_n^m(\cos\theta) \tag{8}$$

$$rS = \sum_{n=1}^{\infty} \sum_{m=0}^{n} \{S_n^{mc}(r) \cos m\phi + S_n^{ms}(r) \sin m\phi\} P_n^m(\cos\theta) \tag{9}$$

Here (r, θ, ϕ) are spherical polar coordinates with origin O, Σ being the surface $r = 1$. Following Bullard and Gellman[4] we adopt the Ferrer normalization of the associated Legendre function P_n^m. Equations (3) and (4) are equivalent to requiring solutions of (5) to obey

$$T_n^m = \frac{dS_n^{mc,s}}{dr} + nS_n^{mc,s} = 0 \quad \text{on } r = 1 \tag{10}$$

Treating the conducting medium as being incompressible, we may introduce expansions like (7) to (9) for **u**; in these s, t, $s_n^{mc,s}(r)$ and $t_n^{mc,s}(r)$ will replace S, T, $S_n^{mc,s}(r)$ and $T_n^{mc,s}(r)$, respectively.

Now, using (7), we operate on (5) by **r**. and **r**. curl, thus reducing the vector equation to two scalar equations in T and S:

$$L^2 \nabla^2 S + R\mathbf{r} \cdot \text{curl}\{\text{curl}^2 \, s\mathbf{r} \times \text{curl}^2 \, S\mathbf{r}$$
$$+ \text{curl}\, t\mathbf{r} \times \text{curl}^2 \, S\mathbf{r} + \text{curl}^2 \, s\mathbf{r} \times \text{curl}\, T\mathbf{r}\} = 0 \qquad (11)$$

$$L^2 \nabla^2 T + R\mathbf{r} \cdot \text{curl}^2 \{\text{curl}^2 \, s\mathbf{r} \times \text{curl}^2 \, S\mathbf{r}$$
$$+ \text{curl}\, t\mathbf{r} \times \text{curl}^2 \, S\mathbf{r} + \text{curl}^2 \, s\mathbf{r} \times \text{curl}\, T\mathbf{r}$$
$$+ \text{curl}\, t\mathbf{r} \times \text{curl}\, T\mathbf{r}\} = 0 \qquad (12)$$

where L^2 is the operator

$$L^2 = \left(x_i \frac{\partial}{\partial x_i}\right)^2 + x_i \frac{\partial}{\partial x_i} - x^2 \nabla^2$$

$$= -\left\{\frac{1}{\sin\theta} \frac{\partial}{\partial\theta}\left(\sin\theta \frac{\partial}{\partial\theta}\right) + \frac{1}{\sin^2\theta} \frac{\partial^2}{\partial\phi^2}\right\} \qquad (13)$$

By introducing the expansions (8) and (9) into (11) and (12), multiplying by Y, the surface harmonic $P_n^m(\cos\theta)\cos m\phi$ (or $\sin m\phi$), and integrating over the surface of the sphere with radius r and centre O, we may reduce the partial differential equations to an infinite set of ordinary differential equations. We write these as

$$r^2 \frac{d^2 S_\gamma}{dr^2} - \gamma(\gamma+1)S_\gamma = R\sum_{\alpha\beta}\{(s_\alpha S_\beta S_\gamma) + (t_\alpha S_\beta S_\gamma) + (s_\alpha T_\beta S_\gamma)\} \qquad (14)$$

$$r^2 \frac{d^2 T_\gamma}{dr^2} - \gamma(\gamma+1)T_\gamma = R\sum_{\alpha\beta}\{(s_\alpha S_\beta T_\gamma) + (t_\alpha S_\beta T_\gamma) + (s_\alpha T_\beta T_\gamma) + (t_\alpha T_\beta T_\gamma)\} \qquad (15)$$

We have here introduced the abbreviated notation T_γ and S_γ for $T_n^{mc,s}$ and $S_n^{mc,s}$, in terms of which the boundary conditions (10) may be rewritten as

$$T_\gamma = \frac{dS_\gamma}{dr} + \gamma S_\gamma = 0 \qquad \text{on } r = 1 \qquad (16)$$

The terms on the right of (14) and (15) arise from the induced electric field $\mathbf{u} \times \mathbf{B}$. Thus, for example, $(s_\alpha S_\beta S_\gamma)$ represents the interaction of an s_α motion with an S_β field to produce an S_γ field; it arises from the term $\mathbf{r}.\text{curl}(\text{curl}^2 \, s\mathbf{r} \times \text{curl}^2 \, S\mathbf{r})$ on the right of (11). These 'interaction terms' have been evaluated* (Bullard and Gellman[4],

---

* Since $\int \mathbf{A}.\text{curl}(\mathbf{u} \times \mathbf{B})\,dV = 0$, when $\mathbf{A}$ is the vector potential of $\mathbf{B}$, it is possible to show that

$$\sum_{\beta,\gamma} N_\gamma F_{\alpha\beta\gamma} = 0$$

where
$$F_{\alpha\beta\gamma} = (t_\alpha S_\beta S_\gamma)T_\gamma + (t_\alpha T_\beta T_\gamma)S_\gamma$$

or
$$F_{\alpha\beta\gamma} = (s_\alpha T_\beta S_\gamma)T_\gamma$$

Also
$$\sum_{\beta,\gamma} N_\gamma \int_0^1 \frac{dr}{r^2} F_{\alpha\beta\gamma} = 0$$

where
$$F_{\alpha\beta\gamma} = (s_\alpha S_\beta S_\gamma)T_\gamma + (s_\alpha T_\beta T_\gamma)S_\gamma$$

or
$$F_{\alpha\beta\gamma} = (t_\alpha S_\beta T_\gamma)S_\gamma \quad \text{or} \quad F_{\alpha\beta\gamma} = (s_\alpha S_\beta T_\gamma)S_\gamma$$

(Here $N_\gamma = \gamma(\gamma+1)K_{0\gamma\gamma}$ is a normalization constant.) These results are occasionally useful in checking the accuracy of a truncated system of equations such as (14) and (15). In the present application, however, for any given specification of γ, T_γ and S_γ are not both present in the solution. They are therefore useless in checking the accuracy of the equations shown in table 1.

TABLE 1 Equations for interaction terms

$$r^2\ddot{S}_1 - 2S_1 = R\left\{-\frac{216}{5}Q_s T_2^{2s} + \frac{432}{7}(Q_s \dot{S}_3^{2c} + \dot{Q}_s S_3^{2c})\right\}$$

$$r^2\ddot{T}_2 - 6T_2 = R\left[-\frac{2}{3}\left(\dot{Q}_T - \frac{2Q_T}{r}\right)S_1 + \frac{12}{7}\left(\dot{Q}_T - \frac{2Q_T}{r}\right)S_3\right.$$

$$-\frac{72}{7}\left\{Q_s \dot{T}_2^{2c} + 2\left(\dot{Q}_s - \frac{Q_s}{r}\right)T_2^{2c}\right\}$$

$$+\frac{360}{7}\left\{Q_s \ddot{S}_3^{2s} + 2\left(\dot{Q}_s - \frac{Q_s}{r}\right)\dot{S}_3^{2s} + 2\left(\ddot{Q}_s - \frac{2\dot{Q}_s}{r}\right)S_3^{2s}\right\}$$

$$\left.+\frac{600}{7}\left\{Q_s \dot{T}_4^{2c} + 2\left(\dot{Q}_s - \frac{Q_s}{r}\right)T_4^{2c}\right\}\right]$$

$$r^2\ddot{S}_3 - 12S_3 = R\left\{\frac{36}{5}Q_s T_2^{2s} - 12(3Q_s \dot{S}_3^{2c} - 2\dot{Q}_s S_3^{2c}) - 60Q_s T_4^{2s}\right.$$

$$\left.+\frac{420}{11}(3Q_s \dot{S}_5^{2c} + 5\dot{Q}_s S_5^{2c})\right\}$$

$$r^2\ddot{T}_4 - 20T_4 = R\left[-\frac{12}{7}\left(\dot{Q}_T - \frac{2Q_T}{r}\right)S_3 + \frac{36}{35}\left\{3Q_s \dot{T}_2^{2c} - \left(\dot{Q}_s + \frac{6Q_s}{r}\right)T_2^{2c}\right\}\right.$$

$$-\frac{36}{7}\left\{3Q_s \ddot{S}_3^{2s} - \left(\dot{Q}_s + \frac{6Q_s}{r}\right)\dot{S}_3^{2s} + 6\left(\ddot{Q}_s - \frac{2\dot{Q}_s}{r}\right)S_3^{2s}\right\}$$

$$-\frac{324}{77}\left\{17Q_s \dot{T}_4^{2c} + \left(27\dot{Q}_s - \frac{34Q_s}{r}\right)T_4^{2c}\right\} + \frac{30}{11}\left(\dot{Q}_T - \frac{2Q_T}{r}\right)S_5$$

$$\left.+\frac{252}{11}\left\{3Q_s \ddot{S}_5^{2s} + 2\left(4\dot{Q}_s - \frac{3Q_s}{r}\right)\dot{S}_5^{2s} + 15\left(\ddot{Q}_s - \frac{2\dot{Q}_s}{r}\right)S_5^{2s}\right\}\right]$$

$$r^2\ddot{S}_5 - 30S_5 = R\left\{\frac{24}{7}(3Q_s \dot{S}_3^{2c} - 4\dot{Q}_s S_3^{2c}) + 24Q_s T_4^{2s}\right.$$

$$\left.-\frac{168}{13}(9Q_s \dot{S}_5^{2c} - 5\dot{Q}_s S_5^{2c})\right\}$$

$$r^2\ddot{T}_2^{2s} - 6T_2^{2s} = R\left[-\frac{2}{3}\left\{3Q_s \ddot{S}_1 + \left(\dot{Q}_s - \frac{6Q_s}{r}\right)\dot{S}_1 + \left(\ddot{Q}_s - \frac{2\dot{Q}_s}{r}\right)S_1\right\}\right.$$

$$+\frac{6}{7}\left\{Q_s \ddot{S}_3 + 2\left(\dot{Q}_s - \frac{Q_s}{r}\right)\dot{S}_3 + 2\left(\ddot{Q}_s - \frac{2\dot{Q}_s}{r}\right)S_3\right\}$$

$$-2Q_T T_2^{2c} + \frac{20}{7}\left(\dot{Q}_T - \frac{2Q_T}{r}\right)S_3^{2s}$$

$$\left.+400\left\{Q_s \dot{T}_4^{4s} + 2\left(\dot{Q}_s - \frac{Q_s}{r}\right)T_4^{4s}\right\}\right]$$

TABLE 1 (contd.)

$$r^2\ddot{T}_2^{2c} - 6T_2^{2c} = R\left[-\frac{6}{7}\left\{Q_s\dot{T}_2 + 2\left(\dot{Q}_s - \frac{Q_s}{r}\right)T_2\right\} + \frac{10}{21}\left\{Q_s\dot{T}_4 + 2\left(\dot{Q}_s - \frac{Q_s}{r}\right)T_4\right\}\right.$$

$$+ 2Q_T T_2^{2s} + \frac{20}{7}\left(\dot{Q}_T - \frac{2Q_T}{r}\right)S_3^{2c}$$

$$\left.+ 400\left\{Q_s\dot{T}_4^{4c} + 2\left(\dot{Q}_s - \frac{Q_s}{r}\right)T_4^{4c}\right\}\right]$$

$$r^2\ddot{S}_3^{2c} - 12S_3^{2c} = R\left\{\frac{2}{15}(3Q_s\dot{S}_1 - 2\dot{Q}_s S_1) - \frac{1}{5}(3Q_s\dot{S}_3 - 2\dot{Q}_s S_3) + 2Q_T S_3^{2s}\right.$$

$$- 56Q_s T_4^{4s} + \frac{1}{11}(3Q_s\dot{S}_5 + 5\dot{Q}_s S_5)$$

$$\left.+ \frac{1512}{11}(3Q_s\dot{S}_5^{4c} + 5\dot{Q}_s S_5^{4c})\right\}$$

$$r^2\ddot{S}_3^{2s} - 12S_3^{2s} = R\left\{\frac{3}{5}Q_s T_2 - \frac{1}{3}Q_s T_4 - 2Q_T S_3^{2c} + 56Q_s T_4^{4c}\right.$$

$$\left.+ \frac{1512}{11}(3Q_s\dot{S}_5^{4s} + 5\dot{Q}_s S_5^{4s})\right\}$$

$$r^2\ddot{T}_4^{2s} - 20T_4^{2s} = R\left[-\frac{3}{35}\left\{3Q_s\ddot{S}_3 - \left(\dot{Q}_s + \frac{6Q_s}{r}\right)\dot{S}_3 + 6\left(\ddot{Q}_s - \frac{2\dot{Q}_s}{r}\right)S_3\right\}\right.$$

$$+ \frac{3}{55}\left\{3Q_s\ddot{S}_5 + 2\left(4\dot{Q}_s - \frac{3Q_s}{r}\right)\dot{S}_5 + 15\left(\ddot{Q}_s - \frac{2\dot{Q}_s}{r}\right)S_5\right\}$$

$$- \frac{6}{7}\left(\dot{Q}_T - \frac{2Q_T}{r}\right)S_3^{2s} - 2Q_T T_4^{2c}$$

$$- \frac{72}{55}\left\{17Q_s\dot{T}_4^{4s} + \left(27\dot{Q}_s - \frac{34Q_s}{r}\right)T_4^{4s}\right\} + \frac{42}{11}\left(\dot{Q}_T - \frac{2Q_T}{r}\right)S_5^{2s}$$

$$\left.- \frac{1512}{55}\left\{3Q_s\ddot{S}_5^{4c} + 2\left(4\dot{Q}_s - \frac{3Q_s}{r}\right)\dot{S}_5^{4c} + 15\left(\ddot{Q}_s - \frac{2\dot{Q}_s}{r}\right)S_5^{4c}\right\}\right]$$

$$r^2\ddot{T}_4^{2c} - 20T_4^{2c} = R\left[\frac{3}{35}\left\{3Q_s\dot{T}_2 - \left(\dot{Q}_s + \frac{6Q_s}{r}\right)T_2\right\} - \frac{18}{770}\left\{17Q_s\dot{T}_4 + \left(27\dot{Q}_s - \frac{34Q_s}{r}\right)T_4\right\}\right.$$

$$- \frac{6}{7}\left(\dot{Q}_T - \frac{2Q_T}{r}\right)S_3^{2c} + 2Q_T T_4^{2s}$$

$$- \frac{72}{55}\left\{17Q_s\dot{T}_4^{4c} + \left(27\dot{Q}_s - \frac{34Q_s}{r}\right)T_4^{4c}\right\} + \frac{42}{11}\left(\dot{Q}_T - \frac{2Q_T}{r}\right)S_5^{2c}$$

$$\left.+ \frac{1512}{55}\left\{3Q_s\ddot{S}_5^{4s} + 2\left(4\dot{Q}_s - \frac{3Q_s}{r}\right)\dot{S}_5^{4s} + 15\left(\ddot{Q}_s - \frac{2\dot{Q}_s}{r}\right)S_5^{4s}\right\}\right]$$

TABLE 1 (contd.)

$$r^2\ddot{T}_4^{4s} - 20T_4^{4s} = R\left[\frac{3}{70}\left\{3Q_s\dot{T}_2^{2s} - \left(\dot{Q}_s + \frac{6Q_s}{r}\right)T_2^{2s}\right\}\right.$$

$$-\frac{3}{70}\left\{3Q_s\ddot{S}_3^{2c} - \left(\dot{Q}_s + \frac{6Q_s}{r}\right)\dot{S}_3^{2c} + 6\left(\ddot{Q}_s - \frac{2\dot{Q}_s}{r}\right)S_3^{2c}\right\}$$

$$-\frac{9}{770}\left\{17Q_s\dot{T}_4^{2s} + \left(27\dot{Q}_s - \frac{34Q_s}{r}\right)T_4^{2s}\right\} - 4Q_T T_4^{4c}$$

$$+\frac{3}{110}\left\{3Q_s\ddot{S}_5^{2c} + 2\left(4\dot{Q}_s - \frac{3Q_s}{r}\right)\dot{S}_5^{2c} + 15\left(\ddot{Q}_s - \frac{2\dot{Q}_s}{r}\right)S_5^{2c}\right\}$$

$$\left.+\frac{54}{11}\left(\dot{Q}_T - \frac{2Q_T}{r}\right)S_5^{4s}\right]$$

$$r^2\ddot{T}_4^{4c} - 20T_4^{4c} = R\left[\frac{3}{70}\left\{3Q_s\dot{T}_2^{2c} - \left(\dot{Q}_s + \frac{6Q_s}{r}\right)T_2^{2c}\right\}\right.$$

$$+\frac{3}{70}\left\{3Q_s\ddot{S}_3^{2s} - \left(\dot{Q}_s + \frac{6Q_s}{r}\right)\dot{S}_3^{2s} + 6\left(\ddot{Q}_s - \frac{2\dot{Q}_s}{r}\right)S_3^{2s}\right\}$$

$$-\frac{9}{770}\left\{17Q_s\dot{T}_4^{2c} + \left(27\dot{Q}_s - \frac{34Q_s}{r}\right)T_4^{2c}\right\} + 4Q_T T_4^{4s}$$

$$-\frac{3}{110}\left\{3Q_s\ddot{S}_5^{2s} + 2\left(4\dot{Q}_s - \frac{3Q_s}{r}\right)\dot{S}_5^{2s} + 15\left(\ddot{Q}_s - \frac{2\dot{Q}_s}{r}\right)S_5^{2s}\right\}$$

$$\left.+\frac{54}{11}\left(\dot{Q}_T - \frac{2Q_T}{r}\right)S_5^{4c}\right]$$

$$r^2\ddot{S}_5^{2c} - 30S_5^{2c} = R\left\{\frac{2}{35}(3Q_s\dot{S}_3 - 4\dot{Q}_s S_3) - \frac{2}{65}(9Q_s\dot{S}_5 - 5\dot{Q}_s S_5)\right.$$

$$\left.+\frac{112}{5}Q_s T_4^{4s} + 2Q_T S_5^{2s} - \frac{432}{65}(9Q_s\dot{S}_5^{4c} - 5\dot{Q}_s S_5^{4c})\right\}$$

$$r^2\ddot{S}_5^{2s} - 30S_5^{2s} = R\left\{\frac{2}{15}Q_s T_4 - \frac{112}{5}Q_s T_4^{4c} - 2Q_T S_5^{2c}\right.$$

$$\left.-\frac{432}{65}(9Q_s\dot{S}_5^{4s} - 5\dot{Q}_s S_5^{4s})\right\}$$

$$r^2\ddot{S}_5^{4c} - 30S_5^{4c} = R\left\{\frac{1}{35}(3Q_s\dot{S}_3^{2c} - 4\dot{Q}_s S_3^{2c}) - \frac{1}{15}Q_s T_4^{2s}\right.$$

$$\left.-\frac{1}{65}(9Q_s\dot{S}_5^{2c} - 5\dot{Q}_s S_5^{2c}) + 4Q_T S_5^{4s}\right\}$$

$$r^2\ddot{S}_5^{4s} - 30S_5^{4s} = R\left\{\frac{1}{35}(3Q_s\dot{S}_3^{2s} - 4\dot{Q}_s S_3^{2s}) + \frac{1}{15}Q_s T_4^{2c}\right.$$

$$\left.-\frac{1}{65}(9Q_s\dot{S}_5^{2s} - 5\dot{Q}_s S_5^{2s}) - 4Q_T S_5^{4c}\right\}$$

p. 225) in terms of 'coupling integrals' of two distinct types, which are often called Gaunt and Elsasser integrals:

$$G_{\alpha\beta\gamma} = \int_0^{2\pi} \int_0^{\pi} Y_\alpha Y_\beta Y_\gamma \sin\theta \, d\theta \, d\phi \tag{17}$$

$$E_{\alpha\beta\gamma} = \int_0^{2\pi} \int_0^{\pi} Y_\alpha \left(\frac{\partial Y_\beta}{\partial \theta} \frac{\partial Y_\gamma}{\partial \phi} - \frac{\partial Y_\gamma}{\partial \theta} \frac{\partial Y_\beta}{\partial \phi} \right) d\theta \, d\phi \tag{18}$$

They are studied by Bullard and Gellman and also in the appendix of this chapter.

The presence of the interaction terms in (14) and (15) implies that, if we start with any field harmonic and any fluid velocity, an infinite* sequence of other field harmonics will be excited. If, following Bullard and Gellman[4], we restrict ourselves to flows which are linear combinations of $\mathbf{t}_1$ and $\mathbf{s}_2^{2c}$ harmonics, and start with the S_1 field harmonic, we find that the harmonics shown in figure 1 are excited. In this interaction diagram,

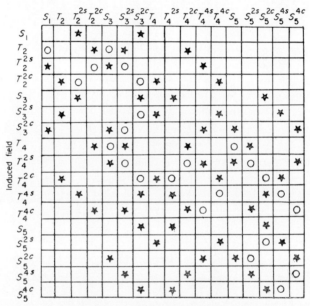

FIGURE 1 The interaction diagram

which should extend to infinity, the rows refer to the indu*ced* field (γ) and the columns to the indu*cing* field (β); a star signifies that the excitation is through the s_2^{2c} flow, a circle that it is through the t_1 flow.

Numerical Results and Inconclusions

As we have mentioned above, we restrict ourselves to flow of the $\mathbf{t}_1 + \mathbf{s}_2^{2c}$ variety. More particularly, following a case considered by Bullard and Gellman[4], we take

$$Q_T t_1 = \varepsilon r^2 (1-r) \qquad Q_s s_2^{2c} = r^3 (1-r)^2 \tag{19}$$

* The only exception to this rule appears to be when the fluid rotates as a rigid body. This 'flow' is, of course, incapable of maintaining a dynamo field.

and generally suppose that
$$\varepsilon = 5 \qquad (20)$$
Since it is clearly impossible to retain the infinite train of harmonics excited by (19) from S_1 we keep only the first s, the remainder being set zero. A value R_s of R is sought for the resulting truncated set of equations. The equations relevant to the fifth harmonic are shown* in table 1. Bullard and Gellman considered only three levels of truncation: $s = 4$, $s = 7$, and $s = 12$. This corresponds to keeping all harmonics up to and including P_2^2, P_3^2, and P_4^4, respectively. We have truncated at the levels $s = 4, 5, 7, 8, 10, 12, 13, 15$, and 17 corresponding to P_2^2, P_3, P_3^2, P_4, P_4^2, P_4^4, P_5, P_5^2, and P_5^4, respectively. We have, therefore, obtained greater information than they about the convergence of R_s, or lack of it!

We note that a reversal of the velocities $\mathbf{t}_1$ and $\mathbf{s}_2^{2c}$ is equivalent to a symmetry operation, so that if R is an eigenvalue so is $-R$. This is also true of the truncated equations as we may see by reversing the signs of $S_1, S_3, S_5, T_2^{2c}, S_3^{2s}, T_4^{2c}, T_4^{4s}, S_5^{2s}, S_5^{4c}$, and R in the equations shown in table 1. This provides a useful check on the numerical work which we made use of at several levels of truncation. A similar check that we used was that R remains unaltered if we replace t_1 by $-t_1$, i.e. if we reverse the sign of ε.

In solving the truncated set of equations, we have made use of Chebyshev collocation[5]. In this, the solution for each defining scalars S and T is represented by a finite sum of $m+2$ Chebyshev polynomials. Derivatives are estimated by applying the recurrence relations for these functions to the series, and the coefficient in the series are evaluated by insisting that the differential equations are satisfied exactly at m collocation points within 0, 1, and that the boundary conditions at $r = 0$ (boundedness) and $r = 1$ (cf. (16)) are obeyed. This results in a system of $s \times (m+2)$ linear algebraic equations for R_s, the smallest positive eigenvalue. For each s, it is clear, by examining the coefficients obtained, whether enough or too many terms have been included in each series, i.e. whether m is too small or unnecessarily large. The results are shown in table 2. By examining the behaviour of R_s for fixed s as m is increased, we obtain a

TABLE 2 The values of R for various s and m

s \ m	7	8	9	10	11	12	13	14	15	16
4		69.4589						66.445		66.46213
5		65.525					62.877	62.90687		62.91218
7		83.376	82.404	83.43368			83.03			83.0908
8	80.164	69.14	69.7	70.67		70.14963				70.27900
10		72.40629		72.85449				72.6929		72.72458
12		75.90126		76.02077	75.7			75.9		75.948
13		63.0168					62.997			
15		112.53779			114.64	120.4				
17		139.48704		143.2						

good idea of the convergence of the Chebyshev method. We see, for example, that, for $s = 7$, R is 82.38 for $m = 8$, 83.03 for $m = 13$, and 83.09 for $m = 16$, indicating a 1%

* Only harmonics up to the fourth are considered by Bullard and Gellman[4]. Bullard and Gellman's table 7 is, however, very useful in going further. Two minor omissions may be noted: on p. 264 of Bullard and Gellman[4], a Q_s should be included on the right-hand side of the penultimate line: on the sixth line of p. 265, the coefficient of $\dot{T}_\beta^m$ should be multiplied by Q_s.

accuracy at $m = 8$. This may be compared with the results of Bullard and Gellman[4] where finite difference methods were employed. Here R was found to be 68·8 for a grid of 0·1, and 83·9 for a grid of 0·05; even at this stage, it was not clear that the eigenvalue had settled down to give the correct eigenvalue of the equations. In what follows we shall assume that the required eigenvalue R_s is well given, for any s, by the largest m entry shown in table 2. Before leaving this point, however, we should record one distressing feature of any numerical method used to determine R_s. If (as appears to be the case in the present work) R_s increases with s, the corresponding eigensolution is increasingly of boundary layer type and requires for its resolution an increasingly fine grid (in a finite difference method) or a larger value of m (in our collocation method). The size of the corresponding matrix ($s \times (m+2)$ in our work) is therefore greatly increased.

Our next objective must be to demonstrate a convergence of R_s as s, the level of truncation, is increased. It is clear, from table 2, that our results are no more convincing here than those of Bullard and Gellman. The change in R from 63 for $s = 13$ to 120 for $s = 15$ and 143 for $s = 17$ is particularly discouraging. It may be conjectured that the closeness of the eigenvalues for eight, ten, and twelve radial functions reflects the fact that it is easy to produce and maintain a toroidal field by the interaction of the t_1 shear and the poloidal field. Indeed as ε (see (19)) becomes large one would expect these eigenvalues to become even closer. In the same light the jump in R in increasing the number of radial functions from thirteen to fifteen probably arises from the difficulty in producing a poloidal field from a toroidal field.

At the final level of approximation that we considered, we found that $S_5/S_1 \sim 0\cdot1$. This once again emphasizes that convergence (or otherwise!)) is very slow, and that neither Bullard and Gellman nor ourselves are close to demonstrating numerically that their model works or not.

Acknowledgements

We are grateful to Miss M. F. Tabrett for assistance in the programming and development of the work. We thank the Science Research Council, both for a special research grant (No. B/SR/150) and for direct computational assistance at the Atlas Computer Laboratory, M. E. Claringbold being a most helpful liaison officer between Chilton and Newcastle. Travel between these areas was supported by the research fund of our University. We thank the director of our own computing laboratory Professor E. S. Page, for his support of this project.

Appendix. The Gaunt and Elsasser integrals

by S. SCOTT

Introduction; definitions

We are concerned with the properties of the Gaunt and Elsasser integrals defined, respectively, by

$$G_{lmn}^{uvw} = \int_{-1}^{1} P_l^u(\mu) P_m^v(\mu) P_n^w(\mu) \, d\mu \tag{A1}$$

$$E_{lmn}^{uvw} = \int_{-1}^{1} P_l^u(\mu) \left\{ v P_m^v(\mu) \frac{dP_n^w(\mu)}{d\mu} - w P_n^w(\mu) \frac{dP_m^v(\mu)}{d\mu} \right\} d\mu \tag{A2}$$

The Bullard–Gellman dynamo

where $P_l^u(\mu)$ denotes the associated Legendre function of degree l, order u, and argument μ; l, m, and n are positive and

$$u+v+w = 0 \tag{A3}$$

These integrals arise in geomagnetic studies[6], in meteorological work, and in molecular structure calculations[7].

Many normalizations of P_l^u have been proffered. (Fletcher and coworkers[8] list six different definitions!) We shall confine our attention to the Ferrer form:

$$P_n^m(\mu) = \begin{cases} (1-\mu^2)^{m/2} \dfrac{1}{2^n n!} \dfrac{d^{n+m}}{d\mu^{n+m}} (\mu^2-1)^n & (m \geq 0) \\ \dfrac{(n-m)!}{(n+m)!} P_n^{|m|}(\mu) & (m \leq 0) \end{cases} \tag{A4}$$

With this definition, we have

$$\int_{-1}^{1} P_l^m(\mu) P_n^m(\mu) \, d\mu = \frac{2}{2n+1} \frac{(n+m)!}{(n-m)!} \delta_{ln} \tag{A5}$$

where δ_{ij} denotes the Kronecker delta symbol.

Analytic properties; recurrence relations

Clearly G and E are symmetric and antisymmetric, respectively, under the simultaneous interchange $m \leftrightarrow n$, $v \leftrightarrow w$:

$$G_{lmn}^{uvw} = G_{lnm}^{uwv}, \qquad E_{lmn}^{uvw} = -E_{lnm}^{uwv} \tag{A6}$$

Under the simultaneous permutations $l \to m \to n \to l$ and $u \to v \to w \to u$, both G and (as can be shown by integration by parts) E are unchanged:

$$G_{lmn}^{uvw} = G_{mnl}^{vwu} = G_{nlm}^{wuv}, \qquad E_{lmn}^{uvw} = E_{mnl}^{vwu} = E_{nlm}^{wuv} \tag{A7}$$

We may, without loss of generality, suppose now that $v \geq 0$ and $w \geq 0$, since two of u, v, and w must be of the same sign (or zero) by (A3) and, if they are both negative, we can reverse the sign of all three and by (A4) the integral is changed only by a constant factor. A permutation, using (A7), then reduces the integral to a form in which $v \geq 0$ and $w \geq 0$.

Let us define g and e by

$$g_{mn}^{vw} = P_m^v P_n^w, \qquad e_{mn}^{vw} = v P_m^v \frac{dP_n^w}{d\mu} - w P_n^w \frac{dP_m^v}{d\mu} \tag{A8}$$

Relations between these quantities are now given. Provided $v+w$ is conserved in these relations, we can, by multiplying through by P_l^u and integrating over μ from -1 to 1, obtain relationships between the integrals G and E themselves. As has been shown by Barnett[9], the g quantities obey the relations

$$\frac{1}{2n+1}\{(n-w+1)g_{m\,n+1}^{v\,w} + (n+w)g_{m\,n-1}^{v\,w}\}$$

$$-\frac{1}{2m+1}\{(m-v+1)g_{m+1\,n}^{v\,w} + (m+v)g_{m-1\,n}^{v\,w}\} = 0 \tag{A9}$$

$$\frac{1}{2n+1}\{g_{mn+1}^{v\ w+1}-g_{mn-1}^{v\ w+1}\} = \frac{1}{2m+1}\{g_{m+1\ n}^{v+1\ w}-g_{m-1\ n}^{v+1\ w}\}$$

$$= \frac{1}{2n+1}\{(n+w)(n+w-1)g_{mn-1}^{v\ w-1}-(n-w+1)(n-w+2)g_{mn+1}^{v\ w-1}\}$$

$$= \frac{1}{2m+1}\{(m+v)(m+v-1)g_{m-1\ n}^{v-1\ w}-(m-v+1)(m-v+2)g_{m+1\ n}^{v-1\ w}\} \quad (A10)$$

By using the standard recurrence formulae between Legendre functions, we can show that

$$e_{mn}^{vw} = \frac{1}{2(2m+1)}\{(m-v)(m-v+1)(m-v+2)g_{m+1\ n}^{v-1\ w+1}$$

$$+(m+v+1)(m+v)(m+v-1)g_{m-1\ n}^{v-1\ w+1}\}$$

$$-\frac{1}{2(2n+1)}\{(n-w)(n-w+1)(n-w+2)g_{m}^{v+1\ w-1}{}_{n+1}$$

$$+(n+w+1)(n+w)(n+w-1)g_{m}^{v+1\ w-1}{}_{n-1}\} \quad (A11)$$

$$(v+w)g_{mn}^{v\ w} = \frac{1}{2n+1}\{(n-w+1)e_{m\ n+1}^{v\ w}+(n+w)e_{m\ n-1}^{v\ w}\}$$

$$-\frac{1}{2m+1}\{(m-v+1)e_{m+1\ n}^{v\ w}+(m+v)e_{m-1\ n}^{v\ w}\} \quad (A12)$$

relations which are clearly consistent with the symmetry of g and the antisymmetry of e, and which provide a means of passing from the Gaunt integral to the Elsasser integral, and vice versa.

The relationships between the e's are more complex than (A9) and (A10); we may note, however, the following simple results:

$$\frac{1}{2n+1}\{e_{m\ n+1}^{v\ w}-e_{m\ n-1}^{v\ w}\} = vg_{mn}^{v\ w}-wg_{m}^{v+1\ w-1}{}_{n} \quad (A13)$$

$$\frac{1}{2n+1}\{(n+w)(n+w-1)e_{mn-1}^{v\ w-1}-(n-w+1)(n-w+2)e_{mn+1}^{v\ w-1}\}$$

$$= (w-1)(m+v)(m-v+1)g_{m}^{v-1\ w}{}_{n}-v(n+w)(n-w+1)g_{mn}^{v\ w-1} \quad (A14)$$

It can be shown[4, 10] that G_{lmn}^{uvw} vanishes unless its indices obey the following selection rules:

(Ga) $l+m+n$ is even

(Gb) l, m, n can form the sides of a triangle

Similarly E_{lmn}^{uvw} vanishes unless its indices obey the selection rules:

(Ea) $l+m+n$ is odd

(Eb) l, m, n can form the sides of a triangle

(Ec) no two superfixes are zero; no two suffix–superfix pairs are equal (e.g. $l = m, u = v$)

The Bullard–Gellman dynamo

The integrals may also vanish when the selection rules are satisfied; particular cases have been noted by Bird[10] and Gjellestad[11]. We have observed the following cases in addition:

$$G^{02\,-2}_{44\ \ 6}, \quad G^{14-5}_{25\ 6}, \quad G^{13-4}_{45\ 5}, \quad G^{12-3}_{53\ 6}, \quad G^{12-3}_{54\ 5}, \quad E^{0-33}_{5\ 64}, \quad \text{and} \quad E^{1-32}_{2\ 54}$$

Numerical results

In evaluating G and E it is convenient to consider non-negative superfixes only, reversing signs where necessary using (A4). Equation (A3) is now written

$$|u| \pm |v| \pm |w| = 0 \tag{A15}$$

It is convenient to divide the Gaunt integrals into two main categories

$$G^+(u, l, w, n, m) \quad \text{and} \quad G^-(u, l, w, n, m)$$

related to definition (A1) by

$$G^+(u, l, w, n, m) = \frac{(m+u+w)!}{(m-u-w)!} G^{u\ -u-w\ w}_{l\ \ m\ \ n}$$

$$= \int_{-1}^{1} P_l^u P_m^{w+u} P_n^w \, d\mu \tag{A16}$$

$$G^-(u, l, w, n, m) = \frac{(n+w)!}{(n-w)!} G^{u\ w-u\ -w}_{l\ m\ \ n}$$

$$= \int_{-1}^{1} P_l^u P_m^{w-u} P_n^w \, d\mu \tag{A17}$$

and similarly to divide the Elsasser integrals into two main categories

$$E^+(u, l, w, n, m) \quad \text{and} \quad E^-(u, l, w, n, m)$$

related to definition (A2) by

$$E^+(u, l, w, n, m) = \frac{(m+u+w)!}{(m-u-w)!} E^{u\ -u-w\ w}_{l\ \ m\ \ n}$$

$$= \int_{-1}^{1} P_m^{w+u}\left(wP_n^w \frac{dP_l^u}{d\mu} - uP_l^u \frac{dP_n^w}{d\mu}\right) d\mu \tag{A18}$$

$$E^-(u, l, w, n, m) = -\frac{(n+w)!}{(n-w)!} E^{u\ w-u\ -w}_{l\ m\ \ n}$$

$$= \int_{-1}^{1} P_m^{w-u}\left(wP_n^w \frac{dP_l^u}{d\mu} + uP_l^u \frac{dP_n^w}{d\mu}\right) d\mu \tag{A19}$$

where, in (A17) and (A19), $w \geq u$. The integrals are related by

$$G^-(u, l, w, n, m) = G^+(u, l, w-u, m, n) \tag{A20}$$

and

$$E^-(u, l, w, n, m) = E^+(u, l, w-u, m, n) \tag{A21}$$

Table 3

ELSASSER INTEGRALS
$E^+(U,L,W,N,M) = 4A/B$

U	L	W	N	M	A	B
0	1	1	1	1	1	3
0	1	1	2	2	3	5
0	1	1	3	3	6	7
0	1	1	4	4	10	9
0	1	1	5	5	15	11
0	1	2	2	2	24	5
0	1	2	3	3	120	7
0	1	2	4	4	40	1
0	1	2	5	5	840	11
0	1	3	3	3	1080	7
0	1	3	4	4	840	1
0	1	3	5	5	30240	11
0	1	4	4	4	8960	1
0	1	4	5	5	725760	11
0	1	5	5	5	9072000	11
0	2	1	1	2	3	5
0	2	1	2	1	3	5
0	2	1	2	3	36	35
0	2	1	3	2	36	35
0	2	1	3	4	10	7
0	2	1	4	3	10	7
0	2	1	4	5	20	11
0	2	1	5	4	20	11
0	2	1	5	6	315	143
0	2	2	2	3	72	7
0	2	2	3	2	72	7
0	2	2	3	4	240	7
0	2	2	4	3	240	7
0	2	2	4	5	840	11
0	2	2	5	4	840	11
0	2	2	5	6	20160	143
0	2	3	3	4	360	1
0	2	3	4	3	360	1
0	2	3	4	5	20160	11
0	2	3	5	4	20160	11
0	2	3	5	6	816480	143
0	2	4	4	5	241920	11
0	2	4	5	4	241920	11
0	2	4	5	6	21772800	143
0	2	5	5	6	299376000	143
0	3	1	1	3	6	7
0	3	1	2	2	36	35
0	3	1	2	4	10	7
0	3	1	3	1	6	7
0	3	1	3	3	12	7
0	3	1	3	5	150	77
0	3	1	4	2	10	7
0	3	1	4	4	180	77
0	3	1	4	6	350	143
0	3	1	5	3	150	77
0	3	1	5	5	420	143
0	3	1	5	7	420	143
0	3	2	2	2	−72	35
0	3	2	2	4	120	7
0	3	2	3	3	120	7
0	3	2	3	5	600	11
0	3	2	4	2	120	7
0	3	2	4	4	4680	77
0	3	2	4	6	16800	143
0	3	2	5	3	600	11
0	3	2	5	5	19320	143
0	3	2	5	7	30240	143
0	3	3	3	3	−720	7
0	3	3	3	5	7200	11
0	3	3	4	4	5040	11
0	3	3	4	6	453600	143
0	3	3	5	3	7200	11
0	3	3	5	5	40320	13
0	3	3	5	7	1360800	143
0	3	4	4	4	−80640	11
0	3	4	4	6	6048000	143
0	3	4	5	5	2177280	143
0	3	4	5	7	39916800	143
0	3	5	5	5	−108864000	143
0	3	5	5	7	598752000	143
0	4	1	1	4	10	9
0	4	1	2	3	10	7
0	4	1	2	5	20	11
0	4	1	3	2	10	7
0	4	1	3	4	180	77
0	4	1	3	6	350	143
0	4	1	4	1	10	9
0	4	1	4	3	180	77
0	4	1	4	5	450	143
0	4	1	4	7	3920	1287
0	4	1	5	2	20	11
0	4	1	5	4	450	143
0	4	1	5	6	560	143
0	4	1	5	8	8820	2431
0	4	2	2	3	−40	7
0	4	2	2	5	280	11
0	4	2	3	2	−40	7
0	4	2	3	4	1800	77
0	4	2	3	6	11200	143
0	4	2	4	3	1800	77
0	4	2	4	5	12600	143
0	4	2	4	7	23520	143
0	4	2	5	2	280	11
0	4	2	5	4	12600	143
0	4	2	5	6	28000	143
0	4	2	5	8	705600	2431
0	4	3	3	4	−3600	11
0	4	3	3	6	151200	143
0	4	3	4	3	−3600	11
0	4	3	4	5	50400	143
0	4	3	4	7	705600	143
0	4	3	5	4	50400	143
0	4	3	5	6	604800	143
0	4	3	5	8	3175200	221
0	4	4	4	5	−3628800	143
0	4	4	4	7	940800	13

Table 3 (contd.)

0	4	4	5	4	−3628800	143	1	1	2	5	5	−10080	11
0	4	4	5	6	−3628800	143	1	1	3	4	4	−2240	1
0	4	4	5	8	101606400	221	1	1	3	5	5	−181440	11
0	4	5	5	6	−399168000	143	1	1	4	5	5	−1814400	11
0	4	5	5	8	127008000	17	1	2	1	3	2	−216	35
0	5	1	1	5	15	11	1	2	1	3	4	−60	7
0	5	1	2	4	20	11	1	2	1	4	3	−20	1
0	5	1	2	6	315	143	1	2	1	4	5	−280	11
0	5	1	3	3	150	77	1	2	1	5	4	−480	11
0	5	1	3	5	420	143	1	2	1	5	6	−7560	143
0	5	1	3	7	420	143	1	2	2	2	3	432	7
0	5	1	4	2	20	11	1	2	2	3	4	120	1
0	5	1	4	4	450	143	1	2	2	4	3	−1080	7
0	5	1	4	6	560	143	1	2	2	4	5	0	1
0	5	1	4	8	8820	2431	1	2	2	5	4	−8400	11
0	5	1	5	1	15	11	1	2	2	5	6	−90720	143
0	5	1	5	3	420	143	1	2	3	3	4	2880	1
0	5	1	5	5	600	143	1	2	3	4	5	120960	11
0	5	1	5	7	11760	2431	1	2	3	5	4	−80640	11
0	5	1	5	9	198450	46189	1	2	3	5	6	2721600	143
0	5	2	2	4	−120	11	1	2	4	4	5	2419200	11
0	5	2	2	6	5040	143	1	2	4	5	6	16329600	13
0	5	2	3	3	−1200	77	1	2	5	5	6	326592000	13
0	5	2	3	5	4200	143	1	3	1	4	2	−80	7
0	5	2	3	7	15120	143	1	3	1	4	4	−1440	77
0	5	2	4	2	−120	11	1	3	1	4	6	−2800	143
0	5	2	4	4	3600	143							
0	5	2	4	6	16800	143							
0	5	2	4	8	529200	2431	1	3	1	5	3	−2700	77
0	5	2	5	3	4200	143	1	3	1	5	5	−7560	143
0	5	2	5	5	16800	143	1	3	1	5	7	−7560	143
0	5	2	5	7	635040	2431	1	3	2	2	4	240	1
0	5	2	5	9	1587600	4199	1	3	2	3	3	1440	7
0	5	3	3	3	1800	77	1	3	2	3	5	7200	11
0	5	3	3	5	−100800	143	1	3	2	4	4	4320	11
0	5	3	3	7	226800	143	1	3	2	4	6	151200	143
0	5	3	4	4	−12600	13	1	3	2	5	3	−3600	11
0	5	3	4	6	0	1	1	3	2	5	5	20160	143
0	5	3	4	8	1587600	221	1	3	2	5	7	151200	143
0	5	3	5	3	−100800	143	1	3	3	3	5	129600	11
0	5	3	5	5	−100800	143	1	3	3	4	4	120960	11
0	5	3	5	7	12700800	2431	1	3	3	4	6	7560000	143
							1	3	3	5	5	5806080	143
							1	3	3	5	7	19958400	143
0	5	3	5	9	85730400	4199	1	3	4	4	6	12096000	13
0	5	4	4	4	403200	143	1	3	4	5	5	130636800	143
0	5	4	4	6	−8467200	143	1	3	4	5	7	838252800	143
0	5	4	4	8	25401600	221	1	3	5	5	7	15567552000	143
0	5	4	5	5	−10886400	143	1	4	1	5	2	−200	11
0	5	4	5	7	−25401600	221	1	4	1	5	4	−4500	143
0	5	4	5	9	228614400	323	1	4	1	5	6	−5600	143
0	5	5	5	5	54432000	143	1	4	1	5	8	−88200	2431
0	5	5	5	7	−1524096000	221	1	4	2	2	3	−240	7
0	5	5	5	9	4000752000	323	1	4	2	2	5	6720	11
1	1	1	2	2	−12	5	1	4	2	3	4	7200	11
1	1	1	3	3	−60	7	1	4	2	3	6	252000	143
1	1	1	4	4	−20	1	1	4	2	4	3	36000	77
1	1	1	5	5	−420	11	1	4	2	4	5	252000	143
1	1	2	3	3	−360	7	1	4	2	4	7	470400	143
1	1	2	4	4	−280	1	1	4	2	5	4	126000	143

TABLE 3 (contd.)

1	4	2	5	6	403200	143
1	4	2	5	8	11642400	2431
1	4	3	3	4	-28800	11
1	4	3	3	6	4536000	143
1	4	3	4	5	4536000	143
1	4	3	4	7	1881600	13
1	4	3	5	4	4032000	143
1	4	3	5	6	21772800	143
1	4	3	5	8	88905600	221
1	4	4	4	5	-36288000	143
1	4	4	4	7	33868800	13
1	4	4	5	6	29030400	13
1	4	4	5	8	279417600	17
1	4	5	5	6	-4790016000	143
1	4	5	5	8	5334336000	17
1	5	2	2	4	-1680	11
1	5	2	2	6	181440	143
1	5	2	3	3	-9000	77
1	5	2	3	5	201600	143
1	5	2	3	7	529200	143
1	5	2	4	4	189000	143
1	5	2	4	6	604800	143
1	5	2	4	8	1587600	221
1	5	2	5	3	126000	143
1	5	2	5	5	504000	143
1	5	2	5	7	19051200	2431
1	5	2	5	9	47628000	4199
1	5	3	3	5	-1814400	143
1	5	3	3	7	907200	13
1	5	3	4	4	-1512000	143
1	5	3	4	6	8467200	143
1	5	3	4	8	69854400	221
1	5	3	5	5	9072000	143
1	5	3	5	7	76204800	221
1	5	3	5	9	285768000	323
1	5	4	4	6	-16934400	13
1	5	4	4	8	101606400	17
1	5	4	5	5	-163296000	143
1	5	4	5	7	609638400	221
1	5	4	5	9	12002256000	323
1	5	5	5	7	-3048192000	17
1	5	5	5	9	240045120000	323
2	2	2	3	4	-1920	1
2	2	2	4	5	-120960	11
2	2	2	5	4	13440	11
2	2	2	5	6	-5443200	143
2	2	3	4	5	-1209600	11
2	2	3	5	6	-119750400	143
2	2	4	5	6	-1437004800	143
2	3	2	4	4	-86400	11
2	3	2	4	6	-3024000	143
2	3	2	5	5	-5443200	143
2	3	2	5	7	-1209600	13
2	3	3	3	5	1296000	11
2	3	3	4	6	3024000	13
2	3	3	5	5	-72576000	143
2	3	3	5	7	-119750400	143
2	3	4	4	6	145152000	13
2	3	4	5	7	43545600	1
2	3	5	5	7	217945728000	143
2	4	2	5	4	-3024000	143
2	4	2	5	6	-9676800	143
2	4	2	5	8	-25401600	221
2	4	3	3	6	9072000	13
2	4	3	4	5	81648000	143
2	4	3	4	7	33868800	13
2	4	3	5	6	14515200	13
2	4	3	5	8	76204800	17
2	4	4	4	7	67737600	1
2	4	4	5	6	8622028800	143
2	4	4	5	8	6401203200	17
2	4	5	5	8	160030080000	17
2	5	3	3	5	-18144000	143
2	5	3	3	7	32659200	13
2	5	3	4	6	33868800	13
2	5	3	4	8	177811200	17
2	5	3	5	5	254016000	143
2	5	3	5	7	2133734400	221
2	5	3	5	9	8001504000	323
2	5	4	4	6	-203212800	13
2	5	4	4	8	4267468800	17
2	5	4	5	7	4267468800	17
2	5	4	5	9	480090240000	323
2	5	5	5	7	-42674688000	17
2	5	5	5	9	11522165760000	323
3	3	3	4	6	-108864000	13
3	3	3	5	7	-65318400	1
3	3	4	5	7	-10059033600	11
3	4	3	5	6	-609638400	13
3	4	3	5	8	-3200601600	17
3	4	4	4	7	135610675200	143
3	4	4	5	8	32006016000	17
3	4	5	5	8	2560481280000	17
3	5	4	4	8	128024064000	17
3	5	4	5	7	102419251200	17
3	5	4	5	9	11522165760000	323
3	5	5	5	9	23044331520000	19
4	4	4	5	8	-2048385024000	17
4	5	5	5	9	414797967360000	19

TABLE 4

ELSASSER INTEGRALS
$E^-(U,L,W,N,M) = 4A/B$

U	L	W	N	M	A	B
1	2	1	5	6	-315	143
1	2	2	5	6	-11340	143
1	2	3	5	6	-30240	13
1	2	4	5	6	-544320	11
1	2	5	5	6	-81648000	143
1	3	1	4	6	-350	143
1	3	1	5	7	-420	143
1	3	2	4	6	-10500	143
1	3	2	5	7	-1680	13
1	3	3	4	6	-16800	11
1	3	3	5	7	-635040	143
1	3	4	4	6	-2419200	143
1	3	4	5	7	-15422400	143
1	3	5	5	7	-18144000	13
1	4	1	4	7	-3920	1287
1	4	1	5	6	-560	143
1	4	1	5	8	-8820	2431
1	4	2	3	6	-700	13
1	4	2	4	7	-15680	143
1	4	2	5	6	-1120	13
1	4	2	5	8	-35280	187
1	4	3	3	6	-84000	143
1	4	3	4	7	-376320	143
1	4	3	5	6	67200	143
1	4	3	5	8	-1058400	143
1	4	4	4	7	-4704000	143
1	4	4	5	6	14515200	143
1	4	4	5	8	-44452800	221
1	4	5	5	6	117210240000	46189
1	4	5	5	8	-635040000	221
1	5	1	5	7	-11760	2431
1	5	1	5	9	-198450	46189
1	5	2	2	6	-3780	143
1	5	2	3	7	-840	11
1	5	2	4	6	-6720	143
1	5	2	4	8	-370440	2431
1	5	2	5	7	-23520	187
1	5	2	5	9	-11907000	46189
1	5	3	3	7	-136080	143
1	5	3	4	6	235200	143
1	5	3	4	8	-10054800	2431
1	5	3	5	7	2540160	2431
1	5	3	5	9	-47628000	4199
1	5	4	4	6	8467200	143
1	5	4	4	8	-12700800	221
1	5	4	5	7	508032000	2431
1	5	4	5	9	-15717240000	46189
1	5	5	5	7	1270080000	221
1	5	5	5	9	-1714608000	323
2	2	2	5	6	2520	143
2	2	3	5	6	60480	143
2	2	4	5	6	1088640	143
2	2	5	5	6	10886400	143
2	3	2	4	6	4200	143
2	3	2	5	7	6720	143
2	3	3	4	6	71400	143
2	3	3	5	7	191520	143
2	3	4	4	6	672000	143
2	3	4	5	7	362880	13
2	3	5	5	7	45360000	143
2	4	2	4	7	7840	143
2	4	2	5	6	-3360	143
2	4	2	5	8	211680	2431
2	4	3	3	6	37800	143
2	4	3	4	7	156800	143
2	4	3	5	6	-362880	143
2	4	3	5	8	635040	221
2	4	4	4	7	1693440	143
2	4	4	5	6	-1209600	13
2	4	4	5	8	12700800	187
2	4	5	5	6	-217728000	143
2	4	5	5	8	190512000	221
2	5	2	5	7	-23520	2431
2	5	2	5	9	6350400	46189
2	5	3	3	7	70560	143
2	5	3	4	6	-329280	143
2	5	3	4	8	4868640	2431
2	5	3	5	7	-9878400	2431
2	5	3	5	9	238140000	46189
2	5	4	4	6	-6585600	143
2	5	4	4	8	59270400	2431
2	5	4	5	7	-35562240	187
2	5	4	5	9	571536000	4199
2	5	5	5	7	-8890560000	2431
2	5	5	5	9	8001504000	4199
3	3	3	4	6	-12600	143
3	3	3	5	7	-30240	143
3	3	4	4	6	-100800	143
3	3	4	5	7	-544320	143
3	3	5	5	7	-5443200	143
3	4	3	4	7	-47040	143
3	4	3	5	6	282240	143
3	4	3	5	8	-1905120	2431
3	4	4	4	7	-439040	143
3	4	4	5	6	6773760	143
3	4	4	5	8	-39372480	2431
3	4	5	5	6	84672000	143
3	4	5	5	8	-444528000	2431
3	5	3	5	7	9313920	2431
3	5	3	5	9	-85730400	46189
3	5	4	4	6	3951360	143
3	5	4	4	8	-20321280	2431
3	5	4	5	7	284497920	2431
3	5	4	5	9	-2000376000	46189
3	5	5	5	7	4267468800	2431
3	5	5	5	9	-2286144000	4199
4	4	4	4	7	62720	143
4	4	4	5	6	-1693440	143
4	4	4	5	8	5080320	2431
4	4	5	5	6	-16934400	143
4	4	5	5	8	50803200	2431
4	5	4	5	7	-116847360	2431
4	5	4	5	9	457228800	46189
4	5	5	5	7	-1422489600	2431
4	5	5	5	9	5143824000	46189
5	5	5	5	7	254016000	2431
5	5	5	5	9	-571536000	46189

Table 5

GAUNT INTEGRALS
$G^+(U,L,W,N,M) = A/B$

U	L	W	N	M	A	B
0	1	0	1	2	4	15
0	1	0	2	1	4	15
0	1	0	2	3	6	35
0	1	0	3	2	6	35
0	1	0	3	4	8	63
0	1	0	4	3	8	63
0	1	0	4	5	10	99
0	1	0	5	4	10	99
0	1	0	5	6	12	143
0	1	1	1	2	4	5
0	1	1	2	1	4	5
0	1	1	2	3	48	35
0	1	1	3	2	48	35
0	1	1	3	4	40	21
0	1	1	4	3	40	21
0	1	1	4	5	80	33
0	1	1	5	4	80	33
0	1	1	5	6	420	143
0	1	2	2	3	48	7
0	1	2	3	2	48	7
0	1	2	3	4	160	7
0	1	2	4	3	160	7
0	1	2	4	5	560	11
0	1	2	5	4	560	11
0	1	2	5	6	13440	143
0	1	3	3	4	160	1
0	1	3	4	3	160	1
0	1	3	4	5	8960	11
0	1	3	5	4	8960	11
0	1	3	5	6	362880	143
0	1	4	4	5	80640	11
0	1	4	5	4	80640	11
0	1	4	5	6	7257600	143
0	1	5	5	6	7257600	13
0	2	0	2	2	4	35
0	2	0	2	4	4	35
0	2	0	3	1	6	35
0	2	0	3	3	8	105
0	2	0	3	5	20	231
0	2	0	4	2	4	35
0	2	0	4	4	40	693
0	2	0	4	6	10	143
0	2	0	5	3	20	231
0	2	0	5	5	20	429
0	2	0	5	7	42	715
0	2	1	1	1	-4	15
0	2	1	1	3	24	35
0	2	1	2	2	12	35
0	2	1	2	4	8	7
0	2	1	3	1	24	35
0	2	1	3	3	24	35
0	2	1	3	5	120	77
0	2	1	4	2	8	7
0	2	1	4	4	680	693
0	2	1	4	6	280	143
0	2	1	5	3	120	77
0	2	1	5	5	180	143
0	2	1	5	7	336	143
0	2	2	2	2	-96	35
0	2	2	2	4	48	7
0	2	2	3	3	0	1
0	2	2	3	5	240	11
0	2	2	4	2	48	7
0	2	2	4	4	640	77
0	2	2	4	6	6720	143
0	2	2	5	3	240	11
0	2	2	5	5	3360	143
0	2	2	5	7	12096	143
0	2	3	3	3	-480	7
0	2	3	3	5	1920	11
0	2	3	4	4	-1120	11
0	2	3	4	6	120960	143
0	2	3	5	3	1920	11
0	2	3	5	5	13440	143
0	2	3	5	7	362880	143
0	2	4	4	4	-35840	11
0	2	4	4	6	1209600	143
0	2	4	5	5	-1451520	143
0	2	4	5	7	725760	13
0	2	5	5	5	-36288000	143
0	2	5	5	7	8709120	13
0	3	0	3	2	8	105
0	3	0	3	4	4	77
0	3	0	3	6	200	3003
0	3	0	4	1	8	63
0	3	0	4	3	4	77
0	3	0	4	5	120	3003
0	3	0	4	7	70	1287
0	3	0	5	2	20	231
0	3	0	5	4	40	1001
0	3	0	5	6	14	429
0	3	0	5	8	112	2431
0	3	1	1	2	-12	35
0	3	1	1	4	40	63
0	3	1	2	1	-12	35
0	3	1	2	3	8	35
0	3	1	2	5	80	77
0	3	1	3	2	8	35
0	3	1	3	4	40	77
0	3	1	3	6	200	143
0	3	1	4	1	40	63
0	3	1	4	3	40	77
0	3	1	4	5	760	1001
0	3	1	4	7	2240	1287
0	3	1	5	2	80	77
0	3	1	5	4	760	1001
0	3	1	5	6	140	143
0	3	1	5	8	5040	2431
0	3	2	2	3	-32	7
0	3	2	2	5	80	11

TABLE 5 (contd.)

0	3	2	3	2	-32	7	0	4	2	4	2	-480	77
0	3	2	3	4	-240	77	0	4	2	4	4	-720	91
0	3	2	3	6	3200	143	0	4	2	4	6	0	1
0	3	2	4	3	-240	77	0	4	2	4	8	117600	2431
0	3	2	4	5	480	143	0	4	2	5	3	-800	143
0	3	2	4	7	6720	143	0	4	2	5	5	-560	143
0	3	2	5	2	80	11	0	4	2	5	7	26880	2431
0	3	2	5	4	480	143	0	4	2	5	9	352800	4199
0	3	2	5	6	2240	143	0	4	3	3	3	1440	77
0	3	2	5	8	201600	2431	0	4	3	3	5	-28800	143
0	3	3	3	4	-1440	11	0	4	3	3	7	33600	143
0	3	3	3	6	28800	143	0	4	3	4	4	-30240	143
0	3	3	4	3	-1440	11	0	4	3	4	6	-67200	143
0	3	3	4	5	-40320	143	0	4	3	4	8	235200	221
0	3	3	4	7	134400	143	0	4	3	5	3	-28800	143
0	3	3	5	4	-40320	143	0	4	3	5	5	-80640	143
0	3	3	5	6	-40320	143	0	4	3	5	7	-1545600	2431
0	3	3	5	8	604800	221	0	4	3	5	9	12700800	4199
0	3	4	4	5	-967680	143	0	4	4	4	4	161280	143
0	3	4	4	7	1478400	143	0	4	4	4	6	-1612800	143
0	3	4	5	4	-967680	143	0	4	4	4	8	2822400	221
0	3	4	5	6	-3628800	143	0	4	4	5	5	-1451520	143
0	3	4	5	8	14515200	221	0	4	4	5	7	-9676800	221
0	3	5	5	6	-7257600	13	0	4	4	5	9	25401600	323
0	3	5	5	8	14515200	17	0	4	5	5	5	14515200	143
0	4	0	4	2	40	693							
0	4	0	4	4	36	1001							
0	4	0	4	6	40	1287	0	4	5	5	7	-217728000	221
0	4	0	4	8	980	21879	0	4	5	5	9	355622400	323
0	4	0	5	1	10	99	0	5	0	5	2	20	429
0	4	0	5	3	40	1001	0	5	0	5	4	4	143
0	4	0	5	5	4	143	0	5	0	5	6	160	7293
0	4	0	5	7	560	21879	0	5	0	5	8	980	46189
0	4	0	5	9	1764	46189	0	5	0	5	10	1512	46189
0	4	1	1	3	-8	21	0	5	1	1	4	-40	99
0	4	1	1	5	20	33	0	5	1	1	6	84	143
0	4	1	2	2	-16	35	0	5	1	2	3	-40	77
0	4	1	2	4	40	231	0	5	1	2	5	20	143
0	4	1	2	6	140	143	0	5	1	2	7	672	715
0	4	1	3	1	-8	21	0	5	1	3	2	-40	77
0	4	1	3	3	8	77	0	5	1	3	4	40	1001
							0	5	1	3	6	56	143
							0	5	1	3	8	3024	2431
0	4	1	3	5	40	91	0	5	1	4	1	-40	99
0	4	1	3	7	560	429	0	5	1	4	3	40	1001
0	4	1	4	2	40	231	0	5	1	4	5	40	143
0	4	1	4	4	360	1001	0	5	1	4	7	12880	21879
0	4	1	4	6	280	429	0	5	1	4	9	70560	46189
0	4	1	4	8	3920	2431	0	5	1	5	2	20	143
0	4	1	5	1	20	33	0	5	1	5	4	40	143
0	4	1	5	3	40	91	0	5	1	5	6	1120	2431
0	4	1	5	5	80	143	0	5	1	5	8	35280	46189
0	4	1	5	7	560	663	0	5	1	5	10	7560	4199
0	4	1	5	9	88200	46189	0	5	2	2	3	80	77
0	4	2	2	2	16	35	0	5	2	2	5	-1120	143
0	4	2	2	4	-480	77	0	5	2	2	7	6048	715
0	4	2	2	6	1120	143	0	5	2	3	2	80	77
0	4	2	3	3	-80	11	0	5	2	3	4	-9600	1001
0	4	2	3	5	-800	143	0	5	2	3	6	-1120	143
0	4	2	3	7	3360	143	0	5	2	3	8	60480	2431

TABLE 5 (contd.)

0 5 2 4 3	−9600	1001	1 2 1 4 2	−16	7
0 5 2 4 5	−1680	143	1 2 1 4 4	720	77
0 5 2 4 7	−6720	2431	1 2 1 4 6	4480	143
0 5 2 4 9	211680	4199	1 2 1 5 3	−480	77
0 5 2 5 2	−3360	429	1 2 1 5 5	1680	143
0 5 2 5 4	−1680	143	1 2 1 5 7	6048	143
0 5 2 5 6	−22400	2431	1 2 2 2 4	96	1 0
0 5 2 5 8	352800	46189	1 2 2 3 3	480	7
0 5 2 5 10	362880	4199	1 2 2 3 5	3840	11
			1 2 2 4 4	2400	11
			1 2 2 4 6	120960	143
0 5 3 3 4	7200	143	1 2 2 5 3	−480	11
0 5 3 3 6	−40320	143	1 2 2 5 5	67200	143
0 5 3 3 8	60480	221	1 2 2 5 7	241920	143
0 5 3 4 3	7200	143	1 2 3 3 5	34560	11
0 5 3 4 5	−40320	143	1 2 3 4 4	26880	11
0 5 3 4 7	−1646400	2431	1 2 3 4 6	2419200	143
0 5 3 4 9	5080320	4199	1 2 3 5 5	1693440	143
0 5 3 5 4	−40320	143	1 2 3 5 7	725760	13
0 5 3 5 6	−2016000	2431	1 2 4 4 6	2419200	13
0 5 3 5 8	−4233600	4199	1 2 4 5 5	21772800	143
0 5 3 5 10	1088640	323	1 2 4 5 7	17418240	13
0 5 4 4 5	483840	143	1 2 5 5 7	191600640	11
0 5 4 4 7	−3763200	221	1 3 1 3 2	192	35
0 5 4 4 9	5080320	323	1 3 1 3 4	960	77
0 5 4 5 4	483840	143	1 3 1 3 6	4800	143
0 5 4 5 6	−29030400	2431	1 3 1 4 3	640	77
0 5 4 5 8	−279417600	4199	1 3 1 4 5	2560	143
0 5 4 5 10	30481920	323	1 3 1 4 7	6720	143
0 5 5 5 6	72576000	221	1 3 1 5 2	−240	77
0 5 5 5 8	−508032000	323	1 3 1 5 4	10320	1001
0 5 5 5 10	457228800	323	1 3 1 5 6	3360	143
1 1 1 1 2	16	5	1 3 1 5 8	151200	2431
1 1 1 2 3	48	7	1 3 2 2 3	−192	7
1 1 1 3 2	−48	35	1 3 2 2 5	1920	11
1 1 1 3 4	80	7	1 3 2 3 4	960	11
1 1 1 4 3	−80	21	1 3 2 3 6	86400	143
1 1 1 4 5	560	33	1 3 2 4 3	8640	77
1 1 1 5 4	−80	11			
1 1 1 5 6	3360	143			
1 1 2 2 3	288	7	1 3 2 4 5	51840	143
1 1 2 3 4	160	1 0	1 3 2 4 7	201600	143
1 1 2 4 3	−160	7	1 3 2 5 4	47040	143
1 1 2 4 5	4480	11	1 3 2 5 6	120960	143
1 1 2 5 4	−1120	11	1 3 2 5 8	604800	221
1 1 2 5 6	120960	143	1 3 3 3 4	−11520	11
1 1 3 3 4	1280	1	1 3 3 3 6	864000	143
1 1 3 4 5	80640	11	1 3 3 4 5	241920	143
1 1 3 5 4	−8960	11	1 3 3 4 7	403200	13
1 1 3 5 6	3628800	143	1 3 3 5 4	645120	143
1 1 4 4 5	806400	11	1 3 3 5 6	2419200	143
1 1 4 5 6	7257600	13	1 3 3 5 8	21772800	221
1 1 5 5 6	87091200	13	1 3 4 4 5	−9676800	143
			1 3 4 4 7	53222400	143
			1 3 4 5 6	0	1
1 2 1 2 2	144	35	1 3 4 5 8	43545600	17
1 2 1 2 4	96	7	1 3 5 5 6	−87091200	13
1 2 1 3 3	48	7	1 3 5 5 8	609638400	17
1 2 1 3 5	240	11	1 4 1 4 2	1600	231

TABLE 5 (contd.)

1	4	1	4	4	14400	1001	1	5	3	4	9	25401600	323
1	4	1	4	6	11200	429	1	5	3	5	4	-806400	143
1	4	1	4	8	156800	2431	1	5	3	5	6	-2419200	187
1	4	1	5	3	10000	1001	1	5	3	5	8	25401600	4199
1	4	1	5	5	2800	143	1	5	3	5	10	76204800	323
1	4	1	5	7	84000	2431	1	5	4	4	5	4838400	143
1	4	1	5	9	352800	4199	1	5	4	4	7	-135475200	221
1	4	2	2	4	-960	11							
1	4	2	2	6	40320	143							
1	4	2	3	3	-4800	77	1	5	4	4	9	355622400	323
1	4	2	3	5	9600	143	1	5	4	5	6	-130636800	221
1	4	2	3	7	134400	143	1	5	4	5	8	-609638400	323
1	4	2	4	4	14400	143	1	5	4	5	10	2286144000	323
1	4	2	4	6	67200	143	1	5	5	5	6	870912000	221
1	4	2	4	8	470400	221	1	5	5	5	8	-21337344000	323
1	4	2	5	3	24000	143	1	5	5	5	10	36578304000	323
1	4	2	5	5	67200	143	2	2	2	2	4	768	1
1	4	2	5	7	2889600	2431	2	2	2	3	5	34560	11
1	4	2	5	9	16934400	4199	2	2	2	4	4	-7680	11
1	4	3	3	5	-518400	143	2	2	2	4	6	1209600	143
1	4	3	3	7	134400	13	2	2	2	5	5	-483840	143
1	4	3	4	4	-403200	143	2	2	2	5	7	241920	13
1	4	3	4	6	-403200	143	2	2	3	3	5	345600	11
							2	2	3	4	6	2419200	13
							2	2	3	5	5	-4838400	143
1	4	3	4	8	11289600	221	2	2	3	5	7	8709120	13
1	4	3	5	5	0	1	2	2	4	4	6	29030400	13
1	4	3	5	7	268800	17	2	2	4	5	7	17418240	1
1	4	3	5	9	50803200	323	2	2	5	5	7	243855360	1
1	4	4	4	6	-3225600	13	2	3	2	3	4	19200	11
1	4	4	4	8	11289600	17	2	3	2	3	6	1728000	143
1	4	4	5	5	-29030400	143	2	3	2	4	5	691200	143
1	4	4	5	7	-130636800	221	2	3	2	4	7	403200	13
1	4	4	5	9	1422489600	323	2	3	2	5	4	-268800	143
1	4	5	5	7	-4790016000	187	2	3	2	5	6	1209600	143
1	4	5	5	9	21337344000	323	2	3	2	5	8	14515200	221
1	5	1	5	2	1200	143	2	3	3	3	6	1728000	13
1	5	1	5	4	2400	143	2	3	3	4	5	12096000	143
1	5	1	5	6	67200	2431	2	3	3	4	7	9676800	13
1	5	1	5	8	2116800	46189	2	3	3	5	6	4838400	13
1	5	1	5	10	453600	4199	2	3	3	5	8	43545600	17
1	5	2	2	3	480	77	2	3	4	4	7	9676800	1
1	5	2	2	5	-26880	143	2	3	4	5	6	958003200	143
1	5	2	2	7	60480	143	2	3	4	5	8	1219276800	17
1	5	2	3	4	-2400	13	2	3	5	5	8	18289152000	17
1	5	2	3	6	0	1	2	4	2	4	4	518400	143
1	5	2	3	8	302400	221	2	4	2	4	6	2419200	143
1	5	2	4	3	-108000	1001	2	4	2	4	8	16934400	221
1	5	2	4	5	0	1	2	4	2	5	5	1209600	143
1	5	2	4	7	100800	187	2	4	2	5	7	8064000	221
1	5	2	4	9	12700800	4199	2	4	2	5	9	50803200	323
1	5	2	5	4	16800	143							
1	5	2	5	6	1209600	2431							
1	5	2	5	8	6350400	4199	2	4	3	3	5	-5184000	143
1	5	2	5	10	1814400	323	2	4	3	3	7	4838400	13
1	5	3	3	4	57600	143	2	4	3	4	6	2419200	13
1	5	3	3	6	-1209600	143	2	4	3	4	8	33868800	17
1	5	3	3	8	3628800	221	2	4	3	5	5	29030400	143
1	5	3	4	5	-1209600	143	2	4	3	5	7	256435200	221
1	5	3	4	7	-2822400	221	2	4	3	5	9	2133734400	323

TABLE 5 (contd.)

2	4	4	4	6	−38707200	13
2	4	4	4	8	474163200	17
2	4	4	5	7	174182400	17
2	4	4	5	9	64012032000	323
2	4	5	5	7	−6096384000	17
2	4	5	5	9	1024192512000	323
2	5	2	5	4	940800	143
2	5	2	5	6	67737600	2431
2	5	2	5	8	355622400	4199
2	5	2	5	10	101606400	323
2	5	3	3	6	−2419200	13
2	5	3	3	8	14515200	17
2	5	3	4	5	−16934400	143
2	5	3	4	7	33868800	221
2	5	3	4	9	1422489600	323
2	5	3	5	6	67737600	221
2	5	3	5	8	711244800	323
2	5	3	5	10	4572288000	323
2	5	4	4	7	−270950400	17
2	5	4	4	9	21337344000	323
2	5	4	5	6	−2438553600	221
2	5	4	5	8	−4267468800	323
2	5	4	5	10	146313216000	323
2	5	5	5	8	−640120320000	323
2	5	5	5	10	146313216000	19
3	3	3	3	6	20736000	13
3	3	3	4	7	9676800	1
3	3	3	5	6	−29030400	13
3	3	3	5	8	609638400	17
3	3	4	4	7	135475200	1
3	3	4	5	8	18289152000	17
3	3	5	5	8	292626432000	17
3	4	3	4	6	67737600	13
3	4	3	4	8	948326400	17
3	4	3	5	7	338688000	17
3	4	3	5	9	64012032000	323
3	4	4	4	8	14224896000	17
3	4	4	5	7	8534937600	17
3	4	4	5	9	2048385024000	323
3	4	5	5	9	2048385024000	19
3	5	3	5	6	3251404800	221
3	5	3	5	8	34139750400	323
3	5	3	5	10	219469824000	323
3	5	4	4	7	−3793305600	17
3	5	4	4	9	1024192512000	323
3	5	4	5	8	512096256000	323
3	5	4	5	10	438939648000	19
3	5	5	5	8	−10241925120000	323
3	5	5	5	10	7900913664000	19
4	4	4	4	8	227598336000	17
4	4	4	5	9	2048385024000	19
4	4	5	5	9	36870930432000	19
4	5	4	5	8	18435465216000	323
4	5	4	5	10	15801827328000	19
4	5	5	5	10	15801827328000	1
5	5	5	5	10	316036546560000	1

Table 6

GAUNT INTEGRALS
$G^-(U,L,W,N,M) = A/B$

U	L	W	N	M	A	B
1	1	1	5	6	-60	143
1	1	2	5	6	-1680	143
1	1	3	5	6	-40320	143
1	1	4	5	6	-725760	143
1	1	5	5	6	-7257600	143
1	2	1	4	6	-80	143
1	2	1	5	7	-84	143
1	2	2	4	6	-1680	143
1	2	2	5	7	-2688	143
1	2	3	4	6	-26880	143
1	2	3	5	7	-72576	143
1	2	4	4	6	-241920	143
1	2	4	5	7	-1451520	143
1	2	5	5	7	-1451520	13
1	3	1	3	6	-600	1001
1	3	1	4	7	-280	429
1	3	1	5	6	0	1
1	3	1	5	8	-1680	2431
1	3	2	3	6	-1200	143
1	3	2	4	7	-2240	143
1	3	2	5	6	1680	143
1	3	2	5	8	-60480	2431
1	3	3	3	6	-9600	143
1	3	3	4	7	-40320	143
1	3	3	5	6	94080	143
1	3	3	5	8	-1814400	2431
1	3	4	4	7	-403200	143
1	3	4	5	6	2903040	143
1	3	4	5	8	-3628800	221
1	3	5	5	6	43545600	143
1	3	5	5	8	-43545600	221
1	4	1	4	6	-40	1287
1	4	1	4	8	-15680	21879
1	4	1	5	7	-560	7293
1	4	1	5	9	-35280	46189
1	4	2	2	6	-560	143
1	4	2	3	7	-4480	429
1	4	2	4	6	1680	143
1	4	2	4	8	-47040	2431
1	4	2	5	7	91840	7293
1	4	2	5	9	-1411200	46189
1	4	3	3	7	-13440	143
1	4	3	4	6	67200	143
1	4	3	4	8	-940800	2431
1	4	3	5	7	2190720	2431
1	4	3	5	9	-4233600	4199
1	4	4	4	6	1075200	143
1	4	4	4	8	-940800	221
1	4	4	5	7	6048000	187
1	4	4	5	9	-101606400	4199
1	4	5	5	7	120960000	221
1	4	5	5	9	-101606400	323
1	5	1	5	6	480	2431
1	5	1	5	8	-5880	46189
1	5	1	5	10	-37800	46189
1	5	2	2	7	-672	143
1	5	2	3	6	1680	143
1	5	2	3	8	-30240	2431
1	5	2	4	7	32480	2431
1	5	2	4	9	-1058400	46189
1	5	2	5	6	53760	2431
1	5	2	5	8	635040	46189
1	5	2	5	10	-151200	4199
1	5	3	3	6	33600	143
1	5	3	3	8	-302400	2431
1	5	3	4	7	141120	221
1	5	3	4	9	-2116800	4199
1	5	3	5	6	134400	221
1	5	3	5	8	4233600	3553
1	5	3	5	10	-5443200	4199
1	5	4	4	7	28224000	2431
1	5	4	4	9	-25401600	4199
1	5	4	5	6	-7257600	2431
1	5	4	5	8	203212800	4199
1	5	4	5	10	-10886400	323
1	5	5	5	6	-1088640000	2431
1	5	5	5	8	3810240000	4199
1	5	5	5	10	-152409600	323
2	2	2	4	6	240	143
2	2	2	5	7	336	143
2	2	3	4	6	3360	143
2	2	3	5	7	8064	143
2	2	4	4	6	26880	143
2	2	4	5	7	145152	143
2	2	5	5	7	1451520	143
2	3	2	3	6	2400	1001
2	3	2	4	7	560	143
2	3	2	5	6	-1680	143
2	3	2	5	8	13440	2431
2	3	3	3	6	2400	143
2	3	3	4	7	8960	143
2	3	3	5	6	-53760	143
2	3	3	5	8	362880	2431
2	3	4	4	7	80640	143
2	3	4	5	6	-1209600	143
2	3	4	5	8	7257600	2431
2	3	5	5	6	-14515200	143
2	3	5	5	8	7257600	221
2	4	2	4	6	-160	13
2	4	2	4	8	15680	2431
2	4	2	5	7	-35840	2431
2	4	2	5	9	423360	46189
2	4	3	3	7	4480	143
2	4	3	4	6	-43680	143
2	4	3	4	8	282240	2431
2	4	3	5	7	-112000	187
2	4	3	5	9	12700800	46189
2	4	4	4	6	-537600	143
2	4	4	4	8	2822400	2431
2	4	4	5	7	-39191040	2431
2	4	4	5	9	25401600	4199

TABLE 6 (contd.)

2	4	5	5	7	−544320000	2431
2	4	5	5	9	304819200	4199
2	5	2	5	6	−40320	2431
2	5	2	5	8	−47040	2717
2	5	2	5	10	604800	46189
2	5	3	3	6	−23520	143
2	5	3	3	8	120960	2431
2	5	3	4	7	−1097600	2431
2	5	3	4	9	8467200	46189
2	5	3	5	6	−188160	2431
2	5	3	5	8	−39795840	46189
2	5	3	5	10	1814400	4199
2	5	4	4	7	−15805440	2431
2	5	4	4	9	8467200	4199
2	5	4	5	6	33868800	2431
2	5	4	5	8	−1244678400	46189
2	5	4	5	10	43545600	4199
2	5	5	5	6	1016064000	2431
2	5	5	5	8	−1778112000	4199
2	5	5	5	10	43545600	323
3	3	3	3	6	−2400	1001
3	3	3	4	7	−1120	143
3	3	3	5	6	13440	143
3	3	3	5	8	−40320	2431
3	3	4	4	7	−8960	143
3	3	4	5	6	241920	143
3	3	4	5	8	−725760	2431
3	3	5	5	6	2419200	143
3	3	5	5	8	−7257600	2431
3	4	3	4	6	19040	143
3	4	3	4	8	−62720	2431
3	4	3	5	7	595840	2431
3	4	3	5	9	−2540160	46189
3	4	4	4	6	188160	143
3	4	4	4	8	−564480	2431
3	4	4	5	7	12983040	2431
3	4	4	5	9	−50803200	46189
3	4	5	5	7	152409600	2431
3	4	5	5	9	−50803200	4199
3	5	3	5	6	−779520	2431
3	5	3	5	8	20603520	46189
3	5	3	5	10	−5443200	46189
3	5	4	4	7	7024640	2431
3	5	4	4	9	−25401600	46189
3	5	4	5	6	−3386880	221
3	5	4	5	8	40642560	3553
3	5	4	5	10	−10886400	4199
3	5	5	5	6	−677376000	2431
3	5	5	5	8	7112448000	46189
3	5	5	5	10	−130636800	4199
4	4	4	4	6	−35840	143
4	4	4	4	8	62720	2431
4	4	4	5	7	−2257920	2431
4	4	4	5	9	5080320	46189
4	4	5	5	7	−22579200	2431
4	4	5	5	9	50803200	46189
4	5	4	5	6	23224320	2431
4	5	4	5	8	−157489920	46189
4	5	4	5	10	21772800	46189
4	5	5	5	6	304819200	2431
4	5	5	5	8	−1828915200	46189
4	5	5	5	10	21772800	4199
5	5	5	5	6	−72576000	2431
5	5	5	5	8	254016000	46189
5	5	5	5	10	−21772800	46189

The selection rules on $E^+(u, l, w, n, m)$ now include the vanishing of the integral when both $u = l$ and $w = n$.

To evaluate $G^+(u, l, w, n, m)$ and $G^-(u, l, w, n, m)$ we have used the formula by Gaunt[12]

$$G^+(u, l, w, n, m) = 2(-1)^{s-l-w} \frac{(l+u)!(n+w)!(2s-2n)!s!}{(l-u)!(s-m)!(s-l)!(s-n)!(2s+1)!}$$

$$\times \sum_t \frac{(-1)^t(m+v+t)!(l+n-v-t)!}{(m-v-t)!(l-n+v+t)!(n-w-t)!t!} \quad (A22)$$

where $v = w+u$, $2s = l+m+n$ and t ranges from the greater of 0 and $n-l-v$ to the smallest of $n-w$, $m-v$, and $l+n-v$.

To evaluate $E^+(u, l, w, n, m)$ and $E^-(u, l, w, n, m)$ we have used the formula (which we believe is new)

$$E^+(u, l, w, n, m) = \frac{(-1)^{n+l}(m+u+w)!}{2^{n+l+u+w}(m-u-w)!} \sum_{r,s} \frac{(-1)^{r+s}(2r)!(2s)!(ws-ur)}{r!s!(n-r)!(l-s)!(2r-w-n)!(2s-u-l)!}$$

$$\times \frac{\Gamma(1/2+q/2)\Gamma(1+q/2)}{\Gamma(1+q/2+u/2+w/2-m/2)\Gamma(3/2+q/2+u/2+w/2+m/2)} \quad (A23)$$

$$(g = 2r+2s-u-w-n-l-1)$$

where Γ denotes the gamma function; r ranges from the integer $(w+n)/2$ or $(w+n+1)/2$ to n, and s ranges from the integer $(u+l)/2$ or $(u+l+1)/2$ to l. (To obtain (23) we have used a result of Barnes[13], p. 182.)

Values of the integrals are presented exactly as integer fractions in the tables 3–6. The tabulation is systematic and includes all integrals for which u, l, w, n are less than or equal to 5 while m is permitted to assume all values for which the selection rules are satisfied. The quantity of tabulation has been reduced by omitting the symmetric and antisymmetric values of G and E respectively. In addition all those values of

$$G^-(u, l, w, n, m) \quad \text{and} \quad E^-(u, l, w, n, m)$$

which can be obtained from the corresponding G^+ and E^+ tables by relations (A20) and (A21) are omitted.

Acknowledgements

The author is indebted to Professor P. H. Roberts for bringing to his notice Barnes' result, used in (A23), and for obtaining (A11) to (A14).

References

1. R. Hide and P. H. Roberts, *Phys. Chem. Earth*, **4**, 25–98 (1961).
2. P. H. Roberts, *An Introduction to Magnetohydrodynamics*, Longmans, London, 1967, Chap. 3.
3. R. D. Gibson and P. H. Roberts, in *Magnetism and the Cosmos* (Eds. W. R. Hindmarsh F. J. Lowes, P. H. Roberts, and S. K. Runcorn), Oliver and Boyd, Edinburgh, 1966, p. 108.
4. E. C. Bullard and H. Gellman, *Phil. Trans. Roy. Soc. London, Ser. A*, **247**, 213 (1954).

5. K. Wright, *Comput. J.*, **6,** 358 (1964).
6. R. Baer and G. W. Platzman, *J. Meteorol.*, **18,** 393 (1961).
7. E. U. Condon and G. H. Shortley, *The Theory of Atomic Spectra*, University Press, London, 1953.
8. A. Fletcher, J. C. P. Miller, L. Rosenhead, and L. J. Comrie, *An Index of Mathematical Tables*, Blackwell, London, 1962.
9. M. P. Barnett, *Univ. Wisconsin Rept.*, No. ONR–30 (1958).
10. J. R. Bird, M.A. Thesis, University of Toronto (1949).
11. G. Gjellestad, *Proc. Natl. Acad. Sci. U.S.*, **41,** 954 (1955).
12. J. A. Gaunt, *Phil. Trans. Roy. Soc. London, Ser. A*, **228,** 192 (1929).
13. E. W. Barnes, *Quart. J. Pure Appl. Math.*, **39,** 97 (1908).

4

G. O. ROBERTS

School of Mathematics
University of Newcastle upon Tyne
England

Dynamo waves

Introduction

The kinematic dynamo problem

This chapter is concerned with the kinematic dynamo problem. This is the study of the excitation and maintenance of a magnetic field by an assumed motion of a uniformly conducting fluid. The equation of motion of the fluid is ignored.

Backus[1] has shown that it is sufficient to confine attention to the equations for the magnetic field. For an incompressible fluid with uniform conductivity σ e.m.u., and with the usual notation, these are as follows:

$$\dot{\mathbf{B}} + (\mathbf{u}.\nabla)\mathbf{B} = (\mathbf{B}.\nabla)\mathbf{u} + \lambda\nabla^2\mathbf{B} \tag{1}$$

$$\nabla.\mathbf{B} = 0 \tag{2}$$

where $\lambda = 1/4\pi\sigma$.

For the case of infinite conductivity, $\lambda = 0$, and the equations show that the magnetic field lines behave as if they are frozen into the fluid. In other words, if the magnetic field at a point at a given instant is parallel to an infinitesimal fluid element $\mathbf{dl}$, then at any subsequent time the magnetic field at the instantaneous position of the same fluid element will be parallel to the fluid element and will have changed in magnitude in proportion to the change in magnitude of the fluid element. This is because the rate of change of the magnetic field at a point moving with the fluid is $\dot{\mathbf{B}} + (\mathbf{u}.\nabla)\mathbf{B}$ and the rate of change of the fluid element is $(\mathbf{dl}.\nabla)\mathbf{u}$ at any instant.

For finite conductivity, the term $\lambda\nabla^2\mathbf{B}$ is a diffusion term. Its effect is to tend to smooth out all irregularities in the magnetic field.

Equation (2) can be regarded as an initial condition. For the divergence of equation (1), using the incompressibility condition $\nabla.\mathbf{u} = 0$, gives

$$\frac{\partial}{\partial t}(\nabla.\mathbf{B}) + (\mathbf{u}.\nabla)(\nabla.\mathbf{B}) = \lambda\nabla^2(\nabla.\mathbf{B})$$

These magnetic field equations (1), (2) usually have to be combined with certain boundary conditions. The external medium can be a conductor or an insulator. In this chapter a fluid of infinite extent is assumed.

Previous work

Roberts[2] has recently reviewed previous work on this problem. Cowling[3,4] and Braginskiĭ[5] showed that an axisymmetric field cannot be maintained or amplified by a fluid motion. The same applies to fields which are functions of only two Cartesian coordinates. The kinematic problem is thus effectively that of choosing motions complicated enough to give dynamo action and simple enough for this action to be proved.

All previous papers proving dynamo action use, explicitly or implicitly, expansions in ascending powers of small dimensionless parameters, keeping only a few terms. The only dimensionless numbers available are the magnetic Reynolds number $R = LU/\lambda$, where L is a length scale and U a velocity scale, and ratios of different velocity scales, length scales, and time scales.

Backus[1] uses 'jerky' motions in his dynamo model. In these, periods of zero motion alternate with rapid motions for very short times, during which diffusion is negligible and the field is practically 'frozen in'. In his limit, the stationary periods become so long that the fields left at the end are almost entirely the slowest decaying modes. The behaviour of these during the 'jerks' is then analysed. The calculation is detailed and exact. Clearly a small parameter limit is involved in the definition of these jerky motions.

In Herzenberg's dynamo model[7] two spheres of conducting fluid rotate uniformly about inclined axes within a large sphere of conducting fluid at rest. The velocity discontinuities simplify the proof; they can be regarded as an implicit limit of a shear layer. Herzenberg also uses explicit limits; the ratio of the radius of the spheres to their distance apart must be small, and so must the ratio of their distance apart to the radius of the large sphere. As in Backus' paper, the calculation is detailed and exact.

Braginskiĭ's[5,6] model is more realistic. His geometry is spherical. He uses a large axisymmetric toroidal motion, in terms of which the large magnetic Reynolds number R is defined, together with poloidal and non-axisymmetric motions smaller by factors of R and $R^{1/2}$. He then expands the magnetic field in ascending powers of $R^{-1/2}$ as far as a term representing dynamo action, to obtain a solution asymptotically valid as R tends to infinity.

The geometry and motions considered here

Two steady two-dimensional motions of a uniformly conducting infinite fluid are proposed in this chapter. They are

$$\mathbf{u} = (2U \cos lz \cos ky, V \sin lz, W \sin ky)$$

and

$$\mathbf{u} = (U \sin(lz+ky), V \sin 2lz, W \sin 2ky)$$

The magnetic field amplified is written as

$$\mathbf{B} = \mathbf{H}(y, z) \exp(\omega t + ijx)$$

where $(\mathbf{H}y, z)$ is periodic with the same periods in the two directions as the motions.

In the above, j, k, and l are real wave numbers in the three directions, $i = (-1)^{1/2}$, and ω can be complex; dynamo action requires that its real part be non-negative. Suffixes are avoided at this stage, and kept for the Fourier components. The factor of 2 in the first motion is introduced for later numerical convenience.

The geometry and motions have been chosen for maximum simplicity. Clearly the Cartesian geometry is a useful simplification. The use of an infinite fluid means that there are no boundary conditions to consider; the application of the results to a real

Dynamo waves

fluid depends on the unproved assumption that fitting the velocities into a large box or sphere and applying suitable boundary conditions has only a small effect on the rate of growth and on the form of the growing field at points away from the edge of the sphere. The forms assumed for **u** and **B** reduce the magnetic field equation to a partial differential equation in two variables, a useful simplification in view of the theorem stated above that a magnetic field which is a function only of two Cartesian variables cannot be amplified indefinitely by a fluid motion. The velocities have been chosen on the basis of the qualitative arguments given in the second section (below); their precise form facilitates the application of Fourier analysis.

Methods of solution

The partial differential equation for $\mathbf{H}(y,z)$, where $\mathbf{H}(y,z)$ is periodic with the same period as the motions, is Fourier analysed. In the resulting equations the dimensionless ω appears as an eigenvalue of a linear operator on the space of Fourier coefficients.

The first method of solution is a small-parameter one, in which an asymptotic form for $\mathcal{R}(\omega)$ is obtained in a limit. It is thus related to the previous work described above. It is shown that in a certain limit attention can be confined to a very small number of Fourier coefficients in obtaining the first term of an asymptotic series for ω. The method is outlined on p. 619.

The second method of solution is numerical, and makes no use of small-parameter methods. The linear operator is represented by a linear operator on a very large finite number of Fourier coefficients; the eigenvalue with the largest real part is then found, with the corresponding eigenvector, using a digital computer. It is shown, theoretically and numerically, that the solution converges as the number of Fourier coefficients considered tends to infinity. Results obtained by this method are described on p. 621.

The Two Dynamo Mechanisms

Illustrative motions

The arguments of this section are qualitative rather than quantitative. They are included to give some physical picture of why the two motions were chosen and of how the dynamo operates.

We consider first the two motions $\mathbf{u} = (\cos z, \sin z, 0)$ and $\mathbf{u} = (\sin z, \sin 2z, 0)$. Both are functions only of z and are in the (x, y) plane. Each fluid plane normal to the z axis moves with a velocity in the plane which is a periodic function of z, with period 2π. The velocities are represented as polar curves in the (x, y) plane, for values of z from 0 to 2π, in figures 1 and 2.

Now we consider the effect of these motions on a single field line initially along the z axis. We neglect the diffusion term $\lambda \nabla^2 \mathbf{B}$ in the equation, so that the magnetic field line can be regarded as 'frozen in' in the manner described in the first section. So the diagrams for the shape of the field line after unit time are precisely the curves of figures 1 and 2 for the two motions, with the values of z at each point of the curves as marked. The line becomes a spiral for the first motion and a double spiral for the second.

Next we consider the effect of these two motions on the initial magnetic field $\mathbf{B} = (0, 0, \cos x)$. The initial field lines are all in the z direction, with the field alternating in sign with increasing x.

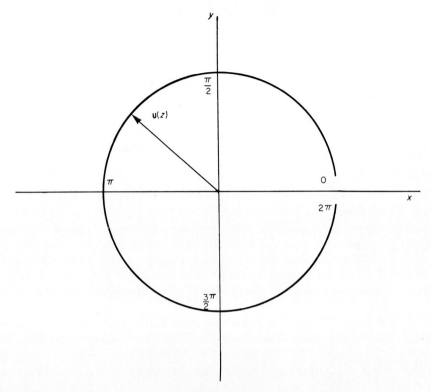

FIGURE 1 Velocity represented as polar curve in the (x, y) plane for the first motion $(\cos z, \sin z, 0)$

We allow for diffusion by regarding the field lines as 'frozen in' for a time T and then examining the effect of diffusion for time T on the resulting field. This approximation is sufficient for the qualitative arguments of this chapter.

With the field lines 'frozen in', then, each one is distorted in exactly the same way as the single field line described above. The effect of the first motion is illustrated in figure 3, which shows the directions and shapes after time $T = \pi/2$ of the field lines initially through the points $x = 0$, π, 2π, and 3π on the x axis. These field lines were initially in alternating directions. The effect of the second motion is illustrated in figure 4, which shows the shape after time $T = \pi/2$ of the field lines initially through the points $x = 0$ and $x = \pi$. For both motions, the field at time T is

$$\mathbf{B}(x, z) = \cos\{x - Tu_x(z)\}\left\{(0, 0, 1) + T\frac{d}{dz}\mathbf{u}(z)\right\} \tag{3}$$

This field is readily shown to satisfy equations (1) and (2), with $\lambda = 0$.

The diffusion term $\lambda\nabla^2\mathbf{B}$ tends to smooth out the variations of the magnetic field. We examine now the effect of a time T of diffusion on the magnetic field given by equation (3).

Clearly, $\lambda(\partial^2/\partial x^2)\,\mathbf{B}(T) = -\lambda\mathbf{B}(T)$. So diffusion in the x direction for time T causes a decay of the magnetic field by a factor $\exp(-\lambda T)$.

Dynamo waves

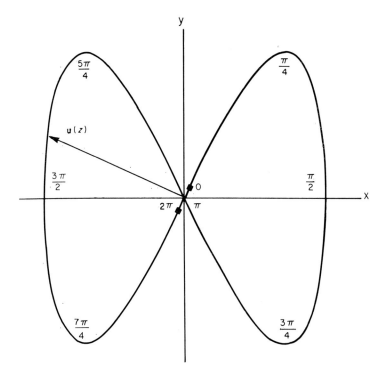

FIGURE 2 Velocity represented as polar curve in the (x, y) plane for the second motion $(\sin z, \sin 2z, 0)$.

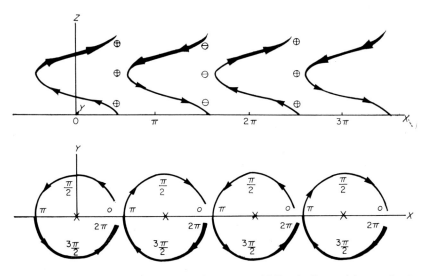

FIGURE 3 For the first illustrative motion, each field line is distorted into a simple spiral. The broader parts of the lines indicate nearness to the reader, that is, larger z and more negative y. The reinforcement of the y components for x near odd multiples of $\pi/2$ is obvious; the sign of the y component is marked at these points in the (x, z) plane. An added shear velocity $u_z = Wy$ has the simultaneous effects of regenerating the original field and causing the field pattern to move in the positive x direction

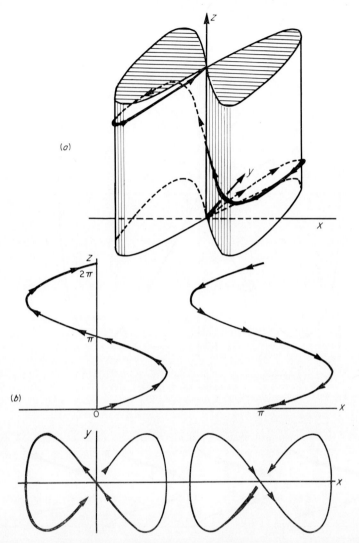

FIGURE 4 For the second illustrative motion, each field line is distorted into a double spiral. (a) This shows how each field line can be regarded as wrapped round a double cylinder. (b) This follows the style of figure 3 with the thicker lines indicating nearness to the reader. The y components reinforce at multiples of π, so that an added shear velocity $u_z = Wy$ can regenerate the original field

Diffusion also tends to reduce the field to its mean with respect to z, that is to

$$\frac{1}{2\pi}\int_0^{2\pi} \mathbf{B}(x,z)\,dz$$

For the two motions this is readily shown to be $(0,\ a_1 \sin x,\ b \cos x)$ and $(0,\ a_2 \cos x,\ b \cos x)$ respectively, where

$$b(T) = \frac{1}{2\pi}\int_0^{2\pi} \cos(T\cos z)\,dz = J_0(T)$$

Dynamo waves

$$a_1(T) = \frac{T}{2\pi}\int_0^{2\pi} \sin(T\cos z)\cos z \, dz = TJ_1(T)$$

$$a_2(T) = \frac{2T}{2\pi}\int_0^{2\pi} \cos(T\sin z)\cos 2z \, dz = 2TJ_2(T)$$

Figures 3 and 4 illustrate these results. They show how for the particular case $T = \pi/2$ the y components of the field reinforce each other at different values of z for certain values of x. For the first motion this reinforcement is for x near to odd multiples of $\pi/2$, corresponding to the y component $a_1 \sin x$ in the magnetic field averaged with respect to z. Similarly for the second motion the reinforcement is for x near to multiples of π, corresponding to the y component $a_2 \cos x$ in the mean magnetic field. This reinforcement is the crucial point of the dynamo mechanism; the y components can be distorted by a shear velocity component $u_z(y)$ to regenerate and to amplify the initial magnetic field.

So the average of the magnetic field with respect to z, after time T, is, approximately for the two motions,

$$\mathbf{B} = \exp(-\lambda T)\,(0,\, a_1 \sin x,\, b \cos x)$$

$$\mathbf{B} = \exp(-\lambda T)\,(0,\, a_2 \cos x,\, b \cos x)$$

respectively.

The fact that $|b| = |J_0(T)| < 1$ indicates that the effect of these two motions is to accelerate the ohmic decay of the initial magnetic field $\mathbf{B} = (0, 0, \cos x)$. Dynamo action depends on the addition to the two motions of z components.

We consider now the added shear velocity component given by $u_z = Wy$. For the qualitative arguments of this section we ignore the modification made by convection by this velocity component to the results of the previous paragraphs. The only other effect is a rate of growth of the z component of the magnetic field equal to the y component multiplied by W.

Thus we can approximate the magnetic field averaged with respect to z after time T by

$$\mathbf{B} = \exp(-\lambda T)\,(0,\, a_1 \sin x,\, b \cos x + c_1 \sin x)$$
$$\mathbf{B} = \exp(-\lambda T)\,(0,\, a_2 \cos x,\, (b+c_2)\cos x)$$

respectively, for the two motions, where

$$c_1'(T) = Wa_1(T) \quad \text{and} \quad c_2'(T) = Wa_2(T)$$

For the first motion, the amplitude of the z component has been multiplied by $\exp(-\lambda T)(b^2 + c_1^2)^{1/2}$, and the phase has been changed so that the field pattern has been shifted in the x direction through a distance $\tan^{-1}(c_1/b)$.

For the second motion, the amplitude has been multiplied by $\exp(-\lambda T)(b+c_2)$; the phase has not been changed.

This completes the demonstration that motions of this type might give dynamo action. For the above results indicate that for suitable W an initial magnetic field $(0, 0, \cos x)$ might have the amplitude of its sinusoidal z component arbitrarily increased. For the first motion the field pattern would also be shifted steadily in the positive x direction.

These qualitative arguments suggest the detailed study of the motions $\mathbf{u} = (U \cos lz, V \sin lz, Wy)$ and $\mathbf{u} = (U \sin lz, V \sin 2lz, Wy)$. The difficulty is that the component $u_z = Wy$ is not sinusoidal in form and not amenable to Fourier analysis. An attempt has been made to prove dynamo action by these motions assuming the following form for the magnetic field:

$$\mathbf{B} = \sum_{-\infty}^{\infty} \mathbf{B}_n(y) \exp(inlz) \exp(\omega t + ijx)$$

But no conclusive results were obtained

The motions and the mechanism of dynamo action

This failure led to the consideration of an alternating shear velocity component in which $u_z = Wy$ is replaced by $u_z = W \sin ky$. This velocity is periodic and amenable to Fourier analysis. But the shear rate $(d/dy)u_z$ is alternating; it is positive near even

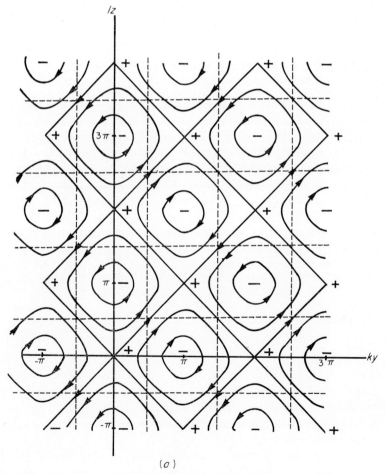

(a)

FIGURE 5(a) Streamlines in the (y, z) plane for the y and z components of the first motion for the case $Vk = Wl$

Dynamo waves

values of ky/π but negative near odd values. So one of the remaining components of each illustrative motion needs to be changed as well to give dynamo action corresponding to the simple pictures above.

It is attractive to use motions with symmetry between the y and z co-ordinates. So we write the two motions as $\mathbf{u} = (u_x, V \sin lz, W \sin ky)$ and $\mathbf{u} = (u_x, V \sin 2lz, W \sin 2ky)$, where the x components are to be chosen as symmetric functions of y and z and so that dynamo action can be expected on arguments like those used with the illustrative motions in the preceding subsection.

The two motions we use, as stated in the first section, are

$$\mathbf{u} = (2U \cos lz \cos ky, V \sin lz, W \sin ky)$$

$$\mathbf{u} = (U \sin (lz+ky), V \sin 2lz, W \sin 2ky)$$

Before analysing the dynamo action of these motions, we endeavour to describe the motions themselves. Both represent flows of incompressible fluid, since $\nabla \cdot \mathbf{u} = 0$. Further, both are functions of y and z only. Thus each motion can be described as

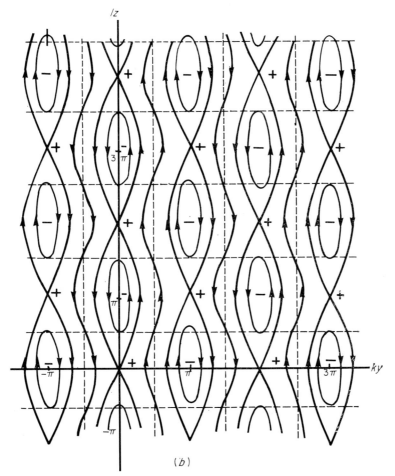

FIGURE 5(b) Streamlines in the (y, z) plane for the y and z components of the first motion for the case $Vk < Wl$

the sum of a motion in the (y, z) plane derivable from a stream function, and a motion in the x direction. For both motions, the streamlines in the (y, z) plane are all closed only if $|Vk| = |Wl|$.

In figures 5(a) and 5(b) are represented in the (y, z) plane the streamlines for the y and z components of the first motion, for the two cases $Vk = Wl$ and $Vk < Wl$. The sign of u_x at each point is also shown. Figures 6(a) and 6(b) represent the same things for the second motion.

The dynamo mechanism for these two motions can be understood on the same basis as in the previous subsection, with the approximation of regarding the field lines initially in the z direction as 'frozen in' for time T, then considering the effect of diffusion and the effect of $u_z(y)$ on the resulting y component. The initial field is $(0, 0, \cos jx)$.

The dynamo mechanisms for the two motions are illustrated in figures 7 and 8 and are described in the associated figure captions.

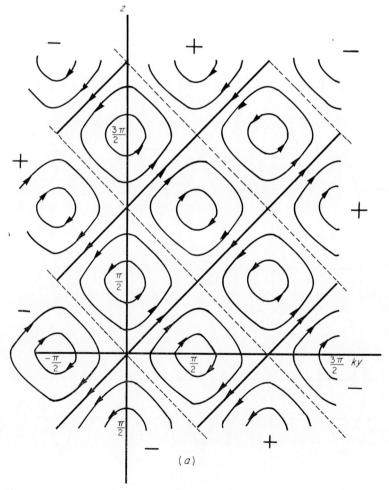

FIGURE 6(a) Streamlines in the (y, z) plane for the y and z components of the second motion for the case $Vk = Wl$

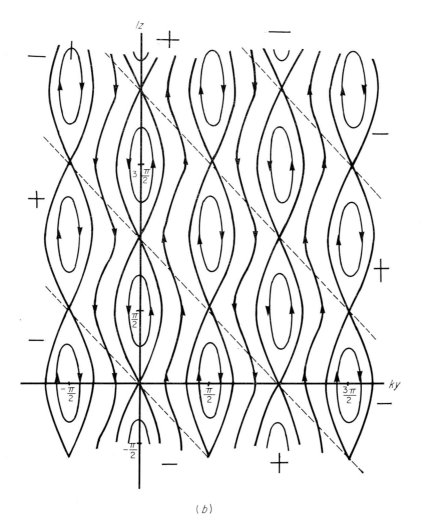

(b)

FIGURE 6(b) Streamlines in the (y, z) plane for the y and z components of the second motion for the case $Vk < Wl$

Analysis of Equations

Equations in dimensionless form

We write the magnetic field equations as

$$\dot{\mathbf{B}} - \lambda\nabla^2\mathbf{B} = -(\mathbf{u}\cdot\nabla)\mathbf{B} + (\mathbf{B}\cdot\nabla)\mathbf{u} \tag{4}$$

$$\nabla\cdot\mathbf{B} = 0 \tag{5}$$

Now we write $\mathbf{B} = \mathbf{H}(y,z)\exp(\omega t + ijx)$, where $\mathbf{H}$ is periodic with the same periodicity as the motion $\mathbf{u}(y,z)$, $i = (-1)^{1/2}$, j is real, and ω is complex in general.

Substituting in equations (4) and (5), we obtain

$$(\omega + \lambda j^2)\mathbf{H} - \lambda\nabla^2\mathbf{H} = -u_x ij\mathbf{H} - (\mathbf{u}\cdot\nabla)\mathbf{H} + (\mathbf{H}\cdot\nabla)\mathbf{u} \tag{6}$$

$$\nabla\cdot\mathbf{H} + ijH_x = 0 \tag{7}$$

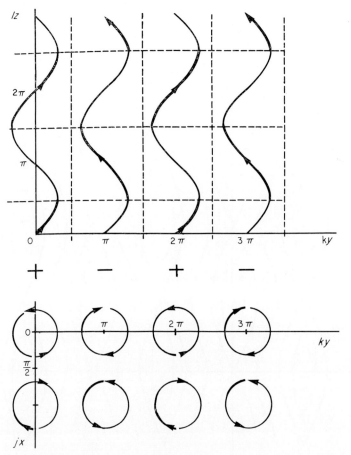

FIGURE 7 The thicker parts of the field lines indicate nearness to the reader, that is, greater x and greater z in the two diagrams. The four field lines marked in the (y, z) diagram were initially in the plane $jx = 0$. The broken lines divide the regions of positive and negative u_x. Each field line is distorted into a simple spiral. The signs between the diagrams indicate the sign of the shear rate $(d/dy)u_z$. Eight field lines are marked in the (x, y) diagram; four were originally in the plane $jx = 0$ and four were in the plane $jx = \pi$ and were in the negative z direction. The field lines are drawn for values of lz from 0 to 2π. The diagram indicates how the y components reinforce for jx equal to odd multiples of $\pi/2$. These y components interact with u_z to shift the original field pattern in the direction of increasing x and maintain or increase its amplitude. It may be seen that, for $jx = \pi/2$ and for each value of y, the reinforcing y components have the same sign as the shear rate $(d/dy)u_z$. Thus the shear produces a field in the positive z direction at
$$jx = \pi/2$$

Equation (7) determines H_x in terms of H_y and H_z, and the y and z components of equation (6) do not involve H_x. So it is sufficient to confine attention to these components and to use (7) to determine H_x. This is in accord with the remark in the first section that the equation $\nabla\cdot\mathbf{B} = 0$ is effectively an initial condition.

Now we make the y and z components of equation (6) non-dimensional. We write $\eta = ky$, $\zeta = lz$, and $\mathbf{u} = (Uv_x, Vv_y, Wv_z)$, where $\mathbf{v} = (2\cos\eta\cos\zeta, \sin\zeta, \sin\eta)$ for the first motion, and $\mathbf{v} = (\sin(\eta+\zeta), \sin 2\zeta, \sin 2\eta)$ for the second motion. We then define the two-dimensional vector function $\mathbf{h}(\eta,\zeta)$ by

Dynamo waves

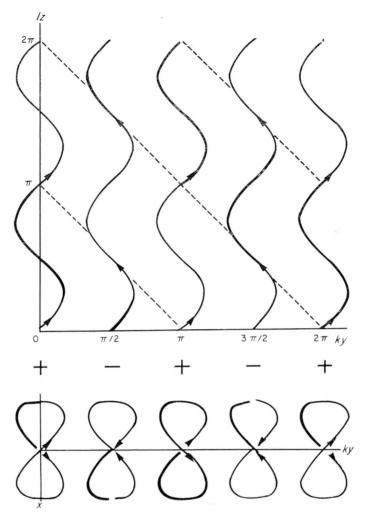

FIGURE 8 The thicker parts of the field lines indicate nearness to the reader, as in figure 7, and the broken lines in the (y, z) diagram divide regions of positive and negative u_x. The five field lines marked in the two diagrams were originally in the plane $x = 0$. Between the two diagrams the sign of the shear rate $(d/dy)u_z$ is indicated. In the (y, z) diagram the direction of the field is marked at the points where the lines cross the plane $x = 0$; at these points the y component is always such as to interact with u_z to regenerate and amplify the original field. This is confirmed in the (x, y) diagram, where the field lines are indicated for values of z from 0 to 2π

$$\mathbf{h} = (h_y, h_z) = (kH_y, lH_z)$$

Then

$$(\omega)\mathbf{h} + (\lambda j^2)\mathbf{h} - (\lambda k^2)\frac{\partial^2}{\partial \eta^2}\mathbf{h} - (\lambda l^2)\frac{\partial^2}{\partial \zeta^2}\mathbf{h} = -(Uj)iv_x\mathbf{h} - (Vk)v_y\frac{\partial}{\partial \eta}\mathbf{h} - (Wl)v_z\frac{\partial}{\partial \zeta}\mathbf{h}$$

$$+ \left\{(Vk)h_z\frac{\partial}{\partial \zeta}v_y, (Wl)h_y\frac{\partial}{\partial \eta}v_z\right\} \qquad (8)$$

The bracketed quantities in equation (8) all have dimension $[T]^{-1}$. They are, respectively, the orders of magnitude of the terms $(\partial/\partial t)$, $\lambda(\partial^2/\partial x^2)$, $\lambda(\partial^2/\partial y^2)$, $\lambda(\partial^2/\partial z^2)$, $u_x(\partial/\partial x)$, $u_y(\partial/\partial y)$, and $u_z(\partial/\partial z)$. We make the equations non-dimensional by dividing through by Uj; this preserves the symmetry between the y and z co-ordinates.

Let us define the dimensionless numbers

$$P = \frac{2\omega}{Uj}$$

$$Q = \frac{2\lambda j^2}{Uj}, \quad A = \frac{2\lambda k^2}{Uj}, \quad B = \frac{2\lambda l^2}{Uj}$$

$$Y = \frac{Wl}{Uj}, \quad Z = \frac{Vk}{Uj}$$

$$X = P + Q$$

Now we write

$$\mathbf{h} = \sum \mathbf{h}_{m,n} \exp\{i(m\eta + n\zeta)\}$$

where for the first motion the sum is over all integers m and n, while for the second motion it is over all pairs of integers m, n with $m+n$ even, corresponding to the form of $\mathbf{v}(\eta,\zeta)$ for the second motion. We then write $\mathbf{h}_{m,n} = (y_{m,n}, z_{m,n})$ to avoid the use of too many suffixes and substitute in equation (8).

Then, for the first motion,

$$(X + m^2 A + n^2 B)\mathbf{h}_{m,n} = -i(\mathbf{h}_{m+1,n+1} + \mathbf{h}_{m+1,n-1} + \mathbf{h}_{m-1,n+1} + \mathbf{h}_{m-1,n-1})$$

$$+ mZ(\mathbf{h}_{m,n+1} - \mathbf{h}_{m,n-1})$$

$$+ nY(\mathbf{h}_{m+1,n} - \mathbf{h}_{m-1,n})$$

$$+ \{Z(z_{m,n+1} + z_{m,n-1}), Y(y_{m+1,n} + y_{m-1,n})\} \tag{9}$$

For the second motion,

$$(X + m^2 A + n^2 B)\mathbf{h}_{m,n} = \mathbf{h}_{m+1,n+1} - \mathbf{h}_{m-1,n-1}$$

$$+ mZ(\mathbf{h}_{m,n+2} - \mathbf{h}_{m,n-2})$$

$$+ nY(\mathbf{h}_{m+2,n} - \mathbf{h}_{m-2,n})$$

$$+ \{2Z(z_{m,n+2} + z_{m,n-2}), 2Y(y_{m+2,n} + y_{m-2n})\} \tag{10}$$

The following points should be noted about equations (9) and (10). The terms on the right-hand sides correspond, respectively, to convection by u_x, i.e. the term $-u_x(\partial/\partial x)B$, convection by u_y, convection by u_z, and to distortion, i.e. the term $(B.\nabla)\mathbf{u}$. A coupling between Fourier coefficients with only the m different involves a factor of Y. A coupling with only the n different involves a factor of Z. Convection couplings are anti-Hermitian. In other words, if the term $\alpha \mathbf{h}_{s,r}$ appears as a convection term on the right-hand side of the $\mathbf{h}_{p,q}$ equation, then the term $-\bar{\alpha}\mathbf{h}_{p,q}$ appears on the right-hand

Dynamo waves

side of the $\mathbf{h}_{r,s}$ equation, where $\bar{\alpha}$ denotes the complex conjugate of α. Distortion couplings have the following property. If the term $\alpha z_{r,s}$ appears on the right-hand side of the $y_{p,q}$ equation, then the term $\alpha z_{p,q}$ appears on the right-hand side of the $y_{r,s}$ equation. The same applies with y and z exchanged throughout.

Finally, it can be shown that there is no loss of generality in the assumption that Uj is positive, and therefore also A and B. In this case, dynamo action requires $\mathcal{R}(P) \geq 0$.

Behaviour of solutions at infinity

In this subsection we analyse the behaviour of solutions $\mathbf{h}_{m,n}$ of the above equations as $|m|$ and $|n|$ tend to infinity.

Let $\mathbf{h}_{m,n}$ be a solution of equation (9) above. We define S_r as the set of points (m,n) with $\max(|m|,|n|) = r$. We define

$$h_r = \max_{(m,n) \in S_r} (|y_{m,n}|, |z_{m,n}|)$$

In other words, h_r is the largest component magnitude on S_r. We can suppose that $h_r = |y_{p,r}|$, where $|p| \leq r$. Now

$$(X + Ap^2 + Br^2)y_{p,r} = -i(y_{p+1,r+1} + y_{p+1,r-1} + y_{p-1,r+1} + y_{p-1,r-1})$$
$$+ pZ(y_{p,r+1} - y_{p,r-1})$$
$$+ rY(y_{p+1,r} - y_{p-1,r})$$
$$+ Z(z_{p,r+1} + z_{p,r-1})$$

Let us take the modulus of each side and use the fact that $|y_{m,n}| \leq h_r$, $|z_{m,n}| \leq h_r$ for $(m,n) \in S_r$. Then

$$|(X + Ap^2 + Br^2)|h_r \leq a_r h_{r-1} + b_r h_r + c_r h_{r+1}$$

for $r \geq 1$, where a_r, b_r, and c_r are numerical constants and are all less than the sum of the moduli of the coefficients on the right-hand side, that is, less than $4 + 2|pZ| + 2|rY| + 2|Z|$.

We now write this inequality in a form independent of which particular y or z component on S_r has modulus h_r. A suitable form is

$$\{\mathcal{R}(X) + r^2 \min(|A|, |B|)\} h_r \leq \{4 + (2r+2)(|Y| + |Z|)\}(h_{r-1} + h_r + h_{r+1})$$

We write

$$C = \frac{\min(|A|, |B|)}{4(|Y| + |Z| + 2)}$$

Then the above equations can be readily shown to imply that there is an r_0 such that, for $r \geq r_0$, $rCh_r < h_{r-1} + h_{r+1}$.

The same applies to the second motion, with H defined slightly differently and with S_r as the set of points (m, n) with $|m| + |n| = 2r$.

For sufficiently large r, rC is large and at least one of h_{r-1} and h_{r+1} must be large compared with h_r. If $h_{r+1} \gg h_r$ then $h_{r+2} \gg h_{r+1}$, etc., and if $h_{r-1} \gg h_r$ then $h_{r-2} \gg h_{r-1}$, etc.

Thus, as $r \to \infty$, h_r must either grow very rapidly or decay very rapidly. Specifically, there are two possibilities. Either there is a positive constant α such that $h_{r+1} > \alpha r! C^r$ and $h_r \to \infty$, or there is a constant β such that $h_r < \beta/(r! C^r)$.

If X, $\mathbf{h}_{m,n}$ satisfy one of the equations above, then the following four conditions, proved equivalent by the preceding calculation, are necessary for physical relevance:

$$\mathbf{h}_{m,n} \to 0 \quad \text{as} \quad m^2 + n^2 \to \infty$$

$$\sum |\mathbf{h}_{m,n}|^2 < \infty$$

the energy density $(1/16\pi)\sum |B_{m,n}|^2$ is finite

there is constant β such that $h_r < \beta/(r! C^r)$

Eigenvector interpretation

Let us consider the infinite dimensional vector space $\mathscr{V} = \{\mathbf{h}_{m,n}\}$ and the Hilbert space $\mathscr{H} = \{\mathbf{h}_{m,n} : \sum |\mathbf{h}_{m,n}|^2 < \infty\}$.

Either of the equations (9) and (10) can be written

$$X\mathbf{h} = T\mathbf{h}$$

where $\mathbf{h} \in \mathscr{V}$ and T is a linear operator on $\mathscr{V}$. Thus X is an eigenvalue of T. It was shown in the last subsection that, for physical relevance, $\mathbf{h} \in \mathscr{H}$. There is a countable set of generalized eigenvectors of T which are in $\mathscr{H}$ and are complete in $\mathscr{H}$. Discussion of this statement is reserved for later publication. Therefore, any given initial small magnetic field can be expressed as a sum of generalized eigenvectors and its behaviour with time thus determined. Either all the eigenvectors decay to zero or there is dynamo action and one or more grow.

Small-parameter Results

Discussion

In order to give an existence proof for dynamo action it is only necessary to show that for some choice of A, B, Y, Z there is an eigensolution X, $\mathbf{h}_{m,n}$ with $\mathscr{R}(X) > 0$. For then we can choose Q, which is independent of A, B, Y, and Z, so that $Q < \mathscr{R}(X)$, and $\mathscr{R}(P) > 0$. For an analytical proof of dynamo action, as opposed to a numerical demonstration, some sort of small-parameter limit is required here.

In this section it is shown for each motion that, if a number of inequalities of the kind $A \ll B$ are satisfied, then the dominant term in $\mathscr{R}(X)$ is positive. Such inequalities have strict mathematical meaning in terms of limiting processes where the ratio of the two terms tends to zero. For example, $A = N^2$ and $B = N^3$ in the limit $N \to \infty$ is a limiting process giving the inequality $A \ll B$.

The results below are expressed in terms of the inequalities, and also of particular limits giving these inequalities is given for each motion.

Method

The method of proof is not given here in full detail, but the approach used is described. We write **h** for the vector $\{\mathbf{h}_{m,n}\}$ in $\mathscr{V}$. We write the equations as

$$(X+m^2A+n^2B)\mathbf{h}_{m,n} = (S\mathbf{h})_{m,n} \tag{11}$$

where S is a linear operator on $\mathscr{V}$ defined by one of the equations (9) and (10).

It can be shown that it is sufficient and consistent to confine attention to fields with $\mathbf{h}_{0,0} = (0, 1)$.

We define $\mathbf{h}_{m,n}(X)$ as follows:

$$\mathbf{h}_{0,0} = (0, 1)$$

$$(X+m^2A+n^2B)\mathbf{h}_{m,n} = (S\mathbf{h})_{m,n} \quad \text{for } (m, n) \neq (0, 0) \tag{12}$$

$$\mathbf{h}_{m,n} \to 0 \text{ at infinity}$$

Then, if X is an eigenvalue, $X\mathbf{h}_{0,0} = (S\mathbf{h})_{0,0}$.

Now, with the assumption $|X| \ll A \ll B$, it is reasonable to suppose that $\mathbf{h}_{m,n}(X)$ is only a very slowly varying function of X. Thus, even if X is not very close to the eigenvalue, $(S\mathbf{h}(X))_{0,0}$ should be very close to being $\mathbf{h}_{0,0}$ multiplied by the eigenvalue.

$\mathbf{h}_{m,n}(X)$ is found by an iterative process; a number of assumed inequalities ensure its rapid convergence so that it is only necessary to consider a few terms in obtaining the dominant term of the eigenvalue X.

The iteration proceeds as follows. Let us define $\mathbf{g}^{(0)}$ by

$$\mathbf{g}^{(0)}_{0,0} = (0, 1)$$

$$\mathbf{g}^{(0)}_{m,n} = \mathbf{0}, \quad \text{for } (m, n) \neq (0, 0)$$

Let us define $\mathbf{g}^{(k+1)}$ for $k \geq 0$ by

$$\mathbf{g}^{(k+1)}_{0,0} = \mathbf{0}$$

$$(X+m^2A+n^2B)\mathbf{g}^{(k+1)}_{m,n} = (S\mathbf{g}^{(k)})_{m,n}, \quad \text{for } (m, n) \neq (0, 0)$$

Then, if $\sum_{k=0}^{K} \mathbf{g}^{(k)}$ converges as $K \to \infty$, then $\mathbf{h}(X) = \sum_{0}^{\infty} \mathbf{g}^{(k)}$ satisfies the conditions (12) on $\mathbf{h}(X)$. The inequality $\max(1, |Y|, |Z|) \ll \min(A, B)$ ensures that $\mathbf{g}^{(k)} \to 0$ rapidly, but it excludes dynamo action. The inequality $\max(1, |Y|, |Z|) \ll B$ is applied, and sufficient weaker conditions on A are applied to ensure the rapid convergence of the iteration. The conditions are given below for the two motions; the detailed proof that they are sufficient is not presented.

For each motion the final inequality is to ensure that the term representing dynamo action dominates over the effect of the x component of the velocity in accelerating the decay.

Result for the first motion

In a limit where

$$B \gg A, \quad B \gg Y, \quad AB \gg 1, \quad AB \gg Z^2, \quad A^2B \gg YZ, \quad \text{and} \quad Y^2Z^2 \gg A^3B$$

the dominant terms in X for the first motion are

$$\mathscr{R}(X) = \frac{56Y^2Z^2}{A^3B^2}, \qquad \mathscr{I}(X) = -\frac{8YZ}{AB}$$

One particular limiting process which gives this result is as follows. For $A = N^2$, $B = N^5$, $Y = N^4$, $Z = N^2$, and $Q = N^{-4}$, as $N \to \infty$,

$$\mathscr{R}(P) \sim 55N^{-4}, \qquad \mathscr{I}(P) \sim -8N^{-1}$$

$$j:k:l = Q^{1/2}:A^{1/2}:B^{1/2} = 1:N^3:N^{9/2}$$

$$U:V:W = 1:\frac{Zj}{k}:\frac{Yj}{l} = 1:N^{-1}:N^{-1/2}$$

$$\lambda = \frac{U}{2j}N^{-4}$$

Result for the second motion

In a limit where

$$B \gg A, \quad B \gg Y, \quad AB \gg 1, \quad AB \gg Z^2, \quad A^2B^2 \gg YZ, \quad \text{and} \quad YZ \gg AB$$

X is real and its dominant term is $X = 3YZ/AB^2$.

One particular limit which gives this result is as follows. For $A = 1$, $B = N^6$, $Y = N^5$, $Z = N^2$, and $Q = N^{-6}$, as $N \to \infty$,

$$P \sim 3N^{-5}$$

$$j:k:l = 1:N^3:N^6$$

$$U:V:W = 1:N^{-1}:N^{-1}$$

$$\lambda = \frac{U}{2j}N^{-6}$$

Discussion of results

These two small-parameter dynamo models are remarkable in that they depend on a small magnetic Reynolds number limit. In other words, the situation is in a sense dominated by diffusion. The field amplified has $h_{0,0}$ as its dominant term, representing the field $\mathbf{B} = (0, 0, 1)\,\mathscr{R}\{\exp(\omega t + ijx)\}$. The distortions of this field by the slow fluid motions against the rapid decay rates associated with the length scales in the y and z directions are very small. But they are sufficient to maintain the magnetic field against the very slow decay rate associated with the length scale in the x direction. The iteration used in the calculation of $\mathbf{h}(X)$ as described on p. 619 makes it clear that the limiting process used is a small magnetic Reynolds number limit, since the diffusion term at the $k+1$ order is balanced with the other terms at the k order.

It remains to compare these results with the earlier dynamo models described in the first section on grounds of typical relevance and of simplicity. These motions share with Braginskiĭ's the property of being continuous functions of position and time. His motions and geometry have greater physical relevance because they do not involve multiple length scales and because the geometry can be spherical. It is felt that the dynamo mechanism is more easily understood for the motions reported here.

Dynamo waves

Numerical Results

Practical values of the dimensionless numbers

The calculations of the previous section establish the possibility of dynamo maintenance by flows of this type. But the motions considered there, with their multiple length scales and velocity scales, are unlikely to occur in physical dynamos. Numerical methods are used to demonstrate and analyse dynamo action for more relevant length and velocity scales.

For an actual flow, U, V, W, k, and l are prescribed. We want to know for what values of j there are eigensolutions with $\mathscr{R}(\omega)$ positive.

It can be shown that it is sufficient to confine attention to positive U, V, j, k, l, and, for the first motion, positive W. Here we confine attention to positive W for the second motion as well; the arguments of the second and fourth sections and some numerical investigation indicate that dynamo action by the second motion requires that Y and Z have the same sign.

We further confine attention to $V = W$ and $k = l$, implying $A = B$ and $Y = Z$. We do not assume $U = V$ since the geometry of the motion singles out the x direction. Thus the problem is defined by the quantities U, V, k, and λ, in terms of which two dimensionless numbers $S = U/V$ and $R = V/\lambda k$, the magnetic Reynolds number, can be defined. We want to find the dependence of the dimensionless ω defined by $G + iH = \omega/Vk$ on the dimensionless length scale in the x direction, defined by $L = k/j$, for a range of values of R and S.

From these definitions and those on p. 616, $Y = L/S$, $A = 2Y/R$, $P = 2Y(G+iH)$, and $Q = 2/RS^2Y$. The numerical results give X as a function of A and Y, and therefore as a function of R and Y. We then have

$$G + iH = \frac{P}{2Y} = \frac{X}{2Y} - \frac{1}{RS^2Y^2}$$

The procedure adopted for each value of R is to find X for each value of Y, and thus to find $G + iH$ as a function of Y and S. It is convenient to use Y as a variable instead of L; they are equivalent since $L = SY$.

For given R, it can be proved that $X(R, Y)$ has the following properties:

$$Y\mathscr{R}(X) \to f(R) \quad \text{as } Y \to \infty, \text{ for both motions}$$

$$\mathscr{I}(X) \to e(R) \quad \text{as } Y \to \infty, \text{ for the first motion}$$

$$\mathscr{I}(X) = 0 \quad \text{for sufficiently large } Y, \text{ for the second motion}$$

$$\mathscr{R}(X) < 0 \quad \text{for sufficiently small } Y, \text{ for both motions}$$

So we have

$$Y^2 G \to \frac{f(R)}{2} - \frac{1}{RS^2} \quad \text{as } Y \to \infty$$

$$YH \to \frac{e(R)}{2} \quad \text{as } Y \to \infty \text{ for the first motion}$$

Thus dynamo action for large Y requires

$$1/S < \Sigma(R)$$

where $\Sigma(R) = \{\mathscr{R}f(R)/2\}^{1/2}$.

The speed with which the field pattern advances in the x direction is determined by the factor $\exp(\omega t + ijx)$ in the magnetic field, and is $-\mathscr{I}(\omega)/j$. It is sensible to make it non-dimensional by dividing by U, the size of the velocity component in the x direction. The resulting dimensionless velocity is $-\mathscr{I}(\omega)/Uj = -\tfrac{1}{2}\mathscr{I}(X)$.

The numerical methods

The basis of the method is to use a digital computer to find the eigenvalue with the largest real part, and the corresponding eigenvector, of the operator on a finite dimensional vector space defined by setting $\mathbf{h}_{m,n} = 0$ for $\max(|m|, |n|) > r$ for the first motion and by setting $\mathbf{h}_{m,n} = 0$ for $|m|+|n| > 2r$ for the second motion, and ignoring the corresponding equations. It has been proved that the solutions obtained converge to solutions of the full problem, but the proof is not presented here.

The numerical results confirm that the solutions obtained converge. For values of Y greater than about 0·1, the value of r needed to give 5% accuracy in the eigenvalue is a function of R and is practically independent of Y. For the first motion R values of 5, 10, and 30 require r values of 3, 4, and 5. For the second motion R values of 4, 8, 16, and 32 require r values of 4, 5, 7, and 11.

The numerical methods used in finding the eigensolutions will not be described here. But the results were checked by adding up the terms in the equations, adding their moduli, and finding the quotient. The results were all of order 10^{-10}, the accuracy of the computer being about eleven significant figures.

The programmes were also checked by finding X for various realizations of the inequalities on pp. 619–621, with L values of 10 and 100; the results confirmed the accuracy both of the programmes and of the small-parameter calculations.

Results for the first motion

Figure 9 shows X as a function of Y, for $R = 2$, in the complex plane. It may be seen that, for $Y \geq 2$, the real part of X is positive. We do not discuss here the behaviour for $Y < 1$, merely pointing out that the eigenvalues marked are genuine eigenvalues and not errors, and that eigenvalues of continuously varying finite dimensional matrices often show such behaviour.

Figure 10(a) shows X again as a function of Y for $R = 2$. Figure 10(b) shows $\mathscr{R}(X)/2Y$ and $1/2Y^2$; the difference between them is G for $R = 2$ and $S = 1$, as a function of Y. It may be seen that G is largest for $Y = L \simeq 6$, and is then approximately 0·02. Further, G is positive for all values of $L = Y$ greater than 4.

Figure 11 shows the various functions of R, as marked. The accuracy estimated for the R values 30 and 100 is not so high as for the other points, but there is no doubt about the reversal in gradient of $\Sigma(R)$. The fact that $\Sigma(R)$ increases with R indicates that dynamo action occurs increasingly readily with increasing R, as might have been expected but was certainly not obvious.

Figure 11(a) shows that dynamo action can only occur for R greater than about 0·85. Figure 11(b) indicates that in the limit $Y \to \infty$ the velocity of the field pattern in the x direction ranges from $0·7U$ to $2U$, and in the limit of large R is apparently exactly $2U$, the same as the largest value of the x component of the motion.

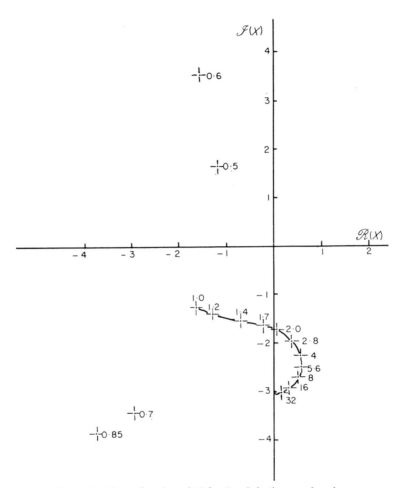

FIGURE 9 X as a function of Y, for $R = 2$, in the complex plane

Results for the second motion

Figure 12(a) shows X as a function of Y for $R = 4$, and also shows $f(4)/Y$. It may be seen that X is positive for $Y > 0.63$.

Figure 12(b) shows $X/2Y$ as a function of Y for $R = 4$, and also gives the curve $1/16\,Y^2$. The difference gives G as a function of Y for $R = 4$ and $S = 2$. It may be seen that the maximum value of G is about 0.045 and that at that point Y is 1·2 and L is 2·4.

Figure 13(a) shows $f(R)$, and figure 13(b), $\Sigma(R)$, for values of R up to 32. All the values are accurate to 1%. Again, $\Sigma(R)$ increases, showing that dynamo action occurs increasingly readily with increasing R. It may be seen that there is no dynamo action for R less than about 2.

Conclusion

Discussion of the dynamo mechanisms

For the first motion the field pattern being amplified moves in the x direction. This property is thought to be a common one; the second motion is the exception rather

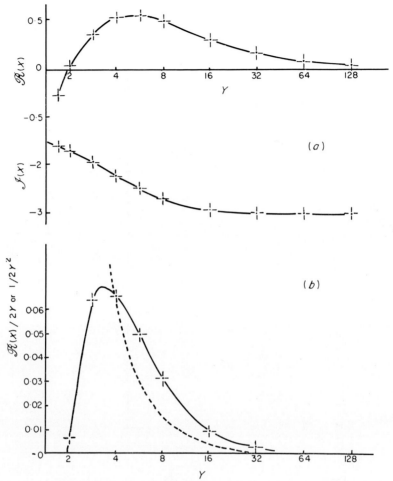

FIGURE 10 (a) X as a function of Y for $R = 2$. (b) full curve, $\mathcal{R}(X)/2Y$; broken curve, $1/2Y^2$

than the rule in that the field pattern is stationary. It is worth noting that in the papers referred to on pp. 603, 604 Braginskiĭ obtains a $\mathbf{v}_{\text{eff}}$ which is analogous to the motion in the x direction for this model.

Clearly there is a special difficulty in fitting the first dynamo model into a box or sphere, since the field pattern cannot move in the x direction indefinitely. This problem can perhaps be circumvented by making the motions transitory or oscillating; if the x component of the motion is reversed, the field continues to grow but the field pattern moves in the opposite direction.

For the first motion it can be shown readily that there is dynamo action with the same growth rate whatever the signs of U, V, and W. For the second motion dynamo action appears to require that V and W have the same sign. Herzenberg's model, described in the first section, has a similar property. But it is possible that with Herzenberg's model an oscillatory growing solution is possible if the direction of rotation of one of the spheres is reversed, and that this property of our second motion, if true, is exceptional.

Dynamo waves

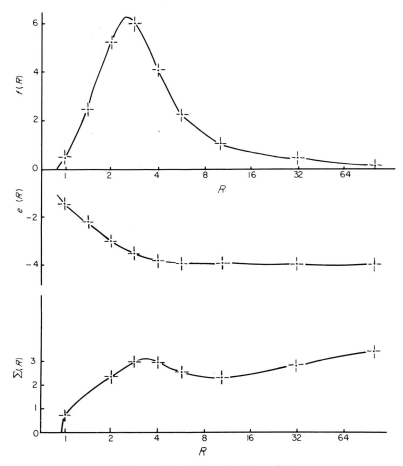

FIGURE 11 Various functions of R

Significance of the results

Clearly these motions can give dynamo action. We now discuss the significance of the results.

It is worth reiterating that conclusive results about the dynamo problem are hard to obtain. Simple motions will not give dynamo action, and with complicated motions and geometry it is impossible to give clear, simple, and conclusive proofs.

It is felt that the second section gives a fairly clear picture of the mechanism of dynamo operation, in terms of the shearing of the distorted field regenerating and amplifying the original field and magnetic diffusion keeping down the irregularities.

The small-parameter approach reported here is fairly short, and is perhaps the simplest small-parameter proof of dynamo action. But any work is bound to seem clearer to its author than to others.

This chapter apparently gives the first report of a successful numerical dynamo model without any use of small-parameter expansions. The results described above are therefore remarkable in their detail, and in the range of magnetic Reynolds number considered.

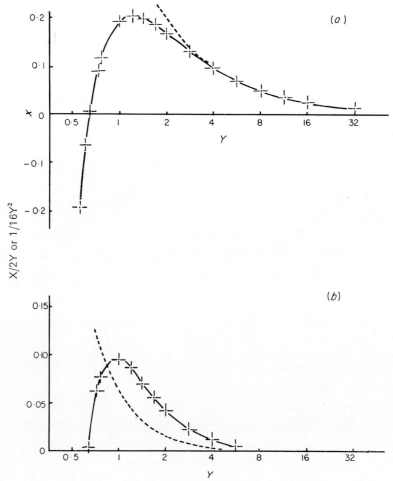

FIGURE 12 (a) full curve, X as a function of Y for $R = 4$; broken curve, $f(4)/Y$. (b) full curve, X as a function of Y for $R = 4$; broken curve, $1/16Y^2$

Application of these methods to other motions

Preliminary investigation indicates that the motion given by

$$\mathbf{u} = (U(\cos ky - \cos kz), V \sin kz, V \sin ky)$$

gives dynamo action far more readily than the two motions described here. This motion satisfies the equation of motion for zero viscosity and for no Coriolis force, Lorentz force, or buoyancy force.

We have confined attention to motions independent of one Cartesian co-ordinate. This was convenient since it effectively reduced the problem to a partial differential equation in two variables. But this property is by no means necessary for dynamo action, not even for reasonably simple proofs of dynamo action. With a reasonable choice of $\mathbf{u}(x, y, z)$, periodic in all three directions and of simple sinusoidal form, and with $\mathbf{B} = \mathbf{H}(x, y, z) \exp(\omega t + ijx)$ where $\mathbf{H}$ is periodic with the same periods as $\mathbf{u}$, dynamo action can apparently be demonstrated in a manner similar to that in this

Dynamo waves

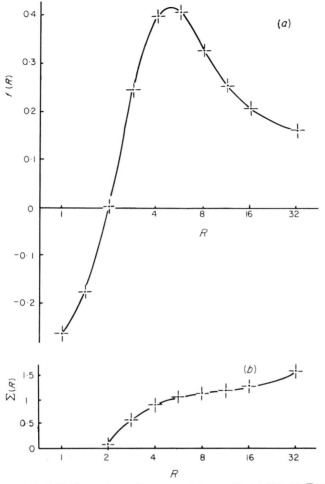

FIGURE 13 This figure shows, for values of R up to 32, (a) $f(R)$, (b) $\Sigma(R)$

chapter. It seems that this has already been done by Childress (see section E II, chapter 5 of this book).

Numerical proofs of dynamo action have been handicapped by the fact that the fields cannot be functions of only two variables. This chapter has shown how it is possible with a motion which is only a function of two variables to obtain a partial differential equation in two variables for the field. It is possible to apply this method to motions which are not periodic; a suggestion is the motion due to convection in a rotating annulus.

Acknowledgements

The author wishes to thank his research supervisor, Dr. H. K. Moffatt, for his constant encouragement and help. The computations were done on 'Titan' in the Cambridge University Mathematical Laboratory, and thanks are due to the Director and staff for making the machine available and for help given. The Science Research Council have supported the author with a Research Studentship during this work.

References

1. G. E. Backus, *Ann. Phys. (N. Y.)*, **4**, 372 (1958).
2. P. H. Roberts, *An Introduction to Magnetohydrodynamics*, 1st ed., Longmans, London, 1967, pp. 65–100.
3. T. G. Cowling, *Monthly Notices Roy. Astron. Soc.*, **94**, 39 (1933).
4. T. G. Cowling, *Quart. J. Mech. Appl. Math.*, **10**, 129 (1957).
5. S. I. Braginskiĭ, *Zh. Eksperim. i Teor. Fiz.*, **47**, 1084 (1964) (*Soviet Phys.—JETP (English Transl.)*, **20**, 725 (1965)).
6. S. I. Braginskiĭ, *Zh. Eksperim. i Teor. Fiz.*, **47**, 2178 (1964) (*Soviet Phys.—JETP (English Transl.)*, **20**, 1461 (1965)).
7. A. Herzenberg, *Phil. Trans. Roy. Soc. London, Ser. A*, **250**, 543 (1958).

5

S. CHILDRESS

New York University
Courant Institute of Mathematical Sciences
New York, U.S.A.

A class of solutions of the magnetohydrodynamic dynamo problem

Introduction

This chapter will be devoted to certain aspects of magnetohydrodynamics which arise in connection with the dynamo theory of the main geomagnetic field. One of the central questions of this theory concerns the existence of kinematic fluid dynamos in a bounded conductor. Under certain broad constraints on the allowable flows, a solution of the kinematic problem provides, in principle, a dynamically self-consistent solution of the complete system of equations with suitable body forces. On the other hand, a given solution of the dynamical equations for a prescribed magnetic field will not in general be a dynamo in the sense of the kinematic problem. This emphasis on the kinematic problem may serve to distinguish dynamo theory from, say, the related dynamical theory of contained rotating fluids.

In the present discussion our object will be to explore the possibility of an explicit construction of a class of kinematic dynamos, and to formulate an approach to the self-consistent dynamo problem, using two basic tools. Firstly, we shall give special attention to the *average* effect of a given fluid motion upon a magnetic field which varies on a relatively larger spatial scale. Secondly, the solutions will be sought in the form of convergent parameter-type expansions.

We shall restrict our study of the existence problem to the steady-state case. This assumption reduces the kinematic problem to an eigenvalue problem, the real eigen-parameter being, physically, the *speed* of the dynamo represented in dimensionless form by the magnetic Reynolds number Rm. In customary notation

$$Rm = U_{max} L_{max} \sigma \mu \qquad (1)$$

It is known that Rm is necessarily bounded away from zero for steady dynamo maintenance in a given bounded domain, so that a parameter-type expansion method must ultimately reduce to the limit process $|Rm| \to \infty$ through a discrete (or continuous) spectrum of values. To apply this idea we must in principle find a family of flows (discrete or continuous), such that for all sufficiently large $A > 0$ the family contains at least one flow with a real eigenvalue whose absolute value exceeds A.

The problems considered below provide examples illustrating the parameter method.

The dynamos are characterized by the existence of two scales of spatial variation of the velocity and magnetic fields. We shall exhibit flows which are uniformly bounded in a small parameter $\varepsilon > 0$, but which give rise to arbitrarily large eigenvalues as $\varepsilon \to 0$. The parameter ε represents the spatial scale of the dominant velocity field relative to that of the magnetic field, and the regenerative induction is due to the average effect of interaction between the small-scale components of the former and the (higher-order) small-scale components of the latter.

This method of analysis, and indeed certain of the results given in the next section, are closely related to the time-dependent dynamo process proposed in 1955 by Parker[1]. The notion of scale separation also appears in various other ways in the existing non-numerical dynamo theory. In Herzenberg's two-sphere steady-state dynamo[2] the ratio of sphere radius to conductor radius is introduced as a small parameter. In Backus' time-dependent dynamo[3] scale separation (with respect to time) occurs as a separation of the dynamo cycle into well-defined periods of motion and arbitrary (but sufficiently long) periods of 'stasis'. More recently Braginskiĭ[4,5] has utilized similar techniques of averaging and expansion to study slightly asymmetric dynamos.

The analysis is divided into three sections. In the next section, the first section of the three, we introduce the problem of induction in an infinite conductor and discuss the steady-state kinematic problem over the trigonometric velocity fields. This leads to the corresponding notion of a spatially periodic dynamo. In the following section the periodic dynamos are utilized in the construction of a class of bounded dynamos in a spherical conductor. Finally, in the last section, we consider briefly the problem of the self-consistent spherical dynamo driven by small-scale body forces. Additional discussion and detailed proofs of results outlined in the second section will be given elsewhere.

Spatially Periodic Dynamos

Formulation

Under the most favorable circumstances we might hope to 'solve' the steady-state kinematic dynamo problem in the following sense. Given an admissible class of normalized motions, necessary and sufficient conditions are found that a given member of the class admit a real eigenvalue. In the present section we shall consider a simpler but related problem in an infinite fluid conductor, for which a 'solution' of this kind is available. In the case to be described the spectrum of eigenvalues is continuous and the dimensionless equations to be solved may be taken to be

$$\nabla \times \mathbf{h} = \mathbf{e} + \mathbf{q} \times \mathbf{h} \tag{2}$$

$$\nabla \cdot \mathbf{h} = 0 \tag{3}$$

$$\nabla \times \mathbf{e} = 0 \quad (\mathbf{e} = \nabla \phi) \tag{4}$$

where $\mathbf{h} = (h_1, h_2, h_3) = \{h_\alpha\}$ is the magnetic field with components h_α relative to a fixed right-handed Cartesian frame, $\mathbf{e} = \{e_\alpha\}$ is the electric field, and ϕ is the electric potential. The vector field $\mathbf{q}(\mathbf{x})$ is the velocity of the conductor at a point $\mathbf{x} = \{x_\alpha\}$ and is to be regarded as a given function.

A class of solutions of the magnetohydrodynamic dynamo problem

The admissible class of motions is defined in the following way. Each **q** is of the form

$$\mathbf{q} = \varepsilon^{1/J-1} \sum_K \boldsymbol{\mu}(\mathbf{k}) \exp\left(\frac{i\mathbf{k}\cdot\mathbf{x}}{\varepsilon}\right) = \varepsilon^{1/J-1} \mathbf{q}' \tag{5}$$

where summation is over the set K of all non-zero vectors having integer coordinates. For convenience we adopt complex exponentials although all fields are understood to be real. In particular the quantities $\boldsymbol{\mu}(\mathbf{k})$ are complex vectors attached to K satisfying $\boldsymbol{\mu}(-\mathbf{k}) = \overline{\boldsymbol{\mu}(\mathbf{k})}$ (complex conjugate). We also assume that only a finite number of the $\boldsymbol{\mu}$ are non-zero, i.e. **q** is a trigonometric field. The number J is a positive integer ≥ 2 to be specified later (see p. 633), and ε is a small expansion parameter. Among the various motions of this class we shall give special attention to the smaller families of solenoidal ($\nabla\cdot\mathbf{q}=0$) fields and Beltrami ($\mathbf{q}\times\nabla\times\mathbf{q}=0$) fields. (A Beltrami field of the form (5) represents a perfect fluid motion, and is of interest here because a dynamo with this property becomes dynamically self-consistent in the limit of vanishing magnetic field.)

We next introduce, for any given real non-zero vector **n**, the complex vector space $V(\mathbf{n})$ generated by linear combinations of quantities like $\mathbf{i}_\alpha \exp\{i(\mathbf{n}+\mathbf{m}/\varepsilon)\cdot\mathbf{x}\}$, where $\mathbf{i}_\alpha$ is a real orthonormal set and **m** is either zero or a vector in K. We shall restrict our solutions of (2)–(4) to lie in this vector space, so that it is sufficient to write the equation for **h** in the operational form

$$\mathbf{h} = -\nabla^{-2}\nabla\times(\mathbf{q}\times\mathbf{h}) \equiv T\mathbf{h} \tag{6}$$

say, where T is an operator defined by an obvious definition of ∇^{-2} on V. If $\mathbf{h} = T\mathbf{h}$ in V and **e** is defined by

$$\mathbf{e} = -\nabla(\nabla^{-2}\nabla\cdot\mathbf{q}\times\mathbf{h})$$

then **h**, **e** solve (2), (4), as is easily shown.

The expansion method

We propose to solve the equation $\mathbf{h} = T\mathbf{h}$ in $V(\mathbf{n})$, for some given admissible **q**, using an operational adaptation of the classical Floquet theory[6]. We first define the projection (or averaging) operator P by

$$P\exp\{i(\mathbf{n}+\mathbf{m}/\varepsilon)\cdot\mathbf{x}\} = 0, \quad \mathbf{m}\neq 0$$

$$= \exp\{i\mathbf{n}\cdot\mathbf{x}\}, \quad \mathbf{m}=0$$

Let

$$\tilde{T} = T - PT, \quad \mathbf{f} = \Gamma(\mathbf{n})\exp(i\mathbf{n}\cdot\mathbf{x})$$

and suppose that $(I-\tilde{T})^{-1}$ exists for sufficiently small ε. It is then a simple matter to show that $\mathbf{h} = (I-\tilde{T})^{-1}\mathbf{f}$ will solve (6) if, and only if, **f** satisfies

$$\mathbf{f} = PT(I-\tilde{T})^{-1}\mathbf{f} \tag{7}$$

That is, (7) appears as an equation of *compatibility*, the compatible **f** being precisely those elements of PV which generate solutions of the form $(I-\tilde{T})^{-1}\mathbf{f}$. Also we note that necessarily $P\mathbf{h} = \mathbf{f}$, i.e. the solution is generated by its projection or mean value. If $\tilde{T}$ is in fact a 'small' operator when ε is small, then the projection $\mathbf{f}$ is the dominant part of $\mathbf{h}$ and a scale separation occurs. It will be evident later that this is indeed what happens here (cf. p. 633).

In Fourier vector form the compatibility equation (7) is equivalent to

$$i\mathbf{n} \times \boldsymbol{\Gamma} - i\mathbf{n}\Phi - A.\boldsymbol{\Gamma} = 0, \quad i\mathbf{n}.\boldsymbol{\Gamma} = 0 \tag{8}$$

where $A = A(\varepsilon^{1/J}, \varepsilon\mathbf{n}) = \{a_{\alpha\beta}\}$ is a complex matrix operator having components $a_{\alpha\beta}$ and $\Phi = \Phi(\varepsilon^{1/J}, \varepsilon\mathbf{n})$ is a complex scalar determined by $\boldsymbol{\Gamma}$. These quantities are defined by

$$A.\boldsymbol{\Gamma} = \sum_{j=2} (i)^{j-1} \varepsilon^{j/J-1} \sum_{(j)} \frac{\boldsymbol{\mu}_j \times \mathbf{m}_{j-1} \times \ldots \times \boldsymbol{\Gamma}}{m_{j-1}^2 \ldots m_1^2} \tag{9}$$

$$n^2 \Phi = i\mathbf{n}.A.\boldsymbol{\Gamma} \tag{10}$$

In (9), $\mathbf{m}_j = \mathbf{k}_1 + \mathbf{k}_2 + \ldots + \mathbf{k}_j + \varepsilon\mathbf{n}$, $\boldsymbol{\mu}_j = \boldsymbol{\mu}(\mathbf{k}_j)$, and $\mathbf{k}_j$ denotes an arbitrary vector from the jth copy K_j of the set K. The symbol

$$\sum_{(j)}$$

is used to denote summation over ordered sets of vectors $(\mathbf{k}_1, \mathbf{k}_2, \ldots, \mathbf{k}_j)$ from j copies of K, such that $\mathbf{k}_1 + \mathbf{k}_2 + \ldots + \mathbf{k}_j = 0$ but $\mathbf{k}_1 + \mathbf{k}_2 + \ldots + \mathbf{k}_i \neq 0$, $i = 2, 3, \ldots, j-1$. We shall refer to this as summation over the irreducible chains of length j. As a final remark concerning (9) we point out that the cross product operations are understood to be applied in succession from right to left, i.e. $\mathbf{a} \times \mathbf{b} \times \mathbf{c}$ means $\mathbf{a} \times (\mathbf{b} \times \mathbf{c})$.

Compatible fields

We may write (8) as a matrix equation $Mv = 0$ for the column four-vector $v = (\boldsymbol{\Gamma}, \Phi)^t$, with

$$M = \begin{bmatrix} a_{11} & a_{12} - in_3 & a_{13} + in_2 & in_1 \\ a_{21} + in_1 & a_{22} & a_{23} - in_1 & in_2 \\ a_{13} - in_2 & a_{23} + in_1 & a_{33} & in_3 \\ in_1 & in_2 & in_3 & 0 \end{bmatrix}$$

The compatible wave-number vectors $\mathbf{n}$ must therefore satisfy the 'dispersion relation'

$$\det(M) = n^4 - \sum_{\alpha,\beta} a'_{\alpha\beta} n_\alpha n_\beta - in^2 \sum_{\alpha,\beta,\gamma} \varepsilon_{\alpha\beta\gamma} a_{\alpha\beta} n_\gamma = 0 \tag{11}$$

where $A' = \{a'_{\alpha\beta}\}$ is the matrix of cofactors of A. The selection of compatible $\mathbf{f}$ then reduces to a study of the surface in $\mathbf{n}$ space determined by (11).

We shall say that a given admissible $\mathbf{q}$ is a *regular periodic dynamo* if the surface determined by (11) contains a real closed analytic branch $\mathscr{B}$, uniformly for $0 \leq \varepsilon \leq \varepsilon_0$ with $\varepsilon_0 > 0$. This definition is based upon the intuitive idea of a self-exciting dynamo. If the surface $\mathscr{B}$ exists, it follows that for each $\mathbf{n}$ on $\mathscr{B}$ there exists a corresponding $\boldsymbol{\Gamma}(\mathbf{n}) = 0$ such that $\boldsymbol{\Gamma}(\mathbf{n})\exp(i\mathbf{n}.\mathbf{x})$ generates a solution $\mathbf{h}(\mathbf{n})$, $\phi(\mathbf{n})$ of

A class of solutions of the magnetohydrodynamic dynamo problem

(2)–(4) in the manner described above. In fact, these solutions will depend continuously on **n** and we may form the surface integral

$$\mathbf{h}(\mathbf{x};\mathcal{B}) = \int_{\mathcal{B}} \mathbf{h}(\mathbf{n})\,ds, \qquad \phi(\mathbf{x};\mathcal{B}) = \int_{\mathcal{B}} \phi(\mathbf{n})\,ds$$

which, by the superposition principle, will also solve (2)–(4). The important point is that for closed analytic $\mathcal{B}$ we have

$$\lim_{\mathbf{x}\to\infty} \mathbf{h}(\mathbf{x};\mathcal{B}), \qquad \phi(\mathbf{x};\mathcal{B}) = 0$$

(see e.g. Lighthill's[7] discussion of asymptotic evaluation of surface integrals). This property of the solutions ensures that the electromagnetic field is not excited at infinity, and hence that the corresponding **q**, although itself periodic in space, is capable of maintaining a nonperiodic magnetic field which is localized in space. Such motions occur here as the natural analog of bounded kinematic dynamos.

Dynamo conditions

In the present paragraph we shall outline a derivation of conditions on **q** which are necessary and sufficient for the motion to have the dynamo property defined above. As a preliminary step we now fix the integer J, which first occurred in (5), to be equal to the value of j marking the first nonzero operator on PV in the sequence $PT\tilde{T}^{j-1}$, $j = 2, 3, \ldots$, and now refer to J as the *degree* of **q**. With this choice the matrix A in (9) has a nontrivial limit as $\varepsilon \to 0$ so we may set

$$A_0 = \lim_{\varepsilon \to 0} A$$

We may now state the first theorem as follows.

Theorem 1: A given admissible **q** *is a regular periodic dynamo if, and only if,* (i) *the degree of* **q** *is even,* (ii) A'_0 *is positive definite, and* (iii) $PT\tilde{T}^{j-1} = 0$ *on PV for all odd* $j \geq 3$.

The proof rests on a natural splitting of the matrix A into two parts, and upon the form of the compatibility equation (11). In particular we shall show that $A = A_1 + iA_2$, where A_1 and A_2 are Hermitian matrices given by

$$A_1 = \sum_{j\text{ even }\geq J} A^{(j)}, \qquad iA_2 = \sum_{j\text{ odd }\geq J+1} A^{(j)} \tag{12}$$

where $A^{(j)}$ is the jth term in the summation (10). It follows that conditions (i) and (iii) above will ensure that A is Hermitian, and hence that the limit A_0 can be represented in a suitable coordinate system by a real diagonal matrix. If in addition (ii) is satisfied (this is equivalent to the condition the eigenvalues of A_0 have one sign), it then follows from (ii) that a suitable $\mathcal{B}$ exists and is arbitrarily close to the closed surface, quadratic in n^{α}/n^2 determined by

$$n^4 - \mathbf{n} \cdot A_0 \cdot \mathbf{n} = 0$$

it is then not difficult to verify that (i)–(iii) are both necessary and sufficient for the dynamo property.

To prove (12), we observe that to any given term in the (j) summation for $A^{(j)}$ there corresponds an ordered set of vectors $(\mathbf{k}_1, \ldots, \mathbf{k}_j)$ which form an irreducible chain of length j. Let $S(i,j) = m_j^{-2} \mathbf{m}_j \times (\boldsymbol{\mu}_i \times\ )$, $U(i) = \boldsymbol{\mu}_i \times\ $, and $S(i) = S(i,i)$. In this notation

$$A^{(j)} = \varepsilon^{j/J-1}(i)^{j-1}\sum_{(j)} U(j)S(j-1)\ldots S(1)$$

Now we observe that (j) summation is unchanged by a permutation of chain $(\mathbf{k}_1, \ldots, =\mathbf{k}_j)$ into $(\mathbf{l}_1, \ldots, \mathbf{l}_j)$, where $\mathbf{l}_1 = -\mathbf{k}_j, \ldots, \mathbf{l}_j = -\mathbf{k}_1$. Thus also

$$A^{(j)} = \varepsilon^{j/J-1}(i)^{j-1}\sum_{(j)} \overline{U(1)}\,\overline{S(2,1)}\ldots\overline{S(j,j-1)}$$

But then

$$(A^{(j)})^* \equiv \overline{(A^{(j)})^t}$$
$$= \varepsilon^{j/J-1}(-1)^{j-1}\sum_{(j)} S^t(j,j-1)\ldots S^t(2,1)\,U^t(1)$$
$$= -\varepsilon^{j/J-1}(-1)^{j-1}\sum_{(j)} U(j)\,S(j-1)\ldots S(1)$$
$$= (-1)^j A^{(j)}$$

where in the next to last step we have used the identity

$$S^t(i+1,i)\,U(i) = U(i+1)\,S(i), \qquad i = 1,2,\ldots j-1$$

Since j is arbitrary, this establishes the decomposition (12).

Examples and remarks

As a simple specification which makes A_2 identically zero we may allow $\mu(\mathbf{k})$ to differ from zero only when $\mathbf{k}$ is one of the vectors $\pm m\mathbf{k}_\alpha$, $m = 1, 3, 5, \ldots$, $\alpha = 1, 2, 3$, $\{\mathbf{k}_\alpha\}$ being any set of real linearly independent vectors. Indeed, in this case it is impossible to form an irreducible chain containing an odd number of vectors $\mathbf{k}$ from K which have $\mu(\mathbf{k}) \neq 0$. Identifying a given value of α with one *wave* of the flow, such fields may be said to consist of three linearly independent waves. If $\mathbf{q}$ is solenoidal, the motion in each wave is normal to $\mathbf{k}_\alpha$, i.e. $\mathbf{q}$ then consists of a sum of three linearly independent shear waves.

We next show that there exist periodic dynamos of degree $J = 2$ consisting of three linearly independent shear waves. If $\mathbf{q}$ is solenoidal, we can introduce for each $\mathbf{k}$ a real orthogonal unit vector $\mathbf{c}(\mathbf{k})$, such that $\boldsymbol{\mu}$ can be expressed as a linear combination of $\mathbf{k}\times\mathbf{c}$ and $\mathbf{c}$

$$\boldsymbol{\mu}(\mathbf{k}) = ia\,\mathbf{k}\times\mathbf{c} - bk^2\mathbf{c} \tag{13}$$

If in addition $J = 2$, then

$$A_0 = i\sum_K k^{-2}(\boldsymbol{\mu}\times\boldsymbol{\mu})^\circ \mathbf{k} \tag{14}$$

Using (13) in (14), we obtain a result which is independent of the choice of $\mathbf{c}$.

$$A_0 = 2\sum_K \mathscr{R}(\bar{a}b)\mathbf{k}^\circ\mathbf{k} \tag{15}$$

A class of solutions of the magnetohydrodynamic dynamo problem

Thus it is sufficient for the dynamo property that either $\mathcal{R}(\bar{a}b)$ or $-\mathcal{R}(\bar{a}b)$ be ≥ 0 for every **k**, and > 0 on at least a linearly independent set. To take a specific example, let us set

$$a_\alpha = \tfrac{1}{2}\sigma_\alpha e^{i\theta_\alpha}, \qquad b_\alpha = \frac{\lambda_\alpha}{2\sigma_\alpha k_\alpha^2} e_n^{i\theta_\alpha}, \qquad \alpha = 1, 2, 3$$

when σ_α, θ_α are real parameters, the σ_α being necessarily nonzero. The θ_α may be removed by a shift of the origin of the coordinate system, so that this motion reduces (for given λ_α, k_α) to the three-parameter family

$$\varepsilon^{1/2} q_\alpha = \sigma_\beta k_\beta \sin\left(\frac{x_\beta}{\varepsilon}\right) + \frac{\lambda_\gamma}{\sigma_\gamma} \cos\left(\frac{x_\gamma}{\varepsilon}\right) \tag{16}$$

where for $\alpha\beta\gamma$ we take the cyclic permutations of 123. The motions (16) are dynamos for arbitrary nonzero k_α, σ_α, and arbitrary positive (or negative) λ_α.

If **q** is a Beltrami field, then there is a real number $\kappa \neq 0$ such that $i\mathbf{k} \times \boldsymbol{\mu} = \kappa\boldsymbol{\mu}$ for all **k**. If $k_\alpha = k = 1$ in (16) the corresponding Beltrami dynamos are of the form

$$\varepsilon^{1/2} q_\alpha = \sigma_\beta \sin\left(\frac{x_\beta}{\varepsilon}\right) \pm \sigma_\gamma \cos\left(\frac{x_\gamma}{\varepsilon}\right) \tag{17}$$

where $\pm$ is assigned in the same way for each α. (In general, if $\mathbf{q}(\mathbf{x})$ is a dynamo, so is $-\mathbf{q}(-\mathbf{x})$ (cf. section on pp. 640–7).)

The motions (17) emerge here as the 'simplest possible' dynamos in that the number of determining parameters is equal to the number of parameters (the eigenvalues $\lambda_\alpha = \pm \sigma_\alpha^2$) needed to determine uniquely A_0. We may clearly produce all possible A_0 by suitable rotations of (17). If one and only one of the σ_α is zero, then $\mathcal{B}$ tends to a plane as $\varepsilon \to 0$ and regeneration is not possible. This shows, incidentally, that these dynamos do not exist in spaces of dimension less than three.

More generally, let us suppose that **q** consists of three linearly shear waves, with the $\boldsymbol{\mu}(\mathbf{k})$ now real numbers. Then $\nabla \times \mathbf{q} + \kappa\mathbf{q}$ is a solenoidal dynamo of degree 2 for arbitrary real κ, and is also a Beltrami field if $k = |\kappa|$ for all **k** occurring in **q**. This follows in the same way as before from (13) and (15). Finally, we remark that there also clearly exist nonsolenoidal spatially periodic dynamos, since the addition of a certain small amount of 'compressibility' to one of the preceding examples cannot, for reasons of continuity, alter the positive definiteness of A_0'.

In these examples the dominant regenerative process has a simple geometrical interpretation. Specifically, let us consider the dynamos (17), with ε small, relative to principal axes. In this case the dominant effect of the motion on a local magnetic field **h** results in a uniform average current $A_0 \mathbf{h}$. The induced current generates orthogonal components of magnetic field according to a right-hand rule, and, if all eigenvalues of A_0 have the same sign, it is seen that the interaction of distinct spheres of fluid of radius $\gg \varepsilon$ becomes regenerative after two steps. The process has the following diagram (figure 1). The symbol $h_\alpha \to \lambda_\alpha h_\alpha$ represents the average effect of a three-step interaction, corresponding to successive operations on **f** by T, $\mathbf{q}\times$, and P. To follow this in detail, let **q** be given by (17) with the plus sign. Then, for example,

$$T\mathbf{i}_1 \sim \varepsilon^{1/2}\{-\sin(x_1/\varepsilon)\mathbf{i}_2 + \cos(x_1/\varepsilon)\mathbf{i}_3\}$$

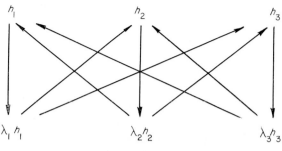

FIGURE 1

so that $\mathbf{i}_1 + T\mathbf{i}_1$ is a helical field which turns clockwise as x_1 increases. From $\mathbf{q} \times T\mathbf{i}_1$ we then obtain a current whose average is $\mathbf{i}_1$. (If we take the minus sign in (17), the sense of the helical field is reversed, as is the direction of mean current.)

The geometry of periodic dynamos is difficult to study in detail, even in the simplest examples (17). The only case where explicit integration has been found can be obtained by setting, for example, $\sigma_1 = \sigma_2 = 1, \sigma_3 = 0$. Then A'_0 has a one in the lower right-hand corner and zeros elsewhere, and the flow can be shown to consist of a pattern of square cells independent of x_3, the boundaries of the cells being nodes for both the vorticity and the velocity component q_3. We can therefore represent (17) not only as a linear combination of orthogonal shear waves, but also as a linear combination of two-dimensional orthogonal cells.

Kinematic Dynamos in a Spherical Conductor

Introduction

We now turn to the physically more relevant question of *bounded* kinematic dynamos. For any given periodic dynamo as defined in the last section, it is possible to construct a related class of motions which 'fit' a bounded conductor, but which retain sufficient properties to suggest that they form a likely class for a perturbational theory as envisaged in the first section. For simplicity we confine attention to the case of a homogeneous spherical conductor of unit radius containing a concentric spherical core of fluid, the radius of the latter being $a < 1$. Retaining the parameter ε in place of Rm, we shall show that the core motions

$$\mathbf{q} = -\varepsilon \nabla \times \omega \mathbf{v}\left(\frac{\mathbf{x}}{\varepsilon}\right) + \varepsilon^\beta \mathbf{w}(\mathbf{x}), \qquad 0 \leq x \leq a, \quad \beta > 0 \tag{18}$$

are dynamos provided ε is sufficiently small. In (18)

$$\mathbf{v}(\mathbf{x}) = (\sin x_2 + \cos x_3, \sin x_3 + \cos x_1, \sin x_1 + \cos x_2)$$

(cf. (17)), $\mathbf{w}$ is an arbitrary continuously differentiable solenoidal field satisfying $\mathbf{w} = 0$ on $x = a$, β is a positive number, and ω is a smooth 'cut-off' function, given, for example, by

$$\omega = \begin{cases} \varepsilon^{-1/2}, & x < a - \varepsilon \\ (\varepsilon\pi)^{-1/2} \int_{-\infty}^{\zeta(x)} \exp(-\xi^2)\,d\xi, & \zeta(x) = \frac{a - \varepsilon/2 - x}{(x - a + \varepsilon)(a - x)}, \quad a - \varepsilon \leq x \leq a \end{cases} \tag{19}$$

A class of solutions of the magnetohydrodynamic dynamo problem

From (18) we observe that the dominant part of **q** for small ε is the small-scale field which, at interior points of the core, behaves locally like a periodic dynamo. As a result we might expect that, as $\varepsilon \to 0$, the large-scale electromagnetic field induced by (18) would tend to a 'compatible' field which, in the core, satisfies the partial differential form of the ε limit of the compatibility equations (8). It turns out to be somewhat more direct to restate this result as a perturbation theorem for a certain operator equivalent of the exact induction equations. Rephrased in this way, the limiting compatibility problem becomes a 'comparison' eigenvalue problem for a simple operator. Moreover, the perturbational theory reveals that the *second* iterate of the exact induction operator (appropriate to a degree *two* periodic field) is close (in an appropriate norm) to the comparison operator. It then becomes possible to develop a perturbation of the comparison operator and to extract a real eigenvalue of the exact operator which is numerically close to a square root of each *positive* comparison eigenvalue. A second class of dynamos, associated with the negative comparison eigenvalues, can be obtained by reflection, since in a sphere $-\mathbf{q}(-\mathbf{x})$ is a dynamo if $\mathbf{q}(\mathbf{x})$ is a dynamo.

One mathematical technicality of the perturbational analysis deserves special mention. The success of the proposed method rests on a restriction of the fields to a Banach space equipped with a maximum norm containing first derivatives of all components (cf. p. 639). We emphasize this point since the topology most commonly utilized in dynamo theory is the topology of the Hilbert space norm (finite magnetic energy). For flows of the form (18) such spaces appear to be too 'wide'. A magnetic field which is bounded in the inner-product norm may interact with a flow (18) to induce a magnetic field which becomes arbitrarily large in the same norm as $\varepsilon \to 0$. In such a topology the exact induction operator becomes unbounded as $\varepsilon \to 0$.

The above assertion can be illustrated by a simple one-dimensional model of the induction operator. Let $f(x)$ satisfy $N_\gamma(f) < \infty$ on $[0, 1]$, where

$$N_\gamma(f) = \left(\int_0^1 f^2 \, dx\right)^{1/2} + \varepsilon^\gamma \left\{\int_0^1 \left(\frac{df}{dx}\right)^2 dx\right\}^{1/2}$$

and consider the operator l defined by

$$lf = \varepsilon^{-1/2} \int_0^1 k(x, y) \sin\left(\frac{y}{\varepsilon}\right) f(y) \, dy$$

with $k(x, y)$ independent of ε, positive, and continuously differentiable on $[0, 1] \times [0, 1]$. Let us consider the function

$$g(x) = \begin{cases} 0, & 0 \leq x \leq \tfrac{1}{2} - \sqrt{\varepsilon}, \quad \tfrac{1}{2} + \sqrt{\varepsilon} \leq x \leq 1 \\ \varepsilon^{-1/4} \sin\left(\dfrac{x}{\varepsilon}\right), & \tfrac{1}{2} - \sqrt{\varepsilon} < x < \tfrac{1}{2} + \sqrt{\varepsilon} \end{cases}$$

Clearly, if $\gamma \geq 1$, $N(g) = O(1)$ as $\varepsilon \to 0$. On the other hand

$$N_\gamma(lg) \geq C\varepsilon^{-1/4}, \quad \gamma > 0$$

for some $C > 0$ and ε suitably small. If, however, we adopt the maximum norm

$$\|f\|_{1/2} = \max_{[0,1]} |f| + \varepsilon^{1/2} \max_{[0,1]} \left|\frac{df}{dx}\right|$$

it can be seen that on the associated Banach space that $\|lf\|_{1/2}$ may be bounded in terms of $\|f\|_{1/2}$, uniformly as $\varepsilon \to 0$.

Operational formulation

The equations governing the dynamo are (3), (4), and

$$\nabla \times \mathbf{h} = \mathbf{e} + \lambda^{-1} \mathbf{q} \times \mathbf{h} \qquad 0 \leq x > a \qquad (20)$$

$$= \mathbf{e} \qquad a < x < 1 \qquad (21)$$

$$= 0, \nabla \cdot \mathbf{e} = 0 \qquad x > 1 \qquad (22)$$

where $\mathbf{q}$ is given by (18), (19) and λ is an auxiliary parameter. We require also

$$\mathbf{e} \text{ continuous, } \mathbf{h} \text{ continuously differentiable if } x = a, 1 \qquad (23)$$

$$\mathbf{h}, \mathbf{x} \times \mathbf{e} \text{ continuous on } x = 0, 1 \qquad (24)$$

$$\lim \mathbf{h} = 0, \lim \mathbf{e} = 0 \text{ as } x \to \infty \qquad (25)$$

Let S denote the core region $0 \leq x < a$, $\bar{S}$ the closure $0 \leq x \leq a$, B the boundary $x = a$. Also let $K(\mathbf{x}, \boldsymbol{\xi})$, $E(\mathbf{x}, \boldsymbol{\xi}) = \nabla_x \Phi(\mathbf{x}, \boldsymbol{\xi})$ be respectively the magnetic and electric tensor field obtained by solving the above problem with $\lambda^{-1} \mathbf{q} \times \mathbf{h}$ replaced by $I\delta(\mathbf{x} - \boldsymbol{\xi})$ in (20), for any $\boldsymbol{\xi} \varepsilon \bar{S}$. By the superposition principle we obtain the formal integral equation

$$L\mathbf{h} \equiv \int_S K(\mathbf{x}, \boldsymbol{\xi}) \cdot \{\mathbf{q}(\boldsymbol{\xi}; \varepsilon) \times \mathbf{h}(\boldsymbol{\xi})\} d\boldsymbol{\xi} = \lambda \mathbf{h}, \quad \mathbf{x}\varepsilon \bar{S} \qquad (26)$$

defined on $\bar{S}$. The corresponding electric potential is then given by

$$\phi(\mathbf{x}) = \lambda^{-1} \int_S \Phi(\mathbf{x}, \boldsymbol{\xi}) \cdot (\mathbf{q} \times \mathbf{h}) d\boldsymbol{\xi} \qquad (27)$$

It can be shown that any solution of the eigenvalue problem (26) which is continuously differentiable in the closure of the core provides, through (27) and an obvious extension of $\mathbf{h}$ to the region $x > a$, a solution of the original problem (2), (4), (20)–(25).

Writing

$$L = HQ$$

where

$$H\mathbf{h} = \int_S K \cdot \mathbf{h} \, d\boldsymbol{\xi}, \qquad Q\mathbf{h} = \mathbf{q} \times \mathbf{h}$$

we define the *comparison* eigenvalue problem on $\bar{S}$ by

$$H\mathbf{h} = u\mathbf{h} \qquad (28)$$

There are several reasons for introducing (28). In the first place, for a spherical conductor (28) is easily solved to obtain a discrete set of real eigenvalues μ (see the appendix for a discussion of the distribution of eigenvalues and explicit formulas for the eigenfunctions). It is the natural operator to associate with the fundamental tensor K. In the second place, the dominant regenerative effect of the small-scale part of $\mathbf{q}$ can be represented by the limit matrix $A_0 = I\lambda^{-2}$. If $\lambda^2 \sim \mu$ it then follows that the eigenvalue problem $H\mathbf{h} = \mu\mathbf{h}$ is equivalent on S to the limiting compatibility problem on p. 632. Recalling that the average current $A_0\mathbf{h}$ results from an interaction between $\mathbf{q}$ and the current induced by the interaction of $\mathbf{q}$ and $\mathbf{h}$, we expect the proper comparison with H to be given by

$$L^2 = H + (L^2 - H) \equiv H + H'$$

That is, (28) is to be a comparison problem for the iterated eigenvalue problem $L^2\mathbf{h} = \lambda^2\mathbf{h}$, with $L^2 - H = H'$ occurring as the perturbation operator. We note that H is not a *positive* operator on $\mathcal{H}$, so that smallness of H' does *not* imply that L is itself 'close' to an equally simple operator. Nor does it imply that QHQ is close to the identity, since such a comparison omits the effect of the final 'averaging' induced by multiplication on the left by H.

The perturbation problem

To make the approximating property precise we carry out the perturbation of H on a normed space of vector fields. For reasons mentioned on p. 636 we introduce the Banach space $C^1(\bar{S})$ of fields $\mathbf{h}$ which have continuous first derivatives in $\bar{S}$, and a family of γ norms defined for any tensor $W = \{W_{\alpha\ldots\beta}\}$ by

$$\|W\|_\gamma = \|W\| + \varepsilon^\gamma \|\nabla W\|, \qquad 0 < \gamma \leq 1$$

$$\|W\| = \max_{\alpha\ldots\beta}(\max_{S} |W_{\alpha\ldots\beta}|)$$

We shall utilize standard O and o symbols to express boundedness or smallness as $\varepsilon \to 0$. Given an operator $T(\varepsilon)$ on C^1, we say that T is bounded (for fixed ε) if there is a number M such that $\|T\mathbf{h}\|_\gamma \leq M\|\mathbf{h}\|_\gamma$ for all $\mathbf{h} \varepsilon C^1$. The least upper bound on the set of all such M is called the γ norm of T and is denoted by $\|T\|_\gamma$.

It can be shown that here, as in the model problem discussed on p. 630, the operator L has the following estimates on C^1:

$$\|L\mathbf{h}\| = O(\varepsilon^{1/2} \log \varepsilon \, \|\mathbf{h}\|) + O(\varepsilon^\beta \, \|\mathbf{h}\|) + O(\varepsilon^{1/2} \, \|\nabla \mathbf{h}\|) \tag{29}$$

$$\|\nabla L\mathbf{h}\| = O(\varepsilon^{-1/2} \, \|\mathbf{h}\|) + O(\varepsilon^{1/2} \log \varepsilon \, \|\nabla \mathbf{h}\|) \tag{30}$$

from which it follows that

$$\|L\|_{1/2} = O(1) \tag{31}$$

It is interesting to note that the proof of (29), (30) (which we omit) can be applied to obtain similar results for motions whose small scale part is a periodic Beltrami field of arbitrary degree J, but that (31) follows *only* if $J = 2$. The conditions we impose on the small-scale motion are therefore closely connected to the methods used in the existence proof.

More involved asymptotic estimates are needed to prove the smallness of H' in a γ norm. In broad outline the argument is summarized by the following theorems.

Theorem 2: With H defined as above, there exists a two-parameter family of operators $\hat{H}(\varepsilon, c)$ such that for any $c \neq 0$, $\|\hat{H} - H\|_\gamma = o(1)$ as $\varepsilon \to 0$ for every γ satisfying $\frac{1}{2} < \gamma < 1$.

Theorem 3: To each positive eigenvalue μ of H having odd multiplicity there corresponds a certain positive eigenvalue $\hat{\mu}$ of $\hat{H}$ defined for $0 < \varepsilon < \varepsilon_0$, $\varepsilon_0 > 0$. Moreover $(\hat{\mu} - \mu) \to 0$ as $\varepsilon \to 0$.

Theorem 4: If $v(c)$ is a positive eigenvalue of $\hat{H}$ with associated eigenfunction $\hat{\mathbf{h}}(c)$, then there is a real number $v(\varepsilon)$ lying in some neighborhood of μ such that $\mathbf{h} = \hat{\mathbf{h}}(v)$ satisfies the exact iterated equation $L^2 \mathbf{h} = v\mathbf{h}$ uniformly for all sufficiently small ε.

We point out that the operator $\hat{H}$ is a modification of the iterated operator L^2 which allows the needed estimate (theorem 2). The eigenvalue problem is then solved by standard means for arbitrary values of the free parameter c (theorem 3). Finally, the parameter c is fixed at a value which makes L^2 operate like $\hat{H}$ on the eigenfunction $\hat{\mathbf{h}}$ (theorem 4). The existence proof is completed by noting that we have $(L - \sqrt{v})(L + \sqrt{v}) \mathbf{h} = 0$ and hence that one of the two numbers $\pm\sqrt{v}$ is an eigenvalue of L.

Dynamically Self-consistent Models

Equations for the large-scale fields

A dynamically self-consistent dynamo will here mean a solution of a coupled system of dynamical-induction equations, satisfying the usual conditions of the dynamo problem, and driven by a certain distribution of body forces. As we have remarked in the first section, it is always possible to construct a trivial example of a self-consistent dynamo from any kinematic dynamo (steady or unsteady), in any dynamical model, simply by the addition of that particular distribution of body forces which will provide a momentum balance in the core. For dynamos involving two or more spatial scales, however, it is possible to allow a nontrivial partial self-consistency, by requiring that the body force distribution on one or more of the scales vanish. In the present section we utilize the results of the preceding section to examine the problem of general time-dependent dynamos driven predominantly by small-scale body forces.

For simplicity let us first assume that the fluid fills all space. We introduce dimensionless equations in a rotating coordinate system, with reference values U the speed of large-scale velocity field, H the magnetic field, L the characteristic dimension of large-scale fields, $E = H/L\sigma$ the electric intensity, $L^2 \sigma \mu$ the unit of time, and with Ω the angular velocity of coordinate system about the x_1 axis. The equations governing the system are then assumed to be

$$\frac{1}{Rm} \frac{\partial \mathbf{q}}{\partial t} + \mathbf{q} \cdot \nabla \mathbf{q} + \alpha \nabla P + \alpha \mathbf{i} \times \mathbf{q} + \beta \mathbf{h} \times \nabla \times \mathbf{h} = \mathbf{F} \tag{32}$$

$$\nabla \cdot \mathbf{q} = 0 \tag{33}$$

$$\nabla \times \mathbf{h} = \mathbf{e} + Rm\, \mathbf{q} \times \mathbf{h} \tag{34}$$

$$\nabla \times \mathbf{e} = -\frac{\partial \mathbf{h}}{\partial t} \tag{35}$$

$$\nabla \cdot \mathbf{h} = 0 \tag{36}$$

where

$$\alpha = \frac{2\Omega L}{U}, \quad \beta = \frac{\mu H^2}{\rho U^2}, \quad P = \frac{1}{\alpha \rho U^2}\{p - \tfrac{1}{2}\rho \Omega^2 L^2 (x_2^2 + x_3^2)\} \qquad (37)$$

and p denotes pressure, ρ the fluid density, and $\mathbf{F}$ a body force. The velocity field is taken to have the same structure as the kinematic dynamos of the previous section ($\mathbf{v}$ and ω are defined as before), but with an arbitrary parameter dependence and a time-varying $\mathbf{w}$:

$$\mathbf{q} = \mathbf{w}(\mathbf{x}, t) + (\lambda\, Rm)^{-1} \varepsilon^{-1/2} \mathbf{v}\left(\frac{\mathbf{x}}{\varepsilon}\right) \qquad (38)$$

We next introduce the averaging operation

$$A_\gamma \mathbf{h} = \lim_{\varepsilon \to 0} \int_{\mathbf{x}' \le \varepsilon^\gamma} \mathbf{h}(\mathbf{x}+\mathbf{x}')\,d\mathbf{x}' = \mathbf{h}^*(\mathbf{x}), \quad 0 < \gamma < 1$$

which serves to extract the dominant average over spatial scales of order ε or smaller. In particular, $A_\gamma \mathbf{w} = \mathbf{w}$ and $A_\gamma \mathbf{v}(\mathbf{x}/\varepsilon) = 0$. By considering the small-scale magnetic field induced by the rapidly varying part of $\mathbf{q}$ we also find that

$$Rm\, A_\gamma(\mathbf{q} \times \mathbf{h}) = Rm\, \mathbf{w} \times \mathbf{h}^* + r_m \mathbf{h}^*, \quad r_m = \lambda^{-2} \qquad (39)$$

the second term on the right of (39) being the usual dominant generation term due to $(\lambda \varepsilon)^{-1/2} \mathbf{v}(\mathbf{x}/\varepsilon)$. Finally, and less obviously, we have

$$A_\gamma [\mathbf{h} \times \nabla \times \mathbf{h}] = \mathbf{h}^* \times (\nabla \times \mathbf{h}^*) \qquad (40)$$

This result follows from the fact that the dominant contribution from the periodic motion is proportional (in the notation of the second section) to

$$\sum_K \frac{(\mathbf{k}\cdot \mathbf{h}^*)^2}{k^4} |\mu|^2 \mathbf{k} = 0$$

since summation is over $\mathbf{k}$ and $-\mathbf{k}$. Using (39) and (40) we obtain the averaged equations

$$\nabla P^* + \alpha\, \mathbf{i} \times \mathbf{w} + \beta \mathbf{h}^* \times \nabla \times \mathbf{h}^* = -\frac{1}{Rm}\frac{\partial \mathbf{w}}{\partial t} + A_\gamma[\mathbf{F} - \mathbf{q}\cdot\nabla \mathbf{q}] \qquad (41)$$

$$\nabla \cdot \mathbf{w} = 0 \qquad (42)$$

$$\nabla \times \mathbf{h}^* = \mathbf{e}^* + Rm\, \mathbf{w} \times \mathbf{h}^* + r_m \mathbf{h}^* \qquad (43)$$

$$\nabla \times \mathbf{e}^* = -\frac{\partial \mathbf{h}^*}{\partial t} \qquad (44)$$

$$\nabla \cdot \mathbf{h}^* = 0 \qquad (45)$$

Our model is now obtained by assuming that

$$A_\gamma[\mathbf{F} - \mathbf{q}\cdot\nabla\mathbf{q}] = \frac{1}{Rm}\frac{\partial \mathbf{w}}{\partial t} \qquad (46)$$

If the physical situation is such that the velocity-dependent terms in (46) are numerically negligible compared with the left-hand side of (41) (this assumption appears to be justified in the core of the Earth), then (46) states that the large-scale body forces are negligible also. The necessary small-scale body forces have, incidentally, the orders $(\lambda\,Rm)^{-1}\varepsilon^{-3/2}$ (inertial), $\alpha(\lambda\,Rm)^{-1}\varepsilon^{-1/2}$ (Coriolis), $\beta(\lambda\,Rm)^{-1}\varepsilon^{-1/2}$ (Lorentz) and for small finite ε these can be negligible compared with terms on the left of (41) whenever $\alpha \sim \beta > \varepsilon^{-1}$, $(\lambda\,Rm)^{-1}\varepsilon^{-1/2} \ll 1$.

Separable solutions

For convenience we now drop the superscript stars in (41)–(45) and let $\delta = \beta/\alpha$. Expansions for δ small may be particularly relevant to study of the Earth's core (typically $\delta \sim 10^{-1}$, cf. the next subsection). Therefore let $\mathbf{h}$, $\mathbf{e}$, $\mathbf{w}$, and P have expansions of the form

$$\mathbf{h} = \sum_{n=0}^{\infty} \mathbf{h}^{(n)} \delta^n$$

and consider the case of rotational symmetry about the $x_1 = \xi$ axis. To order unity we have

$$\nabla P^{(0)} + \mathbf{i}_\xi \times \mathbf{w}^{(0)} = 0, \qquad \nabla \cdot \mathbf{w}^{(0)} = 0 \tag{47}$$

In cylindrical orthogonal $(\mathbf{i}_\xi, \mathbf{i}_\eta, \mathbf{i}_\phi)$ coordinates, the axisymmetric solutions of (47) are of the form

$$\mathbf{w}^{(0)} = U(\eta, t)\mathbf{i}_\xi + W(\eta, t)\mathbf{i}_\phi, \qquad P^{(0)} = \int^\eta W(\eta)\,d\eta \tag{48}$$

where W, U are arbitrary. Writing

$$\mathbf{h}^{(0)} = -\frac{1}{\eta}\frac{\partial(\eta A)}{\partial \eta}\mathbf{i}_\xi + \frac{\partial A}{\partial \xi}\mathbf{i}_\eta + B\mathbf{i}_\phi$$

with A and B functions of ξ, η, and t only, we obtain from (43)–(45) and (48) the lowest-order equations

$$LB - \frac{\partial B}{\partial t} + r_m LA + Rm\left(\frac{W}{\eta}\right)'\eta\frac{\partial A}{\partial \xi} - Rm\,U\frac{\partial B}{\partial \xi} = 0 \tag{49}$$

$$LA - \frac{\partial A}{\partial t} - r_m B - Rm\,U\frac{\partial A}{\partial \xi} = 0 \tag{50}$$

where

$$L = \frac{\partial^2}{\partial x^2} + \frac{\partial^2}{\partial \eta^2} + \frac{1}{\eta}\frac{\partial}{\partial \eta} - \frac{1}{\eta^2}$$

To obtain the simplest equations let us choose

$$W = a\eta + b\eta \log \eta, \qquad U = 0 \tag{51}$$

where a and b are arbitrary constants. Using (51) in (49) we then have solutions

$$A = r_m \exp\{\sigma t + i(\omega t + kx)\} J_1(l\eta)$$

$$B = -(i\omega + s^2)A, \quad s^2 = k^2 + l^2 + \sigma, \quad \omega, \sigma, k, l \text{ real}$$

provided that

$$\omega^2 = s^2(s^2 - r_m^2), \quad 2\omega s^2 = -r_m Rm\, kb \qquad (52)$$

The following results follow easily from (52).

(i) To obtain steady-state solutions we must have $r_m^2 = s^2$ and $b = 0$, i.e. the lowest-order large-scale motion must reduce to a solid-body rotation.

(ii) Otherwise we may set $b = 1$. Then for arbitrary Rm there exist time-periodic solutions ($\sigma = 0$) provided that

$$r_m^2 = r_m^{*2} = \frac{4s^8}{4s^6 + k^2 Rm^2} \qquad (53)$$

(iii) If $r_m^2 \lessgtr r_m^{*2}$ then $\sigma \lessgtr 0$, i.e. amplitudes increase or decay exponentially.

We may now ask whether or not there exists a continuation of the lowest-order solution in the parameter δ which is consistent with (41), (42). If we require the terms of order δ to be periodic in ζ and bounded in both ζ and η, then it is easily seen that necessarily $\omega = 0$ in the lowest-order terms. This follows from the $\mathbf{i}_\phi$ component of the $O(\delta)$ momentum equation

$$w_\eta^{(1)} = \frac{1}{\eta}\left\{\frac{\partial A}{\partial x}\frac{\partial(\eta B)}{\partial \eta} - \frac{\partial B}{\partial x}\frac{\partial(\eta A)}{\partial \eta}\right\}$$

Since $B = -(i\omega + s^2)A$ in complex notation, $w_\zeta^{(1)}$ will have a nonzero ζ mean, and therefore $w_\zeta^{(1)}$ can be periodic in ξ, only if $\omega = 0$. If $\omega \neq 0$, solutions exist but $w_\zeta^{(1)}$ contains the secular term

$$k\omega r_m \frac{\xi}{\eta} \frac{d}{d\eta}\left\{J_1 \frac{d(\eta J_1)}{d\eta}\right\}$$

This term represents a flux towards or away from the plane $\xi = 0$ which balances the radial flux needed to ensure dynamical equilibrium of a cylindrical surface $\eta = $ constant. If $\omega = 0$, this term disappears and the continuation is well behaved to this order.

Solutions in a spherical core

Referring now to the geometry considered in the preceding section, we assume that $a = 1$ (insulating mantle, L the core radius), and require that $\mathbf{w}$ satisfy a tangency condition on the core boundary:

$$\mathbf{w} \cdot \mathbf{n} = 0 \text{ on } x = 1 \qquad (55)$$

Any analysis of self-consistent dynamos based upon a kinematic model rests upon the possibility of 'inverting' the dynamical equations to obtain $\mathbf{w}$, P in terms of $\mathbf{h}$ and derivatives of $\mathbf{h}$. This important question has been discussed by Taylor[8], who finds that, if $\mathbf{w}$ and P are to be a single valued, a necessary and sufficient condition for invertibility in this sense in a spherical core is the vanishing of net electromagnetic torque on all cylindrical surfaces $\eta = $ constant:

$$\int_{\eta \text{ fixed}} \{\mathbf{h} \times (\nabla \times \mathbf{h})\}_\phi \, ds = 0, \quad 0 < \eta < 1 \tag{55}$$

Assuming that (55) holds for all time there is a (nonlinear) operator on $\mathbf{h}$ and its derivatives through second order, $\mathbf{N}(\mathbf{h})$ say, such that

$$\mathbf{w} = W(\eta, t; \delta, Rm, r_m)\mathbf{i}_\phi + \delta \mathbf{N}(\mathbf{h}) \tag{56}$$

where $W(\eta, t)$ is arbitrary. The model induction equations may then be written

$$\frac{\partial \mathbf{h}}{\partial t} - \nabla^2 \mathbf{h} = r_m \nabla \times \mathbf{h} + Rm \nabla \times (W\mathbf{i}_\phi \times \mathbf{h}) + \delta Rm \nabla \{\mathbf{N}(\mathbf{h}) \times \mathbf{h}\} \tag{57}$$

$$\nabla \cdot \mathbf{h} = 0 \tag{58}$$

and we now regard W as *determined* (at least up to a solid-body rotation), for any time, by the invertibility condition (56). Thus (23), (24), (56)–(58), the matching condition on $x = 1$ with a vacuum field satisfying (25), and (if required) an initial or periodicity condition, define the general self-consistent dynamo problem based upon our kinematic model.

We do not know of any exact steady and time-dependent solutions of this problem which are non-trivial in the sense that $\mathbf{w} \times \mathbf{h} = 0$. (It is not difficult to find solutions in an infinite fluid which decay to zero at infinity and satisfy $\mathbf{w} \times \mathbf{h} = 0$.) The present brief discussion will therefore be limited to two simplified models obtained by expansion in the various parameters. We first consider a model suggested by the parameter dependence of the separable time-periodic solutions. We look for time-dependent solutions, symmetric about the rotation axis, in the limit $r_m \to 0$, $Rm \to \infty$, $r_m Rm \to R = $ constant $\neq 0$, $A = A'r_m = O(r_m)$, $B = O(1)$, $\mathbf{w} = O(1)$. With these assumptions we obtain from (41), (42), (46)

$$\mathbf{N}(\mathbf{h}) = r_m(w'_\xi \mathbf{i}_\xi + w'_\eta \mathbf{i}_\eta) + w'_\phi \mathbf{i}_\phi + O(r_m^2)$$

$$w'_\eta = \frac{1}{\eta}\left\{\frac{\partial(\eta B)}{\partial \eta}\frac{\partial A'}{\partial \xi} - \frac{\partial(\eta A')}{\partial \eta}\frac{\partial B}{\partial \xi}\right\} \tag{59}$$

$$w'_\phi = \frac{B^2}{\eta}$$

and from (57)–(59) the $O(1)$ induction equations

$$LB - \frac{\partial B}{\partial t} + \eta R \frac{d}{d\eta}\left(\frac{W}{\eta}\right)\frac{\partial A'}{\partial \xi} + R\delta\left\{\frac{\partial}{\partial \eta}\left(\frac{B}{\eta}\right)^2 \frac{\partial(\eta A')}{\partial \xi} - \frac{\partial}{\partial \xi}\left(\frac{B}{\eta}\right)^2 \frac{\partial(\eta A')}{\partial \eta}\right\}$$

$$- \delta R\eta\left\{w'_\xi \frac{\partial}{\partial \xi}\left(\frac{B}{\eta}\right) + w'_\eta \frac{\partial}{\partial \eta}\left(\frac{B}{\eta}\right)\right\} = 0 \tag{60}$$

$$LA' - \frac{\partial A'}{\partial t} - B - R\delta\left\{w'_\xi \frac{\partial A'}{\partial \xi} + \frac{w'_\eta}{\eta}\frac{\partial(\eta A')}{\partial \eta}\right\} = 0 \tag{61}$$

A class of solutions of the magnetohydrodynamic dynamo problem

We again regard W as determined by the invertibility condition, which in the axially symmetric case can be shown to be equivalent to

$$\int_{-c}^{+c} B \frac{\partial A'}{\partial \xi} d\xi = 0, \quad c = (1-\eta^2)^{1/2}, \quad 0 < \eta < 1 \tag{62}$$

For this model we have a negative result concerning dynamo behaviour as $t \to \infty$. From (62) and the boundary conditions of $\mathbf{w}$, B it follows from (60) that

$$\int_{x<1} \frac{B}{\eta^2} \frac{\partial B}{\partial t} d\mathbf{x} = \int_{x<1} \frac{B}{\eta^2} LB \, d\mathbf{x}$$

so that total magnetic energy in the toroidal field decays to zero as if there were no motion. It follows that A' must also decay to zero, so the dynamo cannot be self-sustaining. We call attention to the fact that this need not be true if (62) is relaxed, and that the missing terms in w'_ϕ are of order r_m relative to the terms retained.

Our second model is based upon expansion in powers of δ and the hypothesis $W = O(\delta)$. We look for steady-state solutions with lowest-order magnetic field determined by

$$\nabla \times \mathbf{h}^{(0)} = \nabla \phi^{(0)} + r_m^{(0)} \mathbf{h}^{(0)}, \quad \nabla \cdot \mathbf{h}^{(0)} = 0 \tag{63}$$

That is, the lowest-order induction problem is identical with the comparison eigenvalue problem of the third section ($r_m^{(0)} = \mu^{-1} = \gamma$), whose solutions are given in the appendix. If we choose a dipole vacuum field with dipole moment vector $\mathbf{m}$, then the eigensolutions take the form

$$\begin{aligned}
\mathbf{h}^{(0)} &= (\gamma^2 \psi - 2)\mathbf{m} + \nabla(\mathbf{m} \cdot \nabla \psi) - \gamma \mathbf{m} \times \nabla \psi \\
\phi^{(0)} &= 2\gamma \mathbf{m} \cdot \mathbf{x} \\
\psi &= \frac{3}{\gamma^2 \sin \gamma} \frac{\sin \gamma x}{x}
\end{aligned} \tag{64}$$

and γ satisfies $\gamma \cos \gamma - \sin \gamma = 0$. It is not difficult to see that (55) is satisfied by (64) for all $\mathbf{m}$, and that if $\mathbf{m} = m\mathbf{i}$ the first-order velocity is

$$\mathbf{w}^{(1)} = 2\gamma m^2 (\mathbf{h}^{(0)} + 2\mathbf{i}) + W^{(1)}(\eta) \mathbf{i}_\phi \tag{65}$$

With

$$\mathbf{h}^{(1)} = -\frac{1}{\eta} \frac{\partial(\eta A')}{\partial \eta} \mathbf{i}_\xi + \frac{\partial A^{(1)}}{\partial \xi} \mathbf{i}_\eta + B^{(1)} \mathbf{i}_\phi$$

it then follows that $r_m^{(1)} = 0$ and that

$$LB^{(1)} + \gamma LA^{(1)} = -Rm \left\{ \eta \frac{d}{d\eta}\left(\frac{W^{(1)}}{\eta}\right) + 4\gamma^2 m^3 \right\} \frac{\partial^2 \psi}{\partial \xi \partial \eta} \tag{66}$$

$$LA^{(1)} - \gamma B^{(1)} = 4Rm \gamma m^3 \frac{\partial^2 \psi}{\partial \xi \partial \eta} \tag{67}$$

To complete the solution we must choose $W^{(1)}$ so that

$$\int_{-c}^{+c} (B^{(1)} + \gamma A^{(1)}) \frac{\partial A^{(0)}}{\partial \xi} d\xi = 0 \qquad (68)$$

with $B^{(1)} = 0$ on $x = 1$ and $A^{(1)}$ matching with a vacuum potential (vertical quadrupole) field. Again, the underlying analytic problem is the functional dependence of $\mathbf{h}^{(n)}$ on $W^{(n)}$ that must be carried before $W^{(n)}$ can be determined up to a solid-body rotation. The solution procedure also involves, at the nth step, the determination of $r_m^{(n)}$ and (since comparison eigenvalues are not simple) terms of order δ^{n-1} in the weighting of linearly independent eigensolutions which appear in $\mathbf{h}^{(n-1)}$. If the scheme fails at any point, the conclusion is that steady-state solutions having the assumed zeroth-order eigensolution do not exist. (In the dipole solutions, symmetry suggests that a north–south or arbitrary equatorial orientation are equilibrium states provided we add a suitable solid-body rotation.)

A few properties of (64), (65) should be noted. In figure 2 we show the parity of

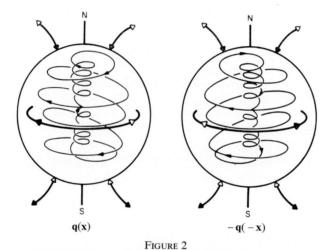

FIGURE 2

poloidal and toroidal magnetic fields, as well as a typical closed particle path relative to the rotating frame for the motion $\mathbf{w}^{(1)} - W^{(1)}\mathbf{i}_\phi$, for the minimum-$\gamma$ dipole eigenstates ($\gamma = \pm 4\cdot 5$). There are four north–south eigenstates maintained by two distinct dynamos. It is interesting that for all eigenstates the particle motion is always to the west. Physical reasoning then suggests that the invertibility condition may require $W^{(1)}$ to be primarily eastwards relative to motion at the core boundary. If $\mathbf{w}^{(1)}$ and $W^{(1)}$ are comparable in magnitude, the net effect could appear (for a suitable solid-body rotation of the core relative to the mantle) as a predominantly westward motion at the core boundary in the absence of significant core angular momentum relative to the mantle. Such a motion would appear to be needed in order to explain within a steady-state model the differences in the rates of westward drift of the various nondipole components of the geomagnetic field. The velocities which are given by the known part of (65) are, however, too small to account for the magnitude of the observed westward drift of the equational dipole, if we assume the regenerative mechanism to be operative over the entire core. We take $L = 3{\cdot}5 \times 10^6 M$, $\sigma = 3 \times 10^5 Q^2 S/M^3 K$, $U = 3 \times 10^{-4} M/S$, $\rho = 10^2 K/M^3$, $\mu = 4\pi \times 10^{-7} KM/Q^2$, $\Omega = 7 \times 10^{-5} S^{-1}$, $\mu H = 10^{-2} K/QS(100\ \text{G})$, to obtain $\alpha = 1{\cdot}6 \times 10^6$, $\beta = 0{\cdot}9 \times 10^5$, and $\delta = 0{\cdot}05$. If the maximum poloidal field at the core boundary is 4 G, then we must take $m = 0{\cdot}02$, so that the maximum value of $\delta(W_\phi^{(1)} - W^{(1)})$ is about 3×10^{-5} cm/sec compared with the observed drift of about 5×10^{-3} cm/sec. It should be mentioned, however, that our model neglects the presence of a central core and it seems likely that this will affect numerical values of the azimuthal fields. The maximum toroidal field in the fluid core would probably be increased from the value obtained from (64). For eigenvalues $\gamma \sim (i+1/2)\pi$, $j = 1, 2, \ldots$ the above numbers give $(h_\phi^{(0)})$ max $\sim 10(j+1/2)$ G, so that typical toroidal fields associated with the dipole eigenstates will be in neighborhood of 50 G.

Acknowledgements

The research reported in this paper has been supported by the Air Force Office of Scientific Research, Office of Aerospace Research, United States Air Force, under Grant No. AF-AFOSR-815-67.

Appendix. Eigenfunctions and Eigenvalues of the Comparison Problem

The eigenvalue problem $H\mathbf{h} = \mu \mathbf{h}$ can be solved for the geometry considered on p. 636 by standard means and we omit the details of the computations. The set of eigenfunctions have the following form ($\gamma = \mu^{-1}$):

$$\mathbf{h} = \nabla \psi_1, \qquad \phi = \phi_1, \qquad x > 1$$

$$\mathbf{h} = \nabla \psi_2 + \nabla \times (\mathbf{x}\psi_2), \qquad \phi = \phi_2, \qquad a < x < 1$$

$$\mathbf{h} = \nabla \psi_3 + \nabla\left(\frac{\partial}{\partial x} x \psi_4\right) + \gamma \nabla \times (\mathbf{x}\psi_4) + \gamma^2 \mathbf{x}\psi_4$$

$$\phi = \phi_3, \qquad 0 < x < a$$

where, in terms of the surface harmonics $Y_{mn}(\theta, \phi)$,

$$\psi_1 = x^{-(n+1)} Y_{mn}, \qquad \phi_1 = \gamma \frac{n+1}{n} \psi_1$$

$$\psi_2 = A(x^n - x^{-(n+1)}) Y_{mn}, \qquad \phi_2 = \frac{\partial}{\partial x}(x\psi_2)$$

$$\psi_3 = B x^n Y_{mn}, \qquad \phi_3 = -\gamma \psi_3$$

$$\psi_4 = C f_n(x) Y_{mn}$$

$$f_n(x) = x^n \left(\frac{1}{x}\frac{d}{dx}\right)^n \frac{\sin \gamma x}{x} = (-1)^n (\gamma)^{n+1/2} \left(\frac{\pi}{2x}\right)^{1/2} J_{n+1/2}(\gamma x)$$

and

$$A = \gamma \frac{n+1}{2n^2+n}, \qquad B = -\frac{n+1}{2n^2+n}(na^{-(2n+1)}+n+1)$$

$$C = \frac{n+1}{2n^2+n}\frac{a^n - a^{-(n+1)}}{f_n(a)}$$

The characteristic equation determining the eigenvalue is found to be

$$(a^{2n+1}-1)\{(n+1)f_n(a) + af'_n(a)\} = (2n+1)f_n(a)\left(a^{2n+1} + \frac{n}{n+1}\right)$$

Each distinct finite real root of the last equation provides a nonzero real eigenvalue μ_n of H having multiplicity $2n+1$. (There are $2n+1$ distinct surface harmonics of order n.) Associated with each order n there is, moreover, a sequence of eigenvalues $\mu_n^{(j)}$, $j = 1, 2, \ldots$ such that $|\mu_n^{(j)}| \to 0$ as $j \to \infty$ and $|\mu_n^{(j)}| \to 0$ as $n \to \infty$. If we restrict attention to the positive eigenvalues, we see that the unperturbed eigenfunctions of the present theory depend upon three integer parameters, two which determine the surface harmonic of the vacuum field, and a third which determines the core field. There should also be added a discrete parameter taking values ± 1 which specifies the direction of the lines of force, if it is assumed that the solutions are renormalized to have core magnetic energy unity. Finally, associated with the dynamos $-\mathbf{q}(-\mathbf{x})$ (cf. the first section) we have another similar family of eigenfunctions of H associated with the negative spectrum.

The above results are easily proved from well-known properties of the spherical Bessel functions $J_{n+1/2}(x)$, using in particular the lower bound $\{(n+1/2)(n+5/2)\}^{1/2}$ for the smallest positive zero. If we let $a \to 1$ (a limit process which is nonuniform in n for large n), we see from the eigenvalue equation that the eigenvalues tend to the zeros of $J_{n+1/2}(\gamma a)$. An interesting special case are the eigenvalues for the vacuum dipole solutions, obtained by setting $n = 1$ in the above equations. We find that γa satisfies in this case

$$\tan(\sigma a) = \frac{\gamma a}{1 + b(a)(\gamma a)^2}, \qquad b(a) = \frac{2}{3}\frac{1-a^3}{1+2a^3}$$

Thus the sequence $\mu_1^{(j)}$ satisfies

$$\frac{a}{(j+1/2)\pi} < \mu_1^{(j)} < \frac{a}{j\pi}, \qquad j = 1, 2, \ldots$$

The maximum positive eigenvalue of H is thus bounded above by a/π.

References

1. E. N. Parker, *Astrophys. J.*, **122**, 293 (1965).
2. A. Herzenberg, *Phil. Trans. Roy. Soc. London, Ser. A*, **250**, 543 (1958).
3. G. Backus, *Ann. Phys. (N.Y.)*, **4**, 372 (1958).
4. S. I. Braginskiĭ, *Soviet Phys.—JETP (English Transl.)*, **20**, 726 (1965).
5. S. I. Braginskiĭ, *Soviet Phys.—JETP (English Transl.)*, **20**, 1462 (1965).
6. E. L. Ince, *Ordinary Differential Equations*, Dover Publications, New York, 1956, p. 381.
7. M. J. Lighthill, *Phil. Trans. Roy. Soc. London, Ser. A*, **252**, 397 (1960).
8. J. B. Taylor, *Proc. Roy. Soc. (London), Ser. A*, **274**, 274 (1963).

E

MAGNETO-HYDRODYNAMICS

III. Aspects of the geomagnetic secular variation

1

R. HIDE*

Department of Geology and Geophysics
Massachusetts Institute of Technology
Cambridge, Massachusetts, U.S.A.

On the dynamics of the Earth's deep interior

Free Hydromagnetic Oscillations of the the Earth's Core and the Strength of the Magnetic Field in the Core

A theoretical study[2] of free hydromagnetic oscillations of a thick rotating spherical shell of incompressible fluid suggests that if the average strength of the toroidal magnetic field in the Earth's liquid core has the reasonable value of 100 G then the interaction of free hydromagnetic oscillations of the core with the Earth's poloidal magnetic field is consistent with present knowledge of the geomagnetic secular variation and of the physical conditions prevailing in the Earth's deep interior. Associated with the slow 'magnetic' modes of free hydromagnetic oscillation of the Earth's core (periods of the order of decades to centuries) are very rapid 'inertial' modes (periods of the order of days)[1]. Variations in the poloidal magnetic field due to these inertial modes would fail to penetrate the weakly conducting mantle, but the eddy currents induced in the lower mantle of these modes might significantly affect the mechanical coupling between the core and mantle.

Although the theoretical treatment of the problem is incomplete, various aspects of it are illuminated by two recent mathematical investigations[2,3].

Electrical Conductivity and Viscosity of the Core

In the 'free oscillations' theory of the geomagnetic secular variation it is assumed that the electrical conductivity of the core, σ, is so high that on the time scale of the geomagnetic secular variation σ can be regarded as infinite. One implication of this assumption is that the total number of lines of magnetic force threading the surface of the core, N, should not change significantly over intervals as short as a few decades. If the magnetic field at the core–mantle interface can be expressed accurately in terms of the extrapolated values of the first four harmonics only of the surface magnetic field, then N has changed by no more than 0.5% since 1900[4]; the corresponding change (a decrease) in the contribution to N due to the first spherical harmonic (the centered dipole) was much bigger, about 3%. (An incidental implication[4] of this result is that the kinematic viscosity of the core is probably much less than 1 m²/sec (assuming that[5] $\sigma \sim 10^6\ \Omega^{-1}\ m^{-1}$).)

* Now at the Meteorological Office, Bracknell, Berkshire, England.

Topography of the Core–Mantle Interface

It has recently been suggested[1] that horizontal variations in the properties of the core–mantle interface that would escape detection by modern seismological methods might nevertheless produce measurable geomagnetic effects. One consequence of this suggestion is that the shortest interval of time that can be resolved in paleomagnetic studies of the geocentric axial dipole component of the Earth's magnetic field might be very much longer than the value often assumed by most paleomagnetic workers. Another consequence[6] is that reversals in sign of the geomagnetic dipole might be expected to show some degree of correlation with processes due to motions in the mantle (for example, tectonic activity, polar wandering).

If the first four harmonics of the regional gravitational field of the Earth (as determined from the motions of artificial satellites) are due to horizontal variations in the level of the core–mantle interface (an admittedly extreme assumption), then the maximum amplitude of these variations about the mean level is about 8 km and the mean amplitude is about[7] 4 km.

The Westward Drift of the Geomagnetic Field and Variations in the Length of the Day

Several workers have suggested that observed variations of the length of the day (i.e. the rotation period of the mantle) of as much as 5×10^{-3} sec can in principle be accounted for in terms of angular momentum transfer between the core and mantle. The chief assumption underlying their work is that U, the speed of the westward drift of the pattern of the geomagnetic field at the Earth's surface (about 0.03 cm/sec), is comparable with or less than V, the average zonal component of the hydrodynamical velocity of the material of the core relative to the mantle. The free oscillations hypothesis of geomagnetic secular variation raises the serious possibility that U could be a measure of the speed of hydromagnetic waves in the core, in which case U may exceed V, possibly by quite a significant factor[1]. Studies of particle motions in the upper reaches of the core along the lines suggested by several recent studies[1, 8, 9] might eventually resolve this apparent difficulty in the theory of the dynamics of the mantle and core.

References

1. R. Hide, *Phil. Trans. Roy. Soc. London, Ser. A*, **259**, 615 (1966).
2. W. V. R. Malkus, *J. Fluid Mech.*, **28**, 793 (1967).
3. K. Stewartson, *Proc. Roy. Soc. (London), Ser A.*, **299**, 173 (1967).
4. J. C. Cain and R. Hide, to be published.
5. R. Hide, *International Dictionary of Geophysics*, Vol. 1, Pergamon Press, Oxford, 1967, p. 358.
6. R. Hide, *Science*, **157**, 55 (1967).
7. R. Hide and K.-I. Horai, to be published.
8. P. H. Roberts and S. Scott, *J. Geomagnetism Geoelectricity*, **17**, 137 (1965).
9. E. H. Vestine and A. B. Kahle, *J. Geophys. Res.*, **71**, 527 (1966).

2

G. SUFFOLK

and

D. W. ALLAN

Department of Mathematics
King's College University of London
England

Planetary magnetohydrodynamic waves as a perturbation of dynamo solutions

Introduction

The new class of magnetohydrodynamic waves which has been discussed by Hide[1] is likely to play an important part in our attempts to understand a number of astronomical problems concerning rotating fluid bodies containing magnetic fields. Of these, the one we are most concerned with here is the homogeneous dynamo process thought to occur in the Earth's core. One very natural way of arriving at the equations discussing possible wave motions in the core is to consider them as an aspect of the *stability* of the terrestrial dynamo. The equations for small disturbances of the fundamental solution should describe wave-type phenomena when it is stable, but these may be of a complicated nature; even the very simple self-regenerating system consisting of several coupled disk dynamos is by no means trivial in its stability properties[2].

We still do not have any satisfactory mathematical solution for the homogeneous dynamo, but there have been real advances in understanding the *general* nature of possible solutions (see section E II, chapter 5 of this book). One qualitative conclusion which seems confirmed is that a dynamo-produced field must involve a strong toroidal field lying within the core, as well as the poloidal field we are able to observe outside. There have been a number of attempts to construct theories of the secular variation in which disturbances of the main velocity field producing the dynamo have outside effects depending on both the toroidal and poloidal main fields[3-6], but without taking proper account of the dynamics of the problem. Hide[1] does do so, for a very simplified situation in which a constant toroidal field is disturbed. The new feature of the waves he considers is to incorporate the effect of rotation in a manner which presents at least a preliminary discussion of the modification of their behaviour by a spherical boundary, through the use of the Rossby β-plane approximation in the equations.

This approximation has been successful in dealing with thin-shell problems connected with the oceanic and atmospheric circulations. Whether it can be applied to the thick-shell problem of behaviour in the core is debatable[7]. The work we present below raises the rather different question of whether the more complicated nature of the field actually present in the core may modify the wave motion considerably.

Perturbation Equations for a Disturbed Dynamo

The induction and momentum equations for the problem are

$$\frac{\partial \mathbf{B}}{\partial t} = \text{curl}(\mathbf{V} \times \mathbf{B}) + \lambda \nabla^2 \mathbf{B} \qquad (1)$$

$$\rho \left\{ \frac{d\mathbf{V}}{dt} + 2(\mathbf{\Omega} \times \mathbf{V}) \right\} = -\text{grad } P + \mathbf{J} \times \mathbf{B} + \mathbf{F} \qquad (2)$$

$$\text{div } \mathbf{V} = 0, \qquad \text{div } \mathbf{B} = 0 \qquad (3)$$

$$\mathbf{J} = \frac{1}{4\pi\mu} \text{curl } \mathbf{B} \qquad (4)$$

where $\mathbf{V}$ is the velocity, $\mathbf{B}$ the magnetic induction, $\mathbf{J}$ the current, P the pressure, and $\mathbf{F}$ the body force driving the dynamo. By leaving $\mathbf{F}$ unspecified to this extent, we shall be neglecting such matters as the effect of perturbations of the temperature field. λ is the magnetic diffusivity.

Let us set $\mathbf{B} = \mathbf{B}_0 + \mathbf{b}$, $\mathbf{V} = \mathbf{V}_0 + \mathbf{v}$, etc., and linearize the resulting equations by neglecting terms such as $\mathbf{v} \times \mathbf{b}$. In virtue of the fact that we are linearizing about a hypothetical dynamo solution, for which

$$\frac{\partial \mathbf{B}_0}{\partial t} = \text{curl}(\mathbf{V}_0 \times \mathbf{B}_0) + \lambda \nabla^2 \mathbf{B}_0 \qquad (5)$$

$$\frac{d\mathbf{V}_0}{dt} + 2(\mathbf{\Omega} \times \mathbf{V}_0) = -\text{grad } P_0 + \mathbf{J}_0 \times \mathbf{B}_0 + \mathbf{F} \qquad (6)$$

we then have the perturbation equations

$$\frac{\partial \mathbf{b}}{\partial t} = \text{curl}(\mathbf{v} \times \mathbf{B}_0) + \text{curl}(\mathbf{V}_0 \times \mathbf{b}) + \lambda \nabla^2 \mathbf{b} \qquad (7)$$

$$\rho \left\{ \frac{\partial \mathbf{v}}{\partial t} + 2(\mathbf{\Omega} \times \mathbf{v}) \right\} = -\text{grad } p + \mathbf{j} \times \mathbf{B}_0 + \mathbf{J}_0 \times \mathbf{b} - (\mathbf{v} \cdot \text{grad})\mathbf{V}_0 - (\mathbf{V}_0 \cdot \text{grad})\mathbf{v} \qquad (8)$$

$$\text{div } \mathbf{v} = 0, \qquad \text{div } \mathbf{b} = 0 \qquad (9)$$

Effect of a Varying Field

A specific example of how these extra terms in the perturbation equations may complicate the solution is furnished by using Rossby's β-plane approximation to analyse the propagation of a disturbance set up on the surface of the core, when the dominant toroidal magnetic field is assumed to vary with latitude.

The spherical fluid core rotates uniformly with an angular velocity Ω, not necessarily equal in magnitude to that of the Earth, but in the same direction. A right-handed set of Cartesian coordinates (x, y, z) is chosen with its origin at the centre of the disturbance and its (x, y) plane as tangent plane to the core (at the origin). The y axis points northwards, the x axis eastwards, and the z axis radially outwards from the core. We neglect the main velocity field; this amounts to disregarding the possible effects of a stratified flow. In lieu of a working geodynamo we assume a simple stationary magnetic field of the form

$$B_0 = (B_0(y), 0, 0)$$

which satisfies the condition $\text{div} \, B_0 = 0$. We further assume that $B_0(y)$ varies slowly and linearly with y for the small values of y we consider. The magnetic Reynolds number is taken to be large.

The perturbation applied to this system consists of both vorticity and current disturbances in the z direction, so that the perturbations of velocity and magnetic field are in the (x, y) plane only. The z component of Ω is taken to vary linearly with y (measured by β). We then obtain from (7) and (8), operated on by curl,

$$4\pi\mu \frac{\partial j_z}{\partial t} - B_0 \frac{\xi}{\partial x} + 2B_0' \frac{\partial v_x}{\partial x} = 0 \tag{10}$$

$$\frac{\partial \zeta}{\partial t} + \beta v_y - \frac{B_0}{\rho} \frac{\partial j_z}{\partial x} = 0 \tag{11}$$

and have a full set of equations when we add

$$\zeta + \frac{\partial v_x}{\partial y} - \frac{\partial v_y}{\partial x} = 0 \tag{12}$$

$$\frac{\partial v_x}{\partial x} + \frac{\partial v_y}{\partial y} = 0 \tag{13}$$

Assuming the four variables are each proportional to $\exp\{i(kx + ly - \omega t)\}$, we obtain the dispersion relation

$$\omega^2 + \frac{\omega \beta k}{k^2 + l^2} - k^2 V_A^2 + \frac{ik^2 l (V_A^2)'}{k^2 + l^2} = 0 \tag{14}$$

which has been written in terms of the local Alfvén velocity

$$V_A = \frac{B_0}{(4\pi\mu\rho)^{1/2}}$$

Provided $l = 0$, both roots of the equation have a non-zero imaginary part. Since we have taken B_0 as slowly varying with y, V_A' is small compared with V_A and the imaginary part of ω is approximately

$$\pm \frac{kl(V_A^2)'}{\{\beta^2 + 4V_A^2(k^2 + l^2)\}^{1/2}}$$

When V_A is zero we have the dispersion relation discussed by Hide[1].

Conclusions

The alteration in the character of the dispersion relation which we have obtained by allowing for variation in the disturbed dynamo field suggests solutions which will increase or decrease in amplitude with latitude, as well as dispersing. Our discussion has done no more than suggest the *possible* importance of one modification of the Hide type of wave, and some detailed solutions must be obtained before this can be assessed in relation to the secular variation problem.

None of this is meant to disparage the novelty and interest of Hide's waves. For example, they seem bound to play a fundamental role in understanding the motions of the Sun's surface layers[8], and indeed in any astrophysical situation where rotation and magnetic fields interact, and disturbances are constrained to move on a curved surface.

References

1. R. Hide, 'Free hydromagnetic oscillations of the Earth's core and the theory of the geomagnetic secular variation', *Phil. Trans. Roy. Soc. London, Ser. A*, **259**, 615–47 (1966).
2. D. W. Allan, 'On the behaviour of systems of coupled dynamos', *Proc. Cambridge Phil. Soc.*, **58**, 671–93 (1962).
3. E. C. Bullard, 'The secular change in the Earth's magnetic field', *Monthly Notices Roy. Astron. Soc., Geophys. Suppl.*, **5**, 248–57 (1948).
4. R. Hide and P. H. Roberts, *Phys. Chem. Earth*, **4**, 25–98 (1961).
5. T. Nagata and T. Rikitake, 'Geomagnetic secular variation and poloidal magnetic fields produced by convectional motions in the Earth's core', *J. Geomagnetism Geoelectricity*, **13**, 42–53 (1961).
6. D. W. Allan and E. C. Bullard, 'The secular variation of the Earth's magnetic field', *Proc. Cambridge Phil. Soc.*, **62**, 783–808 (1966).
7. W. V. R. Malkus, 'Hydromagnetic planetary waves', *J. Fluid Mech.*, in press (1967).
8. H. H. Plaskett, 'The polar rotation of the Sun', *Monthly Notices Roy. Astron. Soc.*, **131**, 407–33 (1966).

3

K. STEWARTSON

Department of Applied Mathematics
University College
London, England

Second-class magneto-Rossby waves

Abstract

The slow oscillations of a rotating fluid in the presence of a uniform toroidal field are discussed, both when the fluid is confined to a thin sheet and when it occupies the region between two concentric spheres. Results are applied to the study of the slow westward drift of the non-dipole components of the Earth's magnetic field. (For further work see Stewartson[1].)

References

1. K. Stewartson, *Proc. Roy. Soc.* (*London*), *Ser. A*, **299**, 173–87 (1967).

F

THE EARTH AS A FUNDAMENTAL PHYSICS LABORATORY

1

H. H. KOLM

Massachusetts Institute of Technology
Francis Bitter National Magnet Laboratory*
Cambridge, Massachusetts, U.S.A.

Search for magnetic monopoles†

Introduction

The laws of electromagnetism, summarized by Maxwell's equations

$$\text{div } \mathbf{E} = \rho_e \qquad \text{curl } \mathbf{E} = 0 - \frac{\partial \mathbf{B}}{\partial t}$$

$$\text{div } \mathbf{B} = 0 \qquad \text{curl } \mathbf{B} = j_e + \frac{\partial \mathbf{E}}{\partial t}$$

(neglecting the pedagogic distinction between B, H and E, D) exhibit a basic symmetry which is not reflected in nature, as evidenced by two missing terms: magnetic charge ρ_m and magnetic current j_m. Elementary particles carrying electric charge, protons, electrons, positrons, exist in abundance, yet no evidence of magnetic charge has ever been observed. All known magnetic fields are dipolar, that is, their lines of force form closed loops. They never diverge from a single pole.

As early as 1931, Dirac[1] pointed out that quantum theory permitted, or in fact required, the existence of magnetic poles carrying multiples of a certain quantum of magnetic charge. He also predicted, from somewhat analogous reasoning, the existence of antielectrons, or positrons. The positron was discovered several years later, but Dirac's magnetic monopole has eluded observation to this day, despite the fact that it has received almost continuous attention. Experimentalists have searched for it with increasingly powerful synchrotrons, while theorists have endeavored to find a law which might prohibit its existence. The Dirac monopole appears equally impossible to find or to explain away. One of the most compelling arguments for its existence is the fact that, without it, the quantization of electric charge seems doomed to remain nothing more than an isolated experimental observation, completely unrelated to any fundamental principle or symmetry. Only the coexistence of electric and magnetic charge can explain the quantization of both. Current trends in elementary particle physics have made the missing monopole even more of a dilemma than it was at the time of Dirac's suggestion. The key speaker at a recent high-energy physics conference even went so far as to describe the search for the magnetic monopole as the only significant experiment on the horizon in high-energy physics today. Although

* Supported by the U.S. Air Force Office of Scientific Research.
† Work supported by the U.S. Atomic Energy Commission.

undoubtedly an oratorical exaggeration, the statement (having been made by a scientist not involved in the search) may be regarded as an indication of how seriously the missing magnetic monopole is taken among high-energy physicists.

In other areas of science the magnetic monopole has attracted very little attention, despite the fact that its existence would have far-reaching consequences. It might, for example, explain the high-energy component of cosmic radiation as well as extensive air showers; it might account for some of the properties of quasistellar objects; it might even be implicated in reversals of the Earth's magnetic field.

It is the purpose of the present chapter to present a concise account of progress in monopole physics, addressed primarily to geologists, cosmologists, and others whose fields would be affected by the existence of magnetic monopoles.

Theory

The basic concept which led Dirac[1] in 1931 to predict the existence of quantized magnetic 'charge' is a purely mathematical one. He observed that the wave function representing a particle is inherently multi-valued inasmuch as differences in its phase have no physical significance. This circumstance permits the existence of singularity lines, Dirac's 'strings' or 'nodal lines', around which a wave function can be integrated to yield a phase change which is a multiple of 2π. All wave functions must exhibit the same phase change when integrated around the same nodal line, and therefore a nodal line, once postulated, becomes 'observable' to all particles. It is this circumstance which endows the mathematical singularity with physical significance. Dirac's nodal line is in fact indistinguishable from a tube containing a quantized amount of magnetic flux; one quantum corresponds to the flux producing a phase change of 2π upon circumintegration. The terminal point of such a nodal line is a singularity in the electromagnetic field, or a source equivalent to a magnetic pole. It follows directly that the strength or 'magnetic charge' of such poles must be quantized in units of

$$g = \frac{\hbar c}{2e}$$

where $\hbar$ is Planck's constant divided by 2π, c is the velocity of light, and e is the quantum of electric charge (electron charge). In terms of the fine structure constant $\hbar c/e^2 = 137$, this condition becomes simply

$$g = 68 \cdot 5e$$

in mixed or Gaussian units.

The incorporation of magnetic poles into electromagnetic theory was further explored by Tamm[2] in 1931, by Grönblom[3] in 1935, by Jordan[4] in 1938, by Fierz[5] in 1944, by Banderet[6] in 1946, again by Dirac[7] himself in 1948, and by Ramsey[8] in 1958. More recently, Schwinger[9] in 1966 concluded that the magnetic charge quantum should have twice the value deduced by Dirac. The general result of all these studies is that the monopole is not capable of forming a bound state with a charged particle (a 'linked orbit' state), at least if point charges are assumed. Nobody has succeeded in solving the problem for charge distributions of finite radius. There appears to be no theoretical basis for predicting the monopole's spin, dipole moment, or mass. A speculative value of the latter has been deduced from the Salam–Tiomno rule[10], based on the observation that the ratio of the proton mass to the electron mass is the same as the ratio of the squares of the strong-interaction coupling constant to the electric coupling constant, which leads to a 'canonical mass' of 2·4 GeV. The same mass would be predicted by arbitrarily assuming equal classical radii for monopole and electron.

The interaction of monopoles with matter has also been studied extensively, notably by Cole[11] in 1951, Bauer[12] in 1951, Malkus[13] in 1951, Ford and Wheeler[14] in 1951, and by Katz and Butts[15] in 1965. When passing through nuclear emulsion, a monopole should behave essentially like an ion of charge 68·5 at relativistic energies. As it slows down, however, a monopole's track should differ dramatically from the track of any known particle. The ionization probability along the track of an electrically charged particle, and hence the number of silver halide grains rendered developable, is proportional to the length of time during which emulsion atoms are exposed to the constant electric field of the particle. Charged particle tracks therefore become very noticeably heavier near their termination. The electric field surrounding a passing monopole, however, is proportional to its velocity, and it therefore decreases at the same rate at which the exposure time increases, resulting in constant ionization probability, or constant track density. A monopole track should therefore maintain constant density to its very end, terminating abruptly when the monopole's electric field has fallen below the minimum value required to ionize emulsion atoms. Thus the identifying criterion in all monopole detection experiments, with one exception, has been an emulsion track of sufficient density (corresponding to charge 20 or greater), and of sufficient length to establish constancy of ionization.

A number of other deductions concerning the behavior of monopoles can be made on the basis of their large charge. Since Coulomb-type binding energy is proportional to the square of charge, it follows that opposite monopoles should be bound with about 5000 times more energy than electrons and protons. This explains why opposite monopole pairs, if indeed they should exist in abundance, are not readily separated and detected. It also explains the lack of magnetically bound atoms, composed of a family of magnetically charged poles analogous to the prevalent family of electrically charged particles. Monopoles should have been created along with protons and electrons at the genesis of the Universe. But, whereas the laws of quantum mechanics permitted protons and electrons to form stable atoms, these same laws would have precluded the formation of magnetic atoms for the simple reason that the ground state for a magnetic atom would involve an orbit (analogous to the Bohr orbit of hydrogen) having a radius smaller than the radius of the monopole is likely to be. Monopoles created during the primordial fireball would therefore have recombined and anihilated each other very quickly, emitting large amounts of high-energy radiation. This process might still be operative in remote galaxies, and might in fact explain some of the peculiar properties of quasistellar objects.

It is inevitable, however, that at least a small number of magnetic monopoles escaped the high-density plasma of their birth and have been cruising through cosmic space for 3×10^{17} sec (Hubble's age of the Universe), protected from recombination by the circumstance of their rarity. More detailed speculations concerning the fate of these monopoles have been made by Goto, Kolm, and Ford[16], by Goto[17], and by Alvarez[18]. In a Universe occupied predominantly by electric charge, all magnetic fields encountered by a monopole are closed (solenoidal), and such fields appear to a monopole like accelerating funnels. The monopole is overwhelmingly more likely to traverse the funnel in the accelerating direction than in the decelerating direction. Quantitative evaluation suggests that in 3×10^{17} sec a monopole should acquire an energy of about 10^{21} ev, which would account for the high-energy component of cosmic radiation for which no adequate accelerating mechanism has yet been devised. Thus very-high-energy cosmic particles, and the extensive air showers they initiate, might possibly be related to magnetic monopoles.

Are monopoles likely to manifest their arrival in any other way? If they reached the vicinity of the Earth with low energies, or if they were created in the upper atmosphere by other incident particles of enormous energy, they would be attracted preferentially by the Earth's magnetic field toward the Earth's opposite poles. An arrival rate of one Dirac monopole per square centimeter per second would neutralize the Earth's magnetic field in about one month. These monopoles may subsequently diffuse through the Earth so that the surface charge would dissipate at a certain rate, but an arrival flux of this order of magnitude would certainly manifest itself in macroscopically observable magnetic charge of surface minerals. It may therefore be concluded that the present arrival rate of monopoles must be substantially lower. It is not unreasonable to speculate, however, that the Earth may have moved through regions of high monopole density during past epochs, and that reversals of the Earth's field may in fact have been initiated by the resulting magnetic 'short circuit' across the Earth's dynamo. Recently presented evidence which correlated the most recent magnetic reversal with an extensive tektite shower seems to suggest the possibility that this reversal was linked to a cosmic event.

Resolution of all these speculations requires that the existence of magnetic monopoles be ruled out by a compelling theoretical argument, or that it be established by experimental observation. We shall now summarize what has been done in the latter direction.

Search for Magnetic Monopoles

New elementary particles are traditionally discovered by allowing the proton beam from a synchrotron to impinge on a suitable target. An abundant number of collisions can be achieved in this way, and, if the beam energy is adequate, any particle capable of existence is bound to be produced along with its antiparticle in due course. The likelihood of producing a particle depends on its production cross section, and the beam energy required to produce it depends on its mass. The production cross section for monopoles is incalculable, and their mass is unknown. One of the first experiments conducted upon the completion of each new, more energetic synchrotron has been a search for magnetic monopoles. The first of these was performed in 1958 at Berkeley with 6 GeV protons[19], the second in 1961 at CERN with 14 GeV protons[20], and the third in 1962 at Brookhaven with 30 GeV protons[21]. The result was negative in each case (on the basis of nuclear emulsion track analysis). The energy available in the latter experiment would have sufficed to produce monopoles having a mass of up to 2·9 GeV, and the number of collisions observed was adequate, even assuming the smallest conceivable production cross section, to provide convincing evidence against the existence of monopoles having a smaller mass than 2·9 GeV. Their 'canonical mass' would be 2·4 GeV assuming Dirac's charge, or 9·6 GeV assuming Schwinger's charge. However, there is no reason for believing that the monopole's mass may not be considerably larger, so that pair production would require considerably higher energies than are available in man-made accelerators.

Particles of higher energy are available in cosmic radiation, but they are much less plentiful than the particles in a synchrotron beam. If a pair of monopoles were formed by a high-energy cosmic-ray particle colliding with an atom in the upper atmosphere, and if the monopoles had reasonably low energies after their creation, they would fall to opposite hemispheres of the Earth along the nearest magnetic line of force, arriving at the surface with a low terminal velocity, and then diffusing through the Earth's interior. In 1951 Malkus[13] attempted to detect such monopoles by aligning a 250 g solenoid, 1 m long, with the Earth's field. His solenoid would have collected flux lines, or monopoles, from an area of 8300 cm^2, accelerated them to 500 MeV in an evacuated tube, and then allowed them to pass through nuclear track emulsions for identification. Malkus observed no monopoles in two weeks, which corresponds to a total cosmic radiation flux, or area–time product of 10^{10} cm^2 sec, thereby setting an upper limit for the arrival rate of north monopoles: 10^{-10} north monopoles/cm^2 sec, or 1 north monopole/cm^2 in 300 years. The total number of primary protons with energy above 1 GeV incident during this time is about 2×10^{10}, and, of these, 5×10^8 have energies above 30 GeV. Considering likely values of production cross section, Malkus' experiment does not provide convincing evidence against the existence of heavy monopoles. Nor, for that matter, does a recent refinement of Malkus' experiment, performed in 1966 by Carithers, Stefanski, and Adair[22], who used a large bubble chamber solenoid at Brookhaven. They would have detected 1 north monopole/cm^2 in 10^6 years. This may sound like an exceedingly rare event to be able to detect, but it is in fact not at all. Meteorite falls, which are not considered exceptionally rare, occur at the rate of one per square centimeter in 10^{16} years. Extensive air showers, caused by very-high-energy cosmic particles, occur at the rate of about one per square centimeter in 10^9 years. If the incidence rate of monopoles is comparable, we are not likely to observe their arrival by a Malkus' experiment. Fortunately a more promising approach is available.

A monopole, once having passed through enough matter to dissipate its kinetic energy by ionization, should not interact strongly with ordinary atoms. It might induce a magnetic dipole moment in nearby atoms by the strong magnetic field in its immediate vicinity, but the resultant attractive forces should be quite negligible. It is for this reason that incident monopoles may be expected to diffuse through the Earth along lines of magnetic force, so that any magnetic charge caused by a temporary accumulation of monopoles near the surface would eventually dissipate.

The situation is quite different, however, if a monopole encounters ferromagnetic material. A single ferromagnetic domain (or a region of aligned atomic dipoles forming a small permanent magnet) is indistinguishable to all outward appearances from a small current loop. A fundamental difference appears, however, as soon as a magnetic monopole enters the picture: the monopole can pass through the current loop, thereby being accelerated at the expense of energy from the source of current, but it cannot be similarly accelerated through the ferromagnetic domain because this would involve the extraction of energy from a quantum system in its lowest energy state. Purely macroscopic considerations of energy conservation therefore require that the magnetic field experienced by a monopole *inside* a ferromagnetic domain cannot be the B field we measure on the outside, but the oppositely directed H field which we have come to regard as a purely pedagogic artefact. The monopole thus provides us with the only basis for assigning a physical meaning to the two magnetic fields inside a permanent magnet: the B field is the field experienced by a monopole if it passes *through* the atomic dipoles which constitute the magnet, while the H field is the oppositely directed field experienced by the monopole if it passes *between* the dipoles. The H field is, so to speak, the average 'return flux' in the immediate vicinity of the atomic dipoles.

The energy paradox was first discovered in 1958 by Goto[23], who interpreted it as compelling proof that a monopole would encounter two opposing forces at the surface of a ferromagnetic domain. Purcell[24] reached the same conclusion on the basis of a convincing microscopic argument. He pointed out that, in order to feel the B field, a monopole would have to pass through the atomic dipoles in a time less than the characteristic Larmor precession time or the time required by an atomic dipole to flip over. This circumstance makes the accelerating process overwhelmingly improbable. The obvious significance of this, as was pointed out by Goto, is that monopoles would remain trapped in ferromagnetic materials with considerable binding energy, and would therefore accumulate during geological periods of time. It should be possible to extract these accumulated monopoles either by destroying the ferromagnetism which binds them (by chemical dissolution or by heating above the Curie temperature), or by exposing them to a sufficiently intense magnetic field.

In 1962 Goto, Kolm, and Ford[16] calculated this extraction field on the basis of a somewhat idealized image charge model. It turned out to be 53 kG for iron (having a saturation magnetization of 22 kG), and 16·4 kG for magnetite (with a saturation magnetization of 6 kG). Even for Schwinger monopoles having twice Dirac's charge the ideal extraction fields would only be raised to 57 and 17·4 respectively. The availability of magnetic fields above 100 kG thus provides a means for searching significant quantities of terrestrial and meteoritic material for accumulated magnetic monopoles. Such fields would suffice not only to remove a monopole from a ferromagnetic surface, but also to drag it readily through any solid material, even assuming that the monopole is loosely bound to the largest conceivable cluster of atoms polarized by its proximity. For example, a sphere of 10 Å diameter containing a single Dirac monopole, in a 50 kG field, would exert a pressure of 2×10^{11} dyn/cm^2 = 3×10^6 lb/in^2, which exceeds the yield strength of any known material. Once removed from the solid material, the monopole would acquire relativistic energies within fractions of a millimeter, and the first collision with a residual gas atom would strip it of any adhering atoms. Falling through a typical 100 kG Bitter solenoid having a length of about 15 cm, the monopole would be accelerated at the rate of 2 GeV/cm, acquiring an energy of 30 GeV, which is equal to the energy acquired by protons making hundreds of circuits around the Brookhaven synchrotron. At such energies the monopole could easily be identified, and then retrapped in an iron target for subsequent reextraction and further experiments. Quite aside from its intrinsic interest, a trapped monopole would be a convenient, reusable projectile for high-energy experiments. It could readily be accelerated to energies several orders of magnitude higher than energies available in any existing machine.

In falling through the Earth's magnetic field, a Dirac monopole would gain an energy of $3 \cdot 8 \times 10^{12}$ ev. The atmosphere is capable of stopping all monopoles with energy less than 8×10^{12} ev. Incident monopoles having moderate initial energies would therefore be thermalized in the atmosphere, reaching the Earth's surface with a low terminal velocity. They would penetrate only a centimeter or two into rock. The subsequent fate of such monopoles depends on the specific mechanism of erosion. If the surface erodes by gradual dissolution, the monopole is likely to remain trapped in the residual solid. If erosion is by the detachment of granules, the monopole is likely to be washed into a riverbed with flakes and sand granules.

Our first search for trapped monopoles was conducted on Brand Pinnacle (44° 44′ N latitude, 74° 6′ W longitude), a rock outcrop in the Owls Head area in the northern Adirondack Mountains. The site was suggested by magnetic anomaly maps and a detailed examination of the area from a light airplane at low altitude. Brand Pinnacle contains veins of pure magnetite which are likely to have been exposed since recession of the glacier; it is reasonably accessible, though sufficiently remote to have remained undisturbed. The apparatus consisted of a pulsed magnet and accelerating chamber (shown schematically in figure 1), along with a specially constructed generator, capacitor bank, high-voltage charging circuit, and high-current discharge circuit, coaxial cables, etc.; it is fully described elsewhere[16]. The entire apparatus was portable

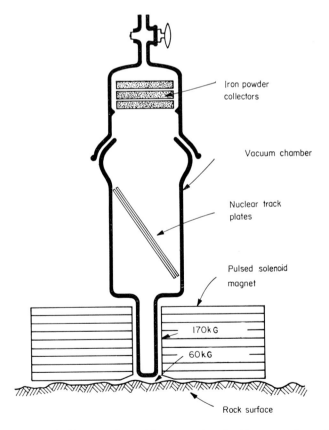

FIGURE 1 Schematic diagram of portable monopole extraction apparatus used in the Brand Pinnacle experiment

in the human sense of the word: it weighed a total of 240 lb, and the heaviest single unit, one capacitor, weighed 35 lb. It was transported on packboards by the three authors in a double portage from a jeep road. An area of about 1000 cm^2 of exposed magnetite was searched for north monopoles by subjecting it to 200 magnetic pulses. These pulses were sufficiently intense to extract monopoles from a depth of several centimeters, and left the area oppositely magnetized, so that the Earth can now be said to possess two south magnetic poles. Conversations with several geologists acquainted with the area have led us to believe, that at the Brand Pinnacle site, monopoles would remain for 100 to 1000 years, the time required to erode 1 cm of rock. Taking an intermediate estimate of 300 years, our search extended over an area–time product of 10^{13} cm^2 sec. Our negative result thus implies an upper limit arrival rate of 10^{-13} north monopoles/cm^2 sec or 1 north monopole/cm^2 in 300 000 years, having an energy less than 8×10^{12} ev (before traversing the atmosphere).

The Brand Pinnacle experiment was supplemented by a search of some meteoritic material. The first specimen was a half-inch thick slab, about 900 cm^2 in area, cut from the Carbo iron meteorite, made available through the courtesy of Professor Frondel of Harvard University. It was searched for north and south monopoles on opposite surfaces, but it is questionable whether the pulsed field of our portable magnet would have sufficed to extract monopoles since the 60 kG surface field was attenuated by eddy currents induced in the conducting material. We therefore exposed some small chips of Carbo, as well as fragments of the Estherville stony meteorite in a continuous 100 kG Bitter solenoid magnet, with accelerating chambers protruding into it from both ends, so as to collect both north and south monopoles. These additional experiments increased the area–time product of the search by about an order of magnitude in view of the age of meteoritic material.

This first search for trapped monopoles is only a preliminary attempt. Before searching more significant amounts of material it was necessary to modify the approach so as to eliminate several disquieting circumstances. The first and most obvious is the fact that, as has been explained, monopoles are likely to have considerable energies. They would therefore be distributed over depths ranging from meters to kilometers and would not have been found on the surface of Brand Pinnacle. A small meteorite, in the absence of a moderating atmosphere, is even less likely to stop energetic monopoles. The most likely terrestrial source of monopoles appears to be sediment from the ocean bottom at great depth. This material accumulates at a slow rate (between 0·1 and 1mm/century), and it contains sufficient quantities of ferromagnetic material in the form of cosmic spherules and magnetite sand to ensure that an arriving monopole would eventually be trapped near the surface, providing it has been thermalized by the ocean water above. To stop a monopole of 10^{16} ev energy takes 8 miles (12·8 km, or 1000 fathoms) of ocean water, about twice the greatest depth from which sediment can be dredged with a reasonable effort. If the distribution of monopole energies really has a maximum in the vicinity of 10^{20} ev, we can expect to find only the monopoles comprising the low-energy tail accumulated in accessible sediment. We continued the search, using a Bitter solenoid to expose deep sea sediments. The first specimens were only minute amounts, some of them collected on the historic Challenger expedition, supplied by the courtesy of Dr. J. Wiseman of the British Museum (Natural History). The second set of specimens were several deep sea cores supplied by Dr. C. Harrison of the Scripps Institution of Oceanography.

However, the experiment had another disquieting flaw, one which involves the confidence level of our detection criteria. In the Brand Pinnacle experiment we used a relatively thin and insensitive nuclear emulsion in order to facilitate processing and alleviate the tedious task of scanning. A monopole, we reasoned, could easily be identified by the spectacular width and the three-dimensional orientation of its track. Scanning turned up one track which proved very difficult to account for. It traversed the entire emulsion within the acceptance angle for monopoles, it was heavier than an alpha-particle track, though not nearly as heavy as the expected track for a relativistic monopole, and it showed too much Coulomb-type scattering (continuous meandering in direction) to be compatible with any known multiply charged particle. Unfortunately the identification of emulsion tracks (unlike the identification of butterflies, for example), is based on characteristics which exhibit considerable statistical variation, and the several emulsion track experts we consulted were able to do little more than to confirm its peculiarity. Detailed comparison with 1200 known alpha tracks in identical emulsion served to convince us that, on the basis of scattering alone, the track could not have been made by even the most unusual alpha particle.

At this juncture in the experiment we were offered the expert help of H. Yagoda, who has probably analyzed more cosmic-ray emulsion tracks than any other person in the world. He felt confident that he could have made a positive identification if the particle had left a substantially longer track in his own, very thick emulsion, processed under carefully controlled and more familiar conditions. All further emulsions were furnished, processed, and scanned by Yagoda's staff at the Air Force Cambridge Research Center at Hanscom Field.

In March 1964 Yagoda called early one Sunday morning to report with considerable excitement that he had found two tracks which were unlike any he had ever seen before. He persisted in this opinion until his tragic death three months later. The two tracks were geometrically compatible with south monopoles, and had been produced during the exposure of two specimens (as evidenced by adjacent control emulsions), one being a chip of the Carbo meteorite, and the other a deep sea sediment from the southern Pacific. Needless to say, the tracks were subjected to every known method of track analysis. They were 1350 and 1850 μm long, slightly heavier than alpha tracks with almost constant width and density over their entire length; both showed excessive continuous scattering and both terminated in the emulsion without any broadening. A Dirac monopole of the expected energy (20 GeV) should have produced twelve times the ionization of an alpha particle and left a track at least three times as broad. The result must therefore be regarded as inconclusive, but more quantitatively so than before.

Could there be a reason why a monopole having Dirac's charge might leave a track substantially lighter than that of a comparably charged electrical particle? The most obvious explanation which suggests itself is that the monopole had substantially less than the anticipated energy. In both the pulsed and continuous magnets, the magnetic field was applied relatively gradually. Our calculation of the extraction field was based on an idealized model, and may represent an upper limit. It is possible that the monopole actually emerges long before the maximum value of magnetic field is reached and thus never acquires the expected energy. The use of a continuous magnet made it relatively easy to eliminate the possibility in further tests by reversing the procedure: the 100 kG field was turned on first, and the specimens were then inserted along its axis. No matter where extraction occurred, the monopole would thus have fallen through at least half of the magnetic potential.

One further change was made in the experimental procedure. We used a sandwich of several emulsions, which remained in superposition only during the actual experiment, and for the shortest possible period of time. This turned out to be of the order of ten minutes; the probability that a cosmic-ray track of any type traversed the emulsion sandwich in a direction within the acceptance cone for monopoles during this time was found to be less than 2% from background observations. It follows that an observed track through the sandwich in the correct direction is likely to have been caused by the experiment, in the sense that it only has 2% probability of having been caused by extraneous particles. Now an event of 2% probability does not in itself constitute overwhelming statistical evidence for the existence of monopoles. However, if the track can be produced again by inserting the collector target into the magnet, it will have a probability of $(2\%)^2$, or only 0·0004. A third observation of the track would constitute an event of probability 8×10^{-6}, which may be regarded as a significant observation, regardless of the characteristics of the track.

Using this improved procedure, we obtained another track traversing the three emulsions of a sandwich (an event of 2% probability). Its geometry was again compatible with a south monopole, and its density was again too low. Unfortunately, however, we were again unable to reproduce the track by reinserting the target material. Thus, if the track had been produced by a monopole, the monopole had not been trapped in the collector target. Can we assert with a high degree of confidence that a monopole traversing the emulsions would have been trapped?

The most obvious question involves the trajectory of the monopole after it leaves the emulsion. The collector target must be located beyond the high-field region to ensure trapping and, therefore, owing to the limited angular aperture of the solenoid, it can subtend only a small fraction of the magnetic flux. If the monopole, upon its extraction at the center of the magnet, happens to be accelerated initially in a slightly diverging region of flux, it will continue to diverge from the magnetic axis and may miss the target. The situation was remediated with the aid of a computer program kindly developed by E. M. Purcell of Harvard University. By calculating the trajectories of monopoles having a wide range of mass values in the actual field of the solenoid, it proved possible to find an absolutely reliable extraction point in a region of converging flux ahead of the center of the solenoid. Placing the specimen within the specified region provides positive assurance that any extracted monopole will converge toward the magnetic axis without crossing it, even it if should be as massive as 100 protons. As a precaution against accidental scattering the accelerating tube was partially evacuated.

The availability of a well-defined monopole beam provided opportunity for a further refinement of the experiment: that of replacing nuclear emulsion by a series of scintillation detectors located sufficiently far beyond the magnet to permit effective magnetic shielding. Failure of track analysis to provide unambiguous detection has modified the experiment to a statistical one, and the need for tedious emulsion scanning would have made it difficult to search significant quantities of material.

A modified apparatus was constructed with the cooperation of F. Villa and A. Odian of the Stanford Linear Accelerator Center. It is shown schematically (not to scale) in figure 2. Deep sea sediment in the form of a slurry is pumped to the extraction

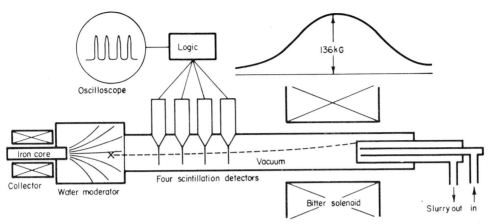

FIGURE 2 Schematic diagram (not to scale) of improved monopole extractor used to search large quantities of deep sea sediment

point along the magnetic axis through two coaxial tubes. After extraction through a thin bronze window, a monopole would be accelerated along the magnetic axis through a partially evacuated tube. It would pass first through four scintillation detectors, and then through an emulsion sandwich which is omitted during the first search and will be used only for confirmation. The monopole then passes through a second thin window and enters a moderating chamber filled with water, in which it is thermalized and allowed to follow a magnetic flux line leading to the collecting magnet. Provisions are made to insert the collecting magnet in place of the pumping tube for reextraction.

The output pulses of the scintillation detectors are monitored by logic circuitry which displays any fourfold coincidence on an oscilloscope and records it photographically. The background due to cosmic-ray showers amounts to 4·5 coincidences per hour, and the recorded pulses are predominantly at minimum-ionizing level, and randomly distributed.

During the approximately two hours required to pump the first 150 gal of slurry through the magnet, the coincidence rate increased to nine per hour. The probability of this increase being a random fluctuation is 1·6%. Two of the recorded coincidences showed approximately equal, high quadruple pulses corresponding to better than twice the alpha ionization, an event which is rarely observed simultaneously in all four detectors. Exposure of the collector target again failed to reproduce the observation.

The result of this refined experiment is thus in exact agreement with all previous results, and just as inconclusive. However, the confidence level has now become so high that it compels us to give serious consideration to a possibility we had previously dismissed: the possibility that magnetic monopoles exist which have a charge lower than the minimum quantum predicted by Dirac. The energy acquired by a monopole in a given magnetic field is proportional to its charge, while the rate at which it loses energy (ionizes) in passing through matter is proportional to the square of its charge. The ionization produced in our scintillators is compatible with monopoles having one-third of Dirac's charge. Such monopoles would easily have passed through our moderating chamber as well as the collecting magnet without being stopped.

We are thus forced to construct a further refinement of the apparatus in order to investigate a new hypothesis which seems absurd within the framework of Dirac's reasoning, and twice as absurd from the viewpoint of Schwinger. However, it appears to be the only hypothesis capable of explaining all of our observations, as well as the negative results of all those who have looked for the monopole before us.

References

1. P. A. M. Dirac, *Proc. Roy. Soc. (London), Ser. A*, **133**, 60 (1931).
2. I. Tamm, *Z. Physik*, **71**, 141 (1931).
3. P. O. Grönblum, *Z. Physik*, **98**, 283 (1935).
4. P. Jordan, *Ann. Physik*, **5**, 32, 66 (1938).
5. M. Fierz, *Helv. Phys. Acta*, **17**, 27 (1944).
6. P. P. Banderet, *Helv. Phys. Acta*, **19**, 503 (1946).
7. P. A. M. Dirac, *Phys. Rev.*, **74**, 817 (1948).
8. N. F. Ramsey, *Phys. Rev.*, **109**, 225 (1958).
9. J. Schwinger, *Phys. Rev.*, **144**, 1087 (1966).
10. A. Salam and J. Tiomno, *Nucl. Phys.*, **9**, 585 (1959).
11. H. J. D. Cole, *Proc. Cambridge Phil. Soc.*, **47**, 196 (1951).
12. E. Bauer, *Proc. Cambridge Phil. Soc.*, **47**, 777 (1951).
13. W. V. R. Malkus, *Phys. Rev.*, **83**, 899 (1951).
14. K. W. Ford and J. A. Wheeler, *Ann. Phys. (N.Y.)*, **7**, 287 (1959).
15. R. Katz and J. J. Butts, *Phys. Rev.*, **137**, B198 (1965).
16. E. Goto, H. H. Kolm, and K. W. Ford, *Phys. Rev.*, **132**, 387 (1963).
17. E. Goto, *Progr. Theoret. Phys. (Kyoto)*, **30**, 700 (1963).
18. L. Alvarez, private communication.
19. H. Bradner and W. M. Isbell, *Phys. Rev.*, **114**, 603 (1959).
20. E. Amaldi, G. Baroni, H. Bradner, L. Hoffman, A. Manfredini, G. Vanderhaege, and H. G. de Carvalho, *Notas Fis.*, **8**, No. 15 (1961).
21. E. M. Purcell, G. B. Collins, T. Fujii, J. Hornbostel, and F. Turkot, *Phys. Rev.*, **129**, 2326 (1963).
22. W. C. Carithers, R. Stefanski, and R. K. Adair, *Phys. Rev.*, **149**, 1070 (1966).
23. E. Goto, *J. Phys. Soc. Japan*, **10**, 1413 (1958).
24. E. M. Purcell, private communication.

Author Index

This author index is designed to enable the reader to locate an author's name and work with the aid of the reference numbers appearing in the text. The page numbers are printed in normal type in ascending numerical order, followed by the reference numbers in brackets. The numbers in italics refer to the pages on which the references are actually listed.

Adadurov, G. A. 452(28), *460*
Adair, R. K. 665(22), *672*
Adams, L. H. 288(10), *297*
Adams, R. D. 292(25), *297*
Ahrens, L. H. 40(32), *45*
Ahrens, T. J. 439(39), *447*, 449(14, 15), 450(20, 21), 452(14), 453(14), 456(21), *460*
Aiyama, Y. 489(5), *496*
Akimoto, S. 218(17), *220*, 456(35, 36), *460*
Alder, B. J. 61(16), *63*, 449(6), *459*
Aldrich, L. T. 36(11), 38(11, 18, 20, 23), 39(11, 18, 20, 28, 29), 40(11, 34), *45*
Allan, D. W. 653(2, 6), *656*
Alley, C. O. 523(1, 2), *529*
Allsopp, H. L. 549(4), *552*
Alsop, L. E. 513(2), *522*
Al'tschuler, L. V. 317(25), *402*, 449(3), 459(42), *459, 461*
Alvarez, L. 663(18), *672*
Amaldi, E. 400(99), *403*, 664(20), *672*
Anakina, L. I. 195(44), *209*
Andel, S. I. van 114(5), 115(5), 116(7, 8), 119(7, 8), *121*
Anders, E. 42(44), *46*
Andersen, F. 140(7), *153*, 155(2, 3, 5), 156(2), 157(2, 3), 158(3), 160(3, 5), 170(2), *172*
Anderson, D. L. 225(6), 231(24), 232(24), *245*, 442(46, 47), 443(46, 47), *447*, 450(17, 18), 456(17, 18), *460*
Anderson, O. L. 61(18), *63*, 426(3, 5, 6, 28, 52, 53, 61), 427(3, 8, 9, 28), 428(3, 5, 6), 429(3, 7), 430(7, 8, 9, 10), 431(3, 8, 13), 432(3, 13, 15, 16), 437(3, 5, 6, 8, 28), 439(13), 440(40), 442(3, 5, 6), *446, 447, 448*, 457(40), *461*, 465(2), 475(32, 34), *476, 477*
Ando, K. J. 486(3), *496*
Andrade, E. N. 257(7), *272*
Andreatch, P. Jr. 426(11, 28), 427(28), 430(11), 437(28), 440(11), *446, 447*
Arkell, W. J. 95(9), *101*
Armstrong, D. S. 108(26), *110*
Armstrong, R. J. 482(4), *484*
Ashkin, J. 322(48), 400(93), *402, 403*
Audley-Charles, M.G. 87(1), *101*

Auzins, P. 181(10), 182(10), *189*

Baadsgaard, H. 39(26), *45*
Babb, S. E. Jr. 202(90), *210*
Backus, G. E. 603(1), 604(1), *628*, 630(3), *648*
Bacon, G. E. 479(1), *484*
Baer, R. 587(6), *602*
Baker, E. 330(61), 345(61), *402*
Balashov, D. B. 452(28), *460*
Balchan, A. S. 184(17), *189*, 195(30, 66), 202(30, 93, 97, 100), 203(30, 93), 204(30), 205(30), 207(30), *209, 210*, 452(31), *460*
Ballhausen, C. J. 194(25), 207(25), *209*
Bancroft, D. 437(26), *446*
Bancroft, G. M. 199(80), 201(84), *210*, 216(8), *220*
Banderet, P. P. 662(6), *672*
Banerjee, S. K. 483(5, 6, 7), *484*, 490(7), 491(10), 493(10, 11), *496*
Barnes, E. W. 601(13), *602*
Barnett, M. P. 587(9), *602*
Baroni, G. 664(20), *672*
Bass, R. C. 452(30), *460*
Bates, C. 196(74), *210*
Bates, C. C. 133(18), *138*
Bauer, E. 663(12), *672*
Bauminger, R. 491(8), *496*
Beck, A. E. 79(3), 82(3), 83(6), *83*
Becker, G. 56(6), *63*
Beloussov, V. V. 241(43), *246*
Belova, E. N. 195(44), *209*
Bemmelen, R. W. van 74(14, 15), *74*
Bender, P. L. 523(1, 2), *529*
Benedek, G. B. 465(4, 6, 8), 466(8), 467(6, 8), 469(6), 470(4, 8), 471(24), 472(4, 8), 474(4, 8), *476, 477*
Benioff, H. 244(47), *246*, 273(6), *283*, 514(3), *522*
Bentley, W. H. 202(99), *210*
Berg, C. A. 273(1), 276(8), 277(8), 278(8, 10, 11), 279(10), 280(8), 281(1), 283(1), *283*
Berktold, A. 128(7), *137*
Berman, R. 203(115), *210*, 248(20, 21), *250*
Bernal, J. D. 439(31), *447*

Bethe, H. A. 317(32), 326(32), *402*
Bethke, P. M. 426(54), *447*
Birch, F. 79(1, 4), *83*, 224(3), *245*, 288(9), 291(9, 20), 292(27), 293(9), 295(27), *297*, 317(29), *402*, 425(1), 433(17), 437(26, 27), 444(49), *446*, *447*, 450(16), 452(27), 455(32), 456(27, 34), *460*
Bird, J. R. 588(10), 589(10), *602*
Birnbaum, G. 509(1), *512*
Blacic, J. D. 232(26), *245*
Black, R. 133(18), *138*
Blackett, P. M. S. 35(2), *45*
Blackhurst, J. 108(25), *110*
Blair, J. S. 108(25), *110*
Blanford, H. F. 48(5), *51*
Bloomfield, P. 216(9), *220*
Bohm, D. 412(7), *422*
Bolt, B. A. 291(23), *297*
Bond, W. L. 248(18), *250*
Bondi, H. 11(15), *18*, 47(3), *51*
Bone, M. N. 169(21), *172*
Boyd, F. R. 108(28), *110*, 191(8), *208*
Brachman, M. K. 319(43), 322(43), 323(52), *402*
Bradley, J. E. S. 195(31), *209*
Bradley, O. 195(31), *209*
Bradley, R. S. 186(22), *189*
Bradner, H. 664(19, 20), *672*
Bragg, W. L. 200(81), *210*
Braginskiĭ, S. I. 555(1), 558(3), 562(1), 564(1), *569*, 603(5), 604(5, 6), *628*, 630(4, 5), *648*
Bramkamp, R. A. 95(9), *101*
Brans, C. 9(7), 10(7), *18*, 22(5), *28*, 527(10), *529*
Brazhnik, M. J. 317(25), *402*
Brewer, R. G. 467(16, 17, 18), *476*
Bridgman, P. W. 317(27), *402*, 441(44), *447*
Brilliantov, N. A. 195(45), *209*
Brillouin, L. 314(9), 353(9), *401*, 465(3), *476*
Brooks, H. 416(15), *422*
Brouwer, D. 303(3), *309*
Brown, E. W. 528(18), *530*
Brown, H. 317(33), *402*
Brueckner, K. 400(95), *403*
Bülow, K. von 61(15), *63*
Bullard, E. C. 233(28), *245*, 442(48), 433(48), *447*, 579(4), 580(4), 584(4), 585(4), 588(4), *601*, 653(3, 6), *656*
Bullen, K. E. 61(17), *63*, 79(2), *83*, 195(67, 68), *210*, 241(41), *246*, 287(1, 2, 3), 288(1, 4, 5, 6, 7, 8, 11, 12, 13), 289(1, 2, 7, 14, 15), 290(3, 8, 14, 17), 291(8, 17, 19, 21), 292(2, 26, 28, 30), 293(2), 294(1), 296(30), 297(2, 36), *297*, 317(28), *402*, 431(14), 437(25), 438(25), 439(14), 440(14), *446*
Burke-Gaffney, T. N. 291(21), *297*
Burns, R. G. 191(9, 11, 13, 15, 17), 192(9, 21), 193(9, 27), 195(11, 13, 21, 59, 63), 197(27, 78), 199(21, 80), 201(11, 84, 85), 204(9, 11, 27), 206(9, 15, 17), *208*, *209*, *210*, 216(8, 10), *220*
Burstein, E. 202(102), *210*
Butler, R. 108(31), *110*
Butts, J. J. 663(15), *672*

Cahen, L. 40(31), *45*
Cain, J. C. 651(4), *652*
Caloi, P. 108(30), *110*
Cameron, A. G. W. 304(12), *309*
Camichel, H. 303(9), *309*
Caner, B. 140(6), 144(10), *152*, *153*, 155(7, 8, 9), 156(9), 167(9), 168(9), 169(7, 8, 9), 170(7, 8, 9), 171(7), *172*
Cannon, W. H. 144(10), *153*, 155(7, 9), 156(9), 167(9), 168(9), 169(7, 9), 170(7, 9), 171(7), *172*
Carey, S. W. 48(8), *51*, 103(4), *110*, 120(14), *121*
Carithers, W. C. 665(22), *672*
Carmichael, I. S. E. 214(4), *220*
Carnegie Institution of Washington 133(19), *138*
Carrison, L. C. 439(37), *447*
Carvalho, H. G. de 664(20), *672*
Casaverde, M. 133(20, 21), 134(21), 135(21), 136(21), *138*, 140(8), *153*
Castillo, J. 133(21), 134(21), 135(21), 136(21), *138*, 140(8), *153*
Chadwick, P. 275(7), *283*
Chamalaun, F. H. 549(4), 551(12), *552*
Chandrasekpar, S. 317(31), *402*
Chandrasekharen, V. 465(12), 473(12, 31), 474(12), *476*, *477*
Chao, E. C. T. 440(41), *447*
Chesnokov, B. V. 195(32, 33), *209*
Chiao, R. Y. 465(5), 467(14), 468(5, 19), *476*
Choubert, B. 87(2), *101*
Christian, R. H. 317(23), *402*, 449(1), *459*
Clark, M. G. 194(27), 197(27, 78), 204(27), *209*, *210*, 216(10), *220*
Clark, R. H. 195(59), *209*
Clark, S. P. Jr. 131(16), 133(16), *138*, 180(6), 183(6), *189*, 191(1, 2, 10), 195(10), 202(10), *208*, 225(8), 226(10), *245*, 439(38), *447*
Cleek, G. W. 426(55), *448*
Clemence, G. M. 303(6), *309*
Clendenen, R. L. 217(12), *220*, 459(41), *461*
Cohen, S. G. 491(8), *496*
Cole, H. J. D. 663(11), *672*
Collins, C. B. 38(21), *45*
Collins, G. B. 664(21), *672*
Compston, W. 40(41), 41(41), *46*
Compton, W. D. 192(20), *209*
Comrie, L. J. 587(8), *602*
Condon, E. U. 587(7), *602*
Coogan, H. M. 486(3), *496*
Cook, A. H. 292(29), *297*, 511(2), *512*
Cook, M. A. 108(29), *110*
Corless, G. K. 417(17), *422*
Corruccini, R. J. 426(60), *448*
Cotton, F. A. 194(29), *209*, 217(15), *220*
Coulson, C. A. 342(70), 345(70), *402*
Cowan, R. D. 322(48), 400(93), *402*, *403*
Cowan, G. R. 452(31), *460*
Cowling, J. G. 603(3, 4), *628*
Cox, A. 58(9), *63*, 549(1, 2, 3, 11), 551(11), 552(11), *552*
Cox, C. S. 130(13), *137*
Creer, K. M. 48(7), *51*, 65(3), 71(13), *74*

Cummins, H. Z. 465(9, 10), 468(20), 469(9, 10, 20), 472(10), 474(10), *476*
Curtis, C. D. 191(16), 206(16), *208*
Cutler, I. B. 186(31), *190*

Dalrymple, G. B. 58(9), *63*, 549(1, 2, 3, 9, 11), 551(9, 11), 552(11), *552*
Darby, J. K. 415(12), *422*
Davies, M. O. 186(27), *189*
Davis, G. L. 36(11), 38(11, 17, 18, 19, 20, 23), 39(11, 18, 20, 28), 40(11), *45*
Davis, L. 416(16), *422*
Davisson, J. W. 202(102), *210*
Day, M. F. 108(27), *110*
Dearnley, R. 69(5), *74*, 87(3, 4), *101*, 103(14), 104(15), 105(14, 15), 106(15, 18), 107(18), 109(18), *110*
Debye, P. 313(4), *401*
DeCarli, P. S. 450(22, 24), *460*
Decker, E. R. 133(17), *138*
Deer, W. A. 180(7), *189*
Dekker, A. J. 181(11), 182(11), *189*
DeLaurier, J. 155(3), 157(3), 158(3), 160(3), 164(15), 165(15), *172*
Demarque, P. R. 22(9), 23(10), *28*
Deribas, A. A. 450(23), *460*
Deutsch, T. 416(15), *422*
Dicke, R. H. 9(5, 7), 10(5, 7), *18*, 22(5, 6), 26(14), *28*, 29(1), *31*, 35(4, 5, 6), 44(6, 48), *45*, *46*, 47(1), 49(1), *51*, 103(5), *110*, 523(1), 527(10), 528(11, 12, 13, 14, 15, 16), *529*, *530*
Dickey, M. 413(8), 416(8), *422*
Dietz, R. S. 230(19), 235(19), *245*
Dirac, P. A. M. 9(4), *18*, 35(1), 37(1), *45*, 47(4), *51*, 70(10), *74*, 314(7), 388(7), 399(7), *401*, 527(8), *529*, 661(1), 662(1, 7), *672*
Dobretson, N. L. 450(23), *460*
Doell, R. R. 58(9), *63*, 549(1, 2, 3, 9), 551(9), *552*
Dollfus, A. 303(7, 8), *309*
Doran, D. G. 449(5, 7), 452(5), *459*
Dosk, J. B. 40(34), *45*
Douglas, T. B. 426(57), *448*
Dremin, A. N. 452(28), *460*
Drickamer, H. G. 184(16, 17), *189*, 195(30, 65, 66), 202(30, 88, 89, 91, 92, 93, 94, 95, 96, 97, 98, 99, 100, 103, 104, 105, 106, 107, 108, 109, 110, 111, 112, 113), 202(30, 88, 89, 91, 93, 105, 106, 107, 108, 109, 114), 204(30, 91, 116, 117, 118), 205(30), 207(30), *209*, *210*, *211*, 217(12, 16), 218(16), *220*, 415(11), *422*, 459(41), *461*
Dugdale, J. S. 415(9, 10), *422*
Duncombe, R. L. 23(11), *28*, 303(4), *309*
Dunham, A. C. 517(5), *522*
Dunham, K. C. 517(5), *522*
Dunn, T. M. 192(23), 207(23), *209*
Duvall, G. E. 449(4), 452(4), *459*
Dvir, M. 195(50), *209*

Y 2

Eaton, J. P. 439(36), *447*
Eckelmann, W. R. 38(14), *45*
Eddington, A. S. 9(1, 2), *17*
Eder, G. 69(9), *74*
Edwards, J. L. 426(54), *447*
Egyed, L. 48(6), *51*, 65(1), 69(6, 7), *74*, 96(10), 97(11), *101*, 103(1, 2, 3), 109(2, 32), *110*, 113(1, 2), 115(2), 116(1), 120(12), *121*
Einstein, A. 10(8, 10), *18*
Elsasser, W. M. 62(19), *63*, 191(6), 207(6), *208*, 225(7), 228(13), 233(29), 240(7), *245*, 291(18), *297*, 316(21), 317(21), *401*
Enderby, J. E. 420(21), *422*
Enikova, M. A. 195(46), *209*
Eppler, R. A. 202(104), *210*
Evans, E. Ll. 426(56), *448*
Evans, T. 247(4, 5, 6), 248(19), *250*
Ewing, J. I. 236(37), *245*
Ewing, M. 236(37), *245*

Fahey, J. J. 440(41), *447*
Faller, J. E. 523(1), *529*
Farquhar, R. M. 38(21), *45*
Farrell, E. F. 183(12), *189*, 195(34), *209*
Faulkner, J. 30(2), *31*
Fényes, I. 314(10), *401*
Fermi, E. 313(2, 3), 384(79), 400(99), *401*, *403*
Feynman, R. P. 316(19), 319(19), 321(19), 322(19), 328(19), 329(19), 355(19), 374(19), 399(19), 400(19), *401*
Fierz, M. 662(5), *672*
Figgis, B. N. 215(6), *220*
Filloux, J. H. 130(13), *137*
Finck, J. L. 323(50), *402*
Fink, V. 465(7), 468(7), *476*
Finnie, I. 267(12), *272*
Fischer, B. 56(6), *63*
Fitch, R. A. 195(65), *210*
Flanagan, T. J. 452(29), *460*
Flawn, P. T. 39(25), *45*
Fleischer, R. L. 499(1, 2, 3, 4), 500(3, 5, 6, 7, 8, 9), 501(6, 10), 502(4, 6, 14, 15, 16), *503*
Fletcher, A. 587(8), *602*
Fleury, P. A. 468(19), *476*
Flynn, K. F. 42(45), 44(45), *46*
Forbush, S. E. 133(20, 21), 134(21), 135(21), 136(21), *138*, 140(8), *153*
Ford, K. W. 663(14, 16), 666(16), 667(16), *672*
Forsyth, J. B. 495(13), *496*
Foster, G. G. 108(25), *110*
Foster, J. H. 549(8, 10), *552*
Foster, M. D. 439(32), *447*
Fowler, R. 323(49), *402*
Fowles, G. R. 449(4), 452(4), *459*
Frank, F. C. 247(1, 10), 248(1, 16), *250*
Franken, P. A. 523(1), *529*
Fred, M. 195(56), *209*
Freeman, G. P. 247(3), *250*
Fritsch, K. 465(4, 8), 466(8), 467(8), 470(4, 8), 472(4, 8), 474(8), *476*

Fritz, J. N. 441(42), 444(50), *447*, 449(10, 11), 452(11), 453(11), 455(11), *459*
Fritze, K. 42(47), 44(47), *46*
Fronsdal, C. 36(10), *45*
Fujii, T. 664(21), *672*
Fujisawa, H. 456(35), *460*
Fuller, P. J. A. 450(19), *460*
Fulton, T. 36(10), *45*
Funnell, B. M. 549(7), *552*
Funtikov, A. I. 317(25), *402*
Fyfe, W. S. 191(9, 15), 192(9), 193(9), 195(59), 204(9, 119), 206(9, 15), *208*, *209*, *211*, 213(2), 214(4), 218(2), *220*

Galdin, N. E. 453(33), *460*
Gammon, R. W. 465(9, 10), 468(20), 469(9, 10, 20), 472(10), 474(10, 30), *476*, *477*
Gamow, G. 50(15), *51*, 56(4), *63*
Garmire, E. 467(15), *476*
Gast, P. W. 39(30), 40(37, 38), 41(38), *45*, *46*
Gastil, G. 103(13), *110*
Gaunt, J. A. 601(12), *602*
Geisler, P. A. 23(12), *28*
Gellman, H. 479(4), 580(4), 584(4), 585(4), 588(4), *601*
Gell-Mann, M. 400(95, 100), *403*
Geological Survey Map 518(6), *522*
Gerharz, R. 303(10), *309*
Ghose, S. 194(28), 199(28), 201(82, 83), *209*, *210*
Gibb, T. C. 483(6), *484*
Gibbs, G. V. 202(87), *210*
Gibson, R. D. 571(2, 3), 572(2), 573(3), 575(2, 3), 576(4), *576*, 578(3), *601*
Giesecke, A. A. 133(20, 21), 134(21), 135(21), 136(21), *138*, 140(8), *153*
Gilbert, C. 16(19), 17(19), *18*
Gilvarry, J. J. 315(16), 316(16, 17, 20), 318(16, 35, 36, 37), 319(39, 40, 41), 320(44, 45), 321(47), 322(44, 47), 323(47, 51, 55), 324(44, 45, 56), 325(51, 58), 328(44, 45, 51), 329(44), 337(16, 37), 341(67), 354(35, 76), 365(16, 65), 374(44), 381(77, 78), 382(78), 389(84), 396(84), 399(47, 89, 90, 91), 401(16, 17, 47), *401*, *402*, *403*
Gjellestad, G. 589(11), *602*
Glashoff, H. 58(12), *63*
Glass, B. 549(8), *552*
Glendenin, L. E. 42(45), 44(45), *46*
Glynn, P. 475(34), *477*
Gniewek, J. J. 426(60), *448*
Goldenberg, H. M. 528(16), *530*
Goldenberg, R. 29(1), *31*
Goldich, S. S. 39(26), *45*
Goldschmidt, V. M. 219(18), *220*
Goldstein, J. I. 501(12), *503*
Gombas, P. 313(5, 6), 315(5, 6), 321(5, 6), 333(5), 388(5), *401*
Gon'shakova, V. I. 453(33), *460*
Goto, E. 663(16, 17), 666(16, 23), 667(16), *672*
Gough, D. I. 141(9), 143(9), *153*

Green, D. H. 195(69), *210*, 230(20), 240(20), *245*
Greene, M. P. 415(13), *422*
Greenwood, N. N. 483(6), *484*
Gregson, V. G. Jr. 449(14), 452(14), 453(14), *460*
Greytak, T. 465(4, 6), 467(6), 469(6), 470(4), 472(4), 474(4), *476*
Griffith, J. S. 177(3), 181(3), *189*, 206(122), *211*, 217(7), *220*
Griffiths, K. 30(2), *31*
Griggs, D. T. 108(23), *110*, 232(26), *245*, 296(34), *297*
Grönblum, P. O. 662(3), *672*
Gross, E. P. 471(25), *477*
Gruen, D. M. 195(56), *209*
Grum-Grzhimailo, S. V. 195(35, 36, 37, 38, 39, 40, 41, 42, 43, 44, 45, 46, 47, 48), *209*
Guggenheim, E. A. 323(49), *402*
Gutenberg, B. 206(123), *211*, 231(23), *245*, 264(9), 265(9), *272*, 291(22), 295(33), *297*, 436(23, 24), *446*

Haddon, R. A. W. 292(30), 296(30), 297(30), *297*
Hamilton, R. M. 186(20), *190*
Hanke, K. 194(26), 197(26), *209*
Hanna, S. S. 493(12), 496(14), *496*
Hansen, K. W. 186(31), *189*
Harrison, C. G. A. 549(7), *552*
Harrison, E. R. 247(11), *250*
Hart, P. J. 425(2), 427(2), *446*
Hartle, R. E. 381(77), 399(90, 91), *403*
Hartmann, O. 129(9), 133(20, 21), 134(21), 135(21), 136(21), *137*, *138*, 140(8), *153*
Hayden, A. W. 38(24), 39(29), *45*
Hays, J. D. 549(8), *552*
Heard, H. C. 108(23, 24), *110*, 232(25), *245*
Heberle, J. 493(12), *496*
Hedley, I. G. 495(13), *496*
Heezen, B. C. 48(9), *51*, 103(6), *110*, 236(36), *245*, 254(4), *272*, 273(5), 274(5), *283*, 549(10), *552*
Hellner, E. 201(82), *210*
Hensley, E. B. 178(5), *189*
Herczeg, T. 305(18), 307(18), 308(18), *309*
Herrin, E. 131(16), 133(16), *138*
Herzenberg, A. 571(1), *576*, 604(7), *628* 630(2), *648*
Herzog, L. F. 40(33), *45*
Hess, H. H. 230(18), 234(32), 244(18, 32), *245*
Hide, R. 106(19), *110*, 577(1), *601*, 651(1, 4, 5), 652(1, 6, 7), *652*, 653(1, 4), 655(1), *656*
Hilgenberg, O. C. 120(15), *121*
Hill, J. E. 318(35), 354(35), 365(35), *402*
Hilten, D. van 58(10), *63*, 113(4), 114(4), 115(4), 120(13), *121*
Hirahira, E. 491(9), *496*
Hirtz, P. 95(10), *101*
Hodge, B. L. 517(5), *522*
Hoffman, B. 21(4), *28*
Hoffman, J. H. 39(26), *45*
Hoffman, L. 664(20), *672*

Holmes, A. 40(31), *45*, 99(13), *101*, 103(7), *110*, 266(11), *272*, 282(14), *283*
Hopkin, L. M. T. 108(26, 27), *110*
Horai, K.-I. 652(7), *652*
Hornbostel, J. 664(21), *672*
Hospers, J. 114(5), 115(5), 116(7, 8), 119(7, 8), *121*
Houten, F. B. van 479(2), *484*
Houtermans, F. G. 35(3), *45*
Howie, R. A. 180(7), *189*, 199(80), *210*
Hoyle, F. 30(2), *31*, 304(11), *309*
Hückel, E. 313(4), *401*
Hughes, D. S. 449(8), *459*
Hughes, H. 186(19), *189*
Hugon, M. 303(9), *309*
Hulme, K. F. 247(14), *250*
Hurley, P. M. 38(16), *45*
Hutson, A. R. 186(24), *189*
Hyndman, D. H. 155(6), 167(6), *172*

Ida, Y. 456(36), *460*
Ince, E. L. 631(6), *648*
Ingard, K. U. 456(13), *476*
Ioffe, A. F. 198(32), *190*
Irving, E. 118(9), 119(9), 120(9), *121*
Isacks, B. 235(35), *245*
Isbell, W. M. 664(19), *672*
Isenberg, I. 408(5), *421*
Ishikawa, Y. 491(9), *496*
Ito, A. 491(9), *496*

Jackson, J. D. 36(10), *45*
Jacobs, I. S. 202(101), *210*
Jaeger, J. C. 242(45), *246*
Jaffe, L. 525(3), *529*
Jamieson, J. C. 191(4), *208*, 450(24), *460*
Jamil, A. K. 186(22), *189*
Jander, W. 186(21), *189*
Japiksie, B. 280(12), *283*
Jeffreys, H. 289(15, 16), *297*
Jenkins, F. A. 469(22), *477*
Jenkinson, E. A. 108(26, 27), *110*
Jennings, D. A. 471(23), *477*
Jensen, H. 399(86), *403*
Johnson, C. E. 483(7), *484*, 491(10), 492(10), 493(10, 11), 495(13), *496*
Johnson, G. A. L. 517(5), *522*
Johnson, M. D. 406(2), 417(18), *421*, *422*
Jones, H. 418(19), *422*
Jones, H. S. 528(17), *530*
Jones, R. B. 231(22), *245*
Jordan, J. 133(18), *138*
Jordan, P. 9(3, 6), 10(6), *18*, 21(3), *28*, 35(3), *45*, 47(2), *51*, 55(1), 56(2), 57(7), 58(2), 60(2, 14), *63*, 527(9), *529*, 662(4), *672*

Källén, G. 36(10), *45*
Kahle, A. B. 652(9), *652*
Kaiser, W. 248(18), *250*
Kanamori, H. 317(26), 398(26), *402*

Kanasewich, E. R. 35(8), *45*
Kapitonova, M. M. 195(48), *209*
Kaplan, S. M. 278(11), *283*
Katz, R. 663(15), *672*
Keating, K. B. 203(114), *210*
Keester, K. L. 183(13), *189*, 191(12), 195(12), 198(79), 199(12), 205(12), *208*, *210*
Kelting, H. 186(30), 187(30), *190*
Kerner, E. H. 353(74), *403*
Kertz, W. 129(8), *137*, 140(2), 146(2), *152*
Khramov, A. N. 58(11), *63*
King, L. C. 103(11, 12), *110*
Kippenhahn, R. 31(3), *31*
Kirkby, H. W. 108(25), *110*
Kittel, C. 192(19), *208*
Klimovskaya, L. K. 195(62), *209*
Knopoff, L. 228(14), *245*, 317(30), *402*
Kobayashi, M. 195(60), *209*
Kobayashi, S. 342(69), 345(69), *402*
Kohn, W. 415(13), *422*
Kolbe, E. 195(49), *209*
Kolm, H. H. 633(16), 666(23), 667(16), *672*
Komada, E. 218(17), *220*
Komissarova, R. A. 58(11), *63*
Konig, E. 216(11), 217(11), *220*
Kormer, S. B. 317(25), *402*
Kouvo, O. 39(30), *45*
Kovach, A. 42(46), *46*
Kovach, R. L. 442(47), 443(47), *447*
Kriess, F. M. 465(7), 468(7), *476*
Krishnan, R. S. 465(11), 471(28), 472(11), 473(11, 29, 31), 474(11, 29), *476*, *477*
Krogstad, R. S. 513(1), *522*
Kronberg, P. 108(31), *110*
Krueger, H. W. 39(26), *45*
Kruseman, G. P. 119(11), *121*
Krutter, H. M. 315(14), 318(14), 324(14), 329(14), 355(14), 374(14), *401*
Kubaschewski, O. 426(56), *448*
Kudinov, V. M. 450(23), *460*
Kuenen, P. H. 65(2), *74*
Kulp, J. L. 40(38), *46*
Kummel, B. 60(13), *63*
Kushiro, I. 218(17), *220*

Lahiri, B. N. 127(2), 132(2), *137*
Lambert, A. 140(6), *152*, 155(8), 169(8), 170(8), *172*
Landau, L. D. 22(7), *28*, 242(44), *246*
Landisman, M. 71(12), *74*
Landsberg, P. T. 400(98), *403*
Lang, A. R. 247(7, 9, 10), *250*
Larson, R. B. 23(10), *28*
Lastovka, J. B. 465(4), 470(4), 471(24), 472(4), 474(4), *476*, *477*
Latter, R. 351(72), 354(72, 75), 363(75), *402*, *403*
Law, L. K. 155(3), 157(3), 158(3), 160(3), 166(16, 17), *172*
Lawrence Radiation Laboratory 452(26), *460*
Lawson, A. W. 191(4), *208*, 216(9), *220*

Lee, W. H. K. 170(23), *172*
Leutz, H. 35(7), *45*
Lewis, H. W. 400(97), *403*
Lieberman, R. L. 465(2), *476*
Lifshitz, E. M. 22(7), *28*, 242(44), *246*
Lighthill, M. J. 633(7), *648*
Linde, R. K. 449(7), *459*
Littlejohn, C. 493(12), *496*
Littler, J. 440(41), *447*
Livingstone, C. E. 144(10), *153*, 155(9), 156(9), 167(9), 168(9), 169(9), 170(9), *172*
Long, L. E. 40(38), *46*
Loomer, E. I. 155(1), *172*
Love, A. E. H. 520(7), *522*
Lovering, J. F. 439(33), *447*
Low, W. 195(50), 197(76), 205(76), *209*, *210*
Lubimova, H. A. 191(3), *208*, 224(1), *245*, 266(10), *272*
Lyttleton, R. A. 436(22), *446*
Lyubimova, E. A. 191(5), *208*

Ma, T. Y. H. 49(12), *51*
McClintock, F. A. 278(11), *283*
McClure, D. S. 177(2), 181(15), 183(15), *189*, 192(22), 195(58), *209*
McCrea, W. H. 11(14), 13(14), *18*, 20(1), *28*
MacDonald, G. J. F. 66(4), *74*, 82(5), *83*, 106(20), *110*, 224(2), 227(11), 228(14), *245*, 317(30), *402*, 526(5), 527(5), 529(19), *529*, *530*
McDougall, I. 549(4, 5, 6), 551(12), *552*
McDougall, J. 108(31), *110*, 326(60), 383(60), *402*
MacGregor, I. D. 108(28), *110*
Machin, D. J. 214(4), *220*
McKenzie, D. F. 227(12), *245*
McMillan, W. G. 316(20), 323(55), *401*, *402*
McMullen, C. C. 42(47), 44(47), *46*
McQueen, R. G. 217(13), *220*, 317(24), *402*, 441(42, 45), 442(45), 444(50), *447*, 449(2, 8, 10, 11, 12, 13), 452(11, 13), 453(11), 455(11), 458(12, 13), 459(12, 13), *459*, *460*
McSkimin, H. J. 426(11), 428(4), 430(11), 440(11), *446*
McVittie, G. C. 23(12), 25(13), *28*
Maddock, A. G. 201(84), *210*
Madeja, K. 216(11), 217(11), *220*
Magnée, I. de 119(10), *121*
Maish, W. G. 202(103), *210*
Major, A. 456(37, 38), *460*
Major, M. W. 514(4), *522*
Malkus, W. V. R. 651(2), *652*, 653(7), *656*, 663(13), 665(13), *672*
Manfredini, A. 664(20), *672*
Mao, H. 456(39), 459(39), *460*
March, N. H. 314(11), 315(12), 323(54), 337(65), 341(67), 342(70, 71), 345(70), 355(65), 388(83), 391(85), 399(90), *401*, *402*, *403*, 406(1, 2), 407(1, 3), 415(12), 417(17, 18), 420(20, 21), *421*, *422*

Marinov, A. 491(8), *496*
Markowitz, W. 525(6), *529*
Marsden, B. G. 303(5), *309*
Marsh, S. P. 217(13), *220*, 317(24), *402*, 441(42, 45), 442(45), 444(50), *447*, 449(10, 11, 12, 13), 452(11, 13), 453(11), 455(11), 457(12, 13), 459(12, 13) *459*, *460*
Marshak, R. E. 317(32), 326(32), *402*
Marussi, A. 511(2), *512*
Mason, B. 196(73), 205(73), *210*, 226(9), *245*
Matsukuma, T. 342(69), 345(69), *402*
Matthews, D. H. 244(50), *246*
Maurette, M. 500(5), 502(14, 15), *503*
Maxwell, A. E. 233(28), *245*
Mayer, J. E. 320(46), 328(46), 333(46), 378(46), 378(46), *402*
Mayer, M. G. 320(46), 328(46), 333(46), 378(46), 381(78), 382(78), 385(80), 386(80), *402*, *403*
Melankholin, N. M. 195(51, 61), *209*
Menard, H. W. 170(22), 171(22), *172*, 230(17), 235(17), 236(38), 237(38), *245*
Metropolis, N. 316(19), 319(19), 321(19), 322(19), 329(19), 355(19), 374(19), 399(19, 87), 400(19), *401*, *403*
Metsger, R. 514(4), *522*
Meyer, A. 413(8), 416(8), *422*
Meyer, M. D. 217(15), *220*
Meyer, O. 127(4), *137*
Miller, J. C. P. 587(8), *602*
Milne, E. A. 11(13, 14), 13(13, 14), *18*, 20(1), 21(2), *28*, 324(57), 331(57), *402*
Milne, W. G. 169(21), *172*
Milton, D. J. 440(41), *447*, 450(22), *460*
Minomura, S. 202(105, 109), 203(105, 109), *210*
Mitoff, S. P. 186(25, 26, 28), 187(28), *189*, *190*
M.I.T. Staff 38(16), *45*
Mizoguchi, T. 489(4), *496*
Moavenzadeh, F. 282(13), *283*
Mössbauer, R. L. 486(1, 2), *496*
Morgan, G. 500(5), *503*
Morgan, W. J. 40(35), *45*, 242(46), *246*
Morin, F. J. 198(34), *190*
Morrish, A. H. 482(4), *484*
Morse, P. M. 466(13), *476*
Moss, R. W. 513(1), *522*
Mott, N. F. 418(19), *422*
Moulin, N. 195(54), *209*
Muir, A. H. Jr. 486(3), *496*
Mullin, J. B. 247(14), *250*
Munk, W. H 66(4), *74*, 526(5), 527(5), *529*
Munro, D. C. 186(22), *189*
Murnaghan, F. D. 431(12), 439(12), 444(12), *446*
Murphey, B. F. 38(15), *45*
Murray, J. D. 108(25), *110*
Muskhelishvili, N. I. 277(9), *283*
Myers, A. T. 439(32), *447*

Nabarro, F. R. N. 251(1), *251*
Naeser, C. W. 502(15), *503*

Nafe, F. E. 71(12), *74*, 432(15), *446*
Nagai, S. 342(69), 345(69), *402*
Nagata, T. 653(5), *656*
Nairn, A. E. M. 119(10), *121*
Neuhaus, A. 195(52, 53), *209*
Newell, H. E. 525(3), *529*
Newnham, R. E. 183(12), *189*, 195(34), *209*
Nguyen Hai 292(24), *297*
Niblett, E. R. 155(1), *172*
Nicholson, W. 56(5), *63*
Nicolaysen, L. O. 40(34), *45*
Nier, A. O. 38(15, 22), 39(26), *45*
Ninkovich, D. 549(10), *552*
Norton, F. H. 257(8), *272*

Oberley, J. J. 202(102), *210*
O'Connor, C. L. 469(21), *477*
Ofer, S. 491(8), *496*
Oliver, J. 235(35), *245*, 514(4), *522*
Ono, K. 491(9), *496*
Opdyke, N. D. 549(8, 10), *552*
O'Reilly, W. 483(5, 6, 7), *484*, 490(7), 491(10), 492(10), 493(10), *496*
Orgel, L. E. 192(24), 194(24), *209*, 215(5), *220*
Orowan, E. 108(21, 22), *110*, 233(27), 239(40), 241(42), 242(27), 244(42), *245*, *246*, 253(1, 3), 254(3), 269(14), *272*, 273(4), 283(15), *283*
Orton, J. W. 181(10), 182(10), *189*
Overton, A. 166(18), *172*

Pabst, A. 196(77), *210*
Pagannone, M. 204(118), *211*
Pakiser, L. C. 169(20), 170(20), *172*
Parker, E. N. 558(2), *569*, 630(1), *648*
Parkinson, W. D. 127(5), *137*, 140(4), 148(4, 14), *152*, *153*
Parsons, R. W. 202(112), *210*
Paterson, M. S. 244(51), *246*
Paterson, W. S. B. 166(16, 17), *172*
Patterson, C. C. 40(40), 42(40), *46*, 224(4), *245*
Paul, W. 416(15), *422*
Pauling, L. 213(1), *220*
Peč, K. 166(19), *172*
Peebles, G. H. 389(84), 396(84), *403*
Peebles, P. J. 44(48), *46*, 528(15), *530*
Peria, W. T. 181(8), 182(8), *189*
Perlow, G. J. 493(12), *496*
Perneva, L. A. 195(47), *209*
Phaal, C. 247(4), *250*
Phillips, D. 415(10), *422*
Pines, D. 400(96), *403*
Pinson, W. H. Jr. 40(39), 41(39), *46*
Plaskett, H. H. 656(8), *656*
Plaskett, J. S. 314(11), *401*
Platzman, G. W. 587(6), *602*
Plotkin, H. H. 523(1), *529*
Pochoda, P. 22(8), *28*, 56(3), *63*
Pozo, S. del 133(21), 134(21), 135(21), 136(21), *138*, 140(8), *153*

Prandtl, L. 249(22), *250*
Press, F. 522(8), *522*
Preston, R. S. 493(12), *496*
Price, A. T. 127(2), 129(11, 12), 132(2), *137*, 145(12), 147(12), 151(12), *153*, 161(14), *172*
Price, J. H. 450(19), *460*
Price, P. B. 499(1, 2, 3, 4), 500(3, 5, 6, 7, 8, 9), 501(6, 10), 502(4, 6, 14, 15, 16), *503*
Purcell, E. M. 664(21), 666(24), *672*
Puttick, K. E. 248(16), *250*

Quate, C. F. 475(35), *477*

Rabe, E. 302(2), *309*
Raff, A. D. 236(39), *246*
Raleigh, C. B. 244(51), *246*
Ramachandran, G. N. 247(2), *250*
Rama Murthy, V. 40(40, 41), 41(41), 42(40), *46*
Raman, C. V. 247(8), *250*, 471(26), *477*
Ramsey, N. F. 662(8), *672*
Ramsey, W. H. 70(11), *74*, 295(31, 32), *297*, 316(22), *402*, 436(21), *446*
Randall, M. J. 292(25), *297*
Rank, D. H. 465(7), 468(7), *476*
Reiling, G. H. 178(5), *189*
Reiss, H. 323(53), *402*
Reitz, J. R. 399(87), *403*
Reitzel, J. S. 141(9), 143(9), *153*
Rendall, G. R. 247(8), *250*
Revelle, R. 233(28), *245*
Rey, C. 216(9), *220*
Rice, M. H. 449(2), *459*
Richartz, W. 195(53), *209*
Richter, C. F. 436(24), *446*
Rieckhoff, K. E. 467(16), *476*
Rijnierse, P. J. 339(66), *402*
Rikitake, T. 127(3), 137(22), *137*, *138*, 139(1), 140(1), 148(1, 13), *152*, *153*, 155(4), 159(4), 160(4), *172*, 653(5), *656*
Rinehart, J. S. 452(25), *460*
Ringwood, A. E. 108(28), *110*, 191(7), 195(7, 70, 71), 196(7, 72), 204(7, 70, 71), 205(7, 71, 72), 206(7, 72), 208, *210*, 229(15), 230(20), 240(20), *245*, 439(34, 35, 38), *447*, 456(37, 38), *460*
Rittmann, A. 100(14), *101*
Roberts, P. H. 576(4), *576*, 577(1), 578(2, 3), *601*, 603(2), *628*, 652(8), *652*, 653(4), *656*
Robertson, H. P. 13(17), *18*, 307(21), *309*
Robertson, W. W. 202(90), *210*
Robie, R. A. 426(54), *447*
Robinson, L. J. 195(64), *209*
Rochow, E. G. 186(29), 187(29), *190*
Roeder, R. C. 22(9), *28*
Rösch, J. 303(9), *309*
Roessler, D. M. 178(4), 181(4), *189*
Rogers, J. 38(13), *45*
Roman, P. 36(9), *45*
Rosenhead, L. 587(8), *602*
Ross, C. S. 439(32), *447*

Rossiter, M. J. 489(6), *496*
Rowley, W. R. C. 511(2), *512*
Roy, R. 195(57), 196(74), *209, 210*
Runcorn, S. K. 49(13), 50(13), *51*, 74(16), *75, 83*(7), *83*, 184(18), *189*, 233(31), *245*
Russell, R. D. 38(21), *45*
Rustamov, A. G. 186(23), *189*

Sadler, D. H. 56(5), *63*
Sakai, T. 319(42), 321(42), *402*
Salam, A. 663(10), *672*
Salgueiro, R. 133(21), 134(21), 135(21), 136(21), *138*, 140(8), *153*
Samokhvalov, A. A. 186(23), *189*
Sandage, A. R. 37(12), *45*
Sandqvist, 108(31), *110*
Satô, Y. 71(12), *74*
Sauvenier, H. 333(62), 341(62), 350(62), 362(62), *402*
Savage, J. C. 35(8), *45*
Saville, G. 181(10), 182(10), *189*
Savino, J. 513(2), *522*
Sawatzky, G. A. 482(4), *484*
Scheidegger, A. E. 244(48), *246*
Schlupf, J. P. 469(21), *477*
Schmidt, O. J. 305(13), *309*
Schmidt-DuMont, O. 195(54), *209*
Schmucker, U. 127(6), 128(6), 133(20, 21), 134(21), 135(21), 136(21), *137, 138*, 140(3, 5, 8), 143(5), 144(5), *152, 153*, 156(10), 167(10), 170(10), *172*
Schnetzler, C. C. 40(39), 41(39), *46*
Scholte, J. G. 318(34), *402*
Schreiber, E. 426(3, 5, 6, 29, 61), 427(3, 8, 9, 59), 428(3, 5, 6), 429(3), 430(8, 9), 431(3, 8), 432(3), 437(3, 5, 6, 8, 29), 438(29), 442(3, 5, 6, 29), *446, 447, 448*, 475(32), *477*
Schrödinger, E. 16(20), *18*
Schüller, K. H. 195(55), *209*
Schwartz, C. M. 439(37), *447*
Schwarzschild, M. 22(8), *28*, 56(3), *63*
Schwinger, J. 662(9), *672*
Sciama, D. W. 10(9), *18*
Sclar, C. B. 439(37), *447*
Scott, J. M. C. 388(82), *403*
Scott, S. 652(8), *652*
Scrutton, C. T. 49(11), *51*
Scuderi, T. G. 426(55), *448*
Seal, M. 247(12, 13, 15), 248(17), *250*
Segal, E. 491(8), *496*
Seitz, F. 192(18), 208, 315(13), *401*
Serrin, J. 269(15), *272*
Shankland, T. J. 175(1), 185(1), *189*
Shapiro, S. M. 465(10), 469(10), 472(10), 474(10) *476*
Shields, R. M. 40(42), 41(42), *46*
Shoemaker, E. 502(13), *503*
Short, J. M. 501(12), *503*
Shortley, G. H. 587(7), *602*
Shulman, J. H. 192(20), *209*
Sibaiya, L. 471(27), *477*

Siebert, M. 129(8), *137*, 140(2), 146(2), *152*
Silver, L. T. 39(25), *45*
Simakov, G. V. 453(33), *460*
Simcox, L. N. 406(1), 407(1), *421*
Simmons, G. 433(18, 19), *446*, 465(1), *476*
Simon, F. E. 248(20), *250*
Skinner, B. J. 426(58), *448*
Slack, G. A. 181(9, 14), 182(9), 183(9, 14), *189*, 196(75), *210*
Slater, J. C. 315(14, 15), 318(14), 324(14), 329(14), 355(14), 374(14), *401*
Slykhouse, T. E. 195(65), 202(92), 204(117), *210, 211*
Smales, A. A. 40(35), *45*
Smith, A. I. 108(26, 27), *110*
Smith, J. V. 202(87), *210*
Soga, N. 426(30, 51, 52), 427(9, 30, 51), 429(7), 430(7, 9), 433(20), 437(30), 442(30), *446, 447*, 475(32), *477*
Sommerfeld, A. 333(63), 341(63, 68), 343(68), 388(81), *402, 403*
Soshea, R. W. 181(11), 182(11), *189*
Speranskaya, M. P. 317(25), *402*
Sprenkel-Segal, E. L. 496(14), *496*
Stamm, W. 186(21), *189*
Staver, T. 412(7), *422*
Stefanski, R. 665(22), *672*
Steineke, M. 95(9), *101*
Stephens, D. R. 184(16), *189*, 202(98, 106, 107, 108, 113), 203(106, 107, 108, 109), *210*
Stewartson, K. 561(3), *652*, 657(1), *657*
Stoicheff, B. P. 465(5), 467(14), 468(5), *476*
Stokes, G. G. 256(6), 258(6), 262(6), *272*
Stone, A. J. 194(27), 197(27), 204(27), *209*, 216(10), *220*
Stoner, E. C. 326(30), 383(60), *402*
Strakhow, N. M. 95(8), *101*
Strauss, E. G. 10(10), *18*
Strens, R. G. J. 195(63), 201(85), 204(120, 121), *209, 210, 211*, 213(3), 216(3), *220*
Stubican, V. 195(57), *209*
Sturtz, J. P. 181(11), 182(11), *189*
Suchan, H. L. 202(93, 94, 96), 203(93), *210*
Suess, H. E. 306(19), *309*
Sukhanova, O. N. 195(45, 48), *209*
Sutton, G. H. 514(4), *522*
Sutton, J. 104(17), *110*
Sviridov, D. T. 195(45), *209*
Sviridova, R. K. 195(48), *209*
Sykes, L. R. 234(33), *245*
Syono, Y. 439(39), *447*, 450(21), 456(21), *460*

Takagi, M. 247(9), *250*
Takeuchi, H. 317(26), 398(26), *402*
Takuma, H. 471(23), *477*
Talwani, M. 236(37), *245*
Tamm, I. 662(2), *672*
Tanaka, M. 489(4, 5), *496*
Tarling, D. H. 549(5), *552*
Tatel, H. E. 425(2), 427(2), *446*
Tatsumoto, M. 224(4), *245*

Taylor, F. B. 103(9), *110*
Taylor, J. B. 643(8), *648*
Teller, E. 50(14), *51*, 316(18, 19), 319(19), 321(19), 322(19), 328(19), 329(19), 355(19), 374(19), 399(19), 400(19), *401*
Ten Seldam, C. A. 407(4), *421*
Termier, G. 95(5, 6, 7), *101*
Termier, H. 95(5, 6, 7), *101*
Thomas, L. H. 313(1), 399(88), *401, 403*
Thomas, T. Y. 10(11), *18*
Thompson, R. W. 38(15), *45*
Thurston, R. N. 426(11), 430(11), 440(11), *446*
Tilton, G. R. 36(11), 37(43), 38(11, 17, 18, 19, 20, 23), 39(11, 18, 20, 28, 29, 30), 40(11), *45*
Tiomno, J. 663(10), *672*
Tischer, R. E. 202(111), *210*, 217(16), 218(16), *220*
Toit, A. L. du 103(10), 105(10), *110*
Tokudo, T. 489(5), *496*
Tolansky, S. 247(11), *250*
Tolman, R. C. 13(16), *18*
Tolstikhina, K. 195(44), *209*
Tomishima, Y. 399(92), *403*
Tomlinson, R. H. 42(47), 44(47), *46*
Tough, J. G. 566(5), *569*
Toulmin, M. S. 426(54), *447*
Townes, C. H. 467(14, 15), *476*
Tozer, D. C. 126(1), *137*, 225(5), 233(30), *245*
Trask, D. W. 525(4), *529*
Trunin, R. F. 455(33), *460*
Tsuchida, R. 195(60), *209*
Tugarinov, A. I. 104(16), *110*
Turkot, F. 664(21), *672*
Turner, F. J. 108(23), *110*
Tuve, M. A. 425(2), 427(2), *446*

Umeda, K. 334(64), 342(69), 345(69), 352(73), 399(92), *402, 403*
Urey, H. C. 302(1), 305(14, 15, 16, 17), 306(20), 307(1, 14), *309*
Uyeda, S. 170(23), *172*

Vacquier, V. 236(39), *246*
Vali, V. 513(1), *522*
Vanderhaege, G. 664(20), *672*
Van Valkenberg, A. 475(33), *477*
Veen, H. J. van 513(2), *522*
Vegos, C. J. 525(4), *529*
Velden, H. A. van der 247(3), *250*
Vening-Meinesz, F. A. 267(13), *272*
Venkateswaran, C. S. 471(26), *477*
Verhoogen, J. 229(16), *245*, 296(35), *297*
Vestine, E. H. 145(11), *153*, 652(9), *652*
Victor, A. C. 426(57), *448*
Vincent, D. H. 493(12), *496*
Vine, F. J. 57(8), 60(8), *63*, 244(49, 50), *246*, 255(5), *272*, 273(3), *283*
Vinogradov, A. P. 104(16), *110*
Vishnevskii, V. N. 195(62), *209*
Vladimorov, L. A. 317(25), *402*

Wachtman, J. B. Jr. 426(55), *448*
Wackerle, J. 441(43), *447*, 449(9), 459(9), *459*
Walker, R. M. 499(1, 2, 4), 500(5, 6, 8), 501(6, 10), 502(4, 6, 14, 15, 16), *503*
Walker, W. C. 178(4), 181(4), *189*
Walsh, J. M. 317(23), *402*, 449(1, 2), *459*
Ward, M. A. 69(8), *74*, 113(3), 115(6), 116(3), 119(3), *121*
Wares, G. 318(38), *402*
Warren, B. E. 200(81), *210*
Warren, R. E. 131(15), 133(15), *138*, 236(39), *246*
Wasserburg, G. J. 38(24), 39(24, 25, 27, 29), *45*
Weaver, J. T. 129(10), *137*
Webster R. K. 40(35), *45*
Weertman, J. 231(21), *245*
Wegener, A. 103(8), *110*
Weger, M. 197(76), 205(76), *210*
Wells, A. F. 217(14), *220*
Wells, J. W. 49(10), *51*
Wenniger, H. 35(7), *45*
Wensink, H. 549(6), *552*
Wertheim, G. K. 481(3), *484*
Wertz, J. E. 181(10), 182(10), *189*
Wetherill, G. W. 36(11), 38(11, 17, 18, 19, 20, 23), 39(11, 18, 20, 25, 28, 29, 30), 40(11, 36), *45, 46*
Weyl, H. 14(18), 15(18), *18*
Whatley, L. S. 475(33), *477*
Wheeler, J. A. 663(14), *672*
White, G. K. 426(53), *447*
White, H. E. 469(22), *477*
White, W. B. 183(13), *189*, 191(12), 195(12), 196(74), 198(79), 199(12), 205(12), *208, 210*
White, W. R. H. 169(21), *172*
Whitham, K. 140(7), *153*, 155(1, 2, 3, 4, 5), 156(2), 157(2, 3), 158(3), 159(4, 11), 160(3, 4, 5, 12), 161(11, 12, 13), 162(13), 163(12), 164(12), 165(13), 166(16), 167(12), 170(2), *172*
Wickens, A. J. 166(19), *172*
Wiederhorn, S. 202(96), 204(116), *210*
Wiese, H. 130(14), *138*
Wiggens, T. A. 465(7), 468(7), *476*
Wigner, E. 314(8), 315(13), 400(8, 94), *401, 403*
Wild, R. K. 247(5, 6), *250*
Wilkins, G. A. 129(11), *137*, 145(12), 147(12), 151(12), *153*
Wilkinson, C. D. W. 475(35), *477*
Wilkinson, D. T. 523(1), *529*
Wilks, E. 248(16), *250*
Williams, R. J. P. 191(14), 206(14), *208*
Williamson, E. D. 288(10), *297*
Wilson, A. G. 10(12), *18*
Wilson, J. 198(33), *190*
Wilson, J. T. 234(34), *245*, 253(2), *272*, 273(2), *283*
Winslow, D. K. 475(35), *477*
Witt, H. 186(30), 187(30), *190*
Wood, J. A. 501(11), *503*
Wright, K. 560(4), *569*, 585(5), *602*

Young, W. H. 413(8), 415(14), 416(8), *422*

Zahner, J. C. 202(88, 95, 110), 203(88), 210
Zartman, R. E. 39(27), 45
Zeldovich, Ya. B. 44(49), 46
Zemansky, M. W. 325(59), 329(59), 402

Ziegler, K. 35(7), 45
Zietz, I. 169(20), 170(20), 172
Ziman, J. M. 411(6), 422
Zussman, J. 180(7), 189, 202(86), 210
Zyuzin, N. I. 450(23), 460

Subject Index

Adams-Williamson equation 288, 296
Adiabatic bulk modulus 426, 428
  at mantle-core boundary 436
Advection, definition 240
Alaska 87
Algae 91
Alkali metals,
  compression 414
  partial waves 414
  self-consistent field 414
Amphiboles, spectra 201
Andradean fluid 257, 282
  distance of diffusion in 257
  plane flow 269, 270
  velocity profiles 262
Angular momentum 314, 316
Apollo (Lunar sample analysis programme) 533–546
  biochemical and organic investigations 545
  chemical and isotopic analysis 543, 544
  field geologic investigations 535
  geologic tools 535, 536
  lunar material 538, 539
  Lunar Receiving Laboratory 539
    gas analysis 540
    low level counting 540
    preliminary examination of sample 540
  physical property measurements 544
  principal investigators 541, 543
  sample containers 535, 536
  sample types 536, 538
Archaeocyathid,
  bioherms 92
  reefs 93
Asthenosphere 231, 234, 239, 241
Asymptotic forms 326, 332–334, 337, 340, 349, 363, 367, 378, 394
Asymptotic series 326–328, 383, 387, 388, 398
Asymptotic solutions 320, 326, 328, 329, 342, 345, 346
  large atom radius 341, 342
  small atom radius 337–340
Atom,
  boundary 330, 331, 342, 379, 399
  compressed 353, 354, 399
  infinite 329, 333, 337, 341, 354, 355, 378, 386, 390
  isolated 341, 342, 347–349, 352, 353
  limiting 388
  magnetically bound 663
  neutral 345, 353, 355
  perturbed 375, 379, 381

  radius 323, 328, 329, 341, 348, 373, 379, 398
  shape 322, 323
  spherically symmetric 320, 321, 323
  unperturbed 375–377, 379, 384–386, 389
  volume 327, 330, 339, 365, 369, 371, 373, 377, 384, 388
Atomic number 316, 317, 318, 321, 323, 324, 331, 383, 399, 400
Atomic orbitals 193
Atomic shell structure 314
Atomic weight, increase in 219
Australasia 87

Backus dynamo 604, 630
Banded iron ores 87–89, 91, 100
  era 87
Bathygenesis 99
Beltrami dynamos 635
Beltrami field 631, 635
Bending of light rays problem 23
Biokinesis 90
Birch's Law 433
Blackett's hypothesis 35
Bohr orbit 326, 383
Boltzman factor 251
Born approximation 414, 417
Boundary conditions 321, 326–328, 330, 338, 347–351, 353, 375, 377–381, 384, 385, 392
Boundary parameters 324, 325, 328, 337, 339, 341, 342, 347, 350, 361, 385, 387, 388, 391
Boundary values 328, 339, 341, 348, 351, 353, 363, 365, 366, 375, 377, 378, 384–387, 389, 390
Bound states, screened point ions 408
Brachman relations 323–325, 376
Braginskii dynamo 555–569, 604, 624, 630
  Chebychev collocation 560
  dynamo in a cylinder 558–561
  effective components 555, 558, 565, 566
  first approximation 555–558
  generating mechanism 556
  generating term 558
  magnetic Reynolds number 555
  second approximation 561–568
Brans-Dicke Theory 23
  empty case 23, 24
Brillouin scattering 465–476
  advantages 467, 468
  elastic constants
    measurement 466, 474
  experimental results 471–474
    solids 472, 473
  stimulated 467

Brillouin spectra,
  detection 468–471
    diffraction grating 468, 470, 471
    Fabry-Perot Etalon 468, 469
    optical superheterodyne spectrometer 468, 471
Bullard-Gellman dynamo 577–601
  Chebychev collocation 585
  Elasser integrals 584, 586–601
    numerical values 590–593
  Gaunt integrals 584, 586–601
    numerical values 594–600
  magnetic Reynolds number 578
  toroidal and poloidal vectors 579–583

Cainozoic era 97
Cambrian era 90–93, 95, 99, 100
Carbon 248
Carboniferous era,
  Mississipian 95
  Pennsylvanian 94, 95
Celestial mechanics 4
Celestial phenomena,
  celestial fire 1
  comet of 1680 4
  freely moving bodies 2, 3
  identity of matter 4
  moon and earth 2, 3
  sun
    magnetic attraction 4
  terrestrial analogies 1–5
Chandler wobble 523, 526, 527
Charge density 323
Chebychev collocation 560, 585
Chemical binding 316
Chemical potential 320, 323, 324, 326, 332, 383
  Fermi-Dirac gas 339
  positive 374
Chondrites, carbonaceous 196, 206
Clipperton fault 236
Coelomata 89, 92
Coesite, shock induced formation 450
Cohesion 399
Compressibility,
  in Earth
    seismic evidence 287–297
  isothermal 412
  olivine 455
Compressibility—pressure hypothesis 288, 290, 293, 297
Compression 294
Condensed phases 318
Conductivity 407, 410–417
  anomalies 127–132, 139, 170
    Andes 125, 133–137
    Canada 155
    Colorado 143, 144, 149
    continental edge 139, 146
    intracontinental edge 139
    North German 128, 130
    Rio Grande 127, 131
  core 651

forsterite 175, 185
high magnetic fields
  production 417
olivine 185
pressure ionized metals 419–421
  liquid 410–412
  solid 412–415
pressure variation 417
relaxation time 420
variable 129
Constitutive relation, shock wave propagation 449
Continental crust, formation 67
Continental drift 48, 72, 269, 526
Convection,
  asymetry 235
  core 228
  currents 75, 105
  definition 240
  energy 227
  generating forces 233
  Mantle 106
    steady state 255
    upper 223–245
  onset 29
  thermochemistry 227
Convective zones 29
Copernican system,
  and Kepler 3
  and Tycho's system 3
  dynamics 3
  infinite universe 3
  planetary motions 3
Corals 50
  Devonian 49
Core 49, 61, 288, 295, 299, 652, 654
  composition 452
  conductivity 651
  convection 228
  dynamo solutions 643, 644
  hydromagnetic oscillations 651
  inner core rigidity 291–293
  outer core 295
  pneumatic viscosity 651
  radius increase 296
  rotation
    relative to mantle 49
  turbulent mixing 228
Core-mantle boundary 293, 294
  relative motion 49
  topography 652
Coriolis force 106
Correlation energy 314, 320, 398, 400, 401
Corresponding states, law of 432, 442
Corundum 426, 433
Cosmic spherules 665
Cosmology,
  ancient cosmogonies 2
  Brans-Dicke 528
  Dirac's principle 9, 35–45, 71
  earth
    formation 5

Subject Index

Newtonian 10, 15
   expanding space 11
   opposition to popular religion 2
   planetary systems
      formation 5
   quasi-Newtonian 20
   relativistic 13–15
   systems of galaxies 5
Coulomb potential 400
Coulomb repulsion 314, 400
Cowling's theorem 556
Creep 225, 230, 231
   chemical factors 232
   steady state 230
      at low temperatures 251
         diffusion 251
         in large grained material 251
Cretaceous era 95, 97, 99
Crystal field,
   spectra
      transition metals 193
   splitting 193
   stabilization energy 204, 206
Crystal structures, stable at high pressures 405

Darwin Rise 237
Day, variations in length 49, 652
De Broglie wavelength 412
Debye-Huckel theory 313, 323
Debye temperature 405, 412, 426, 428, 457
Deerite 214
Degeneracy 319
   factor 320
   limit 320, 374
   locus 318
   temperature 318
   transition 318
Density gradient formula 288
Density–pressure trajectories 439–444
   Murnaghan's equation 440
   shock wave data 442
   static compression data 441
   Voigt-Reuss-Hill values 440
Devonian era 93
Diamonds,
   defects 247
   growth 249
   in upper mantle 247–250
   shock induced formation 450
Dielectric response function 419
   of Fermi gas 411
Dielectric screening 417
Diffusion,
   characteristic distance 261
      of heat 266
      of momentum 263
Diopside 450
Dislocation 251, 502
Ductility 225
Dynamo 629
   Backus 604, 630
   Beltrami 635

   Braginskii 555–569, 604, 624, 630
   Bullard-Gellman 577–601
   comparison eigenvalue problem 638, 647, 648
   compatible fields 632, 633
   conditions 633
   dimensionless parameters 604
   dynamically self consistent 640–646
   frozen-in field lines 603, 605, 612
   Herzenberg 571–576, 604, 624, 630
   homogeneous 653
   kinematic 578, 603, 629, 636–640
   magnetic Reynolds number 604, 620, 625, 629
   motions 610–612
   periodic 630–636
      non solenoidal 635
      regular 632
   scale separation 630
   small parameter models 618–620
   solutions 629–648
      eigenvalue 605, 665
      in spherical core 644
   waves 603–622
   westward drift 646

Earth,
   Age 507
   ancient radii
      Carboniferous 116, 119
      Devonian 116, 119
      Grenville Regime 109
      minimum dispersion method 113
      paleomeridian method 111, 113, 116, 119
      Permian 111
      triangulation method 113
         new 116
         old 116–119
      Triassic 111, 116, 119
   composition 299
   density 293
      central 291, 292
   disequilibrium bulge 227
   expansion 48, 55, 77, 87–101
      energy changes 77–83
      rate 66, 83, 109, 120
      slow 65–75
      tectonic evidence 103
   gravitational field
      regional 652
   gravitational potential energy
      present 79
      primitive 79
   incompressibility 293
   magnetic field—*see* Geomagnetic field
   models 296
      A 289
      B 290, 291
      involving compressibility 289–293
   moment of inertia 83, 288, 292
      artificial satellite observations 292
      change in 49
   radioactive energy 82

regions
  continental 55
  deep sea 55
  internal 287
rigidity 293
rotation 47, 69
  rate 525, 527
temperature distribution
  internal 72
Earthquakes 242
  limiting depth 250
Eclipses, ancient 49
Ediacarian era 89–91, 100
  fauna 91, 92
    Charnian 90
    Coelomata 91
    Cuidarians 91
    Medusae 90, 91
Elastic constants,
  pressure and temperature derivatives 425–428, 430, 444, 466
    adiabatic bulk modulus 426–428
    compression wave velocity 428
    Debye temperature 426–428
    density 426–428
    in lower mantle 444
    in oxide compounds 425–444
    shear modulus 426–428
    shear wave velocity 426–428
Electron,
  cloud 315
  density 313, 314, 354, 381, 382, 399, 400
  distribution 313
    spherically symmetric 315
  mean free path 420
  shells 314
  wave functions 314, 399, 400
  wavelength 314
Electron-electron interaction 321, 341, 399, 400
  distance 323
Electron-nucleus interaction 321, 341
  distance 323
Energy 324, 327, 328, 337, 369, 376, 381, 382, 386–388, 399, 400
  binding 388
  cohesive 315
  electronic 406
  exchange 322, 399
  ground state 400
  internal 315
  kinetic 321–323, 372, 400
  potential 321, 322, 332, 355, 356, 372, 399, 400
  total 322, 327, 328, 332, 354, 355, 376, 400
Enthalpy 323, 325, 331, 332, 376
Entropy 315, 319, 322, 324, 325, 327–329, 331, 376, 381, 382, 386, 388, 397, 401
Eocambrian era 90
Ephemeris time 56
Epeirogenesis 94, 99
  Caledonian 95
  mid-Carboniferous 95
  Taconic 95

Equation of state 313–318, 335, 365, 369, 373, 397, 398, 439
  at high pressures 340, 341
  mantle 439
  metals 449
  olivine 450
  oxides 432
  parameters 432
    density 432
      atomic weight 432, 433
  relativistic effects 314
  temperature perturbation 397
Equatorial bay field 134
Equatorial electrojet 134
Exchange 320, 323, 398–401
Exclusion principle 399

Fayalite, transition pressure 217
Feldspar glass, meteoritic 500
Fermi,
  distribution function 320, 322
  energy 314
  factor 314
  gas
    conductivity 411
    dielectric function 411
    free equation of state 413
    linear response 411
    non-interacting 407
Fermi-Dirac,
  gas 315, 328, 332, 333, 337, 339, 377, 378, 388, 389
  limit 377, 391
  statistics 313, 314, 318, 324
Ferrites 489
Fission track dating 499–503
  chemical processes 502
  mechanical processes 502
    dislocation motion 502
    plastic deformation 502
    plastic strains 502
  thermal processes 502
Fitted function 334, 335, 337, 354, 357–361, 363–366, 369, 371, 373, 374, 390, 394–398
Flandrian transgression 92
Fluxing agent 232
Fold belts 103, 105
  age determination frequency histogram 103
  Grenville Regime 106
  Hudsonian Regime 106
  Superior Regime 104, 106
Forsterite 433, 436
  conductivity 175
  elastic constants
    pressure and temperature derivatives 425
  optical absorption 175
Fractionation 304, 305
  causes
    chemical 305, 306
    magnetic 305–307
    radiative 305–307
    thermal 305

Subject Index 687

Free earth oscillations 288, 521
   spheroidal 296
   toroidal 296
Free electron gas 315, 323, 340, 399, 400
   degenerate 319
Friedel sum rule 414
Fusion curve 316, 318, 365, 401

Garnet 216, 436, 437
   elastic constants
      pressure and temperature derivatives 425
   spectra 202
   spin pairing 205
Gaunt & Elsasser integrals 584–601
Gauss's theorem 315, 321
Geocentric longitude 523, 526
Geomagnetic field,
   age of 577
   depth sounding 126, 139, 148
   dipole 111
      decay time 577
      reversals in sign 652, 664
   external 139, 148
   horizontal disturbances 170
   internal 140, 148
   poloidal 651
   rate of change 111
   separation 145–147
      integral formula 150–152
   toroidal 651
   westward drift 646, 652, 657
Geomagnetism 56
Gibbs,
   Free energy 322, 450
   function 323, 325, 327, 332, 376, 388
Gibbs-Helmholtz equation 322
Gillespite
   spectra 197
   spin pairing 204
Glaciation,
   Congo 60
   Eocambrian 92
   Wurmian 92
Goethite 487
   axis of magnetization 493–495
Gondwanaland 72, 87, 95, 105
Gravitation 36
   and other forces 5, 48
   and thermodynamics 5–6
   constant 19, 71
      change in 9, 19, 47–50
         by lunar retroflector 527–529
   function 19
      hydrodynamical fluid 22
      spherically symmetric body 22
   intensity 293
   quantization 5
   quasi-Newtonian theory 19, 20, 26, 28
      equilibrium 21
      Newton's 2nd Law 20
   universality 4–6
Greenland 87

Grüneisen
   constant 401
   law 315, 401
   parameter 294, 377
Haematite 481, 488
Hall,
   constant 416
   effect 188, 410
      anisotropic relaxation time 416
      pressure dependence 416, 417
Hamiltonian 320
Hartree equation 314
Hartree-Fock equations 314
Heat capacity 322, 324, 327–329, 377
   at high pressure 401
   Dulong and Petit lattice 315
   free electron gas 315, 401
Heat transport, infrared 224
Helium,
   Brillouin zone 407
   equation of state 406
   face-centred cubic 407
Helmholtz function 319, 322, 324, 327, 328, 332,
      376, 386, 388
Herzenberg dynamo 571–576, 604, 624, 630
   basic model 571, 572
   calculus of rigid rotors 572, 573
   laminated rotors 573
High pressure—high temperature regime 430
Homogeneous equation 328
Hornblende, fission tracks 480
Hubble,
   age 37, 44, 603
   constant 36
   effect 57
   law 109
Hugoniot curves 451–453
   elastic limit 453
      shock state 452
   principal 451
Hydrodynamic flow 318
Hydromagnetic oscillations 651
   of core
      inertial modes 651
      magnetic modes 651
Hydrosphere 100
Hyperfine field 487
Hypsographic levels 59

Incompressibility 287
Indonesia 87
Induction anomalies *see* Conductivity anomalies
Induction parameter 130
Inertial time 56
Inhomogeneities, in Earth 290
Initial conditions 321, 326, 330, 338, 341, 342,
      380, 381, 384, 385, 392
Initial parameters 324, 325, 328, 337, 339, 341,
      342, 350, 359, 385, 387, 388
Initial slope 328, 330, 331, 339–342, 345, 349,
      354–356, 359, 361, 363, 366, 375, 377, 379–
      382, 385

Interatomic forces 315
Interionic forces,
  dielectric function 417
  dielectric screening 417
  in metals 405, 407
    at high pressure 417
  pair potential 417
Iron, cosmic abundance 302
Islands, ages 253
Isomer shift 487
Isometric motions 242

Jaramillo event, detection in ocean cores 549–552
Juan de Fuca Ridge 236

Kepler's third law 49

La Brea tar lake
  as Andradean fluid 282
  surface 282
Lamb dip 509
Laser seismometer 513–520
  Earth tides 522
  free Earth oscillations 521
  helium-neon source 507–511
  krypton source 510
  Michelson interferometer 505–507, 510, 514
  strain change measurements 510–512, 517, 522
  undesirable source signals 519
Lattice,
  effect 316, 398, 401
  frequencies 315
  structure 316
Laurasia 72, 92, 95, 105
Legendre-Laplace density relation 289
Lias era 95
Ligand field theory, some aspects 214
Lindemann law 316, 401
Lomonosov ridge 105
Lower Old Red Sandstone 93
Low velocity layer 436–439
  critical thermal gradients 437
  shadow zone 437
Lunar orbit, precession of the nodes 50
Lunar retroflector 523–529
  laser range measurements
    accuracy 525
    Chandler wobble 523
    change in gravitational constant 523
    geocentric longitude 524
    lunar temperature variation 524
    velocity aberration 524

Mach's principle 10, 26, 47
Madagascar 88
Maghemite 482
Magma 248
  rising
    thermal changes 249
Magnetic anomalies 155
  Alert 156–160
  Mould Bay 160–166
  W. Canada and United States 167–170

Magnetic deep sounding 126, 139, 148
Magnetic epochs,
  Brunhes normal 549
  Jaramillo event 549
    oceanic core detection 552
  Matuyama reversed 549
Magnetic monopoles 661
  accumulated 666
  calculated trajectories 670
  charge 662, 671
    Dirac 662, 665, 669
    Swinger 665
  energies 668
  magnetically bound atoms 663
  search for 664–671
    Brand Pinnacle 667
    Carbo iron meteorite 668
    Estherville stony meteorite 668
  track 663
Magnetite 489, 668
  Titanium substituted 493
Magnetohydrodynamics, dynamo theory *see* Dynamo
Magnetohydrodynamic waves,
  as dynamo perturbation 653–656
  Rossby
    β-plane approximation 653
    second class waves 657
Mantle 42, 252, 299
  Andradean fluid 282
  base 290
  Byerly sphere 62
  chemical segregation 225
  composition 195
  conductivity 137
  convection 106, 555
    steady state 255
  density-pressure trajectories 439–444
  equation of state 439
  grey body 191
  heat transfer 207
  induction in horizontal slab 164
  magnetism 206
  material grain size 267
  melting 225
    lamellar 269
  mineralogy 195
  momentum diffusion 268
  non-Newtonian fluid 280–282
    boundary disturbance diffusion 253–271
    momentum diffusion 256, 261
    steady state motion 261
  non-planar flow 254
  permanent deformation
    mechanism 253, 268, 273, 283
      Andradean viscous creep 269
      Newtonian viscous flow 269
    plastic 266–269, 283
  radiation 206, 207
  Repetti sphere 62
  spin pairing 205, 206
  steady flow 264

Subject Index

temperature gradient 223
transition elements distribution 206
transitions 195
upper mantle 223
  aesthenosphere 234
  anomalous structure 156
  composition 425
  conductivity 126, 127, 170
  convection 223–245
  heat balance 207
  melting point gradients 226
  mesosphere 234
  tectosphere 234
viscosity 106, 108
Mass-radius relation 317
Maxwell-Boltzman statistics 318, 319, 324
Maxwell relations 325, 377
Megundulations 72
Melting point relation, reversal 219
Mercury,
  chemical composition 301
  density 308, 309
  diameter 302
  mass 302, 308
  perihelion advance 23, 29
Mesosphere 234, 239
Metals, stable at high pressures 417
Metazoa 91
Meteorites 302, 306, 318, 337, 354, 501
  age 35, 41, 42, 44
  Carbo iron 668, 669
  craters 318
  Estherville stony 668
  impacts 365
  Toluca iron
    cooling rate 501
Mid-Atlantic ridge,
  faults crossing 234
Michelson interferometer, strain measurements 505–507, 510, 514
Mineral magnetism 479
Mobility 187
Molecules 316
Moon 55
  central peaks 61
  secular acceleration 49
  volcanism 61
Mössbauer,
  effect 481, 483
    of magnetic materials 485–495
  spectra 199, 201, 479
    ferrite 489
    iron compounds 487, 488
    magnetite 489
    spinels 489
    ulvöspinel 490
Mountain folding 58
Multinomial coefficient 344
Murnaghans
  equation 440, 446
    aluminium oxide 441
    α quartz 441
    forsterite 441
    garnet 441
    magnesium oxide 441
    spinels 441
    stishovite 441
  Law 431

Negative ion 321
Neutron diffraction, application to rock magnetism 479–481
Nucleus 315, 321, 322
Newtonian fluid 257
  characteristic distance of diffusion 256
  surface motion 276
Non-Newtonian fluid 256, 257, 280
  adjacent to accelerated plate 257, 258
  crystalline material 257
  momentum diffusion 257
  surface instabilities 282

Ocean,
  basins
    formation 69
  floor
    amplification of ripples 275
    spreading 57, 526
    surface features 273, 281
    ridges 72, 236–238, 253, 254
      correlation with
        continental margins 237
        trenches 236
      convective velocity pattern 253
  rifts 57
  trenches 238, 273
    formation 273
    correlation with continental margins 238
Olivine,
  breakdown to oxides 459
  compressibility 455
  conductivity 126
  equation of state 450
  fission tracks 500
  Hugoniot data 453–454
    double shock fronts 453
  in high pressure Regime 456
  phase transitions
    shock induced 453
  spectra 197, 198
  specific heat 457
  spin pairing 205
  Twin sisters 453
Optical absorption,
  charge transfer 179, 184, 192
  colour centres 192
  crystal field 182
  in forsterite 175, 177
  in silicates 191–208
  pressure effect 202, 203
  temperature effect 202, 203
  transitions
    fundamental 178
    strong impurity 179–183

weak impurity 183–184
Orbits, solar system 40
Ordovician era 50, 93–95, 99
Orogenesis 91, 93–95
  Caledonian 93, 95
  Variscan 94, 95
Oxides 425,
  spin pairing 206

Paleogeography, Earth expansion 87–101
Paleomagnetism 58
  dating—see Fission track dating
Pauli principle 313, 314
Periclaze 426, 433
  elastic constants 425
Permian era 87, 95
Perturbation 328, 342, 375, 383–385, 399
  differential equation 328, 381, 387, 393
  function 342, 379, 382
  general method 326, 397
  homogeneous equation 380, 381, 384
  inhomogeneous equation 380
  parameters 382, 397
  temperature 382, 391, 393
    first order 382, 383, 385, 386, 392, 397, 398, 400
  general order 383, 387, 388
Phase,
  high pressure 71
  solid 318
  space 314, 316, 320
  transitions—see Transitions
Phonons 415, 421
  spectrum 419
Planets 301
  composition 301, 302, 307
    iron anomaly 308, 309
  core size 302
  evolution 309
    early 304
    fractionation processes 302, 304
  formation 303, 305
    condensation of gas sphere 304
    dust accumulation 304, 307, 309
  interiors
    cosmogonical aspects 301–309
  mass-radius relation 317
  statistical atom model
    application 317
Plastic strains 267, 502
Pleochroism 194
  haloes 192
Plutonium, fission tracks 501
Pneumatosphere 100
Poisson equation 314, 320, 381
Polar wandering 60, 526, 527, 652
Polycrystalline materials, at high temperatures 256
Polyhedral cell 315
Positive ion 321, 341, 342
  perturbation treatment 388
Potential 314, 316, 320, 321, 332, 399, 355, 356, 383

nuclear 321, 332
Poynting-Robertson effect 307, 308
Pre-Cambrian 50, 103
Pressure 293, 316, 322, 326, 331, 332, 337, 340, 341, 354, 365, 372, 375, 376, 381, 382, 385–388, 394, 399, 401
  critical 316
  effect on spectra 202, 203
Protocaryota 89, 91, 100
P-Wave velocity 169
Pyroxenes,
  spectra 198, 199
  spin pairing 205

Quadrupole splitting 487
Quantum statistical mechanics 319, 320
Quartz,
  α 433
    elastic constants 425

Radiative transfer 219
Radiative transport 184
Radial distribution function 411
Radioactive ages 35
Radioactive decay, heat generated 71
Ramsey's hypothesis 61, 301
  avoidance of 55
Rankine-Hugoniot conservation equations 451
  for solids 451
  olivine 456
Red shift 47
Relativistic theories 21, 22
  Brans-Dicke 10, 22, 27, 28, 47
  Einstein 27, 28
  General 9, 25, 47
    with varying G 14–17, 47
    modified by cosmical constants 25
  Hoffman 21
  Jordan 10, 21, 27, 28, 47
  Milne 21
Resistivity,
  at high pressure
    shock induced state 450
    melting temperature 420
    temperature dependence 420
Ridges—see Ocean ridges
Rift system 60
Riphean era 89, 90, 91, 99, 100
Rossby β-plane approximation 653, 654
Rutherford's equation, of radioactive decay 36
Rydberg 337, 375, 397

San Andreas fault 236
Schroedinger equation 313–315
Secular variation 49, 651, 653
Seismometer—see Laser seismometer
Self-reversal 479, 483
Semi-conductors, low mobility 188
Shock experiments 292
  at high pressure 450
    resistivity 450
  compression

properties of solids 449
induced polymorphs 450
  coesite 450
  diamond 450
  stishovite 450
in iron 450
phase changes,
  in silicates 450
pressure, volume, energy relations 450
wave data 316, 317, 449–459
Silicates 248
  optical absorption 191–208
  phase transitions 450
Silurian era 93, 95
Similarity transformation 322
Skin depth 125
Slater sum 323
Slip band thickness 267
Solar wind 306
Specific heat, at constant pressure 294
Spectra,
  absorption,
    amphiboles 201
    forsterite 177
    garnets 202
    gillespite 197
    intensity 194
    olivines 197, 198
    pleochroism 194
    polarized 195
    pyroxenes 198, 199
    silicates 195
    techniques 195
    width 194
  analysis 4
Spherical cell 315
Spinels 426, 433, 489
  elastic constants 425
Spin 399, 400
  pairing 204, 205
    garnet 205
    gillespite 204
    iron minerals 213–220
    olivines 205
    oxides 206
    pyroxenes 205
  state 314, 320
    equilibrium 215, 218
    transitions see Transitions
Standard rate 328, 354, 355
Statistical atom model 313, 315–318, 323
  boundary conditions 315
  liquid phase 318
  non-relativistic case 318
  planetary applications 317
  pressure ionization 405
  relativistic effects 318
  self consistent field 405
  shell structure 405
  solid phase 318
  white dwarfs 317
Staurolite 216

Steels, deformation 108
Stishovite 440
  shock induced formation 450
Strain,
  energy 294
  finite theory 293–296
    in core 295
    mantle-core boundary 294
  gauge 505, 513
  hardening 267
  rates 251, 522
    steady state 251
  tensor measurement 511, 512
Stress, spreading 244
Stromatolites 50, 88, 91
Strong coupling constant 36
Structure factor 411
Subsidence 235
Sun,
  acceleration 31, 49
  evolution 22
  oblateness 29–31
    perihelion of Mercury 29
    quadrupole moment 29
    rotation 29, 31
    turbulent convection 31
Superior Regime 87, 91, 99, 100

Taylor expansion 330, 337, 382, 393
Tchouktchen Peninsular 87
Techniques, spectra measurement 195
Tectonic activity 652
Tectonic maps 105
Tectosphere 223, 234
  compression 241
  sliding 234, 236, 239, 241
  stress guide 237
  tension 241
Telescope 4
Temperature, effect on spectra 202, 203
Tethys 87, 95–97
Thermal conductivity, radiative transport contribution 184
Thermal dating 499, 500
  cooling rate 500, 501
  stepped cooling 500
Thermal expansion 315
  at high pressure 401
  coefficient 315
Thermal gradients 436
  critical 436
  hydrostatic law 437
  ultrasonic data 436
Thermodynamic functions 313, 315, 318, 320, 321, 324, 325, 327–333, 335, 339, 374, 375, 377, 378, 381, 382, 385–388, 393, 394, 398, 399
  effect of lattice 318, 319, 337, 365, 401
Thomas-Fermi atom model 313, 315, 317, 322, 333, 341, 354, 399, 405, 407
  approximation 313
  boundary conditions see Boundary conditions

conductivity 407
equation of state 318, 319, 326, 332, 333, 337–339, 341, 362, 363, 387, 394
  asymptotic solutions see Asymptotic solutions
  compressed atom 315
  generalized 326–328, 383
  large radius 341, 342, 400
  positive ion 341
  small radius 339, 391, 400
  temperature perturbed 374, 375, 388, 401
    first order 374
    general order 374
  zero temperature 329, 330, 398
  function 326–328, 337, 341, 348, 354, 359, 361, 363, 381–383
  hydrodynamic applications 354, 373
  interatomic forces 407
  ionic screening 408
  planetary applications 318
  semi-classical method 313, 314
  spherically symmetric case 313
  square bracket symbol 343, 344, 346
Tides 49, 522
  frictional torque 50
Tight binding method,
  band width 418
  body centred cubic lattice 418
    lattice vectors 418
  face centred cubic lattice 418
  Hamiltonian 418
  isolated pseudoatom 418
Tillites 90
Titanomagnetites 482
  oxidation 483
Transport properties, of olivines 175–190
Tracheophytes 93
Trenches,
  correlation
    with continental margins 238
    with ridges 236
Transition,
  elements 193, 204–206
    high spin state 204
    low spin state 204
    mantle distribution 206
    spin pairing 204
  pressure
    at mantle temperature 217
  zone
    radiation in 207
Transitions,
  electronic 407
    helium 407
  phase 229, 455, 456
    basalt 230

iron 450
olivine to spinel 456
olivine to oxides 456
peridotite 230
pressure induced 229
silicates 450
spinel to oxides 456
spin state 216
  fayalite 217
  fayalite to spinel 218
  gillespite 216
  geophysical consequences 219
    atomic weight increase 219
    radiative transfer 219
    melting point reversal 219
  olivine-spinel 218, 219
  wustite 217
Tremadocian era 93
Trias era 95
Trilobites 92, 94
Turbulent mixing, in core 228

Ulvöspinel 490
Universe,
  age 21, 35, 48, 50
  De Sitter 25
  empty 26
Uranium, fission tracks 501
Urey's hypothesis 304

Variometer 140–142
Velocity of sound 412
  at high pressure 419
Vendian era 90, 91
Vertebrates,
  agnathids 93
  fishes 93
Virial theorem 322–324, 356, 376
Viscosity 108
Viscous surface,
  ripple growth 275–280
Volcanism 60
  on Moon 61

Weak coupling constant 36
Wentzel-Kramers-Brillouin-Jeffreys approximation 314
Westward drift 646, 652, 657
White dwarf stars 317, 318
  limiting mass 317
  mass-radius relation 317
Wigner Seitz equal volume sphere 406
Wustite, transition pressure 217

Zircon, fission tracks 450